My Mathematical Universe

Universe

People, Personalities, and the Profession

with a Foreword by George Andrews

T0321429

About the Author

KRISHNASWAMI ALLADI is professor of mathematics at the University of Florida where he was department chairman during 1998–2008. He received his PhD from the University of California, Los Angeles, in 1978. His area of research is number theory where he has made notable contributions. He is the Founder and Editor-in-Chief of The Ramanujan Journal (Springer), Founder and Editor of the book series Developments in Mathematics (Springer) and Chair of the SASTRA Ramanujan Prize Committee. He was an Associate Editor of the Notices of the American Mathematical Society. In 2012, he was inducted as an Inaugural Fellow of the American Mathematical Society for his distinguished contributions. In recognition of his research accomplishments and service to the profession, he was awarded an Honorary Doctorate (Honoris Causa) by SASTRA University in September 2022.

Other book by the author:

The Alladi Diary: Memoirs of Alladi Ramakrishnan
Edited By: Krishnaswami Alladi (University of Florida, USA)
ISBN: 978-981-120-287-2

More information on the book can be found on the website:
https://www.worldscientific.com/worldscibooks/10.1142/11346#t=aboutBook

My Mathematical Universe

Universe

People, Personalities, and the Profession

with a Foreword by George Andrews

Krishnaswami Alladi

University of Florida, USA

World Scientific

NEW JERSEY · LONDON · SINGAPORE · BEIJING · SHANGHAI · HONG KONG · TAIPEI · CHENNAI · TOKYO

Published by

World Scientific Publishing Co. Pte. Ltd.

5 Toh Tuck Link, Singapore 596224

USA office: 27 Warren Street, Suite 401-402, Hackensack, NJ 07601

UK office: 57 Shelton Street, Covent Garden, London WC2H 9HE

Library of Congress Control Number: 2022045176

British Library Cataloguing-in-Publication Data
A catalogue record for this book is available from the British Library.

Cover: Photo of Krishnaswami Alladi is courtesy of The American Mathematical Society.
Photo taken at the Annual Meeting of the Society in Baltimore, Maryland, on January 17, 2019.

MY MATHEMATICAL UNIVERSE
People, Personalities, and the Profession

ISBN 978-981-126-305-7 (hardcover)
ISBN 978-981-126-306-4 (ebook for institutions)
ISBN 978-981-126-307-1 (ebook for individuals)

For any available supplementary material, please visit
https://www.worldscientific.com/worldscibooks/10.1142/13046#t=suppl

Desk Editor: Nijia Liu

Typeset by Stallion Press
Email: enquiries@stallionpress.com

Printed in Singapore

Dedication

Dedicated to my wife Mathura, and my daughters Lalitha and Amritha —
companions and ardent supporters in my life's journey

Left to right: Mathura, Lalitha, and Amritha.
Photo taken by Jis Joseph in Cancun, Mexico, November 2016.

My parents who raised me with affection and encouraged me in all of my efforts.

My father Prof. Alladi Ramakrishnan (Aug 9, 1923–June 7, 2008), a visionary scientist who created MATSCIENCE, The Institute of Mathematical Sciences, in Madras, India in 1962. He was my mentor and my inspiration. Photo taken at the G. K. Vale Studios in Madras in 1959, the year he launched his Theoretical Physics Seminar at our family home, the precursor to MATSCIENCE.

My mother Mrs. Lalitha Ramakrishnan (born April 4, 1931), who so graciously hosted the visitors of my father's seminar. Photo taken at G. K. Vale Studios in Madras in 1963.

Contents

Part II: Chairmanship and the Ongoing Ramanujan Mission   **339**

Ekamra Nivas, the home built in Madras by my grandfather Sir Alladi Krishnaswami Iyer. It was the venue of my father's Theoretical Physics Seminar (1959–61). I grew up in this home and as a young boy met the eminent scientists who addressed this Seminar.

Krishna, age 4, with his father Alladi Ramakrishnan, and Nobel Laureate Niels Bohr, at Ekamra Nivas in Madras, January 1960. Professor Bohr spent a leisurely evening having discussion with Alladi Ramakrishnan and his students.

Foreword

This is a grand, unique book. It is an excellent mixture of autobiography and commentary on the world of mathematics. Not surprisingly, many mathematicians are driven people concentrating on narrow specialties; they seldom look up from their calculations. In the very competitive world of mathematics, this is understandable. Some, however, rise above these natural tendencies. Krishna Alladi has done this spectacularly. Fortunately, he possesses the several talents necessary to excel in mathematical research, teaching, administration and publishing.

Alladi's background and upbringing are also unique combining a serious commitment to Hindu culture with a cosmopolitan outlook. His career path is similar to that of his father whom he describes as "...not just a father to me, but my academic mentor and my greatest academic supporter." Alladi Ramakrishnan, Krishna's father, was an eminent physicist whose concern for the good of his profession led him to found the MATSCIENCE Institute in Chennai. It is natural then that Krishna not only became an excellent mathematician but also distinguished himself in academic administration and publishing.

While the book is an autobiography, it is much more. In Part I, we are provided with intimate glimpses of many eminent mathematicians and physicists. These stories extend over decades from boyhood when Krishna met many of his father's famous colleagues, to maturity when his own career brought him in contact with mathematicians from around the world.

Of equal interest is Part II where the ups and downs of chairing a large mathematics department are detailed. One senses vividly how carefully and ardently Krishna worked at building the reputation of the University of Florida Mathematics Department. After an astonishingly successful decade, one feels the anguish produced by the failure of the university to perceive the importance of this achievement. Whether this was owing to financial constraints or the lack of interest on the part of the Administration is unclear. In any event, those of us who have experienced the joys and sorrows of department headship will see much of our own life mirrored dramatically here.

The final portion of the book chronicles the many contributions of Krishna which have served mathematics in particular and society in general. Many of us (especially me) owe Krishna a huge debt for his tireless efforts in research, in publishing, in organizing outstanding conferences, in making the SASTRA Ramanujan Prize a reality, and in much more. Recognizing the great importance of these contributions, SASTRA University awarded Krishna the Honorary Degree of Doctor of Science (honoris causa) on September 18, 2022.

This is a rich book. It is a great story from cover to cover. It is also a book where you can "dip in." If a department headship is in your future, check out Chapters 8 and 9. If you are young and thinking about the possibilities of a career in mathematics, Chapters 2–4 will be of special interest for you. If you are just interested in learning something about the life of a mathematician, this book is filled with stories and adventures that will be illuminating.

In summary, we have here a unique perspective on the mathematical life from a man of many talents and many mathematical pursuits. Enjoy!

George E. Andrews
Member - US National Academy of Sciences
Past-President: American Mathematical Society
University Park, Pennsylvania
September 30, 2022

Author's Preface

I feel extremely fortunate to have had a very unusual and wonderful introduction to the academic world during my boyhood days, and in my career to have moved closely with, and influenced by, several world famous mathematicians and physicists. Being in such great company had a significant influence on my research, and on my contributions to the profession. This is therefore not just an academic autobiography, but one in which I describe some fascinating aspects of the lives, personalities, and works of many distinguished mathematicians and physicists with whom I had an opportunity to interact. Many stories that I narrate about these distinguished scholars are based on my own observations since I got to know them and observe them in close quarters from an early age. I have my observations on Nobel Prize winning physicists Murray Gell-Mann, Richard Feynman, Niels Bohr, Abdus Salam, Hans Bethe, and Subrahmanyam Chandrasekhar, as well as descriptions of Fields Medal winning mathematicians Enrico Bombieri, Atle Selberg, Alan Baker, John Thompson, Manjul Bhargava, among others. These recollections not only convey the genius of these and other luminaries, but also reveal their human side. The book has mathematics in it because I want to highlight major contributions of some of these eminent researchers; there is some mathematics also in descriptions of my research. One can skip these parts and still enjoy the rest of the book owing to the non-technical nature of the narrative; the book is intended to appeal both to members of the academic community and the lay public.

The manner in which I was introduced to academic life was unique and marvelous. My father, the late Professor Alladi Ramakrishnan, a distinguished scientist, conducted a Theoretical Physics Seminar in our family home in Madras, India, during 1959–61. Several world famous physicists and mathematicians lectured in this seminar, and some were our house guests. Thus from the age of four, I was in the company of these makers of modern physics — personalities like Niels Bohr and Abdus Salam, to name just two. Bohr was so impressed with my father's efforts that he talked very highly of this seminar to India's Prime Minister Jawaharlal Nehru. The consequence of this was that in 1962, MATSCIENCE, The Institute of Mathematical Sciences, was inaugurated with my father as the Director. From 1962 until 1975 when I joined graduate school at UCLA, my mother and I accompanied my father annually on his global lecture tours, and thus I had the opportunity to visit several great institutions of higher learning and meet important physicists and mathematicians. In addition, I attended conferences and lectures at MATSCIENCE which, under the directorship of my father, had a vibrant international visiting program. The constant contact with eminent physicists and mathematicians both in India and overseas influenced my decision to take to an academic career.

The manner in which I entered the world of mathematics was equally wonderful. I was conducting research on my own in Number Theory in Madras as an undergraduate student; in those days, there were no special programs to expose undergraduates to research. My father sent my work to many eminent mathematicians to get an assessment. Many of them responded favourably. In particular, Paul Erdős, whose life's mission was to spot talented young mathematicians and encourage them, rerouted his journey from Calcutta to Sydney in early January 1975, and came to Madras to meet me. The meeting with Erdős had a profound influence on my life. During the few days he was in Madras, we discussed several aspects of my research. This not only led to a collaboration over the next several years, but he also wrote a letter of recommendation to Ernst Straus of UCLA which resulted in my getting a Chancellor's Fellowship for a PhD starting in Fall 1975. Since then I have been in the United States except for a brief period in India. I owe much to Erdős, Straus, and several mathematicians for influencing my research and career. Straus was the last assistant of Albert Einstein and I heard several Einstein stories from him, a few of which I will recollect in this book.

G. H. Hardy of Cambridge University who was the mentor of the Indian mathematical genius Srinivasa Ramanujan said that he was very fortunate in having collaborated with Ramanujan and Littlewood in something like equal terms. Although I am no Hardy, I consider myself similarly very fortunate to have collaborated and moved closely with two of the greatest mathematicians of our generation — Paul Erdős and George Andrews, both of whom I consider as my mentors. In discussing my research, I will describe my close interaction and collaboration with Erdős and Andrews.

Since this is an academic autobiography, the narrative is in chronological order starting from 1959. But the emphasis is on commentaries on the academic profession and on eminent scientists with whom I had close association. As a background, I describe my high school and undergraduate experience in India. In reminiscing about my graduate student days at UCLA, or about my first job at the University of Michigan, I discuss the functioning of these two mathematics departments as fine examples of large mathematics departments in top public universities, but then I also talk about the faculty and the great mathematicians who visited UCLA and Michigan when I was there. Similarly, I also describe the faculty and programs at the University of Texas, Austin, The Pennsylvania State University and the University of Hawaii, each of which I visited for a year. In addition, I provide a detailed account of my visit to the Institute for Advanced Study in Princeton, one of the greatest centers of learning in the world, where Einstein was a permanent member. There I had the opportunity to observe several mathematical giants such as Andre Weil, Harish Chandra, S. T. Yau, Freeman Dyson, Enrico Bombieri, and Atle Selberg.

The book also describes some aspects of my research in number theory and those of many mathematical luminaries in a manner that could be understood by a mathematically literate audience, but the anecdotes which spread all over the book should be of interest to lay persons as well. Thus the book has been written to be of interest to academicians, students, as well as those with an interest in the academic world. While most of the sections in the different chapters are thematic, the narrative within Chapters 9 and 10 is chronological. And so, within Chapters 9 and 10, the section titles reflect only the most significant event(s) of each year. My international travels were extensive and frequent, and in these two chapters I talk about the many mathematicians I met on these trips. But I have also included descriptions of the destinations I visited both for work and pleasure, since I felt this would be of interest to the non-academician reading the book. Before taking to a career of research in number theory, I was keenly interested in pursuing aerodynamics, since I was fascinated with commercial aviation owing to my international travels in my boyhood days; this passion for aviation which I imbibed from my father, continues to this day. In view of my continued interest in aviation, I describe some aspects of the aircraft I flew in, and the flights with various airlines that I took; I hope my comments on aviation will interest the general reader.

The zero teaching load that I was given during my term as Chair enabled me to make significant progress in my research even though I was immersed in department administration. Also, since I was free from teaching, I could travel more easily. Like my father who traveled widely when he was Director of MATSCIENCE, I utilized these trips to meet and invite leading mathematicians to the University of Florida.

Whenever I lecture to graduate and undergraduate students, I tell stories about great scientific personalities to make the discussion more interesting and to provide a historical context. Several students have said that I should write up these anecdotes. That was one of the motivations for writing this book. I have also included a description of my experiences as an administrator — as Chairman of the Mathematics Department at the University of Florida for a decade (1998–2008). Many great things were achieved during those ten years, but I have also discussed the difficulties I encountered; in the decade of my Chairmanship, I came to realize that mathematics departments are really doing fine work despite their budget shortcomings, and that their real difficulty is in convincing the upper administration to provide adequate support for mathematics. Finally I describe some of my contributions to the profession, such as the launch of the Ramanujan Journal and the creation of the SASTRA Ramanujan Prize.

In closing, I point out that this book has been written for readers in India and outside India. So, there may be some parts of the narrative that are of interest only to one of these two groups.

Just as my mother and I accompanied my father on many of his international lecture tours, my wife Mathura and my daughters Lalitha and Amritha have been with me often on my worldwide academic trips. I owe a lot to them for urging me to write this book and for their support of all my endeavors. I hope I have succeeded in making the book of interest to a wide audience, and that my recollections about the great figures in the worlds of physics and mathematics will be an inspiration to students aspiring a research career.

Krishnaswami Alladi
Gainesville, Florida
December 2021

Academic/Mathematical Abbreviations and Acronyms Used

The following abbreviations/acronyms have been used in the text and/or in the index.

AMS - American Mathematical Society

AMTI - Association of Mathematics Teachers of India

CERN - Conseil Européen pour la Reserche Nucléaire, Geneva, Switzer-land

CLAS - College of Liberal Arts and Sciences (of the University of Florida)

DEVM - Developments in Mathematics (Springer book series)

FRS - Fellow of the Royal Society (of London)

HRI - Harish-Chandra Research Institute, Allahabad

IAS - The Institute for Advanced Study, Princeton

ICM - International Congress of Mathematicians

ICTP - (Abdus Salam) International Centre for Theoretical Physics, Trieste, Italy

IIT - Indian Institute(s) of Technology

JMM - Joint Mathematics Meeting (AMS, MAA, SIAM)

MAA - Mathematical Association of America

MATSCIENCE - Institute of Mathematical Sciences, Madras, India - now known as IMSc (The city of Madras is now called Chennai)

MIT - Massachusetts Institute of Technology

NIAS - National Institute for Advanced Study, Bangalore, India

PNT - Prime Number Theorem

RH/GRH - Riemann Hypothesis/Generalized Riemann Hypothesis

SASTRA - Shanmugha, Arts, Science, Technology, Research Academy, India

TIFR - Tata Institute of Fundamental Research, Bombay, India

TNAS - Tamil Nadu Academy of Sciences (now Academy of Sciences, Chennai)

UCLA - University of California, Los Angeles

WCNTC - West Coast Number Theory Conference(s)

WSPC - World Scientific Publishing Company, Singapore

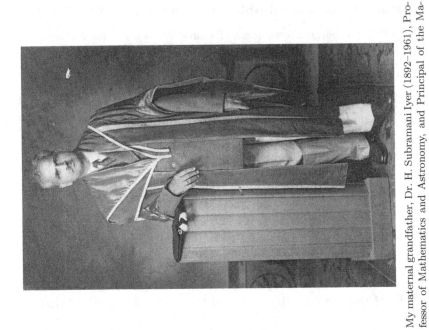

My paternal grandfather Sir Alladi Krishnaswami Iyer (May 14, 1883–Oct 3, 1953), one of India's most eminent jurists who played a key role in drafting the Constitution of India.

My maternal grandfather, Dr. H. Subramani Iyer (1892–1961), Professor of Mathematics and Astronomy, and Principal of the Maharajah's College, Trivandrum, India.

The Alladi family in November 21, 2015 during a celebration of Krishna's 60th birthday. Standing left to right — Aditya Srinivasan (first son-in-law), granddaughter Kamakshi Alladi Aditya, Lalitha Alladi holding grandson Keshav Alladi Aditya, Krishna Alladi, Mathura Alladi, Amritha Alladi Joseph, and Jis Joseph (second son-in-law). Seated — Lalitha Ramakrishnan (Krishna's mother).

Krishna and Mathura with their four grandkids: Standing — Keshav Alladi Aditya and Kamakshi Alladi Aditya. Seated on laps — Shreyas Alladi Joseph and Sahana Alladi Joseph. Jan 2022.

Acknowledgements and Credits

I am deeply indebted to several people for their help, input, and encouragement. A very special thanks to George Andrews not only for writing the Foreword, but also for providing valuable comments related to several sections of the book. I have also benefited from comments by Edward and Alice Bertram, Alfred Hales, Jeffrey Vaaler, Cyndi Garvan, Paul Robinson, G. P. Krishnamurthy, David Bressoud, Doron Zeilberger, Peter Paule, Bill Chen, M. Vidyasagar, Mathura, and my mother, for reading the book (in whole or in part), pointing out errors, and making suggestions. I should also acknowledge the invaluable input of Rochelle Kronzek of World Scientific — for numerous zoom calls and personal meetings in which she critiqued the narrative and gave several constructive suggestions. My colleague Frank Garvan and my son-in-law Jis Joseph gave crucial technical support that helped me work with the WSPC files and prepare them for publication. Max Phua, the Executive Director of World Scientific, his editors Rok Ting in Singapore, and Rochelle Kronzek in Florida, have been most supportive in publishing this book, and most patient as I took more time to finish this project. Rajesh Babu of World Scientific was immensely helpful not only in preparing the manuscript in WSPC format, but patiently and cheerfully guiding me as I worked with the WSPC files. Also of help at WSPC in Singapore, was Nijia Liu, who prepared the book for printing, and Shubei Lou, who handled the marketing. To all of the above at World Scientific, my thanks and gratitude. Finally, I wish to acknowledge the constant support of my family — Mathura, my mother, my daughters Lalitha and Amritha, and my sons-in-law Adi and Jis — without whom I could not have undertaken this task.

Credits:

I am grateful to the American Mathematical Society for permission to reproduce the photograph of me that appears on the front cover.

The photo of Prof. Marshall Stone (on page 44) was given to me by his daughter Phoebe Liebig. I thank her and the Robert S. Cox Special Collections section of the University Archives Research Center of the University of Massachusetts, Amherst, Libraries, for permission to use the photograph.

The photo of Prof. Ernst Straus (on page 96) was given to me by his late wife Louise. My thanks to her for the photograph.

The photograph on the Dedication page v and the two family photographs on page xxi were taken by Jis Joseph. I am thankful for the permission given by him to use the photographs.

The group photograph of the Alladi-60 Conference on page 648 was taken by Ali Uncu. I appreciate his permission to use the photograph. This photograph appeared in the Proceedings of the Alladi-60 Conference published by Springer [B4] in 2018.

The photograph on page 690 of me receiving the Honorary Doctorate is courtesy of SASTRA University, and the permission granted is appreciated.

The photograph at the top of page 646 of Prof. Zeilberger being honored by Prof. Chen, is courtesy of Professor William Chen of Tianjin University.

All other photographs are mine.

I thank the Pacific Journal of Mathematics for permission to reproduce page 1 of my 1997 joint paper with Paul Erdős in that journal.

My thanks to *The Hindu*, India's National Newspaper for permission to reproduce a section on my article on the Erdős Memorial Conference that appeared there in 1996.

I gratefully acknowledge the permission granted by the American Mathematical Society to reproduce certain parts of my article — "The SASTRA Ramanujan Prize — Its origins and its winners", which appeared in the Notices of the AMS in January 2020.

Finally, I appreciate the permission granted by Springer to reproduce certain parts of my article "Royal Society Conference and Publication for the Centenary of Srinivasa Ramanujan's Election as FRS", which appeared as Chapter 33 in my book "Ramanujan's Place in the World of Mathematics - Ed. 2", Springer, New York (2021).

Krishnaswami Alladi

Alladi Ramakrishnan in discussion with Prof. Harish-Chandra (Institute for Advanced Study, Princeton) at Ekamra Nivas, January 12, 1965. Krishna and his mother are keenly listening.

Eleven-year-old Krishna presenting a bouquet to India's Prime Minister Indira Gandhi at MATSCIENCE, January 6, 1967.

Part I
Development as a Mathematician

Krishna with his mentor, the great Paul Erdős, at Ekamra Nivas, Madras, June 1981.

L to R: Mrs. Gell-Mann, Richard Feynman, Murray Gell-Mann, and Krishna's mother, with seven-year-old Krishna, at Gell-Mann's home in Altadena, California, October 6, 1962. Feynman and Gell-Mann won Nobel Prizes in Physics in 1965 and 1969, respectively.

The legendary physicist Richard Feynman carrying on his head the child of his eminent Caltech colleague Murray Gell-Mann, during a poolside lunch party at Gell-Mann's house in Altadena, California, in honor of the Alladi family. Seven-year-old Krishna is next to Feynman - Oct 6, 1962.

Chapter 1

Boyhood contact with eminent scientists

1.1 My father's Theoretical Physics Seminar (1959–61)

I was born on October 5, 1955, in the home of my maternal grandfather Dr. H. Subramani Iyer in Trivandrum, in the state of Kerala in India. The names of my parents are Alladi Ramakrishnan (father) and Lalitha Ramakrishnan (mother). In the Indian custom, a pregnant lady would return to her parents home to give birth under the solicitous care of her mother. I am the only child to my parents. Since I was the first paternal grandson born after the demise of my grandfather Sir Alladi Krishnaswami Iyer, I was given his name. Although I was born in Trivandrum, I grew up in Madras in Ekamra Nivas, the home that my paternal grandfather Sir Alladi built. My name was Alladi Krishnaswami until I was 18 years old, at which time I changed my name to Krishnaswami Alladi to conform to the Western practice of having the family name as the last name since I knew that I will be going to the United States for graduate study.

Right from an early stage of my boyhood, I had the good fortune to have come into contact with world famous scientists who regularly visited our family home in Madras at the invitation of my father Professor Alladi Ramakrishnan. Naturally I did not understand anything about their work at that time, but I heard my father talk in very high terms about these visitors, and that made a deep impression on me.

After completing his PhD at the University of Manchester where he did important work in probability and stochastic processes, my father joined the physics department of the University of Madras in 1952. There he initiated research in the theory of probability. After a year (1957–58) at the Institute for Advanced Study in Princeton, my father changed the focus of his research to theoretical physics. He returned to India full of ambition and vision to create a setting in Madras to emulate the Princeton spirit. He wanted to introduce new material in the physics classes — inspired by the hundred or so lectures on modern physics he had heard at Princeton. But the powers at the University of Madras resisted any change in the syllabus, and so he decided to give lectures on advanced topics in physics at our home *Ekamra Nivas*. This mansion of splendid British colonial architecture was built in 1919 by my grandfather Sir Alladi Krishnaswami Iyer, one of the greatest lawyers of India and Member of the Drafting Committee of the Indian Constitution. As the first paternal grandson born after my grandfather's demise, I was named after him.

Ekamra Nivas has a large hall upstairs, and my father decided to use that for his lectures on modern physics. Eager students of the Masters course (and all his PhD students) gathered to hear his lectures and so this was christened the Theoretical Physics Seminar in early 1959. My father was regularly in contact with leading researchers worldwide, and he took every opportunity to invite them to lecture at his Seminar if they travelled anywhere near India. Thus there was a steady flow of eminent scientists to our home, some of them staying as our house guests. Naturally to have students and visitors at our home so often, my father relied on the support of my mother Lalitha Ramakrishnan, which she gave unhesitatingly. Indeed, my mother arranged lavish dinners at Ekamra Nivas to the visiting scientists and the members of the Seminar.

Initially I was overawed by the presence of these luminaries, and used to watch the lectures by hiding behind one of the doors of the lecture hall. Similarly I watched the group having dinner on our front lawn by sitting at a distance in the long verandah of our house. Later as my shyness wore off, I joined the visitors at dinner and enjoyed their company.

Although I was only four, I remember the visits of Abdus Salam of Imperial College, London, and of Nobel Laureate Niels Bohr from Copenhagen, both in January 1960. Salam was dressed in a full suit and lectured in the Seminar. He then spent a leisurely few hours talking to the students. My father arranged for a photographer to take pictures of all the seminar speakers — both while lecturing and in discussion with the students. These pictures refresh my childhood memories.

Although Salam was a Muslim and my father a Hindu, they were close friends and admired each other for their research, and had mutual respect for their religious beliefs. My father gave a gift to each visitor — a gift that had something to do with Indian culture of which he was very proud. He presented Salam with a portrait of Lord Krishna, which Salam accepted with grace. In return, when Salam said that during his brief stay with us he would like pray (*namas*) at a nearby mosque, my father joined him. Such was the closeness of their friendship and the mutual respect they had.

Salam was already very famous in 1960 for fundamental work in several problems in physics, especially in meson theory. He was Professor at Imperial College, London, and when elected at the age of 33 as Fellow of the Royal Society (FRS) in 1959, was one of the youngest to receive that honor. He was in India then as a principal speaker at the Indian Science Congress in Bombay. My father was organizing a session on theoretical physics at that Congress. He invited Salam to speak in his session and to come to Madras to address his Seminar. Salam accepted both invitations.

Abdus Salam won the 1979 Nobel Prize in Physics (jointly with Steven Weinberg and Sheldon Glashow) and visited MATSCIENCE in 1980 when my father was its Director. In between, Salam invited my father several times to the International Center for Theoretical Physics (ICTP) in Trieste, Italy, of which he was the Founder Director. I have had the pleasure of accompanying my parents to the ICTP in the sixties, and in each such visit, Professor Salam would spent some time talking to me to see how I was growing up and what my new interests were. This early and continued contact with him resulted in the ICTP offering me a handsome scholarship in 1975 to attend the Summer Institute in Complex Analysis there and I was able to do that when I was on my way to California in the summer of 1975 to join the PhD program at UCLA (see Section 2.9).

It is interesting to note that even though a scientist is one who is deeply involved in seeking "rational explanations" for phenomena in nature, he/she could be deeply religious. Both Salam and my father were esteemed men of science but had their strong and separate religious beliefs. After receiving the 1979 Nobel Prize in physics, when Salam had a television interview, he said that he was inspired by the architecture of a mosque in Lahore to come up with certain symmetries in his research.

Another visitor I remember vividly was Nobel Laureate Niels Bohr, who came soon after Abdus Salam. The entire theory of atomic physics rested on Bohr's model of the atom (for which he received the Nobel Prize in 1922), and so he was a most influential figure in the world of science. Professor Bohr was touring India as the guest of Prime Minister Jawaharlal Nehru. My father invited him to meet his seminar students, and Professor Bohr accepted the invitation. Great minds always set apart time, however busy they are, to meet and talk to young and aspiring researchers. To them the advancement of knowledge through discussion is of primary importance and they give preference to this even over official appointments. Even though Niels Bohr had a full plate in terms of his official appointments and was staying at the Raj Bhavan (= Governor's residence), he found time to spend a leisurely evening at Ekamra Nivas having dinner with the seminar students and discussing with them. Since he was a state guest, Bohr was accompanied by a Government aide throughout his trip in India. The aide suggested to Professor Bohr that it was getting late and that he should return to the Raj Bhavan because of a full program the next day. But Bohr continued to stay at Ekamra Nivas and was engrossed in a serious conversation with my father.

While my father talked to Professor Bohr, my mother conversed with Mrs. Bohr. I watched everything from a distance. But then when a group photo was taken, my father picked me up and stood next to the great man. I cherish this photograph with Bohr even today.

While Professor Bohr wore a jacket and tie, my father wore a traditional Indian dress — a dhothi — proud of his Indian heritage. This is something I learnt from my father: while you should admire other cultures, do not give up your own wonderful heritage.

Professor Bohr was so much impressed with my father's Seminar, that during a press conference at the end of his visit, he said that two things impressed him the most: the massive set up of the Tata Institute and the Atomic Energy Commission headed by Homi Bhabha in Bombay, and the small band of students trained by my father in Madras. This statement of Bohr was flashed in the newspapers. It attracted the attention of Prime Minister Nehru and the Minister for Education Mr. C. Subramaniam. The happy outcome was the creation of MATSCIENCE, The Institute of Mathematical Sciences in January 1962 with my father as its Director.

In the summer of 1961, Murray Gell-Mann of Caltech, a brilliant and charismatic physicist, was our house guest, along with Professor Richard Dalitz of the University of Chicago. My father admired Gell-Mann for his pathbreaking work and for his magnetic personality. Gell-Mann had just formulated his quark theory of the atom, and his use of the special unitary group of symmetries $SU(3)$ in quantum physics was creating a revolution. Gell-Mann and Dalitz were the principal speakers at a Summer School in Theoretical Physics organised by the Tata Institute in Bangalore. My father was one of the four organizers of that Summer School and took with him six of his PhD students as participants. My mother and I accompanied him to Bangalore and we joined Gell-Mann, Dalitz and my father on excursions in and around Bangalore. One excursion was to Nandi Hills just outside of Bangalore. Homi Bhabha, Director of the Tata Institute, and

Professor M. G. K. Menon, also of the Tata Institute, came on this excursion. That was when I first met Bhabha, the leader of the atomic energy program of India, about whom my father had spoken so much. After a full afternoon of sightseeing, when we all were seated on the gardens of a cafeteria atop Nandi Hills for tea, Bhabha, Gell-Mann, Dalitz, Menon, and my father, were engrossed in a serious discussion about Indian science.

Gell-Mann exuded an air of confidence and had a very cheerful disposition. He was eager to learn about Indian culture and our way of life. He and Dalitz graciously accepted my father's invitation to be our house guests and lecture at the Theoretical Physics Seminar. The guest room in our house that Gell-Mann and other visitors occupied is next to the Seminar Hall upstairs. Each morning, when he came down for breakfast, Gell-Mann would say that he had come up with an Indian name like Hariharan or Sivaraman having understood how names like Ramakrishnan or Krishnaswami are formed. Gell-Mann had an deep interest in, and good knowledge of, linguistics, and therefore was interested in Indian names.

Gell-Mann was the physicist my father admired the most. My father said that he would win the Nobel Prize — which he did in 1969. My father, among others, had nominated Gell-Mann for the Nobel Prize and he received a nice acknowledgement from the Nobel Committee.

In August 1961, Nobel Laureate Donald Glaser of Berkeley visited Ekamra Nivas. He had won the Nobel Prize the previous year for the discovery of the Bubble Chamber which became a crucial apparatus to observe paths of particles. He had active discussions with my father and the seminar students. We took him on an excursion to Mahabalipuram, about 30 miles south of Madras on the Coromandel coast. Mahabalipuram was a standard excursion that we did for our visitors owing to the grandeur of temple ruins found there as well as the spectacular beaches. In particular, the shore temple partly submerged in the ocean is a fabulous sight. After the sightseeing, Professor Glaser had a desire to take a dip in the ocean. So he took his shirt off and just entered the water! My father had to ward off the fisherman and their children who had gathered there in curiosity to see the foreigners.

In addition to the world famous physicists, there were eminent pure mathematicians from overseas who lectured in the Seminar. One was Laurent Schwartz from France who had won the highest prize in mathematics, the Fields Medal, in 1950 for his fundamental work in the theory of distributions that put the ideas underlying the Dirac delta function on a rigorous mathematical foundation. Another mathematics visitor was

Marshall Stone, Distinguished Service Professor at the University of Chicago. Stone's generalisation of the Weierstrass theorem on the approximation of continuous functions by polynomials is so fundamental, that every graduate student in mathematics learns it. Besides being a great research mathematician, Marshall Stone also contributed to the profession admirably. As Chairman of the Mathematics Department in Chicago, he transformed it so effectively, that his term as Chairman is jokingly referred to as the "Stone Age"! Stone was the son of a Justice of the US Supreme Court. Just as my father got a degree in law initially owing to my grandfather's influence, but later turned to physics, Stone was expected by his family to take to law, but he chose mathematics instead. So he understood my father's family background, and appreciated my father's efforts in running the Seminar. Stone visited India regularly, and was at MATSCIENCE as well.

Every visitor to the Theoretical Physics Seminar expressed great appreciation of my father's efforts. What my father wanted was a new Institute in Madras modeled along the lines of the Institute for Advanced Study in Princeton. Every time he had a distinguished visitor in his Seminar, he would either invite a prominent member of Indian administration to our home to meet the scientist, or take the visitor to meet an important representative of the Government. One important politician who was a regular visitor to Ekamra Nivas to meet the visitors was Minister for Education Mr. C. Subramaniam. He was convinced that a new Institute had to be created in Madras. He arranged for a meeting between my father and the Prime Minister at the Governor's Residence on Oct 8, 1961. It was a dinner with the Prime Minister for my parents, following which the Prime Minister met my father's students. After discussion with the students, Jawaharlal Nehru asked my father what he really wanted. Here was the Prime Minister of India asking what you want! In such an instant you do not ask for a promotion or a salary raise. So my father asked for a new institute. The Prime Minister said that he will give this matter some consideration.

Observing the Prime Minister's interest in the idea, Mr. Subramaniam asked my father to write a proposal that he would show to the Prime Minister. When Jawaharlal Nehru received such a proposal from my father, it was natural for him to refer it to Homi Bhabha, Director of the Atomic Energy Commission and of the Tata Institute, for an opinion. Bhabha reacted negatively and said that another institute was not necessary when the Tata Institute was already there; he also pointed out that the creation of new institutes would split the already weak budget of India. So it looked like my father's proposal was doomed to failure. But then Mr. Subramaniam intervened and convinced both Bhabha and the Prime Minister that a new institute in the southern part of India was needed and in keeping with the competitive spirit that pervades the United States, as emphasised by my father in his proposal. So the Prime Minister gave the nod. It was decided that the new institute to be called The Institute of Mathematical Sciences (MATSCIENCE for short) was to be inaugurated on Jan 3, 1962, with my father as its Director. Thus the enormous effort of my father in the face of opposition both locally from the University of Madras, and externally from the powerful authorities in Bombay, suddenly and dramatically bore fruit. But in the intervening two months, Oct–Dec 1961, my father kept his Seminar going. In particular there were two very distinguished visitors to the Seminar in the last two months of 1961.

One of the last visitors in December 1961 was Professor Sir James Lighthill (FRS), who was then Director of the Royal Aircraft Establishment in Farnborough. Professor Lighthill was a close acquaintance of my father back from 1949–51 when my father did his PhD at the University of Manchester. Lighthill was a great applied mathematician who was especially known for work in aero-acoustics (a field that he founded) and fluid dynamics. The work that he did at Farnborough was subsequently used in the development of the Anglo-French supersonic commercial aircraft, the Concorde. In the 70's he occupied the prestigious Lucasian Professorship of Mathematics at Cambridge University, a chair occupied by Newton himself. I remember the dinner at Ekamra Nivas when Lighthill gave his support to Mr. Subramaniam of my father's proposal for a new institute.

The final visitor to the Theoretical Physics Seminar, who then was associated with the inauguration of MATSCIENCE, was the great astrophysicist Subrahmanyam Chandrasekhar. He was Morton D. Hull Distinguished Service Professor at the University of Chicago. The Chandrasekhar limit, a fundamental idea astrophysics, determines the mass of a star, heavier than the Sun, that will eventually become a white dwarf. The process of a massive star shrinking under its own gravity, is what ultimately leads to black holes. Chandrasekhar received the Nobel prize in astrophysics in 1983 — a belated recognition for his fundamental work several decades earlier. My father had a great admiration for, and interest in, Chandrasekhar's work, and had worked on applications of probability and stochastic processes to certain astrophysical questions. Chandrasekhar arranged for the publication of my father's work (including joint work with his students) in a series of papers in the Astrophysical Journal of which he was the Editor.

When my father heard that Professor Chandrasekhar was visiting India in December 1961, he invited him to lecture at the Theoretical Physics Seminar. Professor Chandrasekhar said he would do so on one condition — that he be given a sit down South Indian style dinner served on a banana leaf! This was of course the easiest thing to arrange at Ekamra Nivas.

Chandrasekhar gave two lectures at Ekamra Nivas of two hours each. He had a very formal style of lecturing. He was in a full suit while lecturing in the Seminar. But he wore a traditional dhothi to enjoy the dinner at Ekamra Nivas served on a banana leaf. My father invited Professor Chandrasekhar to inaugurate the new institute, and he graciously accepted this invitation.

The inauguration of MATSCIENCE on Jan 3, 1962 was a dream come true for my father. It was a miracle not just in Indian science but in the world of science. A few days before the inauguration, telegrams and letters from prominent scientists around the world poured in — from Nobel Laureates Niels Bohr, Werner Heisenberg, P. A. M. Dirac, and others, as well as from all who spoke at the Seminar (see [B2]).

My father, who was an orator par excellence, delivered the finest speech of his life entitled "The miracle has happened". As a six year old boy, I was seated in the front row of the Old English Lecture Hall of the Presidency College in Madras, as my father delivered this speech extempore, as was his custom. That speech made such a big impression on me. I know several parts of it by heart! I decided at that moment, that I would pursue a research career. This decision was reinforced when I had the opportunity to travel internationally with my parents in the next decade, visited several leading academic institutions, and observed the leaders in the world of science at work. This is described in the next section.

In [B2], the inaugural speech of my father as well as a selection of telegrams received for the inauguration of MATSCIENCE are reproduced. MATSCIENCE celebrated its 50th Anniversary with a conference in Jan 2013. As one who had witnessed the birth of the institute, I was invited to deliver the opening lecture at the conference. On that occasion, in my one hour talk, I described in full detail the activities of the Theoretical Physics Seminar, all the visitors to the Seminar, and the efforts of my father to run the Seminar and create the new institute.

1.2   International travel with my parents (1962–70)

My father had travelled overseas on academic assignments while at the University of Madras, but after he became the Director of MATSCIENCE, he made annual trips abroad lecturing at several leading academic centers around the world on his work and those of his students. These trips gave him not only an opportunity to discuss his work with experts worldwide, but also enabled him to contact scientists whom he could invite to his institute. From 1962 until 1970, I had the pleasure of accompanying my parents on these trips, which with the exception of one, were all around-the-world. On these trips I had the opportunity to meet and observe great scientists at work in their home institutions, and get to know some of them well. Here I shall describe some unique experiences. In doing so, I will classify these trips chronologically and in each, describe certain things that made those visits special, and the eminent scientists I met.

1962: This was my first trip abroad and it was around-the-world. Although I was only seven years old, I remember many things vividly.

After briefly visiting Hong Kong, Tokyo, and Honolulu, we went to Los Angeles for a two-month stay. My father was invited by Richard (Dick) Bellman of the RAND Corporation (RAND being an acronym for Research and Development) in Santa Monica, a suburb of Los Angeles.

Dick Bellman was a prolific and reputed applied mathematician. He had become famous for his theory of dynamic programming in 1953, and published the first and definitive book on the subject in 1958. My father and he had interacted in the field of stochastic processes even while my father was doing his PhD in England in 1949–51. In particular there were connections between my father's research and that of Bellman and Ted Harris. That is what prompted Bellman to invite my father to RAND where Ted Harris also was working. Bellman actually started with pure mathematics, having obtained his PhD from Princeton under Solomon Lefschetz. He later moved to applied mathematics where he made notable contributions. He founded the Journal of Mathematical Analysis and Applications published by Academic Press and appointed my father on the Editorial Board.

Bellman was an avid tennis player like mathematician Peter Lax. He loved the good life and dined in the very best restaurants. He took us on a drive through Sunset Boulevard in Los Angeles in his Lincoln Continental, and then to dinner at an exclusive restaurant on La Cienega Boulevard — in a section called Restaurant Row. Bellman was a great friend of my father and asked me and my mother what we would like to see in Los Angeles. While my mother chose the MGM Studios, I said I wanted to see the Douglas Aircraft Factory. I was an aviation enthusiast even then, an interest kindled by my father. Dick Bellman arranged that visit, and I was in awe when I saw the giant Douglas DC-8 in production — an airplane I was soon to fly in. Another highlight of the stay in Los Angeles for me was the visit to Disneyland (in Anaheim) which had opened a few years before. Walt Disney was alive then, and he was a hero to every kid including me.

Murray Gell-Mann who had visited India the previous year and stayed in our home, invited my father to Caltech. There my father met the brilliant physicist Richard Feynman for an afternoon of discussions. Feynman was a legend of modern physics, both for the originality of his work, and for his electrifying lecturing style. Expert audiences and students were spellbound in his lectures. The Feynman Lectures in Physics given during 1961–63 at Caltech are world famous and widely read by students entering the domain of modern physics. Everyone knew that Feynman would get a Nobel Prize, which he did in 1965 for his work on quantum electro dynamics. He shared the prize with Schwinger and Tomonaga.

One of Feynman's ingenious contributions is the explanation of an electron travelling back in time through a technique known now as Feynman graphs or Feynman diagrams. Feynman actually had given a one hour private lecture to my father on this topic back in 1956 when my father first met him. That got my father interested in the problem of trying to understand what it means for an electron travelling back in time. My father used his work on inverse probability to interpret it as tracing back in time in the inverse probability sense. In that process, my father gave a simplified way of classifying the various Feynman diagrams and published this work in the Journal of Mathematical Analysis and Applications.

Gell-Mann hosted a lunch party in honor of my father in his lovely home in the hills of Altadena. Richard Feynman came to this lunch. Feynman had an unusual personality. He was a non-conformist. He never cared for awards and recognitions, and shunned pomp and pageantry. He hated the snobbery associated with awards and honors. Although he attended the Nobel Prize Ceremony and accepted the prize, he resented getting the prize. He had lots of idiosyncracies and inspired the title "Surely You're Joking, Mr. Feynman!" for a book about him.

Feynman was a natural choice for the position of Permanent (Faculty) Member at the Institute for Advanced Study in Princeton. This was a most prestigious and coveted position but Feynman turned it down because he wanted to be in a University so that he could do research and teach as well. At the Institute for Advanced Study there are no teaching duties; the permanent faculty are left alone to ponder on fundamental problems in their fields of expertise. Feynman had openly said that such total freedom without teaching would lead to tremendous pressure especially because everyone goes through periods when research productivity takes a dip. So if there are no teaching duties, the individual might feel a sense of guilt during such dry spells, and owing to this pressure, might not be able to get out of such a dry spell. Feynman loved to interact with students and Caltech attracted the best students from around the world. So Feynman was not inclined to leave Caltech even for a prestigious position without teaching duties in Princeton. Actually even Gell-Mann did not want to leave Caltech, and the combination of Gell-Mann and Feynman made Caltech one of the greatest centers in the world for theoretical physics.

While all the physicists and their spouses were talking near the pool area, Feynman came into the house and sat next to me. I was inside, all by myself, reading a book on dinosaurs. Feynman asked me if dinosaurs interested me and I said they did. He then challenged me to a game: I need to name a dinosaur for every dinosaur he named. He won, but I held on for quite a while. He then picked up the book on dinosaurs that I was reading, turned it upside down, and read the book at normal speed. You must be joking, Mr. Feynman! That was my first and only meeting with Feynman, and I was charmed by his personality.

The dinosaur book that I was reading was not mine. It belonged to the Gell-Mann family. My father later told me that Gell-Mann very much wanted to pursue archeology, but was advised to take to physics because he was so strong in mathematics — as demonstrated by his performance in the Putnan Exam in mathematics as an undergraduate at Yale. Gell-Mann took this advice seriously and went on to do revolutionary work in particle physics. But he continued to have a strong interest in archeology.

It was during our stay in Los Angeles that we heard the sad news that Professor Niels Bohr passed away. My father was much affected on hearing this. He wrote a condolence letter to Niels Bohr's son, Aage Bohr, himself a distinguished physicist, who some years later also won the Nobel Prize.

From Los Angeles we flew to San Francisco by United Airlines DC-8, my first of many flights on that magnificent aircraft. In the Bay Area, my father visited Stanford University and the University of California, Berkeley, for two weeks each. I met several eminent physicists — Leonard Schiff and Nobel laureates Robert Hofstadter in Stanford, and Emilio Segre in Berkeley.

From California we went to Washington DC for a few days before returning to India via Europe. In Washington we stayed at the home of Professor Ugo Fano, our host. When we entered his home, Professor Fano turned towards me and asked, "What would you like to have by way of food?" Without any hesitation I responded saying that I would like to have *Rasam*, which is a spicy South Indian lentil soup. The Fanos graciously offered the kitchen to my mother so that she could prepare rasam for me. My father took me aside and scolded me. He said that it was improper of me to have made such a request when we were guests, and that I was overextending the welcome. As a seven-year-old, I took Professor Fano's offer literally. When Prime Minister Nehru asked my father what he wanted, my father took that offer seriously and said he wanted a new institute. My request was much simpler!

From Washington we moved to Europe and did the trans-atlantic flight to Paris on a Pan American DC-8. After a few days in Paris where my father was the guest of Prof. Claude Bloch at Saclay, we moved to Geneva and to Rome, our final stop, before returning to India.

This round-the-world tour opened my eyes to the scientific and technological advances of the Western World. I saw first-hand several eminent scientists in their home institutions.

1965: My next trip abroad with my parents was in the summer of 1965, to Europe, with the main part being a two-month stay in Trieste where my father was invited to the International Centre for Theoretical Physics (ICTP) founded by Abdus Salam the previous year. Salam's initial desire was to create such an institute in Pakistan, his home country, just as my father created MATSCIENCE in Madras in 1962. But conditions were much more favorable for him to set it up in Europe. Back in 1960, Salam invited a group of visionary scientists including my father to a meeting at the lovely Miramare Castle outside of Trieste to discuss the idea of a center like the ICTP. Subsequently, Salam sent around his proposal to several countries. In the end, Salam succeeded in creating the ICTP, not in Pakistan as he originally intended, but in Trieste, Italy, with significant support from UNESCO. The ICTP had as its mission the development of science in the Third World, and that is still its core mission today.

Salam had great admiration for my father's research and service to the profession. He was such a generous host that he offered my father a First Class air ticket between India and Trieste. My father was a vegetarian and a teetotaler, and so the wines served during the "Maharaja Service" in First Class on Air India were no attraction to him; so on the Boeing 707 flight, for most of the time, he sat in Economy class with my mother and me! Our Air India flight from Bombay to Rome was via Beirut. We arrived in

Beirut at the crack of dawn and at the airport shopping arcade, my father bought me a Tissot automatic watch — I was eight years old then — and I wore it with pride for many years.

Prior to going to Trieste, we visited Rome and Naples where my father gave lectures at the invitation of distinguished Italian physicists Francesco Calogero and E. Caianiello. Besides being an eminent theoretical physicist, Calogero was Secretary General of the Pugwash Conferences on nuclear disarmament; in 1998, he accepted the Nobel Peace Prize on behalf of Pugwash. Caianiello was the Founder and Director of several research institutes in Italy, one of which being the Institute of Nuclear Physics in Naples, where he hosted my father in 1965. I remember the wonderful trip around Naples in his Alpha Romeo car. After the brief but very enjoyable stays in Rome and Naples, we proceeded to Trieste.

The ICTP was located in 1965 in a building in Piazza Oberdan in the center of Trieste. Like MATSCIENCE, the ICTP had an impressive flow of international scientists from the beginning. Salam was a gracious host. He had regular parties at the ICTP some of which I attended with my parents. I remember one party that he hosted for Robert Marshak of Rochester, who was not only an eminent physicist, but a great statesman for the discipline. Like Bellman, Salam loved the good life and entertained his guests at dinners in expensive restaurants. I remember joining my parents for dinners hosted by Salam in Trieste.

During the two months in Trieste, I picked up Italian, not so much as to speak the language fluently, but good enough to help my parents while ordering vegetarian food in restaurants, or when making purchases at stores. I quickly learnt to express numbers in Italian and the names of many vegetarian items. When you are young, you absorb like a sponge.

My mother and I accompanied my father regularly to the ICTP in the mornings. We would relax in the Lounge which had a wonderful collection of newspapers like the London Times and magazines like the New Yorker. The Lounge also afforded a nice opportunity to converse with the families of other visitors. At that time, the ICTP was small, and so one could get know the visitors quite well.

After a very enjoyable stay in Trieste, we flew to Geneva, Switzerland, via Zurich because my father was invited to CERN (Council of European Nuclear Research). Interestingly on the segment from Milan to Zurich, Murray Gell-Mann was a co-passenger! What I remember about this Swissair Caravelle flight was that the cabin was very hot; it was not a serious mechanical problem, but that the crew had set the thermostat incorrectly! It was set right after several passengers including Gell-Mann complained.

CERN is a sprawling experimental lab set up in 1954 with the support of several European nations. It has a gigantic accelerator used to observe collisions of particles. CERN experiments have confirmed important theoretical predictions, and therefore have been responsible for the award of some Nobel Prizes. Even today, CERN is a dominant center for experimental physics. Indeed the elusive Higgs Boson was observed at the Large Hadron Collider in CERN in 2013 and soon led to a Nobel Prize.

What was special about this visit was that our hosts were Professor Victor Weisskopf, Director General of CERN, and Rolf Hagedorn, one the active scientists who was at CERN since its inception in 1954. Hagedorn was previously at the Max Planck Institute, working with Nobel Laureate Werner Heisenberg. When CERN was launched in 1954, Heisenberg recommended Hagedorn from his group to the new experimental set up, and Hagedorn stayed at CERN until his retirement. Weisskopf, who worked under Nobel Laureates Max Born and Eugene P. Wigner, was a great force in physics. He would have received a Nobel Prize in physics had he published his results on the Lamb shift; he did not publish it because he was not entirely sure of its mathematical correctness, but he had done everything correctly as was realized later. Prior to coming to CERN, Weisskopf was at MIT, where he was department head in physics. While at MIT, he had Murray Gell-Mann as his PhD student. So Weisskopf was the academic father of a Nobel Laureate.

Both Weisskopf and Hagedorn had visited MATSCIENCE (Weisskopf in January 1965, and Hagedorn in 1964), and were so impressed that they invited my father to CERN. To me it was a wonder to see the CERN accelerator as we were taken around by Weisskopf and Hagedorn. After the tour of the accelerator, my father gave a talk to a group of experimental and theoretical physicists in CERN. That evening we were entertained to a marvellous slide show of India at Hagedorn's residence after dinner. These included pictures he had taken on excursions with us in South India. To reciprocate the sightseeing trips we made in India together, the Hagedorns took us on an excursion to Mont Blanc. The next day, we had dinner at Professor Weisskopf's home, where we met several scientists with their spouses. Of note was the presence of Dr (Mrs.) Sulamith Goldhaber of Berkeley. She was a world authority on K-mesons and nucleons. It was at that dinner that my father suggested to her that she and her physicist husband Gerson Goldhaber (also a professor at Berkeley) should visit MATSCIENCE which they did in November that year. We received them at the Madras airport and had a wonderful evening together on the night of their arrival. Tragically, within a few days of their arrival in Madras, Sulamith Goldhaber suffered a stroke owing to a brain tumor and died in Madras. In an obituary of Sulamith Goldhaber prepared by physicist Louis Alvarez of Berkeley for the Physical Review, it is stated that the Goldhabers were at the Hebrew University of Jerusalem in November 1965 and preparing for a series of lectures they were to give in Madras. Those lectures were to have been delivered at MATSCIENCE, but could not be given since the visit of the Goldhabers was cut short by the tragedy. Nevertheless, Prof. Gerson Goldhaber graciously sent the notes of their lectures and their 50-page paper was published in Volume 6 of the series *Symposia in Theoretical Physics and Mathematics* edited by my father and published by Plenum Press, New York.

From CERN we moved to Paris where my father had lectures at Saclay and Orsay at the invitations of Professors Michael Gourdin, Philip Meyer, Cirano de Dominicis, and Claude Bloch, each a distinguished physicist.

The flight from Paris to London was memorable. We were in a British European Airways (BEA) Trident aircraft, and they tested the Instrument Landing System (ILS) as we landed at London Heathrow Airport. At my father's suggestion to the cabin crew, I was taken to the cockpit for a few minutes after the crew had set up the ILS procedure. Those were days when flying was a pleasure — no threats of hijacking and no security screening.

The Trident was a three engined short to medium distance jetliner designed by de Havilland, but by the time the Trident came into operation in the early sixties, de Havilland was taken over by Hawker Siddeley. The Trident was actually a precursor to the three engined Boeing 727, but unfortunately the Trident was not a commercial success even though it preceded the 727 and was very similar; the 727 was the best selling airliner of its time. About 100 Tridents were sold for commercial operations. BEA was the primary operator of the Trident. The ILS was termed as Autoland by the British, and the Trident was the aircraft used to test the ILS.

In London, my father took me to a departmental store called Barkers and bought me a fine cricket bat. It was made by Gray Nicholls and autographed by Ted Dexter, an electrifying batsman and a captain of the England team in the late fifties and early sixties. I used that bat effectively and with pride when I played cricket in my school. My classmates used to joke that even my modest success as a batsman was due to the Dexter cricket bat and not due to my talent!

From London, we returned to India after a brief stop in Athens, Greece, for sightseeing. The flight from London to Athens via Rome was equally memorable. We were on board a BEA Comet 4 and flying in delightful weather. We flew over the island of Elba which was the scene of a fateful crash of a Comet 1 of BOAC, and also where Napoleon was exiled.

The Comet 1 which actually ushered in the Jet Age of commercial aviation as early as 1952 with BOAC (British Overseas Airways Corporation), unfortunately suffered crashes in the next two years. Investigations revealed that these crashes were due to metal fatigue under pressurization of the cabin at high altitudes. The defect was rectified and the Comet 4 evolved. But the Comet never recovered from the damage to its reputation, and only a few were sold. BEA and BOAC were the primary operators of the Comet. Both Boeing and Douglas benefited from the investigations on the metal fatigue of the Comet and made design changes to their Boeing 707 and Douglas DC-8 to avoid this problem.

1966 and 1967: I will discuss my two overseas trips in 1966 and 67 together because there many things in common. I will point out some special features of these trips.

Both trips were around-the-world and in the Westbound direction. On each of the two trips we visited more than a dozen universities and research centers where my father was invited, and attended international conferences on high energy physics. In 1966 we had an extended stay in Syracuse where my father was invited by George Sudarshan.

Sudarshan and my father became friends when they met in 1957–58 at the Institute for Advanced Study in Princeton where my father was visiting for that academic year. In addition to discussing the most exciting problems of physics, Sudarshan and my father also discussed the scientific scene in India — focussing on problems faced by scientists in a sterile atmosphere, and how to go about changing that. After my father created MATSCIENCE in 1962, he invited Sudarshan as its first visiting professor in 1963. Sudarshan reciprocated by inviting my father to Syracuse for two months in 1966.

Sudarshan received his PhD from the University of Rochester under the direction of the famous physicist Robert Marshak. In collaboration with his thesis advisor Hans Bethe, Marshak had done fundamental work on fusion, and so was invited to be part of the Manhattan Project (to build the atom bomb) along with his advisor Bethe. In 1957, Marshak and his PhD student Sudarshan proposed a new theory of weak interactions called the V-A theory and published it in the proceedings of a conference in Italy where they presented this work. Six months later, Murray Gell-Mann and Richard

Feynman derived the same result by a different method and published it in a leading journal. The V-A theory is considered extremely important but no Nobel Prize was given for it. It definitely propelled Sudarshan to stardom. The Nobel Prizes awarded later to Feynman (1965) and Gell-Mann (1969) were for their other work. Also the V-A theory paved the way for the theory of the electro-weak force for which Sheldon Glashow, Abdus Salam and Steven Weinberg received the Nobel Prize. I remember several conversations in Syracuse and elsewhere when Sudarshan spoke to my father about the politics of the Nobel Prize.

During our stay in Syracuse, we made two trips by car, one to New York, New Jersey and Washington D.C., and another to Rochester. On these trips, my father's former student Jagannathan (who was then in Syracuse for his graduate studies) accompanied us. While in New Jersey, we were invited to a lunch hosted by President Robert Goheen of Princeton University at the stately Nassau Inn, a Princeton icon. Goheen was born in South India in the state of Andhra Pradesh and attended one of the best high schools in Kodaikanal in the state of Madras. Later under the Jimmy Carter presidency, he served as US Ambassador to India. He had a deep understanding of India both culturally and economically, and so my father and he were able to discuss several aspects of the development of science in India.

While in Princeton, we went to the Institute for Advanced Study for a few hours. My parents showed me the apartment where they stayed in 1957–58, and the office my father occupied at the Institute next to Fuld Hall. Little did I realize that about 15 years later, I will be visiting the Institute for Advanced Study for a year.

During that afternoon at the Institute, we were invited to the apartment of T. S. Bhanumurthy, an Indian mathematician working on the representation of semi-simple Lie algebras. He was the older brother of T. S. Shankara who was doing his PhD in physics under my father at that time. I was surprised to see my elementary school teacher Ms. Kalyanalakshmi, who had now become Mrs. Bhanumurthy! As a school teacher she was a very strict disciplinarian like her twin sister Anandalakshmi. As young children, we were very scared of them. So I was a bit reticent at first on seeing Ms. Kalyanalakshmi in Princeton, but then when she was all smiles and laughs, I got to relax. It was at that lunch that Bhanumurthy told my father that he was returning to India upon completion of his stay in Princeton, and he enquired about positions at MATSCIENCE. Bhanumurthy became the Director of the Ramanujan Institute in Madras in 1969, and his first term lasted until 1975. I did interact with him during my undergraduate days (1972–75) when I began some research investigations in number theory.

Our other trip from Syracuse to Rochester was equally interesting. Professor Robert Marshak hosted us magnificently not only in Rochester, but took us to the Niagara as well. He and his wife visited India in 1963 as our guests and wanted to reciprocate the hospitality in equal measure. We had taken the Marshaks to our village 60 miles North of Madras and shown them our farm lands. To reciprocate this in equal measure, Marshak took us to his farm house. Marshak also arranged a meeting for us with McCrea Hazlett, the Provost of the University of Rochester. Hazlett and his wife had visited Madras twice as our guests in 1961 and 64, and so they took us to dinner at a fancy restaurant.

Marshak was not just a great physicist, but a statesman for the discipline. Starting in 1950, he organized a series of international conferences on high energy physics in Rochester. These conferences are now held once every two years in different parts of the world, but the first seven of them from 1950 to 57 were held in Rochester, and so they are also referred to as the Rochester Conferences. In 1966, the (Rochester) High Energy Physics Conference was held in Berkeley, and my father participated in it. At the conference banquet which I attended with my parents, Marshak spoke and reminisced about the origins of the conference.

In addition to the conference banquet, there was a dinner at the home of Nobel Laureate Emilio Segre to which we were invited as part of a small group. Segre had visited MATSCIENCE in 1962 while we were in the United States. He was impressed with the Institute my father had created, and therefore wanted us to have dinner at his home during the 1966 conference. Nobel Laureate Don Glaser of Berkeley, who had visited Ekamra Nivas in 1961, was at Segre's home for that dinner. Next year (1967), when we visited Berkeley again, Don Glaser invited us to dinner at his home to reciprocate our hospitality to him in Madras in 1961.

Prior to the conference in Berkeley, my father visited Stanford University at the invitation of Professor L. I. Schiff, who was the Chairman of the Physics Department. Schiff had also visited MATSCIENCE in 1963 when Marshak was there. Schiff's book on quantum mechanics is a bible in the field, and my father lectured from it at his Seminar. Schiff was a gentleman to the core and very soft spoken. At a party he hosted for us, his colleague Robert Hofstadter, a Nobel Laureate, was present. Hofstadter was very extroverted. He was extremely jovial, and had a most infectious laughter.

During our stay at Stanford, we were taken around the Stanford Linear Accelerator (SLAC). It was interesting to see sections of this being built and compare it with what I had seen the previous year in CERN in Europe. The Schiffs also took us on a sightseeing journey along the beautiful Pacific coast, covering lovely towns like Monterrey and Carmel. Years later, as a mathematician, I would be back in these towns during my participation in the West Coast Number Theory Conferences at the Asilomar Conference Grounds in the Monterrey peninsula (see Section 3.3).

Even though the High Energy Physics Conferences had moved out of Rochester and were conducted at leading centers worldwide, Marshak and his group were still actively arranging other conferences in Rochester. One such big meeting was the International Conference on Particles and Fields in the summer of 1967 that my father attended during the round-the-world academic tour that year. The conference opened with the scintillating lectures of Richard Feynman and Julian Schwinger who had shared the 1965 Nobel Prize along with Tomonaga for their groundbreaking work in quantum electro-dynamics. I confess that I did not attend these lectures, but my father told me and my mother how exciting they were. I did attend the conference banquet along with my parents and there I saw Feynman, Schwinger and other luminaries. During the day, when the sessions were going on, my mother and I joined the tours that were arranged for the families of the participants. One such tour was to see the stately homes of Rochester and this was organized by Antoinette Emch, wife of Gerard Emch, a young mathematical physicist on the Rochester faculty. Interestingly, nearly two decades later, when I joined the University of Florida in December 1986, Gerard Emch had just assumed the chairmanship of the mathematics department there, and so my wife Mathura and I were welcomed by the Emchs on our arrival in Florida.

Prior to the conference in Rochester, we were at the University of Illinois in Urbana where my father gave a colloquium in the physics department at the invitation of Professor Ravenhall who had visited MATSCIENCE earlier in the year in February. That evening, there was a party at the home of Ravenhall for which he had invited several faculty including Paul Bateman, Chair of the Mathematics Department. Bateman was a noted number theorist, and was Chair at Illinois for many years over the course of which he not only built up the department but expanded the number theory program as well. Later as a number theorist I had several fruitful interactions with Bateman, but that party at the Ravenhall's was my first meeting with him. I was eleven years old then.

Even through the 1966 trip was round-the-world, we did not stop in Europe but went straight to America via Europe. However, in the 1967 trip, we visited Europe and had a two month stay in Trieste where my father was again invited by Abdus Salam to the ICTP. The special feature of our stay in Trieste was the visit of Mr. C. Subramaniam who was touring Europe. He was at that time a minister in the Cabinet of Prime Minister Indira Gandhi. At my father's suggestion, Abdus Salam invited Mr. Subramaniam to the ICTP, and the Minister graciously accepted to come to Trieste for a few days. Leading politicians often speak about their support for science and fundamental research, but rarely do they take the time to visit scientific institutions in a relaxed manner and get to know the researchers. Mr. Subramaniam was of a different mold; he was passionately interested in fundamental scientific research, and used every opportunity to visit the major centers worldwide. During Mr. Subramaniam's stay in Trieste, Salam put his personal car (a lovely Peugeot) along with his chauffeur Joe, at our disposal to show Mr. Subramaniam around Trieste. We took Mr. Subramaniam on an excursion to the Postagna caves in neighboring Yugoslavia; these caves are world famous for their spectacular limestone formations called stalactites and stalagmites.

Mr. Subramaniam was a close friend of our family and so he stayed in our apartment during his visit to Trieste. My mother not only prepared Indian food, but also served him Mellanzane Parmagiana (Eggplant Parmesan as they say in the USA), a baked vegetarian dish of aubergines, cheese, and tomato sauce. This is a favourite Italian dish to many, and Mr. Subramaniam said it was the best European dish that he had tasted. Mr. Subramaniam departed from Venice which is quite close to Trieste, but before his flight, we showed him some of the great sights of this truly marvellous city set on a labyrinth of canals.

After Trieste, we were in Paris, where my father had his usual round of lectures at Saclay and Orsay. This time we were invited to a dinner at the residence of Dr. Malcolm Adiseshiah, Deputy Director General of UNESCO. The discussion there centered on the scientific climate in India. Three years later, after retiring from UNESCO, Adiseshiah settled down in Madras. There he founded the Madras Institute of Development Studies (MIDS) aimed at developing and supporting initiatives in education at all levels. This early contact in Paris led to a close relationship between Adiseshiah and my father. Adiseshiah subsequently became the Vice-Chancellor of Madras University and his term as VC intersected with my father's term as Director of MATSCIENCE.

While Director of MIDS, Adisheshiah held a meeting in November 1971 in which he asked for suggestions to improve scientific research in the state of Tamil Nadu (Madras). At that meeting my father proposed a Science Policy for Tamil Nadu. Adisheshiah liked this proposal very much and the outcome was the creation on the Tamil Nadu Academy of Sciences in 1976 when he was the Vice-Chancellor of the University of Madras. My father played a major role in developing this Academy and served as its first Vice-President. Several of the meetings and lectures of the Academy were held at MATSCIENCE during the initial years and I have attended many of those meetings and talks. In fact I too have given talks on my research at the meetings of the Academy during my annual visits to Madras.

A special feature of the 1967 trip was the visit to Boeing in Seattle, where my father was invited to give talks on his work related to queueing theory. This was the first of three annual visits to Boeing. During the 1967 visit, we were shown the mock-up the Boeing Supersonic Transport (SST). The SST was called the Boeing 2707, because it was to be exactly twice the length of the Boeing 707 which so successfully initiated the jet era in commercial transport. Boeing eventually decided not to produce the SST because it was not commercially viable. At the Boeing factory, I was presented an impressive artist's rendition of the SST. This adorned the wall of my study in Ekamra Nivas along with other aeroplane pictures.

Prior to the visit to Seattle, while we were in Los Angeles, my father made his initial contact with the research group at the Douglas Aircraft Corporation in Long Beach — which had just become McDonnell Douglas. Dr. Dave Pandres of McDonnell Douglas, our host, was also interested in the queueing theory aspect of my father's work, and invited him to Douglas during the next few years. I also had occasion to see the Douglas assembly line at Long Beach, each time we visited McDonnell Douglas. In 1967 we saw the DC-9 and the Super DC-8 in production. These visits to the Boeing and Douglas factories increased my boyhood passion for aviation. Even though I had met so many theoretical physicists, I strongly desired to pursue aerodynamics.

1968/69: The trip during these two years were round-the-world in the easterly direction. I recollect some very special experiences.

On both these trips, we broke our trans-pacific journey in Honolulu, where my father gave talks at the University of Hawaii at the invitation of Professor T. K. Menon, who is an astrophysicist. Construction of the magnificent observatory on the top of the extinct volcano Mauna Kea on the neighboring Big Island of Hawaii had begun in 1967, and this naturally boosted the program in astronomy at the University of Hawaii. Menon had joined the University of Hawaii in 1968, and we had the pleasure of staying at his lovely home in the Aina Haina section of Honolulu in 1968, 69 and 70. After a few years in Hawaii, Menon joined the radio astronomy group of the Tata Institute in India, but finally settled down in Vancouver as a professor at the University of British Columbia.

One of the highlights of the 1968 trip was the visit to Boeing in Seattle where we got to see the prototype Boeing 747 Jumbo Jet in production. Boeing's production of the 707, 727, and 737, were all at Renton, but the 747 required an entirely new set up, and so Boeing built a new plant in Everett — the building being the largest in the world (at that time) under a single roof. The emblems of the various airlines that had ordered the 747 were painted on the fuselage and it was exciting to see this gigantic airliner in production. The 747 made its debut in January 1970 with Pan American which put out a slogan — "First in the world with the 707, first in the world with the 747."

Another special feature of the 1969 trip was a visit to the McDonnell Douglas factory in Long Beach. Douglas which was famous for its line of piston engined aircraft from the DC-3 to the DC-7, had an assembly line for them in Santa Monica. But with the production of its first jet, the DC-8, it moved its assembly line to Long Beach. In 1969 after we were shown the assembly lines of the Super DC-8 and the short range DC-9, we got to see the mock up of the DC-10 trijet. The prototype DC-10 flew in 1970, and was delivered simultaneously to its launch customers American Airlines and United Airlines in July 1971. Douglas decided to close the assembly line of the DC-8 by 1972 in order to focus on the production of the DC-10 on which it had invested so much.

The 1969 trip had another factory visit — the Cadillac plant at General Motors in Detroit, where my father gave a talk. In the audience was mathematician Garrett Birkhoff of Harvard University who was a consultant to GM. He was one of the most influential algebraists, and the son of one of the greatest American mathematicians — George David Birkhoff. In a few years, as a graduate student, I was to read the classic book of (Garrett) Birkhoff and Saunders Maclane on Algebra.

Cadillac is the finest car produced by General Motors and so they marketed it as "The Mark of Excellence". I was so impressed with the Cadillac plant I saw, that when I returned to Madras, I wrote an article for my school magazine entitled "The Mark of Excellence".

These trips had their full share of visits to top universities. In 1968, my father was invited to Cornell by Professor Wolfgang H. J. Fuchs, a well known complex analyst. Fuchs had visited MATSCIENCE earlier that year, liked what he saw, and invited my father to Cornell. Since my father's talk was on mathematical physics, it was attended by Nobel Laureate physicist Hans Bethe. My father invited Bethe to MATSCIENCE, and he accepted the invitation to visit in 1969.

After visiting several universities in America in 1968, we spent a few weeks in Europe. The flight from Washington Dulles to London Heathrow was by a Pan American Douglas DC-8. Pan Am was the first airline to order the DC-8 in 1955, but in 1968, it phased out all of its DC-8s. So it was a rare opportunity for me to fly in this historic aircraft on one of its last flights with Pan Am.

In London, during 1967 and 1969, our host was Sir James Lighthill who at that time was Royal Society Professor at Imperial College, London. Professor Lighthill was the founder of the field of aero-acoustics. He told us that he would often swim in the English Channel, and in that process observe the hydrodynamics of the fishes. Ironically, three decades later, Lighthill died of a heart failure while swimming in the English Channel around the island of Sark off the coast of Normandy.

When Professor Lighthill picked us up at London Heathrow Airport and took us to our hotel, my father asked him whether he liked life in London or in Cambridge where he was previously. In his impeccable British accent, Lighthill replied, "I prefer the tranquillity of Cambridge to the rat race of London." A few years later, Lighthill became Lucasian Professor of Mathematics at Cambridge University.

In 1967 and 69, on the transatlantic segments between London and Montreal, we flew by the VC-10, the British counter to the Boeing 707 and the Douglas DC-8. The VC-10 was produced the British Aircraft Corporation (BAC) in response to a requirement by BOAC of an inter-continental range aircraft capable of take-off from hot and high air fields like Nairobi (elevation 5000 ft above sea level). Although technologically superior to the 707 and the DC-8, the VC-10 was not a commercial success because of high operating costs. But with all its four engines in the rear, it looked extremely elegant, and was very quiet for passengers sitting in the front. On my first flight on the VC-10 in 1967, I joined the *Junior Jet Club* of BOAC. I was invited by the captain into the cockpit, and he entered the miles flown in my logbook and signed it. I still have that logbook of BOAC.

In 1968 in Europe we spent three weeks in Trieste at the invitation of

Abdus Salam. The International Centre for Theoretical Physics had now moved out of the modest offices in Piazza Oberdan to its own imposing buildings in Grignano on the outskirts of Trieste. To mark the occasion of this move, Salam organized a major conference in physics in which my father participated. At the conference party that I attended with my parents, I got to meet Nobel Laureate P. A. M. Dirac.

Dirac, a legend of twentieth century physics, received the 1938 Nobel Prize for his theory of the positron. During his visit to Madras in 1954, at a lecture at the Madras University that my father had arranged, one of the members in the audience remarked during the question period that when it seems so natural to postulate an anti-electron, namely the positron, why did it need a Dirac to develop the idea. Dirac patiently responded by explaining how physical theories evolve, how mathematics helps in the formulation of physical theories, and how such formulations have to be substantiated later by experiment. Dirac's view was that great physical theories are based on beautiful mathematical equations. He often emphasized the role of mathematics in physics.

During the party at the ICTP in Trieste, in a conversation with Dirac, my father casually mentioned that the next week on our way to Moscow, we will be having several hours of transit in Milan. Dirac immediately said that we should utilize the time in Milan to go into the city and see the Last Supper — the famous painting of Leonardo da Vinci. My father said that this was a great idea. I was quite upset at my father because a few days earlier I had suggested exactly this, and my father did not pay attention to my suggestion. He felt it was too much trouble to go into the city of Milan during the transit time. But now that Dirac said that we should see the Last Supper, my father agreed to do this. In my eighth grade class in school in Madras in 1967–68, we studied about Leonardo da Vinci's great paintings, and that is why I wanted to see the Last Supper. In any case, we did get to see this remarkable work of da Vinci.

I now relate a story of Dirac's visit to Madras told to me by my father. Dirac had keen interest in getting to know other cultures and traditions. For example, even though he was no believer in religion, during his visit to Madras in 1954, he went to the Hindu temple of Thirukazhikunram to see the eagles come at a specific time to eat the food offered by the priests. To the Hindus eagles are sacred because of the belief that Lord Vishnu, the Protector, travels through the heavens on the back of an eagle. So at this temple they offer (vegetarian!) food to the eagles — namely, rice. Actually, one has to walk up the hill in Thirukazhikunram to see the eagles. Mrs. Dirac felt that the climb was too strenuous, but Dirac did it since he was all set to see the eagles. On the way back from that temple, Dirac stopped at a village where he was fascinated to see a potter skillfully manipulate his wheel to produce clay pots; right beside the potter was a villager getting his haircut and a shave from a barber right in the middle of a busy intersection. Dirac was amused to see all this — sights he could not get in Cambridge!

After our stay in Trieste, we went to Russia via Milan. My father's book on cascade theory had been translated into Russian, and there was a substantial amount of Russian money (rubles) that had accumulated. In those days one could not take rubles outside of Russia (this rule was enforced to artificially maintain the value of the ruble) and so The Russian Academy of Sciences invited my father for a week to Moscow and Leningrad and it was a unique opportunity for me to accompany my parents to Russia at the peak of communist rule.

The flight from Milan to Moscow was by Alitalia DC-8. Upon arrival at Sheremetvevo airport, the officials collected our passports and said that they would be returned when we departed a week later. We were definitely concerned not to have our passports with us. But we were met at the airport by Dr. I. J. Roysen, of the Lebedev Physical Institute, Moscow, who was with us all the time. He was extremely friendly and helpful, and showed us around Moscow — Lenin's tomb, Red Square, the Bolshoi Ballet, the Russian circus, and museum of aviation. I remember the mile long queue at Lenin's tomb, but when the police saw my mother in a sari, he understood that we were tourists, and so arranged for us to approach the tomb directly without standing in the queue. In India, VIPs bypass queues, and so I felt that this was like the special treatment for VIPs in India.

My father's lecture at the Academy of Sciences in Moscow was arranged by Professors I. M. Khalatnikov of the Institute of Theoretical Physics and by E. L. Feinberg of the Lebedev Physical Institute, the latter having visited MATSCIENCE in 1964 as the leader of a Russian delegation. So the visit to Moscow was hospitality reciprocated.

From Moscow we went to Leningrad for two days. My father wanted me to experience both the air and rail journeys in Russia. So we went by an overnight train to Leningrad and returned by an Aeroflot Tupolev TU-104 aircraft, arriving in Moscow at Domododevo airport. While in Leningrad, we saw the Hermitage museum — comparable to the Louvre in Paris. My father then arranged for us to be on the shores of the Baltic Sea in Leningrad and reminded me there of Churchill's famous words: "From

Stettin in the Baltic to Trieste in the Adriatic, an iron curtain descended on the continent." Just the previous week we were on the Mediterranean coast in Trieste. Even in Leningrad we were accompanied by a representative of the Academy everywhere. Our hosts in Leningrad were Professors V. V. Novoshilov, Vice-Chancellor of Leningrad University and V. N. Gribov of the Physico-Technical Institute with whom my father had discussions about his work on cascade theory.

When I mentioned to my Russian colleague Alexander Berkovich at the University of Florida that in Russia we were always accompanied by a scientific representative, he exclaimed, "Krishna, these were representatives from the KGB!"

When we reached Sheremetvevo airport to depart for India, we were relieved to get our passports back. The Air India Boeing 707 flight from Moscow to Delhi was most memorable. I remember enjoying an excellent Indian breakfast prior to landing in Delhi as we were flying over the snow-capped Pamir Knot, the Hindu Kush, and the Himalayan mountain ranges.

In 1970 as well, I had another round-the-world trip with my parents, but I do not have anything special to report on that trip.

These international trips with my parents gave me an opportunity to observe some of the greatest scientists in their home turf, and to visit some of the leading academic institutions worldwide. So this naturally made me keen to pursue an academic career. But then, my passion for aviation was increased by visits to the Boeing and Douglas factories, and by travel on various international airlines. So during my teenage years, I was torn between aviation and pure science. In the end Mathematics prevailed and I will describe how that happened later. In the next section I will talk about some special visitors to MATSCIENCE during 1962–69.

1.3 Visiting scientists to MATSCIENCE (1962–69)

In terms of the vibrancy of the visiting program at MATSCIENCE, the first decade was perhaps the finest. Almost on a weekly basis, leading scientists from all of the world visited the institute for varying durations and gave lectures. Here I will just provide a selection of these eminent scientists and describe certain special features of their visits.

One of the highlights each year at MATSCIENCE was the anniversary symposium. This was held in January because MATSCIENCE was inaugurated on January 3. For each of the anniversary symposia, very eminent scientists were invited to speak. The anniversary symposia were augmented by Summer Schools for which also there were leading scientists as the main lecturers. The lectures given at the Anniversary Symposia and the Summer Schools during the period 1963–68 were published in ten volumes by the Plenum Press, New York, and I will refer to some of the articles in these publications in the sequel.

The origins of Plenum Press go back to the Consultants Bureau in the late 1940s which started by doing translations of scientific work. Subsequently, its founder Earl M. Coleman, did full scale publishing and changed the name of the company to Plenum Press in 1965. Plenum was bought by the Dutch publishing company Kluwer in 1998, and after Kluwer's merger with Springer in 2004, became part of Springer.

During the sixties, I was only a school boy, but my father asked me to attend the popular lectures by these visitors, some of which I did. I interacted with almost all the visitors at various social gatherings at our home and at MATSCIENCE. Actually MATSCIENCE had its guest rooms only from 1969 when it moved into its own building; until then guests were accommodated in hotels or in our house, and I had close interaction especially with those who stayed with us.

In 1963, the First Anniversary Symposium was opened by Robert Marshak. The Niels Bohr Visiting Professorships had just been created at MATSCIENCE, and Marshak was the first such visiting professor. He spoke about Group Symmetries in Physics and contributed a paper to the first Plenum volume. Leonard Schiff from Stanford was also there when the Marshaks visited Madras. The Marshaks and the Schiffs loved their stay and the many sights they saw. We took the Marshaks to our village. They relaxed in our ancestral home where my grandfather was born, and strolled through our mango orchards.

For the Second Anniversary Symposium in January 1964, my father invited Professor Leon Rosenfeld from Copenhagen to inaugurate it. Rosenfeld was known for fundamental work in quantum electrodynamics that predated Dirac. He was also a collaborator of Niels Bohr. His opening lecture for the symposium was on Bohr's contributions to physics. Rosenfeld was the Second Niels Bohr Visiting Professor, and so the choice of his topic for the talk at the symposium was very appropriate.

Rosenfeld was totally impressed with what he saw at MATSCIENCE. He was amazed that so much was being done in such modest surroundings and with an even more modest budget. He said that wherever he travelled in Europe, there was always someone who had just visited, or planning to visit, MATSCIENCE. In the Visitors Book he wrote : "I am most grateful to an audience composed of young men not afraid of asking questions to the point of cornering the lecturer!" It was actually Professor Rosenfeld who suggested that the lectures delivered at the Anniversary Symposia ought to be published and that is how the Plenum volumes came about.

Victor Weisskopf, Director General of CERN, was the principal speaker for the Third Anniversary Symposium in January 1965. He was so much impressed that he wrote in the Visitors Book that my father had created such a fine institute from almost nothing. It was that visit that led Weisskopf to invite my father to CERN later that year (see previous section).

In view of the steady stream of eminent visitors to MATSCIENCE in the early years, the progress was phenomenal. Homi Bhabha, Director of the Tata Institute and Head of the Department of Atomic Energy, also visited the Institute during that period. Even though Homi Bhabha had initially opposed the creation of MATSCIENCE, he agreed to serve on its Board of Governors once the Institute was created. My father very much wanted him to attend a meeting of the Board of Governors and to meet the visiting professors and the faculty of the Institute. This happened on 18 March, 1964. On that night, there was a special dinner in honor of Bhabha at the roof garden of the Dasaprakash hotel. Bhabha was dressed impeccably — in a dark full suit, with a white shirt and tie. He exuded an air of confidence consistent with the high offices he occupied. But he was very relaxed as he conversed with the faculty, visitors, and the students. At that dinner, Mr. R. Venkataraman, Chairman of the Board of Governors, asked Bhabha his frank opinion of the visiting program at MATSCIENCE. Bhabha was quite complimentary about it and that pleased my father immensely. That was our last meeting with Bhabha, because on 24 January 1966, he tragically died in the Air India Boeing 707 flight (AI 101) that crashed into Mont Blanc as it was descending into Geneva. The flight had left Bombay and made a stop in Beirut before taking off for Geneva. Upon take off from Beirut, the pilot realized that the VOR — an equipment that determines the exact position of the aircraft — was not working properly. So on approach to Geneva, he asked the control tower for the exact position of the aircraft. The control tower told him that after crossing Mont Blanc, he could descend. The pilot mis-heard this instruction, thought that he had crossed Mont Blanc, and therefore began the descent which led to the crash in white-out conditions. Quite coincidentally, that crash was just a few hundred meters away from the Air India Super Constellation crash of November 1950! All eleven Air India Boeing 707s were named after famous Himalayan peaks. It is ironic that one of its Boeing 707s should crash into a very famous mountain.

In addition to physicists, several leading mathematicians came to MATSCIENCE. In January 1965, Professor Harish-Chandra of the Institute for Advanced Study, Princeton, visited MATSCIENCE. He is perhaps the greatest Indian mathematician after Srinivasa Ramanujan. Harish-Chandra, like my father, started working with Homi Bhabha for his PhD, and like my father, he left for England. There he did his PhD work at Cambridge University under the guidance of Dirac. During 1947–48, when Dirac visited the Institute for Advanced Study in Princeton, he took Harish-Chandra with him. It was then that he got deeply interested in the mathematics underlying physical theories and soon made a move to the theory of Lie Algebras in pure mathematics. He made several fundamental contributions to Lie theory and established himself as a leader of this field. During his visit to Madras in 1965, Harish-Chandra spoke at the MATSCIENCE Anniversary Symposium on "Characters of semi-simple Lie groups". A write up of his talk was published in the Plenum Volume V.

In early February 1965, Jayant Vishnu Narlikar of Kings College, Cambridge, visited the Institute. He had completed his PhD at Cambridge University under the guidance of the great astrophysicist Sir Fred Hoyle. Hoyle and Narlikar proposed a new theory of the creation of the universe — the steady state theory. There are three theories that have been proposed for the creation of the universe:

(i) The Big Bang Theory which says that everything started when an incredibly dense object exploded with a big bang; ever since all objects in our universe are flying outward, and the universe is expanding forever,

(ii) The Steady State Theory of Hoyle and Narlikar which says that matter is being constantly created to keep the density of the expanding universe nearly at a constant, and

(iii) The Pulsating Universe Theory which states after a certain period the universe will start shrinking under its own gravitational pull, and become a dense object only to explode again.

Of the three theories, perhaps the pulsating universe theory resonates best with the Hindu philosophy of birth and rebirth in an unending cycle. In any case, the Big Bang Theory is the one which is now most widely accepted, because of the evidence provided of the 3-degree Kelvin background radiation. But at that time the steady state theory was making headlines. Narlikar's name was in the news in India wherever he went. His lecture was arranged in the usual seminar room of the Institute which at that time was in one of the upper floors of a building in the Central Polytechnic Campus. My father announced his lecture in The Hindu, on the theme "A new theory of the creation of the universe". About an hour before the lecture, people started entering MATSCIENCE in large numbers — high school students, college teachers, and even retired judges of the Madras High Court (!) to hear about the creation of the universe. The crowd was several times the number that the seminar hall could accommodate. So an immediate decision was made to move the lecture to the nearby Nehru Auditorium of the Central Polytechnic Campus.

In view of the experience with Narlikar's lecture, the next time a popular lecture was arranged by MATSCIENCE, namely that of the brilliant and mercurial Swedish physicist Gunnar Kallen in January 1966, it was at the Nehru Auditorium. This was for the Fourth Anniversary Symposium, and Kallen spoke about the state of modern physics. To have an appropriate person preside over the lecture, my father invited Sir C. P. Ramaswami Iyer, a great lawyer, an outstanding orator, and a famous statesman. He was a contemporary of my grandfather and a close friend of our family. Sir C. P. was very well read, and for his opening remarks he spoke about discovery of zero as one of India's most fundamental contributions to mathematics. Indeed it was the concept of zero that led the Hindus to invent the decimal system of notation. Also attending Kallen's and Sir C. P.'s lectures was the Minister for Education, Mr. R. Venkataraman who later became the President of India. Several years later when my father and I called on President Venkataraman, he said he remembered Kallen's thought-provoking lecture. Such was the impact that Kallen's lecture had. Sir C. P. passed away a few months later in September 1966. Kallen tragically died in 1968 while piloting his own airplane.

Impressed with the progress of MATSCIENCE, Prime Minister Indira Gandhi accepted the invitation to visit the Institute in January 1967. I had the pleasure of attending that reception for Prime Minister Indira Gandhi and presenting her a bouquet of flowers. Her visit was during the Anniversary Symposium and so there were plenty of international visitors including physicists Michael Gourdin from Orsay, France, Georges Charpak from CERN, Ravenhall from the University of Illinois, and mathematician J. H. Williamson from York, England. Charpak later won the Nobel Prize in Physics in 1992, one year after he retired from CERN. At MATSCIENCE, Charpak presented results from recent CERN experiments on the absorption of pions by a passage of nucleons and discussed its significance in the study of nuclear structure. Charpak's lecture appeared in Volume 8 of the Plenum series edited by my father.

In 1968 there was a spate of very eminent mathematics visitors. At the start of the year, we had the pleasure of hosting the brilliant Indian mathematician Shreeram Abhyankar. He was an algebraic geometer who had received his PhD under the guidance of Oscar Zariski at Harvard. Abhyankar was famous for his contributions to the resolution of singularities problem in algebraic geometry. He was known to work on deep and abstract problems in algebraic geometry without the use of sophisticated techniques, and often emphasized that one can get an understanding of algebraic geometry with just high school algebra. He wrote articles expounding this elementary approach to algebraic geometry. I will have more to say about this later when I will describe a discussion I had with Abhyankar in Florida about the elementary methods Paul Erdős used with miraculous effect in number theory (see §9.1).

Even though Abhyankar had settled in America as a Professor at Purdue University, he had a deep regard for Indian culture and traditions. In particular, he had great respect for the work of ancient Indian mathematicians like Bhaskara and Brahmagupta. In the seventies he started a mathematics institute in his hometown of Poona (now Pune) in Maharashtra, and named it *Bhaskaracharya Prathistama*. Because of his strong ties to India and his deep regard for Indian traditions, he very much admired my father's efforts in creating MATSCIENCE.

Abhyankar had visited MATSCIENCE previously in 1963, but in 1968 he and his wife stayed at our home. During the 1968 visit, at my father's suggestion, he gave a talk at the Sir C. P. Ramaswami Iyer Foundation on the contributions of the Indian mathematical genius Srinivasa Ramanujan. This Foundation was formed soon after Sir C. P.'s demise and the lecture was held in the open air on the sprawling grounds of "The Grove", which was Sir C. P.'s mansion. Abhyankar's talk on Ramanujan at the Sir C. P. Foundation appeared in Volume 10 of the Plenum series.

In the second half of 1968, we had the visit of Professor Robert Rankin of the University of Glasgow, an expert on Ramanujan's work. The role of Rankin in connection with Ramanujan's Lost Notebook is discussed in Chapter 7 when I report on the Ramanujan Centennial Celebrations in Madras in 1987. On this first visit to Madras in 1968, Rankin gave a series of lectures at the Ramanujan Institute which became the basis of his book *The modular group and its subgroups* that was published the next year. Rankin also lectured at MATSCIENCE in 1968 on the famous τ-function of Ramanujan and talk appeared in Volume 10 of the Plenum series.

In January 1968 we had the visit of two top complex analysts — Wolfgang H. J. Fuchs of Cornell University and Lee Rubel of the University of

Illinois. Fuchs' lectures on Nevanlinna Theory and Gap Series, and Rubel's on the two themes of Uniform Distribution on Compact Spaces and on Vector Spaces of Analytic Functions, all appeared in Volume 9 of the Plenum Series. Later in 1973 and 1975 I had interaction with Fuchs when I was beginning to do research in mathematics and was about to join graduate school.

Another eminent analyst to visit MATSCIENCE in 1968 was Walter Hayman from Imperial College, London. It was with the recommendation of Hayman and of Williamson, that Professor Unni was appointed at the Institute and the program in pure mathematics was initiated. It had an emphasis on functional analysis since it was Unni's speciality.

Professor Hayman was very friendly to me. When he came home for dinner, he wanted to play ping pong with me. I was a very avid ping pong player in my school days. Several years later when we met, he remembered the exciting ping pong game we had at our home.

One celebrated mathematician who visited India and Madras almost annually was Professor Marshall Stone of the University of Chicago. He first visited at MATSCIENCE in 1963 as Srinivasa Ramanujan Professor and later gave the first lecture when the Institute moved to its own building in 1969. He was very much involved in mathematics education. So on one occasion he gave a lecture of wide appeal at MATSCIENCE on "The role of mathematics in contemporary education". He spoke slowly and with a drawl, but in a halting tone that revealed his commanding position in the world of mathematics. Professor Stone had a global view of mathematics and so on another occasion he spoke about "Current trends in mathematical research", a write up of which appeared in the Plenum Volume 2.

Like Marshall Stone, the physicist George Sudarshan from Syracuse was a regular visitor to the Institute in the sixties. In the mid-sixties, he proposed a new theory of faster than light particles. According to Einstein's Theory of Relativity, nothing could travel faster that the speed of light, and so Sudarshan's proposal was a radical departure from conventional wisdom. Sudarshan named the faster than light particle as the tachyon — which is yet to be found. My father was open minded and arranged a one day symposium on faster than light particles in March 1968 with Sudarshan as the lead speaker. Sudarshan was very pleased at this and wrote the following in the Visitor's Book of the Institute on March 6: "Appreciation of one's scientific creativity is most cherished when it is critical; I find my visits to Madras and the Institute most inspiring in the sense that it makes me strive my best. May I have many more opportunities to visit and let not the speed of light put a barrier on our thinking." A few years later, my father started working on Special Relativity and proposed a new approach based just on the velocity transformation formula which would imply that nothing travels faster than light. The great mathematician Norman Levinson of MIT published a paper "Ramakrishnan's approach to Relativity" providing a mathematical basis for my father's approach.

The decade of the sixties at MATSCIENCE concluded with the visit of Nobel Laureate Hans Bethe who arrived in late December and stayed until mid-January. His first lecture on nuclear physics attracted a huge audience, but as went on into details, the audience shrunk and only the experts at MATSCIENCE attended his final talk. In his honor the Governor of Madras arranged a lavish dinner at the Raj Bhavan (= the Governor's Mansion). My father wanted to take Bethe to some South Indian classical music concerts because he had arrived at the peak of the music season in Madras. But Bethe politely told my father that he would attend just one concert to get an idea of our music. My father therefore wanted him to hear M. S. Subbulakshmi (MS as she was famously known) — the Queen of Song. But MS was not giving a concert during the dates when Bethe was in Madras. So my father spoke to Mr. T. S. Sadasivan, the husband of MS, and told him that a Nobel Laureate would be interested in listening to his wife's concert. Sadasivan immediately offered to arrange a private concert for Bethe by MS at their mansion in Kalki Gardens, and invited several important persons. It was a unique opportunity for all of us to listen to one of the greatest Carnatic vocalists at her own home, and in the presence of Hans Bethe! This would be like listening to Luciano Pavrotti in an opera arranged at his own home for a private audience! How often does that happen? All in all, Bethe formed such a positive opinion, that he wrote in the MATSCIENCE Visitors Book: "This is an important center for the advancement of science in India free of bureaucratic interference."

During this visit to Madras, and during our visits to Cornell, we got to know the Bethe's quite well. In 1978, when Mathura and I got married and went to Ann Arbor, Hans Bethe sent a lovely gift to us in view of the deep friendship with our family.

Thus both my international travels with my parents and the steady flow of foreign visitors to MATSCIENCE provided me a unique opportunity to get to know so many eminent physicists and mathematicians as a school boy. It was with this background that I was entering college in 1971 and was soon to make a decision between pure science and aerodynamics. But before describing that, I will discuss how my interest in mathematics started in high school and the encouragement I received from my high school teachers.

1.4　Interest in mathematics in high school (1968–71)

It was in high school (9th to 11th grade classes) when I really began to understand the method of mathematics, namely the art of giving proofs of theorems. This happened in Euclidean Geometry which we had for all three years in high school.

In the seventies, high school in India was up to the 11th grade. After this we had one year of Pre-University Classes (PUC) in College followed by three years of Bachelors Degree Classes. PUC has now been abolished in India, and so we now have twelve years in school like in the USA. My comments about school education in India relate only to English medium schools, namely schools where the medium of instruction is in English. A vast majority of students in India attend schools in which the medium of instruction is in the local language.

The school I attended was Vidya Mandir, one of the best English medium schools in Madras. It was founded in 1955 by Mr. M. Subbaroya Iyer, an extremely successful Income Tax lawyer whose "hobby" was to launch educational institutions! He founded Vidya Mandir, was instrumental in creating Vivekananda College (which I attended after high school), and the Madras Institute of Technology (also known as MIT!). Subbaroya Iyer was one the closest associates of my grandfather Sir Alladi, and so our family was well aware of his various successful enterprises. All he touched turned gold, and Vidya Mandir was no exception. Even though it was a newly created school, owing to my family's faith in Subbaroya Iyer's efforts, it was decided that I would join Vidya Mandir. That was the only school I attended from kindergarten to the 11th grade. Unlike the set up here in the USA where Elementary, Middle, and High Schools are separated, in India many schools have all grades from kindergarten upwards together.

By the time students are in high school, they are mature enough to understand detailed logical deductions, and so high school, in my opinion, is the perfect time to initiate students to the art of giving proofs. I also think Euclidean Geometry and Elementary Number Theory are two subjects that could be taught in high school with emphasis on proofs. Of the two, only Euclidean Geometry was chosen for high school students in India, but I am glad it was done thoroughly and in detail over a period of three years.

We had two excellent mathematics teachers in Vidya Mandir — Ms. Alamelu Krishnan and Ms. Alamelu Gopalan. In India, before one can become a high school teacher in any subject be it mathematics or history, one should have a bachelors degree in that subject and one year in a

Teachers College to get a BT degree after completing BA or BSc. With a full fledged Bachelors degree in a subject, one would know that subject well enough to teach in high school. Unfortunately in the USA, one can nowadays become a high school teacher with just a B-Ed degree which teaches mostly pedagogy, and only a sprinkling of knowledge of the subject you are supposed to teach — like mathematics. Such teachers cannot teach a subject like mathematics very well because they do not have a real grasp of it.

Ms. Alamelu Krishnan was my algebra teacher through all three years of high school and Ms. Alamelu Gopalan taught geometry for those three years. One of the things I enjoyed in algebra was *factorization* of polynomial expressions such as

$$a^3 + b^3 + c^3 - 3abc = (a + b + c)(a^2 + b^2 + c^2 - ab - bc - ca).$$

We had to prove or derive each such identity, which in this case is to expand the right-hand side and perform the cancellations to get the expression on the left-hand side. The beauty and symmetry of such identities were emphasized. The above identity implies that if $a + b + c = 0$, then the arithmetic mean of a^3, b^3, and c^3, is equal to their geometric mean!

As I had described earlier, I used to travel internationally with my parents annually, and these trips were from June to August to coincide with the summer session in American universities that my father visited. In India schools close in April and reopen in June. Thus I missed the first quarter of every year including even the years I was in high school. I used to pick up all my books for the next school year and take them with me on my world tours with my parents. While on these three month tours abroad, I would study each one of those books and work out the problems in them. If I had doubts, I would ask my father. Back home in Madras, I also had an excellent private tutor — Mr. A. Srinivasan — and classes with him at my home strengthened my understanding. (He was a Tamil teacher at Vidya Mandir and also the teacher in charge of the Boy Scouts program at the school.) By studying systematically, I was well prepared to take my quarterly exams on return from my trips abroad in September. Since I did very well on these exams, my school permitted me to travel with my parents annually even though I was missing three months of classes each year. Actually I was first in my class through all years in school from the 6th grade, and was awarded a Gold Medal for securing the highest total marks in the Matriculation Examination for the school final in 1971 (my classmate P. R. Shankar also had the same aggregate marks, and so he too was awarded a gold medal). Basically what is needed to be successful in high school is to study and do homework systematically on a daily basis which is what I did even while travelling abroad. Some of my classmates made fun of me saying that I was a book worm studying on my foreign trips and not sightseeing. Actually, I did a lot of sightseeing, but was on top of school work by studying every day for a few hours even on my trips.

In September 1968, when I returned from one such trip abroad, my classmates told me that in this first year of high school (ninth grade), the new subject was Euclidean Geometry, and everything was being taught with proofs. They said that proofs were extremely difficult, and that unlike the previous years, I would now flunk the geometry exams. In contrast, I enjoyed proofs and the subject thoroughly.

The text book for Euclidean Geometry that we had was by Hall and Stevens. After every *theorem*, there were a set of problems which were called *riders*. After proving a theorem, Ms. Gopalan would start the discussion of the riders. She would write down a rider on the board, turn toward the class, and ask if any of us had a proof. My hand would go up, and most of the time she would ask me to come to the board and supply the proof. This was a great encouragement for me. I enjoyed demonstrating the proof on the board for my classmates. This was perhaps my initiation to becoming a mathematics professor in later years!

I am aware that the proofs by Euclid starting from his axioms are not perfect. E. T. Bell in his classic treatise Men of Mathematics, praises Archimedes as a true genius, and considers Euclid as just someone who collected the available knowledge of his day and wrote a text book. He also points out the flaws in Euclid's axioms when discussing other geometries such as Projective Geometry and non-Euclidean Geometry. But Euclidean Geometry as it was taught to me and others before my time, is really a fine way to teach a student the methodology of proofs and the art of logical deduction. This criticism of Euclid by the elite Bourbaki school and the influence Bourbaki had on major academic groups, eventually led to the removal of Euclidean Geometry as the medium to do proofs in high schools in the USA. As a consequence, American high school students are now given a mass of facts and recipes from Euclidean geometry, analytic geometry and calculus and are not trained to think logically. So they have no idea of how these results are derived. The use of advanced calculators has also adversely affected the ability to reason logically. This is why for a majority of high school students entering universities in the USA, their mathematics skills are woefully inadequate, and so they are made to take remedial courses.

What a waste of their time in high school that they are taught mathematics without proofs!

A statement such as *the three perpendiculars from the three vertices of a triangle to the opposite sides are concurrent* is not at all obvious. Anyone hearing this statement must ask why this is true, that is, how this is proved? The point of concurrency is the orthocenter.

A high point of the course on Euclidean Geometry I had was the *Nine Point Circle Theorem* for triangles which states that the nine points consisting of the mid-points of the three sides, the feet of the three perpendiculars from the vertices to the opposite sides, and the mid-points of the three lines connecting the vertices to the orthocenter, all lie on one circle. What was more amazing was that the radius of this nine point circle is one half of the radius of the circumcircle of the triangle, and that the nine point center, the circumcenter, and the orthocenter, are collinear! What could be more beautiful than this? As students we were amazed by these properties. We did not ask what use or application this had. We were happy to have been told this lovely fact, and more importantly, shown how to prove it.

One problem I see with mathematics instruction nowadays in the USA is that every mathematical idea has to be justified with an immediate application. Very few have the patience or interest to grasp a mathematical idea unless its relevance is explained rightaway. This obsession with relevance and application prevents students from appreciating mathematics for its intrinsic beauty. When seeing a painting or hearing music, both of which we enjoy, do we ask for relevance or an application? Mathematics is indeed applicable, but why can't one enjoy it for its intrinsic beauty? I am not against application. I think one should not be obsessed with it.

By emphasizing proofs in high school classes, there is only time to discuss some algebra, geometry, and analytic geometry. Calculus instruction might have to wait until college, but then the foundation laid in high school is so strong in geometry, that calculus with proofs is easy in college. I have first hand experience with this since I learnt Calculus only in College and not in high school. But in high school, I learnt algebra and Euclidean geometry really well. I think that in high school one should be taught algebra, Euclidean geometry, trigonometry, and analytic geometry with full proofs, and that Calculus should be taken up only later after a solid foundation has been laid. At that point, Calculus can be taught with proofs.

The sixties was when "New Maths" (New Math in the USA) was making headlines in India, that is elementary set theory as a means to explain and describe various mathematical concepts at the school level. We were told that New Maths was making a revolution in the United States and changing the way mathematics was being taught. The Bourbaki school emphasised abstraction, but since this was too much for high school students to grasp, they were given a watered down version of set theory without any real proofs of important results. What New Maths did was to eliminate the well tested Euclidean Geometry from schools and provide a mathematics education to high school students without any emphasis on proofs. Thus the revolution initiated by New Maths was disastrous in the long run. When America sneezes, the world follows whether the change is good or bad.

I remember a conversation in 1982 with Heini Halberstam, a British mathematician at the University of Illinois which I will relate now. When I told him that school education in India was modeled according to the British system and so mathematics is taught properly in Indian high schools, he replied that it would be just a matter of time before the defects in the American school education would be copied first in Europe and then transmitted to India. Halberstam was right. I understand that nowadays Calculus is taught in high schools in India in addition to analytic geometry and trigonometry, and owing to the number of topics that are covered, proofs are mostly omitted.

Ms. Alamelu Krishnan, our algebra teacher, was very dynamic. She asked us to do a project on Set Theory and learn some important things. To help us with this project, she invited an outstanding mathematics teacher from Muthialpet High School — Mr. P. K. Srinivasan, who had earned the title 'Master Educator'. Mr. Srinivasan gave us a pep talk and explained some highlights of set theory, such as a hotel with infinite number of rooms that never turns away a guest even if it is full. Another example that he discussed was the one-to-one correspondence between points of two line segments of unequal length. All this was very helpful for the project on Set Theory that we did. New Maths was dreaded by parents who felt that they could no longer help their children with mathematics homework because New Maths was way above their heads. In any case, this project on set theory was publicized by our school, and several parents came to hear us explain these concepts.

One benefit of the New Maths project was my contact with that wonderful human being P. K. Srinivasan. He was a Gandhian and never cared for any material pleasures or possessions. He wore a Gandhi cap, and a traditional white Indian dress made of *khadar* (= home-spun cloth) in the true Gandhian tradition. His sole passion was mathematics teaching. His hero was Srinivasa Ramanujan. He started a Ramanujan Museum in Madras because he a had a fine collection of letters, documents, and photographs pertaining to Ramanujan. He really should have received major national awards for his contributions to mathematics education. Little did I realize at that time, that several years later, I would interact with him again because of our mutual interest in spreading the message of Ramanujan's mathematics to school students.

We had projects on different subjects each year in school. On Project Day, there were several visitors including parents, and we had to explain the topics that were displayed on the charts. Owing to my interest in aviation, one year I suggested to my class teacher that we could do a project on that subject. To my pleasant surprise, the class teacher selected this topic for my class. My classmates enthusiastically worked on this project for which I supplied significant material in the form of books, pamphlets, and pictures, many of which were acquired on my trips abroad.

Although I have focused here on describing my mathematics classes in high school, I should say we had excellent teachers in other subjects as well.

It was in Vidya Mandir that I started to learn Sanskrit, one of the greatest languages, even though it is not spoken now. Vidya Mandir is an English medium school, and so in the primary sections, that is up to Class 5, Tamil, the native language of Madras, and Hindi, the national language of India, were offered as second languages, and all students had to choose one. I chose Tamil. Starting from the 6th grade, Sanskrit was also offered as a second language. My father felt that by taking Sanskrit, I would get a better understanding and appreciation of Indian culture. I am happy that I took Sanskrit in school.

Sanskrit is a very highly structured language and its grammar is almost perfect. Almost every word in Sanskrit has a derivation — its *vigraha vakya*. For example, the original name for Sanskrit is *Samskritam* and it is derived from *Samyak + Kritam*; Samyak means "well" and kritam means "done" or "deed". Thus Samskritam means that which is well done — a perfect name for a complete language! To give another example which will appeal to mathematicians, the name Ramanujan has a beautiful derivation: "ja" in Sanskrit means to be born, and "anu" means to follow. Thus Ramanujan means one who followed Lord Rama in birth — a younger sibling of Rama. Sanskrit grammar and structure is so logical, that mathematicians would appreciate it very much.

With regard to Indian culture, I should mention that when I was just under eleven years of age, I had my *Upanayanam* (sacred thread ceremony). For Brahmin boys, their initiation to the world of learning is the

Upanayanam. In Hindu culture, one's father is the first teacher, and during the Upanayanam, the father whispers the sacred mantra into the boy's ears. In ancient times, after the Upanayanam, the boy would leave his parents home for the home of a Guru to learn the vedas. My parents conducted my Upanayanam in vedic style at our family home on June 24, 1966.

During my last two years in high school, we were taught physics by the venerable Ramamurthy Iyer, a retired teacher from the P. S. High School, who had taught my father there three decades before! This was a unique experience for us in many ways. Ramamurthy Iyer identified some of us in class on the first day and said that he had taught our fathers. He wore a *dhothi* in the traditional *panchagajam* style along with a western style jacket and tie — a strange combination that was quite common among the educated classes in India in the first half of the twentieth century. My grandfather wore the same type of dress. That style was a mixture of the Indian and British traditions. Ramamurthy Iyer always carried a meter stick, not to punish the errant boys, but to point out important things being written on the board in his beautiful handwriting. He insisted on very precise statements of definitions and laws; this emphasis on precision was very helpful to those of us who followed his example. In fact, when I teach undergraduate classes, I often have one question on an exam which asks the students to give definitions and statements of theorems in precise form. I find that such a question helps identify the strong students, because only they are able to write down precise statements.

My extra-curricular scholastic activities in school included frequent participation in oratorical competitions and debates as well as in essay competitions. All this was related to my English classes. We were asked to recite famous passages such as Lincoln's Gettysburg Address or Marc Anthony's speech in Shakespeare's Julius Caesar. Oratorical competitions stress intonation and emotion in delivery. Definitely participation in oratorical competitions and debates removes stage fear. These competitions also help in moulding an individual to be an effective speaker.

All in all, I had a most enjoyable educational experience at Vidya Mandir. But the bane of the Indian educational system is the University Examinations which for high school is to be taken at the end of the final year. This is a common exam given to all students completing high school but it depended on the program in which your school was part of. Mine was the Matriculation Examination. It did not matter how well you did in your high school exams that are given periodically. All that mattered in graduating from high school to go to college was your performance on these University Examinations that is given at the end of the final year. I and many of my classmates started preparing for them about three to four months in advance since our whole future depended on our performance in that intense one week of the University Examinations. My effort paid off but I hated these University Exams both in high school as well as in college. I suppose my dislike of the University Exams was like the way the great mathematician G. H. Hardy of Cambridge University, Ramanujan's mentor, loathed the Tripos Exams. In India the University Exams were introduced because it was felt that otherwise schools and colleges would resort to unfair grade inflation to get better placement for their graduates. Thus I can understand the need for a common University Examination as the yardstick to measure the ability of students.

Although I enjoyed mathematics immensely in school, my fascination was for aerodynamics. But this changed in College when I delved deep into mathematics to experience the joy of research in Number Theory to such an extent that I chose mathematics for my profession and not aerodynamics. Aviation just remained a lifelong hobby. I discuss this transition next.

Lalitha Ramakrishnan with physicists Murray Gell-Mann of Caltech (in full suit) and Richard Dalitz of the University of Chicago, at Ekamra Nivas, July 16, 1961. Four decades later, when I met Gell-Mann in Florida, he said he remembered the beautiful sari my mother wore in Madras.

Prof. Marshall Stone (University of Chicago), one of the most influential mathematicians of the 20th century, visited Alladi Ramakrishnan's Theoretical Physics Seminar in 1959, and MATSCIENCE on a regular basis during the sixties. Prof. Stone gave Krishna valuable advice about PhD work shortly before Krishna left Madras for UCLA. Photo credit - Phoebe Liebig (Prof. Stone's daughter), and the Special Collections on the University of Massachusetts, Amherst, where Prof. Stone was after retiring from the University of Chicago.

Noted mathematician Wolfgang Fuchs (first from left), with Nobel Laureate Hans Bethe (second from left) and other physicists at Cornell University, June 3, 1968. Picture taken by Alladi Ramakrishnan. See letter of Prof. Bethe at the end referring to Prof. Fuchs' comment on Krishna's undergraduate research.

Chapter 2

Decision to pursue mathematical research

2.1 The Pre-University Course in College (1971–72)

After completing my high school, I joined the Pre-University Course (PUC) at the Vivekananda College of the University of Madras in June 1971. At that time, school was for eleven years, followed by one year of PUC in College, and then three years of Bachelors classes. Indian universities are modeled after the British system in that there are a number of affiliated colleges to a major university. Vivekananda College, close to my home, was affiliated with the University of Madras. So all final exams were given by the University and not by the colleges.

All colleges had three quarterly terms, starting in late June and ending in early April. Whereas the first term, second term, and third term exams each year in September, December, and March, were college exams, all year-end exams for the PUC, BS and MS courses, each year in April were given by the University. Performance in the university exams was what really counted for promotion.

For the PUC, I chose Mathematics-Physics-Chemistry (MPC) as the combination. For all students choosing MPC, the mathematics courses were in Algebra, (Euclidean) Geometry, Trigonometry, Analytic Geometry, and Calculus I. This was the same in all affiliated colleges of the University of Madras. Thus Calculus was taught only after a good grounding of the basic subjects. Since PUC is equivalent to the 12th grade, this means Calculus 1 was taught in the final term of school by present standards.

The Geometry course was a fine extension of what we learnt in school. In the PUC the emphasis was on concurrency and collinearity. Our enthusiastic teacher was Prof. Ramakarthikeyan. We were taught the proofs of the famous theorems of Ceva, Menelaus, and Desargues — the piece-de-resistance of that course. It was delight to learn that the converse of the

45

Desargues theorem is its dual! Our teacher emphasized the duality between points and lines in geometry — that is, for every theorem involving the notions of points and lines, there is a dual theorem where the notions relating to points and lines are interchanged. Thus concurrency and collinearity are dual notions. The teacher said that even though we are proving all these theorems in Euclidean geometry, this duality is best approached and understood from the point of view of Projective Geometry, which unfortunately was not offered for the BSc in India at that time.

For trigonometry, we followed the classic text of S. L. Loney — a book that interested Ramanujan. Our physics teacher Prof. Lakshminarayanan had to use trigonometric identities in his discussions. He said emphatically, "My dear boys, Loney's trigonometry is worth its weight in gold!". By the bye, very few high schools and colleges were co-educational in India at that time. This has changed now with western influence on conservative Indian society. Vivekananda College was an all boys college. One of the co-educational institutions in Madras was the famous Presidency College which boasted a prime location on the Marina Beach. Our Sanskrit teacher Prof. S. Viswanathan used to say: "In the Presidency College you can enjoy both sea breeze and she breeze!"

In Algebra, one of the topics discussed was geometric progressions and the convergence of infinite geometric series. Although geometric progressions (sequences) are defined by the rule that each term is a fixed multiple of the preceding term, I tried to characterize geometric progressions in terms of the sums of the first n terms. I defined a *cumulative progression* to be one for which each term is the sum of all preceding terms. So if the first term is 1, the cumulative progression is $1, 1, 2, 4, 8, 16, \ldots$. All terms after the first generate a geometric progression of common ratio 2. By considering more generally progressions in which each term is r times the sum of all its predecessors, one gets $1, r, r(r+1), r(r+1)^2, r(r+1)^3, \cdots$, where the terms after the first now form a geometric progression with common ratio $r + 1$. A convergent geometric series with common ratio $1/2$ such as

$$\frac{1}{2} + \frac{1}{4} + \frac{1}{8} \cdots$$

is also characterized by a cumulative property, namely that each term is the sum of all terms that follow. In the first cumulative progression, instead of having two ones in the beginning, if we interpret the first 1 to be the value of the infinite sum above, then we could replace the cumulative progression with the doubly infinite sequence

$$\cdots, \frac{1}{8}, \frac{1}{4}, \frac{1}{2}, 1, 2, 4, 8, \cdots.$$

This double infinite sequence has the property that each term is twice the previous term AND that each term is the sum of all the terms that precede it. Thus as a PUC student (= 12th grade in the USA) I wrote a paper on Cumulative Progressions as an alternate way to look at geometric progressions and series, and published it in early 1972 in the Mathematics Teacher, a journal brought out by the Association of Mathematics Teachers of India (AMTI).

On the last day of our regular university examinations for the PUC, two of my classmates — G. Ramesh and M. Ramesh — told me as we were leaving for our summer vacation, that there is a very interesting sequence of numbers called the *Fibonacci Numbers* which are generated by the formula that each term in the sequence is a sum of the two previous terms with initial values 0 and 1. The first few terms of the Fibonacci sequence are

$$0, 1, 1, 2, 3, 5, 8, 13, 21, 34, 55, 89, 144, \cdots.$$

The Fibonacci numbers possess many beautiful properties and they occur in mathematics and in nature in several settings. Thus I came home on the last day of our university exams with a sheet of paper containing the Fibonacci numbers.

A few days after my PUC exams, I left for the USA with my parents in April 1972. I was planning on applying for admission to BSc Physics after my return from the trip. In India one had to wait for the University Exam results before applying for the next level course. Since the Final Exams for the PUC were in April, and the results would be announced only in May, I could apply for admission to the BSc only in early June after I returned from my trip. This process is very different from admission to US Universities where you apply several months in advance and supply the results on the final exams after securing admission but prior to joining; that is admission is contingent on full transcripts being supplied before joining.

It was decided that I would leave the USA on May 19 from New York with my mother and return to India so that I could apply for admission to the BSC in time. My father was to return a few weeks later. On the morning of my departure from New York to India, my father took me to Princeton University to meet Professor Hazen of the Department of Aerospace and Mechanical Sciences, an aerodynamist who had ties with the Hindustan Aeronautics Limited (HAL), Bangalore, India. Hazen was also instrumental in the formation of the aeronautical engineering program at IIT Madras. Hazen had been a Visiting Professor at IIT Kanpur and a Member of the Kanpur Indo-American Program Steering Committee in the sixties. Hazen's name was suggested to us by our friend Dr. Sankaranarayanan of HAL. Prof. Hazen said that much of aerodynamics was really fluid dynamics, for which one needed a good background of mathematics. He suggested that after the PUC, I should enroll in BSc Mathematics instead of BSc Physics. Following Professor Hazen's advice, I enrolled in the BSc Mathematics course at Vivekananda College.

The BSc classes were to start a few weeks later and so in the meantime I started looking closely at Fibonacci numbers at my home in Madras. I soon "discovered" several of its charming properties. The identities I found were of course well known, but they were new to me. One question that intrigued me was whether the sequence can be extended to the left using the same recurrence, and if so, what happens? If we let F_n to be the n-th term of the Fibonacci sequence with $F_0 = 0$ and $F_1 = 1$, then this yields $F_{-1} = 1$, which in turn yields $F_{-2} = -1$, $F_{-3} = 2$, $F_{-4} = -3$, and so on. I was excited to note that by extending the Fibonacci sequence to the left, one gets the same sequence but with alternating minus signs. Thus the bilateral Fibonacci sequence is,

$$\cdots, 89, -55, 34, -21, 13, -8, 5, -3, 2, -1, 1, 0, 1, 1, 2, 3, 5, 8, 13, 21, 34, 55, 89, \cdots.$$

The Fibonacci numbers occur as the rising diagonal sums of the Pascal triangle of binomial coefficients. The Pascal triangle is usually written as a triangular array of numbers in the lower right quadrant, if we think of the location of the apex of the Pascal triangle as representing the origin $(0, 0)$. I then extended the binomial coefficients to negative indices and consequently the Pascal triangle to the entire plane — that is into every quadrant. I was pleased with these results, and so I wrote a paper entitled "A new extension of the Fibonacci sequence" in the summer of 1972 just before entering the BSc mathematics course. This paper and some other early papers of mine were typed by our dedicated secretary R. Ganapathi.

When my father returned from America, I showed him these results and the paper I had written. He said that he would like to get the opinion of certain experts about the quality and originality of my results. Although I had cut to short my USA trip, the good thing was that I had a few weeks of leisure in Madras to think about Fibonacci numbers. I could not have had done this with a packed schedule of travel in the United States. This one month in solitude in Madras changed the course of my life as we shall see. It was bye-bye to aerodynamics and welcome to Number Theory.

2.2 The joy of Number Theory revealed (1972–73)

On seeing the work on Fibonacci numbers I did in the summer of 1972, many persons told me that I should study Number Theory. The first number theory book that I actually read was by Ralph G. Archibald, which had appeared in print just two years before. It was in the MATSCIENCE library, and it provided a leisurely and solid introduction to various topics in number theory at the undergraduate level. Interestingly, for my birthday in 2019, my daughters and sons-in-law presented this book (purchased through Amazon since it was out of print) to kindle memories of my infancy in number theory! Next I started reading the classic by Hardy and Wright "Introduction to the Theory of Numbers", and later the first volume of the two volume book by W. J. LeVeque called "Topics in number theory". One unusual aspect of Hardy and Wright is that after the first two chapters on divisibility and prime numbers, the next chapter is a discussion of *Farey Fractions*. I think that Hardy's interest and emphasis on Farey fractions is due to the fact that in the revolutionary *Circle Method* that he and Ramanujan initiated to get an asymptotic series for the number of partitions of an integer, they made crucial use of dissections of the unit circle into arcs determined by Farey fractions.

The Farey sequence $\mathcal{F}_n$ of fractions of order n is the set of reduced fractions with denominator not exceeding n and arranged in increasing order. For example, the Farey sequence of order 5 in the interval [0,1] is

$$\frac{0}{1} < \frac{1}{5} < \frac{1}{4} < \frac{1}{3} < \frac{2}{5} < \frac{1}{2} < \frac{3}{5} < \frac{2}{3} < \frac{3}{4} < \frac{4}{5} < \frac{1}{1}.$$

Simple as this is, the arrangement in increasing order leads to two striking properties:

If $\frac{h}{k} < \frac{h'}{k'} < \frac{h''}{k''}$ are three consecutive members of $\mathcal{F}_n$, then

$$\frac{h + h''}{k + k''} = \frac{h'}{k'} \quad \text{(medaint property) and} \quad kh' - hk' = 1.$$

These two properties can be verified with the example given above. On seeing this in Hardy and Wright, I immediately asked myself the following question: Instead of arranging ratios of all integers with a bounded denominator in increasing order, why not take ratios of members of a strictly increasing sequence $\{a_k\}$ of integers and arrange them in increasing order? Since the Fibonacci numbers mimic the integers in terms of divisibility properties and since I knew several properties of the Fibonacci sequence, I went about considering the *Farey sequence of Fibonacci numbers* and its properties. It turned out that the medaint property was valid except when the middle fraction was $1/F_m$, and consequently second property was also valid almost everywhere in the sequence. The Farey sequence of Fibonacci numbers which I denoted by FF_n, had several other nice properties and so I wrote up a paper on this topic in 1972 while I was in the first year of the BSc class, and this paper was published in 1975 in the Fibonacci quarterly just as I was completing my BSc.

I read other chapters of Hardy and Wright and was fascinated by the topic of arithmetic functions. One result due to Hardy and Ramanujan that interested me very much was that the number of prime factors of an integer n had average value loglog n and was also almost always asymptotically loglog n. Hardy and Ramanujan wanted a mathematical explanation of the phenomenon *round numbers are rare* and in order to understand this, they studied the number of prime factors as a measure of roundness or compositeness of a number. They pointed out that even though the values of most arithmetic functions like $d(n)$, the number of divisors of n, and $\sigma(n)$, the sum of the divisors of n, went up and down significantly, yet they had a well behaved average order, but the average was not necessarily the most commonly occurring asymptotic size (which they called the *normal order*), because the average could be influenced by sporadically occurring large values. This was the case with $d(n)$, but what was most interesting was that the number of prime factors of n has average order AND normal order loglog n. From the point of view of average and normal orders, they showed that it did not matter whether one considered $\omega(n)$, the number of distinct prime factors of n, or $\Omega(n)$, the number of prime factors of n counted with multiplicity. Thus they pointed out that the number of prime factors of an integer is a better measure of roundness than the number of divisors of an integer. These fundamental observations of Hardy and Ramanujan on $\omega(n)$ and $\Omega(n)$ eventually led to the creation of Probabilistic Number Theory in 1939 with the great theorem of Erdős-Kac. I did not know about Probabilistic Number Theory at that time as an undergraduate, but I felt that it was surprising that even though prime numbers have been investigated since the Golden Age of Greece, the first significant results on the number of prime factors of integers was as recent as Hardy and Ramanujan — recent in comparison with the long history of number theory. Another thing also struck me as unusual. There was a significant literature on $\sigma(n)$ (the sum of the divisors of n) dating back to the perfect number problem of Greek antiquity, and on $d(n)$ (the number of divisors of n) due to the famous Dirichlet Divisor Problem, and quite a bit known about the number of prime factors of n since Hardy and Ramanujan, but I found nothing in the number theory books on the sum of the prime factors of a number. So I defined $A(n)$ to be the sum of the prime factors of n with multiplicity, and $A^*(n)$ to be the sum of the distinct prime factors of n. I immediately noticed that $A(n)$ had two very fundamental properties:

$$A(mn) = A(m) + A(n),$$

for all positive integers m and n and for any integer $m > 0$, the number of solutions to $A(n) = m$ is the number of partitions of m into primes:

$$|\{n|A(n) = m\}| = p_\pi(n),$$

where $p_\pi(n)$ is the number of partitions on n into primes. These two properties convinced me that $A(n)$ (which I called a new logarithmic function) was worthy of study, and so I set about investigating its properties. One thing I showed rightaway in the spirit of the Hardy-Ramanujan observation on $\omega(n)$ and $\Omega(n)$ was that both $A(n)$ and $A^*(n)$ had the same average order $\pi^2 n / 6\log n$. But I could not determine their normal order. This dilemma on the normal order of $A(n)$ and $A^*(n)$ was resolved only in January 1975 when I met Paul Erdős in Madras (see §2.7).

Being an academic, my father wanted to get the opinion of top mathematicians about this early work of mine. He wrote to Helmut Hasse, a giant in the field of Algebraic Number Theory. Hasse is credited for several fundamental ideas, one of the most significant of which is called the *Hasse principle* which he enunciated in the 1920's. The principle gives necessary and sufficient conditions for a homogeneous polynomial in several variables of degree not exceeding 2 (that is the linear and quadratic polynomials) to have integer solutions. This had a profound effect in both algebraic number theory and in the study of Diophantine equations.

It was a great pleasure to receive a letter from Hasse saying that he felt my idea of a Farey sequence of Fibonacci numbers was original, and that while I should read books and papers, he said, "let not excessive reading spoil your innate originality." This was very different from the type of advice I received from most others, namely that I should broaden my knowledge by reading a number of excellent books in number theory, and not be seduced by the charm of Fibonacci numbers.

My father also wrote to the great physicist Freeman Dyson at the Institute for Advanced Study in Princeton asking for an opinion. Dyson actually started out in number theory, and as an undergraduate at Cambridge University, England, he gave in 1944 a beautiful and important combinatorial explanation of two of Ramanujan's famous divisibility properties of the partition function by introducing a statistic he called the *rank of a partition*.

This early work of Dyson has had immense impact and the study of ranks and similar partition statistics became a major area of investigation in the theory of partitions. My father chose to write to Dyson not only because of this outstanding work in number theory he did as an undergraduate, but also because my father had interacted with Dyson while visiting the Institute for Advanced Study in 1957–58. Dyson was known to be frank and forthright, and my father wanted a critical assessment from him. Dyson did respond promptly. He said my work was very interesting, but he felt that number theory was a recreational pursuit; a talented youngster should pursue a more significant subject like physics. This was perhaps the reason that Dyson shifted from number theory to become a physicist. This loss to number theory was a gain to physics. Several years later, in 2013, I had a chance to host Dyson at the University of Florida, and then I had an opportunity to discuss his views on number theory versus physics.

I had opportunities to discuss my work with mathematicians in Madras, especially those at the Ramanujan Institute of the University of Madras. In particular, I met Professor Bhanumurthy, the Director of the Ramanujan Institute, several times at his residence on weekends for discussions on Number Theory. He extolled the contributions of the Russian mathematicians to analytic number theory, most specifically I. M. Vinogradov. Bhanumurthy explained to me how Vinogradov initiated the method of trigonometric sums to prove his great theorem that every sufficiently large odd number is a sum of three primes. Observing my passion for number theory, he presented me his notes for the number theory course he would often give; I still have that hard bound note book filled with his calligraphic handwriting. Professor Bhanumurty invited me to give a talk at the Ramanujan Institute while I was a undergraduate student at Vivekananda College, and I appreciated his encouragement.

While visiting the Ramanujan Institute, I got to know Prof. Rangachari. The Ramanujan Institute had a tradition of research in classical topics like summability of series and Prof. Rangachari work was in such areas. He is extremely orthodox in appearance and practice. He always wears a dhothi in traditional panchagajam style, but with a full sleeved shirt. He is an Iyengar and has the full *namam* on his forehead. The front part of his head was shaven and he had long hair curled into a tuft like the Hindu priests. He never ate food outside — that is at restaurants or at people's homes. He ate only home cooked food. I interacted with him quite a bit on convergence and divergence of series. During our discussion in his office, he offered me coffee and cool drinks from a nearby shop, even though he would never drink these himself since they were not from his home.

There were also opportunities for me to get an assessment of my work from top mathematicians visiting MATSCIENCE from abroad. In particular, in the first week of January 1973, as part of the anniversary celebrations, a big International Conference on Functional Analysis was organized at MATSCIENCE by Professor Unni. That conference attracted about 50 leading analysts from around the world and gave me an opportunity to interact with them. One of the stars at the conference was Louis Nirenberg from Courant Institute, and I remember his lecture on the Laplacian. He was very soft spoken and modest in spite of his greatness. Several years later I had the pleasure of hosting him at the University of Florida when he came to deliver the Ulam Colloquium in November 2005.

Another very eminent analyst with whom I had the pleasure of interacting at that conference was I. J. Schoenberg from the University of Wisconsin. He was famous for the theory of splines. One of the most important results in analysis is the Weierstrass Approximation Theorem which tells you that continuous functions can be nicely approximated by polynomials. Splines are polynomials pieced together to provide better approximations. Splines have been used extensively in aerodynamics especially in the study of stability of aircraft in tactical warfare. They have also been used significantly in the automobile industry for modeling automobile bodies. Garrett Birkhoff referred to earlier as a consultant to General Motors, has explained the use of splines in his GM technical reports.

I attended the talk that Schoenberg gave at the conference on cardinal spline interpolation. Schoenberg stayed back in Madras for a few days after the conference and so I had the opportunity to give a talk at MATSCIENCE with Schoenberg in the audience. Schoenberg actually had started out in Number Theory by working on a topic suggested by the great German mathematician Issai Schur. Only later did Schoenberg shift his research to functional analysis. Thus it was good for me to give a talk in the presence of Schoenberg and get his reactions.

Besides Niremberg and Schoenberg, I remember my conversation with the great Jean-Pierre Kahane from Paris. He made some suggestions regarding the style of presentation for my paper on Fibonacci numbers. I also had the pleasure of renewing contact with Kahane several decades later — at the Erdős Centennial Conference in Budapest in 2013.

While it was very beneficial for me to listen to leaders in the field of analysis at this conference, they were not number theorists and so could not exactly say what part of my work was well known, and what really was new. So I needed the critical opinion and guidance of expert number theorists, and this was to happen in a few months.

2.3   The Number Theory Summer Institute in Ann Arbor (1973)

The next natural thing to do was to attend a major international conference or workshop not only to meet several leading researchers in number theory, but also to get to know the recent advances in the field. In the summer of 1973, a two-month Institute in Number Theory was held at the University of Michigan in Ann Arbor featuring several top mathematicians as speakers. The University of Michigan has an outstanding program in number theory, and it was natural therefore for it to be the venue of a major summer institute. This summer institute was organized by Professor Donald Lewis of the University of Michigan, a noted algebraic number theorist who had done important work in collaboration with the great Harold Davenport of Cambridge University. Lewis used to visit Cambridge frequently, and likewise, Davenport visited Michigan often, and as a consequence, the University of Michigan enjoyed strong ties not only with Cambridge, but with the number theorists in England as well. It was this connection with Davenport that led Lewis to recruit the brilliant young number theorist Hugh Montgomery to Michigan after Montgomery completed his PhD under the direction of Harold Davenport. Davenport died in 1969, and so when Montgomery submitted his thesis, Alan Baker was his official advisor in place of Davenport. Montgomery had won the Adams Prize for his PhD thesis and so was a rising star. Seeing the announcement of the Summer Institute in the Notices of the American Mathematical Society at MATSCIENCE, I wrote to Professor Lewis in December 1972 expressing my interest in attending it. I also sent him the papers in number theory I had written as an undergraduate and provided him names of some mathematicians as references. It was a great gesture on the part of Lewis to accept me as an undergraduate participant in that summer institute. It was an eye opener for me and influenced my academic life is many ways.

My annual visits to the USA each summer with my parents continued, and in so in 1973 I could be in Ann Arbor. The dates of the Summer Institute were June 18–August 10, but I could be there only for one month because classes at College were starting in early July. Even with this short stay in Ann Arbor, I missed the first few weeks of classes for the II BSc in College. It was generous of Vivekananda College to give me leave to attend the Summer Institute. My college teachers realized how important this was to my research efforts in number theory. My father left me and my mother in Ann Arbor for a month while he travelled around the USA on his lecture assignments. In between, he visited us in Ann Arbor a couple of times.

From the book "Topics in Number Theory" by William LeVeque, Volume I of which I read in Madras, I thought that LeVeque would be on the faculty at the University of Michigan. So I wrote to him expressing my desire to discuss with him in Ann Arbor. He responded saying that he liked my work, but that he was no longer in Michigan; he had moved to the Claremont Graduate School in Claremont, California. My letter was forwarded to him at Claremont. He enquired whether I could meet him at Claremont since he would not be at the Summer Institute in Michigan while I would be there; he would be at the Summer Institute only in August. I quote from his letter of 9 April 1973: "Thank you for sending me your preprints. You have a number of very interesting ideas and have worked out the details very nicely. Your work shows a great deal of promise, and I urge you to develop your talents. I am sure you will find the conference in Ann Arbor this summer most stimulating and that you will come away from it with a number of new ideas. ... I would of course be quite happy to have you come here..." Thus prior to arriving in Ann Arbor, my father took me to California, mainly to meet Professor LeVeque at Claremont. There LeVeque introduced me to a very bright undergraduate student by name Ted Chinburg who was doing research under LeVeque's guidance and was at Harvey Mudd, one of best among the various Claremont colleges. LeVeque mentioned that Chinburg would be an undergraduate participant at the Summer Institute and that we should interact more closely in Ann Arbor. Thus Chinburg and I were the two undergraduate participants at that Summer Institute in Ann Arbor. I did interact with Chinburg, but his interests were in algebraic number theory, while I was more attracted to analytic number theory. Chinburg had a successful career as a mathematician and became a professor at the University of Pennsylvania.

In addition to doing important research in transcendental number theory and Diophantine approximations, LeVeque wrote and edited several books, and was also a competent administrator. Besides his well known text books in number theory, he also produced a six volume *Reviews in Number Theory,* collecting the reviews of all the papers in number theory that appeared in Mathematical Reviews from 1940 to 1972. This was published in 1974 and I found it extremely useful as I was starting research in number theory. LeVeque was Chair of the Mathematics Department at the University of Michigan before moving to Claremont. After Claremont, he took up the post as Executive Director of the American Mathematical Society in 1997 and held this position until his retirement in 1988. Among the many things he accomplished as Executive Director was to make Mathematical Reviews available electronically.

Ann Arbor is very close to Detroit and so after meeting LeVeque in California, we flew into Detroit. Prof. M. S. Ramanujan of the Michigan Mathematics Department, who knew my father, picked up us at Detroit airport and took us to Ann Arbor. He helped us get settled.

All the participants were housed in two large apartment complexes — University Towers and Tower Plaza — both of them high rises. We stayed at University Towers. Angell Hall, the most impressive building on campus, was the home of the Mathematics Department. I was in awe as I entered the corridors of Angell Hall and saw the names of famous mathematicians etched on the glass of their office doors. It was inspiring and daunting to see Hugh Montgomery pace up and down the corridors of Angell Hall in his shorts, with a serious look and in deep thought.

On the evening prior to the opening of the Summer Institute, there was a welcome party at the home of Professor Lewis to which I was invited. I was only an undergraduate participant and so was touched by the care shown by Professor Lewis. My mother and I attended this reception.

Unlike the opening ceremonies of conferences in India with their fanfare, the Summer Institute opening was simple. Don Lewis walked in with a cup of coffee, said Hi, and informed us that apart from a certain scheduled talks, there was plenty of unalotted time when participants could get together for discussions or arrange lectures. As it turned out, there were a good number of wonderful lectures.

One of the impressive talks was by Professor Ian Richards of the University of Minnesota. Richards and his student Douglas Hensley had just established a startling result: the famous Hardy-Littlewood conjecture was incompatible with the prime k-tuples conjecture. Let me explain this.

The sequence of prime numbers, namely those integers greater than 1 which cannot be decomposed into a product of smaller numbers, is

$$2, 3, 5, 7, 11, 13, 17, 19, 23, 29, 31, \ldots.$$

One of the most intriguing aspects of prime numbers is that one can see certain patterns with some regularity, but it is very difficult to prove results about these. It is easy to show that the sequence of primes is infinite and the most well known proof is due to Euclid. One can see that there are several instances of prime twins, namely pairs of primes like (5,7) or (17,19) that differ by 2. But the prime twins conjecture, asserting that there are infinitely many such pairs of primes, is unsolved to this day. Generalizing the prime twins conjecture is the conjecture about k-tuples of primes, not just that there are infinitely many of certain types of k-tuples, but on the number of such k-tuples below a given magnitude. A very famous conjecture due to Hardy and Littlewood is that

$$\pi(x + y) \leq \pi(x) + \pi(y)$$

for all x and $y \geq 2$, where $\pi(x)$ is the number of primes up to x. Basically what this says is that the greatest concentration of the primes is at the beginning. What Hensley and Richards showed is that the prime k-tuples conjecture is not compatible with the Hardy-Littlewood Conjecture, that is both cannot be true. Expert opinion is that the prime k-tuples conjecture is probably true.

G. H. Hardy and J. E. Littlewood of Cambridge University formed the greatest partnership in mathematical history. As part of their collaboration, they wrote a sequence of papers under the title *Some problems in partitio numerorum* in which they stated the strong form of the k-tuples conjecture on the number of k-tuples below a given magnitude.

After his talk, Ian Richards asked me to sit next to him, and wrote in my notebook in large block letters various important theorems and conjectures of number theory. One of the results he emphasized was that of Montgomery on gaps between primes.

The Prime Number Theorem is the statement that

$$\lim_{x \to \infty} \frac{\pi(x)}{x/\log x} = 1.$$

This fundamental result conjectured independently by Gauss and Legendre, was proved toward the end of the nineteenth century simultaneously and independently by Hadamard and de la Vallee Poussin using complex variable theory following a program outlined by Riemann. From this it follows that the n-th prime p_n is asymptotically $n\log n$ and so the average gap $d_n = p_{n+1} - p_n$ between the n-th prime and the next is asymptotically $\log n$. An important and deep problem is to determine upper bounds for d_n of the form

$$d_n < n^{c+\varepsilon},$$

for integers n sufficiently large (depending on ε), where c is a non-negative constant less than 1, and ε arbitrarily small. It was a tremendous achievement by Hoheisel who showed in 1930 that the above inequality holds with $c = 32999/33000$. In the next few years, the exponent in Hoheisel's result was significantly improved first by Heilbronn in 1933 to $c = 248/250$, then by Chudakov in 1936 to $c = 3/4$ and finally by Ingham in 1937 to $c = 5/8$. There matters stood for several decades until Montgomery broke the impasse in 1969; using new methods he showed that $c = 3/5$ holds. The Riemann Hypothesis, the greatest unsolved problem in number theory and analysis, would yield $c = 1/2$. It is actually conjectured that $c = 0$.

Ian Richards patiently explained the significance of these theorems and conjectures. I was touched by his kindness and care. I never met Richards after that conference in Ann Arbor. But I did get to know his student Douglas Hensley very well. Although Hensley and I never collaborated, our research intersected heavily in the eighties when we both worked on integers without large prime factors.

The seventies was a period of intense activity in the area of Transcendental Number Theory following the revolutionary work of Alan Baker of Cambridge University on linear forms in the logarithms of algebraic numbers. Baker received the Fields Medal for this in 1970 at the International Conference of Mathematicians in Nice. Baker's methods enabled one to show that certain Diophantine Equations had at most a finite number of solutions, and provided effective bounds for these solutions. Among the most striking applications of Baker's method was the theorem of Robert Tijdeman from Leiden, Netherlands, that there are at most a finite number of consecutive powers like 8 and 9. Note that 8 is a cube and 9 a square. Tijdeman proved this theorem in 1976, but by 1973 he was known for several important results in Transcendental Number Theory. He was one of the participants at the Summer Institute along with his student Cijsouw. At the Summer Institute, there were several lectures on transcendence. One lecture I remember was by Dale Brownawell of Penn State University. Brownawell and Michel Waldschmidt (Paris) had established certain major transcendence results relating to the exponential function and this was the theme of Brownawell's talk.

Observing my interest on irrational numbers and on Farey sequences, Professor Richard Bumby of Rutgers University walked into my office and explained to me the Markov spectrum of constants in Diophantine Approximations. Yes, I was given an office, not just a cubicle; the office was not in Angell Hall but in one of the neighboring buildings. The fact that I was given an office, showed that I was viewed seriously as a participant even though I was just an undergraduate student. I was also really impressed that senior professors took time to explain things to me, knowing that my interest was great, but that my knowledge was modest.

Hugh Montgomery gave lectures as well — on his seminal work on the large sieve, as well on the zeros of the Riemann zeta function. I did not have enough background at that time to understand his lectures in depth.

One of the lectures that I understood was by Harald Niederreiter from Southern Illinois University. He spoke about uniform distribution of sequences. His presentation was thorough and his boardwork exceptionally beautiful. He and Professor Kuipers of Southern Illinois (also a participant at the Summer Institute) were writing a book at that time on Uniform Distribution of Sequences. This was published a year later and has become a standard reference in the field. It was the lecture of Niederreiter that got me interested in uniform distribution. Two years later, as a graduate student at UCLA, I took a year long course of Niederreiter on uniform distribution, by which time his book with Kuipers had appeared. Someone told me later that there was not a single typo in the book of Kuipers and Niederreiter because they were so thorough in proof reading!

There were also lectures by leading algebraic number theorists that I attended — such as those by K. Iwazawa (Princeton), H. Heilbronn, and Olga-Taussky Todd. But due to my lack of background in algebraic number theory, I did not much understand those lectures compared to those in other parts of number theory.

I was given an opportunity to speak about my work in one of the seminars in July. I spoke about the function $A(n)$ and certain generalizations of the Euler function I was investigating. I had lectured on my work previously in Madras but here I was giving my talk for the first time in front of so many of the world's experts in number theory. Needless to say that I was very nervous, but I survived. After my talk, several professors came up to me and said very encouraging things. So I have pleasant memories of that first conference lecture of mine. If that had been a disaster, I might have quit pursuing research in number theory. I therefore appreciated the support from several experts in the audience.

Since I had to depart for India in mid-July, I missed meeting some outstanding mathematicians like Paul Erdős, Wolfgang Schmidt and Hans Zassenhaus who attended only the second half of the eight week session. But the Summer Institute in Ann Arbor exposed me to several areas of number theory and gave me an opportunity to meet many leading researchers and learn about their important work.

The 1973 Summer Institute in Ann Arbor was one of the best organized and important summer institutes. It was at this conference that I got to meet Hugh Montgomery and Don Lewis, and I think that might have played a role later in the decision by Michigan to offer me a Hildebrandt Assistant Professorship in 1978 upon completion of my Phd. Definitely my acceptance of that position was due to the positive experience and impressions I had during at that Summer Institute.

2.4   Visit to Australia to work under Kurt Mahler (1973)

Back in Madras from Ann Arbor, I had to get back to work on my college curriculum having missed three weeks of classes. But I continued to learn number theory on my own. In particular, I started reading introductory books on Diophantine Approximations. Within a month of my return from Ann Arbor, I saw a paper in the Journal of Number Theory by Kurt Mahler that interested me.

Kurt Mahler was one of the leaders in the field of transcendental number theory. He received his PhD under the direction of Carl Ludwig Siegel, one the greatest mathematicians of the twentieth century. In the 1920s, and 30s, Mahler made several fundamental contributions to transcendental number theory. In particular, his classification of transcendental numbers into three types — S, T, and U — is now at the core of the subject.

Kurt Mahler was of Jewish descent. With Hitler's rising power in the 30's and with the persecution of the Jewish community becoming stronger day by day, Mahler decided to leave Germany. He spent the bulk of the period from 1933 to 1962 at the University of Manchester where he was invited by Louis Mordell. In 1963 Mahler took up a position at the Australian National University (ANU) in Canberra, where he was until his death in February 1988 at the age of 84, except for the period 1968–72 when he was at the Ohio State University. So in 1973, when Mahler's paper attracted my attention, he was back in Australia.

I told my father that I liked the work of Mahler and would like to interact with him. He was very happy to hear that. Actually, when my father was at the University of Manchester during 1949–51 for his PhD, Kurt Mahler was one of the leading professors in the mathematics department there. Even though my father was in the statistics group which was part of the mathematics department at Manchester, he had no occasion to interact with Mahler, but had heard great things about him. My father suggested I should contact Professor B. H. Neumann, a famous group theorist, who was the head of the mathematics department of the Australian National University. Professor Neumann knew my father well and had visited MATSCIENCE in 1970. I sent Professor Neumann some of my preprints and expressed my desire to work with Professor Mahler. Professor Neumann spontaneously responded and offered me a Student Scholarship at ANU to start in October 1973 so that I could work under Professor Mahler. What a fantastic opportunity to learn from one of the central figures of number theory! Once again I would be missing classes in the second year of my

BSc, but again within the same year, my teachers at Vivekananda College gave me the leave to enable me to accept the assignment in Australia to work under Professor Mahler. So I planned to go to Australia for six weeks from the middle of October. This way I would be back on December 1 and would have enough time to catch up on class work to take my half-yearly exams in the third week of December. My parents decided to join me on the trip because my father had lots of academic connections in Australia, and also because my father's younger brother and younger sister were in Australia with their families in Melbourne and Sydney respectively. So this would also provide an opportunity to visit them.

When we arrived in Canberra, Professor Neumann came to the airport to pick us up. He knew my father well, but still that was a very kind gesture. Even though Professor Neumann had a car, he rode a bicycle to campus. I was surprised to see the Head of the Department arrive for work on a bicycle. That showed his simplicity and informality.

After spending a few days in Canberra, my parents departed because my father had assignments in various universities in Australia. Of particular note was his visit to Adelaide where he was the guest of the great physicist Sir Mark Oliphant who was then the Governor of South Australia. So my father actually stayed at the Governor's mansion in Adelaide and was picked up in a Rolls Royce at Adelaide Airport! While in Canberra, my father gave two talks in the physics department. It was a pleasure for him that his former PhD advisor Professor M. S. Bartlett from the University of Manchester who was visiting ANU, attended his talks.

At the mathematics department in Canberra, there was a professor from India by name K. M. Rangaswamy. He was a group theorist who had done his PhD at the University of Madurai in Tamil Nadu, India, under the supervision of Prof. M. Venkataraman, a great friend of my father. In the 1950s when the University of Madras started a branch campus in Madurai, my father and Venkataraman were sent to Madurai to start the programs in physics and mathematics respectively. That branch campus grew significantly and became Madurai University. While my father returned to Madras after a year in Madurai, Venkataraman continued in Madurai and trained several PhD students there over the decades. He once told me proudly when he came to UCLA where I was a graduate student, that he had produced about 20 PhDs, each in a different area of mathematics!

Professor Rangaswamy was very helpful in getting me settled and took me home from time to time for Indian meals cooked by his wife. I was put up in one of the student dormitories at ANU called Bruce Hall, very close to the Mathematics Department. Having attended an all-boys college in Madras, it was a strange feeling to be in a Co-ed dorm. Not only was the dorm Co-ed, but it turned out that the bathrooms and showers (with doors, of course) were also common for boys and girls! I was taken aback by this. In India we thought of America as a very permissive society, but Australia was one step ahead! But all this was of secondary importance. I had come to Australia to learn from Professor Mahler and had to make the best of my six week stay there. To describe what I learnt from him, I need to provide a bit of a background of transcendental number theory.

A transcendental number is one which cannot be obtained as a root of a polynomial equation with integer coefficients. An algebraic number is one which can be obtained as the root of a polynomial equation with integer coefficients. The degree of an algebraic number is the smallest degree among all such polynomials with integer coefficients for which it is a root. Thus rationals are algebraic numbers of degree 1, quadratic irrationals are algebraic of degree 2, and so on. For many years it was not known whether transcendental numbers existed.

The theory of transcendental numbers was born when Liouville in 1844 found the first transcendental number, namely

$$\lambda = \sum_{n=1}^{\infty} 10^{-n!} = \frac{1}{10} + \frac{1}{100} + \frac{1}{1000000} + \cdots.$$

The way Liouville arrived at λ was to first observe that algebraic numbers can be approximated by rationals only to a certain degree of precision. More precisely, the theorem that Liouville established is that if α is algebraic of degree $k > 1$, then there exists a constant c depending on α such that

$$|\alpha - \frac{p}{q}| > \frac{c}{q^k}, \quad \text{(Liouville's theorem)}$$

for all rationals p/q. The rationals p_m/q_m obtained by adding just the first m terms of the series for λ at position m, would be such strong approximations that they would violate Liouville's theorem for sufficiently large m. Thus λ cannot be algebraic, and so it is transcendental. More generally, transcendental numbers having rational approximations that are so strong as to violate Liouville's theorem, are called Liouville numbers.

After Liouville's discovery of transcendental numbers, it was shown that certain well known and important numbers like e, the natural base of the logarithm, and π, the ratio of the circumference of a circle to its diameter, are transcendental, but are not Liouville numbers. The transcendence of π was confirmed by Lindemann toward the end of the nineteenth century and it settled the famous problem of squaring the circle, in the negative. The three problems of antiquity are:

(i) *Trisection of an angle:* Trisect an angle using only ruler and compass.

(ii) *Doubling the cube:* Given a cube, construct another cube of twice the volume using only ruler and compass, and

(iii) *Squaring the circle:* Given a circle, construct a square using only ruler and compass that is equal in area to the circle.

All three problems are impossible to solve. The impossibility of (iii) came out of the transcendence of π. The numbers which are constructible using ruler and compass are certain types of algebraic numbers. To square the circle means to construct $\sqrt{\pi}$ and this is impossible using ruler and compass because the transcendence of π implies the transcendence of $\sqrt{\pi}$.

While Liouville's theorem for algebraic numbers enabled him to construct the first transcendental number λ, his result was significantly improved over the years. First the Norwegian mathematician Axel Thue improved the exponent k in Liouville's theorem to $(k+2)/2$ in 1909. This improvement had very significant consequences in the theory of Diophantine Equations, such as in showing that equations like

$$x^n - ay^n = b \quad \text{(Thue Equation)}$$

where a, b are integers, have only a finite number of integer solutions in x, y when $n > 2$. Next Siegel in 1921 improved Thue's result by replacing $(k+2)/2$ by $2\sqrt{k}$. The story came to a grand conclusion in 1955 when K. F. Roth showed that for all algebraic numbers, Liouville's holds with any exponent $\mu > 2$ in place of k. This result is referred to as the Thue-Siegel-Roth Theorem and for this Roth received the Fields Medal at the International Congress of Mathematicians in Edinburgh in 1958.

Cantor's theory of infinite sets shows that almost all complex numbers are transcendental in the Lebesque measure sense. Even though there is a preponderence of transcendental numbers, it is generally very difficult to prove a given number to be transcendental. The way Lindemann deduced the transcendence of π was to prove that for every non-zero algebraic α, the value e^α is transcendental. Since $e^{i\pi} = -1$ is algebraic, it follows that $i\pi$ is transcendental, and so π is transcendental.

Following the work of Lindemann, there was considerable interest on functions f which satisfied $f(\alpha)$ is transcendental for every non-zero algebraic α. During discussions in his office, Professor Mahler showed me the proof of a classical result of his, namely that

$$f(z) = \sum_{k=0}^{\infty} z^{2^k} = z + z^2 + z^4 + z^8 + z^{16} + z^{32} + \cdots$$

is transcendental of every algebraic α satisfying $0 < |\alpha| < 1$. Mahler showed me other examples of functions given by series which were transcendental at algebraic values of the argument within the circle of convergence.

Mahler (1953) established effective irrationality estimates for π such as

$$|\pi - \frac{p}{q}| > \frac{1}{q^{42}}$$

for every rational p/q. A few years later, I was involved in research at the University of Michigan on effective irrationality estimates for logarithms of certain algebraic numbers, and k-th roots of certain rationals.

Mahler's classification alluded to earlier, applies to all complex numbers which he decomposed into four classes, namely the A-, S-, T-, and U-numbers. The A-numbers are just the algebraic numbers. In his classification scheme, Mahler considered the quality of approximation of complex numbers by all algebraic numbers, not just by rationals. From this classification it turned out that the Liouville numbers are U-numbers of degree 1. Other noted mathematicians had suggested different classification schemes for transcendental numbers, but Mahler's is the one most widely used.

In addition to work in transcendental number theory, Mahler was an authority on p-adic numbers and functions, and on the popular subject of normal numbers. A normal number to the base 10 is a number in whose decimal expansion, every block of k digits occurs with the expected frequency $1/10^k$. That is the digit 1 occurs with frequency $1/10$, as does 2, and 12 occurs with frequency $1/100$, and so on. One can define a normal number to other bases b using expansions in base b instead of the decimal expansion. A normal number ν is one which is normal to all bases. It is known that almost all numbers are normal but no single example of a number normal to all bases has been constructed! The famous Champernowne number 0.12345678910111213... obtained by writing the integers successively in decimal form, is normal to base 10, but the same construction when used in other bases will not yield the same number. Mahler proved that the Champernowne number is transcendental, and his method yields the transcendence of other similarly constructed numbers.

Kurt Mahler was perhaps the most dominant figure in transcendental number theory until the arrival of Alan Baker in the sixties. For his fundamental contributions, he received numerous awards and recognitions such as the Fellowship of the Royal Society (1948), the Senior Berwick Prize of the London Mathematical Society (1950), and the DeMorgan Medal of the London Mathematical Society (1971). He was honored in Australia with the Thomas Ranken Lyle Medal in 1977. It was a privilege for me to learn mathematics from one of the towering figures of the twentieth century.

Mahler had two hobbies — photography and yachting. He proudly showed me several beautiful pictures he had taken on his voyages on the South Seas while riding on his own yacht.

Mahler had a great veneration for his PhD advisor Carl Ludwig Siegel who had made fundamental contributions to several parts of number theory. In his strong German accent, Mahler often referred to Siegel as "Kaaarl Loodwig Zeeegal", and always in very high terms.

The visit to Australia also gave me an opportunity to meet George Szekeres, a famous Hungarian number theorist who had settled in Sydney. In my investigations on the function $A(n)$, I needed an asymptotic formula for the number of partitions of an integer into primes. Such an asymptotic formula was obtained by K. F. Roth and Szekeres. When I was in Sydney with my parents, we visited Szekeres one evening at his home in Turramurra on the outskirts of Sydney. There I met his wife Esther Klein who was also a Hungarian mathematician. Szekeres actually started out as an analytical chemist and only later became a mathematician. He was very active in encouraging students to pursue advanced mathematics by conducting high school Olympiads and undergraduate competitions in Australia. He was therefore very appreciative of my attempts in research. After discussions that evening, Professor Szekeres invited me to speak in the Number Theory Seminar at the University of New South Wales in Sydney. Actually I gave two talks at UNSW, one on Farey-Fibonacci fractions, and another on the function $A(n)$. When I visited the Mathematics Department of the University of New South Wales, Szekeres introduced me to Alfred van der Poorten (Alf as he was known to friends), a former PhD student of Kurt Mahler who was on the faculty there, and John Mack, one of Szekeres' PhD students. They both attended my talks. That was my first meeting with van der Poorten and Mack. Over the next several years, I met van der Poorten at various conferences and we used to find out who between the two of us had travelled farther to come to the conference — he from Australia or me from India. I got to know Alf quite well and so I invited him in 1996 to the Editorial Board when I launched the Ramanujan Journal. He graciously accepted the invitation. I also got to know John Mack quite well when he later visited the University of Michigan in 1980–81 to work with Hugh

Montgomery. I was a Hildebrand Assistant Professor in Michigan during 1978–81. Actually, that contact with Szekeres in 1973 later led to one of my papers being accepted in the Journal of the Australian Mathematical Society, of which he was the Editor. Paul Erdős had a role to play in that (see Section 2.7). In summary, the visit to Australia was fruitful in many ways, and I cherish most the opportunity of learning some transcendental number theory from Kurt Mahler, the great master of that fascinating field.

2.5   A special summer in sunny California (1974)

Summer 1974 was special. I was on a trip to the United States with a two-fold purpose: (i) to visit different universities where I was planning on applying for graduate studies, and (ii) to work for six weeks with the Fibonacci expert Professor Verner Hoggatt of San Jose State University.

In 1973–74, with the exposure I gained from visits to Michigan and Australia, I went through several books in number theory. This broadened the areas of my own investigation. I had a few more papers written up and I lectured on these at MATSCIENCE. I thus had more material that I sent to mathematicians abroad for critical opinion. Since I received favourable comments, I decided by the end of the academic year 1973–74, that I would apply in Fall 1974 to various American universities for admission to graduate school starting Fall 1975. Some of the mathematicians from Europe and USA who had gone through my work, agreed to write letters for me.

It is quite common nowadays for undergraduate students in the USA to visit various universities before deciding where exactly to apply for graduate studies. It is very uncommon for students in India to make such a trip of exploration to the USA because the costs are prohibitive. Fortunately in my case, a trip to the USA in the summer of 1974 was just another of my trips accompanying my father, and so I could explore various American universities and meet number theorists there before deciding where to apply. In addition, since I had submitted my work on Fibonacci numbers to the Fibonacci Quarterly, Prof. Verner Hoggatt, its Editor, invited me for six weeks to work with him. In view of these programs I had, my mother did not accompany my father and me to the United States on this trip.

The major part of the trip was a six week stay in Santa Clara, California, where Prof. Hoggatt lived. As in the case of my stays in Ann Arbor and in Canberra, my father dropped me in Santa Clara and left on his lecture assignments at various American Universities.

Verner Hoggatt was the world's leading authority on Fibonacci numbers. He was at San Jose State University and invited me as a Visiting Scholar.

He lived in Santa Clara near San Jose and so he arranged accommodation for me in the home of a neighbor Mr. Speisock, but I was to have all meals at the home of the Hoggatts. The Spiesocks were very kind to me, and the Hoggatts took parental interest in me. Mrs. Herta Hoggatt, a gracious lady, cooked the most delicious European style vegetarian meals for me.

Hoggatt was the founder and the Editor-in-Chief of The Fibonacci Quarterly, the official research publication of the Fibonacci Association. Fibonacci numbers occur in nature quite often, and in various mathematical settings as extremal solutions. For example, the Euclidean Algorithm which is used to compute the greatest common divisor (gcd) of two given numbers, works least efficiently when the two numbers are consecutive Fibonacci numbers. Owing to the frequent occurrence of Fibonacci numbers in a wide variety of problems, the Fibonacci Quarterly, a journal devoted primarily to the study of Fibonacci numbers, was launched.

Hoggatt and I went to the University every morning after breakfast at his home. There he worked with me and attended to his departmental duties. We then returned home for lunch. After lunch we worked again in his lovely study. Often in the late afternoons, we played tennis. I was an avid tennis player trained in the copy-book style by my seniors in Madras who admired the British tradition in tennis. Hoggatt was all craft and guile. In the beginning, my youthful power play was blunted by his vicious spins, but toward the end of my stay I got the better of him.

During the six weeks, Hoggatt and I wrote several joint papers and so the stay was productive. I asked him when my paper on the Farey Sequence of Fibonacci Numbers which I submitted in 1972 would appear. He said that there were several papers in the pipeline ahead of mine. My paper on the Farey sequence of Fibonacci numbers and some of my other papers appeared in The Fibonacci Quarterly in 1975.

In addition to collaborating with Hoggatt, I also attended a course on number theory at San Jose State University given by Professor Hugh Edgar who was on the faculty. He was of genial disposition, and was very kind to me. Some years later Edgar published a nice introductory book on number theory which we have used at the University of Florida for undergraduate courses from time to time. Over the years I have met Edgar at several conferences, and he always enquired about my academic progress.

Owing to multitude of elegant properties that Fibonacci numbers have, they have attracted the attention of both experts and non-experts. Hoggatt was not only the world's leading authority on Fibonacci numbers, but also its greatest enthusiast and advocate. Fibonacci numbers were his life, body, and spirit. Many sophisticated mathematicians agree that Fibonacci numbers have beautiful properties, but feel that there are more important and deeper aspects of mathematics that one ought to pursue. This is of course correct. Since Hoggatt worked exclusively on Fibonacci numbers, he was not taken seriously in the higher mathematical circles. Hoggatt was a very cheerful personality, but over time, this lack of recognition bothered him, and so he fell into deep depression. In the end he committed suicide in 1979 by hanging himself from a tree in the backyard of his house when his family was away. I was at the University of Michigan as an assistant professor at that time, and was much hurt on hearing this news. I had known him as a cheerful spirit who enjoyed life and so this news came as a shock. In December 1979 when I attended the West Coast Number Theory Conference in Asilomar, I drove to Santa Clara to pay my condolences to his wife. She told me that Hoggatt had become very depressed with the lack of recognition of his work.

Hoggatt told me that the mathematician who appreciated and encouraged his work on Fibonacci numbers the most was Howard Eves, who was a noted geometer and an authority on the history of mathematics. There were several noted mathematicians who appreciated Hoggatt's work, and many of them — Paul Erdős and Leonard Carlitz, for instance — contributed to a special issue of the Fibonacci Quarterly dedicated to Hoggatt's memory that appeared in 1981. I too contributed to that Special Issue and I will describe that paper now.

It is a well known theorem of Chebyshev, that if two positive integers m and n are randomly chosen, then the probability that they are relatively prime (that is have no common factors > 1), is $6/\pi^2$. In my paper for the Hoggatt Memorial Issue, I showed that the probability that n and $\Omega(n)$ are relatively prime is also $6/\pi^2$, where $\Omega(n)$ is the number of prime factors of n counted with multiplicity. The same result holds for $\omega(n)$, the number of distinct prime factors of n. Thus from the point of view of relative primality, $\Omega(n)$ behaves randomly with respect to n. Results like this for a class of additive functions were established by R. R. Hall, but his theorem did not apply to $\Omega(n)$. So I investigated this problem.

Hoggatt and I corresponded often for some time after my stay with him. We wrote a very nice joint paper with the incomparable Paul Erdős while I was a graduate student at UCLA. I will describe this joint work now.

When I was at UCLA, Hoggatt informed me about the following lovely observation of Joseph Silverman: *The set of positive integers can be partitioned uniquely into two (disjoint) subsets such that no two members from the same subset will sum up to a Fibonacci number.* The first few members of the two subsets S and T (by placing 1 in S) are:

$$S = \{1, 3, 6, 8, 9, \ldots\}, \quad \text{and} \quad T = \{2, 4, 5, 7, 10, \ldots\}$$

Silverman did not have a proof of his observation, and so Hoggatt asked if I could prove this, which I did. To understand this problem better, consider the following general situation: We are given an increasing sequence of integers $\mathcal{A}$. We say that $\mathcal{A}$ produces an *additive partition* of the positive integers $\mathcal{Z}^+$, if $\mathcal{Z}^+$ can be decomposed into two (disjoint) subsets such that no two members of the same subset sum up to an element of $\mathcal{A}$. In proving Silverman's observation, we showed more generally that any integer sequence $\mathcal{A}$ satisfying the Fibonacci recurrence, and with first term 1 and second term > 1 would generate a unique additive partition of $\mathcal{Z}^+$. When I mentioned this to Erdős who was visiting UCLA at that time, he formulated the concept of an additive partition in terms of two-colorability of certain graphs. Thus Erdős brought a graph theoretic perspective and so it became a triple paper which appeared in the journal Discrete Mathematics in 1978. There we note that certain sequences $\mathcal{A}$ growing faster than a Fibonacci sequence generate more than one additive partition, while certain others that grow slower, do not generate an additive partition. Thus the Fibonacci sequence growth is ideal to generate a unique additive partition.

It was during my stay with Hoggatt that I met G. L. Alexanderson of the University of Santa Clara. Alexanderson is independently wealthy and pursued mathematics purely for pleasure. He was very active in organizing the Putnam Exam for mathematics undergraduates, and more generally in activities encouraging student research in mathematics. Hoggatt said that I should meet him and took me to his home. Alexanderson was impeccably dressed in a three piece suit as he chatted with us. This is how he is always dressed. But beneath that formal style of dress, was a most charming personality and a very caring human being, whose main desire was to serve the mathematical community and to encourage talented youngsters. I have met Alexanderson several times since then, especially at the Joint Annual Meetings of the AMS and the MAA, and he always took time to ask how I was progressing and what mathematics was engaging my attention. Alexanderson has immense knowledge of the history of mathematics, and so, he often gives comments on the historical charts and objects that appear on the cover or the frontspiece of the Bulletin of the AMS.

As I mentioned earlier, one of my two goals on this trip to America was to visit certain universities where I wanted to apply for graduate studies.

One of places I was considering was the University of California, Berkeley. I had visited the physics department at Berkeley in the sixties with my father, but now I wanted to see the mathematics department. Berkeley's pride, and rightfully so, was the fact that it is the only public university competing in equal terms with private universities like Harvard, Princeton, Stanford, and others in the top ten. It was Hoggatt who took me to Berkeley to meet Professor Max Rosenlicht of the mathematics department there who showed me around the department and the campus.

After my stay in Santa Clara with Hoggatt, my father took me to various universities in America to meet mathematicians whose work interested me.

I had begun studying the book of Lang on Diophantine Approximations and there I read about the work of Bill Adams his former student. Adams was at the University of Maryland, where he and Larry Goldstein had an active program in number theory. So I visited the University of Maryland to talk to them. I was impressed not only with their number theory program, but also with their friendliness. During our conversation, Goldstein was waxing eloquent about the great John Tate at Harvard, one of the most influential figures in algebraic number theory. So my father and I decided to fly to Boston to meet both Tate at Harvard and Harold Stark at MIT.

The mathematics department at Harvard University had an impressive address — 1 Oxford Street, Cambridge, Massachusetts! It was a very modern building in a campus filled with older buildings covered with ivy creepers. So it definitely stood out. Both Harvard and MIT are located in Cambridge, Massachusetts, a town that is across the Charles River from Boston. For those of us in India, Cambridge means Cambridge, England. Most people in India identify Harvard and MIT with Boston, and are not aware of Cambridge, Massachusetts.

While my father and I waited in the lobby of the mathematics department, Tate arrived riding a bike. He was very informally dressed in jeans and a t-shirt. He told me about the mathematics graduate program in general and the graduate number theory courses in particular. He showed us around the department and took us to the office of his colleague Garrett Birkhoff (another famous algebraist), whom we had met earlier at the General Motors plant in Detroit in 1969. After the meeting with Tate and Birkhoff, we walked "across the street" to the campus of MIT where we met Harold Stark.

Stark was at Michigan until the late sixties. He solved one of the most famous problems in number theory, namely the determination of all imaginary quadratic fields with class number 1 That propelled him into stardom, and he was offered a permanent position at MIT which he accepted. He was actually a participant at the Summer Institute in Number Theory in Ann Arbor in 1973 and spoke about this great work there. I attended his talk, but I did not interact with him then.

Stark is a very shy person. In our conversation at MIT, he said very little. I thought that his silence was because he had formed a poor opinion of me, or perhaps he was in deep thought over a fundamental problem. I was later told that he is extremely shy and therefore I should not interpret his silence as a poor impression of me.

The last university I visited was the University of California at Los Angeles (UCLA) to meet Professor Ernst Straus. I was totally impressed by his kindness. He came to the airport to pick me and my father up, and took us to the campus of UCLA. Straus had begun his academic career as an assistant of Albert Einstein at the Institute for Advanced Study in Princeton, and collaborated with Einstein on the general theory of relativity and on the grand unification program Einstein was working on. In fact the last three papers of Einstein were written jointly with Straus. So Straus had rubbed shoulders with the greatest of the greats — with Einstein himself. But Straus was very simple and unassuming. For him to come to the airport to pick up a potential graduate student showed his simplicity and kindness, as well as his genuine interest in guiding graduate students. This definitely was one of the reasons I decided to work under Professor Straus. After Einstein died, Straus moved to UCLA. It was then that he shifted the focus of his research to number theory. In Chapter 3, I will relate fascinating Einstein stories that Straus told me and several others.

In any case the 1974 summer trip to the USA enabled me to visit several leading mathematics departments and to meet noted number theorits there. This helped me in deciding where to apply in Fall 1974 for graduate studies to begin in Fall 1975. But before I discuss that, I would next like to describe my undergraduate college experience in India and what made that special.

2.6 College student life in India (1972–75)

I thoroughly enjoyed the three years of BSc Mathematics at Vivekananda College, in Madras, and learnt a lot of classical mathematics. The mathematics department at Vivekananda College was one of the best among the colleges in Madras, and several distinguished mathematicians had studied either the BSc or MSc there. In India, the Bachelors course is for three years unlike the four-year program in the USA.

Whatever I say here about undergraduate education in India is with regard to colleges affiliated with universities and my comparison is with large state universities in the United States. I am not discussing the undergraduate and graduate education at the exclusive academic institutions in India like the IITs geared to only the brightest students, nor am I making comparisons with elite universities in the USA like Harvard and Princeton.

There are two essential difference between undergraduate education in India and the USA: (i) In the USA, you have to have a certain number of total credits over a four year period (one credit being equivalent to one hour per week in a course), but each year you can choose the courses to suit your needs. Thus there is flexibility built into the American system. In India, you have to declare your major before admission, and the courses that you take each year, as well as the sequence in which you take them, is pre-determined. You do not have a choice of what to take when. The only choice is in the ancillary subjects. I chose statistics and numerical analysis as ancillary instead of physics and chemistry. (ii) The second difference is that in India, every subject in mathematics was taught (until the end of the twentieth century) with full proofs, the way mathematics should be taught, starting from the basics.

The sequence of courses is pre-determined in India because the prerequisites for a higher level course will be automatically covered in an earlier semester or earlier year. I really like this system and feel fortunate that I had undergraduate education in India and graduate education in America. The main drawback in the Indian system is that if in mid-stream you find out that BSc physics is not your cup of tea and you want to change to BSc chemistry, this is not possible. The flexibility of the American system permits such changes, but the undergraduate training is not as rigorous as in India. So if you are clear as to what your major is, then the Indian system is better. The contents of the courses had also remained more or less the same over several decades, with one change being that textbooks were by British authors when India was under British rule, whereas the textbooks with similar content were by Indian authors during my time as a student. But the style and form of treatment was the same — theorem, proof, examples, and problems in succession. Our textbooks did not have many pictures, just enough diagrams or figures to explain the proof.

My mathematics teachers in the BSc were all very good. Prof. K. V. Parthasarathy taught analytic geometry of the conic sections in great detail for a full year. It was amazing to me that so much could be proved by just using the formulas for distance and angle, and some trigonometry. It was a pleasure to learn and to prove that the equation of the tangent at the point (x_1, y_1) to the general conic given by

$$ax^2 + 2bxy + cy^2 + 2dx + 2ey + f = 0$$

where $a, b, ..., f$ are constants, is

$$axx_1 + b(xy_1 + yx_1) + cyy_1 + d(x + x_1) + e(y + y_1) + f = 0.$$

This is so beautiful! When I teach undergraduate classes in America, I often exclaim that a certain idea or equation is beautiful. The idea of beauty in mathematics is alien to most American students who in school are never told that mathematics is beautiful.

Analytic geometry in two and three dimensions was done in detail both in Cartesian and polar coordinates. This was a good grounding for Calculus in two and three dimensions which was done over a two year period. In later years while studying complex variable theory as a graduate student, or while teaching complex analysis as a professor, I found this detailed study of Euclidean geometry and analytic geometry to be very useful. I noticed that many American graduate students taking complex variables have to be reminded of certain fundamental properties of conics because they have not had a thorough study of analytic geometry. Here I would like to tell you Abhyankar's opinion on this matter.

When I was Chair of the Mathematics Department, I arranged for Abhyankar to give a talk to $\pi\mu\epsilon$, the Undergraduate Mathematics Club. He gave a talk entitled "An introduction to algebraic geometry". He started by saying that algebraic geometry is really analytic geometry but in more abstract form. He lamented that students in America are not given enough exposure to analytic geometry (with full proofs) in high school or in undergraduate classes. He said that as a student in India he had two years of analytic geometry in college and that his father, who was also a mathematician, had three years of instruction on that subject in college. He was critical that the time spent on analytic geometry was steadily decreasing.

In America there are many reasons why analytic geometry is not treated in depth with proofs, and why proofs are not emphasized until one gets to linear algebra. Firstly, there is a desire to have more students get college (undergraduate) degrees, and so standards have to be lowered to accomplish this. Secondly, more advanced and newer subjects are taught early on and so not enough time is spent covering the fundamentals like analytic geometry with proofs. In India, in trigonometry, the addition formulas for sine and cosine functions were proved geometrically — at least a quarter century ago. I simply cannot imagine how one can grasp mathematical subjects without a thorough understanding of the fundamentals. I think it is too late to wait until linear algebra to introduce proofs.

There is a misconception that in countries like India, students are asked to memorize a lot and not encouraged to think, and that in the United States, students are asked to think. Having studied up to BSc in India and having taught undergraduate classes in the USA at many universities, I know that the opposite is true at least in mathematics up to the undergraduate level. By teaching mathematics with proofs, students are made to think. By teaching mathematics without proofs as in schools and lower undergraduate classes in the USA at many universities, students tend to memorize rather than understand the how and why. But then, as was told to me by Heini Halberstam, in the educational sphere, for better or worse, in about a decade, England will incorporate the changes in the educational system adopted by the United States, and India would follow a decade later. During a visit to Chennai (formerly Madras) in December 2015, my former analytic geometry teacher Prof. K. V. Parthasarathy (now retired) called on me at my home for a chat. When I asked him how analytic geometry was being taught nowadays, he said that proofs were largely omitted, and most of the time, theorems were just illustrated with examples.

We had a good course in astronomy that used a lot of spherical trigonometry. All the required theorems from spherical trigonometry were proved before they were used. The course included a study of precession and nutation and a detailed treatment of eclipses. By the bye, my maternal grandfather, the late H. Subramani Iyer, was a Professor of Astronomy and the Principal of the Maharajah's College in Trivandrum. He wrote a fine textbook in Astronomy for college students. Had he been alive in the seventies, he would have been happy to see that I had astronomy in my undergraduate classes.

Our enthusiastic astronomy teacher, the late Prof. Ranganatha Rao, was also the faculty advisor to the Astronomy Club of which I was a member. Our equipment was primitive, but our interest in the subject was high. Once a month, the Astronomy Club members gathered at night on the terrace of the mathematics building to view the heavens with a simple telescope. Most students spent the entire night there as Ranganatha Rao would show us the rotation of the constellations around the pole star.

Prof. P. R. Vittal, who taught statistics, is a teacher par-excellence. His voice would raise to a high pitch as he stressed the importance of a big theorem. Years later, when I did research in Probabilistic Number Theory, I found the undergraduate statistics course that I took in India to be very useful. Prof. Vittal is such an outstanding teacher that he is most sought after as a private tutor even today.

Three dimensional calculus was taught by Prof. K. S. Ramachandran, admired by students for his brilliance and affectionately called KSR. He grew up as a student at the Ramakrishna Mission Students Home. My father said that when KSR was of middle school age, he would come to our family home Ekamra Nivas along with other students of the Home to recite the vedas before my grandfather Sir Alladi. KSR could have taken to a career as a research mathematician but out of gratitude decided to serve the Ramakrishna Mission by being a teacher at its Vivekananda College.

Two of the subjects that were missing in my undergraduate mathematics curriculum were number theory and abstract algebra. The latter was taught in India only at the Masters level. I learnt abstract algebra at UCLA from one of the great Masters — Professor Robert Steinberg. I was studying number theory on my own in India and so it did not matter that I had no undergraduate course on that subject. I have so far talked about the merits of the undergraduate instruction in India. So if there was a defect in the mathematics curriculum for the BSc in India, I think it was the absence of abstract algebra and number theory. Instead of Statics and Dynamics, we could have had Abstract Algebra and Number Theory. Or we could have been given a choice between Statics/Dynamics and Algebra/Number Theory just as we were given a choice in our ancillary subjects. In this regard, the flexibility in America to choose the courses one needs is definitely better. My main complaint of mathematics instruction in America nowadays in high school and early college is that the subject is being taught without proofs. This is useless because one never understands mathematics without going through the proofs. This lack of understanding is the main reason why so many students in America hate mathematics.

Our calculus teacher, Prof. A. K. Rajappa (recently deceased) was an enthusiastic rebel. He took permission from the Department Head K. Subramaniam, to do an introduction to analysis (= point-set topology of the reals) before taking up calculus with proofs. I now realize how good this was, but initially I fumbled with point-set topology. I got only 56 out of 100 in my first test with Prof. Rajappa — the first (and only) time I ever scored so low in mathematics. I was determined to master point-set topology and the epsilon-delta method. Seeing my genuine interest, Prof. Rajappa suggested that I read the book of R. R. Goldberg, which I did over the break. Coming back from the break, I did so well, that Prof. Rajappa pointed to me and told the class in a thundering voice —

"There is one among you who has mastered the techniques and gone far ahead." That was the kind of encouragement I received from my college teachers.

Another of my college teachers who encouraged me and who also could be termed as an enthusiastic rebel, was Prof. Ramakarthikeyan. He taught us Statics in the BSc class. He said that even though the class met five days a week, all the material could be comfortably covered in four classes per week. So I immediately suggested that one day a week could be used for a Student Seminar. I offered to start by lecturing on number theory related to problems I was working on. I said in a few weeks other students would volunteer to lecture. Prof. Ramakarthikeyan applauded my idea, got the permission of Prof. K. Subramaniam, the Department Head, and gave the go ahead for the Student Seminar. This Seminar I organized, was a great success. About half a dozen students gave talks during that semester on various aspects of current research in mathematics.

Observing the interest with which this Seminar was being conducted, K. N. Visweswaran, one of the brilliant students of the MSc class, offered to hold a Talent Exam for the BSc and MSc students of my college. I got the First Prize on this talent exam. It really was very gracious and generous on the part of Visweswaran to conduct such a talent test and give prizes. Such an act is unusual even by American standards.

I also took part in the Mathematics Talent Exams conducted by the Association of Mathematics Teachers of India (AMTI). These exams were state-wide. I received prizes every year from the PUC to the III BSc for my performance on these talent exams, culminating in the First Prize when I took it in the final year of the BSc. There were several questions on number theory on this exam since number theory was not on the curriculum of the BSc at that time. My studies in number theory helped me. Other than what the AMTI conducted, there was no Mathematics Olympiad at a national level in India for undergraduates at that time like the Putnam Exam in America. The Indian National Mathematics Olympiad (INMO) for high school students conducted by the National Board for Higher Mathematics (NBHM), was launched only in 1989.

In addition to mathematics, I also had classes in Sanskrit and English until the second year of the BSc. The courses in the third (last) year of the BSc were exclusively in the major which in my case was mathematics.

Vivekananda College had the best Sanskrit department among the (affiliated) colleges in the State of Madras. One of our Sanskrit teachers Prof. S. Viswanathan, taught us the great epic *Kumarasambhava* of the legendary poet Kalidasa (often referred to as the Shakespeare of India). Professor Viswanathan was actually one of the two head priests at the *Kapaleeswar Temple* in Mylapore. Early in the mornings, he could be seen bare topped and clad in a traditional dhothi as he performed the *puja* at the Kapaleeswar Temple. But then, by 10:00 am, he would change into a full suit, wear a turban to hide his tuft of hair, and drive his Ambassador car into Vivekananda College to lecture in his Sanskrit classes. I also had a fine course on the history of Sanskrit literature by Professor Rajagopalan, the Head of the Sanskrit department at Vivekananda College. Two other faculty members of the Sanskrit Department were Drs. Narasimhachari and Ramaratnam. Both were highly knowledgeable and I had them as my teachers for classes on the great Sanskrit epics such as *Kadambari* by the poet Bhatta Bhana and *Nala Charitha* — both these epics in abridged form. I enjoyed the Sanskrit classes immensely. Since I already had Sanskrit in high school, I did extremely well and received the first prize in each of the three years that I had Sanskrit in College; these prizes were given during College Day near the end of the academic year.

One of the unique features of Vivekananda College was that the first period on Mondays was a religion hour — that is Hindu religion. This period started with a vedic chanting led by Prof. Viswanathan and his colleagues in the Sanskrit department. Attendance was not mandatory for this religion class, but most students including me attended it every Monday. Only the PUC class students attended it in the main hall where Prof. Viswanathan and his colleagues were seated on stage as they recited the Vedas. All other students heard the chanting in their classrooms through the loudspeakers. Over a four year period, by listening to the chanting every Monday morning, I learnt by-heart all the vedic verses that were recited such as the *purushasuktham* and certain *upanishads*. The vedic chanting lasted 30 minutes. After that there was a lecture/discourse on the teachings of Swami Vivekananda by one of the monks of the Ramakrishna Mission. I did not find the one hour per week religion hour an infraction into our learning process. It certainly was not forced on us. I enjoyed it and profited from it. It is sad that due to political pressure, this one hour religion class no longer exists in Vivekananda College!

The Head of Department of English at Vivekananda College at that time was Prof. Srinivasan. He wore a traditional Indian dress of a dhothi in *panchagajam style* topped by a *jibba*, and had the red line mark on his forehead that orthodox *Iyengars* wore. Most Indians admired English language and literature — even those who in personal life adhered to orthodox Hindu practices. I remember Prof. Srinivasan's classes in which he repeatedly pointed out the elegance of various English phrases. He was aware of my interest in mathematical research, and so during one conversation I had with him after a class, he told me that his daughter Parimala was doing her PhD in mathematics at the Tata Institute in Bombay under the direction of Professor R. Sridharan. Actually Sridharan did his BSc and MSc Mathematics at Vivekananda College in the 1950s. Parimala became a professor at the Tata Institute and after her retirement from there, joined Emory University. Her younger brother Soundararajan was my classmate in the Pre-University at Vivekananda College. It is a small world! In summary, my College experience was very enjoyable and rewarding. I not only interacted closely with my classmates and seniors, but with my teachers as well, so much so, that my professors became close friends. Even today, whenever I visit Madras, Professors Vittal and Parthsarathy call on me at my house and spend a leisurely few hours with me.

Having pointed out positive aspects of the Indian undergraduate education and the defects of the American system, I must say that at the Masters and PhD level, American Universities leap way ahead of their Indian counterparts. In the USA, it is at the graduate level that the real excellence in education is seen. I realized this, and therefore decided to do my graduate studies in America. That transition was facilitated by my contact with Paul Erdős, one of the most influential figures of twentieth century mathematics, and I describe this next.

2.7 My first meeting with Paul Erdős and its impact (1975)

As I entered the final year of my BSc Course (1974–75) after my return from the USA where I had visited universities to get an idea of the graduate programs, I knew where I was going to apply for graduate admission in the USA. So I spent the bulk of Fall 1974 filling out these applications. I continued communicating with leading number theorists about my work. Some of them said that the person I should really contact is Paul Erdős whose life's mission was to spot and encourage very young mathematicians. My contact with Erdős was the most significant event of my academic life, for it gave me an opportunity to collaborate with one of the most gifted minds of our generation, and completely determined my career path.

Paul Erdős was one of the most eminent mathematicians of the twentieth century. There were several stories — indeed legends — about his unusual life that I had heard. He was Hungarian and had an honorary position at the Hungarian Academy of Sciences in Budapest, but he travelled all the time around the world seldom spending no more than two weeks in one place. This is certainly not easy because in order to be invited with that frequency to academic centers across the globe, not only should your research be first rate, but you should be able to collaborate or interact with mathematicians with varying backgrounds and tastes. Erdős was superbly suited for this because he was the greatest collaborator in the world. Of his papers which numbered more than 1000 at that time (an enormous number for a mathematician), more than half were in collaboration and that too, several of them with young people. Erdős often joked that he sought young collaborators because he believed he would be alive as long as at least half of his collaborators are alive! Erdős was also single and so he could travel more easily with that frequency than someone with a family. In addition, he was so kind and helpful that people simply loved to have him as a guest. He was also unrivalled in communication — writing about a dozen letters a day from wherever he was, to mathematicians across the world. His letters would start as follows: "Dear ..., Let n be an integer satisfying ..." It is not that he did not make personal references in his letters; he always made kind enquiries usually at the end — such as "How are your parents?" or "My enquiries to your wife and children" in letters he wrote to me, for instance. In any case, in those days when there was no e-mail, the best way to communicate mathematical ideas was often in a letter to Erdős. He was the hub through which mathematical ideas in number theory and combinatorics were channeled. He also had an encyclopedic knowledge of the work done in his areas of expertise which was broad — number theory, combinatorics, graph theory — and so writing to Erdős was sometimes better than going to the library! He was one of the most influential mathematicians of the 20th century owing to his fundamental contributions, and his numerous collaborations through which he molded the careers of many. He also was one of the most lovable scientific personalities because of his genial nature and almost childlike simplicity. There are two well known biographies about him (by Hoffman [B10] and Schechter [B13]).

With this background information about Erdős that I had, I sent him a letter to the Hungarian Academy of Sciences in October 1974, since I did not know where he would be, and hoped that it would be forwarded to him suitably. In that letter I mentioned several results that I had obtained on (i) the function $A(n)$, (ii) some generalizations of the Euler and other arithmetical functions, and (iii) Fibonacci numbers. My letter was forwarded, and it was hard to believe that he responded in November. In his response he said that he liked my results very much and would like to meet me. In December 1974 he was going to be in Calcutta at the Indian Statistical Institute (ISI) for an international conference in memory of its founder Professor P. C. Mahalanobis who had died in June 1972. He enquired whether I would be at this conference. I replied saying that my paper on the additive arithmetic function $A(n)$ was accepted for presentation in the student section of the conference, but I could not go because of my half-yearly exams in December. These exams were extremely important because this was my final year, and in my applications for graduate admission, I had to provide information about my grades in the college exams. But I informed Erdős that my father who was invited to deliver one of the main lectures on probability at that conference, would present my work in that student session on my behalf.

Shortly before my father departed for Calcutta, I gave him a summary of my results to him. He understood it of course, but said that because he did not have expertise in number theory, he would not be able to answer questions. I thought that was fair enough. What a surprise it was to my father that just as he was about to start his presentation of my work at the student session, Paul Erdős walked in and sat in the front to hear the talk! That shows how genuinely interested he was in finding out about the work of an aspiring student. After the talk, Erdős walked up to my father and said, "While I am very pleased to meet you, I would be much happier to meet your son." Knowing that I am in Madras, Erdős continued and said, "From Calcutta I need to go to Sydney for my next assignment, but I have a few days free in-between my assignments in Calcutta and Sydney. So I could come to Madras on my way to Sydney to meet your son." That a mathematician of such eminence would be willing to reroute his international journey to meet a potential student, speaks volumes of his generosity, and his passion to encourage young mathematicians! My father was stunned when he heard this. He felt that the proper thing to do would be to arrange some lectures for Erdős in Madras during his brief visit. My father was the Director of MATSCIENCE and Professor Bhanumurthy was the Director of The Ramanujan Institute in Madras. So my father contacted Bhanumurthy upon return from Calcutta and two lectures were arranged — one at MATSCIENCE and one at The Ramanujan Institute.

Erdős arrived from Calcutta on Sunday, January 5, by a night flight and I went to the airport to receive him. Needless to say that I was very nervous since I was receiving a legend of twentieth century mathematics. I spotted Erdős descending the stairs of the aeroplane — yes, there were no vestibules at Madras airport then — and he was wearing his overcoat as usual even though the temperature in Madras was around 75°F in December at night! As soon as I greeted him, he said: "Hello, how are you? Do you know my poem about Madras?" I replied that I did not. So he recited the poem:

This to the city of Madras
The home of the curry and the dhal
Where Iyers speak only to Iyengars
And Iyengars speak only to God

He said that this poem had been modeled along the lines of a similar well known poem about Boston:

This to the city of Boston
The home of the bean and the cod
Where Lowells speak only to Cabots
And Cabots speak only to God

Erdős had understood that Madras was the seat of Brahminical culture in India. In the Hindu religion there are several Gods but among them the premier Gods are *Brahma* (the Creator), *Vishnu* (the Protector), and *Shiva* (the Destroyer), and their consorts *Saraswathi* (Goddess of Learning), *Lakshmi* (Goddess of Wealth) and *Parvathi* (Goddess of Physical Strength). The Brahmins of the Hindu religion can be broadly divided into two classes — the *Iyers* who worship all Hindu Gods but hold Lord Shiva as their premier deity, and the *Iyengars* who primarily worship Vishnu. Iyengars are much more selective when offering their prayers. For example they will not enter the temples of Shiva. The Nobel Laureate physicist Sir C. V. Raman — known for the Raman effect in spectroscopy — was an Iyer, whereas the mathematical genius Srinivasa Ramanujan was an Iyengar.

As soon as Erdős started our conversation with a poem, I was at ease, because I realized that this great man is easily approachable. During the next few days of his stay in Madras, I was with him all the time, attending his talks at MATSCIENCE and the Ramanujan Institute, and taking walks with him along the beautiful Madras Marina beach, and discussing mathematics with him everywhere. This story as how Erdős rerouted his trip to see me in Madras is mentioned in Chapter 1 of the biography [B13] of Erdős by Bruce Schechter.

This was Erdős' first visit to Madras. So in his lectures he emphasized the work of mathematicians from South India, especially from Madras. He said that his very first paper was on a new proof of Bertrand's postulate and this was related to Ramanujan's proof of it. Bertrand's postulate is the simple statement that for each integer $n > 1$, there is always a prime between n and $2n$. Bertrand's postulate was first proved by the great Russian mathematician Chebyshev in the course of obtaining estimates for the sum of the reciprocals of the primes. The proof that Erdős found in 1932 involved ingenious use of properties on the central binomial coefficient. It is Erdős' proof that is given in all number theory books today, not Chebyshev's proof. Erdős was only 19 when he discovered his proof and it propelled him into world fame. News of Erdős' proof spread like wildfire and was accompanied by the rhyme "Chebyshev said it and I'll say it again, that there is always a prime between n and $2n$."

Soon after Erdős published his proof, his attention was drawn to Ramanujan's proof of Bertrand's postulate and he noted the similarities. This was how he first got to know about Ramanujan. Soon he got interested in the beautiful result of Hardy and Ramanujan that *almost all* integers n have about loglog n prime factors, more precisely that the number of prime factors of n has average order and normal order loglog n. He thought that there ought to be a general distribution result for additive functions. Here I focus on strongly additive functions which satisfy

$$f(n) = \sum_{p|n} f(p)$$

where the sum is over primes p, especially because in 1934 his friend Paul Turan (another very great Hungarian mathematician) had proved that for additive functions for which $f(p)$ is bounded,

$$\sum_{n\leq x} |f(n) - A(x)|^2 \leq CxB(x), \quad \text{(Turan's inequality)}$$

where C is a constant, and

$$A(x) = \sum_{p\leq x} \frac{f(p)}{p} \quad \text{and} \quad B(x) = \sum_{p\leq x} \frac{|f^2(p)|}{p}.$$

Here $A(x)$ is asymptotically the mean of f, and the variance about this mean is bounded by $B(x)$. Turan's inequality is easily proved by simply expanding the square on the left. This gave a much simpler proof of the Hardy-Ramanujan theorem that the normal order for $\nu(n)$ is log log n. Turan's result was the first hint that there would be major probabilistic underpinnings to the Hardy-Ramanujan result. All this was clarified and put in a grand setting in 1939 by the celebrated Erdős-Kac Theorem:

Theorem (Erdős-Kac)

Let $f(n)$ be a real valued strongly additive function with $f(p)$ bounded and $B(x) \to \infty$ with x. Let

$$F_\lambda(x) = \frac{1}{x} \sum_{p \leq x, f(p) - A(x) < \lambda\sqrt{B}(x)} 1.$$

Then

$$\lim_{x \to \infty} F_\lambda(x) = \frac{1}{\sqrt{2\pi}} \int_{-\infty}^{\lambda} e^{-u^2/2} du.$$

(This theorem holds more generally for additive functions with $A(x)$ and $B(x)$ suitably defined.)

The manner in which this theorem came about is itself a fascinating story and Erdős described this in his lectures in Madras: Mark Kac, a great probabilist, was giving a lecture in Princeton on how probabilistic models could be applied to problems in analysis and number theory. Erdős attended this lecture. One of the problems Kac discussed was the distribution of additive functions. In that connection, he noted that if p and q are distinct primes, then in an asymptotic sense, the probability that an integer is a multiple of p is "independent" of the probability that it is a multiple of q. Since the probability that a randomly chosen integer is a multiple of p is $1/p$, one could define independent random variables ρ_p for each prime p by

$$\rho_p = 1 \quad \text{with probability} \quad \frac{1}{p}, \quad \text{and} \quad \rho_p = 0 \quad \text{with probability} \quad 1 - \frac{1}{p}.$$

He suggested that one can think of

$$X = \sum_{p|n} f(p)\rho_p$$

as a model for the additive function $f(n)$. If this model would work, then the Central Limit Theorem on sums of infinite random variables would yield the above theorem. Kac did not know how to show that this approximation would work. Erdős realized immediately how to make this model work, and so he walked up to Kac and showed him how to do it. The key idea was to employ Brun's sieve result that $\Phi(x, y)$, the number of integers up to x all of whose prime factors is $\geq y$ is given asymptotically by

$$\phi(x, y) \sim x \prod_{p < y} \left(1 - \frac{1}{p}\right), \quad \text{if} \quad \alpha := \frac{\log x}{\log y} \to \infty, \quad \text{as} \quad x \to \infty.$$

This enabled Erdős to deduce from the Kac model that the above theorem would hold for the truncated function

$$f_y(n) = \sum_{p|n,\, p \leq y} f(p)$$

with $A(x)$ and $B(x)$ replaced by $A(y)$ and $B(y)$, for any choice of y such that $\alpha \to \infty$. Erdős then noted that with the growth conditions on f and B as in the Theorem, y could be chosen suitably to close the gap between $f_y(n)$ and $f(n)$ using Turan's inequality. Thus the Erdős-Kac Theorem was proved and that was the birth of Probabilistic Number Theory. As Erdős said, "Kac did not know number theory, and I did not know probability. Neither of us knew what the other was doing. But two brains are better than one, and together we proved the theorem." Erdős said that he was very pleased to have worked on a problem begun by Hardy and Ramanujan. To me it was a privilege to learn about the celebrated Erdős-Kac theorem from the creator himself.

As was customary, in his Madras lectures Erdős spoke about a number of problems he had raised over the years and discussed their current status. Erdős was the greatest problem proposer in history. What was unique was that he offered prize money for each of these problems depending on the nature of their difficulty. Even to solve a $100 problem of Erdős is not easy; his problems have taken years to solve and so the prize is less than the minimum wage if one thinks of the number of hours one has to invest in working on such problems! But mathematicians, or for that matter serious researchers in general, are not working on problems for the prize money. It is simply for the challenge and the importance of the question in the landscape of their subject. Erdős was often asked what would happen if his problems were solved all at once. Does he have the funds to pay for all of them? He replied jokingly as follows: "What would happen to the strongest bank if all the depositors withdraw their money at the same time? This is more likely to happen than my problems getting solved simultaneously!" The famous mathematician Ron Graham took care of all of Erdős' finances; he would send the cheques to those who had solved problems posed by Erdős. In many instances, if the prize money is not too big, the mathematician who solved the problem would not cash the cheque; instead the cheque would be framed and displayed! In 2014, the $10,000 problem of Erdős concerning large gaps between primes was solved simultaneously and independently by James Maynard by Green-Ford-Konyagin-Tao, and this created a sensation in the mathematical world. Ron Graham presented the cheque for the solution of this problem (see Section 10.8) at a conference at the University of Florida in March 2016 for my 60th birthday.

Erdős was such a celebrity, and his lifestyle so unique, that newspaper reporters loved to interview him. The Indian newspaper reporters asked him why he never married. In Western society, and especially in British society, one does not ask any personal questions unless one knows the individual very well. That is why when two Englishmen meet for the first time, they talk about the weather. In contrast, when two Indians meet for the first time, it is very common for them to find our details about their personal life. Are you married? Do you have children? Are you looking to get your children married? Who are your parents? What kind of a job do you have? Are you happy in your job? These are some of the many questions that will be asked and answered (without discomfort) by both parties. It is as though by asking personal questions, you are showing your care and concern, not your inquisitive nature! I was at this meeting that Erdős had with the newspaper reporters. So when asked why he never married, he replied, "because I am wedded to mathematics" and shrugged. When the newspaper reporters persisted, Erdős looked at me with surprise, and so I told the reporters in Tamil (the native language of Madras) that he had answered their question and they should not pursue this further.

The reporters were also much amused and impressed that Erdős did nothing but mathematics, without any distractions. They asked him what he did for relaxation, or whether he relaxed at all. Erdős gave his favourite reply — "There is plenty of time to rest in the grave!"

Erdős also had his own peculiar (and charming!) terminology that he often used in lectures as well as in conversations. He referred to children as "epsilons", because in mathematics the Greek letter ϵ (epsilon) is used to denote a quantity that can be made arbitrarily small. He valued the ability to do research so highly, that when he said that a certain mathematician had "died", he meant that the mathematician had stopped doing research! When some actually died, he would say that person was "cured", meaning cured of the ills of life! This certainly resonates with the desire of Hindus to attain *moksha* (= salvation) through which you reach God and are never reborn to face the ills of earthly existence. The reporters were fascinated by Erdős philosophy and his unique terminology.

Erdős was an agnostic. He often jokingly referred to God as the "Supreme Fascist". In India, one often associates goodness in a human being with belief in God. You often can hear an Indian refer to another Indian as "he is a very nice person, a very God fearing individual".

Agnostics and atheists are viewed with some skepticism in Indian society, but this is changing in a modern world. Naturally Erdős' feelings about God were also of interest to the newspaper reporters. Everyone knew, including the reporters, that whatever Erdős' feelings on God were, he was the kindest and most loving human being.

Even though Erdős did not believe in God, he often jokingly said that if God existed, he would have a book of the most beautiful proofs of the most important theorems, and that after he died, he would like to take a peek into this Book of Proofs. Interestingly, in 1998, Martin Aigner and Gunter Zeigler published a book entitled "Proofs from The Book" in which they have collected the most beautiful proofs of some of the most important theorems in various branches of mathematics. It is hardly surprising that several of Erdős' proofs are there in this book, including his proof of Bertrand's postulate.

In his Madras lectures, Erdős also spoke about the work of other Indian mathematicians, most notably that of Sivasankaranarayana Pillai (S. S. Pillai). In his Hungarian accent, Erdős said the name of Pillai in full — *Sheevaa Shankaraa Narayaaana Peellaai* much to the amusement of the Madras audience. In the course of describing Pillai's work on iterates of arithmetical functions and on Waring's problem, Erdős said he was looking forward to meeting him at the International Congress of Mathematicians at Harvard in 1950, but sadly Pillai died in an air crash on his way to that conference. Erdős paused for several moments and looked at the ground in silence. He was deeply affected by this tragic accident to Pillai. Erdős was a man of deep human feelings, and of great concern for the suffering of others.

S. S. Pillai (1901–50) was perhaps India's greatest number theorist soon after Srinivasa Ramanujan. He did his undergraduate studies at the Maharajah's College in Trivandrum in India where my maternal grandfather H. Subramani Iyer was Professor of Mathematics and Astronomy, and also the Principal. Pillai was a student in one of my grandfather's classes. Later Pillai was a lecturer at Annamalai University from 1929–41 and following that he spent the year 1941–42 at the University of Travancore in Trivandrum. That year he lived in a house next door to my maternal grandfather's house, and so the family members on my mother's side all got to know him. In 1946, soon after my parents got married, in one of the conversations about the academic life in Trivandrum, my maternal grandfather told my father that he considered Pillai to be perhaps the most brilliant student to have graduated out of the Maharajah's College. Naturally, my grandfather was also affected on hearing the news of S. S. Pillai's untimely death. Pillai died in a TWA Lockheed Constellation crash near Cairo on Aug 31, 1950.

The Ramanujan Institute where Erdős gave his second lecture in Madras, is located close to the famed Marina Beach — not fronting the beach, but about a block inside. So after his lecture there, I took Erdős to the Marina Beach — the pride of Madras — for a walk. The Madras Marina Beach is one of the finest city beaches, second only to Rio de Janiero's Copacabana. All along the three to four mile stretch of the beach, are very impressive and important buildings, including those of the University of Madras. After asking me a few questions about these buildings, Erdős started talking mathematics with me. I told him that while I was able to determine that the average order of $A(n)$ was $\pi^2 n/6 \log n$, I could not determine its normal order. Erdős then said that $A(n)$ would not have a normal order because almost always it is asymptotically the size of the largest prime factor $P(n)$ which dominates the sum of the prime factors of n, and that $P(n)$ does not have a normal order. When I asked him why $P(n)$ does not have a normal order, he said it is because the quantity $P(n)/n^u$ has a density for each u between 0 and 1. He said I should read the papers of N. G. de Bruijn on the function $\Psi(x, y)$ which enumerates the number of integers up to x all of whose prime factors are $\leq y$. This was how I got introduced to de Bruijn's work and to $\Psi(x, y)$. Over the next several years I did a lot of work on $\Psi(x, y)$ and related sums, and made some improvements over certain results of de Bruijn.

Having understood the dominance of the largest prime factor in $A(n)$, I asked Erdős if $P(n)$ is subtracted from $A(n)$, whether the difference be dominated by $P_2(n)$, the second largest prime factor of n, and that if $P(n)$ and $P_2(n)$ are subtracted from $A(n)$, whether the remaining sum is dominated by $P_3(n)$, the third largest prime factor, and so on. He said that this is very likely and that it was a very good question that I had asked. He said I should try to prove such a result. So in the next few months, I worked hard on the problem, and corresponded with Erdős. We wrote a joint paper (my first with Erdős) entitled "On an additive arithmetic function" in which the main result is:

Theorem (Alladi-Erdős)

For each fixed positive integer k, the functions $P_k(n)$ and $A(n) - P_1(n) - P_2(n) - \cdots - P_{k-1}(n)$ have the same average order, namely $c_k n^{1+1/k}/\log^k n$, where c_k is a rational multiple of $\zeta(1 + 1/k)$.

Here ζ is the famous Riemann zeta function, and $P_k(n)$ is the k-th largest prime factor of n, with $P_1(n) = P(n)$ as the largest prime factor. It is to be noted that while there is no ambiguity in the definition of $P_1(n)$, the second largest prime factor $P_2(n)$ could be defined in two ways — either as the largest prime factor of $n/P(n)$, or as the second largest distinct prime that divides n. We show in our paper that such distinctions in the definitions of $P_k(n)$ for $k > 1$ do not affect our result, because almost always the large prime factors occur without repetition. While the genesis of this joint paper with Erdős was during the walk at the Madras Marina beach in January 1975, I wrote it up later that year only after I entered UCLA for my PhD; the paper was published in the Pacific Journal of Mathematics in 1977. $A(n)$ is sometimes called the Alladi-Erdős function.

Besides $A(n)$, I had been working on other arithmetical functions during my undergraduate days. In particular I had a complete manuscript entitled "Arithmetic functions and divisors of higher order" which I showed Erdős. In this paper I defined iteratively an infinite hierarchy of divisors of an integer — divisors of order 1 (which are simply the ordinary divisors), divisors of order 2 (which are the well known unitary divisors), divisors of order 3, and so on. The notion of divisors of order k for $k > 2$ was new. What was very interesting was that if $\mathcal{D}_k(n)$ denotes the set of k-th order divisors of an integer, then we have the remarkable inclusion relations given by

$$\mathcal{D}_2(n) \subset \mathcal{D}_4(n) \subset \mathcal{D}_6(n) \subset \cdots \subset \mathcal{D}_5(n) \subset \mathcal{D}_3(n) \subset \mathcal{D}_1(n),$$

for each integer. These are like the beautiful inequalities satisfied by the convergents of a continued fraction. Unexpectedly, the Fibonacci numbers arose in the characterization of these divisors of higher order. Using the concept of a divisor of higher order, I studied arithmetical functions like the number of k-th order divisors of an integer, the sum of the k-th order divisors of an integer, and the k-th order analogue of the Euler function, and discussed their properties and inter-connections. Erdős liked this paper very much. He said he would show it to his long time friend George Szekeres in Sydney (whom I had met earlier in 1973 during my visit to Australia), to see if it could be published in the Journal of the Australian Mathematical Society. As it turned out, Szekeres was impressed with the paper and so it appeared in that journal in 1977.

Shortly before departing Madras for Sydney, Erdős had a meeting with the Governor of Madras, His Excellency K. K. Shah. The Governor was a mathematics enthusiast. When he heard from my father about the eminence of Erdős, he said he would like to meet him. I was with Erdős and my father at that meeting at the Governor's residence. The Governor said that to encourage talented youngsters to take to mathematics, he had established a fund to provide scholarships to very bright high school students in mathematics. As soon as Erdős heard this, he put his hand in his coat pocket, drew a load of cash he had — a sizeable sum in Indian rupees of the honoraria he had received from MATSCIENCE and the Ramanujan Institute — and handed them all to the Governor for his scholarship fund. The Governor was touched by Erdős' generosity and simplicity.

While in Madras, Erdős asked me what my plans for higher studies were. I told him that I had applied for admission to graduate studies in America and I hoped to join there in Fall 1975 soon after getting my Bachelors degree in June 1975. I asked him if he would write me a letter. He said he would gladly write, but it would be just to one university because he was sure that based on his letter, I would be admitted there with full financial assistance. He therefore asked me where all I had applied. As soon as I mentioned UCLA, he exclaimed that it would be the best place for me, because his long time friend Ernst Straus (whom I had met earlier in the year) would be the ideal advisor for me since Straus would teach me a lot of good mathematics and at the same time allow me to work on problems of my own. Erdős also said that he is an annual visitor to UCLA and so we could meet there to continue our collaboration. So right there at MATSCIENCE as we were having this conversation in January 1975, Erdős asked for a sheet of paper, wrote a letter to Straus, and mailed it immediately. Two months later, I received a letter from UCLA offering me the Chancellors Fellowship (their most prestigious fellowship for graduate students) for a period of four years for my PhD studies. I accepted it and that changed my life completely. I have to thank Providence for the opportunity to meet Erdős, to interact and collaborate with him during the final year of my bachelors degree course.

I consider my meeting with Erdős as one of the most significant aspects of my academic life. For the next twenty years, I interacted with him very closely and hosted him in almost every place I visited on an extended basis. There will more on Erdős later, but next I describe my departure to UCLA and the three wonderful years I spent there as a graduate student, meeting great mathematical personalities during that time.

2.8 Departure to UCLA for my PhD (1975)

In January 1975 I heard from a few universities that I would be admitted to their graduate program in mathematics and that the decision on financial assistance would soon follow. One of the first such letters I received was from UCLA in mid-January that gave me admission. Then in mid-March, UCLA offered me the Chancellor's Fellowship. Such was the effect of the letter of Erdős. So I accepted it rightaway and informed the other universities accordingly. I still had to take the university exams in April for my third and final year of the BSc, but with decision to go to UCLA made, I could fully concentrate on my BSc examinations without the distraction of the graduate application process.

What I appreciated most about American Universities was their flexibility in admission, even though they had very high standards. In the United States, by the time one finishes undergraduate studies, one would have completed 16 years of education. In India with the BSC being only a three-year program instead of four, students completing their Bachelor's degree have only 15 years of education. Thus normally American universities expect the student from India to either apply for graduate study after getting a Master's degree or after completing 16 years of education, which means one year into the Masters program. In my case an exemption to this 16 year rule was made because (i) I had a research record as an undergraduate (modest as it might have been), and (ii) I had letters not just from my college teachers but from some leading mathematicians overseas. Such exemptions would not be possible in the more rigid Indian system.

In early 1975, I saw an announcement that there was to be a Summer Institute on Complex Analysis at the International Centre for Theoretical Physics (ICTP) in Trieste, Italy, starting on May 21. This would be perfect for me to attend prior to joining UCLA in September, and so I wrote to Professor Abdus Salam, Director of ICTP, expressing my interest. He passed on my letter of interest to the organizers of the Summer Institute. What a gracious gesture on the part of the organizers who replied promptly offering me not only all expenses for my stay there for two months, but my round-trip airfare between Madras and Trieste! Actually when I informed the ICTP that I would stop in Trieste on my way to America, they permitted me to convert the roundtrip ticket Madras-Trieste-Madras to Madras-Trieste-Los Angeles (one way). This was a pleasant surprise.

After completing my BSc final exams in April, I started preparing in early May for my departure to USA via Trieste; the trip was to start in the second half of May. It was at the time that my father received a call one morning from Professor Marshall Stone who had just arrived in Madras. He requested some of the best mangoes from us. Professor Stone was on various committees pertaining to the improvement of mathematics education in India, and he was in Madras at that time in connection with such work. The month of May is the hottest and driest in Madras and in South India, but it is also the best season for mangoes. We had over

100 acres of mango orchards in our village 60 miles north of Madras, and several gunny (= jute) bags of the tastiest mangoes from our orchards had arrived at our home a few days earlier. So my father called Professor Stone and told him that a dozen of the choicest mangoes will be delivered in the morning to his room at the Woodlands Hotel near our house where he was staying, and that he would be calling on him later in the day. My father said that I should accompany him to see Professor Stone and that I should get his advice since I was about to begin my graduate studies at UCLA. So I accompanied my father that afternoon to see Professor Stone.

When we knocked on the door, Professor Stone said, "Come in, the door is open." Upon entering the room, we saw him enjoying the mangoes. He thanked my father for the mangoes and asked us to sit down. My father told him that I was going to UCLA for my PhD and that I had attempted research in number theory as an undergraduate student. He said that was really good, but his advice was that my PhD work should be among my best. He said that for a number of great mathematicians, their PhD work was among their finest. Professor Stone advised me to read the recent book [B5] of Alan Baker on Transcendental Number Theory. Baker had won the Fields Medal in 1970 for his pathbreaking work in this field — much of this being work in his PhD thesis under the direction of Harold Davenport. The book was an exposition of the latest advances on transcendence. I did get this book a few months later as a gift from my father.

On the eve of my departure to America for graduate work, it was inspiring for me to get the encouragement and advice from Professor Stone, one of the most dominant figures in twentieth century mathematics. That was my last meeting with him. Professor Stone did visit Madras several times in December in later years and even attended the concerts of the Madras Music Academy with my parents. I was in the United States from 1975 onwards and visited India only in the summers in the seventies and eighties and so did not get to meet him on his visits in December. In January 1989, during a visit to India, Professor Stone died at the Madras Woodlands Hotel after attending the December concerts of the Music Academy, next door to the Woodlands. He loved India, its music, its culture, and its people, and he breathed his last in Madras soon after the music festival.

2.9 The Complex Analysis Summer Institute in Trieste (1975)

The International Centre for Theoretical Physics (ICTP) created by Salam in 1964, had grown significantly within a decade, and had become a leading center in the world for research in theoretical physics. Owing to this growth, the ICTP moved out of its modest home at Piazza Oberdan in the center of Trieste, to its magnificent permanent home in Grignano, an enchanting seaside resort outside of Trieste, facing the azure waters of the Adriatic sea. Since theoretical physics rested on mathematical foundations, it was realized that the ICTP would need to have an active mathematics division. It was with this goal that Salam decided to have a few Summer Institutes on various mathematical topics as a preparation for the launch of a mathematics division. The Summer Institute in Complex Analysis was held from May 21 to August 8, 1975. The lead organizers were Professors James Eells of the University of Warwick, A. Andreotti of the University of Pisa, and G. Gherardelli of the University of Florence.

There were 35 main speakers who gave courses or mini-courses on various aspects of the theory of one and several complex variables. There were more than 100 participants (other than the main speakers) and several of them came from the under-developed countries since it is the mission of the ICTP to help the advancement of science in the Third World. The participants from India included Professors Padmanabhan, Geetha, and Parvatam from the Ramanujan Institute in Madras, Sundararaman from the Tata Institute in Bombay, and R. Narasimhan from the Madras Institute of Technology Sundararaman subsequently joined the Ramanujan Institute. I was the only student participant at the conference.

Since my father knew Salam and the physics group at ICTP very well, he too was invited to ICTP (not to the Summer Institute), and so my parents were there as well. After spending a few days in Trieste, my father visited various centers in Europe and returned to Trieste to pick me up at the end of the Summer Institute. My mother stayed with me in Trieste for the entire two months. The ICTP had arranged for us a most lovely apartment facing the sea in Sistiana, another beautiful seaside resort a bit further away from Trieste than Grignano.

The conference opened with an electrifying lecture by James Eells who gave a panoramic view of the state of complex analysis. His enthusiasm for the subject was infectious. No wonder, when the mathematics division at ICTP was launched in 1986, Eeels was appointed as the first Director of this division. During the time of the Summer Institute in Trieste, he was on the mathematics faculty the University of Warwick, England, where he joined in 1969 at the invitation of Christopher Zeeman who was starting the mathematics department there. In Warwick, Eells was the main organizer of the Warwick Symposia. He was a superb organizer, and that was clear from the way in which this Summer Institute was conducted. Eells was succeeded in 1992 as the Director of Mathematics at ICTP by M. S. Narasimhan of the Tata Institute. M. S. Narasimhan gave a course on Riemann surfaces at this Summer Institute.

One of the most interesting discussions I had at the Summer Institute was with Professor Raghavan Narasimhan of the University of Chicago (who passed away in 2015). He was at the Tata Institute until the mid-sixties and started his career working in analytic number theory. In the late fifties, he and his thesis advisor K. Chandrasekharan established very significant results on Dirichlet series satisfying certain functional equations. Raghavan Narasimhan was a genius and one of the quickest thinkers.

Although Raghavan Narasimhan did significant work in analytic number theory, the thing that propelled him to fame was his solution of Levi Problems I and II in several complex variables that appeared in *Mathematische Annalen* in 1961 and 62. He was immediately invited to speak at the 1962 International Congress of Mathematicians (ICM) in Stockholm. To deliver an invited lecture at the ICM is a great honor, and he received such an invitation after barely completing his PhD. He was made full professor overnight and was the youngest at this rank. In 1965, his mentor K. Chandrasekharan, who was the Head of the School of Mathematics at TIFR, left for a position at ETH in Zurich, one reason being that he had differences on matters of policy with Homi Bhabha, the Director of TIFR. So Raghavan Narasimhan also left TIFR in 1965; he first went to the University of Geneva, and soon after moved permanently to the University of Chicago.

My father had told me about Raghavan Narasimhan's brilliance, and so I requested him to give me some time for a discussion. At the Summer Institute he was giving a mini-course on the Weierstrass Preparation Theorem. He asked me to come to his office at the ICTP for a discussion. When I told him that I was interested in analytic number theory, he said: "All that is used in analytic number theory is Cauchy's theorem on rectangular and circular contours! You should study several complex variables which is more challenging." I replied humbly that I should first understand single complex variable theory before starting on several complex variables. He smiled. He then suggested that I should read Enrico Bombieri's paper on the Large Sieve that had just appeared in *Asterisque* — a newly formed French journal. He then asked me to describe some of the results I had obtained as an undergraduate. I described to him my work on higher order divisors and said that it would appear in the Journal of the Australian Mathematical Society. He immediately said: "Why did you send it there?

You should have sent it to the Proceedings of the AMS where it would gain a wider readership." I told him that it was Erdős who took my manuscript and handed it to George Szekeres, who as Editor of the Journal of the Australian Mathematical Society, accepted it rightaway. I then described to him my work on $A(n)$ and said that I was using my time in Trieste to write up these results. He said that I should send the paper on $A(n)$ to a journal with wider readership; that paper with Erdős was submitted to, and accepted by, the Pacific Journal of Mathematics.

Raghavan Narasimhan had a great regard and love for number theory. That is how his research career began. His advice that I should study several complex variables was like Dyson's advice that I should study physics. Dyson, like Raghavan Narasimhan, started out in number theory, but later moved into physics. Like Dyson, Raghavan Narasimhan was frank — brutally frank — in expressing opinions. However, such comments could sometimes have a negative impact.

Raghavan Narasimhan was a chain smoker. He spent a lot of time in the ICTP cafeteria in discussion with participants, and puffing away all the time. Mathematicians were in awe of him because of his vast knowledge and due to his quick thinking. I saw him sitting in the cafeteria explaining to groups of eager listeners, the key ideas underlying various major problems in different areas of mathematics.

Another leading analyst with whom I interacted in Trieste was Fred Gehring from the University of Michigan in Ann Arbor. He was a world authority on quasi-conformal mappings and was invited twice to give hour lectures at the ICM. His lectures were very clear and his board work, impeccable. He had a very pleasing personality. He recalled seeing me in Ann Arbor in 1973 at the Summer Institute. When I told him that I was joining UCLA for my PhD, he asked why I did not consider the University of Michigan which had a great tradition in number theory. When I told him why I was going to UCLA, he said that I should come to Michigan after my PhD. This actually happened because after my PhD, I took up the Hildebrandt Assistant Professorship at Michigan in 1978, and Gehring was the Chair of the mathematics department there.

It also attended the lectures of Professor Wolfgang Fuchs of Cornell University, whom I had met in Madras in the sixties when I was a schoolboy. Fuchs was an authority on Nevanlinna theory, and in Trieste he spoke about the Viman-Valiron theory pertaining to the growth and values of entire functions represented by Dirichlet series. He had obtained his PhD at Cambridge in 1941 under Ingham, a well known analytic number theorist.

Thus Fuchs had a genuine interest in number theory, and he appreciated my undergraduate research on arithmetical functions.

ICTP Director Abdus Salam was a gracious host. There were several parties that the ICTP arranged for the Summer Institute participants. There was even a musical evening in which participants of the Summer Institute from around the world rendered classical songs of their countries. Professor R. Narasimhan of MIT Madras and my mother rendered South Indian Carnatic music songs. Salam attended the event and graciously wrote personal letters of thanks to everyone who contributed to the cultural program and made it memorable. Salam invited me to his office and wanted to know how beneficial the Summer Institute was to me — both in terms of my mathematical education and research. He was a warm and caring person and carried his scientific fame lightly on his shoulders.

The Summer Institute concluded in August, and my father returned from his European academic tour to pick me up. On his trip he visited Cambridge University at the invitation of his long time friend, the eminent applied mathematician Sir James Lighthill who was Lucasian Professor of Mathematics there. My father requested Lighthill to arrange a meeting with Professor Alan Baker since Marshall Stone had advised me to read Baker's book. When my father met Professor Baker and mentioned Marshall Stone's advice, Baker said, "That is a very good advice indeed. Let me walk you to the Cambridge Bookstore where you could purchase a copy of my book." When my father returned to Trieste, he presented me Baker's book [B5] on Transcendental Number Theory with the inscription: "To Krishna, in the hope that he will attain transcendence in number theory."

From Trieste on our way to UCLA, we stopped in New York for a few days. While in the New York area, we met Professor Harish-Chandra at the Institute for Advanced Study in Princeton. where he was von Neumann Professor in the School of Mathematics.

Professor Harish-Chandra was dressed in a grey full suit with a tie, as he was working in his room, overlooking a spacious lawn. He had met me in 1964 at Ekamra Nivas and was pleased to see me as a teenage entrant in the world of mathematical research. He said that one should develop a theory slowly and systematically starting from the foundations and reach the greatest heights. He compared this to trying to unwind a ball of twine; if you pull the string slowly and steadily, you can unwind the ball without difficulty, but if you rush and pull too quickly, you will get entangled. When he enquired about my research interests, I mentioned to him my joint work with Paul Erdős on $A(n)$. He immediately exclaimed, "This work is very interesting, but I must say that Erdős writes too many papers, some of them very simple. Take my colleague Selberg for instance. He publishes occasionally, and only when the results are really important."

Selberg and Erdős, both of them great in their own ways, had very different views about how to do mathematics and what to publish. Harish-Chandra was a close friend and admirer of Selberg. Like Selberg, Harish-Chandra published papers only if the work represented a significant advance. He wrote very long papers, but that was his style. Erdős had a very different style, but that did not make him less impressive. Both Erdős and Harish-Chandra were awarded the Cole Prize of the AMS for their seminal contributions — Erdős in 1951 and Harish-Chandra in 1954.

The legendary mathematician Paul Erdős at MATSCIENCE with Alladi Ramakrishnan and Krishna (who was 19 years old then) in Jan 1975. Erdős rerouted his trip from Calcutta to Sydney to fly via Madras to meet Krishna. That meeting with Erdős had a profound influence on Krishna's academic life.

My thesis advisor at UCLA, Prof. Ernst Straus. He was the last student and assistant of Albert Einstein. Picture courtesy of Mrs. Louise Straus.

Chapter 3

Graduate student days at UCLA (1975–78)

3.1 The UCLA graduate program and faculty

One of the wonderful aspects of graduate study in mathematics in the United States is that it provides a three-fold experience: (i) graduate students take courses on various subjects, often from professors who are leaders in the field and thus experience an enlightening learning process, (ii) as teaching assistants, graduate students gain a valuable teaching experience, so useful for them when they apply for jobs after obtaining their degrees, and (iii) graduate students get experience in research by preparing a PhD thesis of original work. The principal difference is that PhD work in many countries is focused on research, because all course work is completed for the Masters degree which has to be finished before being admitted as a PhD student. In contrast, in the USA, one is admitted to the combined Masters/PhD program right after an undergraduate degree, and it is the choice of the student to leave with a Masters degree in two years, or stay in the program to complete an original research dissertation and receive a PhD degree. Most well established American universities have a nicely structured set of courses, but it is left to the student to choose the courses subject to certain minimum requirements. Some courses are mandatory because of the PhD qualifying examinations; these exams are separate from the term exams given in each course, and the student must get a high pass in the qualifying exams in order to advance to candidacy and be allowed to write a thesis. I was informed of all these requirements as I entered the graduate program at UCLA. In my case, the Chancellor's Fellowship was a four-year contract that provided me a fellowship in the first year without teaching duties, a teaching assistantship in the second and third years, and a Dissertation Fellowship in the fourth year without teaching duties so that I could focus fully on my thesis. The Chancellor's Fellowship was the most prestigious at UCLA and the best in terms of what it offered, and the mathematics department had a few of these each year. There was another fellowship called the Regents Fellowship, which was quite good, but not as attractive as the Chancellor's Fellowship.

I received the offer of the Chancellor's from the UCLA Chancellor Charles E. Young himself and another letter about it with details of the Fellowship from Professor Henry Dye, Chairman of the UCLA Mathematics Department. Although I did not get to meet Chancellor Young at UCLA, I did meet him several years later at the University of Florida during my term as Chair when he served as the President of the University of Florida. What a coincidence that I would interact with him three decades later!

Henry Dye was a highly accomplished mathematician and an authority in the fields of Operator Theory and Ergodic Theory. He was very soft spoken and a perfect gentleman. He addressed the entering graduate students on the first day of the Fall Term. Sometimes he would talk to me in the corridor and enquire about my progress.

Since my first year was free of teaching duties, I was advised to take as many courses as I could handle and get the Qualifying Exams out of the way. The Qualifying Exams in Algebra, Real Analysis and Complex Analysis were mandatory, but for the fourth, I had a choice and I selected Differential Geometry and Algebraic Topology.

I told Professor Straus that I had no courses in abstract algebra as an undergraduate in India. Straus informed me that Professor Robert Steinberg was scheduled to teach a year long (= three quarter) graduate course on algebra, and that I should take it.

Robert Steinberg was one of the most eminent faculty members in the mathematics department, and revered by students and colleagues alike. He was a world authority on Algebraic Groups. Several things were named after him such as the Steinberg groups, the Steinberg symbol, and the Steinberg cocycle. He had a beard and a far off look, as if he was lost in thought on a problem, and he often was, and so most students like me were in awe of him. There was no light talk with Steinberg (at least for the graduate students) such as what happened last night at the UCLA basketball game. You talked only mathematics with him, and you had to talk sense. But he had a fine sense of humor, and once you got to know him, you could relax and talk non-mathematical topics with him. My thesis advisor Ernst Straus was one of Steinberg's closest friends.

Steinberg was a perfect gentleman and very dedicated to his work. He joined the UCLA faculty in 1948 upon completing his PhD and was there until his retirement; as Emeritus Professor he continued to come to the Department until his death in 2014 at the age of 92. In America, one way to get a significant salary raise is to get a job offer from outside, and use this to leverage a higher salary and/or a better arrangement such as a lower teaching load. Owing to his eminence, he could have gotten a job at any top university, but he never played this "game" to negotiate better terms for his position at UCLA. Awards and recognition came to him naturally; he never changed his practices to seek such recognitions. For his seminal contributions, he was invited to speak at the International Congress of Mathematicians in 1966, was awarded the Steele Prize of the AMS in 1985 and elected Member of the National Academy of Sciences that year. In the important field of algebraic groups, he was considered to be on par with Armand Borel and Claude Chevalley, two of its greatest practitioners.

Steinberg was an outstanding teacher. For the three quarter course, he taught group theory in the first quarter, rings and modules in the second, and fields and Galois theory in the third. Steinberg's lectures were so crystal clear, that the subject appeared deceptively simple. He often broke the result into components which were easy to prove. I worked really hard since these were my first courses in algebra, and got an A in his abstract algebra course during each of the three quarters. His course definitely helped me perform well in the Algebra Qualifying Exam which I took toward the end of my first year at UCLA in Spring 1976.

After the qualifying exams, there was an evening party in the mathematics lounge called the Post-Qualifier Hash-Bash, where alcoholic beverages were served with snacks. Steinberg was there at this party in Spring 1976. One of the graduate students had a few too many glasses of booze, and was involved in a "serious" mathematical conversation with Steinberg. I was next to them hearing and watching. The graduate student told Steinberg that he had a very nice idea of how to prove the famous Poincaré conjecture in topology and had worked out the details the previous night. Steinberg smiled and realized that the student was a little bit drunk. He played along and asked the student to outline the main ideas. The student proceeded to describe his "proof" of the Poincaré conjecture. After about five minutes, the student realized that his proof would not work. Steinberg smiled and said it was a good attempt and that you learn a lot by attempting to solve big problems. Since the student was "under the influence", he had shed his inhibitions and talked frankly with Steinberg. After he had returned to normalcy, the student was embarrassed that he had made such claims, and of all persons — with Steinberg! When the student and I passed Steinberg in the corridor the next day, Steinberg just nodded and smiled — indicating that he understood the background for such tall claims — and this put the nervous student at ease.

For complex analysis, I had the year long course by the brilliant and eccentric Paul Koosis. He chose the great book by Saks and Zygmund as the text for the course, and recommended the classic by Lars Ahlfors for additional reading. All faculty knew of the brilliance of Koosis, but were well aware of his idiosyncrasies. For example, Koosis had a harpsichord in his office that he had himself assembled, and would play it quite loudly. So they moved his office from the main floor to the basement, where he was near the cubicles of the graduate students, and could play his harpsichord without disturbing the peace on the main floor. During the first year, when I was not a TA, I had a cubicle in the basement near Koosis' office, and have heard his musical concerts!

Koosis often gave new proofs or variations of the proofs in the text. On a few occasions he got stuck, but returned the next class to complete the proof in a most elegant fashion. Many students including me enjoyed his treatment of complex analysis, but some students were unnerved by his unconventional approach. He gave an exam at the end of the first quarter in which he had five problems and asked us to solve four. To get an A one needed four problems answered correctly, whereas one would get a B with three solved, and a C if only two were solved. I started working on a problem and soon found there was something wrong with the statement. I thought at first I had made a mistake, but the closer I looked at it, the more I realized that it was stated incorrectly. I therefore showed that the problem was wrongly stated, and suggested a way to fix it. I had spent a lot of time on this, and was not sure whether my solution was correct. I looked around the room, and no one but me seemed bothered by anything on the question paper. Having spent so much time on this problem, I could do only two more. Thus I turned in the solutions for just three, and thought I had only two correct answers. This would give a C and that would mean a disastrous start for me at UCLA. Graduate students are expected to get at least a B, and a C could mean termination of my Chancellor's Fellowship. I spent several sleepless nights and talked to various persons on what the consequences would be in getting a C. To my pleasant shock, when Koosis posted the grades outside his office door, I found out that I got an A! When I went to talk to him, he said that I was the only one who realized that the problem was wrong and had fixed it. So he gave me double credit for that, which meant that I had the equivalent of four correct solutions and therefore received an A. Such a thing could never have happened in India, where I would have just received regular credit for the problem. In India, I would have been told that it was too bad that this problem had taken up so much time on the exam, and that I should have just skipped it and gone to the other problems. This episode also shows that Koosis was a very fair person who realized the mistake he had made and did not want to penalize students for it. I enjoyed the remaining two quarters of complex analysis with Koosis and learnt a lot from his beautiful lectures. These helped me get a high pass the Qualifying Exams in Complex Analysis which I took along with the Algebra Exam toward the end of Spring 1976.

Having passed two of four qualifying exams, I had the exam on real analysis, and the exam on the elective (differential geometry and algebraic topology) remaining. I decided to take them in Fall 1976 in my second year. But I took the courses for these exams in the first year itself.

The all-year course on Real Analysis was given by Ronald Miech. He did his PhD in number theory in the sixties under the direction of Paul Bateman in Illinois, but shifted his research to p-groups after moving to UCLA. His lectures were very thorough. He lectured on measure theory from the fine book of Royden, but switched to Rudin's classic book for Hilbert and Banach space theory. In addition as graduate students we consulted the massive treatise by Hewitt and Stromberg to make sure we had studied all topics for the real analysis qualifying exam.

Among all mathematics qualifying exams at UCLA, the exam in real analysis had the reputation of being the toughest. I was told that many of the best students had to take it twice to pass it. In fact several graduate students said that I should not expect to pass it on my first try. When I mentioned this to Professor Straus, he got upset and said, "Why are you talking to students who had failed it on their first attempt? Instead talk to students who passed it on their first try. That will give you confidence." In any case, I felt it was a good decision that I made to take the real analysis exam along with the one on differential geometry and algebraic topology in 1976–77. That gave me the entire summer to prepare for them.

The program in real analysis was being run by John Garnett and Theodore Gamelin, two of the leaders in the field. It was they who made these exams tough, and only those who did really well could hope to do their PhD in real analysis with them. I did get a high pass in the real analysis qualifying exam on my first attempt. I passed the exam on differential geometry and algebraic topology, but not in flying colors. The Graduate Committee knew that my PhD would be in analytic number theory, and so were not insistent on a high pass on that fourth exam.

Besides Gamelin and Garnett, I interacted also with Steven Krantz, then an assistant professor in analysis. I really got to know him well only in the past few years when he was Editor of the Notices of the AMS and I was an associate editor. He did a fine job as Editor of the Notices and it was a pleasure to work with him, especially on a feature article for the birth centenary of Paul Erdős. Besides his important research contributions, Krantz is widely known as the author of several excellent books.

The high standards set by Garnett and Gamelin on the Real Analysis Qualifying Exam paid off and I was a direct witness to that. One of my classmates, and perhaps the most brilliant of all, was Peter Jones. He was a "drifter" and did not really know what he wanted to do in life. He joined the UCLA graduate program in 1975, the same year I did. He just had a regular teaching assistantship, nothing special. However, contact with Garnett and Gamelin transformed him. He became serious about mathematics, and the genius in him blossomed under their tutelage. He wrote an outstanding thesis in real analysis, accepted a Dickson Instructorship in Chicago, and became one of the top analysts of our generation. He served as Director of the Mittag-Leffler Institute for a few years and is now a professor at Yale.

Mathematics graduate programs in America are built around the qualifying exams to make sure that students get a solid and broad foundation before proceeding to pursue a field of specialization for the PhD thesis. I felt fortunate to have received this rigorous training from highly qualified faculty. At UCLA the rule in the seventies was that one should take at least two qualifying exams (out of the four) each time, until there is only one left to pass. Such a requirement was to ensure that only the strong students would make it through, and also to ensure that students do not take too much time in finishing their qualifying exams by taking one at a time. In the last two decades, for various reasons, several PhD program requirements have been relaxed all over the USA except perhaps in the very best universities. One reason for a relaxation in the requirements is the desire to meet the "time to degree" criterion that both funding agencies and university administrators use to evaluate mathematics programs. In other words, mathematics departments are required to show data or demonstrate that among the students admitted to their PhD program, a good number of them do complete the PhD, and that too in five years. In the experimental sciences, graduate students work in labs on a project of a professor and are able to write collaborative papers based on their findings. This helps in getting the PhD degree without delay, thereby satisfying the "time to degree" requirement. In mathematics, students are not working in labs, nor are they always collaborating with their professors. Often they are working on their own but under the guidance of their advisor. Mathematics is known to maintain the highest standards and so if a student needed more time to make the thesis worthwhile, the student was normally allowed to continue in the program with support. Now owing to the "time to degree" requirement, as well as for budgetary reasons, students are generally not supported beyond five years and expected to finish by then. Whether fortunately or unfortunately, the fact is that both the funding agencies as well as the university administrators are making mathematics departments function like the lab sciences in certain ways, thereby changing the culture and practices of mathematics departments.

Having finished all the courses required for the qualifying exams during my first year at UCLA, I spent the second year taking more advanced courses and pursuing my thesis problem. UCLA had a first rate faculty and I got to know several professors very well either by taking their courses, or by discussions with them in the lounge.

My thesis advisor Ernst Straus had a warm and genial personality and I moved with him like a family friend. He was the last assistant of Albert Einstein at the Institute for Advanced Study in Princeton and had the privilege of collaborating with Einstein. Indeed the first few papers of Straus were on unified field theory and these were the last few papers of Albert Einstein! Straus received his PhD from Columbia University with F. J. Murray as his first advisor, and Einstein as his second advisor, and his thesis was on unified field theory. When Straus was at the Institute for Advanced Study, he was influenced by the work of Erdős, Selberg, and other number theorists, and so after he left the Institute and joined UCLA, his research turned to number theory. Straus told several Einstein stories to friends and colleagues, and I have heard a good number of them at UCLA some of which I will share with you now.

When Straus was interviewed by Einstein for the position of "assistant", Straus said that he knew no relativity. Einstein immediately responded — "You don't have to. I do." Einstein said that he chose Straus for his mathematical knowledge, and required Straus to infuse mathematical rigor into the physical theories they would work on together. With regard to mathematics versus physics, Einstein felt that in mathematics there were too many beautiful problems, and he would be seduced by their elegance without knowing their importance. In physics, Einstein said that he had the "nose" to smell an important problem, and pursue that with a single minded purpose whether it was elegant or not.

The question of what is elegant and important varies from discipline to discipline. What mathematicians consider beautiful or important might not seem significant to physicists. On their very first day of work together, Einstein asked Straus to provide an example of a theorem he had recently proved. Straus responded by stating the following elegant result: *Given any closed curve C in the plane and any triangle T, there exist three points P, Q, R, on the curve such that the triangle PQR is similar to T.* On hearing this Einstein asked what was so nice about this result? He said it was unnatural because similarity of triangles is a metric concept, whereas closedness of curves is a topological concept, and the two should not be mixed because they have different transformation groups. To mathematicians this is a very elegant theorem, and also exciting because of the interplay of invariants under different groups; but Einstein, the physicist, was not charmed by it. This is not an isolated example of a physicist not appreciating the beauty or significance of a mathematical theorem. After Fermat's Last Theorem was proved, several of my friends in other departments at the University of Florida including some physics faculty, asked me what was so great about showing that a certain equation has no solution!

Straus said that once Einstein was focused on something, nothing could distract him. When they were writing one of their joint papers, Einstein was looking for a paper clip. When he could not find one, he took a piece of wire nearby and starting bending it to form a paper clip. As Einstein was converting the wire into a paper clip, Straus was searching for a paper clip in the office. He found one and gave it to Einstein. But Einstein put that paper clip aside, continued working on the wire, and after successfully making a paper clip out of it, he used that instead!

Faculty and students alike were charmed by the Einstein stories that Straus would tell in the UCLA lounge during tea time. I too enjoyed these wonderful recollections. I felt privileged to be a grandstudent of Einstein.

After Straus moved to UCLA, he shifted the focus of his research to the study of analytic functions, and also to number theory and combinatorics. Straus was a close friend and collaborator of Erdős and arranged for Erdős to visit UCLA every year in the Spring.

Straus' published papers are very significant, but he is greater that what they reveal, because he generously shared his ideas with many. Also he was laid back and did not go about claiming his territory in the subject. His research interests were broad and his contributions range over number theory, combinatorics, extremal graph theory, arithmetic properties of analytic functions, and of course, general relativity.

The combinatorics program at UCLA was run by Professors Basil Gordon and Bruce Rothschild. Gordon was quite distinguished in algebraic number theory and the theory of partitions, having established a beautiful extension of the celebrated Rogers-Ramanujan partition theorem to all odd moduli. He was single, and was a descendant of the family that made the Gordon Gin. So he was independently wealthy and did not care what salary he had at UCLA or what annual raises he would get. He was really a mathematical genius and had an almost encyclopaedic knowledge. He was a renaissance man and had a great knowledge of the classics, poetry, and western classical music. His mind was always in the renaissance period. Once on a trip to Russia, he took a very expensive tour because it gave him an opportunity to play on the very piano of Tchaikovsky! Like Straus, Gordon pursued mathematics for the sheer pleasure of doing creative work, and never sought recognition in any form. He once told me almost philosophically, that the pursuit of mathematics is the most gratifying for the soul. Gordon was an outstanding teacher as well and would treat his students not only to beautiful mathematics, but would recite poetry and passages from great books to enliven his lectures. My office mate Bob Miller was his PhD student, and he had a gleam in his eyes as he told stories that revealed Gordon's genius or personality. I really got to know Gordon well only in 1987 during the Ramanujan Centennial when he came to India as my guest (see Section 7.2)

Bruce Rothschild is a highly accomplished combinatorialist specializing in Ramsey Theory. He and Ron Graham had established a central result in Ramsey Theory for which they both received the 1971 Polya Prize along with three other mathematicians, one of whom was Alfred Hales, also in the UCLA mathematics department. Rothschild's office door was always open. Faculty, students, and staff often stopped by his office to have a chat with him, as I did quite regularly. Whenever Erdős visited UCLA, he was either with Straus or Rothschild. Both Gordon and Rothschild were the Managing Editors of the Journal of Combinatorial Theory, Ser. A, and made it one of the leading journals in the world. Rothschild would handle all the correspondence with the authors, editors, and referees, and would consult Gordon for expert opinion on various manuscripts. Rothschild would not bother Gordon with routine matters — important nonetheless — that had to be addressed to ensure the smooth running of the journal.

Across the corridor from Rothschild's office was that of Professor Richard Arens, a leading figure in functional analysis. Arens was always dressed in a full suit and often he would walk out of his office, smoking a pipe, to have conversation with colleagues and students in the corridor. Arens had visited MATSCIENCE in 1966 in the summer while my parents and I were away on a world tour, and so we missed hosting him in Madras. But he enjoyed his visit to MATSCIENCE — he told me this at UCLA — and very kindly invited me to his home when he and his wife had other colleagues and their spouses for dinner. Arens was an Editor of the Pacific Journal of Mathematics at that time when I submitted my first joint paper with Erdős to that journal at the suggestion of Straus. The Pacific Journal was founded in 1951 by Edwin Beckenbach of UCLA and Frantisek Wolf of UC Berkeley. UCLA had always played a strong role in the Pacific Journal, and all the operational work of the journal at that time was managed superbly by Elaine Barth, the Head Secretary of the UCLA Mathematics Department. (Elaine also managed the operations of the Journal of Combinatorial Theory, Ser. A, since Rothschild and Gordon as Managing Editors of the JCT (A) were at UCLA.) The Pacific Journal came into the limelight when the monumental 255 page paper of Walter Feit and John Thompson on the solvability of groups of odd order appeared in it in 1963. Arens told me that it was he who convinced the editorial board to devote an entire issue of the Pacific Journal to that paper alone! With a twinkle in his eyes he said he also made sure that the number of pages of the paper were adjusted (mildly) to be an odd number as well! I felt happy and proud that my first paper with Erdős was accepted by the Pacific Journal.

Close to the office of Arens, in the same corridor, was Professor Alfred Hales, who is very well known for the Hales-Jewett theorem that is now a standard part of Ramsey Theory. He had won the Polya Prize in 1971 along with his colleague Bruce Rothschild, and three others. Hales is very pleasant, and like Arens, would often step out of his office into the corridor for conversation. He later became Chair of the UCLA Mathematics Department and was also Chairman of the Board of Trustees of the Institute of Pure and Applied Mathematics (IPAM) at UCLA. Nowadays I see him at the Annual Meeting of the American Mathematical Society and we reminisce about our UCLA days.

I also got to know many other professors quite well either because I took courses from them, or due to the fact that I was their Teaching Assistant.

Harald Niederreiter with whom I had corresponded since meeting him at the 1973 Summer Institute in Ann Arbor, visited UCLA for a couple of years when I was a graduate student. I took his course on uniform distribution of sequences. By that time his classic book on Uniform Distribution of Sequences co-authored with Lawrence Kuipers had been published by John

Wiley, and so he prescribed that as the text for the course. His lectures were perhaps the most thorough that I had heard, and I enjoyed the subject very much. To this day I use the book of Kuipers-Niederreiter on uniform distribution for the systematic treatment of the subject, and for the wealth of information in it in the form of notes and references.

One of the nice things about the UCLA graduate program was that topics courses at an advanced level could be run with as few as two students provided there were a few others who would attend the course, that is audit it. This may not look financially viable, but it actually helps advanced students in the final stages of the PhD program. I took advantage of this more than once at UCLA and I adopted this convention while I was Chair at the University of Florida.

I very much wanted to learn sieve theory, and bought the outstanding book of Halberstam and Richert on Sieve Methods [B9] which had just appeared. Until then, one learnt sieve theory by reading papers; the book by Halberstam-Richert was the first on sieves. I wanted a course on sieves to be given out of this book, and since I knew that Ronald Miech had done his PhD work in sieve theory, and because I got to know him well by taking his real analysis course earlier, I approached him with a request to give a course on sieves using Halbertstam-Richert as a basis. (Actually Miech's fundamental work on almost primes in polynomial sequences was prominently discussed in Halberstam-Richert.) Miech graciously agreed to give the course provided I could get one other person to enroll in the course. That was not at all difficult because one of my classmates — Eugene Ng — was passionately interested in sieves and he enthusiastically enrolled in the course. Miech gave a masterly treatment of sieve methods with Halberstam-Richert as a basis, but referred to research papers as well such as the seminal works of Ankeny-Onishi and Jurkat-Richert. The course had two consequences: (i) I got Miech to agree to serve on my PhD Committee, and (ii) Eugene Ng became Miech's doctoral student and wrote a thesis in Sieve Theory in 1981. Interestingly one of the persons who was attending these lectures on sieves was S. T. Yau, a brilliant differential geometer who visited UCLA for a few years. Every one knew of Yau as a student of S. S. Chern of Berkeley who was considered as the king in the field of differential geometry. Yau's reputation was that he had solved the Calabi Conjecture and that he was publishing papers at an astonishingly high rate — each one of them in journals as prestigious as the Annals of Mathematics. The joke among graduate students at UCLA was that if one ordered the letters written by Chern on various mathematicians, then the maximal element as given by Zorn's Lemma would be Chern's letter for Yau! In any case, we were all struck by the fact that a differential geometer was attending the course on sieves and asking very sensible and subtle questions! We then found out that he was attending other courses and seminars well outside his research speciality and asking probing questions in those seminars as well. Chern's high opinion of Yau was well founded — Yau went on to win Fields Medal in 1982. I was to interact with Yau later at the Institute for Advanced Study in Princeton in 1981–82 (see §5.6).

Just as I wanted to learn sieve theory, I also had a keen interest in learning Transcendental Number Theory out of the book [B5] by Alan Baker that my father had presented me. So I approached Straus with a request to give a course on Transcendence with Baker's book as a basis, and he too, like Miech, graciously agreed.

The regular graduate course in number theory that was offered during my three years at UCLA, was on algebraic number theory by Murray Schacher, which I took. He used the classic text by Borevich and Shaferevich. It turned out that no course in analytic number theory was offered. This was the area of my research, and basically I learnt it on my own by reading books and papers.

The UCLA Mathematics Department had a Graduate Reading Room with a fine collection of books and journals. All graduate students and faculty had keys to this reading room and so we could enter it at any time. Thus it was most conducive to the learning process. Everything in the graduate reading room and much more on mathematics both in terms of journals and books were available in the central library. What this meant was that the books and journals in the graduate reading room were actually ordered in duplicate by UCLA, and obviously UCLA did not worry about the extra cost that this entailed. Such costs are marginal, and universities should allow mathematics departments to have either such graduate reading rooms, or mathematics libraries in the department building. The mathematics librarian was a charming lady by name Sharon Marcus. She and her husband Stan became close friends of me and my parents. They visited our family home in Madras during a trip to India they made while I was a graduate student. Sharon worked in the mathematics reading room on a voluntary basis. When I asked her why she preferred that instead of having a regular job, she said she enjoyed being in an intellectual atmosphere.

In addition to books and journals, the Graduate Reading Room had the collection of papers of all its faculty and graduate students. Thus a student could get an idea as to who is doing what, and this is very helpful in determining who to have on your Committee and as your PhD advisor.

3.2   The Mathematics Colloquium and Seminars

All mathematics departments in major research universities have an active weekly Colloquium. The Chair or Organizer of the colloquium informs the speakers to make the talk not only accessible to non-specialists but to graduate students as well. I attended the UCLA colloquia regularly. I understood barely 10% of what was presented in the talks, but what I gained from attending the colloquia was an idea of the landscape of mathematics. There were some colloquia that I vividly remember and I will describe a few here.

Philip Griffiths, then at Harvard University, gave a lovely talk on the Poncelet Problem. Poncelet's theorem is that given two ellipses one inside the other, if it is possible to draw a closed polygon that is inscribed in the outer ellipse and circumscribes the inner ellipse, then it is possible to draw infinitely many such polygons using other points on the outer ellipse. Philip Griffiths with Joe Harris (his PhD student) extended this theorem to polyhedra inscribed in and circumscribed about ellipsoids. Not only was the mathematics Griffiths presented beautiful, so was his lecturing style.

Philip Griffiths is a world class geometer — a leader in both algebraic and differential geometry. Between 1991 and 2003 he served as Director of the Institute for Advanced Study in Princeton. During his tenure as Director of the Institute, he visited Madras and on that visit came to Ekamra Nivas to meet my father and see the lecture hall that was the womb of MATSCIENCE. In 2010, during the 80th Anniversary of the Institute for Advanced Study that I attended with Mathura, I met Griffiths. I said that I remembered the beautiful talk that he gave at UCLA and he responded saying that he remembered his visit to my family home in Madras.

Another very exciting colloquium that I attended was by Christopher Zeeman entitled "Catastrophe theory and buoyancy of ships". Rene Thom, the French mathematician, had received the Fields Medal in 1958 for his work on catastrophy theory. Zeeman, who at that time was consultant for the Lloyd's shipping line of England, had become famous for applications of catastrophy theory. The huge auditorium on the ground floor of the mathematics building was packed to capacity for Zeeman's lecture. I could get a seat only in the last row, but I noticed that Straus was sitting in the front row near the center. Within a few minutes of Zeeman starting his lecture, Straus got up and walked out. I was stunned when I saw this. After the talk, I asked Straus why he walked out. He said that he heard Zeeman's talk a few months earlier at the University of Michigan, and since the first few sentences of Zeeman's UCLA lecture were identical to those at the talk at the University of Michigan, he concluded that the entire lecture would be identical. So he got up and left the hall.

Straus was a very genial person, but he could be blunt on occasion. Once at a UCLA conference, Straus was charing a session. The last speaker before lunch was going over time and the audience was getting impatient. The speaker was also taking about odd perfect numbers whose existence is yet to be confirmed. Straus could take it no more. He got up, asked the speaker to stop and said, "I think we better adjourn for lunch because for all we know, you could be talking about the null set!"

The UCLA Mathematics Colloquium featured talks by several luminaries. One of the most inspiring for me was the lecture by Fields Medallist Alan Baker of Cambridge University on advances in transcendental number theory. His lecture was polished and the results were breathtaking. Starting with the illustrious history of the subject, Baker traced the development through the decades, stating one spectacular theorem after the other, like the torrent of the Niagara! I also had attended his colloquium at Caltech a few days earlier when Professor Straus took me there.

The tradition at UCLA was that after each colloquium, the faculty would meet with the speaker in the Lounge to drink sherry — not just any wine, but sherry. In fact the colloquium chair proudly announced at the end of each talk that "Sherry will be served in the Lounge." Only faculty were invited to this, not students. In many instances, the host of the colloquium speaker had a party in his or her home to which both faculty and graduate students were invited. I used to attend some of these parties because it gave me an opportunity not only to converse with the speaker but also with my professors in a relaxed setting.

Besides the weekly colloquia, there were several active weekly seminars. One of the most prominent was the Caltech-UCLA Logic Seminar. UCLA had a very strong program in mathematical logic with Alonzo Church, C. C. Chang, and Donald Martin on its faculty. I actually got to know Chang quite well since I was a TA for his calculus class. Chang had done very important work in Model Theory in the sixties, but by the time I arrived in UCLA, he had become very spiritual. In particular he had become a devotee of the Indian saint Sai Baba; whenever I went to his office, after discussing the calculus assignments, he would talk to me about Sai Baba and his visits to India to see the saint.

Another leading logician at UCLA was Yiannis Moschovakis (now retired) who was co-organizer of the Caltech-UCLA Logic Seminar along with his former student Alexander Kechris on the Caltech faculty. The Seminar alternated between UCLA and Caltech. Moschovakis wrote a definitive text on Descriptive Set Theory that was widely used. Like Arens, he smoked a pipe and would come out of his office from time to time to converse with faculty and students in the corridor. He has a commanding personality and I felt he would be a good administrator as well; indeed he became chairman of the mathematics department at UCLA a few years later.

The logicians at UCLA were also active in managing the production of the Journal of Symbolic Logic. Thus at that time, at least three major mathematics journals were managed by the UCLA mathematics department — JCT (A), The Pacific Journal, and the Journal of Symbolic Logic.

There was no scheduled weekly number theory Seminar at UCLA at that time, but the number theory Seminar would be arranged any time there was a visitor in number theory, and this happened often. Paul Erdős was an annual visitor to UCLA. Even though Erdős travelled around the world constantly, rarely spending more than a week at a place, he returned to a few universities regularly each year — UCLA, University of Colorado, University of Calgary, and the University of Florida, besides Bell Labs in New Jersey. Thus Erdős visited UCLA each of the three years I was there. During those visits to UCLA, we collaborated. We wrote a second joint paper on $A(n)$ and large prime factors of the integers, and this too appeared in the Pacific Journal. It was also during one his visits, that we collaborated on the paper on additive partitions of integers that appeared in Discrete Mathematics. Thus by the time I left UCLA with my PhD, I had three papers with Erdős.

Erdős was paid out of Straus' NSF grant, but he always used Rothschild's office. Starting from the sixties, Erdős worked more in combinatorics than number theory. He interacted a lot with Rothschild on Ramsey Theory. During his visits to UCLA, he would speak in the number theory seminar, the combinatorics seminar, or the set theory seminar.

Whenever Erdős was at UCLA, there were visitors to see him. I believe this was true wherever he went. Mathematicians senior and junior, came to see him in good numbers. These could range from leading mathematicians like Ron Graham to amateur mathematicians in community colleges. In fact the first time I met Persi Diaconis was in Rothschild's office at UCLA when he came to see Erdős. Interestingly, a few years later, in a paper on the Fast Fourier Transform (FFT), Diaconis used the theorem Erdős and I had proved on $A(n)$ and the large prime factors of n.

One of the rituals was "lunch with Erdős". In fact Erdős would routinely invite various people to join him for lunch. It was an occasion to discuss mathematics and world affairs in a relaxed setting. Erdős was very generous. He always paid for the lunch for all guests.

Erdős was always concerned whether as a vegetarian I was getting enough to eat. In these lunch outings, he would often ask me whether the food was to my liking. Guests around the table naturally picked up on the conversation and would ask me what level of restriction I have as a vegetarian; for example, whether I would eat dairy products. Erdős would sometimes interject and answer these questions by jokingly saying — "Alladi does not eat carcasses!" Erdős understood my vegetarian requirement: I would eat dairy products because the animal is not killed for it.

Erdős' visits to UCLA were usually in December or January just before or after the West Coast Number Theory Conference that took place one week before Christmas each year. He attended this conference regularly. I describe my wonderful experiences at these conferences next.

3.3 The West Coast Number Theory Conferences

Within a few weeks of my arrival at UCLA, Charles Grinstead, a fellow graduate student one year my senior, noting my passionate interest in number theory, drew my attention to the following problem that Erdős and Straus had posed at the 1973 West Coast Number Theory Conference: *Write n! as a product of n prime powers, and among all such decompositions, choose one in which the smallest prime power is as large as possible. How big is that smallest prime power in that special decomposition?*

This question was motivated by a similar question (also due to Erdős and Straus) whether one could write $n!$ as a product of n numbers each asymptotically about the size of n/e? That question had been answered recently in the affirmative by Straus himself and so it motivated the problem on prime powers.

Grinstead was very much interested in number theory and so this question sparked his curiosity. But he had "moved out" of number theory to graph theory to do a PhD under Rothschild; so Grinstead felt that someone like me in number theory was better suited to handle this problem. I told Grinstead that we should work on this together, and so we did. Within a few weeks we solved the problem. We showed that the minimal prime power had the maximal value

$$n^{\alpha(1+o(1))}, \quad \text{where} \quad \alpha = \exp\left\{-1 + \sum_{k=2}^{\infty} \frac{1}{k}\log\left(\frac{k}{k-1}\right)\right\},$$

We actually solved a more general decomposition problem involving prime power decomposition of $n!$, wrote this up, and sent it to the Journal of Number Theory where it appeared in 1977. The constant α is now known as the *Alladi-Grinstead constant* and I am thankful to Charles Grinstead for drawing my attention to this problem.

The West Coast Number Theory Conferences had an interesting origin. D. H. Lehmer of the University of California, Berkeley, was a powerful and prolific number theorist with a special strength on the computational side. He was a great mentor and always had a sizeable group of talented PhD students. Two of his most illustrious students are Ronald Graham (a great combinatorialist who headed the mathematics division at AT&T Bell Labs for many years and was President of the AMS), and Harold Stark (at MIT at that time and now in UCSD). In the sixties, Lehmer decided that it would be nice to have a "retreat" for his current and past students before each Christmas and discuss progress on the problems they were working on. Lehmer and his students started meeting in Asilomar on the Monterey Peninsula south of San Francisco. Soon this gathering expanded, and the West Coast Number Theory Conferences were born, their first venue being the California Conference Grounds in Asilomar. Subsequently, as more participants from various universities on the Pacific coast of USA began to participate in the West Coast Number Theory Conference, the venue changed annually, returning to Asilomar every other year. In December 1975, the venue was Asilomar.

Straus advised that it would be profitable for me to attend the conference and that he would take me in his car to Asilomar. Since the conference had an air of informality, and since it was held close to Christmas, many participants came with their spouses. Straus' wife Louise, and Erdős were also in the car and so we had a most interesting conversation all along.

The California Conference Center in Asilomar consists of a number of log cabins strewn over a wooded landscape overlooking spectacular cliffs on the Monterey coast. It is an idyllic location, much like Oberwolfach, but on an oceanside setting. The conference opened with a banquet during which D. H. Lehmer, the organizer, chose the chairs of the various sessions such as on elementary number theory, analytic number theory, algebraic number theory, computational number theory, irrationality and transcendence, etc., and asked the participants to contact the sessions chairs right at the banquet with the titles of their talks. Every one was given 15 minutes, whether one was a giant in the field, or a student. By the end of the night, the session chairs made up the schedule which was then posted rightaway. The atmosphere was informal, but the quality of the conference was high. There were no parallel sessions and so one could attend all the talks. I liked this format of just having 15 minute talks, and deciding the program right at the banquet — a tradition that has not changed to this day.

Since Paul Erdős was at the conference, I spoke about my work with him on the function $A(n)$. It was a delight and honor for me that Kurt Mahler was at this conference and attended my lecture. He very kindly came up to me after my talk and said, "This is very nice work. I could not have done those estimates."

One thing that struck me as unusual was the number of mathematical couples at the conference. Emma Lehmer of Berkeley, wife of D. H. Lehmer, and a well known mathematician on her own right, was there. And then there was John and Olga-Taussky Todd from Caltech. Olga Taussky-Todd was a world authority in algebraic number theory and considered to be the "torch-bearer" of Matrix Theory. I had met her at the Summer Institute in Ann Arbor in 1973. She was a very gracious lady; she remembered seeing me as an undergraduate in 1973 and was pleased that I was a graduate student in number theory at UCLA. Another eminent mathematical couple at this meeting were Raphael and Julia Robinson from Berkeley, the latter having become notable for her contribution to Hilbert's tenth problem which asks whether there is a procedure which can be applied to any Diophantine equation to decide whether it is solvable or not. The problem was resolved in the negative by Yuri Matijasevic in the seventies and his method relied on the pioneering work of Julia Robinson. The regular presence of the Lehmers, the Todds, and the Robinsons, gave a touch of class to the West Coast Number Theory Conferences.

In view of Lehmer's interest and significant role in computational number theory, this session at the conference was especially strong. One of Lehmer's former PhD students who was a regular at the West Coast Number Theory Conference was Harold Stark of MIT. Besides the large number of number theorists from the western part of the USA, one group from the east that came in significant numbers were the number theorists from the University of Illinois at Urbana. Paul Bateman was the Chair at Illinois, and had over a period of a decade, built up the number theory program there. He arrived in Asilomar with a full entourage from Illinois. Perhaps the Illinois crowd wanted to get away from their dreadful winter!

The conference concluded with a Problem Session conducted by John Selfridge (Northern Illinois). In that session, updates on problems from earlier conferences were discussed and new problems were posed. The scribe for the Problem Session was Richard Guy from Calgary. Guy was born in 1916 and lived for 103 years! He died in 2016. He was in perfect shape until his death and participated in conferences despite his advanced age. He kept a careful log of all the problems posed not only at the West Coast Number Theory Conferences, but at various other conferences he attended regularly. In 2004 he published a book entitled "Unsolved problems in number theory" (Springer) which contains a vast collection of problems in number theory posed at conferences worldwide. This book is very valuable not only for the mathematical content, but also for the wealth of references.

At the conclusion of the Problem Session, a decision was made regarding the venue of the next conference — and this tradition still holds.

The West Coast Number Theory Conference in December 1976 was on the campus of the University of California at San Diego (UCSD). The campus located in La Jolla is quite close to the coast but not on the coast. We were all put up at the lovely Torrey Pines Inn on the Torrey Pines State Park that was right on the Pacific Coast.

Torrey Pines Inn was on a cliff near the coast, and right below it is Black's Beach, a famous nude beach. On the opening evening of the conference, as participants were arriving and getting settled, Paul Erdős went out on a walk with Carl Pomerance of the University of Georgia. They were totally immersed in mathematics during their walk. Suddenly they found themselves in the nude beach! Erdős was in his usual suit, without a tie, but with an overcoat, and Pomerance in a wind breaker. But everyone around them were totally naked! Erdős turned to Pomerance and said: "Isn't it strange that nobody is wearing any clothes?!" Just as he said this, one of the conference participants who was playing volleyball in the nude shouted, "Hi Paul"!! By the time Erdős and Pomerance finished their walk, the conference banquet had started. So Erdős and Pomerance walked into the banquet room and Erdős told D. H. Lehmer: "Don't you think it is crazy to play volleyball in the nude? It is completely crazy!"

As I said, all talks at the West Coast Number Theory Conferences were of 15 minute duration, but there were opportunities to give longer lectures by special arrangement. Hugh Montgomery from the University of Michigan was there in San Diego, and Harold Diamond requested him to give an hour lecture on the latest advances pertaining to the Large Sieve. This special lecture was arranged in a cozy fire side chat room of the Torrey Pines Inn at night after dinner, and an eager audience of about 50 (including me) assembled to hear Montgomery. Diamond, who had a great regard for Montgomery, sat on the floor in the front — squatting, like the way people in India do, and taking notes of Montgomery's talk. It was at this talk that Harald Niederreiter introduced me to Bruce Berndt because Berndt was getting deeply interested in Ramanujan's mathematics and wanted to meet someone who could tell him how to procure copies of Ramanujan's Notebooks. I told Berndt that the Tata Institute had brought out very nice photo copies of the Notebooks, and that he should write to the Tata Institute to get a copy. Berndt did that, received the copy he wanted, and from then on he has totally consumed by the work of Ramanujan.

My talk at this conference was on a new Duality Identity I had just found connecting the smallest and largest prime factors of the integers. This duality had some striking consequences such as a new series evaluation involving the Möbius function that was related to the Prime Number Theorem for Arithmetic Progressions. It also led to my thesis work on sums of the Möbius function over integers with restricted prime factors.

In December 1977, during my third and final year at UCLA, the West Coast Number Theory Conference was held at UCLA itself. This was very fortunate for me because in Fall 1977 I had applied for post-doctoral positions at various universities, and during this conference I could talk to many number theorists about my application at the opening banquet which was held at the fabulous Sunset Canyon Recreation Center on the fringe of the beautiful UCLA campus. At the banquet I had a conversation with Patrick Gallagher of Columbia University who told me that I was on the short list for the Ritt Assistant Professorship at Columbia. But Gallagher said that if I would be offered the Hildebrandt Assistant Professorship at Michigan, then I should accept that instead of seeking an offer from Columbia because he felt that Hugh Montgomery in Michigan was doing more fundamental work in analytic number theory than he was. Gallagher was a leading analytic number theorist, and I was struck by his humility when he said this. D. H. Lehmer said that he had put in his recommendation for a position for me at Berkeley. This put me in a great mood because the prospects that I would have a nice job after my PhD seemed very bright. My talk at the 1977 Conference was on my thesis work "Sums of the Moebius function and integers with restricted prime factors."

Basil Gordon on the UCLA faculty graciously hosted a party at his home for the conference participants. My officemate Bob Miller who was finishing his PhD under Gordon said, "Given Gordon's high standards and tastes, the party will offer the finest in food and wine." And indeed it lived up to everyone's expectations.

In addition to giving talks at the West Coast Number Theory

Conferences, as a graduate student I also had opportunities to give seminar talks at various universities on my way to, or back from, India each year.

In 1976, I met Hyman Bass of Columbia University in Madras when I was there during the summer break. Bass gave a talk at MATSCIENCE that I attended and after that I showed him around town. He was genuinely interested in my research attempts. During a walk at the Marina Beach, I described my own research as well as my joint work with Erdős. He immediately said that I should stop in New York on my return to the USA, and meet Patrick Gallagher their senior number theorist. Bass was Chair at Columbia and he arranged this visit. That was the first time I met Gallagher and was impressed that so eminent a mathematician could be so modest. Both Gallagher and Bass took me to dinner at the Woodlands (a South Indian Vegetarian Restaurant) close to the United Nations. I was touched by their warmth and hospitality; I was just a graduate student. They both suggested that I should apply for the Ritt Assistant Professorship at Columbia when I complete my PhD, which I did.

On my second trip to India from UCLA in 1977, I was invited by Bill Adams and Larry Goldstein to give a number theory seminar at the University of Maryland. I had met Adams and Goldstein in 1974 when I was visiting various universities in the USA trying to decide where I should apply for graduate admission. I did get admission into Maryland with a graduate fellowship, but that letter came after I had accepted the Chancellor's Fellowship at UCLA. The most interesting aspect about my visit to Maryland was that my seminar was arranged at 5 pm immediately after the Colloquium by Charlie Fefferman of Princeton, a phenomenon in the mathematical world! Fefferman was a whiz kind who by the age of 17 had earned both a degree in mathematics and physics at the University of Maryland, and by the age of 20 had a PhD from Princeton. At the tender age of 22, he was made a Full Professor in Chicago, but he returned to Princeton as a Full Professor at the age of 24. Fefferman won the Fields Medal in 1978, but was a superstar even before that. In 1977 when he visited Maryland, he was also welcomed back as a local hero. He is an electrifying speaker and so for those who attended both talks, mine paled in comparison with Fefferman's. But the number theorists were very kind to me.

On my third and last trip to India from UCLA in April 1978, I was invited to give a talk enroute at the University of Colorado at Boulder, an invitation that was facilitated by Paul Erdős. Boulder was one of the regular stops for Erdős on his global travels. In an article I wrote for his 80th birthday, I said that "like migrating birds, Erdős always hovered around the isotherm $70^\circ F$. He could be found in Boulder and Calgary in the summers, in Florida during the Spring, and at UCLA in December." Erdős arranged for me to stay at the University Club at Boulder where he stayed. At Boulder I interacted with Wolfgang Schmidt, a giant in the field of Diophantine Analysis, and Peter Elliott, a world authority in Probabilistic Number Theory. After my talk, Erdős took me along with him for dinner at the home of the Stan Ulam, who was a brilliant mathematician known for his contributions to both pure and applied mathematics. From 1974 until his death in 84, Ulam was Graduate Research Professor in the mathematics department at the University of Florida where I joined in December 1986. Thus I did not meet him in Florida, but only in Boulder.

These contacts at the West Coast Number Theory Conferences and at various universities where I gave talks, put me in a position of advantage when I applied for post-doctoral positions. Next I will describe my PhD work and my post-doctoral experience.

3.4   My PhD Committee, thesis, and consequences

After finishing all my Qualifying Exams in 1976–77, the next step was to form my PhD Committee. In India, at least in the universities, the practice is that for the PhD there are always external examiners; that is the thesis is sent to experts overseas. This practice holds in many countries to ensure the quality of the thesis. But in advanced countries like the USA, the PhD thesis is judged internally. However, PhD Committees are required to have one or two members outside of the department where the student is working. At UCLA, in the seventies, the rule was that out of the five PhD Committee members, two must be from other departments.

The PhD Committee is extremely important for the candidate. The letters written by the members of the Committee are crucial for the candidate to get job offers. Thus it is not just the quality of the doctoral work, but also the letters attesting to it that ensures success in job placement. Needless to say that the stature of the letter writers plays a major role as well. Thus the formation of the PhD Committee and its endorsement will determine the orbit for the candidate in the academic world. Straus advised me that of the two members from the mathematics department on the Committee besides himself as Chair, one would be Miech, but the other should be a non-number theorist. Straus suggested that I should request Steinberg to be on my Committee. Straus' reasoning was that, number theorists at UCLA like Basil Gordon or David Cantor, would always be willing to write a letter for me even if they were not on my Committee because they were familiar with my work due to presentations I made at the UCLA seminars or at the West Coast Number Theory Conferences. But someone like Steinberg, not in number theory, would write a letter only if he served on my Committee. Straus also said that the Search Committees at the various universities will be composed of faculty in different areas of mathematics, and Steinberg's letter would attract the non-number theorists on those search committees because of his stature in algebra. This was a fantastic suggestion, but I was unsure whether Steinberg would agree to be on my Committee. Straus assured me that Steinberg would agree because he had already spoken to Steinberg about this! Straus had thought of everything and had taken action as well. But he wanted me to formally request Steinberg to be on my Committee. I confess that I was a bit nervous when I entered Steinberg's office to make this request. With his typical sense of humor and a wry smile he said: "So you want me to be on your Committee. I know of your performance in my classes but I am not familiar with your research. But I agree to be on your Committee assuming that you will provide me enough background material about your research." I was thrilled to hear this, said that I was honored, and left his office. Steinberg's letter was crucial for me in more than one instance as will seen in the sequel.

The selection of the members outside the Department is a formality since they rarely come up with serious questions on the PhD defense, or write letters evaluating the candidate's thesis. The first external member I chose was Professor S. V. Venkateswaran of the Department of Atmospheric Sciences (formerly Department of Meteorology). "Venki" as he was affectionately known, was a family friend. I first met him in 1962 when I accompanied my parents to USA on my first world tour. Actually when I joined UCLA for my PhD, I was only 19, and since I was less than 21 years of age, I had to list a Guardian because my parents were not in America. Venki was my Guardian, and indeed he took a parental interest in me. In fact, on several weekends, I used to visit Venki at his home in Santa Monica. He was very glad to serve on my PhD Committee as I was to have him as a member. Venki suggested the name of a F. Coroniti, an Associate Professor in Astronomy, and so Coroniti was the fifth and final member of my Committee. With the Committee having been formed, my next job was to make progress on my thesis work.

In 1976, at the start of my second year at UCLA, I found a very elegant "Duality" identity connecting the smallest and largest prime factors of the integers via the Möbius function. More precisely the pair of identities I found were the following: *Let $p(n)$ and $P(n)$ denote the smallest and largest* *prime factors of an integer $n > 1$, and let f denote any function defined on the primes. Then*

$$\sum_{1<d|n} \mu(d)f(p(d)) = -f(P(n)) \quad \text{and} \quad \sum_{1<d|n} \mu(d)f(P(d)) = -f(p(n)).$$

I was always interested in the Möbius function $\mu(n)$ defined by $\mu(1) = 1$, $\mu(n) = 0$ if the square of a prime divides n, and $\mu(n) = (-1)^r$ if n is a product of r distinct primes. The Möbius function is one of the most important functions in analytic number theory and my Duality identities generalized the well-known fundamental formula

$$\sum_{d|n} \mu(d) = 0 \quad \text{if it is rewritten as} \quad \sum_{1<d|n} \mu(d) = -1.$$

It was clear that the Duality had to have significant consequences and I found one rightaway.

It is a well-known and famous result due to Landau, that the Prime Number Theorem is equivalent to the statement that

$$LANDAU: \quad \sum_{n=1}^{\infty} \frac{\mu(n)}{n} = 0, \quad \text{which I rewrote as} \quad \sum_{n=2}^{\infty} \frac{\mu(n)}{n} = -1.$$

By equivalent is meant that each result can be derived from the other elementarily. My Duality yields a series evaluation of the Möbius function that is related to the Prime Number Theorem for Arithmetic Progressions (PNTAP), which is the statement that for each integer $k > 1$, the prime numbers are asymptotically uniformly distributed in the reduced residue classes $\ell \pmod{k}$, for $1 \le \ell \le k$ with $\gcd(\ell, k) = 1$. Using the PNTAP, I showed that if we consider the distribution of the sequence $P(n)$ of the largest prime factors in the reduced residue classes, then uniform distribution is retained. Using this uniform distribution property of $P(n)$, and my Duality, I deduced that

$$\sum_{n \ge 2, p(n) \equiv \ell \pmod{k}} \frac{\mu(n)}{n} = \frac{-1}{\phi(k)},$$

for each reduced residue $\ell \pmod{k}$, and this generalized Landau's theorem. This generalization is remarkable because it says that if the convergent series in Landau's theorem is cut up (sliced) into $\phi(k)$ subseries using $p(n) \equiv \ell \pmod{k}$, then all these subseries will converge to same value $-1/\phi(k)$!

In Fall 1976, I wrote a paper entitled "Duality between prime factors and an application to the Prime Number Theorem for Arithmetic Progressions" and submitted it to the Journal of Number Theory. It was quickly accepted and appeared in 1977 before I finished my PhD. That paper was the starting point for my PhD thesis. In the last few years, my generalization of Landau's theorem to arithmetic progressions has been significantly extended in the setting of algebraic number theory by a host of young researchers following a suggestion by Ken Ono. It is gratifying that what I proved more than forty years ago is sparking interest now.

Back in Madras in December 1974, at the Madras Marina beach, during a discussion on connections between my function $A(n)$ and $P(n)$, Erdős introduced me to the fundamental function $\Psi(x, y)$ which enumerates the number of integers up to x all of whose prime factors are $\leq y$. The integers enumerated by $\Psi(x, y)$ are called *smooth numbers*. The function $\Psi(x, y)$ has a long and rich history and Erdős told me about various important asymptotic results of de Bruijn. If we let $y = x^{1/\alpha}$, it can be shown that for each fixed $\alpha \geq 1$, we have $\Psi(x, y) \sim x\rho(\alpha)$, where $\rho(\alpha)$ satisfies a certain difference-differential equation. de Bruijn's great work was to show that this asymptotic estimate for $\Psi(x, y)$ holds uniformly as α varies with x for long ranges of α. This is important because $\rho(\alpha)$ decays very rapidly as $\alpha \to \infty$. de Briujn also obtained asymptotic estimates for the "dual" function $\Phi(x, y)$ which enumerates the number of integers up to x all of whose prime factors are $\geq y$, namely the number of uncancelled elements in the sieve of Eratosthenes.

Having studied the two papers of de Bruijn on the functions Ψ and Φ, I was motivated to study the sum

$$M(x, y) = \sum_{n \leq x, p(n) > y} \mu(n)$$

due to my interest in the Möbius function. $M(x, y)$ is a generalization of the sum $M(x, 1) = M(x)$ made famous by Landau who showed its connections with the Prime Number Theorem. To my pleasant surprise I found that even though $M(x, y)$ is a weighted version of $\Phi(x, y)$, in view of the Duality between large and small prime factors effected by the Möbius function, it was intimately connected with $\Psi(x, y)$. For instance, $M(x, y) \sim \frac{xm(\alpha)}{\log y}$, where $m(\alpha) = \rho'(\alpha)$, the derivative of $\rho(\alpha)$. In my thesis I established an asymptotic formula for $M(x, y)$ for long ranges of α and this is useful because when $\alpha \to \infty$, $m(\alpha)$ decays as rapidly as $\rho(\alpha)$.

I also began estimating the dual sum

$$M^*(x, y) = \sum_{n \leq x, P(n) < y} \mu(n),$$

which even though is a weighted version of $\Psi(x, y)$, is intimately connected via Duality to $\Phi(x, y)$. The asymptotic analysis of $M^*(x, y)$ is more complicated than that of $M(x, y)$. So Straus asked me not to rush through it. He said that my results on the Duality and on $M(x, y)$ were sufficient for the thesis, and that I could investigate $M^*(x, y)$ in detail after my thesis is completed. So my thesis entitled "New results on the Möbius function and applications of a Duality principle" was ready by the Spring of 1978.

Straus' office was right across the corridor from mine, and so I knew when exactly he was free since his door was always open. I would literally barge in whenever I had an interesting idea and he was always ready to hear it. He never said that I should work on a specific problem or area; instead he encouraged me to pursue my own ideas, and owing to his vast knowledge, was able to provide me the references I needed or could explain some subtle features of the analytical tools I required. Thus he gave me complete freedom in what I investigated and how I went about exploring it. I think this was why Erdős told me in Madras that Straus would be the perfect thesis advisor for me, and indeed he was.

In 1976–77, in my second year, I had finished all my PhD Qualifying Exams, and so starting from that time, I was working on my thesis. The results on $M(x, y)$ were actually obtained in my family home in the summer of 1977 when I was visiting my parents in Madras, and I showed Straus my results in September 1977 when I returned to UCLA from India. Straus approved my thesis submission in Spring 1978 even though it was only my third year of my four year Chancellor's Fellowship. He said that I should start applying in Fall 1977 for academic positions to start in Fall 1978.

In the United States, the tradition is that after receiving the PhD degree, one must seek positions elsewhere. To prevent inbreeding, universities do not generally employ their own PhDs. The kind of cross-fertilization that occurs when PhDs from one university go to be employed elsewhere is very beneficial to the discipline. Only in certain exceptional cases would a person with a PhD from a certain university be employed on a tenured position in the same university. Even then, it is expected of the individual to get a post-doctoral experience outside before coming back to his alma-mater for a permanent position. In the UCLA number theory group, an exception was made in the case of David Cantor, a former PhD student of Gordon and Straus. Cantor was a leading authority in Diophantine Approximations, well known for his generalization of the notion of transfinite diameter.

In Fall 1977 I started writing my thesis and also a paper on $M(x, y)$ which I submitted to the Journal of Number Theory. I selected about a dozen universities to send my job applications to, far fewer that the hundred or so applications my classmates were sending out. My choices were based on specific individuals with whom I had interest in working. I wrote to these mathematicians drawing attention to my work and enclosing my papers. I think this is very important, because no matter how strong your file is, in order to actually get a job offer, someone in that department must speak to the search committee in high terms about your work.

In mid-January 1978, within a span of 24 hours, I received two job offers; (i) the Hildebrandt Assistant Professorship at the University of Michigan, (a three year position) and (ii) a two year lectureship at UC Berkeley. Lehmer's support led to the offer from Berkeley. I called Gallagher to enquire about the Ritt Assistant Professorship at Columbia. He said that I was on their short list, but it would be two more weeks before any offers would be made. But he reiterated what he told me at the West Coast Number Theory Conference, that I should accept the offer from Michigan to be with Montgomery. So even though Berkeley was a more highly ranked mathematics department, I accepted the Hildebrandt Assistant Professorship. In April 1978 I received an offer from Institute for Advanced Study for a one year Visiting Membership. I wrote to the Institute saying that I had accepted the offer from Michigan, and hoped that after completing that assignment, I could visit Princeton in 1981–82, and that actually happened.

There were a variety of factors that led to the offer from Michigan. Most important was that Hugh Montgomery liked my work. I had sent him my papers and drew attention to my application. He remembered me from the Summer Institute in 1973 when I was an undergraduate participant. Next Fred Gehring was the Chairman at Michigan. He remembered me from the 1975 Summer Institute in Complex Analysis in Trieste. Thirdly, Straus spent most of Fall 1977 at the University of Michigan on a sabbatical, and he spoke to Montgomery, Lewis and others about my work. Finally, I must acknowledge the support of my PhD Committee members in their letters, as well as the letter of Paul Erdős. I cannot thank Straus adequately for the way he encouraged and supported me. I give you one more example.

I had completed my thesis defense in spring 1978 at the end of my third year at UCLA. The Chancellor's Fellowship I had was a four year offer with the final year being the Dissertation Fellowship which had no residency requirement. So Straus recommended to the department that the entire amount of my fourth year Dissertation Fellowship be given to me as a (dollar) gift at the end of my third year(!) since I had spent the third year on my dissertation on top of my duties as a Teaching Assistant.

I did not ask for this. It was Straus' idea and a very fine gesture. I could not believe that UCLA accepted this suggestion of Straus. So when I left UCLA in 1978 to take my job at the University of Michigan, I actually had some funds with me owing to the gift of the Dissertation Fellowship that I received. Such a thing could never happen in India. Perhaps this was a unique instance of such a gift even in the USA!

Straus also did one more act of support. He recommended my thesis for the Alumni Medal which was a university-wide honor. To my pleasant surprise I was awarded the Alumni Medal in the summer of 1978 for one of five best PhD thesis (all subjects) at UCLA. In between the submission of my thesis and the receipt of the Alumni Medal, a very significant event happened for me: I went back to India to get married to Mathura! She has been my partner in life ever since. I will describe that story next.

3.5   Social life at UCLA and my marriage to Mathura

As a graduate student at UCLA, my focus was my work. I never dated, and so in the opinion of most American students, I did not have a life! I was brought up in a traditional Indian society whose practices I valued. One could think of marriage after completing one's education, and one could get married by meeting girls introduced by your family or close friends. So dating was rare in Indian society, at least in Madras in those days. It is much more common in India now with strong Western influence. Also, the seventies was a time when American society was extremely permissive. It was not just dating in the ordinary sense; people took pride in saying that they had been intimate with dozens of individuals in short periods of time. So a number of my UCLA classmates felt that I was not enjoying life.

During my first year at UCLA, I stayed in the dorms. I had a very sheltered upbringing, and so my parents felt that instead of living in an apartment and commuting to and from campus, it is much safer to reside on campus during the first year. After I get acclimatized, I could move into an apartment, which is what I did in the second year.

Since I was only 19 years old when I joined UCLA, even though I was a graduate student, I could only stay in the undergraduate dorms. This was because in the graduate dorm (called *Mira Hershey*), they would regularly serve wine at dinner. The liquor law in California was quite strong: no serving of liquor to persons under 21. I told the authorities that I was a vegetarian and a tee-totaller, but they would not take any chances. I was given a room in one of the four undergraduate dorms called *Sproul Hall*, located along with the three other undergraduate dorms in a most beautiful section of campus near Sunset Canyon.

Since real estate was very expensive around UCLA, all undergraduate dorm rooms had to be shared in order to maximize the number of dorm occupants in limited space. My roommate was an undergraduate by name Chad Roche, a very friendly chap. He was of course very much into dating. Since I had travelled extensively in America in the sixties with my parents, I had several Indian friends in the Los Angeles area. On weekends I used to go to their homes, especially the homes of Professor Venkateswaran of the Department of Meteorology, and that of Dr. A. Sankaranarayanan, a grandstudent of my father. The UCLA undergraduate dorms would go crazy on weekends with parties and dancing. My roommate was happy that I was away on weekends, because he could enjoy the privacy of his dorm room with his girlfriends! The dorms were noisy even during weekdays with many students playing loud music at nights. As a graduate student, I helped many of the students in the dorms with their calculus homework; thus I enjoyed a certain level of respect from them. The most important effect of this respect was that I could ask my neighbors to turn down the volume of the music at night and they would oblige with a smile!

Besides some graduate students in the mathematics department like Charles Grinstead and Bob Miller with whom I used to go out to dinners, I had a few close friends among the graduate students and post-docs from India. Nadadur Sampath Kumar (Sampath as he is known to friends) was a Rotary Scholar from Hyderabad doing his MBA at UCLA, and he and I became very close friends owing to family connections and our common passionate interest in Carnatic (= South Indian classical) music. Sampath moved on to get a law degree from the University of Southern California and is now one of the leading immigration lawyers in California. He lives in a magnificent mansion in fashionable Brentwood. His house is the Ekamra Nivas of Los Angeles! Our friendship that started at UCLA has grown in strength as a bond between our two families.

When I moved out of the dorms after my first year at UCLA to an apartment in Santa Monica, there were two post-docs from the Indian Institute of Science who lived near me — Srivatsan (who was single then) and Sivaraman and his wife Ponni. They used to invite me regularly to their apartments for dinners in the evening. I found out later that Ponni was a classmate of Mathura whom I was to marry soon!

My main "recreation" was tennis. I was an avid tennis player and had represented my college in Madras while I was an undergraduate student. UCLA had an outstanding record in tennis with the pride that Jimmy Connors was on their team and he lifted it to the NCAA Championship

Title. I was no match to the UCLA team players but I enjoyed watching them practice in the Sunset Canyon courts next to the dorms. I too played regularly on those courts. In fact the showdowns between me and Professor Andy Majda of the mathematics department were quite famous — sort of the graduate students vs faculty rivalry in tennis; after each of our matches, my classmates wanted to know who won the encounter. Andy Majda is world-renowned applied mathematician who had strong ties with the Courant Institute. He waxed eloquent about Courant Institute being a pre-eminent center for applied mathematics with Peter Lax, Louis Nirenberg, Fritz John and others on its faculty. He said that Peter Lax was a fine tennis player. Majda and I became very good friends. I meet him from time to time during my travels and we compare notes about our tennis experiences. My latest meeting with him was at the NYU campus in Abu Dhabi in the UAE in November 2015; I had gone there to evaluate their mathematics program, and he was there giving a course of lectures.

Paul Erdős of course understood why I never dated because he was wedded to mathematics! But he felt that I was spending too much time on tennis. He expressed that view one day when he wanted to meet me at UCLA to discuss mathematics. I said I could do that a few hours later after finishing a tennis game that I had already committed to; it was with Majda who was a very competitive player.

My pastime other than tennis at UCLA was watching the basketball games at the famous Pauley Pavilion. It was Charles Grinstead who introduced me to this exciting game, and I used to watch the UCLA games with him. UCLA created a sensation in college basketball by winning ten national championships in a eleven year period (1965–75) — a record that is unlikely to be surpassed! Just as Erdős felt that I was spending too much time on tennis, Straus felt that watching basketball was a waste of time. According to him, competency in basketball is an increasing function of height as the only parameter. That is, if A is taller than B, then A is the better player! This is obviously incorrect. Straus also felt that it suffices to watch the last few minutes of a basketball game because that is when the real action takes place. I have the highest regard for Straus and Erdős, but I needed to have some form of recreation in the midst of my total focus on studies. Watching UCLA basketball and playing tennis provided me much needed relaxation. Tennis actually played a major role in my marriage. Indeed it was due to tennis that I met the love of my life — Mathura!

I used to go to India every summer during my graduate student days at UCLA. Similarly my parents visited me for three months in Los Angeles during the two years I lived in an apartment. My classmates felt that since my parents were staying with me for three months, I would have no privacy. They said that three months is way too long. Privacy from what? I was not dating anyway. That is why they felt that I did not have a life. But I enjoyed having my parents for three months. After all they had taken me on world tours when I was a school boy. Now it was my turn to host them, and it was a pleasure. Actually, during these three months, my father would go away from time to time on his academic assignments. My mother stayed back with me. Throughout the three months, she prepared delicious Indian meals, a refreshing change for me from the bland cafeteria food that I had on weekdays at UCLA.

On my trips to India, I played tennis almost daily at the Mylapore Club close to my house. One day in 1977, Mr. K. Venkatesan, a family friend of ours who played tennis with us at the Club, asked me and my father whether he could introduce a gentleman by name N. C. Krishnan to us. Venkatesan said that Mr. Krishnan had a charming daughter by name Mathura (Krishnan) who would be a perfect match for me. We agreed to meet Mr. Krishnan and his family over Tea at the home of Venkatesan.

Mr. N. C. Krishnan was a leading Chartered Accountant (= CPA) and the senior partner of Viswanathan & Co., one of the major accounting firms in Madras. He had risen to the level of President of the Institute of Chartered Accountants of India, and thus was a well known throughout India. He and my father had met at a few public functions earlier as it became apparent during our conversation at Mr. Venkatesan's home.

Mathura was exactly like Mr. Venkatesan had described — very charming and elegant. Her grace and modesty impressed me equally. She also shared values similar to mine. She was highly accomplished in *Bharathanatyam*, one of the premier classical dance forms of India. She started learning this great art form at the age of eight, and by the age of ten had her *Arangetram* (full graduation performance). It is quite remarkable that in two years she had mastered this intricate art to give a full two-hour solo performance in her Arangetram. I felt that her interest and expertise in Bharathanatyam nicely complemented mine in mathematics. We had a most pleasant conversation. When we came home, both my parents and I said unhesitatingly that she would be a fine partner for me. I understand that when the Krishnan's went back home, Mathura and her parents felt equally positive. We conveyed our mutual positive feelings through Mr. Venkatesan — the catalyst for the union. From then on the Krishnan family and my family met a few times at their home and our home. With each meeting, I got to know Mathura better, that it was agreed that we would get married next year, that is in the summer of 1978 after I finish my PhD and after she would finish her undergraduate degree. Mathura was in a Bachelor of Commerce (BCom) degree course and 1977–78 would be her final year. So everything fitted perfectly. The summer of 1977 was an outstanding one for me — I made significant progress on my thesis problem in my family home in Madras, and I met Mathura, my future partner in life. So I returned to UCLA in very high spirits.

Just before leaving Madras, I requested the Krishnans to send me a beautiful enlarged photograph of Mathura that I could have with me in Los Angeles. In Madras, the most reputed photo studio is G. K. Vale. Mathura's father arranged for a lovely portrait photograph of her to be taken at G. K. Vale and sent it to me. Among all the photographs of Mathura taken over the years, this is the loveliest, and has been my favourite. I put it my briefcase and therefore she was in my thoughts every moment during the year 1977–78 when we were separated by 10,000 miles.

UCLA was on a quarter system — three quarters in an Academic Year. The Winter Quarter ended in late March and the Spring Quarter was from early April to mid-June. I defended my thesis at the end of the Winter Quarter of 1978 and so I returned to India in April to get married. In India, the final university exams each year are in April. So Mathura took he final BCom final exams in April 1978 and was ready to get married shortly afterwards. Our marriage was fixed on June 29, in the middle of the Madras summer — a date picked by our priests as most auspicious. The priests' choice must have been correct, for we have a most happy married life in all respects! Since the wedding day was two months away, and since I had arrived in Madras early, I requested Mr. Krishnan permission to go out with Mathura on a regular basis BEFORE the wedding day. This was a form of dating that was frowned upon in orthodox Hindu society. But Mr. Krishnan was a progressive man. So he happily agreed but on one condition — that his aged father (Mathura's paternal grandfather) should not come to know of it! I promised that I would not trumpet this dating, but keep it hush-hush. So everyone was happy.

Although I said earlier that I did not succumb to the temptation of dating while I was at UCLA, I do agree that dating is a most enjoyable experience. In these outings with Mathura, I remember every sari she wore on the nearly two dozen dates we had. We dined at some of the finest restaurants in Madras and had a most wonderful time.

Two weeks before the wedding, Mr. Krishnan arranged a special two-hour Bharathanatyam program by Mathura for me and all our relatives. Over the years I have enjoyed Mathura's dance performances in India, Hawaii, Japan, and Florida, but this program on the eve of our wedding is my favourite and I remember every moment of it.

Indian weddings are elaborate and many are conducted over two full days. My wedding with Mathura was a three-day ceremony. This was because, my father-in-law, Mr. Krishnan, a great benefactor of Sanskrit, arranged for an assembly of Sanskrit scholars (*Vidwat Sadas*) on the third day, to discuss the sanctity of the institution of marriage as laid down in our Hindu scriptures. Also, there was a chanting of the vedas by a group of Vedic scholars led by our family priests Nataraja Sastri and Pattabhirama Sastri. Finally, my mother's music teacher Mannargudi Sambasiva Bhagavatar, gave a musical discourse on the marriage of Lord Krishna.

It has generally been a custom, that on the evening of the first day of the wedding, the groom goes out on a grand procession — the *janavasam* — followed by friends and relatives. But in the seventies, many grooms declined the Janavasam since they felt it was ostentatious, but I agreed to have it since it was part of the festivities.

Just a few days before our wedding, I received a letter from Professor Straus that I would be awarded the Alumni Medal. He knew that I could not attend the Award Ceremony, and so he most graciously agreed to accept the Medal on my behalf at the Ceremony. I was delighted to receive this message just before my wedding, as was Mathura.

Mathura and I had a honeymoon in Kodaikanal, a lovely hillstation in the state of Madras. We stayed at the English Club where my father-in-law was a member. The "season" had just ended, and so we were the only occupants at the English Club. Thus we were treated royally by the butlers and the English Club felt like our own house!

The honeymoon in Kodaikanal was arranged by my father-in-law for us. But I wanted to offer Mathura a second honeymoon on my own. On our way from India to Ann Arbor, Michigan, I chose to travel via the Pacific. So we halted in Singapore, Hong Kong, Tokyo and Hawaii for a splendid vacation. I consider the Hawaiian islands as paradise on earth, my first introduction to those isles being in 1962 on my first round-the-world tour with my parents. I had visited Hawaii several times since then and I wanted to introduce Hawaii to Mathura. She was impressed by the beauty of Honolulu and its surroundings, but was totally mesmerized by the tranquil beauty of Hilo on the Big Island of Hawaii, and the awesome grandeur of Volcanoes National Park. After Hawaii, we were ready to leave for Michigan but I had warned

Mathura about the dreadful winters of the American mid-west — a far cry from the heat of Madras she was accustomed to, or the celubrious climate of Hawaii that she had just experienced.

Enroute from Hawaii to Michigan, we stopped in Los Angeles. We met Straus and he gave me the Alumni Medal. We then proceeded to Ann Arbor where Mathura and I began life as a married couple.

(Left) Mathura in 1977, when I first met her in Madras. This picture, taken in G. K. Vale studios, is my favourite of hers. I had it in my briefcase throughout my last year (1977–78) at UCLA before getting married to her on June 29, 1978 (Right picture).

Chapter 4

First job at the University of Michigan (1978–81)

4.1 The Michigan Mathematics Department and its faculty

The pride of the University of Michigan in general, and of the Department of Mathematics in particular, was that apart from UC Berkeley, the only universities more highly ranked (NRC Rankings) are all private universities like Harvard, Princeton or Chicago. Indeed, in many subjects, Berkeley is within the top five, and the public universities that come just after Berkeley and the top private universities are usually Michigan and UCLA. Thus I felt honored to join the esteemed faculty at Michigan.

The Hildebrandt Research Assistant Professorship that I was offered, was named after Theophil H. Hildebrandt who joined the department in 1909 and served as Chair from 1934 to 1957. Mine was a three-year position without possibility of an extension of any type. In previous years, the Hildebrandt Assistant Professorship was a two-year position, but there was a possibility that the individual could continue in a tenure track assistant professorship. Not all Hildebrandt assistant professors were "selected" for the tenure track position, and so this led to some disappointment among those who did not continue beyond their two-year terms. In order to avoid such delicacy of making a choice among the Hildebrandt assistant professors, the Department decided that starting from 1978 (the year I joined), all Hildebrandt assistant professors would have to leave at the end of their terms, but then the Department extended the duration from two years to three. Don Lewis referred to such terminating positions as "folding chairs", a term that is not very encouraging! But I appreciated the rationale behind the decision to make it a three-year non-renewable position. Several years later, as Chair at the University of Florida, I instituted the John Thompson Assistant Professorship (see §8.5), but made it a three-year non-renewable position inspired by the Michigan model.

The home of the Michigan Mathematics Department was Angell Hall, the most imposing building on campus, and definitely one of the grandest campus buildings to be found anywhere in the USA. Angell Hall is composed of white stone, and has huge pillars at its entrance giving it a Roman or Greek grandeur, like the Parthenon! Upon ascending the steps of Angell Hall and entering the building, one is bound to be impressed by the very long corridors and high ceilings, and with the names of its eminent faculty etched on the glass doors of the offices.

The third floor of Angell Hall was where the main mathematics administrative offices were. Most mathematics faculty had offices in Angell Hall, but a few had offices in a portion of the West Engineering Building, not far away. One entire wing of the third floor of Angell Hall was the Mathematics Library, one of the best in the world. The main University Library is acknowledged as one of the most comprehensive, and it had a mathematics section as well. Having a mathematics library or a comprehensive reading room in the mathematics building is very conducive to research. All the major research universities understand this and therefore have such a library or reading room in the mathematics building. I already have spoken about the graduate reading room at UCLA. The Penn State Mathematics Department had a library in the first floor of its home, the McAlister Building. After the recent renovation of McAlister, the mathematics library moved just across the street to a building no more that 50 yards away. The University of Illinois at Urbana has the mathematics library in Altgeld Hall which is where most faculty of the mathematics department have offices. Universities which are not highly ranked, want to avoid either the duplication of journals and books, or do not want to assign large departments the extra space needed to house a library. Therefore these universities just prefer a central library for all. I cannot over emphasize the usefulness of having the mathematics library in the mathematics building.

At the University of Michigan, the practice was that only Full Professors would get private offices; assistant and associate professors had to share offices, but these were quite spacious. In 1973 when I first visited Ann Arbor, Montgomery was an Associate Professor and so he shared an office on the third floor of Angell Hall with Peter J. Weinberger. But by the time I joined the University of Michigan in 1978, Montgomery had been promoted to the rank of Full Professor and so he had his own office on the third floor of Angell Hall. My office was on the fourth floor, and owing to the design of the building, the fourth floor of Angell Hall was not very big like the other floors. So there were very few offices on the fourth floor, and this provided a certain amount of privacy.

My office mate during the first year was K. Y. Shih an assistant professor who previously had a Hildebrandt Assistant Professorship like me, but was in his final year at Michigan. He had received his PhD under the direction of Goro Shimura of Princeton, one of the most influential figures in algebraic number theory. After completing his term at Michigan, Shih took up a job in industry. My office mate during the next two years was Jeffrey Rauch, an associate professor who was in Paris on a sabbatical during my first year in Michigan. He is quite a well known applied mathematician and had received his PhD under the guidance of Peter Lax of the Courant Institute.

The office of Robert Greiss was next door to mine. He used to walk out of his office frequently and pace the corridor of the fourth floor in deep thought. When he saw you, he would smile in a most charming way, but I doubt he really took note of you because he was thinking of something fundamental. How did I know this? Greiss was a group theorist and he was working on the great problem of classification of finite simple groups. More specifically, he was trying to construct an exceptionally large group that the German mathematician Bernd Fischer had predicted in 1973. Whenever Greiss' office door was open, one could see the calculations on his blackboard pertaining to the orders of various groups that he was constructing. On the board was the instruction "DO NOT ERASE" to the janitors who obliged most willingly. In January 1980, we heard from Greiss that he had succeeded in constructing the group predicted by Fischer. It was gigantic and had about 8×10^{53} elements! In a massive 102 page paper that appeared in *Inventiones Mathematicae* (one of the most prestigious and exclusive journals), Greiss gave the construction of this group that he dubbed the *Friendly Giant*, thereby helping the classification that mathematicians had been seeking for decades. I was told the F and the G in the Friendly Giant also stood for Fischer and Greiss! In the end however, this group was called *The Monster*, a name given by John Conway, who simplified Greiss' construction. Greiss was not happy with this because he felt the discoverer ought to have the privilege of naming the group. I agree with Greiss' sentiment. But Conway was one of the giants of the field and so the name he gave was widely used by the mathematical community. Greiss received the 2010 Steele Prize of the AMS for seminal contributions to research.

Robert Greiss received his PhD under the direction of John Thompson, arguably the most eminent group theorist of our generation. Thompson himself made very significant contributions to the Classification Problem and in the end the Monster was discovered by one of his former students — Greiss. In view of Greiss' connection to Thompson, I had occasions to host Greiss at the University of Florida during my term as Chair.

The Chairman of the Mathematics Department at that time was Fred Gehring. He was a world authority on quasi-conformal mappings. He was invited twice to the International Congress of Mathematicians to deliver Plenary Addresses, a rare honor. The subject of quasi-conformal mappings was the brain child of the Finnish mathematician Lars Ahlfors, who had received the First Fields Medal (along with Jesse Douglas) in 1938. Ahlfors was perhaps the premier figure in complex analysis for the next several years. From the 1940s, Ahlfors was a professor at Harvard. Gehring had close ties with Ahlfors, and owing to that, he turned the focus of his attention to quasi-conformal mappings. I remember a beautiful lecture that Gehring gave at a meeting of the Mathematics Club in which he presented more than thirty ways to characterize quasi-conformal mappings.

In view of their close connections, Gehring visited Ahlfors regularly in Harvard and Finland, and Ahlfors visited Ann Arbor quite often. I remember the visit of Ahlfors during my stay there. His office was on the fourth floor of Angell Hall, close to mine.

Fred Gehring was always formally dressed. But what was unique was that he only wore bow ties, never a tie. This was the way he dressed even when he was not a chairman. He was an extremely friendly person and took great interest in my progress.

There was an Indian faculty member in mathematics — Professor M. S. Ramanujan (now retired). He was no relative of Srinivasa Ramanujan; but he was an Iyengar, the same Brahmin sect to which Srinivasa Ramanujan belonged. M. S. Ramanujan and his wife Chitra were our "God-parents" in a certain sense; Mathura and I consulted them on almost everything as we were beginning life as a married couple in the United States, and they assisted and guided us most graciously. M. S. Ramanujan actually taught Mathura how to drive, and for this she is most grateful to this day. He said that to avoid fights, a husband should never attempt to teach his wife how to drive! He therefore volunteered to teach Mathura how to drive, and he was indeed a perfect instructor, as she was an attentive pupil.

Ramanujan was extremely meticulous in everything he did. I could see that from the way he taught driving to Mathura. I watched everything quietly from the back seat. Ramanujan was the Associate Chair for Graduate Studies in mathematics at that time. Fred Gehring said how lucky he was to have Ramanujan as Associate Chair, and that he couldn't manage the Department without Ramanujan's help.

In the field of analysis, two of the leading figures in Michigan were Peter Duren and Allen Shields. In addition to his research contributions to several areas of analysis, Duren is especially known for the many excellent textbooks and monographs he wrote. I myself have benefited by using his book "Invitation to classical analysis" published by the AMS in 2012. In 1993, Duren co-edited the three volume "A Century of Mathematics in America" [B6] for the Centennial of the AMS.

Allen Shields was brilliant, and so friendly, that faculty and graduate students adored him. He served as an Editor of the Proceedings of the AMS. He was also Chair of the mathematics department at Michigan.

There was considerable excitement among the faculty at that time in the appointment of Mel Hochster from Purdue. Hochster is a first rate algebraist, specializing in commutative algebra. Soon after arriving in Ann Arbor, he received the 1980 Cole Prize of the AMS (along with Michael Aschbacher of Caltech). So the enthusiasm of the Michigan faculty about him was well founded.

Many top mathematics departments publish their own journal, which is often ranked among the leading journals in the discipline. The Michigan Mathematics Journal founded in the 1920s is a very good example, another being the Illinois Journal of Mathematics. The premier example is of course Annals of Mathematics published by Princeton University. In his reminiscences of mathematics at Michigan, Raymond Wilder ([B6], p. 197) says, "There is no question, however, that the establishing of the journal has enhanced the reputation of the mathematics department and that it has justified whatever it has cost to run such a journal." Many of the senior mathematicians in Ann Arbor take turns serving on the Editorial Board of the Michigan Mathematics Journal. In addition, Ann Arbor is where the office of Mathematical Reviews (published by the AMS) is located. Many mathematics faculty like M. S. Ramanujan devoted a certain amount of time working for Mathematical Reviews. When I was there, the Executive Editor of Mathematical Reviews was John Selfridge and he used to attend the weekly number theory seminars.

For a major department to function smoothly, the administrative staff must be very efficient, as was the case with the staff in Ann Arbor. I mention two outstanding staff members whom I got to know very well.

The Administrative Assistant was Leon (Lee) P. Zukowski. He was in charge of administering all research grants, and more generally, assisted the Chair in all financial matters. Lee received a Masters Degree in mathematics, and soon after joined the department as Administrative Assistant. He was therefore more mathematically adept than typical administrative assistants/associates, and Michigan was lucky that he took this position.

He was a most charming and humorous person, and everyone loved him. He would sign his name $L_p(z)$, to indicate his mathematical background!

In the same office as Lee, was Ethel Rathbun, the Chairman's Secretary. She literally identified herself with the Department. The Department Chair changed every few years, but Ethel was invariant. Hence her great loyalty to the department which was her body and soul. During the time Selfridge was Executive Director of Mathematical Reviews, he would come to the department from time to time, and so he interacted with Ethel Rathbun. She was so efficient, that after she retired, John Selfridge sought her help in managing the Number Theory Foundation that he founded. Mathura and I were quite close to Ethel and Lee, and we had them over for dinner in our apartment a few times.

Just as I had Andy Majda at UCLA as a formidable tennis opponent, in Ann Arbor my tennis rivalry was with Professor James (Jim) Kister of the mathematics department. He was a well known topologist of the R. H. Bing school. Kister was a very intelligent tennis player mixing power with craft and guile. Initially I could not deal with his tricky play, but in a few months I managed to meet his game with power and consistency. Lee Zukowsky with whom I had friendly informal games regularly, told me that Kister would read books by authors like Bobby Riggs who told you how to outsmart players and destroy their confidence! Kister had won the faculty tennis championship a few times. I was in my twenties and he was twenty years older; yet he could battle younger tennis players fiercely. I could only imagine how much stronger he would have been in his twenties.

Besides Kister, I played singles with Rudy Ong, a physics professor, who was just as good as Kister, but played a serve and volley game instead. We split sets every time we played, Rudy winning the first and me the second. After my marriage to Mathura, I began playing tennis early in the mornings, and this was good because it would not interfere with things Mathura and I planned on weekends later in the day. But in America, people do not like to wake up early on weekends. So my tennis buddies agreed to play with me only once during the weekend which meant that I had to battle one opponent on a Saturday and another on a Sunday on a regular basis. At the University these opponents were Jim Kister and Rudy Ong.

Since one could not play outdoors in the winter in Ann Arbor, we decided to play indoors throughout the year at the Men's Gym. The surface there was wooden and therefore very fast for someone like me trained on the slower clay courts in Madras. But soon I got the hang of it and I was able to mix serve and volley with the baseline game. An interesting feature of the Men's Gym was that the floor was used for a variety of games besides tennis, such as basketball, volley ball, and badminton. Thus there were lines of different colors on the floor, with lines of a certain color pertaining to the boundaries of the court of a specified game. The white lines were for tennis. In the beginning, the variety of lines seemed confusing, but soon, I could see nothing but the white lines of the tennis court. That was due to total concentration which I needed in playing Kister and Ong!

Besides Kister and Ong, I also played tennis with Sriram, a friend of our family from Madras. Sriram's older brother Sridevan, a well known lawyer in Madras, was a fine tennis player who had the most fluid style, and with whom my father and I had played regularly at the Mylapore Club in Madras. Sriram was just as good as Sridevan, and while I gave him a good fight, I could never beat him in singles. Sriram was an executive at the Ford Motor Company, and we played tennis at his indoor club in Dearborn on the outskirts of Detroit where he lived not far from the Ford Assembly Plant. Mathura and I got together often with Sriram and his wife Akhila at their home. Just as Ramanujan and Chitra helped Mathura and me in many ways, so did Sriram and Akhila in equal measure.

Finally, I should mention that Mathura and I often got together with N. Sivaramakrishnan (Sivaram Narayanan as he is known now). He was a mathematics graduate student in Michigan at that time. He was one year my senior at Vivekananda College in the mathematics program, and so I knew him very well. In Ann Arbor we often had him at our apartment for dinner and used to listen to South Indian classical music in which we had a common passionate interest. Sivaram is now a professor at Central Michigan University. I see him regularly at the AMS Annual Meetings.

4.2 The Number Theory Program and Seminar

In the 1970s and for many years thereafter, the outstanding number theory program at the University of Michigan was led by Don Lewis as its leader and with Hugh Montgomery as its star. In addition there was James Milne, a reputed algebraic number theorist who had just been promoted to the rank of Professor. These permanent faculty were supplemented by assistant professors like me, and some long term visitors. In view of the strong program in number theory that Michigan had, many leading researchers opted to spend their sabbatical year, either in full or in part, at Michigan. I already mentioned that my advisor Straus spent the Fall of 1977 in Michigan during his sabbatical. While I was in Ann Arbor, Paul Bateman, who had completed his long term as Chair at Illinois, and who had a year's research leave following that, spent that entire academic year 1980–81 in Ann Arbor. That was very beneficial to me.

Don Lewis systematically built up the number theory program in the sixties using his contact with Harold Davenport of Cambridge University, and other noted British number theorists like Brian Birch at Oxford. Lewis was ably assisted in his efforts by William LeVeque who even served as Chair for a few years. After LeVeque left Ann Arbor for California, Hugh Montgomery arrived and with his appointment, Michigan became a major center for analytic number theory.

Both Lewis and Montgomery made sure to have visitors who were either world leaders in number theory or rising stars who were doing groundbreaking work at that time. In 1978–79, Lewis invited Takuro Shintani, a brilliant young number theorist from Tokyo. Among other things, Shintani had introduced a new type of zeta function that included several well known zeta functions as special cases, and Lewis arranged for Shintani to give a series of lectures on this topic. Shintani was extremely shy and introverted. It was not easy to maintain a conversation with him. It was sad to hear in 1980, that he had committed suicide. He was known to be quite generous in helping other mathematicians. It is not clear what got him into the depression that took his life.

Also during 1978–79, Lewis had another eminent visitor — Professor Peter Roquette of the University of Heidelberg. One of Roquette's major achievements was his joint work with Abraham Robinson applying methods of non-standard analysis to the theory of Diophantine equations. To help us understand Roquette's lectures, Don Lewis gave a preparatory series of talks. It was efforts like this that contributed to the high quality of the number theory program in Michigan.

Another visitor of Lewis was Rainer Schulze-Pillot from Saarbrucken, Germany, an authority on quadratic forms. Owing to this early contact, I invited Schulze-Pillot to the University of Florida in 2009, when in partnership with Manjul Bhargava, I conducted a conference on quadratic forms. A few years later, when I needed a quadratic forms expert on the editorial board of The Ramanujan Journal, I invited Schulze-Pillot, and he graciously accepted my invitation.

Montgomery invited Roger Heath-Brown of Oxford University in 1978–79. Heath-Brown had just completed his PhD at Cambridge under the supervision of Fields Medalist Alan Baker. Heath-Brown was a rising star in analytic number theory, and I remember Montgomery saying that he expected big things from him. Indeed Heath-Brown lived up to everyone's expectations by proving fundamental results in various parts of analytic number theory in rapid succession. He received the Smith's Prize (1976), the Senior Berwick Prize (1996), and Fellowship of the Royal Society (1993). The visits of both Roquette and Heath-Brown had beneficial effects on my research and publications.

The Number Theory Seminar met once weekly after dinner at 7:30 pm. All persons attending the seminar would then go to a bar or pub and discuss over a glass of beer. This was the "ritual" whether the speaker was from outside or one of us local number theorists. With regard to this ritual, I had an interesting conversation with Sid Graham, who in 1977 received his PhD from Michigan under Montgomery's direction. So Graham was academically one year my senior, and he held the prestigious Bateman instructorship at Caltech during 1977–79. We therefore interacted during my final year 1977–78 at UCLA since he came to UCLA for certain number theory seminars. When he heard that I had received the Hildebrandt Assistant Professorship, he mentioned this ritual to me and asked how I would deal with it since I was a tee-totaller. I said I would go to the bar with everyone else but drink only non-alcoholic beverages. Montgomery poked fun at me saying that the soft drinks I would order are more harmful to health than the beer that he and others were having!

It turned out that during my second year in Michigan (1979–80), both Montgomery and Lewis were away in England on their sabbaticals, and so the Number Theory Seminar was left to me to organize. It met infrequently, but whenever it did, the ritual continued; I did not want to change a long held tradition. During the year that Lewis and Montgomery were away, they asked me to teach the graduate course in analytic number theory as well the undergraduate number theory course. Even though I was initially disappointed that during my three year stay there, both Lewis and Montgomery were away one full year, looking back I realize now that I gained considerable experience in teaching both the graduate and undergraduate courses in number theory. So rather than feel lonely, I decided to conduct the number theory courses as best I could. Even today I use those detailed notes I prepared for both the number theory courses. Actually during that academic year, I suffered a wrist fracture owing to fall on the tennis court; so Mathura wrote some of the notes for my lectures as I dictated them.

Montgomery took his senior student Brian Conrey with him to Cambridge University in 1979–80. But other graduate students of Lewis and Montgomery remained in Ann Arbor. David Leep, a PhD student of Lewis, and Todd Cochrane a PhD student of Montgomery, took my graduate course in analytic number theory. Both are professors of mathematics now — Leep in Kentucky and Cochrane in Kansas. That year, Mathura and I stayed in the comfortable town house of the Conreys in Ann Arbor while they were away in Cambridge. Conrey is now the Director of the American Institute of Mathematics (AIM) in Palo Alto, California.

When Montgomery returned in 1980 from England after his sabbatical, he gave a year long course (1980–81) on advanced analytic number theory. He was at that time revising Harold Davenport's classic book *Multiplicative Number Theory* by incorporating the latest developments. In particular he began his course by discussing how Robert Vaughan had obtained an improvement and smoother treatment of Vinogradov's method of exponential sums. Vinogradov introduced this method to prove his great theorem that every sufficiently large odd number is a sum of three primes. His method is deep and complicated. Vaughan's approach was much smoother. Montgomery referred to this as V^3M for Vaughan's version of Vinogradov's method, and incorporated it in his revised version of Davenport's book. Davenport was a master expositor and his writing style was smooth as silk. Heini Halberstam once told me, "Davenport's first draft is better than most people's tenth!" In the Preface to the revised edition, Montgomery says that he has tried to preserve the style of Davenport.

John Selfridge who took over as Editor of Mathematical Reviews in 1978, was a regular attendee of the Number Theory Seminar. In 1980–81, my final year at Michigan, Paul Bateman visited Ann Arbor for the whole year on his research leave, after he completed a long service as Chair at Illinois. Bateman's presence added a touch of class to the Number Theory Seminar because he knew so much of the history of number theory and had met so many eminent number theorists of earlier eras. He provided illuminating comments at the end of the talks as well as at discussions in the bar following the seminar. There were several reasons that Bateman chose to spend the entire year of his research leave in Michigan. Firstly, Michigan had an outstanding program in number theory. Secondly, Montgomery was previously an undergraduate in Illinois and so Bateman knew Montgomery from his student days. Thirdly, Ann Arbor was geographically close to Urbana, Illinois — a one day's driving distance — and so the Batemans could visit Urbana in between without difficulty.

Another visitor in 1980–81 was John Mack from the University of New South Wales, Australia, who spent his sabbatical in Ann Arbor working with Montgomery. I had met Mack in 1973 when I gave a talk at the University of New South Wales in Sydney, and it was a pleasure to renew contact with him. He had brought with him a fine collection of opals and we were impressed by their colors and brightness. So we bought two large opals from him, one for Mathura and one for my mother. During the summer of 1981 when we returned to India for a visit, my mother and Mathura made necklaces using these opals, and they wear these with pleasure even today.

A wonderful thing happened during my first year in Michigan. The Office of Research decided to put the spotlight on the number theory program, and bring out an entire issue of their "Research News" in Spring 1979 describing the work of the number theory group. Thus even though I was a fledgling member of the group, I was roped in by Lewis and Montgomery as one of the editorial consultants of this special issue. Representatives from the Office of Research met with us collectively and individually several times, but the bulk of the work was done by Lewis and Montgomery. Fortunately, my first year in Michigan was very productive and so my PhD work as a background, and my work at Michigan, were both reported in that special issue. It was also in 1978–79 that word came from the NSF that new mathematics institutes would be launched. The NSF sought proposals for such mathematics institutes. Michigan was very enthusiastic and submitted a comprehensive proposal. There were faculty meetings as well as meetings of the research groups such as the number theory group, for input into this proposal. Even though mine was only a three year appointment, the Department graciously asked me to take part both in the faculty meetings and the number theory group meetings for this proposal. I learned a lot by attending those meetings and in observing how the Self Study Report was prepared. Decades later, that experience paid off during my term as Chair when I prepared the Self Study Report for the External Review of the mathematics department in Florida (see Chapter 8).

The University of Michigan submitted an excellent proposal to the NSF for a Mathematics Institute, but did not get it. The winners in the first round were UC Berkeley which got the Mathematical Sciences Research Institute (MSRI) and the University of Minnesota which became the home of the Institute for Mathematics and its Applications (IMA).

4.3 The Mathematics Colloquium and the Ziwet Lectures

The University of Michigan had a very active Colloquium that featured world famous mathematicians as speakers. During my first year, my UCLA classmate Peter Jones, a Dickson Instructor in Chicago, addressed the colloquium, and it was a pleasure to meet up with him again. One of the most memorable colloquium talks I heard was by Irving Kaplansky also of the University of Chicago. Every one knew that he was an expositor par-excellence and so his talk was arranged in the large auditorium on the ground floor of Angell Hall, and it was packed to capacity. His eminence as a mathematician combined with his power of expression and charming personality made him an obvious choice for the Directorship of MSRI in 1984 when S. S. Chern, one of the founders of MSRI, stepped down as Director.

Another colloquium talk I remember very well was by Michael Aschbacher of Caltech. In the seventies he had written a series of fundamental papers in group theory and introduced several new ideas and techniques. With Aschbacher's breakthrough work and the discovery of the Friendly Giant (Monster) by Greiss, the great problem of the classification of finite simple groups was seen to be nearly completed. In 1980, Aschbacher was awarded the Cole Prize of the American Mathematical Society for his path-breaking contributions to group theory along with Mel Hochster of Michigan (for work in commutative algebra). Thus Aschbacher was a natural choice to address the Colloquium in 1980–81. Years later, I had the pleasure of hosting Aschbacher and many other eminent group theorists when they came to the University of Florida in 2002–03 for a conference in honor of John Thompson for his 70th birthday.

During 1979–80 when Lewis and Montgomery were away, I had the opportunity to suggest speakers in number theory for the colloquium. Bruce Berndt who had procured a copy of Ramanujan's Notebooks by writing to the Tata Institute following our conversation at the 1976 West Coast Number Theory Conference, sent me three of his first papers based on his investigations of Ramanujan's Notebooks. So I suggested to the Colloquium Committee that he be invited. I think the Colloquium in Ann Arbor that he gave in 1979–80 should been among the first few colloquium talks on Ramanujan's Notebooks that he must have delivered. Although Ramanujan never gave proofs of the "entries" in his notebooks, he organized them somewhat thematically, and grouped them into chapters. So what Berndt did was to discuss the entries chapter by chapter by (i) providing a reference if the entry was known already, (ii) provide a proof if the entry was a new result, and (iii) comment on the relevance of the entries to existing literature, and their implications. He did not edit these chapters in the order in which they were in the notebooks; instead he chose to edit a chapter when he felt comfortable doing that. Once he completed editing a chapter, he wrote a paper on it and published it. Then he would combine a certain collection of edited chapters and publish a volume entitled "Ramanujan's Notebooks". All in all, over a span of two decades, he completed editing

Ramanujan's Notebooks, and published it in five volumes through Springer (1985–2005). For this monumental contribution, he was awarded the Steele Prize of the AMS in 1996, well before the fifth volume was published! In 1985 he sent me a complimentary copy of Volume 1.

The mathematics department has a distinguished lecture series called the *Ziwet Lectures*, named after Alexander Ziwet, a former faculty member who as Chair from 1888–1925, did much to raise the stature of the Department. The Ziwet Lectures were created in 1934 by a bequest from Alexander Ziwet, and they have a rich history. They are held once every two or three years, and there were two while I was there — one in 1978–79 and another in 1980–81. The Ziwet Lecturers gave three talks, the first of wide appeal followed by two more technical talks. All Ziwet speakers were mathematicians of world renown. Irving Kaplansky whose mathematics colloquium in Michigan I had enjoyed hearing during 1978–79, had actually delivered the Ziwet Lectures way back in 1962–63. Again I quote Raymond Wilder from his reminiscences on Michigan mathematics ([B6], p. 196) where he says, "Another factor which I believe had a very beneficial influence on the evolution of the Department was the Ziwet lectures."

In 1978–79, the Ziwet Lectures were given by Fields Medallist Enrico Bombieri of the Institute for Advanced Study, Princeton. In view of Apery's sensational proof of the irrationality of $\zeta(3)$ in the summer of 1978, there was a resurgence of interest in field of irrationality, and so Bombieri chose to lecture on this exciting topic. Indeed Bombieri himself worked on irrationality measures for the next few years and obtained some very significant results especially on the Thue equation, partly in collaboration with Jeffrey Vaaler of the University of Texas. In fact, it was Bombieri's Ziwet Lectures that inspired me to briefly work in the field of irrationality.

Even though Bombieri did not address the number theory seminar on that visit, he did join the number theory group one evening for dinner and for discussions at a bar following dinner. I then spoke to Bombieri about the possibility of going to the Institute for Advanced Study at Princeton in 1981–82 after completing my three year term as Hildebrandt Assistant Professor. He knew that I could not accept the offer from Princeton in 1978 because I had committed to the University of Michigan already. Bombieri said that the offer of 1978 could not be renewed automatically; I would have to reapply for a visiting position at the Institute in 1981–82, and survive the competition that year. But he said that the Institute would keep in mind that an offer to me was made earlier. I then asked Bombieri who the Visiting Members in Number Theory at Princeton in 1978–79 were. Instead of just giving their names, Bombieri mentioned also one important theorem that each visitor was known for! This was amazing because he gave a list of more than a dozen visiting members in number theory.

In 1980–81, the Ziwet Lectures were by Fields Medallist Bill Thurston of Princeton University. He was leading a revolution in the field of topology. By the bye, Thurston succeeded Kaplansky as Director of MSRI in 1992.

The Ziwet Lectures in Michigan impressed me greatly. So as soon as I became Chair in 1998 at the University of Florida, I launched two series of Distinguished Colloquia — The Erdős Colloquium in pure mathematics and the Ulam Colloquium in applied mathematics, both colloquia featuring luminaries as speakers. Subsequently, when George Andrews graciously offered to sponsor the Ramanujan Colloquium in Florida, I suggested they be modeled along the lines of the Ziwet Lectures — the first talk being a public lecture of wide appeal, followed by two more technical talks. Thus my three year stay in Michigan influenced me in many ways. Next I will describe my research at the University of Michigan.

4.4   Going beyond my PhD thesis

The three years I spent at Michigan were just as productive in research as my three years at UCLA. The Hildebrandt Research Assistant Professorship that I held, had a reduced teaching load (like all other named instructorships in America for young researchers) and that gave me ample time to conduct research and write papers. I pursued problems that were natural outcomes or extensions of my thesis work, and at the same time embarked in new directions inspired by lectures I heard at Michigan and the contacts I made there.

One of the first things that I did was to complete the investigation of the function $M^*(x, y)$ that I had begun at UCLA by establishing an asymptotic formula for long ranges of $\alpha = \frac{\log x}{\log y}$. This analysis was more complicated than the one I did for its dual $M(x, y)$ in my thesis, just as $\Psi(x, y)$, the number of integers $\leq x$ all of whose prime factors are $\leq y$ is much more complicated to analyze than its dual $\Phi(x, y)$, the number of integers $\leq x$ with prime factors all $\geq y$.

In the early twentieth century, Edmund Landau had established an important result for the Möbius function, namely that $M(x)/x \to 0$ as $x \to \infty$, and that this is equivalent to the celebrated Prime Number Theorem (PNT). Here $M(x)$ is the sum of the Moebius function $\mu(n)$ for positive integers n up to x. So among the square-free numbers up to a large number x, roughly half have an even number of prime factors and half have an odd number of prime factors, and this is equivalent to the PNT (a square-free number is one for which none of its prime factors repeat). Thus the Möbius function behaves like an "unbiased coin" (to use a term from the theory of probability) because it takes values 1 and −1 with equal frequency among the square-free numbers. It is also important that the smallness in the size of $M(x)$ relative to x is closely tied to the error term in the PNT. Another way to understand the Möbius function is to consider a random walk on the real line by a mathematical drunkard. That is the drunkard takes a unit step to the right at time n if $\mu(n) = 1$, a unit step to the left at time n if $\mu(n) = -1$, and stays motionless if $\mu(n) = 0$. Now if this is truly like random walk in the theory of probability, then after x units of time, the drunkard should be no more than $x^{1/2+\varepsilon}$ distance from the origin, for every $\varepsilon > 0$. It is to be noted that this is yet unproven, and that it is equivalent to the Riemann Hypothesis, one of greatest unsolved problems in all of mathematics concerning the complex zeros of the Riemann zeta function. In my paper on $M^*(x, y)$, one of the results I established was an extension of Landau's theorem on $M(x)$ to smooth numbers for long ranges of y, namely:

$$\max_{\exp\{(\log x)^{2/3}\} \leq y \leq x} \frac{M^*(x, y)}{\Psi(x, y)} << \frac{1}{\log x}.$$

I conjectured that this upper bound will hold uniformly for all values of y, and this was later established by Adolf Hildebrand.

My paper on $M^*(x, y)$ appeared in the Transactions of the AMS (TAMS), whereas my paper on $M(x, y)$ appeared in the Journal of Number Theory (JNT). These two papers constituted the first systematic study of the Möbius function over integers with restricted prime factors. I will now relate an interesting story why the two papers, although related, but written two years apart, were published almost simultaneously.

I had submitted my paper on $M(x, y)$ to the Journal of Number Theory in December 1977 while I was at UCLA. This was one of my papers that impressed Montgomery. I know this because soon after I arrived in Michigan, he told me that he liked my results on $M(x, y)$. More than a year had passed and I heard nothing from the journal about this paper. Montgomery was on the Editorial Board of the JNT, and so I went to his office in March 1979 and complained that the referee was taking too much time with the paper; I asked if he could help in speeding this up. Montgomery looked at me, smiled, said that he will talk to Hans Zassenhaus, the Editor-in-Chief of the JNT, and tell him that he will handle the paper.

The next academic year (1979–80), Montgomery went to Cambridge University on a sabbatical. He took my paper with him and said that he would send his comments from there. And sure enough he did, but in a letter he said that I had to cut my 30 page paper to half its size and indicated how to do so. Basically he wanted me not to give detailed proofs for certain theorems, but just indicate the key ideas. In his letter he said that I should not be disappointed that he was asking me to do a major revision, because I should know that he would not take so much interest in just anybody's work. I thanked him for his comments, shortened the paper as he suggested, and submitted the revised version in early 1981. He took some more time to accept the revision, and the paper appeared in February 1982 when I was at the Institute for Advanced Study. I was later told that Montgomery would take time to referee papers because he is so thorough, but in the end, he would offer valuable comments, as was the case with my paper, and I appreciated that very much.

With regard to my paper on $M^*(x, y)$, just as I was completing it in 1980, Michael Artin of MIT, son of the famous Emil Artin, and himself a noted mathematician, was in Michigan to give a colloquium. He was an Editor of the Transactions of the AMS, and I had a chance to talk to him after his colloquium. As a consequence of that conversation, I submitted my paper to the Transactions of the AMS, and it appeared in the normal time frame of two years.

Based on my thesis work on $M(x, y)$, I was led to the study of the more general sum

$$S_z(x, y) = \sum_{n \leq x, p(n) \geq y} z^{\nu(n)}$$

for complex z, motivated by some fundamental results on the number of prime factors established in the first half of the twentieth century.

Hardy and Ramanujan had proved that $\nu(n)$, the number of prime factors of n, is almost always about the size $\log\log n$. To establish this result, they made the ingenious observation that it suffices to establish certain uniform upper bounds for the quantity $\nu_k(x)$, the number of integers up to x having precisely k prime factors. Shortly thereafter, using the Prime Number Theorem, Landau obtained asymptotic estimates for $\nu_k(x)$, if k is arbitrary but fixed. In view Landau's asymptotic estimate for $\nu_k(x)$ for fixed k, Hardy raised the problem whether a uniform asymptotic estimate for $\nu_k(x)$ could be obtained for k varying with x. This problem was solved by the Indian mathematician L. G. Sathe who painstakingly used induction to confirm this in four long papers that appeared in the Journal of the Indian Mathematical Society in 1955. Upon seeing Sathe's results, the great Atle Selberg pointed out that the best way to approach the problem of estimating $\nu_k(x)$ is to study the sum

$$S_z(x) = \sum_{n \leq x} z^{\nu(n)},$$

as a function of the complex variable z and use the Cauchy Residue Theorem to obtain uniform asymptotic results for $\nu_k(x)$ sharper than Sathe's. Motivated by Selberg's paper and my earlier study of integers with restricted prime factors, I decided to investigate the sum $S_z(x, y)$ which specializes to $S_z(x)$ when $y = 2$. Also $\Phi(x, y)$ is a special case of $S_z(x, y)$ when $z = 1$.

By using difference-differential equations more general that what was employed by de Bruijn in the study of $\Phi(x, y)$, and in combination with the analytic method of Selberg, I was able to estimate $S_z(x, y)$ asymptotically for all values y up to x. This enabled me to obtain a uniform extension of the celebrated Erdős-Kac Theorem as well as the Sathe-Selberg results to the set of integers without small prime factors.

When Heath-Brown visited Ann Arbor in 1978–79, I discussed my results on $S_z(x, y)$ with him, and felt encouraged by his positive reactions. So I submitted my paper on this topic to the Quarterly Journal of Mathematics (Oxford) of which he was an Editor, and it was accepted.

The two papers in the Transactions of the AMS and the Oxford Quarterly Journal concerned generalizations and extensions of my thesis work. But my stay at Michigan also took me in a different direction of research involving irrational numbers.

4.5 The irrationality of $\zeta(3)$ and the aftermath

One of the sensational developments in 1978 was Roger Apéry's proof that the value of the Riemann zeta function at 3 is irrational. To explain the significance of this result, we need to first recall that in Calculus classes we are shown that the infinite series

$$\zeta(s) := \sum_{n=1}^{\infty} \frac{1}{n^s}$$

converges (= has finite value) for each real $s > 1$ and diverges (= has infinite value) for each real $s \leq 1$. In particular the series converges for $s = 2$, but in Calculus classes we are not told the value of series when $s = 2$, or if we are told what the value is, we are not shown how to prove it because that requires more sophisticated methods.

The great German mathematician Riemann realized that it would be particularly important to study the series for complex values of s, in order to understand the distribution of prime numbers. It was he who used the notation $\zeta(s)$ for the above infinite sum, which actually converges when $\text{Re}(s)$, the real part of the complex number s, is larger than 1. The function $\zeta(s)$ is called the Riemann zeta function. It is one of the most important functions in mathematics since it plays a fundamental role in the study of prime numbers because when $\text{Re}(s) > 1$, $\zeta(s)$ admits a product representation involving primes.

Prior to Riemann, Euler had investigated $\zeta(s)$ for real values $s > 1$ and obtained some important results on prime numbers using the product representation for real numbers $s > 1$. But then he surprised everyone by evaluating the zeta function at all positive even integral values of s. More precisely he showed by ingenious arguments that the sum of the reciprocals of the squares, namely, $\zeta(2)$, has value $\pi^2/6$, thereby solving the *Basel problem* posed by the Bernoullis. More generally, Euler proved that for each even positive integer $2k$, the value of $\zeta(2k)$ is a rational multiple of π^{2k}. Since π is a transcendental number, it follows that all the values $\zeta(2k)$ are transcendental, and hence irrational. In contrast, nothing was known about the value of the zeta function at odd integers $3, 5, 7, \ldots$ until Apéry stunned the world by proving that $\zeta(3)$ is irrational in value. It was not just the fact that Apéry broke an impasse of more than two centuries, but the manner in which he established his theorem also startled everyone.

Apéry announced his great result at a conference in Marseilles in the beginning of the summer of 1978. He was known to be an unorthodox mathematician, highly idiosyncratic, and was not well regarded by the French who were steeped in the rigorous style of doing mathematics. When Apéry started his lecture by stating a few incredible identities, members of the audience asked how he obtained them. He responded by saying that he found them in his garden! After stating one incredible identity after another, he concluded his talk by saying that the irrationality of $\zeta(3)$ followed from them. There were howls and cat calls from the angry and surprised audience! But Henri Cohen who was in the audience during Apéry's talk, starting checking the identities as he was lecturing, and concluded that they were true! Shortly thereafter, Cohen and van der Poorten got together with Don Zagier and established the truth of all of Apéry's claims. The story spread like wildfire throughout the world in the next few days, and Alfred van der Poorten wrote a charming article entitled "A proof that Euler missed" describing the incredible proof; van der Poorten's delightful article in his calligraphic handwriting came to Michigan in Fall 1978 and we all began reading it with great interest. The significance of Apéry's result was so immense, that it was decided that there should be a talk on it at the International Congress of Mathematicians in Hensinki, Finland, in August that year. But Apéry was too crazy a mathematician to be invited to speak at such a prestigious Congress, and so Henri Cohen was asked to make the presentation instead.

After the International Congress, we heard that an ingenious proof of the irrationality of $\zeta(3)$ was given by Frits Beukers, a young PhD student of Tijdeman in Leiden. Beukers cast the identities of Apéry as expressions involving integrals of Legendre polynomials, but in this form, the proof is much smoother.

Most irrationality proofs go by way of the famous criterion of Dirichlet: *A real number θ is irrational if and only if there exist an infinite sequence of integer pairs p_n, q_n such that*

$$|q_n\theta - p_n| \neq 0, \quad \text{and} \quad |q_n\theta - p_n| \to 0 \quad \text{as} \quad n \to \infty.$$

In general it is very difficult to construct these integer pairs (p_n, q_n). Apéry brilliantly achieved the construction of these pairs of integers in the case of $\zeta(3)$. Beukers' keen insight was to observe that the sequence of integer pairs of Apéry for $\zeta(3)$ can be obtained from a certain three dimensional integral involving Legendre polynomials. In the Apéry-Beukers approach, one has to study the rational approximations to $\zeta(2)$ first and in this case, the Beukers integral is simpler and of dimension 2. The advantage of the Apéry-Beukers approach is that it yields irrationality measures, namely inequalities like

$$\left|\zeta(2) - \frac{p}{q}\right| > \frac{c_1}{q^{11.86}}, \quad \text{and} \quad \left|\zeta(3) - \frac{p}{q}\right| > \frac{c_2}{q^{26.2}}$$

for all rationals p/q, with some explicitly determinable constants c_1, c_2.

Beukers' paper also circulated as widely and rapidly as van der Poorten's handwritten manuscript on Apéry's proof. van der Poorten was a fabulous and humorous lecturer. Consequently he was invited to speak at several conferences and universities on Apéry's proof of the irrationality of $\zeta(3)$. I had the pleasure of listening to his prominently announced "final lecture" on this topic at the West Coast Number Theory Conference at the University of California, Santa Barbara, in December 1978. He sent the audience roaring in laughter as he described how Apéry taunted, and was ridiculed by, the audience in Marseilles. van der Poorten had an infectious laughter. His Santa Barbara lecture is fresh in my memory.

The origins of the theory of irrational numbers go back to the proof of the irrationality of $\sqrt{2}$ in Greek antiquity. The subject has remained active in spite of the passage of time and several long standing open problems still engage some of the most gifted minds. But Apéry's proof led to a resurgence of activity in the area of irrationality.

One of the things that mathematicians try to do is to study extensions to higher dimensions. Just as Beukers' two and three dimensional integrals gave rational approximations to $\zeta(2)$ and $\zeta(3)$ respectively, one can construct higher dimensional integrals connected to expressions of the form

$$q_n \zeta(k) - p_n$$

with p_n, q_n integers, but when $k > 3$, such an expression tends to zero. So the problem of showing that $\zeta(k)$ is irrational for any odd $k > 3$ is tantalizing and remains unsolved. Thus Apéry's achievement is truly remarkable.

Instead of going higher in dimension, I asked myself the question as to what happens if one "goes down" in dimension? More precisely, I felt that it would be worthwhile to investigate integrals

$$I_n(z) := \int_0^1 \frac{P_n(x)dx}{1 + zx},$$

for complex z in general. This is related to $\log(1 + z)$ and I felt it would yield irrationality measures not just for $\log 2$ when $z = 1$, but for a class of numbers of the form $\log(1 + z)$. The procedure worked extremely well for $\log 2$ and provided a good irrationality measure. I wanted similar irrationality measures for other logarithms as well. Montgomery at that time had a PhD student by name Mike Robinson who had expressed interest in doing his PhD work in irrationality and transcendence. Montgomery told me that Robinson was interested in the work of Apéry and Beukers, and suggested that I talk to him. Hence I approached Robinson and told him that we could work on this problem together. As anticipated, we obtained good irrationality measures for a class of logarithms, but to establish these results, we exploited orthogonality properties of Legendre polynomials. One of our striking results was that

$$\left| \frac{\pi}{\sqrt{3}} - \frac{p}{q} \right| > \frac{c}{q^{8.33}},$$

which falls out of investigating $I_n(z)$ with $z = e^{2i\pi/3}$, a cube root of unity. From the Apéry-Beukers irrationality measure for $\zeta(2)$, it follows that

$$\left| \frac{\pi}{\sqrt{k}} - \frac{p}{q} \right| > \frac{c(k)}{q^{23.62}}, \quad \text{for all positive integers} \quad k$$

because $\zeta(2) = \pi^2/6$. But result for just $\pi/\sqrt{3}$ was superior. The other benefit of considering the one dimensional version of the Beukers integral was that one could also deal with integrals of the type

$$\int_0^1 \frac{P_n(x)}{(1 + zx)^{\ell/k}}$$

for rationals ℓ/k. Expansion of the above integral gave us irrationality measures $< k$ for certain k-th roots, and this had important consequences in the theory of Diophantine equations as I will explain briefly now.

The theory of Diophantine equations deals with equations for which we seek integer or rational solutions. The name "Diophantine" stems from the Greek mathematician Diophantus who always sought integer or rational solutions to equations. As we already saw in §2.4, the Norwegian mathematician Axel Thue showed that

$$x^m - dy^m = k \quad \text{(Thue equation)}$$

with d not being an m-th power, can have at most a finite number of integer solutions if $m \geq 3$. He deduced this by obtaining an irrationality measure $< m$ for the m-th root of d, when $m \geq 3$.

Our one-dimensional approach using Legendre polynomials yielded irrationality measures $< m$ for m-th roots of certain rations when $m \geq 3$. This established the finiteness of the number of solutions of some Thue equations. Already in the sixties, Alan Baker had employed hypergeometric functions and established irrationality measures for $d^{1/m}$ that were superior to what Robinson and I obtained, but our method was simpler.

I gave a talk in the Number Theory Seminar at the University of Michigan in early 1979 on this work. At that time Professor Peter Roquette of the University of Heidelberg was visiting Michigan. He was the Editor-in-Chief of the prestigious *Journal für die Reine und Angewandte Mathematik* (= Journal for the Pure and Applied Mathematics), also known as Crelle's Journal. He was impressed by my joint work with Mike Robinson. So he suggested that we should submit our paper to Crelle's Journal, and that is where our joint work appeared in 1980.

Crelle's journal is the oldest journal among all active mathematics journals today and has an illustrious history. It was founded by Leopold von Crelle in the first half of the 19th century (hence it is known as Crelle's Journal), a great philanthropist and benefactor of mathematics. The first papers in Crelle's Journal were those of the genius Abel. Abel's contact with Crelle was one of the most fortuitous events of his life. Abel wrote a series of fundamental papers on elliptic functions and integrals and these were published in the first few issues of Crelle's journal. Thanks to a large part due to Abel's papers, Crelle's journal soon became one of the leading journals, a distinction that it still holds today. I was very pleased that my paper with Robinson was to appear in a journal with such a distinguished history. Robinson and I were well aware that several experts would be working on the ramifications of the Apéry-Beukers method, and so we had to complete our work and write it up quickly. We often worked together in my apartment late into the night. Mathura who understood the importance and urgency of this project, provided us late night refreshments to keep our energy level high. I should point out, the besides this joint work, Robinson had other results of his own on transcendence in his PhD thesis.

Prior to the full paper appearing in Crelle's Journal, a shorter announcement appeared in the Proceedings of the 1979 Number Theory Conference (published as Springer Lecture Notes) that was held at Southern Illinois University, Carbondale. Mel Nathanson who was at that time in Carbondale, conducted a fine conference and invited me to give a talk. Further exposure of my work with Robinson, and the opportunity to learn about the work of others in the field of irrationality and transcendence, came in Germany in the summer of 1979, and I describe this next.

4.6   Visits to Germany and my first Oberwolfach conference

In the summer of 1979, Mathura and I were planning to go to India via Europe. In view of this, Professor Roquette invited me to visit the University of Heidelberg enroute and give a colloquium on my work with Robinson on irrationality. Also during that summer, the Mathematisches Forschungsinstitüt in Oberwolfach, was having a conference on Diophantine Approximations. Hugh Montgomery felt that it would be profitable if I would speak at that conference as well. So he very generously wrote a letter to the conference organizer Professor Schneider, who was one of the giants in the areas of irrationality and transcendence. So it was an honor for me to be invited to the Oberwolfach conference by Schneider himself, followed by a letter with specifics about arrangements by Dr. Martin Barner, the Director of the Institute.

Theodor Schneider shot into prominence when he and A. Gelfond of Russia settled the famous Seventh Problem of David Hilbert on the transcendence of numbers like $2^{\sqrt{2}}$. Hilbert, a father figure in mathematics, was perhaps the most prominent member of the elite group of mathematicians at the University of Göttingen during the end of the nineteenth and the beginning of the twentieth century. In 1900, at the dawn of the new century, at the International Congress of Mathematicians in Paris, Hilbert gave a list of 23 problems that he said would influence the development of mathematics for the next one hundred years or more. Subsequently, in a lecture in Göttingen, Hilbert predicted the future of three of the most famous problems in number theory, namely (i) the *Riemann Hypothesis*, (ii) *Fermat's Last Theorem*, and (iii) the transcendence of numbers of the form α^β, where $\alpha \neq 0, 1$ is an algebraic number and β is an algebraic irrational. Hilbert said that complex analysis had undergone such a rapid development, he expected that even he and the older members of the audience would see the Riemann Hypothesis solved in their time. While there was some progress in algebraic number theory, more would be needed to attack Fermat's Last Theorem, and so he said only the younger members of the audience would witness its solution. With regard to the transcendence of α^β, Hilbert noted that no one has any idea how to approach the problem, and so none in the audience will live to see that solved. Ironically, exactly the opposite happened chronologically! The Riemann Hypothesis is still unsolved, and Fermat's Last Theorem was proved in 1995. A few years after Hilbert's lecture, Theodor Schneider, a young German mathematician, and Gelfond in Russia, independently developed techniques that established the transcendence of α^β as conjectured by Hilbert, and that is the famous Gelfond-Schneider theorem of 1934!

The Gelfond-Schneider theorem was the PhD thesis of Schneider. Later Schneider became the assistant of Carl Ludwig Siegel in Gottingen, and the two, along with Kurt Mahler, led the development of transcendental number theory in Germany. Whereas Mahler left Germany fearing the rise of Hitler, Schneider and Siegel stayed on. After the Mathematisches Forschunginstitüt in Oberwolfach was launched in 1944, Schneider served as its Director from 1959 to 63.

I visited the University of Heidelberg before proceeding to Oberwolfach. Professor Roquette received me and Mathura at the railway station and showed us around the lovely city with its stately buildings and castle overlooking the river Neckar. It was springtime weather at its best, and the Neckar looked especially lovely with beautiful swans gliding on its calm waters. I was told that Heidelberg was one of the cities that were spared by Allied bombing in World War II. So the historic buildings of Heidelberg remain intact in their original form.

I was getting worried that the owing to the car tour of Heidelberg, we might be late for my talk by a few minutes. But Professor Roquette informed me that in Germany there is the "academic quarter", which means that lectures begin 15 minutes after the announced time, and that was reassuring. My talk was actually for $1\frac{1}{2}$ hours, split into two 45-minute presentations with a short break in-between, and so I was able to present my joint work with Robinson with all details. In the evening, after my talk, Professor Roquette took me and Mathura to his lovely home where his gracious wife had so thoughtfully prepared a delicious vegetarian dinner for us. I was touched by the hospitality shown by one of the great figures of European mathematics.

I was only a recent entrant to the subject of irrationality. I had started out in elementary and analytic number theory and so my original contacts in Germany were Professors H.-E. Richert and Eduard Wirsing in Ulm and Wolfgang Schwartz in Frankfurt. They had invited me earlier to the Oberwolfach Conference on Analytic Number Theory in October 1977, but I could not attend that owing my course work at UCLA. This time in May 1979 I wrote to them that I would be visiting Germany, and so I was invited to the University of Ulm for a colloquium two days before the start of the Oberwolfach conference. With the presence of Richert and Wirsing, Universtät Ulm became one of the great centers for analytic number theory not just in Germany, but in Europe. Richert and Wirsing were previously at the University of Marburg and moved to Ulm in the early seventies a few years after the Universität Ulm was founded.

Ulm is the birth place of Albert Einstein, and the home of the tallest church spire in the world. Mathura and I were in our twenties then, and so we easily walked up to very top of the Ulm cathedral, a feat we may not be able to do now.

My talk in Ulm was on the sum $S_z(x, y)$ and its applications, an apt choice because on the one hand it was related to the sieve on which Richert was a world authority, and on the other it concerned the study of additive functions using multiplicative techniques, an area where Wirsing had done pioneering work. Following my talk, the Richerts hosted a dinner for me and Mathura in a fancy restaurant on the Danube river, and ordered tender asparagus with cream sauce. It was a German speciality that Richert said we should not miss especially because it was the best season for asparagus.

The Oberwolfach Institute is hidden in the Bavarian mountains and getting there is tricky. Fortunately, Professor Wirsing was going to the Diophantine approximations conference as well, and so he very kindly offered to take us in his car. On the morning of our trip, we stopped at his house for lunch. On entering his house, we were struck by the fact that right next to the flight of stairs going to the upper floor, there was a pole going from the floor to the ceiling. Professor Wirsing called for his wife saying that we were home, and she answered the call by gliding down the pole from upstairs to say hello to us instead of using the stairs! Professor Wirsing then said that he too uses the pole, not just to slide down from the upstairs, but to lift himself up as well, and he demonstrated this as Mathura and I watched his performance in amazement!!

The Oberfolwach Institute was created in 1944 just before World War II ended. Its initial home was a castle in the Bavarian mountains and that is where the lectures were held originally. The Institute holds weekly conferences throughout the year; participants arrive on a Sunday and depart the next Saturday. The number of participants is limited to about forty to facilitate intimate mathematical discussions. Participation is by invitation only from the Director at the suggestion of the conference organizers. The Institute provides room and board to all participants and railway fare within Germany. For the 1979 Diophantine Approximation conference, there were thirty eight speakers and a few more participants like Wirsing, Schwartz, and Erdős who attended but did not give talks.

In the early seventies, with a generous gift from the Volkswagen Siftung, a beautiful new guest house was built for the Institute, as well as another new building containing the main lecture hall and a magnificent library. The new guest house had a few double rooms to accommodate participants like me who were accompanied by their spouses.

Mathura and I enjoyed the breathtaking scenery of the Bavarian mountains and the black forest during our drive to Oberwolfach in Wirsing's car. The Institute has one of the most fabulous locations in the world — on a hill slope surrounded by mountains, valleys, and meadows. Mathematicians attending the Oberwolfach conferences take walks along the beautiful mountainside discussing mathematics in heavenly surroundings. For me it was an opportunity to meet the leading lights of irrationality and transcendence — ranging from senior mathematicians like Theodor Schneider to young stars like Frits Beukers.

One of the nice features of the Oberwolfach conferences is that at dinner and at lunch, every one is assigned a seat and you have to sit there, but then the arrangement is different each time so that you sit with different groups of people during the conference. This helps in making new contacts and enriches the experience of participation.

Alfred van der Poorten with his fine sense of humor and infectious laughter, was the one who interacted with the greatest number of participants.

He was able to hold a meaningful mathematical conversation with almost anyone seated at his dining table, and after dinner, he would continue the discussions in the lounge for several hours, and would smoke several packs of cigarettes in that process.

One of the most inspiring and thrilling lectures I heard was by Gregory Chudnovsky — assisted by his brother David Chudnovsky — on the topic "Transcendental number theory and methods of mathematical physics". The talk dealt with irrationality measures as part of a vast general theory. His talk as announced in the conference program was to have been on "Deformation theory of linear differential equations: complete integrability", but he changed it in view of all the excitement on irrationality measures following Apéry's proof of the irrationality of $\zeta(3)$.

Gregory Chudnovsky is a genius, but unfortunately he suffers from *Myasthenia gravis*, a neuro-muscular disease. So he is confined to a wheel chair, like the great astrophysicist Stephen Hawking of Cambridge University. His older brother David Chudnovsky is an accomplished mathematical physicist who assists Gregory in daily life as well as in lectures. They also collaborated on several projects. The Chudnovskys wanted to leave the USSR in the seventies, and my thesis advisor Ernst Straus was one of those who played a role in getting them out of Russia. Since New York has a large Jewish population, the Chudnovskys settled down in New York where they have been ever since they emigrated to America. In the seventies, Gregory had a special research professorship at Columbia University and David, a teaching position there.

Gregory Chudnovsky shot to prominence when he showed that the value of the Gamma function at $1/4$ is transcendental. The Gamma function is one of the most important functions not just in mathematics, but in science in general. It "extends" the factorial function $n!$ to values n that are not positive integers. It is well known that $\Gamma(1/2) = \sqrt{\pi}$, and therefore it is transcendental. Gregory Chudnovsky's result that $\Gamma(1/4)$ is transcendental stunned the world. The exact value of $\Gamma(1/4)$ is still not known. For his fundamental contributions, Gregory Chudnovsky was awarded the MacArthur Fellowship (known as the genius award) in 1981. The Chudnovskys are also very famous for having built a computer by themselves in their apartment in New York to compute several billion digits of π.

The lecture by the Chudnovsky brothers was thrilling both in content and style. The spoke alternately — one would speak a sentence and pause, and other would come out with the next sentence. Even jokes were delivered in this fashion! There was not even a single instance when they faltered in this style of delivery! They presented a multitude of new and powerful theorems, and the transparencies that contained them were flipped with breakneck speed, that it was difficult even for the seasoned experts to keep up with them. They too had noticed like Robinson and me that $\pi/\sqrt{3}$ was special among the numbers of the form $\pi/\sqrt{k}$ with regard to provable irrationality measures, and indeed the Chudnovsky's obtained the same irrationality measure for $\pi/\sqrt{3}$ that we had established. Gregory Chudnovsky then made a claim in his lecture that it is the best irrationality measure for $\pi/\sqrt{3}$ that can be obtained with the methods available at that time. He subsequently improved it by more sophisticated methods. I became good friends with the Chudnovskys who later invited me to several conferences they conducted in New York owing to our common interest on irrationality. But that was my only paper on irrationality, and so in the conferences in New York, I spoke about my work in other areas. The Chudnovskys have always been exceptionally kind to me and often said that I should return to the subject of irrationality.

It was at this Oberwolfach conference that I met the French transcendence specialist Michel Waldschmidt. He was sporting a long beard and looked like one of the great Hindu sages who retreat to the Himalayas to perform austere penances to seek union with God! He is a charming personality and we became very close friends. I had the pleasure of hosting him in Madras and Florida and enjoying his hospitality in Paris. Many years later, I invited him to serve on the Editorial Board of the Ramanujan Journal, and on the SASTRA Ramanujan Prize Committee, and he graciously accepted both invitations.

Another leader in transcendence I made acquaintance with in Oberwolfach was David Masser (then at the University of Nottingham), a former PhD student of Alan Baker. Both Montgomery and Lewis had spoken very highly of Masser saying that he knew the ramifications of Baker's methods better than anyone else, and was opening up new avenues of exploration in transcendence theory with G. Wüstholz, who was also a speaker at this conference. Masser subsequently joined the University of Michigan faculty, but then settled down in Basel, Switzerland where he is now emeritus.

One of the senior European mathematicians whom I had the pleasure of listening to was Edmund Hlawka of the University of Vienna who was the doyen of the subject of Diophantine approximations in Austria. He was one of the few who spoke in German, but I was told that his was an Austrian version of German. Hlawka's most eminent student is the great Wolfgang Schmidt of the University of Colorado, who was also at the conference. Schmidt, like his PhD advisor Hlawka, also spoke in German. For me it was a privilege to present my work on irrationality in the presence luminaries like Schneider, Schmidt, and Chudnovsky.

Frits Beukers was there in Oberwolfach along with his PhD advisor Robert Tijdeman. Just as Robinson and I obtained irrationality measures for several values of $\log(1 + z)$, Beuker's extended the method of his proof of the irrationality of $\zeta(2)$ and $\zeta(3)$, to obtain irrationality measures for the di-logarithms and tri-logarithms for certain values near 1. The Oberwolfach Conference on Diophantine Approximations not only gave exposure to my work, but also gave me an opportunity to meet leading researchers senior and junior and learn about important recent advances.

From Oberwolfach, Mathura and I proceeded to India, but we had to spend a night in Frankfurt before boarding our flight. It was very kind of Professor Wolfgang Schwartz to take us by car from Oberwolfach to Frankfurt and reserve a room for us at a modest hotel near the Haupt Bahnhof (= central railway station). When he dropped us at our hotel, he cautioned us to be "very careful" in walking at night because he said in his charming German accent, "You will be approached by ladies who will offer *luf for munny*." (love for money) — an interesting way to let us know that the Bahnhof area was teeming with prostitutes at night! Mathura was very uneasy after hearing this and she told me privately that in the future while staying at major European cities, we should never stay near the Haupt Bahnhof. So when we returned to Frankfurt in 1983 at the invitation of Professor Schwartz, we stayed at the fabulous Islamic Guest House near the University because I requested Professor Schwartz to get us more comfortable accommodation in a safer neighborhood! In any event, the visit to Germany and to Oberwolfach in 1979 had a significant influence on my academic career.

The trip to India in the summer of 1979 after Oberwolfach was also academically significant. I was in India in June, July and August, and it was during that visit that I made contact with Professor Ramachandra of the Tata Institute. There was a Number Theory Conference organized by MATSCIENCE during August 19–22 in Mysore, and for that conference Ramachandra was invited as a lead speaker to give four lectures in analytic number theory, more specifically on the Riemann zeta function and Dirichlet series. I was also asked to give three lectures and I spoke about my work in Michigan on multiplicative sums over numbers without small prime factors. The conference venue was the lovely Cheluvamba Palace of the Royal family of Mysore. Prof. Ramachandra, who hails from Karnataka (formerly called the Mysore State) was thrilled by the choice of this venue and the emphasis on classical analytic number theory in the conference. He was the father of analytic number theory in India and always strongly voiced the opinion that classical analytic number theory needs to be more emphasised in India. He thus congratulated MATSCIENCE for conducting the conference. But he had also established significant results in Transcendental Number Theory that were well quoted in Alan Baker's famous book on the subject that came out in 1975. Since I had come to India after speaking at the Oberwolfach conference on Diophantine Approximations, he requested me to give a fourth lecture on the last day on my recent work on irrationality measures. He was delighted on hearing my talk on this theme and said that on my way back to the USA, I should break journey in Bombay for a day and have lunch with him at his apartment at the Tata Institute. Thus Mathura and I departed Madras for Bombay on August 29 morning, and after checking into our hotel, we took a taxi to the Tata Institute. On that visit, Ramachandra presented me a book — "Transcendence Theory - Advances and Applications", which was the refereed proceedings of a 1976 conference at Cambridge University edited by Baker and Masser. In the Preface to that book, Baker and Masser say that if that book is read in conjunction with Baker's 1975 book, then that would put the reader at the forefront of knowledge in the theory of transcendence. So it was a valuable gift to me. Prof. Ramachandra had an inscription in the book, but he dated it 22 December 1887, which was the day Srinivasa Ramanujan was born! This showed his admiration for Ramanujan.

The 1979 summer in India was significant in other ways as well. MATSCIENCE had advertised several new positions and I applied for one. It was in July that I was interviewed and given an associate professorship which I accepted. But I returned to India only in 1981 after finishing my term at the University of Michigan. There is more on this in Section 5.1.

I was in India in the summer of 1980 as well, and as in 1979, I halted in Germany in the forward direction. In 1980, I was invited to the give a colloquium at the University of Bonn by Professor Don Zagier. He is a legendary figure in number theory who is hailed as a prodigy. He wrote his PhD thesis at the age of 20 under the direction of Friedrich Hirzebruch at the University of Bonn. When appointed as a Professor at the Max Planck Institute in Bonn at the age of 24, he was the youngest to hold that rank. Zagier is one of the fastest thinkers. He also talks very fast and with enormous energy.

In Bonn, I also had the pleasure and privilege of meeting Prof. Hirzebruch, Director of the Max Planck Institute, who was a world authority in algebraic topology. I also made acquaintance with Professor Gunther Harder. Both Zagier and Harder had visited MATSCIENCE in January 1980 at the invitation of my father, but I was not in Madras then. During that visit to MATSCIENCE, they came to know that I had just accepted a position there. So upon return to Germany from India, Zagier invited me to visit Bonn that summer and give a Colloquium. When Zagier and Hirzebruch offered me a student scholarship in 1975, I could not avail that opportunity, and so it was nice to make contact with them on that 1980 trip to Germany. Professor Zagier graciously agreed to my invitation to be one of the Founding Editors of The Ramanujan Journal. He also accepted my invitation to serve of the First SASTRA Ramanujan Prize Committee in 2005. I am most grateful for his support of my efforts.

4.7   Ann Arbor and Notre Dame AMS Meetings (1980–81)

In August 1980, during my three-year stay in Ann Arbor, the AMS Summer Meeting took place at the University of Michigan. Although not of the size of the AMS Annual Meeting which takes place each January, the Summer Meetings were quite large, and very beneficial to graduate students and young mathematicians. Like the Annual Meetings, the AMS conducted the Summer Meetings in conjunction with the MAA. Shortly after the Ann Arbor meeting, the AMS pulled out of the summer meetings, but then the MAA decided to continue on its own with the summer meetings (MathFests), and they have been a great success. Three decades later, when George Andrews was President of the AMS, he tried to resurrect the summer meetings, but was not successful.

There are benefits in holding a summer meeting instead of having just one huge Joint Annual Meeting. The dates of the Annual Meetings in January often coincide with the start of classes for the Winter/Spring term in certain universities, and so some faculty are unable to attend the annual meetings regularly. Most faculty do not teach in the summer, and graduate students are not necessarily talking courses then, and so the summer meeting could be more easily attended.

At the Ann Arbor Summer Meeting I gave my very first talk in an AMS Special Session. It was organized by Bruce Berndt whom I had invited earlier in Fall 1979 for a colloquium at the University of Michigan. It was a first hand experience for me to see how AMS Meetings are organized, and how effective their format of one hour talks and special session talks are for both senior and junior mathematicians. One of the most inspiring lectures I heard at that conference was by Richard Askey of the University of Wisconsin on "Ramanujan and some extensions of the beta and gamma functions". He was delivering the J. Sutherland Frame Lecture of Pi-Mu-Epsilon ($\pi - \mu - \epsilon$), a national undergraduate mathematics society. I was much impressed by Askey's lecturing style — informal, yet very engaging. He was exhorting everyone to read Ramanujan's works. Keep in mind that this was seven years before the Ramanujan Centennial which was when the focus of attention on Ramanujan's mathematics really started in a big way. Shortly after the Summer Meeting, Askey communicated with mathematicians worldwide soliciting funds to get busts of Ramanujan made by the American sculptor Paul Granlund. These busts were all ready by the time of the Ramanujan Centennial.

Richard Askey was the doyen in the world of Special Functions. His wide ranging scholarship, his fundamental research on orthogonal polynomials, and his guidance of young mathematicians, made him one of the most influential figures in his field. He received his PhD in Princeton in 1961 under the guidance of Solomon Bochner, and joined the faculty of the University of Wisconsin at Madison where he was ever since. He was very easily approachable, and loved to discuss mathematics at all levels. He took a big interest in Mathematics Education and has written and lectured extensively on this topic. He along with George Andrews and Bruce Berndt had led the Ramanujan revolution in the last few decades. My association and friendship with Askey grew considerably every since the Ramanujan Centennial in 1987 which is when the focus of my attention turned to the enchanting world of Ramanujan. But in Ann Arbor in 1980, I could not have imagined that just a few years later, I would interact with Askey in significant ways. By the bye, at the Ann Arbor Summer Meeting, George Andrews was delivering the Earl Raymond Hedrick Lectures of the MAA on partitions and Ramanujan's Lost Notebook, but my contact with him was to start a year later.

The AMS Special Session participation at the Summer Meeting in Ann Arbor was the start of several such special sessions for me in the next few years. In 1980–81, I had two more AMS conference participations — in Knoxville, Tennessee in November 1980 at the invitation of Carl Pomerance, and in March 1981 at the University of Notre Dame, Indiana. The Knoxville meeting was where I presented my results on $M^*(x, y)$ as acknowledged in the paper that appeared in the Transactions of the AMS. The Notre Dame meeting was very special for me as I now describe.

As mentioned earlier, Professor Paul Bateman spent the year 1980–81 in Ann Arbor on his research leave following his long term as Head of the Mathematics Department at the University of Illinois. Bateman was a Secretary of the AMS in charge of the Eastern Section. One day in Fall 1980, he approached me and said that he would like to invite me to organize a Special Session at the AMS regional meeting in Notre Dame in March 1981. I was very surprised at this, and responded saying that I was too young and had no experience organizing any meetings. He reacted immediately and said, "Well, organize this Special Session and get the experience!" So I went to Montgomery and asked his advice. He smiled when he heard about my conversation with Bateman and said: "I think it would be good for you to organize this Special Session, but make sure to invite strong mathematicians, because you will be judged by the quality of the session you organize." Immediately I asked him if he would be willing give a talk because I knew that his presence would assure the quality of the Special Session. He agreed without any hesitation. So with the one speaker confirmed, I set about contacting other speakers for the session.

I called Professor Patrick Gallagher at Columbia University, a top analytic number theorist, who among other things, had done fundamental work on the Large Sieve. He too said yes without hesitation. So the Special Session in Analytic Number Theory was assured to be a success with two of the leaders of the area — Montgomery and Gallagher as speakers. I was able to get a total of 15 speakers — all active researchers in analytic number theory, and this was adequate for the session in the two day conference.

One of the best ways to scout for young speakers would be to go through the list of visiting members for the year at the Institute for Advanced Study, Princeton. So I contacted a few of those Visiting Members, and two of them accepted my invitation — R. Balasubramaniam of the Tata Institute, Bombay, and Carlo Viola of the University of Pisa, Italy. Balasubramaniam had received his PhD in 1979 under the direction of Prof. K. Ramachandra of the Tata Institute. Viola was a former PhD student of Enrico Bombieri. He and I shared the hotel room in Notre Dame. Since I was going to Princeton in a few months, we talked about Bombieri and what it was like to work with him. Viola said that Bombieri is very intense at anything he does, and this intensity extends beyond mathematics to hobbies.

The speakers in my special session besides those mentioned above were Bruce Berndt, Brian Conrey (then a PhD student in Ann Arbor under Montgomery), Sid Graham, Grigori Kolesnik, Mel Nathanson, Andrew Odlyzko, Carl Pomerance, Don Redmond, Jan Turk (a visitor at Michigan from the Netherlands), and Jeffrey Vaaler.

An AMS Regional Meeting has four one-hour speakers and several special sessions of twenty minute talks, with at least one connected with the area of speciality of each hour speaker. In Notre Dame, Harold Diamond (University of Illinois) was an hour speaker, and my special session was associated with his talk. Diamond is a world authority on elementary methods in prime number theory and that was the title of his lecture in Notre Dame.

My talk at the Notre Dame conference was on a new Turan-Kubilius inequality for integers without any large prime factors. The inequality of Turan alluded to earlier was for strongly additive functions whose values on primes are bounded. Kubilius then extended Turan's inequality to all additive functions. The Turan-Kubilius inequality is a bound for the variance of additive functions and is a fundamental tool in Probabilistic Number Theory. For the additive function $\nu(n)$, the number of prime factors of n, we have $f(p) = 1$ for all primes p, and so

$$A(x) = B(x) \sim \operatorname{loglog} x, \quad \text{for} \quad f(n) = \nu(n).$$

Also Turan's inequality becomes an asymptotic relation, namely

$$\sum_{n \leq x} (\nu(n) - \operatorname{loglog} x)^2 \sim x \operatorname{loglog} x.$$

Thus in the classical theory, the mean and the variance are asymptotically the same for the function $\nu(n)$. Here I was discussing the Turan-Kubilius inequality for integers $\leq x$, all of whose prime factors are $\leq y$, I noticed that for the function $\nu(n)$, in certain ranges of y, the variance is much smaller compared to the mean. This was definitely a deviation of the classical theme and the first natural example where for $\nu(n)$, the mean and variance are NOT asymptotically the same. My paper on this topic which I presented in Notre Dame appeared in Crelle's Journal in 1982.

The AMS Meeting in Notre Dame gave me the first experience in organizing a conference session, an experience that led me to organize several conferences both in India and in Florida in the ensuing decades, and I appreciate Paul Bateman for giving me that first opportunity in my youth.

The conference in Notre Dame was a fine conclusion of my three-year stay at the University of Michigan where I was able to do some nice research, gained experience as a teacher, and made contact with a number of distinguished mathematicians. I was also pleased that the Institute for Advanced Study renewed the offer made to me in 1978. With the three year experience at Michigan, I was in a perfect position to make the best of my stay in Princeton. Another pleasant conclusion to my stay in Ann Arbor was that Mathura was pregnant with our first child and the delivery was to be in early September. Since the term at the Institute was to start only late September, we decided to go to India for the summer and for Mathura to have her delivery there in the care of her mother.

Professors Donald J. Lewis (University of Michigan), Paul Bateman (University of Illinois, Urbana), and Atle Selberg (Institute for Advanced Study, Princeton) at the International Conference on Number Theory, Oklahoma State University, July 1984.

Prof. Hugh L. Montgomery (University of Michigan) was Krishna's post-doctoral mentor during 1978–81. This picture was taken during the International Number Theory Conference for Krishna's 60th birthday in Gainesville, Florida on March 18, 2016, when Montgomery delivered the Erdős Colloquium.

Chapter 5

Visits to Princeton and Austin (1981–83)

5.1 A summer in India (1981) as a prelude to Princeton

In 1979, MATSCIENCE had advertised new positions in mathematics and physics. My father was the Director of the Institute. He told me that even though I was attracted by academic life in the USA, I should apply for one of these positions and should not say no to academic life in India without trying it for a few years. The chief mathematician on the Selection Committee was Professor M. S. Narasimhan of the Tata Institute of Fundamental Research (TIFR). On hearing this, I agreed to apply, knowing fully that the TIFR mathematicians have high standards.

M. S. Narasimhan is known for fundamental work on Riemann surfaces. He was at that time, and until his retirement, the leader of the mathematics group at TIFR whether or not he was the department head (called Dean at TIFR). He sat on various mathematical committees and his opinion always mattered the most when decisions were made. Thus he was very influential in shaping the development of mathematical research in India. He was always referred to in India as MSN in awe and admiration.

The Selection Committee met in the Summer of 1979 just in time when I was visiting Madras from Ann Arbor. So I was interviewed by the Committee. Whenever TIFR mathematicians are present on a committee in India, and the issue under consideration is about mathematics, they take the lead in the discussions and decisions. It should be pointed out that the TIFR mathematicians focus only on mathematics, and do not care about decisions made in other disciplines, such as physics. In fact they are totally detached when discussions about other disciplines take place. But when it comes to decisions relating to mathematics in India, the TIFR mathematicians hold the sway. I found this out first hand later during my brief stay at MATSCIENCE.

During the interview, M. S. Narasimhan asked me to explain the significance of my results on irrationality that I had recently obtained with Mike Robinson. He liked my brief presentation and said that he was also impressed by the letter Professor Robert Steinberg at UCLA had written about my PhD thesis of 1978. Not only was Steinberg world famous, but Narasimhan also knew him quite well because Steinberg had visited TIFR and Narasimhan had visited UCLA at the invitation of Steinberg. A few days after the interview, while I was still in Madras, I received an offer from MATSCIENCE of a tenured Associate Professorship. Straus was wise in suggesting that even though Steinberg was not a number theorist, I should have him on my PhD Committee, because his letter would have effect in certain places. Narasimhan's reaction to my file on seeing Steinberg's letter was a perfect example of what Straus anticipated! But I also found out that among the approximately 80 applicants for various positions, only two offers made, both of them associate professorships — one to me in pure mathematics and the other to Dr. Bhimsen Shivamoggi (a mathematical physicist) in physics. I said that I could join only in 1981 after completing my term in Michigan, and that was accepted by the Selection Committee. My hope was that in two years, other appointments in mathematics would be made, but that was not the case because Narasimhan with his high standards did not approve any more appointments in mathematics. In any case I joined in 1981 as I said I would.

Shortly before leaving Michigan in May 1981, I learnt that Professor Erdős would be travelling through India, and so I asked him if he would be interested in visiting MATSCIENCE in the summer of 1981, and perhaps be the principal speaker in a conference that I would like to organize. After the Notre Dame experience, I felt bold and enthusiastic to organize a conference upon joining MATSCIENCE. Professor Erdős readily agreed and so it was a great opportunity for me to convene this conference within a few weeks of joining the Institute. Erdős arrived straight from Waterloo, Canada, where he had just accepted an Honorary Doctorate.

Madras summers are unbearably hot; in fact the joke is that Madras has only one weather with three variations — hot, hotter, and hottest! And Erdős arrived in Madras during the hottest period. He immediately started writing letters and postcards to friends saying that "the only reality here is the heat!" His previous visit to Madras in 1975 was in the month of January when the weather was much cooler and quite pleasant. In any case, knowing that Madras would simmering in the summer heat, I arranged the conference in Mysore, where the weather is salubrious. Erdős not only loved the Mysore weather, but also everything that charming city and its environs had to offer — temples, palaces, waterfalls, and a bird sanctuary.

I invited Professor K. Ramachandra of the Tata Institute as the other lead speaker and he graciously accepted the invitation. Thus began a long and rewarding friendship with Professor Ramachandra whom I consider as the father of analytic number theory in India. The TIFR is known for having made great strides in various abstract areas of mathematics. The mathematics program at TIFR is influenced very strongly by the Bourbaki approach. Ramachandra who worked on classical aspects of mathematics, felt unappreciated amidst colleagues who used high powered abstract methods; in fact Ramachandra went about saying that elementary and classical analytic number theory were not valued at TIFR. As a reaction to what he perceived as the lack of regard for the area of his work, he started the *Hardy-Ramanujan Journal* in the late seventies, along with his student R. Balasubramanian emphasizing classical analytic number theory and published it privately with his own funds. He was very happy that I was conducting this conference on elementary and analytic number theory and expressed his joy in no uncertain terms in his lecture and in conversation with the participants. Since there were several excellent talks at the meeting by Erdős, Ramachandra, and other participants, I approached Catriona Byrne of Springer Verlag, Germany, to publish the refereed proceedings in their famous Lecture Notes Series. They readily agreed and thus began my long and fruitful relationship with Springer.

I must relate to you a humorous anecdote about Erdős at the conference. One of the great attractions of Mysore city is the palace of the Maharaja of Mysore. So we took Erdős to see the palace. Upon arriving at the gate, an enthusiastic guide volunteered to show us around and Erdős said yes. One of the questions that Erdős asked the guide was when the palace was built, but the guide could not give a precise answer. Tourism in India was not much developed at that time and so there were no brochures available at most tourist spots including the Mysore Palace. Erdős was obviously disappointed. The guide took us through several chambers, and at one point said, "This is where the queen slept. The rooms on the other side are where some of the Maharaja's concubines slept, ..." Erdős, who had no interest in the romantic exploits of the Maharaja, exploded at that point as follows: "I do not want to know where these women slept. I want to know *when* the palace was built!"

MATSCIENCE in those days used to organize two conferences each year outside of Madras. In 1981, their second conference in late August was in Ootacamund — popularly known as Ooty, a lovely hillstation a few hundred miles from Madras. When the British ruled India, they shifted to hillstations in the summer and conducted official business from there. Two of the most famous hillstations are Simla in the north and Ooty in the South. During the conference in Mysore, there was discussion about the next conference in Ooty. Erdős expressed a desire to see Ooty on his next visit having heard how lovely it was and how cool the weather there was throughout the year. So for Erdős' 70th birthday, I arranged a conference in Ooty in January 1984 (see Section 6.2).

Although I joined MATSCIENCE in the summer of 1981, I had the offer from the Institute for Advanced Study for a Visiting Membership for the Academic Year 1981–82 starting in September. MATSCIENCE generously gave me a leave to enable me to visit Princeton. The second MATSCIENCE conference in Ooty was in late August and so I decided to attend that before departing for Princeton. Mathura was to deliver our first child in early September, and so I asked her if I could go to the Ooty conference. She was confident that nothing would happen in August and so she gave me the green signal. After the conference was over, on our return to Madras from Ooty by car, my parents and I decided to spend two nights in the Mudumalai Wild Life Sanctuary. I am a wildlife enthusiast and so I utilize every opportunity to visit wildlife sanctuaries, especially Mudumalai. In Mudumalai we stayed at the Kargudi Guest House which has a secluded location in the middle of the dense jungle. There is a view point near the guest house from where one gets a breathtaking view of the entire Mudumalai sanctuary and the Nilgiri mountains surrounding it. This is my favorite spot on earth. On August 28, as we were relaxing at the view point, a uniformed employee of the postal service brought a telegram to us there; the telegram said that Mathura had just delivered a girl — one week before the projected date! Naturally I was sorry not to have been by Mathura's side. We left immediately for Madras and reached there the next day. We named our daughter Lalitha after my mother, but we call her Vanitha at home because I heard the news of her birth when I was in a forest — *vanam* in Sanskrit means forest.

The Institute for Advanced Study (hereafter in this chapter referred to as the Institute) was opening for the Fall Term during the third week of September. So I was planning to leave for Princeton by the middle of September with Mathura to follow me a month later with the baby after taking some rest in Madras following the delivery. It was only a one day process to get the USA Visa and so I went to the consulate a few days before my departure to apply for the visa for me and my family to go to Princeton. I was shocked to hear that I could not be granted the visa because I had gone to Michigan on an Exchange Visitor Visa (J1), had spent the full term of three years on that visa, and therefore had to remain in my home country for two years — a residency requirement — before I could apply for a visa to re-enter the USA. This two year Residency Requirement was exempt only for specialists in certain fields of activity that the USA had classified as "areas of need". The consulate told me that the only way that a new visa could be issued is if I get a "Waiver" of the two year residency requirement. But getting such a waiver would take months because it involves clearance from the police in India, from the IRS in the USA, and so on. I told the Consul that this was not a visit to just any university in the USA, it was to the Institute for Advanced Study, one of the greatest centers of learning in the world made famous by the presence of Albert Einstein as its Permanent Member. The Consul was receptive to what I said and understood how significant the visit to the Institute would be at the early stage of my career. So she said that instead of insisting on the waiver, she would write to Richard L. Fruchterman of the United States International Communication Agency (USICA) in Washington asking permission to issue a new J-1 visa without the waiver. The final waiver is anyway issued by USICA, and so she was asking Fruchterman to waive the waiver! The Consul told me that it would take a week for Fruchterman to respond, but everything was being done by telex to speed up the process. What this meant was that I had to postpone my departure by a week to the end of September, but I was thankful that the US Consulate in Madras went out of the way to help me. Also, the delay in departure gave me some extra time with Lalitha who was just a few weeks old. What a pleasant news it was to hear from the US Consulate that Fruchterman gave the nod. So I was issued a J1 visa, not just for one year, but for three years! This longer period of the visa enabled me to accept a one year assignment at the University of Texas right after my stay at Princeton without going through the visa application again. In any case, with the new J1 visa, I was in a great mood to depart for Princeton and for a productive year there in the company of some of the most eminent mathematicians in the world.

5.2 The Institute for Advanced Study

I arrived at New York John F. Kennedy International Airport by Pan Am in the evening on October 1, 1981 and wanted to proceed to the Institute without any further delay. I was concerned to take the train with my luggage and so even though it was expensive, I took a New York cab (taxi) to Princeton which is in New Jersey. The cab driver told me that since his is a New York taxi, he was not permitted to transport customers in New Jersey back to New York, and so he wanted me to pay double the fare to cover the return journey. I agreed to this without hesitation. When I arrived at the Institute, it was late (around 9:00 pm), but the door to my apartment was left unlocked so that I could get in. The staff of the Institute had made perfect arrangements. There I was finally in 56 Einstein Drive, on the campus of the Institute for Advanced Study. As a grand student of Albert Einstein, I felt overjoyed to stay on Einstein Drive.

Everything was working in the apartment except the telephone. I could get the connection only two days later. So I could not call home rightaway to inform Mathura and my parents that I had reached safely.

I got up early in the morning and went to the Institute to get "registered" in the office of the School of Mathematics. Caroline Underwood, the Administrative Officer, was very warm to me. One of my goals for visiting the Institute was to interact with Professors Atle Selberg and Enrico Bombieri, both Permanent Members there, and two of world leaders in number theory. I first made a courtesy call on Selberg in his office just to say that I had arrived and he said very politely that after I get settled he would be glad to talk to me about my work.

Selberg's office was in Fuld Hall, the imposing main building. My office was in the adjacent Building C, and I shared it with Amit Ghosh who had received his PhD a few months earlier in Nottingham under the direction of Heini Halberstam. My office on the second floor of Building C was adjacent to the offices of Professors Harish-Chandra and Bombieri. Professors Armand Borel and Robert Langlands had their offices on the first floor, and so Building C had its share of mathematical giants as well.

The next day I saw Bombieri in his office. He smiled and said that my father had called him the previous night concerned whether I had reached safely and he had assured my father that I was fine. After getting my telephone connection in my apartment, I called my home in Madras. My father said that to every question about me that he asked, including the question whether I reached Princeton safely, Bombieri replied "he is fine"!

The idea to create the Institute for Advanced Study crystallized in the 1920s with Louis Bamberger and Felix Fuld agreeing to make generous donations. The Institute was founded in 1930 by which time Felix Fuld had died, and so his wife Carrie Fuld, the sister of Louis Bamberger, stepped in as the principal donor along with her brother. Abraham Flexner was appointed as the first Director. The guiding principle of the Institute was "pursuit of knowledge for its own sake" (see [B8] by Flexner). The idea was to have a small group of exceptionally accomplished Permanent Members (referred to as the Faculty), and a large number of Visiting Members who would come there for a year or two, free of teaching duties, to make advances in their research. The Permanent Members also did not have any teaching duties; they were to lead important research programs and also help the younger visitors by discussing with them. Even though the Institute had no systematic courses that it offered, there had always been plenty of research seminars given by both the Permanent and Visiting Members. It was the Seminar which was the moving spirit of the Institute.

To maintain the highest standards in research, the Institute only appointed the most distinguished academicians as its Permanent Members, and focused on a few subjects instead of broadening out like the big universities. Right across the lawn was Princeton University, one of the elite universities in the world, and the Institute has always enjoyed a harmonious and fruitful relationship with the University. With the University and the Institute side by side, Princeton can be regarded as perhaps one of the two greatest intellectual centers in the world, rivalled only by Cambridge, Massachusetts, with Harvard and MIT in close proximity.

The main building of the Institute, Fuld Hall, named after Felix and Carrie Fuld, was opened only in 1939. It is an imposing colonial style building with spacious lawns in the front and rear. The administrative offices as well as the main offices of the School of Mathematics are in Fuld Hall on opposite sides of the central stairway. The Institute has an outstanding library the mathematics part of which was housed in the upstairs of Fuld Hall. On the ground floor, in the center, is a spacious lounge where Tea is served on weekdays at 4:00 pm. It is a place to meet the distinguished faculty of the Institute in a relaxed setting, and the visiting members as well. The informal discussions at the afternoon tea provide a fine opportunity to get know other members and the research they were engaged in, as well as to discuss one's own work with other experts.

Albert Einstein was one of the very first appointments at the Institute as a Permanent Member and he was there from 1933 until his death in 1955. My father in his Diary [B11] said that "the spirit of Einstein was everywhere in Princeton". Einstein's appointment set the tone and standard for appointments to follow. The salary offered to Einstein at the Institute for a special professorship was $20,000 per year, which indicates that since then, salaries have increased by a factor of 10 to 15. Oswald Veblen, a leading topologist at Princeton University, was appointed at the Institute and Flexner gave him the charge of developing the School of Mathematics, which he did admirably. At that time there was no separation between the faculties in physics and mathematics and so they were all (including Einstein the physicist) in the School of Mathematics, but the two groups functioned separately. The first faculty in the School of Mathematics at the Institute were Albert Einstein, Oswald Veblen, John von-Neumann, Hermann Weyl, James Alexander, and Marston Morse — all legends in their respective fields of research.

Within a few days of joining the Institute, I approached Caroline Underwood, the Administrative Officer, and asked her where Einstein's office was, and whether I could see it. She responded saying that there were, and are, so many academic faculty of excellence at the Institute, that they do not separate Einstein from the rest and make a museum of his office. She said that Einstein's office was being occupied by Arne Beurling. So my admiration for Beurling became greater from that point on! Beurling was an Emeritus Professor in 1981, but I did see him walking through the corridors from time to time. I should point out that Einstein's home on Mercer Street was not made into a museum at his request. It was not even prominently advertised as a "place to see" in Princeton in order to ensure the privacy of Einstein's step-daughter who was living there. My thesis advisor Ernst Straus, who had been Einstein's assistant at the Institute, kindly arranged for Mathura and me to meet Einstein's step-daughter.

Even though I communicated mostly with Bombieri in connection with my application for a visiting membership at the Institute, and I was paid out of Bombieri's NSF grant, the letter of offer came from the Director Harry Woolf, who was on the History Faculty. It was during Woolf's period that the Achives section of the Institute was launched. When my father visited the Institute in 1957–58, Robert Oppenheimer, the leader of the Manhattan Project that made the Atom Bomb, was the Director. Oppenheimer's period saw the most intense activity in theoretical physics in the history of the Institute.

In 1965, during the final year of Robert Oppenheimer as the Director of the Institute, a major confrontation between him and the mathematics faculty took place over certain appointments in mathematics. The confrontation was not about the qualification of the mathematicians being considered for appointment, about which there were no doubts; the issue was whether one could, or should not, consider professors at Princeton University for appointments at the Institute. Oppenheimer was of the opinion that the

Institute should not cause the depletion of talent from the University, and his view led to friction with the mathematicians and a worsening of relations between the mathematics and physics faculties. This resulted in the two faculties separating into their own schools in 1965 with the physicists forming the School of Natural Sciences.

The permanent faculty at the Institute have their own independent views about research and the academic world. They are absorbed in their research and generally appear to be lost in thought. For example, I must have crossed Armand Borel a hundred times near Building C, but we never exchanged greetings. He would, of course, acknowledge my presence with a mild nod. Armand Borel, like Marshall Stone, was deeply interested in South Indian classical music, and therefore attended the Madras Music Season concerts in December every time he came to the Tata Institute for their conferences in January. Borel's passion for Carnatic music was due his deep interest in Jazz; Carnatic music and Jazz are similar in the sense that there is a significant amount of improvization in both. One year, Borel and his wife were seated close to us at the Madras Music Academy. So I went up to him and introduced myself. He smiled and said that he remembered me visiting Princeton some years ago, and was pleased that I remembered and recognized him in his advancing years. So if you go and talk to the permanent faculty at the Institute, they will be pleasant to you. We talked about the Madras Music Season in December. He said he enjoyed coming to Madras during that time. In a memorial article on Borel that appeared in the AMS Notices in 2004, K. Chandrasekharan says: "He (Borel) personally might have seemed dour to those who did not know him well; they could not sense the soft core underneath."

Mathura arrived with our infant daughter Lalitha only in late November after resting for two months following the delivery. Since it would be strenuous for her to travel with a child by herself, Mathura was accompanied by her father Mr. N. C. Krishnan and her uncle Mr. C. N. Srinivasan. They were with us in Princeton for a month and enjoyed their stay immensely. My parents came only in April 1982, but then it was springtime at its best with flowers everywhere. In view of long term guests that I knew we would have, I asked, and was given, a comfortable two bedroom apartment.

A few weeks before Mathura arrived, Nobel Laureate Hans Bethe was at the Institute to give the Einstein Memorial Lecture on "The Energy Problem". Soon after Mathura and I got married, Bethe sent a very thoughtful gift to us. When I chatted with Bethe after his lecture, he made kind enquiries about my parents and Mathura.

The year in Princeton was one of the most productive of my academic life when I made significant progress on a new approach to Probabilistic Number Theory using sieves. One feels inspired by simply breathing the air of Princeton and being in the company of so many movers and shakers of the academic world. I had heard so much about the Institute from my father whose stay in 1957–58 inspired him to launch his Theoretical Physics Seminar and create MATSCIENCE. I was excited to be in the hallowed grounds of the Institute and experience first hand many things I had heard from him. Next I will share with you my observations on some of the Institute faculty like Atle Selberg, Enrico Bombieri and Andre Weil, and then describe briefly some aspects of the research I was able to complete during my stay at the Institute.

5.3 Observations on Atle Selberg

Atle Selberg is universally viewed as one of the greatest mathematicians of the 20th century. He made revolutionary contributions to the theory of the Riemann zeta function, sieve theory, and the spectral theory associated with automorphic forms. It was a sensation when he and Paul Erdős produced an "Elementary Proof" of the celebrated Prime Number Theorem (PNT), but then a bitter dispute between the two arose after which they were not on speaking terms. I will say more about this dispute below.

Selberg published occasionally, but every paper of his has had a deep and lasting impact. He was awarded the Fields Medal in 1950 for his contributions to the theory of Riemann zeta function, his work on sieves, and for the elementary proof of the Prime Number Theorem (PNT). His very influential work on what is now called the Selberg Trace Formula came a few years after the Fields Medal.

Selberg was born in 1917 in Langesund, Norway. His interest in mathematics was aroused when as a boy of thirteen, he read in his father's library the works of the Indian genius Srinivasa Ramanujan. Selberg received his PhD from the University of Oslo in 1943 with the famous mathematician Harald Bohr (brother of the Nobel Laureate physicist Niels Bohr) and Thoralf Skolem as examiners. The Norwegian mathematician Viggo Brun, father of the modern sieve method, also influenced Selberg who went on to create a more powerful sieve known now as the Selberg sieve.

Selberg was a Visiting Member at the Institute in the late forties which is when the interaction with Erdős on the elementary proof of the PNT took place. After receiving the Fields Medal, Selberg was widely viewed as one of the most powerful mathematicians and promoted to a Professorship at the Institute in 1951, a post he held until his retirement.

Selberg had such high standards that he published only if the work represented a major advance. He also frowned on those who published too much and in some instances discouraged mathematicians from publishing their work if it was not going to be highly influential. For example, when Barkley Rosser obtained a significant improvement of the Combinatorial Sieve of Brun, Selberg dissuaded Rosser from publishing it because it did not solve the Goldbach or Prime Twins Conjectures — the sieve was invented to tackle these difficult problems. But in his famous lectures on Sieve Methods delivered in Stony Brook in 1969, Selberg did acknowledge that Rosser's unpublished work was important. Years later, Rosser's work was brought to light by Henryk Iwaniec who published a paper in Acta Arithmetica entitled "Rosser's Sieve".

Selberg never made tall claims about his results. In fact, in his papers, he discussed the limitations of the methods he used. For example, he said that the celebrated Goldbach and Prime Twins problems could not be solved by sieve methods alone, although the sieve was devised to solve these long standing, simple to state, but notoriously difficult problems. Similarly, toward the end of one of his papers on automorphic functions that he wrote in the sixties, he stated that the famous *Ramanujan Hypothesis* concerning the function $\tau(n)$ could not be solved by purely number theoretic methods, but would require the use of ideas from algebraic geometry. This prediction was proved correct in the mid-seventies when the Belgian born mathematician Pierre Deligne proved the Ramanujan Hypothesis and other deep results utilizing techniques from algebraic geometry.

There is a story about Selberg and his attitude towards publication that I heard which I will relate to you now: Once a young Visiting Member was describing his work in detail in Selberg's office. The discussion was proceeding well and Selberg was pointing out ways in which the work could be improved. The young visitor then asked Selberg where he could publish his work. At that point Selberg opened one of the drawers of his desk, pulled out some sheets of papers and said, "Some years ago I had obtained your results in a somewhat stronger form but I decided not to publish them. But standards have declined now and so you may proceed with the publication of your theorems."

I had discussions with Selberg in his office once every two months giving him an update on my progress. He was extremely helpful in our meetings, giving me suggestions how certain estimates could be improved, and why some estimates were best possible. Since I was aware of his views on publication, and knew the story in the preceding paragraph, I never asked him whether I should publish my work or where I should publish. So my discussions with Selberg went smoothly.

The way I would get an appointment with Selberg was to approach him at the afternoon tea. He would be there reading a newspaper but seated with his back to the large gathering of people at tea. Whenever I asked for an appointment, he would put the newspaper down, think for a few moments and say something like "How about Tuesday afternoon next week at 2:00 pm?" And when I went to his office at the appointed time, he gave me a full hour without getting distracted; he would have instructed his secretary to hold all phone calls. As soon as I entered his office, he would say: "Please go to the board and explain your work." I really appreciated that he genuinely set apart time to discuss with the visitors, and that was indeed one of the duties of the faculty that he carried out so well.

There is a joke associated with Selberg's view about publications and his assessment of mathematicians. It is said that Selberg held the view that you are only as good as your worst paper! That is, if among all your publications, the least important one is significant, then you must be a great mathematician. No wonder he and other believers of this philosophy did not hold Erdős in high esteem because among Erdős' hundreds of papers, there are some which are very simple. The mathematical version of this philosophy is that Selberg measured mathematicians by the "inf norm" applied to their publications. We all know that the supremum is a norm, and not the infimum, and so this is really a joke.

Selberg never really cared where a paper is published. He only cared whether a result was worthwhile to be in print. In fact some of his own early papers were in obscure Norwegian journals. He felt that if a result is important, mathematicians will read the paper no matter where it is published. I really like this view because otherwise only the very best journals will get the most important papers giving other journals no chance to enhance their reputation by publishing outstanding papers.

Selberg also did not believe in collaborative work in mathematics. Here his philosophy was that *a joint paper is credit to neither*! He wrote only one joint paper — with S. Chowla because Chowla approached him with a question. This negative attitude towards joint papers was one of the reasons he and Erdős had a dispute in connection with the elementary proof of the Prime Number Theorem (PNT), the story of which I now describe.

My report of the dispute between Erdős and Selberg is based primarily on what my thesis advisor Ernst Straus told me, but I have drawn from other reports as well. Erdős was Straus' house guest at the Institute then, so

Straus was a direct witness to the events (see [A8] by Spencer and Graham in which there is a complete account by Straus). I have talked to Erdős about this, but never with Selberg. Dorian Goldfeld has given a masterly account [A7] of the controversy after speaking to Selberg and Erdős. See also Schechter [B14; pp. 142–151] for an equally thorough account.

The PNT was first proved towards the end of the nineteenth century by Hadamard and de la Vallee Poussin using complex variable theory and properties of the Riemann zeta function as a function of the complex variable. No proof avoiding complex variables was known and this was frustrating because prime numbers are defined in such a simple way. Following the complex analytic proof of the PNT, it was noticed that the PNT was "equivalent" to the statement that the Riemann zeta function has no zeroes on the line with real part 1, that is $\zeta(1 + it) \neq 0$ for every real number t. Here by two statements being "equivalent" we mean that each can be easily derived from the other. This equivalence motivated G. H. Hardy, a towering figure in number theory at Cambridge University, to challenge the mathematical world to find an "elementary proof", namely a proof that used only real variables and no complex variables; Hardy went on to make the prophecy that if such an elementary proof were to be found, then the books would have to be rewritten because it would change our view of the way the subject hangs together! That propelled the problem into prominence.

Atle Selberg arrived in Princeton in 1947 as a Visiting Member at the Institute. One of his significant achievements was his elementary derivation in March 1948 of a Fundamental Lemma involving the von-Mangoldt Λ-function, which we state here as

$$\sum_{p \le x} \log^2 p + \sum_{pq \le x} \log p \log q = 2x \log x + O(x). \quad \text{(SELBERG'S IDENTITY)}$$

The first sum is over primes p up to x, and the second sum is over primes p, q such that their product pq does not exceed x. Selberg's identity was immediately recognized as a major step towards the elementary proof of the PNT, but one did not know how to derive the PNT from it. The final solution was to come dramatically at the Institute in the same year.

Paul Turán, a great Hingarian mathematician, was visiting the Institute in 1948. During Turán's discussion with Selberg on Selberg's elementary proof of Dirichlet's theorem, Selberg indicated that he had obtained the above identity. Turán asked Selberg to give a talk, but Selberg was busy contacting various universities to get a permanent position and about to leave town. So he declined to speak. After Selberg left town, Paul Erdős arrived at the Institute. Turán then gave a talk on Selberg's elementary proof of Dirichlet's theorem, following which there was a discussion on Selberg's identity. Immediately, Erdős realized that he could go one step closer to the PNT as follows.

In §2.7 we had mentioned how Erdős catapulted to fame by his ingenious proof of Bertrand's postulate, namely the statement that there is always a prime number between x and $2x$, for $x \geq 2$. This is equivalent to saying that if p_n is the n-th prime, then

$$\frac{p_{n+1}}{p_n} < 2. \quad \text{(BERTRAND'S INEQUALITY)}$$

On seeing Selberg's identity, Erdős realized that he could significantly improve Bertrand's inequality as follows: Given $\epsilon > 0$,

$$\frac{p_{n+1}}{p_n} < 1 + \epsilon, \quad \text{for} \quad n > N(\epsilon). \quad \text{(ERDŐS' INEQUALITY)}$$

This could be recast in the form

$$\lim_{n \to \infty} \frac{p_{n+1}}{p_n} = 1.$$

Actually Erdős could prove something stronger, namely, for each $\epsilon > 0$, there exists a $\delta = \delta(\epsilon) > 0$, such that there are at least $\delta x / \log x$ primes in the interval $[x, (1 + \epsilon)x]$ (ERDŐS PRIME DENSITY RESULT). This is weaker than, and follows from, the PNT, which is equivalent to the statement that $p_n \sim n \log n$.

After Selberg returned to Princeton from his trip, Erdős approached him and said that he could obtain the prime density result from Selberg's identity. Selberg responded by saying that this could not be possible because he knew how to derive the PNT from such a density theorem, and he had convinced himself that the PNT could not be obtained in that fashion from his identity. Later Selberg admitted that he said this mainly to throw Erdős off the track and to dissuade him from working on the problem further. But Erdős persisted. A few days later, Erdős and Selberg checked their claims; indeed Erdős' derivation of his density result was correct, and Selberg's deduction of the PNT from that was also accomplished. This indeed was a great moment and Straus was one of the fortunate few who witnessed this derivation [A8] of the PNT by Selberg and Erdős. The result was of such significance, that even though it was around 10:00 pm, it was suggested that a seminar be arranged immediately to announce the result. But Straus' wife was arriving by train at Princeton junction at that time. Thus the seminar was scheduled two hours later — at midnight(!) — so that Straus could pick up his wife and be back for the seminar. Everyone who attended that midnight seminar at the Institute knew that they were witnessing a historic moment and so there was great jubilation.

Selberg was travelling for the next several days in connection with his job search. Meanwhile, Erdős, the ever enthusiastic communicator, sent several carefully worded postcards announcing the newly found Elementary Proof of the Prime Number Theorem giving Selberg due credit for what he had done. Erdős was already quite famous in 1948 for a variety of results including the Erdős-Kac Theorem, but Selberg was relatively unknown then. What Selberg heard while he was away was the news that Erdős and some other mathematician had found an elementary proof of the PNT. This naturally upset him and so when he returned to Princeton, he told Erdős that they will not be writing a joint paper. Selberg was able to circumvent the argument of Erdős and so he decided to write his paper without appeal to Erdős' intermediate result, but acknowledging Erdős' contribution to the original proof. Selberg also suggested that Erdős publish his prime density result first and that he would publish his paper after that. Selberg said that he never had any idea to collaborate with Erdős and perhaps felt that Erdős had infringed on his territory. Selberg said that while he did not mind that Turán gave a talk on his proof of Dirichlet's theorem, he did not realize that Turán would discuss his identity and also that Erdős would be at the lecture. But once an idea is discussed in a seminar, then anyone can pursue that. And surely Erdős did, but not with the intention of infringing on Selberg's territory.

What happened after this breakdown in relations between Erdős and Selberg was quite disturbing. Both Selberg and Erdős wrote their separate papers on the Elementary Proof of the PNT, with Erdős describing exactly how things unfolded, and Selberg alluding to the first derivation using Erdős' density result but choosing to give a proof without using it. Both submitted their papers to the Annals of Mathematics. Hermann Weyl had a key role in the acceptance Selberg's paper and the rejection of Erdős' paper! Erdős then tried to get it published in the Bulletin of the AMS, but the Bulletin also rejected Erdős' paper. The only way for Erdős to get it published in a timely fashion was to have it communicated to the Proceedings of the National Academy of Sciences (PNAS), a very prestigious journal, and this fortunately happened. So the papers of Erdős and Selberg appeared simultaneously and separately in 1949. Selberg was offered a Permanent Membership at the Institute for Advanced Study in 1949, which he of course deserved, and in which he did his finest work in the fifties. As Straus said [A8], senior mathematicians like Weyl, instead of mediating and effecting a settlement, took sides on the issue and destroyed a relationship between two of the greatest number theorists. This is because some senior mathematicians admired Selberg's way of doing mathematics and never appreciated Erdős practice of working on several problems simultaneously and with different authors. Selberg and Erdős clashed because they had completely opposite views of how mathematics is to be done. What should have been an event for celebration turned out to be a bitter dispute that left everyone unhappy. In the end, even though the Elementary Proof was a tour-de-force, it did not have the significant impact as predicted by Hardy; in contrast the Hadamard-de la Vallee Poussin analytic proof revolutionized the subject.

At the International Congress of Mathematicians in Harvard in 1950, Selberg was given the Fields Medal along with Laurent Schwartz. This was only the second time that the Fields Medals were awarded, because after 1938 when the first Fields Medals were given, the International Congress was interrupted by World War II. In 1951, Selberg was promoted to a Full Professorship at the Institute where he was ever since. Erdős was awarded the Cole Prize of the American Mathematical Society in 1951 for his many influential papers. Starting from 1949, with no permanent job or home, Erdős travelled around the world constantly, collaborating with hundreds of mathematicians young and old, in leading universities and in lesser academic centers. With sheer volume and breadth of contributions, and his numerous collaborations, he became one of the most influential mathematicians of the twentieth century. On my part I felt fortunate to have interacted with both these giants of number theory.

After the fallout with Selberg and the harsh treatment he faced due to that, Erdős never came to the Institute even though he would be a regular visitor to Bell Labs in nearby Murray Hill, New Jersey, where Ron Graham headed the mathematics division. However, during mid-October 1981, when Erdős was at Bell Labs, I met him there and invited him to spend a couple of nights at my apartment at the Institute. He gladly came a few days later enroute from Philadelphia to Rutgers. He relaxed in my apartment, saw pictures of my infant daughter Lalitha, and said she was a "sweet boss child" (Mathura arrived in Princeton from Madras with Lalitha only in mid-November). In conclusion, I must say that even though Selberg knew that I was very close to Erdős, he was very kind to me and did not let his feelings towards Erdős influence the way he treated me. Although I was saddened that Erdős and Selberg were not on speaking terms, I benefited by interacting with both of them.

5.4   Observations on Enrico Bombieri

Enrico Bombieri has made fundamental contributions to many areas of mathematics such as number theory, univalent functions, and partial differential equations. For his far reaching contributions, he has received several prestigious international prizes including the Fields Medal in 1974. He is a genius who excels in everything he attempts to learn within and outside mathematics such as chess, stamp collection, and semi-precious stones, to name a few. He also demands and expects the very best in everything he experiences, and he has always enjoyed the finest in all aspects of life. As a mathematician, he is "ultra-careful", meaning that he will use the work/theorem of another researcher only after he has checked the method/proof himself.

Bombieri was born on 26 November 1940 in Milan, Italy. He was a prodigy and published his first mathematics research paper when he was only 16. He received his PhD at the age of 22 under the direction of Giovanni Ricci in Milan. Soon after, he came under the influence of Harold Davenport at Trinity College, Cambridge University and that is when his powerful mean value theorem for prime numbers in arithmetic progressions was proven using techniques of the *large sieve*.

The Prime Number Theorem for Arithmetic Progressions, (already encountered) is the statement that for each integer k, the prime numbers are uniformly distributed in the reduced residue classes $\ell(\mod k)$. Bombieri's celebrated Mean Value Theorem makes this precise for primes up to a given magnitude x and for k up to a certain large value depending on x. To state his result, define

$$E(x,k) = \max_{2 \le y \le x} \max_{0 \le \ell < k, (\ell,k)=1} |\pi(y,k,\ell) - \frac{1}{\phi(k)} li(y)|,$$

where $\pi(y,k,\ell)$ is the number of primes up to y which are $\equiv \ell(\mod k)$, and

$$li(y) = \int_2^y \frac{dt}{\log t}.$$

Then Bombieri's theorem is that for each $B > 0$, there exists an $A > 0$ such that

$$\sum_{k \le X^{1/2}/\log^A X} E(x,k) <<_B \frac{X}{\log^B X}. \quad \text{(BVT)}$$

Here $<<_B$ means less than a constant multiple of, where the constant depends on B. Simultaneously and independently, A. I. Vinogradov had established a slightly weaker result with $X^{\frac{1}{2}-\epsilon}$ replacing $X^{1/2}/\log^A X$, and so the above is also referred to as the Bombieri-Vinogradov theorem (BVT). The yet unproved *Generalized Riemann Hypothesis* (GRH) on the zeros of *L*-functions implies the BVT. One reason that the BVT is so important is that for a large class of problems, it suffices to use the BVT instead assuming the GRH. In a famous 1971 paper on Sieve Methods, Selberg has emphatically said that after his Harvard Lecture of 1950 (when he received the Fields Medal), there have been only a handful of influential papers on sieves, and he considered Bombieri paper on the Mean Value Theorem to be very significant. Coming from Selberg who is not easily given to hyperbole, this is high praise indeed!

It is said of Bombieri that every few years the focus of his research turns to a totally different area, and in making the change, he effects a fundamental advance. After the sensational proof of Apéry on the irrationality of $\zeta(3)$, Bombieri turned his attention to effective irrationality measures and applications to Diophantine equations. This is what he was working on when I visited the Institute. Whenever Bombieri is pursuing a problem, he works on it with furious intensity and total concentration. I heard the following about Bombieri's amazing working habits: Bombieri feels that more progress can be made by working non-stop on a problem for 72 hours than working intermittently on it for several months. He sleeps when he is tired, gets up when he is rested, and eats when he is hungry. He does not have specific times for sleeping, getting up, or eating. At home, when he wants to focus on a problem, he would shut himself up in his study and not want to be disturbed for days together. Food would be left on a tray outside his study. Whenever he is hungry, he would open the door, take the food inside and eat. After he is finished, he would leave the tray outside and close the door of his study. I heard this story in amazement. When I was at the Institute, Julia Mueller, who had recently received her PhD from Columbia University under the direction of Patrick Gallagher, was Bombieri's assistant. She worked into the wee hours of the night for days together on irrationality just to keep up with Bombieri. I worked on irrationality only during my first academic year in Michigan, but after that I reverted back to analytic number theory. So at Princeton, I interacted mathematically more with Selberg than with Bombieri.

One day during a social event at the Institute, Mathura was wearing a necklace which had one of the opals we had bought from John Mack when he visited Michigan in 1980–81. That year Mack had briefly visited Princeton and got to know Bombieri. So in conversation with Bombieri at that event, I expressed great appreciation for the fine collection of opals that Mack had brought with him. Bombieri said that he saw Mack's collection. He said they were beautiful, but he preferred solid opals which are rarer, and Mack did not have them with him on his visit to the USA. Bombieri then told me that whenever he goes to Australia, he spends some time digging for solid opals in the Australian desert. So that is an example of how much of an expert Bombieri is at anything that he becomes interested in learning.

There is a fascinating story about Bombieri and the Riemann Hypothesis that I would like to relate now:

One day during the International Congress of Mathematicians (ICM) in Hensinki, in August 1978, Bombieri and Montgomery were taking a walk. During the walk Bombieri said that he expected the Riemann Hypothesis to be solved in three years. Hugh Montgomery had spent some time thinking deeply about the Riemann Hypothesis, and he responded saying it would not be solved in three years. So Bombieri and Montgomery had a bet. The wager was that the loser would host a very special dinner for the winner. Montgomery mentioned this wager between him and Bombieri in Fall 1978 during discussions after a number theory seminar in Ann Arbor.

I was in Princeton in Fall 1981 and so three years had passed by then without any news about the solution of the Riemann Hypothesis. I got a phone call from Montgomery saying that he will be at the Institute not only to give a seminar on his work but also enjoy the dinner that Bombieri was going to host for him. After Montgomery's seminar, we were all having lunch at the Institute cafeteria which had a reputation as being one of the best places to eat in Princeton. I was at this lunch with Bombieri, Montgomery and a few others. Naturally the conversation turned to fine dining and how good the Institute cafeteria was. At this point Montgomery said that one of the advantages of living in Ann Arbor is that Detroit is less than an hour away and there are some really good restaurants in Detroit. On hearing this, Bombieri immediately responded as follows: "I do not think there is any good restaurant throughout the United States." After saying this, he got up, left the table, and went to get coffee. While Bombieri was away getting coffee, I looked at Montgomery and said, "Hugh, how can it be that there are no high class restaurants in all of America? Surely New York City has some world class restaurants." To this Montgomery smiled and replied: "Krishna, you should understand that Bombieri has exceptionally high standards and is used to spending in the range of $1000 per person for a high class dinner!" When I got this response from Montgomery, I asked him if no restaurant in the United States is good enough, then where is Bombieri hosting this special dinner? To this Montgomery replied: "Oh, the special dinner will be at Bombieri's house!" (By the bye, Bombieri himself is an expert in cooking, especially Italian dishes.)

5.5 Luminaries galore at the Institute

In 1981–82, there was a special year-long program on Algebraic and Arithmetic Geometry at the Institute jointly organized with Princeton University. Fields Medalists Pierre Deligne from Paris and David Mumford from Harvard were the main organizers of the special program. This perhaps was the beginning of Deligne's long association with the Institute because three years later he was appointed as a Professor there.

Owing to the special program, the atmosphere of the Institute was charged with the presence of several mathematical giants. One day at Tea I saw seated on adjacent sofas in the lounge, about half a dozen of the greatest mathematicians in the world — Selberg, Deligne, Bombieri, Mumford, Weil and Langlands. I felt that so much brilliance was concentrated in so small an area that the Earth would cave in under the sheer weight of their intellect! I wish I had a camera and taken a picture of that ensemble of masterminds.

One of the visitors for the year owing to the special program in algebraic geometry was young Michael Rapoport from Paris. His office was in Building C and so I would see him regularly. Three decades later, in 2013 December, I had the pleasure of hosting him in India when he came to attend a conference in Ramanujan's hometown of Kumbakonam where his former PhD student Peter Scholze received the SASTRA Ramanujan Prize Another brilliant young visiting member in algebraic geometry was Shigefumi Mori from Japan. He too was in Building C. Mori's PhD advisor was the great M. Nagata, known for ingeniously constructing counter-examples to deep conjectures in commutative algebra. A few years later (in 1990) Mori won the Fields Medal at the International Congress of Mathematicians in Kyoto. Some of the other visiting members in algebraic geometry were David Gieseker of UCLA (whom I knew when I was in graduate school there), James Milne of Michigan, and Philip Griffiths of Harvard.

Owing to the special program, most of the visitors were in algebraic geometry and related areas. So there were just a few of us in number theory. Andrew Wiles from the neighboring campus of Princeton University was a visiting member. He too occupied an office in Building C. Besides Julia Mueller and Amit Ghosh, the three number theory visitors I interacted with were Professor K. G. Ramanathan (now deceased) from TIFR, Mark Sheingorn from Baruch College, CUNY, and Eira Scourfield from London.

Sheingorn lived in a lovely apartment in Manhattan which commanded a splendid view of the Empire State building, and he used to come to the Institute in the morning by train from New York and get back the same night. I had met Sheingorn previously at the 1973 Ann Arbor Number Theory Institute and at the West Coast Number Theory Conferences, but during the year at the Institute, I got to know him really well. He is a former PhD student of Marvin Knopp and works in the area of automorphic functions. Sheingorn is a charming personality and known among the number theory community as someone who would discuss controversial topics and get into long (friendly) arguments. He and I had our share of such discussions, and this only strengthened our friendship. He was keen on learning about Indian culture and so we had long conversations not only at the Institute but in New York City where we met when I would go to Manhattan from time to time. I even took him to a South Indian classical music concert at Columbia University and introduced him to that great art form. He expressed interest in visiting India, but that never happened even though I invited him for conferences that I organized there.

It was K. G. Ramanathan's third visit to the Institute, his first visit having been during 1948–51, and his second in 1961–62. He actually received his PhD from Princeton University in 1951 under the direction of Emil Artin. When he saw me at Tea soon after I had arrived, he came up to me, introduced himself, and the first thing he said was — "From where did you apply for a Visiting Membership at the Institute?" I understood the import of his question: The senior TIFR mathematicians generally keep abreast of the development of mathematics at centers of advanced research in India, and are usually aware of what is happening at these centers with regard to the work of various mathematicians. So if I had applied to the Institute for Advanced Study from MATSCIENCE, K. G. Ramanathan probably would have been aware of it. I responded to his question by saying that I had applied from the University of Michigan and not from India. But I told him that I had arrived in Princeton on leave from MATSCIENCE where I had joined in June 1981 after completing my term at Michigan. By the bye, while I was at Michigan, I was told by a senior Indian mathematician that each year the TIFR would send one person as a Visiting Member to the School of Mathematics of the Institute. The TIFR always had strong ties with the Institute for Advanced Study.

Besides the plethora of lectures in algebraic geometry, one of the main events that year was a course of lectures on the History of Number Theory by Emeritus Professor Andre Weil. He was a legend in mathematics and one of its leaders who commanded universal respect for his brilliant contributions and his vast knowledge of the subject. He was one of the founders of the famous *Bourbaki* group, whose mission was to write a series of self-contained textbooks on various branches of mathematics, starting from scratch, but reaching the high level of research work, with complete proofs and explanations. Weil became a permanent member at the Institute in 1958. His arrival at the Institute was much cause for celebration. In recalling the heady atmosphere at the Institute in 1957–58, my father had told me that many visiting members there used to talk about Andre Weil in great admiration. Weil commanded that respect from one and all throughout his illustrious career.

Weil started his history course by talking about the *Arithmetica* of Diophantus. Weil's lectures were accessible to all and everyone knew that the master was unmatched in thoroughness. In his study of the history of mathematics, Weil not only discussed how the results were proved chronologically, but he also emphasized how ideas evolved and influenced later generations. So the hall was packed and overflowing for his lectures. One day I saw Deligne sitting on the floor in the front since he was unable to find a seat. The course of history lectures Weil gave at the Institute became the basis for his book *Number Theory - an approach through history: from Hammurapi to Legendre*. After one of Weil's talks in his course, I had a conversation with him at Tea about mathematics in ancient India, and especially about Hindu mathematics because he was quite knowledgeable about Hindu religion and Sanskrit literature. I asked him how in India, the Hindu religious scholars are able to precisely fix the dates of eclipses and other astronomical occurrences. Weil replied that these Hindu calendars are prepared using Western calendars! He was very nice to me in conversation and this was a pleasant surprise because I had heard stories as to how Weil would cut you down in size. Later I was told that he was always nice to Indian mathematicians because of his great interest in, and regard for, Hindu philosophy, and Sanskrit literature.

Weil was deeply interested in Indian culture and studied Sankrit in Paris right from his youth. He enthusiastically accepted his very first job at the Aligarh Muslim University in India in 1930. It gave him also an opportunity to see the length and breadth of India. In connection with Indian philosophy and culture, I must point out that Weil had the greatest regard for the *Bhagavat Gita* which is the teaching of Lord Krishna to the warrior prince Arjuna in the great epic, the *Mahabharatha*. During World War II, Weil was imprisoned for not joining the French army for being a conscientiously objecting to war. He spent time in prisons in Finland and France, and even then he had the Bhagavat Gita with him, from whose teachings he drew courage in his convictions. Of course what is remarkable is that, even though he was in prison, his mathematical productivity and excellence remained unabated. Weil was both a world class mathematical researcher and mathematical historian, and I felt fortunate to have heard his lectures at the Institute.

Another mathematical luminary at the Institute with whom I had contact was Professor Harish-Chandra. He very kindly invited Mathura and me to dinner at his house. Professor K. G. Ramanathan and his wife were also invited and I gave them a ride to the Harish-Chandra home. Professor Harish-Chandra was pleased to hear about my academic progress and that I was having regular discussions with Selberg whom he admired so much. He said that he was very pleased with the appointment of Robert Langlands as Professor. One of Langlands' initial great achievements was to adapt methods of Harish-Chandra in representation theory to the theory of automorphic forms.

Harish-Chandra is widely regarded as one of the most eminent mathematicians of the twentieth century and the greatest Indian mathematician after Srinivasa Ramanujan. He was IBM von-Neumann Professor at the Institute. It seems he was considered for the Fields Medal in 1958. The story as to why he did not get it is told by Robert Langlands in an obituary article that appeared in the Biographical Memoirs of the Fellows of the Royal Society: "He was considered for the Fields Medal in 1958, but a forceful member of the selection committee in whose eyes Thom was a Bourbakist, was determined not to have two. So Harish-Chandra, whom he placed in the Bourbaki camp, was set aside."

Harish-Chandra, like all other permanent members of the Institute, worked with great intensity, but it took a toll on his health. He had heart problems and died in 1983 at the age of 60. In 1983 there was a conference at the Institute in honor of Armand Borel and for that conference, Harish-Chandra and his wife graciously hosted a party at their home. After the guests left, Harish-Chandra went out for his customary walk but he collapsed while walking and died!

Another distinguished senior Indian mathematician I got to know well during my stay in Princeton was Sarvadaman Chowla. He was retired by then, but was in Princeton in 1981–82, although not as a Visitor in the formal sense. Chowla was born in England in 1907 but grew up in Lahore which is now in Pakistan, but was then in United India under the

British rule. After getting a Masters degree in Lahore, he went to Cambridge University to get his PhD under the direction of J. E. Littlewood in 1932. Although he returned to Lahore after his PhD, he left for the USA in 1947 when the partition of India took place and there was a lot of sectarian violence. He was in the USA since then and held faculty positions at the University of Kansas, the University of Colorado, and Penn State University, from where he retired in 1976. He made notable contributions to several parts of number theory and held the distinction of being the lone collaborator of Selberg!

Chowla visited the Institute many times and was there in 1957–58 when my father was a Visiting Member. Thus my parents got to know Chowla very well. One common topic for discussion for them was the sterile atmosphere in Indian universities. My father wanted to change this in Madras and the Princeton experience inspired him to do just that.

My conversations with Chowla were about mathematics since we were both number theorists. He used to ask how my discussions with Selberg went. He attended my talk in the Visiting Members Seminar and made positive comments. Chowla was a very genial person with a self-effacing attitude. He was perhaps overly modest about his accomplishments. He was extremely generous and hosted dinners regularly at a French restaurant in downtown Princeton where the wine flowed freely much to the delight of his guests. Mathura and I have enjoyed this hospitality, but we were the least expensive guests for him since we were tee-totallers.

Freeman Dyson, a legend in the world of physics, was Professor Emeritus in the School of Natural Sciences. My father had interacted with Dyson at the Institute in the 1950s when they both were very young, and he had spoken very highly to me of Dyson. Professor Dyson used to come to Tea regularly and I had a few conversations with him there. He remembered discussing with my father in the 1950s and also writing a letter to me advising me to take to physics instead of number theory. Mathura and I interacted with Dyson non-mathematically as well. Dyson's daughter Rebecca who was 14 years old then, did some baby sitting in her spare time. So when Mathura and I had to attend a dinner such as at the Harish-Chandra's, or a social event such as a party at the Institute or the University, we utilized Rebecca's help to babysit our infant daughter Lalitha. On such occasions, Professor Dyson would himself drop Rebecca in the evening at our residence on Einstein Drive and pick her up later at night.

Each year at the Institute there is the Visiting Members Seminar and this is usually organized by one of the Permanent Members. It met regularly to accommodate talks by all visiting members. In 1981–82, the organizer of this Seminar was S. T. Yau. I had known Yau as a visitor at UCLA when I was a graduate student there, and now he was a Professor at the Institute. It was a pleasure to renew contact with him and he remembered the sieve theory course at UCLA that was given by Miech at my request — a course that he attended. At the Visiting Members Seminar, I spoke about work I was doing at Princeton on a new multiplicative generalization of the sieve, and I will briefly describe this next.

5.6   My research at the Institute and seminar presentations

During my final year in Ann Arbor, I purchased the comprehensive two-volume book [B7] on "Probabilistic Number Theory" by P. D. T. A. Elliott that was published in the prestigious Grundlehren Series of Springer. After Erdős, Kac, Turan and Kubilius, the four founders of Probabilistic Number Theory, Peter Elliott of the University of Colorado, Boulder, is the premier expert on the subject, and so this book was an authoritative treatment. After working on integers with restricted prime factors at UCLA and at Michigan, my interest broadened more generally to the distribution of additive functions and certain multiplicative functions derived therefrom. So I was lucky that just as I was turning my attention to Probabilistic Number Theory, the two-volume book by Elliott appeared. I was also very much attracted by a paper of Elliott that had appeared in 1980 in the Canadian Journal of Mathematics in which he had obtained a uniform upper bound for higher moments of additive functions, thereby extending the fundamental Turan-Kubilius inequality. I took Elliott's book and his paper with me to India in the summer of 1981 and decided to study them in detail as a preparation for my research in Princeton.

On going through Elliott's book, I noticed that while there was considerable literature on the distribution of additive functions on all integers, comparatively little was known regarding distribution is subsets. It occurred to me that by using a multiplicative generalization of the sieve, one could discuss the distribution of additive functions in subsets, and extend Elliott's inequality to such subsets as well. While these ideas first occurred to me in India in the summer of 1981, I worked out the details only while I was at the Institute in 1981–82.

In a standard sieve, one starts with a finite set of integers $\mathcal{A}$, and successively deletes members of $\mathcal{A}$ which are multiples of primes p taken from a set of primes $\mathcal{P}$. If one performs the sieve process on an infinite set $\mathcal{S}$, then one truncates $\mathcal{S}$ at x, and this truncation is the set $\mathcal{A}$ under consideration.

A typical sieve problem is to estimate $S(\mathcal{A}, \mathcal{P}, y)$, the number of integers in $\mathcal{A}$, which have no prime factors $p < y$ from $\mathcal{P}$. Ideally one would want an asymptotic estimate for this quantity, but one often settles for good upper and lower bounds. Many classical sieve questions like the Goldbach and Prime Twins problems can be cast in terms of $S(\mathcal{A}, \mathcal{P}, y)$ by suitable choice of $\mathcal{A}$, $\mathcal{P}$, and y.

To get the multiplicative generalization of the sieve, I began by considering a strongly multiplicative function $g(n)$, namely one that satisfies

$$g(n) = \prod_{p|n} g(p)$$

but also with $0 \leq g \leq 1$. I attached to this g its *dual* g^*, namely the strongly multiplicative function $g^*(n)$ defined by

$$g^*(n) = \prod_{p|n} g^*(p), \quad \text{where} \quad g^*(p) = 1 - g(p).$$

My multiplicative generalization of the sieve is as follows. Start with a set $\mathcal{A}$, and for each $n \in \mathcal{A}$, begin with weight 1. Instead of removing all multiples of a given prime p from $\mathcal{A}$, replace the weight 1 by weight $g(p)$ for the multiples of p. So the multiples of p have not been removed, but $g^*(p)$ has been shaved off from their weight. Next consider multiples of another prime q. Now shrink all weights by a factor $g(q)$, and so on. My general sieve problem is to estimate

$$S_g(\mathcal{A}, y) = \sum_{n \in \mathcal{A}} g_y(n), \quad \text{where} \quad g_y(n) = \prod_{p < y} g(p).$$

This reduces to the classical sieve problem by choosing $g(p) = 0$ for $p \in \mathcal{P}$ and $g(p) = 1$ for $p \notin \mathcal{P}$.

Sieve theory began with the pioneering work of the Norwegian mathematician Viggo Brun who showed that the sum of the reciprocals of the twin primes is convergent. In contrast, the sum of the reciprocals of the primes is divergent. Atle Selberg was influenced by Brun's work on sieves.

By using the Moebius function, one can express the typical sieve sum in the form

$$S(\mathcal{A}, \mathcal{P}, y) = \sum_{n \in \mathcal{A}, (n, \mathcal{P}_y) = 1} 1 = \sum_{n \in \mathcal{A}} \sum_{d | (n, \mathcal{P}_y)} \mu(d),$$

where $\mathcal{P}_y$ is the product of the primes in $\mathcal{P}$ up to y. While this is an exact identity, it is not of real use because the number of error terms generated it so large that it becomes unwieldy. Brun's brilliant idea was to choose functions χ_1 and χ_2 that would satisfy

$$\sum_{d|n} \mu(d)\chi_1(d) \leq \sum_{d|n} \mu(d) \leq \sum_{d|n} \mu(d)\chi_1(d), \quad \text{(BRUN INEQUALITIES)}$$

and then use these inequalities with n replaced by $(n, \mathcal{P}_y)$ to get upper and lower bounds for $S(\mathcal{A}, \mathcal{P}, y)$. The idea was to choose the χ_i to take the value 0 often enough to keep the errors terms under control, and he was very cleverly able to come up with such χ_i. Brun's sieve method yielded asymptotic estimates for $S(\mathcal{A}, \mathcal{P}, y)$ as long as $\log |\mathcal{A}|/\log y$ was large.

Analogous to the exact identity for $S(\mathcal{A}, \mathcal{P}, y)$, there is the more general exact identity

$$S_g(\mathcal{A}, y) = \sum_{n \in \mathcal{A}} \sum_{d|(n, \pi_y)} \mu(d) g^*(d),$$

where π_y is the product of all primes less than y. In connection with Brun's sieve, I noticed the following very interesting and useful *monotonicity principle*. If χ_1 and χ_2 are any pair of Brun sifting functions satisfying the Brun inequalities, then for ALL multiplicative functions g that satisfy $0 \leq g \leq 1$, we also have

$$\sum_{d|n} \mu(d)\chi_1(d)g^*(d) \leq \sum_{d|n} \mu(d)g^*(d) = g(n) \leq \sum_{d|n} \mu(d)\chi_2(d)g^*(d),$$

where the dual g^* also satisfies $0 \leq g^* \leq 1$. What this implied was that the same sifting functions χ_1, χ_2 of Brun could be used to get asymptotic estimates for the multiplicative sieve sum $S_g(x, y)$ when $\log |\mathcal{A}|/\log y$ is large. Moreover, the asymptotic estimate would hold uniformly for ALL $0 \leq g, g^* \leq 1$. This uniformity had significant consequences. By writing

$$g(n) = e^{-u(f(n)-A(x)/B(x)},$$

I was able to estimate the bilateral Laplace transform of the distribution of a positive additive function about its mean $A(x)$ with variance $B(x)$. The bilateral Laplace transform enabled me to estimate the moments of the distribution and from this I was able to get the Erdős-Kac theorem (and the Kubilius theorem in certain cases) by the method of moments.

The first person to prove the Erdős-Kac theorem by the method of moments was Halberstam, but his method involved painstaking calculations. The asymptotic estimation of the moments was subsequently simplified by Delange by using generating functions. My approach utilized the monotonicity principle. Because I used sieves, my results were valid for a wide class of sets on which the sieve could operate. I was also able to the more general Kubilius type result with non-Gaussian limiting distributions and felt pleased to have completed this work at the Institute.

I used to go to Selberg's office to discuss my research with him, and this multiplicative generalization of the sieve was one of the topics of our discussion. Selberg had formulated a *monotonicity principle* for the classical sieve, and it was Montgomery who drew my attention to Selberg's monotonicity principle toward the end of my stay in Ann Arbor. It was daunting for me to tell Selberg that I had my own monotonicity principle, but I survived! In fact he was quite encouraging.

Although Selberg did not study the distribution of general additive functions, he had actually done very fundamental work in the fifties on the *local distribution* of the number of prime factors and provided a more natural and powerful approach to the problem than Sathe, as I had mentioned briefly in §4.4. Now I provide some details as to how Selberg took up the study of Hardy's problem solved by Sathe.

In four long papers, Sathe used Landau's approach to evaluate $\nu_k(x)$) (the number of integers up to x with precisely k prime factors) by induction on k; this inductive process was excruciatingly complicated to carry out for k varying with x, but Sathe was brave enough to face the difficulties and go through all the details. Sathe submitted his paper to the Transactions of the AMS and Selberg was the referee for Sathe's paper. On going through Sathe's work, Selberg realized that the natural way was to attack the problem analytically. Selberg noted that one should first estimate the sum $S_z(x)$ (see Chapter 4) asymptotically for complex z. Once this is done, then $N_k(x)$ is to be evaluated using the Cauchy integral formula

$$\nu_k(x) = \frac{1}{2i\pi} \int_C \frac{S_z(x)dz}{z^{k+1}},$$

by choosing the contour C optimally. In doing so, Selberg not only obtained a sharpening of all of Sathe's results much more swiftly, but also was able to explain why such asymptotic estimates for $\nu_k(x)$ hold only for $k \leq C \log\log x$ with $C < 2$ in the case $\nu = \Omega$.

Selberg's referee's report was five pages long. In that report he outlined his analytic approach and the results that can be derived therefrom. Somehow, Sathe's paper intended for the Transactions of the AMS, was published in the Journal of the Indian Mathematical Society in 1955. I do not know how this was arranged. The Editors of the Indian journal decided to also publish Selberg's five page referee's report as a paper entitled "Note on a paper of L. G. Sathe", in the very same issue where Sathe's four long papers appeared! The decision to publish Selberg's referee's report as a separate paper was immensely beneficial to the mathematical world because of the beautiful and powerful approach he discovered, but it definitely was a blow to Sathe (see paragraph below). Referee's reports are confidential and the author is not told who the referee is. The referee's name is divulged to the author in the rare occasion when the referee's remarks improve the paper so substantially that the author requests the referee to be a coauthor. While Selberg's observations provided a new approach to Hardy's problem, Selberg and Sathe were not coauthors, and so the publication of Selberg's report as a paper did not help in boosting Sathe's spirits even though he had solved a major problem proposed by Hardy.

In my investigations of the Erdős-Kac theorem among the uncancelled elements in the Sieve of Eratosthenes, I had discussed the local distribution of the number of prime factors. So in my meetings with Selberg in his office, I discussed with him his note on Sathe's work. It was then that Selberg told me the story about him being the referee for Sathe's paper. I asked Selberg what happened to Sathe after he solved Hardy's problem? Did he publish anything else of significance? Seated at his work desk, Selberg looked out of his office window at the sprawling lawn outside Fuld Hall, and said laughingly, "After seeing my simple and natural approach to the problem, Sathe must have given up doing mathematical research!"

I presented my work on the multiplicative generalization of the sieve and its applications to probabilistic number theory in the Visiting Members Seminar at the Institute in Fall 1981. In April 1982, toward the end of my stay in Princeton, I was invited to give a talk on this work at the New York Number Theory Seminar which met at the Graduate Center of the City University of New York (CUNY) in Manhattan. That seminar was launched in January 1982 and so I was a speaker in its very first semester of operation. The seminar was organized by Gregory and David Chudnovsky of Columbia University, Harvey Cohn of the Graduate Center of CUNY, and Mel Nathanson, who in Fall 1981 moved to Rutgers University in New Jersey from Southern Illinois University in Carbondale. The collection of lectures given each year in that seminar from 1982 to 88 were published in Springer Lecture Notes in a series of volumes, and so my paper appeared in the Proceedings of the 1982 New York Number Theory Seminar. Interestingly, since my last name starts with the alphabet A, my paper is the opening article in the very first proceedings of that famous seminar! The Chudnovskys told me that while they were very pleased to have my paper on moments of additive functions and sieve methods in their proceedings, they would be also happy to see me return to Diophantine Approximations.

During the year in Princeton, I made several visits to Bell Labs in nearby Murray Hill for discussions with Andrew Odlyzko and Jeff Lagarias who are highly reputed number theorists. I benefited immensely by discussions and correspondence with Odlyzko on the behavior of the solutions to classes of difference-differential equations that I was investigating.

I also had an opportunity in Fall 1981 to present my research work in a Colloquium at Penn State University. Next I will describe that visit because it was the start of a great lifelong association with George Andrews.

5.7   My first meeting with George Andrews — Fall 1981

In Fall 1981, while I was in Princeton, I had an opportunity to visit The Pennsylvania State University to give a colloquium there at the invitation of George Andrews. That was the start of a friendship which has grown in strength over the years and eventually led to some very significant collaborative work. It is also an association which has transformed my academic life comparable to the way in which my contact with Paul Erdős shaped my career. So I will describe how that association with Andrews began.

George Andrews has been the unquestioned leader in the theory of partitions for over half a century. A partition of a positive integer is a representation of that integer as a sum of positive integers, two such representations being the same if they differ only in the order of the parts (= summands). Thus 2+2+1 is the same partition of the integer 5 as 2+1+2. When we go to a bank and ask for cash for $1000, the clerk asks in what denomination we would like that. What the clerk is asking is the partition of 1000 into hundreds, fifties, twenties, tens, fives and ones, that we want. Thus partitions occur in daily life, but the layman does not recognize that. The theory of partitions was founded by Euler who showed how generating functions can be used to establish identities connecting partition functions defined in very different ways. The next major advance was more than a century later by Sylvester who showed how combinatorial methods can be used to prove various partition identities. Then in the early twentieth century, under the magic touch of the Indian genius Ramanujan, the subject underwent a glorious transformation; Ramanujan established important and surprising connections between partitions and the theory of modular forms. Starting in the sixties, Andrews systematically studied different classes of partition functions, their analytic representations as series and products, their connections with different fields within and outside of mathematics, and in a decade, established himself as the world leader in the subject. I have often said that the theory of partitions can be broadly divided into four eras — (i) the Era of Euler, the Founder, (ii) the Era of Sylvester, (iii) the Era of Ramanujan, and (iv) the Era of Andrews that began in the 1960s showing how rich the study of various types of partitions is and how they are connected with fundamental questions in other areas such as physics. In 1976, Andrews published his "magnum opus" — his book on partitions [B3] as Volume 2 of the Encyclopedia of Mathematics (John Wiley). It is really an encyclopedia, for it contains everything that was known on partitions at that time, and therefore is widely viewed as The Bible in the field. In 1976 Andrews also "discovered" Ramanujan's Lost Notebook at the Wren Library in Cambridge University while going through papers in the Watson Estate in that hallowed library. Andrews had actually done his thesis work on Ramanujan's mock theta functions and therefore could instantly recognize certain loose sheets in the Watson Estate as the Lost Notebook of Ramanujan since they contained incredible results on mock theta functions that the Indian genius had communicated to G. H. Hardy in January 1920, a few months before his death. (For a brief description of Ramanujan's life and contributions, see Section 7.2 on the Ramanujan Centennial.) Thus Andrews was the right person to "discover" Ramanujan's Lost Notebook, and certainly the most appropriate expert to investigate the identities in the Lost Notebook which he did in the next decade. I consider it as Providential that Andrews was the one destined to discover the Lost Notebook and would even go further to say that the Goddess of Namakkal who inspired Ramanujan in his dreams with formulae, must have had a role in this! While Andrews was very well known in the 1960s among number theorists as a partition expert, his discovery of Ramanujan's Lost Notebook in 1976 and the series of papers that he wrote on the identities contained therein, catapulted him to a celebrity status the world over. In 1981–82 he was made Evan Pugh Professor of Mathematics at Penn State, and also was appointed Chair of Mathematics that year for a two year term. Penn State held him in such high esteem, that during the half-time of their football games, they had clips of George Andrews explaining the Rogers-Ramanujan identities to television audiences!

The Evan Pugh Professorship is the highest honor that can be bestowed on a faculty member at Penn State University. These professorships were launched in 1971. As per the charter, "the number of active Evan Pugh Professorships University-wide shall not exceed one percent of the total number of standing academic appointments". So it was an exclusive and fitting honor that was bestowed on Andrews.

In early August 1981, while I was in Madras preparing for my departure to Princeton, my father told me one day, that he had received a letter from Professor G. P. Patil of the Department of Statistics at Penn State, that George Andrews would be coming to India in October for a few weeks and would be interested in visiting Madras. Andrews' visit was being coordinated by Professor Abdi, Head of the Department of Mathematics at the University of Cochin. That was Andrews' first visit to India and it was triggered by the invitation to speak at a conference of the Indian National Science Academy in Cochin, in the southernmost state of Kerala, adjacent to state of Tamil Nadu where Madras is. Andrews also utilized the trip to visit several academic centers in India, but was particularly interested in visiting Madras to meet Mrs. Janaki Ramanujan who was about 80 years old. When I heard from my father about Andrews' proposed visit, I immediately wrote to him saying that he should visit MATSCIENCE, give one or two lectures, and that my father, as Director, would help in making arrangements to meet his requirements. I also said in my letter that in October I will be in Princeton and not in Madras, but I had spoken to my father and others at MATSCIENCE concerning arrangements for him.

Andrews gave two talks at MATSCIENCE which were attended by faculty and PhD students from other institutions as well, such as the Ramanujan Institute and the IIT. Andrews is a brilliant and charismatic speaker, and so he had a tremendous impact on the Madras audience. There were reports in the local newspapers about his Madras lectures and about his research work, especially because it related to Ramanujan. My father took him to our family home *Ekamra Nivas* and showed him the lecture hall where the Theoretical Physics Seminar was held and how it was responsible for the creation of MATSCIENCE. My father also arranged for Andrews to meet Janaki Ramanujan, and for this, Andrews is most grateful.

The visit to MATSCIENCE, our family home, and meeting Janaki Ramanujan, made a big impression on Andrews. So upon return from India, he called me at Princeton, said that all arrangements in Madras were perfect, and invited me to a Colloquium at Penn State in mid-November. The colloquium invitation could not have come at a better time, because Mathura was arriving from India with our infant daughter Lalitha in late November, and I was planning on going to Ann Arbor in my car to pick up all our household goods we had left there at the home of Sriram and bring them to Princeton before her arrival; so I said that I would stop in Penn State on my way back from Ann Arbor to Princeton in mid-November.

Andrews is a gracious host anytime, but as Chairman, he rolled out the red carpet for me. He put me up at the stately Nittany Lion Inn on campus, and he and his wife Joy hosted a special party for me at their house after my colloquium that was attended by several of his colleagues. Penn State University is in State College, but Andrews lives out in the country, in an area called "Happy Valley", more than half an hour away. His beautifully decorated house had interesting wall hangings relating to Ramanujan.

My colloquium at Penn State was on my new work on the multiplicative generalization of the sieve. Little did I realize then, that about a decade later, influenced by Andrews' lectures at the Ramanujan Centennial, I would shift my area of research to the theory of partitions and collaborate with him in a major way. During my brief visit to Penn State, Andrews told me several times how happy he was that he met Mrs. Ramanujan, and how impressed he was on seeing our family home — the womb of MATSCIENCE. He mentioned specifically the architecture of our family home *Ekamra Nivas* that my grandfather built, and the gigantic teak beams that supported the ceiling — these teak beams are from Burma, and elephants were used to transport them! Thus began a most rewarding friendship and association with Andrews which eventually led to a significant collaboration. The manner in which this collaboration began in Fall 1990 when he visited Florida at my invitation is a fascinating story (see Section 7.4). In summary, my visit to Princeton in 1981–82 not only resulted in one of the most productive academic years of my career, but also put me in close touch with some of most gifted mathematicians of our generation.

5.8   A year at the University of Texas, Austin (1982–83)

Although I was supposed to get back to MATSCIENCE after my one-year stay in Princeton, a wonderful opportunity arose for me to visit the University of Texas in Austin for the year 1982–83, and so I availed this opportunity and returned to India only in 1983. The University of Texas had made a strong commitment to analytic number theory with appointments in 1979 and 80 in rapid succession of Jeffrey Vaaler, Sid Graham, and Gregory Kolesnik. Vaaler had received his PhD from the University of Illinois in 1974 under the direction of Harold Diamond. Sid Graham had received his PhD in 1977 at the University of Michigan under the direction of Hugh Montgomery. Soon after joining the University of Texas, Graham and Vaaler forged a major collaboration studying certain extremal functions in Fourier analysis and their implications pertaining to the large sieve; this was inspired and influenced by the seminal work of Arne Beurling and Atle Selberg. Kolesnik, who had the strongest results on the famous Dirichlet Divisor Problem, had emigrated to the USA from USSR, and had a visiting position at UCLA with the help of my advisor Straus, before moving to Texas. In the summer of 1982, Graham, Kolesnik, and Vaaler, conducted an eight week Summer Institute in Austin, similar in spirit to the Summer Institute in Ann Arbor that I attended in 1973. I was invited to this Summer Institute in Austin as one of three post-doctoral participants will full support for the eight week period, the other two being Amit Ghosh my office mate in Princeton, and Brian Conrey, who had recently received his PhD from Michigan under the direction of Hugh Montgomery. During the academic year 1982–83, Jeffrey Vaaler was going to be away from Texas on a Visiting Membership at the Institute for Advanced Study in Princeton, and so the University of Texas offered me a Visiting Associate Professorship for the year 1982–83 utilizing the salary made available by his absence. Thus I accepted the offer to spend the period May 1982–July 1983 at the University of Texas. MATSCIENCE generously extended my leave. Since the new J-visa that I had obtained with great difficulty to get to Princeton had a three-year duration, there was no problem continuing to remain in the USA for another year on that visa.

My parents arrived in Princeton in April toward the end of my stay there and planned to come to Austin and be with us through the summer. My father had always dreamt of a sightseeing trip by car across the United States. Since we had to drive from Princeton to Austin, his desire was partly realized — I say partly because we travelled not across all of USA but through a broad section of the USA. His desire was fully realized, when we made trip from Austin to the Grand Canyon area by car seeing Zion and Bryce National Parks as well during the two week period between the Summer Institute and the start of classes for the Fall Term in 1982. I felt happy to quench the thirst of my father for such a trip because he had taken me around the world so many times and this was one small way for me to reciprocate and show him around.

For the Number Theory Institute there were three main speakers who gave a course of ten lectures each. They were Henryk Iwaniec of the Polish Academy of Sciences, Heini Halberstam of the University of Illinois, and Hugh Montgomery of the University of Michigan.

Henryk Iwaniec was a new phenomenon in analytic number theory. He had received his PhD in 1972 in Warsaw under the direction of the famous number theorist A. Schinzel. Iwaniec had proved a number of impressive results in analytic number theory and sieve theory in rapid succession and was transforming the field. So he was offered a distinguished position at Rutgers University that he accepted in 1983.

Sieve theory is a technically difficult topic, and very few can bear to go through the terse details. But Heini Halberstam was an expositor par-excellence. His delivery was like that of an orator, and so not only did he make sieve theory palatable, but very enjoyable. Having benefited immensely by his monumental book on sieves co-authored with H.-E. Richert, it was a real pleasure for me to listen to the master deliver a course of lectures on sieves. I got to know Halberstam really well during the Summer Institute, and was able to convince him that he should visit MATSCIENCE in the year 1983–84 when I would be back in India after my term in Austin.

One of the leading participants was Dorian Goldfeld of Columbia University. He had actually received his Bachelors degree from Columbia, as well as his PhD there under the direction of Patrick Gallagher. So he was a Columbia boy all along. He brought to the Summer Institute a brilliant young number theorist — Daniel Bump, who after receiving his PhD from the University of Chicago that year, was living in the mountains of Oregon and working in seclusion on fundamental questions in number theory. Bump's contact with Goldfeld transformed his life. Goldfeld encouraged him, and introduced him as a new force in number theory at many conferences. Bump eventually became a professor at Stanford University, but it was Goldfeld who brought him out of seclusion into prominence. Goldfeld's presence in Austin attracted the attention of the mathematics department, and so he was offered an Associate Professorship with tenure at the University of Texas, which he accepted in 1983. But then when Columbia University offered him a tenured Full Professorship in 1985, Goldfeld returned to Columbia where he has been since then.

It was also at this Summer Institute that I got to know Peter Sarnak quite well because of several discussions I had with him. He was there for a couple of weeks and gave two lectures. He is a world class mathematician, but what impressed me the most was his radiant optimism and confidence — not arrogance, but confidence! By just talking to him, you feel enthused to excel in your research. He and Dorian Goldfeld were quite close to Selberg, by which I mean, that they regularly discussed with Selberg. They both had lots of amazing Selberg stories to tell, and simply hearing those stories makes you want step up your own research work. When you hear Sarnak imitate Selberg, not only with the accent and intonation, but also with hand gestures, you feel that Selberg is talking to you. I can attest to that, having had discussions with Selberg at the Institute. I do a pretty good imitation of Erdős, but Sarnak's imitation of Selberg is better! At the time when I met Sarnak in Austin, he was perceived as a rising star, just a few years after his 1980 PhD at Stanford under the direction of Fields Medalist Paul Cohen, with a great future ahead of him. Indeed as expected, he has become one of the most influential number theorists of our generation with a panoramic view of number theory and related areas. His eminence falls lightly on his shoulders; he is easy to talk to and very encouraging. No wonder, so many of the most brilliant graduate students in number theory gravitate towards him to do their PhD.

After the conference, I spent the full academic year 1982–83 at the University of Texas, where I taught two courses each semester like a regular faculty member, and continued the research that I had begun in Princeton. The experience of working in a large public university in southern United States that was steadily rising in reputation and performance was very helpful for me later when I joined the University of Florida and was Chairman there for a decade.

The mathematics department at the University of Texas in Austin was housed in the tallest building (RLM) on campus named after Robert Lee Moore, a very influential topologist and a distinguished faculty member of the University of Texas for many years. Owing to its size, RLM actually housed the departments of physics and astronomy besides mathematics. One would therefore think that there were strong ties between the mathematics and physics faculties. Not really. The joke was that the mathematicians and physicists only met in the restrooms!

The physics department was proud to have on its faculty two Nobel Laureates — Steven Weinberg who won the Nobel Prize in 1978 when he was at Harvard, and Ilya Prigogine, a Belgian who received the Nobel Prize in Chemistry in 1977. Besides these two stars, the physics department was also proud to have George Sudarshan as a distinguished faculty member. Sudarshan had moved to Texas from Syracuse in 1969. By offering high salaries, the University of Texas attracted very eminent scholars from ivy league level institutions and steadily rose in reputation.

During the year I was there, a very wealthy lady donated $8 million to the University of Texas to create distinguished professorships. The University matched this with $8 million and then the State of Texas matched that with $16 million! So instantly there was $32 million to form endowments of $1 million each for 32 distinguished professorships to be distributed to several departments. There was a lot of excitement in the University about this endowment — an excitement shared by the mathematics department. It was with such an endowment that the mathematics department in Austin lured the great John Tate away from Harvard in 1990, Karen Uhlenbeck from Chicago in 1988, and more recently Luis Caffarelli.

In Austin, I worked on extensions of the method I had begun studying at Princeton on a new approach to the moments of additive functions. I investigated how the method would apply to the distribution of additive functions over the set of shifted primes, namely the set S_a of integers of the form $p + a$, where a ia an arbitrary but fixed non-zero integer and p runs over the primes. Previously, asymptotic estimates for the moments were established in the case when the limiting distribution is Gaussian, but no such results for non-Gaussian limiting distributions were known for S_a. I noticed that it was crucial to utilize an inequality of the type

$$\sum_{d|n} g(d) \le c \sum_{d|n, d < \sqrt{n}} g(d),$$

where the implicit constant c is uniform for ALL multiplicative functions g satisfying $0 \le g \le 1$. Clearly the inequality holds with $c = 2$ when $g(n) = 1$ for all n, because then it is simply an inequality for the number of divisors of n, and we know that half of the divisors of n are $\le \sqrt{n}$. But then my monotonicity argument showed that if the inequality holds for $g = 1$, then it holds with the same c for all $0 \le g \le 1$. Even though I did this work in Austin, I wrote it up only in 1983 after I returned to India, and had it published in the Straus Memorial Volume of the Pacific Journal of Mathematics in 1984.

During my stay in Texas, I visited Canada briefly in Fall 1982 in response to invitations from Professor M. V. Subbarao for a Colloquium at the University of Alberta in Edmonton and from Professor David Boyd for a Colloquium at the University of British Columbia in Vancouver.

Professor Subbarao was a contemporary and a good friend of my father. Subbarao was two years older to my father and also had graduated from the Presidency College in Madras but with a degree in mathematics instead of physics. Subbarao was born in Andhra Pradesh, and so he and my father conversed quite freely in Telugu, the language of that state. Subbarao research spanned two areas: (i) the theory of arithmetical functions and (ii) the theory of partitions. He visited MATSCIENCE in 1972 and gave a talk on partitions which I attended. Indeed, that was long before I met George Andrews and other partition theorists, and so Subbarao's talk was the very first lecture on partitions that I heard.

There is a very famous conjecture on partitions due to Subbarao that bears his name, namely that in every arithmetic progression $An + B$, the partition function takes even and odd values infinitely often. This apparently simple assertion is very difficult to prove. Ken Ono famously proved that even values are taken by the partition function infinitely often in every arithmetic progression. For odd values, Ono showed that if in an arithmetic progression, there is at least one odd value, then there are infinitely many! The odd case was finally settled by Silviu Radu.

Subbarao was a very kind person. While in Madras in 1972, he invited me to his residence and patiently explained the significance of many fundamental results and conjectures in number theory such as the Jacobi Triple Product Identity for theta functions, and the Riemann Hypothesis. When he heard that I was visiting the University of Texas for a year, he kindly invited me to the University of Alberta for a colloquium. Upon arrival at Edmonton airport, I was told by the immigration officials, that I needed a visa to enter Canada but did not have one. In the sixties and even when I was in Ann Arbor (1978–81), there was no difficulty entering Canada with an Indian passport and a temporary US visa such as the J-1 that I had. I was unaware that rules had changed that required Indian passport holders without a US Green card to get visas to enter Canada. The immigration officer realized that I was coming to Canada to give lectures at the invitation of their professors. So he kindly agreed to give me a visa on the spot, but he said that required paperwork which would take about an hour to complete. Subbarao who had come to the airport to pick me up, patiently waited for more than an hour until I received the visa, even though it was nearly 10:00 pm by the time I got out of immigration. That was how kind he was. Even though he put me up in hotel in Edmonton, he took me home both evenings and his wife cooked delicious South Indian vegetarian meals for me. I was touched by the kindness that he and his wife showed. In a small way, I was able to reciprocate the hospitality when Subbarao attended a number theory conference in Florida during my tenure as Chair.

From Edmonton I went to Vancouver for a Colloquium at the University of British Columbia. There my host David Boyd was equally gracious. As a graduate student at UCLA, I met him there when he came to deliver a colloquium at the invitation of David Cantor. I remember the beautiful lecture Boyd gave on Pisot-Vijayaraghavan (PV) numbers and Salem numbers. PV numbers are algebraic integers > 1 and have all their conjugates within the unit circle. An immediate consequence is that the sequence of positive integral powers of PV numbers will tend arbitrarily close to the integers, and so they are very interesting from the point of view of Diophantine Approximation. Salem numbers are like PV numbers except that at least one of the conjugates must have absolute value 1. Salem numbers are important in analysis, and Boyd is an authority on Salem numbers. Boyd was one of the stars of the UBC mathematics department.

I was also invited in Spring 1983 to give a Colloquium at Oklahoma State University in Stillwater. Just as the University of Texas made a big move in number theory with the appointments of Vaaler, Graham and Kolesnik within a span of two years, the Oklahoma State University decided to make a plunge in number theory by appointing four young number theorists — all to start in Fall 1983; they were Amit Ghosh, Brian Conrey, Alan Adolphson and Chuck Yager. Bus Jaco was the mathematics chairman at Stillwater and he believed in making strong commitments like this and straightaway have a viable group. This is not easy to do for a variety of reasons. Firstly, it is very difficult to get faculty to support of allocating all positions in a given year to the same field, when different faculty groups would have their own favorite candidates. Secondly, it is more difficult to convince the higher administration to support such a move. Finally, it is most difficult to get three of four young candidates to simultaneously accept the positions being offered. But Jaco was able to pull it off. He was a successful administrator who later became the Executive Director of the AMS.

In January 1983, I had the opportunity to present my new work on sieve methods and probabilistic number theory at the Annual Meeting of the AMS in Denver in a special session organized by Peter Elliott. The two senior number theorists at the University of Colorado in Boulder at that time were Wolfgang Schmidt and Peter Elliott. It was very gracious of Schmidt and his wife to host a party at their lovely home in Boulder for all the special session participants. Boulder is two hours away from Denver and so a special bus was arranged to take the participants to the party at Schmidt's home and bring them back to Denver. Schmidt is so distinguished, that he could have received an appointment at any ivy league level university. He told me that he and his wife loved the mountains where they enjoy hiking. With the Rocky Mountains in the backdrop, Boulder was perfect for the Schmidts.

At the AMS Meeting in Norman, Oklahoma, in March 1983, I presented the research I did at the University of Texas when Harold Diamond invited me to give a talk in a Special Session he was organizing there. It turned out that Paul Erdős was turning 70 a few days after that conference, and he was speaking at that special session as well. Thus an informal dinner was arranged to celebrate his 70th birthday and it was a pleasure to be part of that celebration. I was returning to Madras a few months later, and so I got the idea in Norman to organize a conference in India for Erdős' 70th birthday in January 1984, before he would turn 71. I planned to have that conference in the hill station of Ootacamund (Ooty) because Erdős had expressed his desire on his visit to India in 1981 to see Ooty on his next visit. Erdős graciously accepted my invitation to come to India in December 83–January 84.

Just as the special session I organized in Notre Dame in March 1981 was associated with the one hour talk of Harold Diamond, the Special Session that Diamond organized in Norman was connected with the one hour talk that Jeffrey Vaaler gave there. Vaaler spoke about his joint work with Graham on extremal functions in Fourier analysis. One of the other hour lectures in Norman was given by Michael Starbird of the University of Texas with whom I interacted not mathematically, but on the tennis court!

During my year in Texas, I played a lot of tennis. Michael Starbird on the mathematics faculty was a fantastic tennis player and the two of us played from time to time at a club where Starbird was a member. Although we had close matches, I never beat Starbird even once. The best I did was to split sets. Besides being an accomplished research mathematician, Starbird excelled as a teacher and expositor.

It was through tennis that I got to know quite well the famous Indian civil servant and diplomat Mr. C. V. Narasimhan — CVN as he was affectionately called — who after his retirement as Under Secretary General at the United Nations, came to the University of Texas in Fall 1982 as a Distinguished Visitor in the department of political science. CVN was a good friend of my father and so as a boy I had met him in Madras in my father's company, and in New York in the sixties when my parents and I called on him at the UN during our trips to the USA. But in Austin, I really became a friend of his. CVN was an avid tennis player and he came to know that I was in town and that I played tennis regularly. So he contacted me and said that he would like to play tennis with me once a week, but he had a requirement that I needed to be aware of. He said that he was no longer as agile as he was in his younger days. So, no matter where I am on the court, from every position I should always hit the ball to within two feet of him so that he need not run! He was confident I could play by this rule. I took up the challenge and met his expectations, and so he was very pleased. CVN was a great connoisseur of Carnatic music, the classical music of India, and so Mathura and I interacted with him closely in view of our deep interest in Carnatic music. We became such close friends, that he asked Mathura and me to stay in his apartment during May–July 1983 while he was away in India. This was very convenient because, during the academic year 1982–83, we stayed in Vaaler's home while he was away in Princeton, but when he returned in May 1983, we had to shift to another residence for a few months before our return to India in July 1983.

I conclude this chapter by describing how I wound up my stay in Texas to return to India permanently — at least that was the intention then!

My father was scheduled to retire on August 9, 1983, his sixtieth birthday. We heard that George Sudarshan on the physics faculty in Austin was to succeed my father as the Director. My father and Sudarshan were old acquaintances; Sudarshan was the first visiting professor at MATSCIENCE in January 1963, and had visited there regularly since then. In view of Sudarshan's eminence in physics, and his strong interest in the development of science in India, he was a natural choice to succeed my father. But the question remained whether he would relinquish his position in Texas to come to India. The terms of that arrangement between MATSCIENCE and Sudarshan were finalized only later in 1983, and so while I was there in Austin, all I heard was that the Director's post was offered to Sudarshan and that negotiations were going on. Since Sudarshan knew me from my childhood, he invited me and Mathura to his home in Austin for dinner. There the conversation was in general about the scientific scene in India, but not specifically about MATSCIENCE. To develop the mathematics program at MATSCIENCE, Sudarshan was actively considering getting the assistance of C. S. Seshadri of the Tata Institute, a very distinguished algebraic geometer, who was a close associate and collaborator of M. S. Narasimhan, and of the same age as Narasimhan. Thus Narasimhan and Seshadri were considered as the twin leaders of the TIFR mathematics group, with Narasimhan having had a more active role in administration. It turned out that Seshadri was visiting the USA in 1983 and so Sudarshan invited him to Austin to discuss the move to MATSCIENCE. Sudarshan told Seshadri that I was visiting Austin, and so Seshadri called me to have a discussion with him over lunch about MATSCIENCE. It was clear from the discussion that Seshadri primarily admired the Tata Institute tradition and practices, and took no cognisance of the enormous effort that went into the creation of MATSCIENCE. Thus I felt uncomfortable, but thought that perhaps after I return to MATSCIENCE, we could work together smoothly. So with that hope, I returned to MATSCIENCE in July 1983. The incredible events that followed will be told in the next chapter.

Krishna and Mathura (with two month old daughter Lalitha all bundled up) on the grounds of the Institute for Advanced Study, Princeton (November 1981). Fuld Hall of the Institute is in the background.

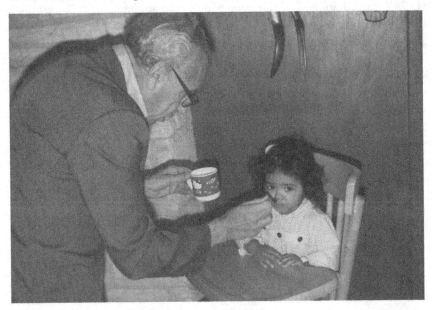

Paul Erdős feeding Lalitha Alladi at Jeffrey Vaaler's home in Austin where Mathura and Krishna stayed during 1982–83.

Chapter 6

To be or not to be

6.1 Return to India and the Straus Memorial Conference (1983)

After a full year at the University of Texas (1982–83), I returned to India in July 1983. My term in Austin ended in June, and on my way back I visited Germany briefly to give talks at the University of Frankfurt at the kind invitation of Wolfgang Schwartz and at the University of Ulm where I was invited by H.-E. Richert and E. Wirsing. From Frankfurt we flew to Nairobi to visit the game parks in Kenya — a desire that I had nursed for a long time as a wildlife enthusiast — and enjoyed the hospitality of Mathura's aunt Gigi and her husband Dr. Ramakrishnan who was working for the United Nations in Nairobi. We watched in amazement more than one hundred thousand wildebeest in migration with lions, wild dogs, and hyenas in their midst, and were thrilled to experience one of the great wonders of life on earth.

Back in Madras, I plunged into my work. I was the only number theorist at MATSCIENCE. My father was to retire on August 9, his 60th birthday. No one knew who was to succeed him as the Director. Strangely, on August 9, my father received notification that his term was extended to October 31, 1983 — an extension he neither asked for, nor was discussed with him.

Upon returning to India in July, I heard the sad news that Professor Straus had passed away. He was a diabetic since childhood, and had fought the disease with courage, but eventually succumbed to it. He was a great mathematician, a noble soul, and a wonderful teacher. My heart sank when I heard the news, and I felt that a void had been created in my academic life. Within a few weeks I received a letter from David Cantor of UCLA, that the Pacific Journal of Mathematics was planning on bringing out a volume in memory of Straus, and inviting me to contribute to that volume. While in Texas, I had completed some work on the Erdős-Kac-Kubilius theorem for the set of shifted primes that required some new ideas. I had planned on writing that paper on return to Madras, and so I decided to submit that to the Straus Memorial Volume. My early papers with Erdős were submitted to the Pacific Journal at Straus' suggestion and it was ironic that my paper in Straus' memory was to appear in the same journal!

I also received a letter informing me that the West Coast Number Theory Conference in December 1983 would be dedicated to the memory of Straus, and inviting me to speak at that conference. While I was a graduate student at UCLA, it was Straus who introduced me to the West Coast Number Theory Conference in 1975. That conference was in Asilomar, and now the Straus Memorial Conference was to be in Asilomar as well.

Before leaving for the Straus Memorial Conference, I started preparing in right earnest in Madras for the Erdős 70th Birthday Conference to be held in the hill station of Ootacamund in January 1984, since Erdős had graciously agreed to attend the conference. First I contacted the Hungarian mathematicians Andras Sarkozy, Vera Sos, and Imre Katai, who were very close to Erdős, and they all accepted my invitation. They were to come under the Indo-Hungarian Exchange Program which meant that Hungary would pay the international airfare and India all the local expenses. Among all mathematicians who had collaborated with Erdős, Sarkozy had the most number of papers with Erdős, and so it was fitting to have him attend the conference. Vera Sos was the wife of the late Paul Turán (Erdős' collaborator and great friend), and she had collaborated closely with Erdős on various combinatorial problems. Katai was an authority in the area of additive functions where Erdős had made very significant contributions. Thus my efforts for the conference was off to a great start, and it was easy to get other stalwarts once a solid Hungarian component was confirmed. The other number theorists from Europe and USA who accepted the invitation to attend the conference were H.-E. Richert and Eduard Wirsing from Ulm, Germany, Bruce Berndt and Heini Halberstam from Urbana, Illinois, Carl Pomerance from Athens, Georgia, and George Andrews from The Pennsylvania State University.

With my father retiring on October 31, and with no person having been appointed as the Director, the administration of MATSCIENCE was handled by the Secretary for Education, Mr. T. D. Sunder Raj, with oversight by Dr. Raja Ramanna, Chairman of the Atomic Energy Commission in Bombay. Both of them were pleased that I was single handedly managing all arrangements for the Erdős 70th birthday conference. So they offered full support for it and also agreed to sponsor my trip to the Straus Memorial Conference in Asilomar. Since there was no Director at MATSCIENCE from November 1 onwards, I had to go regularly to the Secretariat building in Madras, the seat of the legislature, to get signatures of approval from Mr. Sunder Raj for various budgetary items relating to the conference. Unlike a typical bureaucrat in India, he readily agreed to every request of mine for conference funds and visitor support, and also would promptly see me without making me wait outside his office for a long time.

For the Asilomar conference in memory of Straus, I decided to travel via the Pacific. I wrote to my friend Edward Bertram of the University of Hawaii in Honolulu enquiring whether he planned to attend the Asilomar conference. He responded that he would not be in Asilomar, but his colleague Ken Rogers, would be attending the conference; Rogers was an assistant professor at UCLA in the late fifties when Straus was a tenured associate professor there. Bertram enquired whether I could break my journey in Honolulu to give a colloquium suggesting December 9 as the date. This turned out to be a life changer for me as will be seen soon.

Edward Bertram had received his PhD in 1968 at UCLA under the direction of Theodore Motzkin in the area of combinatorial group theory, and joined the faculty at the University of Hawaii that year. In 1975–76, he was back at UCLA on a sabbatical and used to attend the number theory seminar. His wife Alice was in Steinberg's graduate course on algebra that I also took. So I got acquainted with the Bertrams in 1975–76 and it was a nice opportunity to renew the contact.

Edward Bertram also very kindly invited me to stay in his lovely house in Hawaii Kai during my two night stay in Honolulu in December 1983. He and his gracious wife Alice hosted a party in my honor at their house after my colloquium — a party that was attended by several members of the mathematics department. My talk "A new application of the sieve to probabilistic number theory" at the University of Hawaii had a significant impression on the department because it led to an unexpected invitation for me to visit there for a year starting August 1984 (see next section).

It was sad to be back in Asilomar without the genial presence of Straus who was such an integral part of the conference. One evening at the conference, Straus' wife Louise arranged a gathering in which several mathematicians paid memorial tributes to Straus. Louise asked me speak on that occasion and I recalled my days at UCLA under his able guidance, and the kindness he showed me. My conference lecture was on "A new monotonicity principle for the sieve" on which my approach to Probabilistic Number

Theory was based. One notable absentee at the conference was Paul Erdős who had been Straus' close friend and collaborator. Later I found out that Erdős did not attend memorial conferences because they depressed him.

On the academic side, there were two significant developments at this conference I would like to relate.

It was at this conference that Carl Pomerance made the transition to computational aspects of number theory — a transition that propelled him to stardom. The West Coast Number Theory Conferences had a strong computational component owing to Lehmer's strength and interest in computation. At the 1983 Asilomar conference, there were talks on computational number theory in which the role of $\Psi(x, y)$ was emphasized. Pomerance had studied $\Psi(x, y)$ in great depth, and so he perked up. He realized how he could contribute to several aspects of computational number theory with his expertise on $\Psi(x, y)$. So after this conference Pomerance moved into computational number theory, and there was no turning back!

In Asilomar we heard that Adolf Hildebrand of the University of Illinois, Urbana, had devised a method to significantly extend the range of validity of the asymptotic evaluation of $\Psi(x, y)$ by attaching logarithmic weights. In trying to establish an Erdős-Kac theorem for integers counted by $\Psi(x, y)$, I needed asymptotic estimates for a certain multiplicative generalization of $\Psi(x, y)$. Harold Diamond from Urbana who was at this conference, gave me a copy of the important paper of Hildebrand (at that time only in handwritten form), and that was precisely what I needed.

The Asilomar conference concluded as usual with a problem session. I proposed the following problem which was a generalization of the inequality I had established earlier for small multiplicative functions:

(i) *For each positive integer k, show that there is a constant $c = c_k > 0$ such that for all strongly multiplicative functions g satisfying $0 \leq g(p) < c$ for all primes p, we have*

$$\sum_{d|n} g(d) <<_c \sum_{d|n, d \leq n^{1/k}} g(d).$$

(ii) *Show that $c_k = 1/(k-1)$ works in (i).*

The above inequality is of interest because it could be used to establish the Erdős-Kac-Kubilius theorem for polynomial sequences. This problem generated considerable interest over the next several years.

I utilized the Asilomar conference and my brief stay in the USA to also talk to the American number theorists who were planning to attend the Erdős 70th birthday conference in Ooty in January 1984. Unfortunately,

George Andrews who had accepted my invitation, told me that he was unable to come because he had just lost his dear father.

I returned to Madras in late December from the Asilomar conference ready to welcome international visitors to MATSCIENCE and for the Erdős 70th birthday conference in Ooty.

6.2   Visitors to MATSCIENCE and the Erdős-70 conference (1984)

Among the eight international visitors, Heini Halberstam (Illinois) and H.-E. Richert (Ulm) came as Visiting Professors for one month, and utilized their stay at MATSCIENCE to work together on sieves. So they arrived in the latter part of December 1983, ahead of the other visitors, to start their work. Both Halberstam and Richert were accompanied by their spouses and had planned sightseeing in Madras and in South India during their one month stay. Since December was the peak of the music season in Madras, we arranged for them to attend some fine Carnatic music and Bharathanatyam (South Indian classical dance) performances which they enjoyed immensely. Mathura and I accompanied them to these concerts and also showed them the historic buildings of Madras.

Both Halberstam and Richert gave a series of lectures on sieves. Besides the faculty and PhD students of MATSCIENCE, these lectures were attended by some members of the Ramanujan Institute. Halberstam's lecturing style was oratorical, and he gave a glorious panoramic view of sieves. Richert on the other hand was a master of details, and he gave a meticulous treatment of several aspects of sieve theory. I had enjoyed reading the classic book of Halberstam-Richert [B9] on sieves and now it was a pleasure for me to hear both of these masters of the field.

Carl Pomerance arrived on January 1 and gave a lecture at MATSCIENCE the very next day on the topic "How to factor a number", which confirmed that he had moved into computational number theory as I had observed a few weeks earlier in Asilomar. The Halberstams, the Richerts, Pomerance, Wirsing, and us (my parents, me and Mathura), all left for Ooty via Mysore in three cars on January 3 morning. Erdős and his Hungarian colleagues did not arrive in Madras until later in the afternoon of January 3, and so I arranged a car to bring them to Ooty. Bruce Berndt who had been lecturing elsewhere in South India, arrived in Ooty by bus accompanied by Balasubramanian of the Tata Institute.

In December 1983, Erdős was awarded the Wolf Prize of Israel and so it was a privilege to welcome him just after he had received the award. He informed us that the entire prize money had been given away to various worthy causes in mathematics, as could be expected of him. I was pleased that we could arrange this conference in the lovely hillstation of Ooty that Erdős had wanted to see.

The inauguration of the conference was at the magnificent Aranmore Palace, and Erdős gave the Opening Lecture on the theme "Some of my favorite problems in number theory I would like to see solved." Both Halberstam and Richert spoke about combinatorial sieves, a topic very relevant to me in view of my own work on approaching probabilistic number theory utilizing a multiplicative generalization of the sieve.

There was good participation in the conference by number theorists from India including K. Ramachandra and his former students R. Balasubramanian and S. Srinivasan of TIFR. Kumar Murty (from Canada) who was visiting TIFR at that time, also participated in the conference. Since I had published the Proceedings of the MATSCIENCE 1981 Number Theory Conference (Mysore) in the Springer Lecture Notes series, I approached Catriona Byrne of Springer again, and she also accepted the refereed proceedings of the Erdős-70 conference in Ooty for Springer Lecture Notes. Almost all the participants (including Erdős) contributed to the Proceedings which appeared as Volume 1122 in 1985.

My father who attended the conference, arranged accommodation for all the participants at the comfortable Hindustan Photo Films (HPF) Guest House. The friendly and efficient butlers there provided excellent Indian and Western dishes to the satisfaction of the participants Indian and foreign. Owing to the British influence and its cool climate, Ooty is famous for "English" (= non-native) vegetables like brussels sprouts, cabbage, and cauliflower, that are ideally suited for westernized cuisine that the butlers were so good at. In India tipping is expected everywhere and the HPF Guest House was no exception to this. The only person who got into problem with tipping was Halberstam because he looked imperious, smoked a cigar with an air of assurance, and spoke like an orator; so the butlers assumed he was Royalty and therefore expected much higher tips from him!

Ooty has a lot to offer by way of sightseeing. My parents, Mathura and I were pleased to take our guests on excursions in and around Ooty — to the beautiful Botanical Gardens that has been a favourite for romantic scenes in Indian movies, the lovely Glenmorgan Tea Estates, and Dodabetta, the highest point in the Nilgiri mountain range, from where you get an awesome view of an 8000 foot drop into the valley below. We also took the visitors to the see the Radio Telescope that had been set up by the Tata Institute, and on that visit the staff there asked my father to speak about his recent work in special relativity.

On our way back from Ooty by car to Madras, we halted at the Mudumalai Wildlife Sanctuary. We went on a jeep ride and saw plenty of animals — elephants, wild boars, wild dogs, spotted deer, and the magnificent gaur (the largest of all wild oxen), but not the tiger, the king of the Indian jungles. Even if one spots a tiger, it is often only for a fleeting moment, because the elusive big cat will disappear into the dense jungle. This is very different from the open savannas of Africa where you can see lions roaming freely. Since we did not see a tiger, Erdős kept saying throughout our jungle drive: "I don't think there are tigers in this forest!"

Erdős was invited as Srinivasa Ramanujan Visiting Professor, and so his entire trip including international airfare was provided by MATSCIENCE. He spoke about some of his unsolved problems including his $10,000 problem on large gaps between primes (that was solved recently by Maynard and by Ford-Green-Konyagin-Tao). The picture I took of Erdős giving this talk at MATSCIENCE in 1984 with the $10,000 problem on the blackboard, has now been used in several settings and in articles that have appeared in the Notices of the AMS and elsewhere. All the international visitors gave lectures at MATSCIENCE in January.

My father-in-law, the late Dr. N. C. Krishnan, who among his many roles was a CPA for the Taj group of hotels, hosted a magnificent lunch for our international visitors at the (Taj) Fisherman's Cove Hotel on the beach about 20 miles south of Madras. As we enjoyed a fabulous Indian buffet watching the waves of the Bay of Bengal crash on the silvery sands of the Coromandel Coast, Bruce Berndt declared that it was the finest meal in his entire life!

My father arranged for three public lectures by the distinguished visitors under the auspices of the Alladi Centenary Foundation that he had launched in 1983 for the centenary of my grandfather's birth. The first of these was by Halberstam on the theme "How common is common sense - the role of mathematics in contemporary education", followed by talk by Bruce Berndt on "Editing Ramanujan's Notebooks", the same evening. The third was a lecture at our home Ekamra Nivas on January 14 by Erdős entitled "Reminiscences of a mathematician", for which the Governor of Madras, His Excellency H. L. Khurana, presided. This was followed by dinner in honor of the Erdős and the Governor at Ekamra Nivas attended by more than hundred guests. The Governor was so much impressed with Erdős' fame and humility, that he readily agreed to our invitation to preside over the lecture and stay for dinner. A few days later, on January 18, the Governor invited the international visitors and my family for a dinner at the Raj Bhavan (= Governor's Residence) in honor of Erdős. That was a fitting finale for a glorious one month of intense activity due to the presence of eminent number theorists from Europe and the USA. Just as Erdős in 1975 gave all the remaining Indian currency he had to the Governor for his fund to encourage mathematically gifted students, this time Erdős gave most of money remaining from his Ramanujan Visiting Professorship to Mrs. Ramanujan whom he could not meet during his visit. I collected the money from Erdős and gave it to Mrs. Ramanujan when I met her later on February 2. On that occasion I told her that Erdős was a Founding Father of Probabilistic Number Theory, a subject whose origins could be traced to the Hardy-Ramanujan paper of 1917 on round numbers.

After the international visitors left, I felt a sense of void, and to get over this feeling, I plunged into my work. One important fallout of the presence of Erdős was that he and I were able to solve the problem I posed in Asilomar by showing that $c_k = 1/(k-1)$ is admissible. I had some correspondence with Jeff Vaaler on this who showed how to generalize the inequality to sets and measures, and so it became a triple paper. I also studied the new method of Hildebrand and adapted it to make it work for a multiplicative generalization of $\Psi(x, y)$ and was able to establish an Erdős-Kac theorem for integers devoid of large prime factors I also offered a series of 20 lectures at MATSCIENCE on the theory of partitions and q-hypergeometric series. I always had wanted to learn this subject, and the best way to learn a subject is to lecture on it! In late 1983, I had written to George Andrews requesting his papers on partitions. He responded by sending me two large packets containing more than one hundred reprints of his! With those in hand, and his book on partitions [B3] that had appeared as Volume 2 of the Encyclopedia of Mathematics Series in 1976 (John Wiley), I prepared systematic notes for a three month course of lectures (Feb–Mar–Apr, 1984) on partitions at MATSCIENCE. I use these notes even today at the University of Florida! I did not realize then that within a decade I would shift the area of my research to partitions and q-series and begin a major collaboration with Andrews.

6.3   Some unexpected developments (1984)

In April 1984, I received an invitation from Amit Ghosh to speak at International Conference on Number Theory that he and his colleagues were organizing at the Oklahoma State University during June 24–July 3 that year. Almost simultaneously, I was also invited by Don Lewis and Wolf-gang Schmidt (Organizers) to an AMS Summer Conference (July 14–21) on "Diophantine Problems — including Diophantine equations, Diophantine approximations and transcendence" at Bowdoin College, Brunswick, Maine; this was followed by a formal letter of invitation from Willam J. Leveque, Executive Director of the AMS. I decided to attend both conferences during a single one month trip to the USA. Since MATSCIENCE had just sup-ported me fully for the Straus Memorial Conference in Asilomar, I applied to the Department of Science and Technology in Delhi, and they agreed to fully sponsor my trip to America in June–July, on one condition that I had to travel by Air India on the international sectors.

Shortly before my departure for the Oklahoma and Bowdoin confer-ences, Bruce Rothschild from UCLA visited Madras for a few days in early June. It was a pleasure to host him for he was one of my former professors at UCLA and also such a kind and genial person. He is a world authority on Ramsey Theory in the area of Combinatorics, and so he gave two talks at MATSCIENCE on Ramsey Theory.

I was all set to depart Madras on June 20 night by Air India. I went to MATSCIENCE in the morning and came home in the afternoon to rest. While I was resting, I received a phone call from Edward Bertram invit-ing me to the University of Hawaii for one year on a Visiting Associate Professorship starting August 1984! Did I mis-hear the start date in my sleep? I asked him whether the start was August 1985. He confirmed it was August 1984. I told him that I was leaving for the United States in a few hours and so would talk to him regarding arrangements after reaching America. Bertram said that my talk in December 1983 in Hawaii enroute to the Straus memorial conference had made a positive impression, that one of their faculty members had suddenly decided to go on leave with-out pay, and that prompted his department to offer me this visiting posi-tion. I wanted to accept it because there had been no development yet at MATSCIENCE after my father's retirement for any faculty appointments. I called MATSCIENCE and informed the administration about this sud-den unexpected development, that I did not even apply for this assignment, and that after return to Madras in July from my trip to the USA, I would discuss with the authorities about the Hawaii assignment.

I left Madras on the night of June 20 for Bombay by Air India Boeing 707, and connected to an Air India Boeing 747 there for New York via Dubai and London. I reached Stillwater, Oklahoma, and stayed with Professor N. V. V. J. Swamy of the Physics Department of the Oklahoma State

University for the duration of the conference. He was a close friend of my father and I had gotten to know him quite well.

The four number theorists — Alan Adolphson, Brian Conrey, Amit Ghosh, and R. I. Yager — who had all joined the Oklahoma State Mathematics Department in Fall 1983, decided to conduct this major international conference on analytic number theory and Diophantine problems to jump start and bring visibility to their program. They had succeeded in getting several leaders in the field, and among them Enrico Bombieri, Dorian Goldfeld, David Masser, and Bob Vaughan, each gave three one-hour lectures. Goldfeld spoke about analytic number theory in $SL(2, \mathcal{Z})$, while Masser lectured on recent progress in transcendence theory. Bombieri spoke about his recent work on effective estimates in irrationality, including some joint work with Jeffrey Vaaler. Vaughan gave talks on the circle method.

A special feature of the conference was that Atle Selberg was a principal speaker. He rarely attended conferences (only those of special significance). So it was a fantastic opportunity (the first time for me) to hear him lecture. Somehow the organizers were able to convince him to speak at the conference. But he did not contribute to the Proceedings. Selberg spoke about the (Selberg) Trace Formula and consequences (one lecture). I remember how he started the lecture. He was facing the blackboard, not the audience, and his opening sentence was "Number theory is full of idle questions posed by idle questioners, but I hope you will not find the subject of this talk to be idle." A participant who was seated next to me nudged me and said, "Krishna, that is a dig at Erdős." But maybe it wasn't about Erdős. Maybe Selberg was just emphasizing his view about research, that one should only work on fundamental questions.

During the conference, the two great collaborators, Hugh Montgomery and Bob Vaughan, presented a neat solution to an old and famous problem of Erdős relating to the second moments of gaps between numbers relatively prime to a given number, and generalized this to higher moments. They were to receive a prize from Erdős after the publication of their paper.

My talk at the conference was on the solution to the problem I posed in Asilomar, and my joint paper on this with Erdős and Vaaler appeared in the proceedings of the conference published by Birkhauser.

During my stay in Stillwater in Professor Swamy's residence, I made several phone calls to Bertram who helped make all arrangements for me to visit Hawaii for one year with Mathura and my three year old daughter Lalitha. All this was done on the assumption that I would get the necessary leave sanctioned by MATSCIENCE, and that was to be decided only after my return to India from this trip.

After the Oklahoma conference, I spent a few days in Austin to work with Jeffrey Vaaler. We then proceeded together to Boston from where we picked up a rental car and drove to Bowdoin College for the AMS Conference on Diophantine Problems. David Masser (who had moved to the University of Michigan) and Gisbert Wustholz from Bonn, Germany, had made significant advances in transcendental number theory, and so that was one of the main aspects of this conference. After my work on irrationality in 1979–80 at the University of Michigan, I had not worked in that area, and so I did not give a talk in Bowdoin. The conference however afforded me an opportunity to learn about the latest advances in irrationality and transcendence since the leaders in these areas like Wolfgang Schmidt, Gregory Chudnovsky, and Frits Beukers, were giving talks.

After the AMS meeting, I went to New York JFK airport on July 21 via Boston and boarded Air India Boeing 747 named *Emperor Kanishka* for London and Delhi. It was a pleasant surprise that Mathura's aunt Ganga and uncle Dr. Sankaran, a distinguished entomologist working on United Nations projects in South East Asia, were my co-passengers. This Boeing 747 Emperor Kanishka was the one that was blown up a year later (23 June, 1985) over the Irish Sea by Sikh terrorists as an act of revenge because Prime Minister Indira Gandhi had sent troops into the sacred Golden Temple of the Sikhs in Amritsar to capture the terrorists who were seeking to separate Punjab from India. The insistence of the Department of Science and Technology that I should travel by Air India, led me to fly in the ill-fated *Emperor Kanisha*, but fortunately a year before its doom!

6.4 Mathematics in paradise and a turning point in my career (1984–85)

When I returned to Madras in late July 1984 after attending the conferences at Oklahoma State University and Bowdoin College, I learnt that George Sudarshan of the University of Texas, Austin, who had accepted the offer of the Directorship of MATSCIENCE would be joining duty in a few weeks. Sudarshan never resigned his position at UT Austin or took an unpaid leave of absence from there; he continued to draw his full salary from Texas and therefore accepted only a token salary from MATSCIENCE of one rupee per month! He arranged for all his expenses to be paid such as the rent and maintenance of his residence in Madras, his travel between Austin and Madras, etc. He also had to spend at least 180 days each year in Austin and so he negotiated with the authorities for a Joint Director be appointed who would act on his behalf while he was away. The Joint Director chosen was G. Rajasekharan, a physicist who was formerly at the Tata Institute, but had since moved to the University of Madras. Rajasekharan actually joined MATSCIENCE in the first quarter of 1984, a few months before Sudarshan arrived. Sudarshan had asked Seshadri of the Tata Institute to develop the Mathematics program at MATSCIENCE. Seshadri arrived also in 1984; he too had not resigned his job at the Tata Institute since he had arranged to be at MATSCIENCE on leave (deputation) from TIFR for five years. I remember greeting him as he entered MATSCIENCE, and his first words were "We should change many things here." I understand that any new administrator would like to make changes, but Seshadri need not have opened his conversation with me so abruptly. I felt that he could have greeted me more warmly. I had returned to MATSCIENCE fully aware that there would be major changes.

George Sudarshan arrived in early August and he sanctioned my leave to accept the assignment in Hawaii. So I applied for a visa to go to Hawaii. When I went to the US Consulate, the Vice-Consul in charge of visas asked me why I wanted to go there with my wife and daughter. Why could I not go alone leaving my family in Madras? In response I said, "Your country has prominently advertised Hawaii as a paradise, and you are suggesting that a young man leave his wife and child and go to paradise by himself! How does this fit in with America's emphasis on family values?" She smiled and said that she would give the visas for all three of us. But on checking the paperwork, she said that since I had gone to Princeton and Texas on a J-Visa, she had to get permission from USIA (formerly USICA) in Washington DC to waive the two year residency requirement and give permission to issue the visa. So the story of the permission from USIA repeated itself, and for a second time the US Consulate in Madras had to contact Richard L. Fruchterman of the USIA in Washington for the required permission! Fortunately I applied for the visa in early August, a few weeks before the start of classes in Hawaii, and so there was time to get the permission. But this was not easy going. It required great effort on the part of Professor Bill Lampe, Chairman of Mathematics at the University of Hawaii, who spoke to Fruchterman several times to get the permission. In fact, during this period of negotiation, Lampe was even prepared to change the start date of the visiting position to January 1985, but Fruchterman gave the nod on Thursday, August 23, and so I could leave for Hawaii on Saturday, August 25 as planned.

During this period of waiting for the permission, two pleasant things happened. First was the visit of 1970 Fields Medalist John Thompson to

MATSCIENCE. He gave a beautiful lecture on August 11 on the History of Group Theory. The same day he graciously accepted my father's invitation to visit Ekamra Nivas since he recognized the enormous effort my father made to create MATSCIENCE. Later that night, my father and I joined the MATSCIENCE group for a dinner at the Chola Sheraton Hotel in honor of Thompson. Little did I realize that in two years, Thompson and I would be colleagues at the University of Florida!

The second event was the wedding engagement of Mathura's younger brother Chella Srinivasan in mid-August. His marriage was fixed to take place in mid-September. So Mathura decided to stay back for this and said she and my daughter Lalitha would join me in Honolulu soon after that. Hence I departed for Honolulu by myself on August 25.

I left Madras by Air India for Singapore where I boarded a Japan Airlines Boeing 747 overnight flight for Tokyo. Not only was the flight on JAL very smooth, but the inflight service was outstanding. I was especially impressed to see the JAL cabin crew pay so much attention to elderly Japanese passengers and attend to all their needs. We could learn a lot from the Japanese. Since I had almost the entire day in Tokyo, JAL graciously accommodated me at the stately Nikko Narita (Airport) hotel, and so I could rest before taking a second successive overnight flight from Tokyo to Honolulu by a JAL 747.

The Hawaiian islands are incredibly beautiful and the panoramic view one gets from the air is spectacular. As I landed in Honolulu on a bright sunny morning, I could see the entire southern coast line of the island of Oahu — the turquoise blue of the ocean water accentuated by the sunlight on the coral beds, the towering skyscrapers of Waikiki against a lush mountainous backdrop, and the mighty US naval fleet at Pearl Harbor right next to the airport.

I arrived in Honolulu on Sunday, August 26 morning. Bertram picked me up at the airport and took me straight to the mathematics department to show my office and give my course textbooks since classes were starting the next day! He then drove me to the lovely apartment he had arranged for us at Kilauea Gardens in the exquisite Kahala area of Honolulu.

The lovely campus of the University of Hawaii is in the verdant Manoa valley that is nestled between the mountains that form the backbone of the island of Oahu. The campus is botanical garden in itself for it has a fine collection of tropical trees and flowering plants. The northern part of the campus which is at the foothills of the Manoa valley, gets a lot of rain, in contrast to the southern part which is quite dry. Thus the weather in the northern part of campus is dramatically different from the southern section. Hawaii, known as the Rainbow State, is famous for double rainbows, one can see these often as we gaze towards the northern part of the campus.

The University of Hawaii had a quite a strong mathematics department. Their main problem was recruiting and retaining good graduate students. This is because, unlike in the mainland where graduate students could travel to conferences by road to save on costs, from Hawaii one has to fly to the mainland, and this is expensive. The Department enjoyed a sudden expansion in 1968 and 69 when the great analyst Paul Halmos was appointed as the Chairman with a promise that he would be provided funds for several positions and to launch a research institute. Thus Hawaii successfully lured Halmos away from the University of Michigan in Fall 1968. Almost immediately, more than a dozen appointments were made including those to begin in Fall 1968 that the previous Chair had authorized but Halmos had to approve, and several appointments to begin in Fall 1969 that Halmos himself authorized. Halmos also brought several distinguished visitors to the University of Hawaii in 1968–69 — Allen Shields of the University of Michigan, Paul Erdős, Fields Medallist Paul Cohen from Stanford, and Leo Moser, to name a few. The promise of an Institute did not come through, and that was one of the reasons Halmos left Hawaii in 1969 having just spent only a year there, to take up a position at Indiana University. It is very common for university administration in general to go back on promises. Many faculty he appointed also left, but a good number remained in Hawaii such as Edward Bertram, Larry Wallen and George Csordas, and I interacted with all three during my stay in Hawaii. Indeed many of the faculty appointed by Halmos who remained, helped steer the department by serving on the Executive Committee that works closely with the Chair. One of them is Larry Wallen, who is very friendly, but brutally frank in his speech. Bertram told me that Larry attended my colloquium in December 1983, and it was he who urged the mathematics department to offer me a visiting associate professorship.

All members of the department administration were extremely kind and helpful. Chairman William (Bill) Lampe who had patiently put through several calls to Fruchterman in Washington DC to get my visa approved, had not lost his patience; instead he offered help anytime I needed assistance. The graduate chair Tom Craven and the undergraduate chair Ruth Wong made sure that my teaching assignments were appropriate, and Pat Goldstein, the Chair's secretary, was a model of efficiency and politeness.

During my year at the University of Hawaii, I gave a series of lectures in the weekly Number Theory Seminar. In Fall 1984, I lectured on two topics: (i) the theory of partitions and q-hypergeometric series, and (ii) sieve methods. The Notes I had prepared for the lectures on partitions I had given earlier in 1984 at MATSCIENCE proved immensely useful. For the talks on sieves, I prepared notes for the first time in Hawaii, and I use them even today. In the Spring of 1985, I lectured in the Number Theory Seminar on (iii) Probabilistic Number Theory, and (iv) integers with restricted prime factors, both pertaining to my research. The four faculty members who attended my seminars regularly were Ed Bertram (combinatorial group theorist), Larry Wallen (analyst), Ken Rogers (number theorist), and Ron Brown (algebraic number theorist). I used to spice up my lectures with anecdotes of famous mathematicians, and this would always get Ken Rogers excited; he often responded with humorous comments and observations. Ron Brown is one of the most pleasant and polite persons I have ever met. I was not surprised that he later served as the Chair in Hawaii. Another faculty member who attended some my talks in the number theory seminar was the complex analyst George Csordas. His interest in analytic number theory was stimulated by the Polya-Jensen conjecture which is equivalent to the Riemann Hypothesis. Csordas was starting an active collaboration on this theme with Richard Varga of Kent State University, Ohio, and so he arranged a colloquium of Varga in Hawaii during the academic year 1984–85.

The big news in the mathematical world in Fall 1984 was the proof of the famous Bieberbach conjecture by Louis De Branges of Purdue University. The University of Hawaii had an active program in complex analysis that was begun by Halmos. The analysts there perked up on hearing about the work of deBranges, and so George Csordas, Larry Wallen, and David Stegenga of the analysis group ran a seminar on the Bieberbach conjecture in Fall 1984 that I attended.

I had a fruitful period of research in Hawaii. The University had a very good library and it was convenient to have it next door to Keller Hall where the mathematics department is located, so that I could walk to the Library often each week. Nowadays with the internet and the arXiv, you can download most papers you need for your research, but back then one had to go to the library, and so the proximity was helpful.

Mathura and my daughter Lalitha arrived in the third week of September. Lalitha who was three years old, went to a Montessori pre-school just a block away from our apartment, and so Mathura could walk her to school and bring her back. I used to go to campus early in the morning every week day and work there till about 4:00 pm. Then every evening on week days, and on weekends, we would go around the island of Oahu enjoying the breathtaking scenery from every angle, and strolling in the bustling Waikiki beach area teeming with tourists. I noticed that almost all the faculty members of the mathematics department were embarrassed to mention the word Waikiki, since Waikiki was associated with tourists, and they did not want anything to do with it. This strong dislike of Waikiki was taken to ridiculous extents: when I asked some of my colleagues in the mathematics department to join us for dinner at Waikiki — and Waikiki has some good restaurants — their reaction would be "Waikiki is too touristy. Let us go some place else." Yes, Mathura and I dined in various parts of Honolulu, but we did not avoid Waikiki like plague! Ed Bertram and Alice joined us for dinner in Waikiki from time to time, and they confessed they did that for my sake. The one faculty member who enjoyed Waikiki was Larry Wallen who would go to the beach there often on weekday afternoons. On my part, I felt that there were several things attractive about Honolulu — the mathematics department, the idyllic environs of the island of Oahu, and the exciting Waikiki beach area. One distinguished mathematician who shared my view was Patrick Gallagher of Columbia University. During a conversation I had with him at the Oklahoma conference prior to my visit to Hawaii, he said that he enjoyed every aspect of Honolulu including Waikiki. Gallagher visited the University of Hawaii in 1980–81, and the entire department fell in love with him and his mathematics. He was even offered the Chairmanship of the mathematics department, but he declined the offer.

My tennis was also active in Hawaii. David Bleecker on the mathematics faculty, a reputed topologist who had written some fine text books, was my tennis mate — or should I say my tough opponent?! He played a slow and steady game — almost error free — that would frustrate all but the most experienced players. As a student, he was a practice partner for Bob Lutz, one of the world's greatest doubles players who partnered Stan Smith. Bleecker and I had several memorable encounters at the Diamond Head courts and at the Ala Moana beach park courts, my pace and power pitted against his steady and slow game. I also enjoyed playing doubles on the campus, and for this George Csordas joined Bleecker and me.

I had taken a two bedroomed spacious apartment in Honolulu so that we could accommodate guests in the second bedroom — and we had plenty of guests throughout the year. Mathura's parents and her kid brother Sriram visited us in the Fall. They stayed with us for the entire month of November.

Two days before their arrival, on 30th October, we heard the shocking news that India's Prime Minister Indira Gandhi was assassinated at her residence by one of her own body guards — a Sikh who was seeking revenge for her action to send troops into the Golden Temple of Amritsar, the holiest shrine of the Sikhs. Bertram called me to give the news, and almost simultaneously I received a call about the assassination from Mathura's parents who were in Tokyo and about to leave for Honolulu.

All the islands of the Hawaiian chain have wonderful things to offer by way of sightseeing. Mathura and I decided to visit the other islands in the company of our guests. So with Mathura's parents we visited *Hawaii*, the Big Island, and enjoyed the awesome grandeur of the Volcanoes National Park. My parents visited for four months starting in January, and with them we visited *Kauai* which in my opinion is the most idyllic of all the Hawaiian islands, and *Maui* which is known for the majestic Haleakala volcano and fabulous beaches.

In Shakespeare's Richard II, John of Gaunt describes England as "This precious pearl set amidst the silver sea ... this fortress built by Nature against infection and the hand of war, ..., against the envy of less happier lands, ..., this blessed plot of earth, ..., this other Eden, demi paradise..." I think these words describe the Hawaiian islands even more accurately. The only change I would make is to replace "silver sea" by "azure sea"!

During the Erdős 70th birthday conference, both Paul Erdős and Carl Pomerance said that I should return to the United States for good. I too began to feel that way despite my attachment to Madras and to MATSCIENCE. So I decided to use my stay in Hawaii to apply for permanent positions in the USA, a decision I reached only in the Fall of 1984 in Hawaii and not earlier. That decision changed my life completely.

Instead of applying to several Universities, I chose just three where positions in Number Theory were advertised, and where I knew the number theory group. The three universities were — The University of Arizona at Tucson, The University of Colorado at Boulder, and the University of Florida in Gainesville. The University of Florida was one of the annual stops for Erdős — he always visited there for two weeks each Spring — and so even though Florida advertised a position only in combinatorics, Erdős said he would talk to the department to convert that to a position in number theory. And that actually happened. Since the position advertised was in Combinatorics, Erdős advised me that I should send my file to Jean Larson (combinatorics and set theory), and request her to draw the attention of the Chair and the Search Committee about my case, which she did. The response from Florida to my file was quite enthusiastic.

I was invited for colloquia at all three universities, with Arizona and Colorado wanting to schedule my talks in mid-January itself. So I decided to make a trip to the mainland in January to attend the Annual Meeting of the AMS (Jan 9–14) in Anaheim, California, and to go from there to Tucson and Boulder. My parents arrived in Honolulu on January 8 for a four-month stay, and I left for the mainland the same day but after their arrival so that Mathura and Lalitha would not be alone while I was away. At the Annual Meeting of the AMS, Dorian Goldfeld was organizing a Special Session in Analytic Number Theory and he invited me to speak in that session. My talk there was on "The Erdős-Kac theorem for integers without large prime factors" that I had completed in Madras shortly before leaving for Hawaii. There were several leading analytic number theorists at the session Goldfeld organized — Enrico Bombieri, Patrick Gallagher, Peter Sarnak, Harold Stark, Piatetski-Shapiro, Andrew Odlyzko and Heini Halberstam — to provide a sample list.

One of the most thrilling experiences at the Anaheim meeting was to hear the lecture of Louis De Branges to an audience of about a thousand at the Convention Center. De Branges did not give a talk on the Bieberbach Conjecture; instead he spoke about the Riemann Hypothesis for Hilbert spaces of entire functions.

De Branges had been working on the Bieberbach Conjecture for many years and had made claims of the proof previously, but there had been gaps in his proofs. So he was not taken seriously and therefore was not funded by the NSF. When he finally cracked the problem, he still was not taken seriously by the American mathematicians. It was the Russian mathematicians who recognized that he had something going, invited him to lecture in Russia, and confirmed the correctness of his proof. So when he started his lecture at Anaheim, his opening slide had the line — "Research NOT supported by the National Science Foundation!"

Danny Gorenstein of Rutgers gave the Colloquium Lectures at the Anaheim meeting on the theme of the Classification of Finite Simple Groups. It was nice to hear Gorenstein's survey after hearing Ashbacher's colloquium in Ann Arbor on the Classification a few years earlier.

From Anaheim, I left for Tucson and Boulder. My interest in applying to the University of Colorado was because that would give me an opportunity to interact with Peter Elliott, a world authority in Probabilistic Number Theory, the area in which I was working at that time. I also felt that I could profit by the presence there of Wolfgang Schmidt, a giant in the area of Diophantine Approximations. Just before my colloquium at Boulder,

Peter Elliott drew me aside and said: "Do not make everything too clear, because if the audience understands everything you say, they may think your work is not deep. So leave some stuff to be mysterious and that will make them think that your stuff is deep. Remember that you are giving a job colloquium." Well, I cannot change my lecturing style just to impress an audience. I do give clear lectures, but I decided to skip the details in some essential parts following the advice of Elliott.

On the night before my departure from Boulder to Honolulu, I received a phone call from David Drake, Chair of the Search Committee in Florida, inviting me for a talk in Gainesville, and enquiring whether I could come there from Boulder. Since I was scheduled to fly back to Honolulu for my classes, I requested that my talk in Gainesville be scheduled in late February. This turned out to be beneficial because Erdős would be in Florida during my visit and could talk to the authorities there about my case.

On Friday February 15, the day before my departure to Florida from Honolulu, Professor Walter Hayman, an eminent complex analyst from Imperial College, London, gave a talk in the Analysis Seminar in Hawaii that my father and I attended. The next night there was a party in honor of Hayman at the home of Professor Jake Bear who had been the Mathematics Chair in Hawaii from 1969 to 74. I could not attend the party because I departed for Florida earlier that day. But my father attended the party to which he was taken by Bill Lampe. It was nice that my father and Hayman could reconnect because it was on Hayman's suggestion that MATSCIENCE began a program in pure mathematics in 1966.

My visit to the University of Florida, Gainesville, was one of the most pleasant experiences. My father told me that at the University of Florida, there was a young faculty member by name R. Charudattan (Charu as he is called by friends), specializing in plant pathology, and that he was the son of the late Sanskrit professor Raghavan of the University of Madras who was our family friend. My father had visited the University of Florida statistics department in 1980 and during that visit had met Charu. My father said that Charu had been very hospitable and that I should contact him. Charu received me at the airport on Sunday, February 17, and took me to his home for dinner that night. He and his wife Dharini (Judy as she is called) also invited Erdős for dinner at their home. I was put up at the Rush Lake Motel near campus where Erdős was also staying. The next morning Erdős walked with me through campus to the mathematics department. The campus was beautiful, with magnificent oak trees, expansive lawns, and azalea bushes, just as my father had described.

My colloquium on Monday was on my new approach to probabilistic number theory using a multiplicative generalization of the combinatorial sieve. The mathematics department had active programs in probability and in combinatorics, and so my talk resonated well. There were at least three occasions on Monday and Tuesday when I was taken out to lunch and dinner, and for each of these several faculty members and Erdős attended. Naturally I was pleased at the hospitality and friendliness of the mathematics faculty. I had a very fruitful meeting with Dean Charles Sidman of the College of Liberal Arts and Sciences (a historian), and with the long time department chairman Al Bednarek and the Associate Chair Zoran Pop-Stojanovic, a probabilist. One of the Indian faculty members in the mathematics department was Professor Arun Varma, who was goodness personified. My flight back from Gainesville on Wednesday, February 20 departed very early. It was extremely kind of Professor Varma to take me to the airport before sunrise, and for his wife Manju to have packed some Indian food for me to eat on the long air journey back to Honolulu. Of course Erdős who was also staying at the Rush Lake Motel came to the airport with Varma to see me off in the wee hours of the morning.

Back in Honolulu, I told my parents and Mathura that I was touched by the warmth and hospitality shown to me at the University of Florida, and by the caring presence of Paul Erdős. The very next day I received a phone call from David Drake (another very genial person!), the Chair of the Search Committee. He said that my research complemented the research expertise of the department quite well and that I would soon be receiving a call from Al Bednarek, the department chairman, offering me an associate professorship. The University of Florida had only advertised for an assistant professorship in combinatorics. But the University was willing to convert that to an associate professorship in number theory, and Erdős played a key role in this decision.

During my correspondence with Erdős in January, I enquired whether he could visit the University of Hawaii while I was there. He responded saying that he could visit in March after his stay in Florida. I spoke to Bill Lampe, who spontaneously agreed to arrange the visit; Lampe actually got the Department of Geology to co-sponsor the visit, with the understanding that Erdős would give a public lecture in the Geology department!

Erdős arrived in Honolulu on Wednesday, March 6 for a ten-day visit. Since my parents were staying with me and occupying out guest bedroom, I arranged an apartment for him next door to ours in the same apartment complex where we were staying — Kilauea Gardens. This way, as a family we could attend to his needs promptly.

Erdős gave three lectures — (i) a talk in the Number Theory Seminar on the Erdős-Kac Theorem on Thursday, March 7, (ii) a Colloquium on Problems and Results in Number Theory, the next day and (iii) a Public Lecture in the Geology Department on Child Prodigies, the following week. After the colloquium, Ken Rogers and his wife Choleng hosted a dinner party in honor of Erdős at their lovely home in the Niu Valley subdivision of Honolulu that we enjoyed attending. A reporter of the Honolulu Star Bulletin newspaper interviewed Erdős after his public lecture and was charmed by his brilliance, simplicity, and idiosyncracies. She published a half page article on Erdős with input from me and Bill Lampe since Erdős was too modest to talk about his own achievements.

My family did a good bit of sightseeing with Erdős. He chatted with my father about Indian history and the scientific scene in India, and played with my three-year old daughter Lalitha. He appreciated the Indian food that Mathura prepared. Even though Erdős always looked frail, he was an energetic walker. He came on an early morning hike up the Diamond Head volcano to enjoy the magnificent view of Waikiki at sunrise.

While Erdős was in Honolulu, we did a very nice piece of research. We showed that the inequality I had posed at the 1993 Asilomar conference could be resolved even in the boundary case $c_k = 1/(k-1)$, with which I had been struggling. Erdős suggested that I should use a powerful theorem of Baranyai on hypergraphs, and when I applied it, I found a surprising combinatorial argument involving binomial coefficients to complete the details. Since Vaaler had sent his ideas on the problem, it became a triple paper which appeared in the Journal of Number Theory after I joined the University of Florida.

Interestingly, the formal written offer from Florida came while Erdős was visiting me in Honolulu. I accepted the offer but told the University of Florida that I would go back to India and wait there until my green card was approved. I did not want to risk losing my tenured position at MATSCIENCE while waiting for my green card in the USA.

After a brief but very eventful visit, Erdős left Honolulu on March 15. My parents stayed on for another six weeks. My father played tennis with me and joined me in doubles matches with my colleagues in the Department. One very enjoyable tennis game was with Jake Bear and his wife Ruth Wong in the beautiful Hawaii Kai Tennis Club nestled in the valley with a spectacular mountainous backdrop. Following that game, we had dinner at the lovely home of the Bears up in the mountains from where we enjoyed a fabulous view of Hawaii Kai and the marina glistening at night. In late

March, my father left on a short trip to the mainland in response to lectures at various universities. After his return from that trip in mid-April, he gave two talks on Clifford algebras in Hawaii, one as a Colloquium, and another in the algebra and lattice theory seminar, arranged by Professor Jerry Yeh.

My parents left on April 30. We still had one more month in Honolulu because final exams for my classes at the University were in mid-May and Lalitha's school was closing on May 31. So we spent most of May packing and preparing for our return to India. By a strange coincidence, during our final period of our stay in Honululu in May, we ran into an Indian gentleman at the Ala Moana shopping center. A conversation was struck, and we learnt that he was Dr. Ramanathan, a professor in Pharmacology at the University of Hawaii, whom I had met in Honolulu 1966 as a ten-year old boy travelling with my parents! We had lost touch with him because he returned to India shortly thereafter, and then came back to Hawaii a few years later. It was a chance meeting in May 1985 that brought us together. The renewed contact resulted in a close friendship between our entire family and the Ramanathans. Both my family and my parents have enjoyed the hospitality of the Ramanathans over the years at their lovely home in Kalama Valley during our annual visits to Honolulu since 1986.

The Hawaiian visit was like a wonderful dream, satisfying in every aspect. Ed Bertram and Alice took solicitous care of us, and Chairman Bill Lampe was most supportive. The way in which the Hawaii offer came, the events unfolded, and the manner in which Erdős played a role in the Florida offer, convinced me that Providence was guiding my destiny. So I made the difficult decision to leave India and settle down in America. But I would still have a year to wait in India for my green card, and during that period, several interesting events took place that I describe next.

6.5   Decision to immigrate to America (1985–86)

Back in Madras from Hawaii in June 1985, my family and I were mentally getting prepared for the biggest change in our life — the decision that we would leave India for the United States on a permanent basis. This was not easy, because we would be leaving our dear family home *Ekamra Nivas* — a home of a thousand memories and with a great history — only to return annually for short visits. I also was going to relinquish my position at MATSCIENCE, a unique institution that was built by my father in Ekamra Nivas and whose origins I had witnessed. This decision was especially hard on my parents but they supported it because they understood my reasons. I was not comfortable with the type of arrangement that Sudarshan and

Seshadri had — administering MATSCIENCE with tenure elsewhere. I felt that it would eventually lead to trouble. But then, the academic set up and opportunities in the USA had always attracted me.

Although I had made this decision to immigrate to America, I did not inform the authorities at MATSCIENCE until my green card was processed, and this took more than a year. The Immigration and Naturalization Service (INS) of America, expects people to apply for the green card from their home country. But almost everyone who is in the USA on a temporary visa, applies for change of status for the green card by continuing to reside in the USA without returning home. I was an exception, but the fact that I did it as per the rule did not speed up the issuance of the green card. The Labor Certification Process went through quickly, but the process of getting the J-visa Waiver from Richard L. Fruchterman of USIA Washington (for the third time!) took about six months. Once again, Bill Lampe, the Chair in Hawaii, was gracious enough to help me without losing his patience. He put through several calls to Fruchterman assuring him that when I arrived in Hawaii, I had no plans to settle down in Florida — and that was certainly true. Lampe told Fruchterman, that Paul Erdős, one of the greatest mathematicians of the twentieth century, was impressed with my work, and convinced Florida to make the offer. This persuaded Fruchterman to give the waiver, but he could do so only after various preliminary clearance letters were obtained — such as from the Indian tax authorities that I owed no taxes, from the police in India that I was not needed in connection with any criminal case, from the Government of India that they had no objection to my leaving India, etc. Although these appear to be routine, getting such clearance letters in India requires effort and often one has to know the right people to get such obvious things processed without delay. Both my father and my father-in-law, with their vast web of contacts, helped me in getting such clearance letters in India, following which Fruchterman issued the waiver. But then there was a further period of waiting because there was a long queue of applicants for the immigrant visa and the United States would release a new set of visas in a block from time to time to process the persons in the queue. During this period, Al Bednarek, the longtime Chair of the University of Florida Mathematics Department, was very helpful and extremely patient. I must have called him about one hundred times over a period of a year to discuss the visa delays and my late arrival in Florida. Also these phone calls were at 7:00 am Florida time, an unacceptable hour in America to make phone calls! But Bednarek wanted me to call him early before he went to work, and he discussed various matters with patience and kindness. It is due to this inordinate delay in processing the green card, that almost everyone applies for the change of status to the green card while continuing to reside in the USA, and requests the INS to treat the case as "exceptional".

During this long period of waiting, we decided to go about our life in Madras as if we were going to be there permanently. My four-year daughter Lalitha joined the kindergarten class of the P. S. Senior Secondary School. There was a dual significance to this — (i) this school was born out of the famous P. S. High School my father attended, and was located within the campus of the old school, but separated by a wall, and (ii) the Principal of the school was Mrs. Alamelu Krishnan, now Alamelu Ganapathy, who was my mathematics teacher in Vidya Mandir. Mathura continued with her dance and music lessons, and I went about my work at MATSCIENCE with the commitment of a permanent member. I even became a member of the Rotary Club of Madras. This Rotary membership had a beneficial effect as will be seen soon.

During the academic year 1985–86, the mathematics group at MATSCIENCE was being built up by Seshadri. I had no problem with this, because I wanted to see the growth of the mathematics program at the Institute. However both Sudarshan, the Director, and Seshadri, the Head of Mathematics, retained their positions at their home institutions — Sudarshan at the University of Texas in Austin, and Seshadri at the Tata Institute. Both came to MATSCIENCE on five year contracts. This gave us the indication that they were unsure of their commitment to the institute and could leave if things do not go as planned. It so happened, that in 1988, there was a turmoil at MATSCIENCE, following which both Sudarshan and Seshadri left the institute; Seshadri did not return to TIFR but joined the SPIC Science Foundation in Madras as the Head of a newly formed mathematics unit.

During 1985–86, Seshadri invited several members of the Tata Institute for lectures and for advice. I attended many meetings with Seshadri in his office in the company of his TIFR colleagues. In these meetings, there was reference to work being done at the School of Mathematics at TIFR as serious mathematics. So I realized that any major decision in mathematics would have to be vetted by TIFR.

In 1986, plans were being made for the Ramanujan Centennial to be celebrated in December 1987. M. S. Raghunathan of the TIFR, the Chairman of the National Board for Higher Mathematics (NBHM) of India, was the chief organizer of the centennial celebrations because the NBHM was providing significant funds for the event. Raghunathan visited MATSCIENCE frequently and there were meetings with him in Seshadri's office to which I was invited. Raghunathan said that the plan was to have a Public Function in Madras on Ramanujan's 100th birthday, December 22, 1987, when Ramanujan's wife (widow) Janaki Ammal would be honored, and there would be just two lectures that day — one by mathematician Selberg and another by the astrophysicist Chandrasekhar. Following this there would be a conference for a few days in Madras featuring lectures by various experts on Ramanujan's work, most notably the "gang of three" of the Ramanujan world — George Andrews, Richard Askey, and Bruce Berndt. These would be lectures of wide appeal. Raghunathan said that after this, there would be a conference at the TIFR in Bombay in early January 1988, where it will be a more expert audience and the lectures would be more technical. Raghunathan said that he would like to hear proposals from others. I said that the Ramanujan Centennial must be used as an opportunity to showcase mathematics being done in various centers in India, that besides the two main conferences, the NBHM ought to support smaller (satellite) conferences in other centers like the Indian Statistical Institute in Calcutta which had strength in Probabilistic Number Theory whose origins can be traced to Hardy-Ramanujan, and in Mysore where there was activity in the area of hyper-geometric series, to name two. But the NBHM preferred to focus on the conferences in Madras and at TIFR. In the end, different universities in India had their own Ramanujan Centennial celebrations and many of the leading researchers from around the world who came for the Centennial, spoke at these other meetings as well (see §7.2).

On August 31, 1985, I received a letter from Professor Schinzel, Editor of Acta Arithmetica, inviting me to submit a paper for a Special Volume in honor of Paul Erdős to be brought out for his 75th birthday. I had just completed all the details for the paper "An Erdős-Kac Theorem for integers without large prime factors", and so I decided to submit that to the Erdős 75th birthday volume of Acta Arithmetica where it was accepted.

In September 1985, I had the pleasure of hosting Michel Waldschmidt from Paris, who visited MATSCIENCE for one week (September 4–10). At my request, he gave a Public Lecture at the Alladi Centenary Foundation on "Diophantine Equations - a walk through time" that was appreciated by representatives from different professions. At MATSCIENCE he gave three lectures on transcendental number theory.

Michel Waldschmidt loves India — its people, its culture, and its scenery, and so he visits India regularly. He told me that he prefers to visit Asia, whose ancient culture he appreciates so much. Waldschmidt is an avid jogger, and he wanted to jog around Madras. I was nervous that in doing so he might run into the danger of being bitten by stray dogs, or might inadvertently enter certain areas of Madras which are either unsafe or dirty. My good friend G. P. Krishnamurthy (GP as he is known), who enjoys the company of academicians, was an active jogger in those days, and so I introduced GP to Waldschmidt and suggested that GP might be Waldschmidt's jogging companion. This arrangement worked really well, and on one occasion, after the jog, GP invited Waldschmidt to his apartment to enjoy dosas for breakfast prepared by his grandmother. (GP who was two years my senior in Vivekananda College, became my dearest friend. Our acquaintance began when we were doubles partners in various city tennis tournaments. But what strengthened the bond was GP's high regard for my grandfather Sir Alladi and for the many accomplishments of my grandfather and father.)

The next number theory visitor to MATSCIENCE was Matti Jutila from Turku, Finland, in December. He is a powerful analytic number theorist who had done significant work on Linnik's constant related to the problem of the least prime in an arithmetic progression. He gave a beautiful lecture that was thorough in all the details.

In early January 1986, it was a pleasure to receive Springer Lecture Notes 1122, the refereed proceedings of the Erdős-70 Conference that was held in Ooty in January 1984. The Proceedings edited by me, contained contributions from Erdős, and from leading mathematicians who spoke at the conference as well as those who were invited but could not come. My paper in these Proceedings was a detailed account, complete with proofs, of my work on the study of moments of additive functions using Laplace transforms and sieve methods.

During May 7–15, 1986, there was a Workshop on the Mathematics of Computer Algorithms at MATSCIENCE organized by the energetic Veni Madhavan of the Computer Science Department of the Indian Institute of Science, Bangalore. I gave a series of four lectures on "An Introduction to Multiplicative Number Theory". One of the bright Masters students of Veni Madhavan who interacted with me was Tetali Prasad. He was interested in pursuing graduate studies in America and I was happy to support his application. He got his PhD in 1991 at NYU under the direction of Joel Spencer and became a Professor at Georgia Tech. In 2021 he moved to the University of Pittsburgh as Chair of Mathematics.

During this period of waiting in India, I was in active correspondence with leading number theorists in Europe and the United States and kept them informed of progress in my research. In particular, I studied the multiplicative generalization of the sieve (that I had employed in probabilistic number theory), as a sieve for its own sake, and obtained generalizations of various classical sieve results in a multiplicative setting. So I wrote a paper entitled "Multiplicative functions and Brun's sieve" which I submitted to Acta Arithmetica in 1986, where it appeared two years later. Owing to all this correspondence, I received a number of invitations for lectures in Europe and America and so I decided to make a round the world trip from August to October, 1986. The main difficulty was to enter the United States, and this would not usually be permitted because I was waiting for a green card. Here is where the membership at the Rotary Club helped.

One of the members of the Rotary Club was U. S. Consul General Mr. John Stempel. I got to know him really well. He would ask me to sit at his table during the weekly Rotary lunch meetings, and even invited me to his magnificent mansion to play tennis with him. When I informed him about the invitations I had received in America for lectures at various universities, and my desire to attend the International Congress of Mathematicians (ICM) at UC Berkeley in August, he told me to apply at the US Consulate for permission to enter the USA briefly, and asked me come upstairs to his room after submitting my application downstairs in the visa office. I did that following his advice. When I was in his office, he called the consulate office downstairs and said that he stands guarantee that I will not remain in the United States on this trip but would return to India! What a gracious gesture of support!! I went downstairs, and the visa office had my permission letter ready in a sealed envelope, to be opened by the immigration officer at the port of entry into the USA on that trip. So all was set for me to go on a round-the-world academic tour.

I decided to make the trip eastbound via Singapore, so that my entry point in the USA would be Honolulu where I had a colloquium scheduled, following which I would go to Berkeley to attend the ICM. With that in mind, I went to the Singapore Airlines office to arrange my ticket and planned to purchase it a week later. That night there was party at the Taj Coromandel Hotel hosted by Northwest Airlines, and my father and I were invited to this party. It was a pleasant surprise at the party for me to meet my high school classmate V. S. Subramaniam (alias Subbu) who was at that time a senior representative at KLM, the Royal Dutch Airlines. When I told him of my trip in a few weeks, and that I was planning on purchasing the ticket from Singapore Airlines, he said that if it is agreeable to me, he would "take over" the reservation from Singapore Airlines and issue the ticket. He said that if I would pay the regular round-the-world economy fare, he would upgrade me to Business Class for the entire trip! The only requirement was that I had to travel on KLM on the transatlantic, and from Amsterdam to Delhi. He also said that he would keep the Singapore Airlines flights from Madras to Tokyo unchanged. What is more, he said he would provide free hotel accommodation in Singapore, Tokyo, Amsterdam and Delhi. What a deal?! How could I pass this? So I readily said YES!

The full fare economy class round-the-world that I paid was only (Indian) Rs. 25,660, which was about US $2,000, because the exchange rate was a bit less than Rs. 13 for $1. Not only were airfares much less expensive then, but the rupee also was much higher against the dollar than it is now. Even though I was attending two conferences on this trip — the International Congress of Mathematicians in Berkeley in August and the Oberwolfach Conference in Analytic Number Theory in late September — I bought the ticket from my personal funds. All I asked MATSCIENCE was leave to make the trip and this was granted.

I departed Madras on July 27. From Madras to Tokyo via Singapore, I enjoyed the renowned Raffles Class service of Singapore Airlines which advertised it as "a service that other airlines talked about", and it lived up to that high claim. It was a pleasure to be back in sunny Honolulu and give a talk in the Mathematics Department of the University of Hawaii arranged by Ed Bertram. After a two-night stay with the Ramanathans, I left by a United Airlines Boeing 747 for San Francisco. For the first time I was attending an International Congress of Mathematicians, and the campus of the University of California, Berkeley, was teeming with about 3000 participants. It was a thrill to attend the Prize ceremony on the opening day; three Fields Medals were given to Simon Donaldson of Oxford University for his path-breaking work on the topology of four dimensional manifolds, Gerd Faltings of Germany (but at Princeton University in 1986) for his proof of the Mordell Conjecture, and Michael Freedman of U.C. San Diego for his proof of the four dimensional Poincaré Conjecture. The Nevanlinna Prize in Information Sciences was given to Leslie Valiant. Several years later, I would have the pleasure of receiving a letter from Faltings accepting one of my papers for the very prestigious journal *Inventiones Mathematicae*. When I left the auditorium after the Fields Medal ceremony, I found former Fields Medallists Atle Selberg and Lars Hormander (differential equations) — both Scandinavians — engrossed in a conversation on the steps outside, and I quickly snapped a picture of them which I cherish to this day.

The ICM offered me an opportunity to meet several leading mathematicians from around the world. In particular, it was a pleasure for me to meet Professor Jonas Kubilius of Lithuania, one of the four founders of Probabilistic Number Theory (along with Paul Erdős, Mark Kac, and Paul Turan), the area in which I was working then. He was a gentleman to the core and had very kind things to say about my work. He said that I should visit the University of Vilnius in Lithuania where he was the Rector since 1958. Indeed, a few years later I received an invitation to a conference at the University of Vilnius, which regrettably I could not attend.

Paul Erdős, as was to be expected, was there at the ICM, and we took long walks in the Berkeley campus talking about the career at the University of Florida that I was soon to take up.

From Berkeley I proceeded south to spend a few days in Los Angeles visiting UCLA and my old friends there, and then to San Diego, to visit my paternal cousin Professor V. S. Ramachandran, a brilliant neuroscientist with a worldwide reputation. My father was sure that he would rise to the top of his profession which he did, and he in turn always acknowledged that my father inspired him and encouraged him to pursue research when he was studying for his medical degree in Madras in the early seventies.

After the ICM, my next academic visit was to the University of Colorado at Boulder for three days (August 18–20), where I gave a colloquium at the invitation of Professor Peter Elliott. Following my talk, I was treated to a dinner at the home of the Elliotts, where Jean Elliott had so thoughtfully prepared delicious Indian vegetarian food. The Elliotts took me on a trip to the Rocky Mountain National Park. The scenery was spectacular.

From Denver I left for Atlanta where I took a limousine to the University of Georgia in Athens. My host Carl Pomerance arranged my talk in the Number Theory Seminar where I spoke on "Multiplicative functions and small divisors". My good friend Andras Sarkozy from Budapest was visiting there to work with Pomerance. Sarkozy is a fine tennis player, and the two of us had a memorable match on the har-tru (greyish clay) courts on the campus of the University. Pomerance hosted a dinner in my honor at his home and invited a few of his colleagues, one of them being Kannan, a probabilist, who considered my father as one of his gurus. This is because Kannan had received his PhD under the direction of Barucha-Reid whose book on probability dealt with my father's fundamental work in detail.

After my brief but very pleasant visit to Georgia, I went to the

University of Florida for two weeks (late August to mid-September). Al Bednarek, who had been Chairman for 17 years had stepped down, and Gerard Emch, a mathematical physicist from Rochester had been appointed as the new Chair. He had been promised 25 positions in mathematics over a five-year period. So there was great excitement in the Department over this sudden expansion. The excitement was even greater that Fields Medallist John Thompson had accepted a Graduate Research Professorship at the University of Florida, but for only in the Fall of each year starting in 1986, because he had not yet relinquished his position at Cambridge University. So I arrived in Florida at the time of this great development. It was very kind of Gerard Emch to invite me to attend the faculty meeting when these new positions were discussed, even though I had not joined the Department because I was still waiting for the green card.

While in Florida, I visited Bhimsen Shivamoggi at the University of Central Florida (UCF) in Orlando on a weekend. He was the only other appointment at MATSCIENCE in 1979 beside me that M. S. Narasimhan had approved, but he left MATSCIENCE two years after my father retired for a permanent position at UCF.

After the two-week stay in Florida, I went to New Hampshire to visit my father's cousin, Professor K. Sivaprasad and his wife Indira, before departing for academic assignments in Europe. At that time there was no tradition in number theory at the University of New Hampshire, so the visit there was mainly social. (Now the University of New Hampshire is known the world over for the revolutionary work of Zhang — his stunning result on bounded gaps between primes.) While I was visiting New Hampshire, my father called me on September 17 to say that he heard from the US Consulate in Madras that the queue for the immigrant visa had moved significantly since a new quota had suddenly been released. So I was invited to an interview in Madras on the 14th of October to be given the visa. Thus the long wait was over. Also the timing was perfect because after completing my lecture assignments in Europe, I would be back in Madras on October 9. With this positive news about the US immigrant visa, I was in an ecstatic mood as I departed for Europe.

My high-school classmate Subbu had given special instructions to KLM, the Royal Dutch Airlines, and so I was treated royally at JFK airport New York, on board the KLM 747 on the flight from New York to Amsterdam, and in Amsterdam, where KLM very graciously put me up at the Grand Hotel Krasnapolsky. KLM is the oldest airline in the world operating under its original name, and is one of the few airlines — may be the only one —

that gives its premium passengers a sensible and thoughtful gift, namely a porcelain model of a Dutch house. There are several such exquisitely beautiful model houses, and after many flights, one can have an impressive collection of these lovely souvenirs.

After a quick day of sightseeing in Amsterdam, I left for Frankfurt, from where Professor Wolfgang Schwartz took me in his car to Oberwolfach for the Conference on Analytic Number Theory (September 21–27). Schwartz was a very kind and genial person, but on the autobahn, he drove at breakneck speed. When I told him that I was impressed with his skill in driving so fast, he modestly responded by saying that his speed is nothing compared to what could be achieved by a Mercedes or a BMW.

The Oberwolfach Conference on Analytic Number Theory was held once every two years in the Fall and its three organizers were H.-E. Richert, Wolfgang Schwartz and Eduard Wirsing. I spoke at this conference on "Brun's sieve and bounded multiplicative functions". My half-hour talk was the second after lunch on one of the days. The first speaker after lunch on that day was Professor Hubert Delange from Paris, a senior mathematician (after whom the Seminaire Delange is named) who had done significant work in analytic number theory, especially on moments of additive functions using generating functions. He was to speak on "An old theorem of Kubilius" and my work intersected with the theme of his talk. So it was natural that Delange and I were speaking back-to-back. When I entered the lecture hall after lunch, I found that Delange had filled six black boards — yes, not one black board, but six — in his calligraphic handwriting, and had done this by skipping lunch! So I thought that since everything he wanted to say was on the blackboard, he would have no problem finishing on time. But that was not to be. He went 10 minutes beyond his 30 minute allotted time, and out of respect for his seniority and age, the chair of the session did not intervene to stop him after 30 minutes. Thus I had only 20 minutes left for me, and I concluded my talk in those 20 minutes. Several members of the audience including the chair of the session thanked me for bringing the conference program back on time!

The Oberwolfach conference had an especially strong attendance by European number theorists and it was a nice opportunity for me to meet old friends like Paul Erdős and Vera Sos Turan and make new acquaintances such as with Janos Pintz of the Alfred Renyi Institute in Budapest, and Martin Huxley of Cardiff, to name just two. Huxley became famous by improving Montgomerey's exponent 3/5 to 7/12 for the problem of gaps between consecutive primes (see Section 2.3). Even though I was meeting

Huxley for the first time, he was so friendly, that I felt I had known him for years; in fact, before I left Oberwolfach, he had composed a limerick about me! He composes limericks about mathematicians all the time. It was also at this conference that I met Gerald Tenenbaum of Nancy for the first time, and I was to visit his department a few days later.

From Oberwolfach, I returned to Frankfurt from where I flew to Paris, where at the invitation of Michel Waldschmidt and Hedi Daboussi, I gave two talks on Monday, September 29, first in the morning for the Groupe de Travail en Theorie Analytique for which Daboussi was the Maitre de Conference, (= organizer) and then in the afternoon for the Seminaire de Theorie des Nombres, Paris (STNP) — formerly called Seminaire Delange-Pisot-Poitou. Both talks were arranged at the Institut Henri Poincaré. The talks were attended even by some mathematicians outside of Paris, like Gerald Tenenbaum of Nancy. My talk in Paris was published in the Proceedings of the Paris Number Theory Seminar (1986–87) which appeared as Volume 75 in the Progress in Mathematics Series of Birkhauser. By the bye, Waldschmidt had written to me in July saying that the STNP typically starts only in early October for the Fall term, but to accommodate my schedule, they made the first meeting of the STNP on September 29. I definitely appreciated this gesture of support.

The Seminaire de Theorie des Nombres, Paris (STNP) was founded in the late fifties by Hubert Delange. He organized the seminar first along with Charles Pisot and later with George Poitou, and so the seminar was also known as the Seminaire Delange or Seminaire Delange-Pisot-Poitou (DPP) in the French mathematical community. The Seminar was instrumental in reviving the tradition in analytic number theory in France. The Seminar always met on Monday afternoons at the Institut de Henri Poincaré. Delange continued to be active in research even in his retirement and attended the STNP regularly. He attended my talk at the STNP and joined me and my hosts for dinner after my talk. At that time, the STNP had a contract with Birkhauser for the publication of its annual proceedings in the Progress in Mathematics series. By the bye, Delange also founded the mathematics department of the Université de Paris-Sud in Orsay, and Hedi Daboussi, one of my two hosts in Paris, was his former PhD student.

Weather in Paris was gorgeous, and so the charming city looked even more spectacular. On the Sunday before my talks, I walked almost all day to see the magnificent sights of Paris, and did not feel tired at all.

From Paris I took the train from Gare de l'est (Train Station East) to Nancy where I was warmly received by Gerald Tenenbaum. I stayed for two nights at the home of the Tenenbaums where he and his wife hosted me magnificently. The University of Nancy is named after the great mathematical physicist Henri Poincaré, and the Faculty of Sciences after the eminent mathematician Elie Cartan. So great is the respect for mathematics throughout France, that several things are named after mathematicians — streets, buildings, and universities. This high regard for mathematics is one reason that square mile by square mile, France has more mathematics than any other country in the world!

Gerald Tenenbaum is one of the world's experts in the study of divisors and on work relating to $\Psi(x, y)$, the number of integers up to x that are free of prime factors greater than y. Paul Erdős thought very highly of him, and it was Erdős who put me in touch with Tenenbaum owing to our common interests in $\Psi(x, y)$. The visit to Nancy began a lifelong friendship with Tenenbaum, and over the years I have profited by reading his many wonderful papers in analytic number theory.

It was not all mathematics at Nancy. Tenenbaum and I spent an enjoyable afternoon at his tennis club where we played on the slow red clay. In those days, I always travelled with my tennis racquet and never missed an opportunity to play.

I took the train back from Nancy to Paris, from where I flew to Bordeaux. Jean-Marc Deshouillers, my host in Bordeaux, had arranged for his eminent colleague Henri Cohen to receive me at the airport. But when I arrived in Bordeaux, there was no one there to meet me. So I took a taxi to the comfortable hotel where Deshouillers had arranged my stay. That night, when Deshouillers picked me up from the hotel for dinner, he was surprised to learn that Henri Cohen did not show up at the airport. The next day when I met Cohen at the University during my seminar, he apologized and said that he forgot to pick me up — an act of absent mindedness!

After picking me up at my hotel, enroute to the restaurant for dinner, Deshouillers took me to the magnificent mansion — a villa — of his colleague Michel Mendes France, because Mendès France and his wife were joining us for dinner. Michel Mendès France was the Chairman of the Mathematics Department of the University of Bordeaux. His father Pierre Mendès France was Prime Minister of France in 1954–55, but Michel was so modest, that he never wanted to be recognized or identified in public as the son of the former prime minister. Michel Mendès France had a fine sense of humor and I found his company to be most enjoyable.

My talk the next day (October 3) at the University of Bordeaux was very well attended and received. I was invited to submit a written version of my talk for publication in the Seminaire de Theorie des Nombres du Bordeaux, which I did. That night after my talk, Deshouillers and his wife hosted a wonderful dinner in my honor at their lovely home, attended by some of his colleagues and their spouses. Deshouillers is a connoisseur and has a superbly stocked wine cellar in his basement. So the finest of wines were flowing freely at dinner, but for me Deshouillers had the best grape juice he could get in Bordeaux!

Deshouillers was planning his first visit to India in January 1987 which included a stay at MATSCIENCE, and so he wanted tips and advice from me about India in general. By the time I was in Bordeaux, I knew that I would not be in Madras in January 1987 because my green card interview was fixed for mid-October 1986 and so I would leave for Florida before the year ended. Thus I would miss Deshouillers' visit, but he was most pleased that I helped him plan his trip, and in later years has expressed his appreciation of my help on numerous occasions. I said that in my absence, my father would be glad to provide any help he needed. My father arranged a public lecture by Deshouillers on January 23, 1987, at our home under the auspices of the Alladi Foundation. The title of the lecture was "Summing integers - a promenade through arithmetic". That lecture was well attended by academicians, school teachers and students. In fact, one of the students who attended the talk was thirteen year old Kannan Soundararajan who later became one of the leading number theorists of our generation. In subsequent years, every time Deshouillers visited Madras, he made it a point to call on my father.

I had a full free day in Bordeaux before departing and so Deshouillers took me on an excursion to the countryside visiting several vineyards in that process. He met with the owner of each vineyard where we stopped, visited the wine cellar of that vineyard, and purchased wine. He was a true connoisseur and he devoted time to make such trips and purchase the best wines Bordeaux offered. That is why he had an outstanding cellar. Since I was a tee-totaller, I did not taste the wines being offered at these vineyards. Instead I had a picture of me taken posing beside the grape vines!

I departed Bordeaux for Stuttgart via Paris and this was my final academic stop in Europe on this trip. In Stuttgart my host was Professor Bodo Volkmann, whose two beautiful lectures I had heard at MATSCIENCE in 1972 when he visited Madras as the guest of the Max Mueller Bhavan. He was a gracious host. After my talk on October 6, he showed me around Stuttgart which is the home of the Mercedes Benz.

I left Stuttgart for India via Amsterdam in high spirits because this academic trip around the world was so fruitful. On the final portion of the journey from Amsterdam to Delhi, I enjoyed once again the superb hospitality of KLM arranged by Subbu both on ground and in the air. In Delhi, KLM graciously arranged accommodation for a night at the stately Maurya Sheraton Hotel (which combined western amenities with Indian opulence) before my departure for Madras.

Back in Madras on October 9, with the US Green Card interview set for mid-October, we planned to leave soon after so as to reach Florida before the end of December to be in time for classes to begin for the Spring (Winter) term in January. We decided to travel the longer route via the Pacific with stops in Singapore, Honolulu, Los Angeles and Houston, visiting friends enroute. Singapore Airlines which was disappointed losing my ticket to Subbu of KLM a few months earlier, said that they would match "the deal" offered by Subbu. Thus Mathura and my daughter Lalitha enjoyed the transpacific air journey as much as I did a few months earlier.

Our green card interview was most pleasant. The Madras US Consular Office was pleased that we did it the "right way" by staying in India during the waiting period. I have heard from several Indian friends in the USA, that even though their immigrant visas were approved in their interviews, the INS officers were quite rude to them because they stayed back in America and requested a change of status there.

With the immigrant visa in hand, I informed MATSCIENCE about my acceptance of the job in Florida. I appreciated that I was granted one year's leave without pay. I also informed M. S. Raghunathan that I could not serve on the Local Arrangements Committee for the Ramanujan Centennial Conference in Madras organized by the NBHM because I was leaving for Florida, and he understood my reasons. However, a remarkable set of events were to take place in 1987 which gave me a role in the Ramanujan Centenary Celebrations in Madras, and these will be described in the next chapter.

Alice and Edward Bertram with Krishna and Mathura Alladi, at the top of Diamond Head volcano to watch the sunrise over Honolulu, Feb 24, 1985. The visit to the University of Hawaii in 1984–85 was a life changer for Krishna (see §6.4).

Paul Erdős delivering a talk on "Reminiscences of a mathematician" at Ekamra Nivas, the Alladi family home in Madras, India, on January 14, 1984. Seated next to Erdős is His Excellency S. L. Khurana, the Governor of Madras, who presided over the lecture.

Chapter 7

Major change in my life and in my research

7.1 The Florida chapter of my life begins (1986–87)

We arrived in Gainesville during the second half of December 1986, and so we had a couple of weeks to settle down before classes for me began for the Spring Term of 1987 at the University of Florida, and for Lalitha to join school in January. In Florida, they do not call it the Winter Term, but the Spring Term! We had a two bedroom apartment in the "Gardens" apartment complex about 20 minutes away from campus, and close to the home of Professor R. Charudattan. The Charudattans were our first Indian friends in Gainesville, and during our first year, we consulted them on practically everything. For instance, it was at their suggestion that we admitted Lalitha to Martha Manson Academy, a private school near our apartment, which the Charudattans recommended highly because their son had studied there and they had observed the school's programs and activities very closely. Before admission to Martha Manson Academy, five year old Lalitha had to undergo certain tests by a school chosen psychologist that included a general aptitude test and a test of fluency in English, and she did very well in these tests.

Since accepting the offer from the University of Florida in March 1985, and during the period of our waiting in Madras for a year and a half for our immigrant visas to be processed, several changes had taken place in the mathematics department in Gainesville — hereafter to be referred to as the Department. Long time chairman Al Bednarek stepped down at the end of the academic year 1985–86 for various reasons. Over the years, he had asked for the expansion of the Department faculty not only to meet the higher enrollment of students in the mathematics courses, but also to enhance the research profile of the Department. But the College of Liberal Arts and Sciences (CLAS) never fully gave the positions the Department requested each year, and so due to retirements and attrition, the Department did not grow in size, but student enrollment had risen significantly. So Bednarek decided to step down, and the Department requested the Dean for an external chair to be appointed in the hope that a person coming from outside would be in a much better position to leverage with the College for several new positions. This strategy worked; the Dean approved an external search for a new Chair to start in Fall 1986. Gerard Emch, a mathematical physicist from the University of Rochester, came in as the new Chair and with a "promise" from the College for 25 new positions to be spread over the next 5 years. This was indeed remarkable because the Department had at that time a little over 50 faculty members. I say "promise" because as I found out after many years of living in the United States, that these are not binding, and that Universities often go back on their promises much to the aggravation of the departments, as it happened in Florida two years later. In any case, when I arrived in Gainesville, there was great excitement in the Department that Dean Sidman had authorized five new tenure-track positions to begin in Fall 1987. So there were several potential candidates who delivered colloquia in Spring 1987. Thus during my first term (Spring 1987), I witnessed a very active colloquium schedule in the Department.

One thing Bednarek accomplished before he stepped down, was to get 1970 Fields Medallist John Thompson of Cambridge University appointed as Graduate Research Professor for a five-year period, but only for the Fall each year. This was a miracle and so I will tell you how this happened.

Since I did not join the Department in Fall 1985, my salary for 1985–86 was used to offer a visiting position to Chat Ho, a group theorist who was a former PhD student of Thompson. Ho was able to join the Department in Fall 1985 and his position was converted to a tenure-track position the next year. Ho drew the attention of the Department that Thompson might be "available". Since Bednarek wanted to enhance the research profile of the Department, he pursued this idea. In the end, when Thompson agreed to come to Florida, he did not want to relinquish his position at Cambridge University, and so he accepted to be in Florida each year in the Fall Term, for a period of five years. And he did arrive in Florida during Fall 1986, which was his first term in Gainesville.

John Thompson at that time was working on the Inverse Galois Problem, that is to determine which groups arise as the Galois groups of field extensions. He was collaborating with Michael Fried of UC Irvine on this fundamental question, and so at Thompson's suggestion, Fried was appointed. Fried joined in Fall 1986 with leave from Irvine. I had known

Fried when I was a graduate student at UCLA, because he was the organizer of the Irvine Number Theory Seminar, and Straus had taken me a few times to attend that seminar. One of the appointments as a tenure-track assistant professor during this "first wave", was Helmut Voelklein, an algebraist from Germany. He collaborated in a major way with Fried and Thompson and became a world authority on the Inverse Galois Problem.

Some of the other senior appointments made in that first year were (i) Jeffrey Remmel from UC San Diego, who had done significant work in Combinatorics and in Set Theory, and had a major collaboration with Adriano Garsia, (ii) Paul Ehrlich in Differential Geometry, (iii) Greg Gallaway, a mathematical physicist from the University of Miami, and his wife (iv) Michelle Wachs, a reputed combinatorialist, also from Miami. Of these senior appointments, only Ehrlich remained in the Department. The others left after a year or two for reasons that will be described later. In any case, it was exciting for me to join the Department in that period of growth. The one comment I would like to make here is that, while it might be necessary to commit to a major expansion in order to get an outside chair, it should not be necessary to get an external chair in order to expand a department in a major way. Almost always, external searches seem to be the only way to get the university administration to commit to a significant influx of funds. This is a problem with the way the academic set up works.

In view of Thompson's interest in the Inverse Galois Problem and his lectures along with Fried's on that topic in Fall 1986, it was decided to run a seminar on modular forms. So I participated in this seminar and spoke on elliptic and theta functions first in Spring 1987, and then in Fall 1987 when Thompson was back in Florida. Actually, in Fall 1986, during my round-the-world trip, I spent two weeks in Florida in September, and had the occasion to hear the lectures of Thompson and Fried.

As the only number theorist in Florida, and with an interest to activate the program in that area, I started teaching the graduate course in number theory in Spring 1987 (on analytic number theory). The notes I had prepared on analytic number theory when I taught it in Michigan and later at MATSCIENCE, proved useful. I even had a graduate student in physics take my number theory course.

My parents visited us in Gainesville in 1987, as they did every year since then, but that first year was special for two reasons: (i) they had the pleasure of meeting and getting to know my colleagues at the University, and (ii) since Mathura was pregnant with our second child and the delivery scheduled for October, my mother could be of assistance to Mathura in the first months of pregnancy. Since we were new to Florida ourselves, we did a lot of sightseeing around the state during the first year and my parents were with us on many of these trips. In anticipation of the new arrival, we moved to a three bedroom apartment at Millhopper Village in July 1987 while my parents were there, and Lalitha shifted to Littlewood Elementary School next door, in Fall 1987.

Shortly after moving in to Millhopper Village Apartments, Mathura had to undergo a procedure to ensure that the delivery would not be premature. It was most kind and considerate of Mathura's obstetrician Dr. James Lukowski to come to our apartment to check on her progress.

John and Diane Thompson returned to Florida in September 1987 for the Fall Term a few days before my parents departed. They took my parents out to dinner to reciprocate the warm hospitality my father had extended to John Thompson in Madras in 1986.

After a six month stay, my parents left in mid-September, but only after Mathura's mother Mrs. Gomathi Krishnan arrived from Madras to be with Mathura. It is customary for the girl's mother to help her daughter before, during, and after the delivery, and in India, usually a girl will to go to her mother's house a few months before delivery. Mathura did that in 1981 when Lalitha was born because we were in Madras then. Since we had just settled down in Florida, Mathura's mother came to Gainesville instead to help Mathura with the delivery of our second child.

Our second daughter was born on October 4, 1987, one day before my birthday. We named her Amritha, since *Amrith* is the sweet nectar that the Hindu Gods drink that make them immortal! And she, along with Lalitha, have been like that nectar in our lives.

Mathura's father Dr. N. C. Krishnan, and her youngest brother Sriram, arrived in mid-October and spent a month with us. Mathura's parents and her brother left for Madras in mid-November after she had recovered from the strain of the delivery.

One other visitor for us in 1987 was my college teacher Prof. P. R. Vittal, who had come to the USA for a workshop in the area of probability. He spent a few days with us while my parents were visiting.

1987 was not only the year of Amritha's birth, but also the Ramanujan Centenary, and so it was doubly special for me.

In 1987, both during the Spring and Fall Terms, I gave several lectures in Florida on Ramanujan's work, and some of these were in my graduate course in number theory. This caught the attention of M. Prakash, a reporter from *The Hindu*, India's National Newspaper, based in Madras. In

Spring 1987, Prakash had arrived at the University of Florida to get a degree in journalism. After interviewing me, he wrote a fine article entitled "Mathematics Department teaches theorems of Indian genius" which appeared on November 17 in the *Gainesville Sun* the local newspaper. Prakash also interviewed Gerard Emch, the Chair of Mathematics, and Bruce Edwards, the Associate Chair, who was in charge of course assignments, and they said very positive things about Ramanujan's mathematics as being part of my course. By the bye, Prakash was my college mate at Vivekananda College the same years I was there (1971–75)! It is a small world.

In October there was a one-day conference on Ramanujan at Framingham State University near Boston arranged by Thomas Koshy. There were just four speakers — Bruce Berndt, M. Ram Murty, Koshy and me — and we spoke for one hour each, two lectures in the morning and two in the afternoon. It was a well organized event and was attended by a large group of students and faculty. This was a precursor to my participation in the Ramanujan Centennial in Madras in December 1987 which I describe next.

7.2 The Ramanujan Centennial (1987–88)

Srinivasa Ramanujan (1887–1920) is one of the greatest mathematicians in history. He was born on December 22, 1887 in a poor Hindu family in the town of Erode in a rural part of the state of Tamil Nadu in South India, and grew up in a neighbouring town of Kumbakonam where his father worked. Ramanujan did not have any formal mathematical education, yet he produced spectacularly beautiful results that revealed surprising connections between seemingly disparate areas of mathematics. There is a legend that the Hindu Goddess Namagiri, the deity in the temple in the neighbouring town of Namakkal, used to appear in his dreams and give him mathematical formulae. This added to the aura of mystery surrounding his discoveries. Often, Ramanujan would get up in the middle of the night and write down formulae on a piece of slate lest he would forget them when he got up in the morning. He recorded his findings in Notebooks which he would show to his teachers and other educated persons. While some of them realised that Ramanujan was unusually talented and that his Notebooks contained important new results, none of them could properly evaluate his work. Thus Ramanujan wrote a letter in 1913 to G. H. Hardy of Cambridge University, England, stating dozens of mathematical results, and said that he was approaching Hardy because "the local mathematicians are unable to understand me in my higher flights." The letter contained startlingly beautiful formulas, many of which were correct, a few were incorrect, and there were others that Hardy could neither prove nor disprove because they were so deep. So Hardy in consultation with his colleague J. E. Littlewood came to the conclusion that Ramanujan was a true genius in the class of Euler or Jacobi. Hardy felt that Ramanujan should not waste more time in India, but should come to Cambridge so that his untutored genius could be given a sense of direction. But Ramanujan belonged to an orthodox Brahmin family and their belief was that it was a sin to cross the seas. Thus Ramanujan's mother did not give him permission to go, and so Ramanujan declined Hardy's invitation. There is a story that one night the Goddess of Namakkal came in his mother's dream in which she saw him being honored in a large assembly of "white men" and the Goddess ordering the mother not to stand in the way of her son's recognition. So when Ramanujan's mother got up in the morning, she gave him permission to go to Cambridge. Thus Ramanujan sailed to England in 1914.

During the five years that Ramanujan was in England, he did path-breaking work, some in collaboration with Hardy. In particular, the paper he wrote with Hardy in 1917 on the series representation for the partition function, received wide acclaim, and the *circle method* that Hardy and Ramanujan initiated in that paper was subsequently developed by Hardy and Littlewood, and has become the principal technique to attack a wide class of problems in Additive Number Theory. Another paper of Ramanujan in collaboration with Hardy was on the normal number of prime factors of integers, which eventually led to the creation of Probabilistic Number Theory. Although Ramanujan's mathematical productivity was phenomenal, the rigours of life in England during World War I combined with his eating habits led to rapid decline in his health. Fearing the worst, Hardy made great effort to get Ramanujan elected Fellow of the Royal Society (FRS) first, and then Fellow of Trinity College in 1918, because he wanted these honors to be given to Ramanujan while he was alive. In 1919, Ramanujan returned to India a very sick man. Even though he was mortally ill, his mathematical faculties were as active as ever. He wrote one last letter to Hardy in January 1920, outlining his most recent discovery of the *mock theta functions*, which he classified as those of orders 3, 5, and 7. These are now considered among Ramanujan's deepest contributions.

Ramanujan died on April 26, 1920. But before he died, he used to ask his wife Janaki for sheets of paper to write down his new found formulae. After Ramanujan died, his wife had the good sense to collect these loose sheets of paper and hand them to the University of Madras from where they were despatched to Hardy. Although these sheets of paper contained several new results, including many on mock theta functions, Hardy put them aside and did not investigate them. On the other hand, G. N. Watson of the University of Birmingham, a leading authority in the field of special functions, investigated the third order mock theta functions that Ramanujan had communicated in his letter to Hardy, and gave a talk on this topic as his Retiring Presidential Address to the London Mathematical Society in November 1935. The title of Watson's talk was "The Final Problem - an account of the mock theta functions", and he justified the choice of this surprising title for his talk by saying that he compared himself to Dr. Watson who was trying to solve the mystery as to how Sherlock Holmes extricated himself from the grip of Moriarty as they plunged over the waterfalls in Switzerland!

After Watson died, these loose sheets of papers of Ramanujan, along with many of Watson's own papers were filed in the "Watson estate" at the Wren Library of Trinity College. Thus Ramanujan's last writings were buried inside the Watson Estate and so the world remained unaware of what had happened to these papers. (See [B1], Ch. 17 for details.)

In 1976, George Andrews who had worked on mock theta functions for his PhD, went to Cambridge University to look through the Watson Estate. To his surprise, he located there a set of papers in Ramanujan's handwriting containing hundreds of incredible identities, many of which were on mock theta functions. In view of his knowledge of mock theta functions, he realized that he had unearthed a treasure — that he was holding in his hands, what he named as "The Lost Notebook" of Ramanujan! Since the mid-seventies, Andrews has been analyzing the identities contained therein, and for many of them, he has provided proofs and explained them in a number theoretic setting. We owe primarily to Andrews our present understanding of the significance of the contents of Ramanujan's Lost Notebook.

Although Hardy admired Ramanujan's mathematics greatly, he said that it did not have the quality of the discoveries of Euler and Gauss. Since Hardy's time we have gained a much better understanding of Ramanujan's contributions and how they are related to mainstream developments in various branches of mathematics. Indeed, Ramanujan's discoveries have led to major progress in several fundamental areas of mathematics, and he is now regarded as one of the greatest mathematicians in history. The Ramanujan Centennial was an occasion for leading researchers from around the world to come to India, and to Madras (now called Chennai) in particular, to pay homage to Ramanujan. It was an appropriate time to take stock of the influence his work has had, and to assess its impact in shaping future developments. Although the NBHM prioritized its funding to two main conferences, one in Madras and another at the Tata Institute in Bombay, several universities and major educational institutions had strong interest in conducting their own Ramanujan Centenary conference or event, and were successful in getting support from their administration and other sources. The gang of three as they were humorously referred to (or as the Great Trinity in the Ramanujan World, as I would like to refer to them), namely George Andrews, Richard Askey, and Bruce Berndt, were invited to ALL such Centenary Conferences, and they attended as many as they could squeeze into their one month visit.

The NBHM Conference in Madras was scheduled to start on December 22, 1987, Ramanujan's 100th birthday. I will soon provide an exciting report of this conference and the incredible drama that took place, but first I describe the Ramanujan Centenary Conferences at the University of Illinois in the summer of 1987, and another at Anna University that I helped organize in Madras, just before the NBHM conference.

The very first, and indeed the largest and most comprehensive, Ramanujan Centenary Conference, was the one at the University of Illinois in Urbana with Bruce Berndt and Adolf Hildebrand as the lead organizers. The conference attracted more than one hundred participants, and was the only one in which Freeman Dyson of the Institute for Advanced Study, Princeton, and Rodney Baxter, the eminent Australian mathematical physicist, participated. The most exciting development at the Urbana conference was the resolution of Dyson's famous conjecture pertaining to Ramanujan's partition congruences, which I will briefly describe now.

In order to test the asymptotic formula for partitions that Hardy and Ramanujan had proved, Major MacMahon, a noted combinatorialist, prepared a table of values of the partition function $p(n)$ at the request of Hardy. On seeing the table, Ramanujan immediately wrote down three relations:

5 divides $p(5n + 4)$, 7 divides $p(7n + 5)$, and 11 divides $p(11n + 6)$.

Hardy was simply stunned because partitions represent an additive process, and Ramanujan had observed a divisibility property of partitions! Neither MacMahon who prepared the table, nor Hardy who checked this, noticed these divisibility properties. Ramanujan had the eye for the unexpected! Not only that, Ramanujan gave startlingly beautiful formulas for the generating functions of $p(5n + 4)$ and $p(7n + 5)$ from which his divisibility relations involving 5 and 7 followed.

Since partitions are combinatorial objects, it is natural to ask whether there is a combinatorial way to understand these relations, that is if there is a natural way to split the partitions of $5n + 4$ into five equal sets of partitions, and similarly for $7n + 5$ and $11n + 6$. In 1944, Freeman Dyson, then a young undergraduate student at Cambridge University, introduced a statistic called the *rank* of a partition, which splits the partitions of $5n + 4$ and of $7n + 5$ into five and seven equal sets of partitions, respectively. The rank of a partition is the largest part minus the number of parts. Dyson noticed that the rank does not explain the divisibility relation involving 11, but he conjectured the existence of a statistic he named as the *crank* that would! Dyson published his paper in *Eureka*, an undergraduate mathematics journal of Cambridge University. He humorously remarked the crank is probably the first instance in mathematics when an object had been named before it had been found!

Dyson's problem remained unsolved for more than 40 years. But then in 1986, Frank Garvan, a PhD student of George Andrews at Penn State University, introduced the *Vector Crank*, namely a crank for vector partitions. Each vector partition of $11n + 6$ had an integer weight, the sum of which was equal to $p(11n + 6)$. Using the crank Garvan split the vector partitions of $11n + 6$ into eleven subsets such that the sum of the weights over the partitions in each subset were all the same, and hence equal to $p(11n + 6)/11$, thereby "solving" Dyson's problem. What is more, the vector crank of Garvan explained all three of Ramanujan's congruences in a similar fashion. This was the talk that Garvan gave at the Urbana meeting and indeed it was a highlight of that conference. But Garvan's combinatorial proof of Dyson's conjecture used vector partitions. So the question remained whether one could find a statistic for ordinary partitions that would do the job. This happened miraculously in Urbana.

When the conference ended, most participants had left, but Garvan and Andrews stayed on. When everything quietened down, the two got together and found a way to convert the vector crank into a crank for ordinary partitions, and thereby completed the resolution of Dyson's conjecture. The Goddess of Namakkal must have played a role to ensure that this conjecture pertaining to Ramanujan's congruences would be resolved during a Ramanujan Centennial Conference!

The proceedings of the Ramanujan Centenary Conference in Urbana which appeared in June 1988, is the most comprehensive among the proceedings of all Ramanujan centenary conferences that were published. One of the most quoted articles there is the charming talk of Dyson entitled "A walk through Ramanujan's garden". The Proceedings also contain Garvan's important paper describing both his Vector Crank and the Andrews-Garvan Crank of ordinary partitions in the solution of Dyson's problem.

During the summer of 1987, I was approached by Dr. V. C. Ku-
landaiswamy, Vice-Chancellor of Anna University in Madras, to help conduct a Ramanujan Centenary Conference. Anna University was officially
formed in 1978 by combining the Madras Institute of Technology, the College of Engineering, the AC College of Technology, and the College of Architecture and Planning, and so it was, and is, an engineering and technology
university. But it does have an active mathematics department and the
Vice-Chancellor wanted to do something special for the Ramanujan Centenary. I readily agreed to help them. Since the NBHM was planning its
conference in Madras to start on December 22, Ramanujan's birthday, it
was decided that the Anna University Conference will run for three days
(Dec 19, 20, 21) with the first two days on various aspects of mathematics
related to research at Anna University to be conducted by Dean of Sciences, Dr. S. Ramanaiah, and the final day Dec 21, for a Number Theory
Symposium to be conducted by me. I was given a free hand to invite
speakers. I wrote to a several leading number theorists whose work was
related to Ramanujan, and got acceptances from George Andrews (Penn
State), Richard Askey (Wisconsin), Bruce Berndt (Illinois), David Bressoud (Penn State), Basil Gordon (UCLA), M. V. Subbarao (Edmonton),
and Imre Katai (Budapest). I had invited Paul Erdős, but he could not
come due to an eye surgery. I knew he did not like memorial conferences
(like the Straus memorial conference that he missed). But he kindly sent
me a paper entitled "Ramanujan and I" for me to read at the conference
on his behalf and to be included in the conference proceedings.

With my participation and role in the Ramanujan Centenary determined, I left Florida for India in early December with Mathura, daughters
Lalitha (age 6) and Amritha who was just two months old. We travelled
by Singapore Airlines from Los Angeles to Singapore, spent two days there,
and then went to Madras.

Since Basil Gordon was very absent-minded, and since I did not want
him to have any unpleasant experience on his travel to India, I advised
him to buy a Business Class ticket on Singapore Airlines, Los Angeles —
Singapore — Madras round-trip, and made sure that he would be on the
Singapore-Madras flight with us. Upon entering the Boeing 747 of Singapore Airlines for our flight to Madras, I went to the upper deck and found
Gordon working on his talk with his notes spread out on the table, oblivious
to what was going on around him! My father had made arrangements for
us to be received by the Madras airport staff, and so all of us with Gordon
were given a quick passage through customs and immigration. We arrived

in Madras on December 13, but Askey and his wife had arrived earlier and so my father took care of them.

Gordon was staying at the comfortable South Indian style Woodlands Hotel near our house. I had arranged accommodation for all speakers of my symposium at the Woodlands, and some of them including Gordon were to leave for a Ramanujan Centenary conference in Chidambaram the next afternoon (Dec 14). On Dec 14, prior to his departure to Chidambaram, I took Gordon to lunch at the Chola Sheraton Hotel right across from the Woodlands. Gordon had ordered "Chole-Bhatura" a very popular North Indian dish consisting of a large puffed up bread (Bhatura) accompanied by a spicy chickpea curry (Chole). We were deeply engrossed in a discussion of the Rogers-Ramanujan identities, for which Gordon had obtained a significant partition generalization in the sixties. When the waiter brought the Chole-Bhatura, he had decorated the chole curry with hot green chillies on the border to give it an attractive color contrast. Before I could warn Gordon, he took a large spoonful in his mouth with a hot green chilli! He had just started saying: "Krishna, the Rogers-Ramanujan identities which are among the most beautiful in all of mathematics, ...", but as soon as he bit on the chilli, he froze, because the heat from the chilli just hit him, and he was in a trance looking upwards to the ceiling! The heat had taken him to another world, but he controlled himself, and after several seconds of silence, completed the sentence "are extremely significant due to their connections with modular forms", without mentioning anything about the chilli. I asked him if he was OK, and he assured me that he could survive the heat from the chillies.

Just prior to the Anna University conference, there was the Ramanujan Centenary Conference at Annamalai University, an old university with a great tradition, and located in Chidambaram about 200 miles south of Madras in a rural area. Askey, Berndt, Gordon, and few others who were speakers there, were taken to Chidambaram in a van on Dec 14 afternoon. Andrews' arrival flight was delayed, and so he missed the van to Chidambaram. Therefore he stayed at our home for one night before leaving by a taxi that we had arranged to take him to Chidambaram the next morning. Just as he was leaving, I told him that in India many things can go wrong, but the people are very friendly. It was also a time when the Chief Minister of Madras, M. G. Ramachandran, was very ill, and situation in the state was tense. Andrews then held my hand and said that he had told himself that he was here for the Ramanujan's Centenary and so would not let anything upset him! I also said that travel by rural roads in India could be hazardous, and I did not want any unforseen delays preventing him and others getting back to Madras for the start of the Anna University Conference. He reassured me that he, Askey, Berndt and Gordon will be back in Madras on time.

One of the benefits for those attending the conference at Annamalai University was that they were taken to Kumbakonam, Ramanujan's hometown, and shown Ramanujan's humble family home, the temple next door where he worshipped, and the Town High School where he studied. This memorable excursion was not arranged for any of the other conferences. On the way back to Madras by car, Andrews, Askey, Berndt, and Gordon were caught in a political procession and their cars surrounded by political activists not letting them pass through. Andrews who had the presence of mind like James Bond, rolled down the car window, and threw a bunch of rupee (Indian currency) notes into the air! The delighted political activists happily let their car go through and they arrived in Madras as scheduled at 4:00 pm on December 18. That night there was a party in honor of the international speakers of my conference hosted by the U. S. Consul General John Stempel at his magnificent home overlooking the Adyar River.

The Anna University conference was inaugurated on December 19 morning by Richard Askey with an inspiring lecture on Ramanujan's work, and presided by Mr. C. Subramaniam, who had a very distinguished career as a politician, and had done great things for the cause of science in India, including helping my father in creating MATSCIENCE. It was particularly appropriate to have Mrs. Janaki Ammal, age 87, the widow of Ramanujan, at the inaugural ceremony. In her short but very effective and emotional speech, she said that just before Ramanujan died, he told her, "Do not worry. You will live to see my Centenary celebrated in glorious style." And indeed that prophecy came true. She also thanked Richard Askey for commissioning the Ramanujan bust made by Paul Granlund. Following Janaki's speech, George Andrews, in a voice choked with emotion, thanked her for preserving the pages of the Lost Notebook that has been a source of ideas for him and many mathematicians of our generation. On the morning of the inauguration of the Anna University Conference, *The Hindu*, India's National Newspaper based in Madras, not only gave a detailed description of the conference program, but also published my article "Ramanujan - an Estimation" in the center page. My father had told me many times that in addition to conducting first rate research, I should also spread the message of mathematics to the public at large. I have followed his advice and this article was the first of many articles I wrote for The Hindu on Ramanujan every year around Ramanujan's birthday, December 22.

My talk at the Anna University Conference was on my new approach using the multiplicative generalization of the sieve to the study of additive functions in probabilistic number theory. I also gave the talk "Ramanujan and I" on behalf of Erdős who had sent his manuscript to me. In this paper, Erdős discusses several connections between his work and Ramanujan's, such as the similarities in their proofs of Bertrand's postulate, and how the Erdős-Kac theorem in Probabilitic Number Theory emerges from the foundational work of Hardy-Ramanujan on round numbers.

Besides talks by international delegates, there were more than 20 lectures at the Conference by various active researchers from different parts of India. Thus the conference had a fine collection of lectures covering a broad section of number theory influenced by Ramanujan. The Master Educator Mr. P. K. Srinivasan, a great Ramanujan enthusiast, gave a talk entitled "Ramanujan as a gifted student - a case study". We were all ready for the celebrations the next day — Dec 22, Ramanujan's birthday.

M. S. Raghunathan of the Tata Institute, the head of the NBHM, had planned a grand public function on Ramanujan's 100th birthday. India's Prime Minister Rajiv Gandhi opened the festivities that day, released the Lost Notebook of Ramanujan that was brought out in printed form by Narosa of New Delhi (affiliated with Springer), and handed over the first copy to George Andrews. The beautiful Narosa volume contained a photostat copy of Ramanujan's Lost Notebook as well as copies of his other unpublished papers. Andrews had written a marvellous foreword for this volume which was urgently prepared for release on Ramanujan's 100th birthday.

The public function was held at the Kalaivanar Arangam, a huge auditorium that could accommodate two thousand people, and it was packed to capacity. Since the Prime Minister was present for the public function in the morning, there was heavy security, and only VIPs were given reserved seats and car passes. There was a break after the morning function, and in the afternoon, there were just two lectures — one by the Nobel Laureate Astrophysicist Subrahmanyam Chandrasekhar of the University of Chicago, and another by Fields Medallist Atle Selberg of the Institute for Advanced Study Princeton. During the lunch break, I went to the Rotary Club of Madras where as a Guest Speaker, I delivered a brief talk on Ramanujan at their lunch meeting. David Bressoud and his wife Jan very kindly attended my talk at the Rotary Club. I was at the Kalaivanar Arangam for the afternoon session for the talks of Chandrasekhar and Selberg.

During our planning meetings in Madras in 1986 which I attended before leaving for Florida, Raghunathan emphasized that on December 22, 1987, there will be only two talks — by Chandrasekhar and Selberg, because these two stood high above the rest. Chandrasekhar's connection with Ramanujan is that he collected Ramanujan's passport photograph from Janaki Ammal after Ramanujan's death and handed it to Hardy.

Chandrasekhar was not only a world class astrophysicist, but a phenomenal writer of books. At that time he was editing Newton's Principia and so he decided to talk about it at the Ramanujan Centenary. The original version of the Principia was published in 1687 and so 1987 was the 300th Anniversary of Principia. Dressed in his typical dark suit with a white shirt and plain tie, he had a meter stick in his hand, and like the teachers of the British colonial India, used the meter stick to point to equations on the screen. He started his lecture as follows: "To every great man there is a greater man. Ramanujan is great, but Newton is greater. So I will talk about Newton's Principia."

Great as Chandrasekhar was, one might question whether this was the best way to begin. After all, the audience was there to celebrate and honor Ramanujan. Perhaps he could have said something like: "I really have no mathematical connection with Ramanujan and cannot claim any real understanding of his mathematics. I an honored to have been asked to speak at the Ramanujan Centenary, and I will pay tribute to him by speaking about Newton, possibly the greatest figure in the history of science."

Chandrasekhar did give an outstanding lecture, but he spoke for one and a half hours even though the time he was given was one hour. So high was his stature that every one had to keep quiet when he went beyond time! We were later told that he was used to giving one and a half hour lectures and he did that in Madras. No one criticized his opening sentence nor his going beyond time. The newspapers reported the events of December 22 in great detail, and praised his masterly lecture on Newton's Principia.

After Chandrasekhar's magnificent lecture, there was a break for Tea. At that time I approached him and introduced myself as Alladi Ramakrishnan's son. I alluded to his visit to our home Ekamra Nivas in 1961 at the invitation of my father, and his lecture at my father's Theoretical Physics Seminar when I met him as a boy of six. The conversation was very brief because so many people were waiting to talk to him.

The one hour lecture by Atle Selberg after Chandrasekar's was equally inspiring. Selberg used transparencies, and he was so focused on the slides, that he fixed his eyes on them and never looked towards the audience. Selberg said that he was inspired to pursue mathematics after coming across

Ramanujan's work in the library at his father's home. Selberg spoke at length about the Hardy-Ramanujan asymptotic formula for the partition function, which is an amazing series representation for $p(n)$, the number of partitions of n, with the property, that if one sums the terms of the series to about $\sqrt{n}$ terms, then the nearest integer to the sum is $p(n)$. The Hardy-Ramanujan series when summed up to infinity diverges. A few years later, Hans Rademacher made some crucial modifications in the Hardy-Ramanujan series (one of them being the replacement of the exponential function by hyperbolic functions) and obtained an infinite series that converged to the value $p(n)$. In this connection, Selberg mentioned in his lecture that he wanted to see if he could derive the asymptotic formula for partitions by himself. Selberg said that whenever he saw a major theorem that sparked his interest, he would try to derive it by himself and then see how his approach compared with those of others. In any case, he emphasised that in trying to prove the Hardy-Ramanujan formula, he actually ended up with the convergent infinite series involving hyperbolic functions, which he felt was somewhat more natural. Since Rademacher had published his convergent series representation, Selberg did not care to publish his version even though it differed from Radamacher's in some respects. Selberg stressed that Ramanujan had insisted that there ought to be an infinite series which converges to the value $p(n)$, but Hardy who felt it was too good to be true, settled for less, namely the asymptotic series. Indeed, in Ramanujan's letter to Hardy in 1913, there is claim that the value of a certain partition function related to $p(n)$ has an exact formula involving hyperbolic functions. Selberg blamed Hardy for not trusting Ramanujan's intuition and missing the convergent infinite series for $p(n)$.

One other point emphasised by Selberg in his lecture was that sometimes it is more important to raise deep question than even to prove theorems, and Ramanujan's work definitely raised several deep questions in analysis and number theory. The contents of Selberg's lecture in Madras on Dec 22, 1987, and his talk at the Tata Institute in January 1988, are contained in his article "Reflections around the Ramanujan Centenary" that is included in his Collected Papers published by Springer in 1989.

After Selberg's lecture, I went up to him and said that I would like to take him to dinner one evening during his stay in Madras. He graciously agreed that we could do that on December 24 night, but due to a stunning development that will be described below, this did not take place.

Only the December 22 event of the NBHM was at the Kalaivanar Arangam. The NBHM conference for Dec 23–27 were scheduled on the campus of the Indian Institute of Technology (IIT). The main speakers of the NBHM Conference were accommodated at the Adyar Gate Hotel (5 star) starting from Dec 21 night, and so Andrews, Askey, and Berndt moved from the Woodlands to the Adyar Gate Hotel. Other participants of the NBHM Conference were provided accommodation at the IIT.

My father arranged two lectures at our home under the auspices of the Alladi Centenary Foundation (ACF). The first was on December 19 evening by Richard Askey entitled "Thoughts on Ramanujan". The second talk was on December 23 evening by George Andrews entitled "An introduction to Ramanujan's Lost Notebook", for which the U. S. Consul General John Stempel presided. Both lectures were for the general public and were well attended by representatives of different professions — lawyers, judges, civil servants, school teachers, students, and mathematicians. After Andrews' lecture, my family and all the international visitors attended a fabulous dinner at the Taj Coromandel Hotel, graciously hosted by N. Ram, Editor of The Hindu. After we finished dinner and were having conversation over dessert and coffee, word got around in whispers that the Chief Minister of Madras, Mr. M. G. Ramachandran, or MGR as he was affectionately known, had just passed away. Everything was expected to shut down in the city as soon as the word would get out, and so Ram advised all of us to quickly and quietly return to our hotel rooms or homes.

MGR was a local cinema hero like no other in South India. He held the fascination of the entire state of Tamil Nadu, and thus became the Chief Minister. By daybreak of December 24, the city came to a standstill. All shops were closed and there were no buses or taxis operating. Yet, more than a million of MGR's followers converged into the city of Madras from all over the state in lorries and vans chanting slogans that his name will live for ever over Tamil Nadu. In those two days of mourning, December 24 and 25, Ramanujan who was honored two days earlier, was forgotten by the public, and the MGR adulation took the city by storm! Most people were scared to get out of their homes, but Basil Gordon who was at the Woodlands, went up to the gate to watch the funeral procession and the thousands of MGR fans shouting their slogans.

MGR's passing severely disrupted the NBHM Conference schedule because no lectures could be held on December 24 and 25. The Gang of Three (Andrews, Askey and Berndt) all had spoken on December 23, but then the talks scheduled on December 24 and 25 had to be squeezed along with those that were scheduled for December 26 and 27. The speakers and participants were most cooperative to accommodate this change. Fortunately for me, the events I was organising concluded with Andrews' lecture on December 23 evening at my house, but unfortunately, I was denied the honor and pleasure of taking Selberg out to dinner on December 24. I called him at the Adyar Gate Hotel where he was staying, and he said he regretted that we are unable to meet because neither of us could move out. Public transportation and normal movement by cars resumed on December 26, but for those participants scheduled to fly out on December 24 or 25, it was a nightmare. The senior mathematician R. P. Agarwal, the Vice-Chancellor of Rajasthan University, and an expert on Special Functions, was scheduled to fly out on December 25 early morning, and he managed to do this most daringly. He left the Adyar Gate Hotel for the airport before daybreak in a taxi, but lay down on the floor of the taxi in the backseat so that he would not be seen by those outside as a client of the taxi! Empty taxis were allowed to ply, but no passengers were allowed to go in taxis.

Immediately after the NBHM Conference, there was a Ramanujan Centenary Conference at Kandy in Sri Lanka in late December sponsored by the National Institute of Fundamental Studies of Sri Lanka, and organised by Srinivasa Rao of MATSCIENCE, a former PhD student of my father. Don Zagier, Atle Selberg, Robert Rankin and Bruce Berndt went to the Kandy conference. Since Basil Gordon, Subbarao, David Bressoud and his wife Jan stayed back in Madras, my father-in-law Dr. N. C. Krishnan treated them to a lavish lunch over looking the sea at the fabulous Taj Fisherman's Cove Hotel, just as he hosted my conference guests in 1984. I took the Bressouds and Gordon to concerts at the Madras Music Academy since December is the peak of the music season in Madras. Gordon had a great knowledge of Western Classical Music and was himself quite an accomplished pianist; so at the Music Academy concerts, he was making comparisons between Carnatic music and Western Classical Music.

After an exhilarating Ramanujan Centenary in Madras, all international participants left, but there was drama even in that. The absent minded Basil Gordon lost his airline ticket at the Woodlands Hotel and noticed this just the day before his departure! Those were the days of paper tickets and so a paper ticket had to be reissued. We rushed to the Singapore Airlines office before it closed for the day. Since he was travelling in Business Class, Singapore Airlines reissued his ticket without imposing a penalty, and he departed as scheduled on December 30 for Los Angeles via Singapore.

With the Ramanujan Centenary celebrations in Madras completed, I could relax with my family. Since it was Amritha's first time in Madras (she was just two months old), we had a get together of family and friends in Ekamra Nivas to see the baby.

Although the Ramanujan Centenary was completed in Madras, there was still the conference at the Tata Institute during January 4–11, 1988 for which Professor Raghavan kindly invited me, but I could not go since classes at the University of Florida were resuming. Just as the International Congress of Mathematicians (the ICM) is held once every four years (on years which are 2(mod 4)), the Tata Institute's Bombay Colloquium is also held once every four years, during leap years. The first one was in 1956. But the Bombay Colloquia are not large conferences; each is focused on a specific area. In fact they are "closed conferences", which means to be a participant, you need to be invited. The Tata Institute decided that their 1988 Bombay Colloquium would be a Ramanujan Centenary Conference. For the TIFR conference, Jean-Pierre Serre, a mathematical luminary in the same class as Selberg, participated. The proceedings of this conference was published by Oxford University Press.

The Ramanujan Centennial in India was a proud moment for the country. One realized how much Ramanujan's work had influenced various branches of mathematics, and one wondered how much more his influence would be. Two things happened to me:

(i) Inspired by the lectures of Andrews and others at the Centenary, I wanted to do research in the theory of partitions and q-hypergeometric series. A number of favourable circumstances helped me shift the focus of my research from analytic number theory to partitions and q-series, but it took three years to make this transition.

(ii) While the Ramanujan Centennial was magnificent, it was an event that would not repeat. I dreamt of creating a permanent memorial for Ramanujan. This dream became a reality ten years later when I launched *The Ramanujan Journal.* Thus the Ramanujan Centennial profoundly influenced my academic life.

My work for the Ramanujan Centennial wasn't over yet. I still had to edit the Proceedings of the Anna University Conference to be published by Springer in their Lecture Notes Series. Also, before I left Madras in January 1988, The Hindu sent me a copy of "The Lost Notebook and other Unpublished Papers of Ramanujan" that was released on Ramanujan's 100th birthday, for me to review. I wrote the review upon return to Florida and it appeared in The Hindu on February 20, 1988.

In March 1988, there was the Spring Meeting of the Mathematical Association of America (MAA) on the beautiful campus of Rollins College in Orlando. My colleague Bruce Edwards who was an office bearer for the Florida Chapter of the MAA, invited me to deliver a one hour address at that meeting on a theme related to Ramanujan, to which I gladly agreed. My parents had arrived in Florida in early March and so we all went to Orlando and stayed near Rollins College at the elegant Langford Resort Hotel which had a lovely garden setting. I was glad that my father who had encouraged me all along, attended my talk at that MAA Meeting.

7.3 Research, teaching and service at the University of Florida

Anytime one gets appointed to a tenure-track or tenured position at a university, the letter of appointment says that the individual's progress will be measured by contributions to Research, Teaching, and Service. Here by Service one means service to the Department or the University. My letter of appointment was no exception to this standard letter. I shall describe my contributions to Teaching, Service, and Research in that order in general terms in this section, and provide more specifics about these in subsequent sections as I describe various events chronologically.

Teaching: I have always loved to teach whether it be undergraduate or graduate (= post-graduate as in England and in India) classes. My teaching load at the University of Florida has always been the standard one — two courses each in the Fall and Spring terms, which is typical of most mathematics departments in large state level universities. (The Department is now moving to a 2+1 teaching load which is what I have currently.) The only time I did not teach was when I was Chair, but then the administrative duties were considerable; I was also having a vibrant research program, and so the Dean exempted me from teaching. I have been fortunate that almost always I have taught one graduate course and one undergraduate course each semester, and so have contributed to both the graduate and the undergraduate programs of the Department.

I feel that mathematics is both an art and a science, and that we practitioners of the subject are drawn by its intrinsic beauty. Especially when I teach undergraduate classes, I emphasize the aesthetic aspects of the subject. Almost everyone is aware that Mathematics is the Queen of all Sciences, but few really consider mathematics as a beautiful subject — an art in itself. Most undergraduate students perceive mathematics as a calculational device, and think that mathematical techniques are created mainly for the purpose of accurate or speedy calculations. Mathematics is a study of structures and of relationships. While working on an applied problem, a mathematician understands the underlying structure and then extracts certain general principles and methods which usually have much wider application. Therefore when teaching mathematics to undergraduates, I emphasize not only its usefulness, but also draw attention to its intrinsic beauty and the underlying structure. I do not have to emphasize that mathematics is beautiful to the graduate students.

The concept of beauty in mathematics is alien to most undergraduates in America, but after attending my classes, they do accept that mathematics is beautiful. I also believe that *proof* is of utmost importance in mathematics. Sadly, nowadays in America, the treatment of mathematics is not rigorous in lower level classes and proofs are not emphasized until late in the undergraduate curriculum. Without proofs one will not understand how the subject is developed, and so even in the lower level undergraduate classes, I try to prove various results. A question I ask my undergraduate classes is the following: "Would a Chemistry professor teach the course without a lab? No, because lab experiments are crucial for the understanding of the subject. If that is the case, then why should mathematics professors be expected to teach their subject without proofs when proofs are essential for the proper understanding of mathematics?"

Many of my undergraduate students appreciate that I develop the subject rigorously. One reason proofs are dropped is due to the desire to expose the students to more advanced material quickly, such as rushing to teach calculus in high school without covering Euclidean geometry, trigonometry, and analytic geometry rigorously. Thus most students entering the university arrive with a half-baked knowledge of mathematics, and we at the university have to remedy this defect. I am amazed that a majority of undergraduate students nowadays have not the faintest idea as to what a proof is or how to go about giving a proof by logical reasoning. A very humorous anecdote on this was related to me by my friend, the late P. V. Rao.

P. V. Rao was a professor in the Statistics department. One day a student came to his office hour and said: "Professor, I enjoyed your lecture today, but there was one thing I did not understand. Why is e^x the derivative of itself?" PV was pleased that the student wanted to clear this fundamental doubt. So he spent about half an hour patiently proving that the derivative of e^x is e^x. At the end of the proof, the student exclaimed: "Wow, now I understand why e^x is its own derivative. So what is the second derivative of e^x?" At this point PV decided that there would be no benefit (to him or the student) to provide further explanation. He said that he had a meeting to attend and therefore had to leave his office.

I tell my students that mathematics is like fine cuisine for which one takes time to prepare and time to savor it. Mathematics has to be developed slowly and systematically, and students allowed time to assimilate the concepts before proceeding to the next steps. If one rushes through mathematical ideas without proofs or without understanding the inter-relationships, this is like eating fast food without chewing, and one will be then in a permanent state of mathematical indigestion (= aversion to mathematics), as is the case with most human beings!

In the early nineties, at the request of the College of Engineering, the mathematics department introduced a one semester course called "Mathematics for scientists and engineers". In this one semester course one had to cover BOTH linear algebra AND complex analysis (up to the Cauchy residue theorem and its applications to contour integration). My feeling is that Linear Algebra has be taught over one year, and not just one semester as is common, and definitely complex analysis has to be taught over one academic year instead of just one semester. Here we were asked by the College of Engineering to teach both subjects in the same course in one semester!! I was asked to teach this course. I agreed much to my dislike. I did give it a good try, and in spite of all my efforts in class and during office hours, the students had a very hard time digesting the material in such short time. I think this course is ideal for a student who has already had the regular linear algebra and complex analysis courses, went over to industry for a few years, and was back in college; for such a student, this course will be a quick refresher. But it is definitely not a course to be taken to learn the subjects for the first time. After me, a couple of my colleagues taught this course, but it eventually was cancelled because of steadily declining enrollment. I discussed this course in this section to illustrate that there is an unhealthy tendency to rush the learning process in mathematics without proper treatment of the subject.

Prior to becoming Chairman in 1998, for the graduate classes, I used to teach Complex Variable Theory and Analytic Number Theory in alternate academic years. The two subjects are closely related. Indeed the attempt to solve various questions in number theory by analytic methods led to the development of complex analysis in the nineteenth century, and so it was natural to teach these two subjects back-to-back and many graduate students took both my courses to gain a more complete understanding.

After completing my term as Chair in 2008, when I resumed teaching regularly, for graduate classes I have done a rotation of the following subjects over a three year period: (i) analytic number theory, (ii) irrationality and transcendence, and (iii) partitions and q-hypergeometric series. This is not only refreshing for me, but it also gives graduate students an opportunity to learn three different areas of number theory. And many students took all three of my courses

If a graduate class does not have the minimum required number of students, then the class is cancelled; fortunately my classes were, and are, always well attended, and so I have not faced this unpleasant situation.

Most classes at the University of Florida meet three days a week. A few classes like Calculus and Linear Algebra meet four days a week. I never ask for a four credit course, because I like to keep Tuesdays free for the number theory seminar, and Thursdays to attend to my varied academic assignments. It is a custom that if you teach a graduate course, then you are not given a four credit undergraduate class on top of this unless you ask for this. Since I have been teaching graduate classes almost always, I have been given only three credit undergraduate courses, except perhaps on rare occasions when I did not teach a graduate course.

In order for students to understand any subject in mathematics, whether it be one taught in lower level or upper level undergraduate classes, or in graduate classes, it is vital to work on a good number of problems related to the material covered in the course. In the past — certainly until the seventies — even for the upper level undergraduate classes, the professor teaching the class would have a TA; the professor would give the lectures on Mondays, Wednesdays, and Fridays, and the TA would meet the class on Tuesdays and Thursdays to work out several problems. When I was a graduate student at UCLA in the seventies, in my third year, I was a TA for upper level undergraduate courses like number theory and set theory. Back then, for lower level undergraduate classes, there was a grader in addition to a TA because sufficient funds were provided for mathematics departments to provide such support for undergraduate classes. When I was a TA for Calculus, I would collect the homework and give it to the grader. Nowadays, due to lack of funding, TAs are not provided for upper level undergraduate classes unless a faculty member has a "special arrangement". Even if I do not have a TA, I still spend time working out several problems but this slows me down in covering the material.

Even though new technology is being employed in teaching, I believe that a blackboard lecture is the best for students because it gives time to assimilate the ideas being explained. I think power point presentations are suitable in a conference lecture due to time constraints or when graphical/pictorial representations are required. Students nowadays depend on the calculator too much — indeed to such an extent that they are unable to reason logically. I would say that even though the average IQ of human beings has not changed with time, I have noticed a steady decline in mathematical skills among undergraduate students annually. This phenomenon must be true nationwide, and even worldwide. I put the blame on two things: (i) the over dependence on calculators, and (ii) on the lack of emphasis on proofs when teaching mathematics.

With regard to the use of the blackboard, I have a somewhat unique style. I start at the upper left hand corner, write what I say in complete sentences, and when I conclude the lecture, I have usually finished my writing in the lower right hand corner. As far as possible, *I do not erase anything.* This is possible only if the blackboard space is large. I also number the equations during a lecture (like we do when we write papers), and refer to equation numbers in the lecture. Students like my style, because everything is on the blackboard and nothing is erased. My hope is that my style will influence students to develop the habit of writing clearly and expressing their arguments systematically.

Even though research, teaching, and service, are emphasized in the appointment letter, in the academic world it is the research contributions that are used as the yardstick for promotion and recognition. Good teaching is applauded, but never used as a criterion for promotion. On the other hand, poor performance in teaching can lead to denial of promotion or tenure. In view of the fact that excellence in teaching was not really rewarded, but parents of students want their sons and daughters attending the university to get quality instruction, President John Lombardi of the University of Florida had the radical idea to introduce the Teaching Incentive Program (TIP) in 1992–93. Under this program, university faculty were given $5000 base salary raises. Although it was the University of Florida President John Lombardi who proposed the TIP Program, when the State of Florida Legislature approved funds for it, ALL state supported universities in Florida received funds for the TIP Program. Up until this program was introduced, Teaching Awards only provided a one time cash payment and not a salary raise. The mathematics department has always been known for its commitment to teaching, and several faculty including me received TIP Awards. Whenever I mentioned these TIP awards to academicians outside of Florida, they were in disbelief that salary raises were actually given in recognition of teaching. It was a good program and it was adopted by all Florida universities. These TIP awards were offered only for a few years, and were discontinued due to budget reasons.

Research: When I arrived at the University of Florida in December 1986, the main research groups in the Department were in combinatorics, probability, analysis, and topology. By research groups I do not mean large groups, but about three or four in a group having an active seminar. Gerard Emch who had just taken over as Chair in Fall 1986 had been promised a number of positions and so the department began building programs in differential geometry and mathematical physics, dynamical systems, and group theory. I was asked to help the department create a program in number theory. In December 1986 I was still working in analytic number theory, and so when the Department asked me to teach a graduate course in Spring 1987, I lectured on analytic number theory.

In December 1987, I was in India in connection with the Ramanujan Centenary Celebrations and conducted a number theory conference in Madras on December 21, one day before Ramanujan's 100th birthday (Dec 22). Although I spoke about my research in probabilitic/analytic number theory, I was very much influenced by the lectures I heard on Ramanujan's work on partitions and q-hypergeometric series. So I decided to learn this area in the hope of attempting research in this exciting field. At the University of Florida, I was the only number theorist, and so until other number theorists were appointed, I felt it was pointless to begin a number theory seminar. So I decided to speak in the combinatorics seminar on topics in number theory with a combinatorial flavor. Partitions were ideal for such a presentation, and so I lectured on partitions knowing that this will help me make the transition to research in the area. It actually took me two full years to learn the subject well enough to attempt research, and the visit to Florida of Basil Gordon, my former professor at UCLA, was crucial in enabling me to transition from analytic number theory to the theory of partitions and q-hypergeometric series (q-series for short). Details of Gordon's visit and my first papers with him on partitions and q-series can be found in the next section.

I was the last appointment by Al Bednarek during his long term as Chair. During Bednarek's term, I believe no start-up funds were given for new faculty — such start up funds in mathematics were available only from the Emch chairmanship onwards, but even these were not sizeable. In the eighties, there were grants given out by the College of Liberal Arts and Sciences (CLAS) called CLAS Research Awards ranging from \$10,000 to \$25,000 to be used over a three year period. The funds could be used for research related travel, invite visitors for collaboration, and to buy books, and equipment, but NOT for one's own salary. These CLAS Research

Awards came in three categories, one of which was for new faculty. There was no guarantee that every new faculty member who applied would get it, because there was an internal review process to select the awardees. I applied in the year 1987–88 and was given the CLAS Research Award for new faculty. I was at that time working in analytic number theory.

The second category for the CLAS Research Awards was for transition to another area of research. Since I was moving from analytic number theory to the theory of partitions and q-series in the early nineties, I applied for a CLAS Research Award under this category and was successful a second time. In 1994 I received a three year Individual Investigator Grant of the NSF for work in the theory of partitions and q-series, and so the university administration viewed the CLAS Research Award as well justified. Throughout my tenure in Florida, the University was very supportive of my research efforts and I appreciate such support given in various forms by the College and by the Office of Research.

I was asked to serve on the Search Committee to recruit a number theorist, and in two consecutive years, I helped the department identify very strong candidates; also at conferences on number theory that I was regularly attending, I was on the lookout for active researchers in number theory and urged them to apply to the University of Florida. We interviewed Andrew Granville and made him an offer, but he accepted a tenure-track position at the University of Georgia because they had a well established research group in number theory. The Department also made attractive offers to Jeffrey Vaaler (analytic number theory, irrationality and transcendence) and his soon to be wife Leslie Federer (algebraic number theory), but they decided to stay at the University of Texas at Austin. These unsuccessful hiring efforts in number theory were during my first two years. By the time I completed three years in Florida, I was firmly entrenched in partitions and q-series, and so I decided to find candidates in this area, and succeeded in getting Frank Garvan to join the Department.

Garvan was a former PhD student of Andrews. As for me, I was very much influenced by Andrews' book [B6] on partitions (Encyclopedia of Mathematics, Vol. II) as well as his many papers. So I arranged for Andrews to deliver the Frontiers of Science Lecture in November 1990 and get him to visit the University of Florida. That visit of Andrews had a transformative effect in the second phase of my research career, namely in the theory of partitions, just as the visit of Erdős to Madras in 1975 had on the first phase of my career. My major collaboration with him began with that visit when he gave a remarkable proof of an incredible multi-parameter

q-hypergeometric series identity that Gordon and I had found for a generalization and refinement of a deep partition theorem of Göllnitz. My entry into the theory of partitions was firmed up during my year long sabbatical visit to Penn State University in 1992–93 where I closely interacted and collaborated with Andrews, and my frequent short visits to UCLA in the 1990s to work with Gordon. As a consequence of my interaction with both Andrews and Gordon, I worked in the theory of partitions and q-series for a quarter century, that is until 2015, when I returned to analytic number theory to guide PhD students in that area.

I did not find my 2+2 teaching load as an impediment to my research. I was quite productive in the theory of partitions and q-series, and provided new insights because I was approaching the subject with the freshness of a student. Even though I did not directly work on Ramanujan's Notebooks, various aspects of my research were related to Ramanujan's work. My work on partitions and q-series is in the theory of Rogers-Ramanujan type identities. In view of this aspect of my research, and what I witnessed at the Ramanujan Centennial in 1987, namely the deep and wide impact of Ramanujan's work in mainstream mathematics, I launched *The Ramanujan Journal* in 1997, and have played the role as Editor-in-Chief since its inception. This journal was published by Kluwer until 2004, after which it is published by Springer when Kluwer and Springer merged.

The role of Chief Editor of the journal entailed a lot of work but that did not affect my research productivity either. In 1998, I accepted the position as Chair of the mathematics department. This was work of a different order of magnitude. The Dean knew that I was an active researcher, and to help me maintain my research productivity during my term as Chair, he agreed to my request to release me completely from teaching. This zero teaching load was crucial for my research productivity to remain high. It also enabled me to go frequently to research conferences without worrying as to how to deal with my classes during such visits.

Just as I was about to start my term as Chair, George Andrews informed me in the summer of 1998, that he had a post-doctoral associate with him — Alexander Berkovich — a mathematical physicist who had done very impressive work with Barry McCoy and members of his group on Rogers-Ramanujan type identities arising from models in conformal field theory in physics. Andrews said that Berkovich was completing his term at Penn State University and was looking for a job; Andrews added that Berkovich would be a fine addition to the University of Florida. I met Berkovich at the Andrews 60th birthday conference in Maratea, Italy, in the early Fall of 1998, and arranged a visiting position for him to start in Spring 1999. We started collaborating immediately and attacking the difficult problem of finding a four parameter extension of the three parameter Alladi-Andrews-Gordon generalization and refinement of Göllnitz's (Big) theorem. We succeeded in resolving this problem in Spring 2000, and Andrews, Berkovich and I published our results in a paper in *Inventiones Mathematicae*. To support Berkovich, I applied for a CLAS Research Award and was given this award for a third time. The award funds were not used for my salary, but to temporarily provide partial support for Berkovich with whom I was collaborating. But after the award funds ran out, the Number Theory Foundation (NTF) very generously provided partial support for one semester for Berkovich. The College viewed the CLAS Research Award as worthwhile because my NSF grant proposal was approved for funding to start in 2000, and the NTF felt justified of the support because of the acceptance of our joint work in *Inventiones Mathematicae*. In a few years, we were able to make Berkovich's appointment as permanent, and he and I went on to write at least half a dozen substantial papers.

With Frank Garvan, I organized a number of conferences in the theory of partitions, q-series and modular forms, and so Florida became a hub of activity in these areas. George Andrews was always one of the main speakers at these conferences. From 2005 onwards, an arrangement was made which enabled Andrews to spend the Spring term each year at the University of Florida by doubling his duties at Penn State University. With his presence, and with the work of the three of us (Garvan, Berkovich, and me), Florida became one of the world's major research centers for work in partitions, q-series, modular forms, and the mathematics of Ramanujan.

Service: My service to the University of Florida since 1987 January has been confined to the Department. I did not serve on any College or University committees, but my departmental service was substantial especially because I was Chair for ten years. It was also after to coming to Florida that I served the profession in several capacities — (i) as Editor-in-Chief of The Ramanujan Journal since its inception 1997, (ii) as Editor of the book series Developments in Mathematics (DEVM) which I also founded in 1998, (iii) as Chair of the SASTRA Ramanujan Prize Committee since the inception of the prize in 2005, (iv) as an organizer of the annual SASTRA Ramanujan Conferences in Ramanujan's hometown Kumbakonam in India since 2003, and (v) as organizer along with Frank Garvan of conferences in Florida. Thus my service to the profession was just as substantial as to the

Department, and it also brought increased visibility to the University of Florida. Here I summarize first my departmental service and then briefly describe my service to the profession.

A major committee in the Department is the Steering Committee which is advisory to the Chair. Membership on the Steering Committee is by election. An important duty of the Steering Committee is to evaluate the performance of each faculty member based on the Annual Activities Report. The Chair uses this evaluation by the Steering Committee to decide the annual salary raises. The Steering Committee also meets regularly with the Chair to discuss and help address various departmental issues. I served on the Steering Committee in the early nineties and this experience was useful when I took up the position as Chair in 1998.

I have served regularly on the Tenure and Promotion Committee. I consider this to be important because the evaluation and report by this committee of the performance of our colleagues coming up for promotion is crucial to the department decision/vote on the tenure/promotion of the candidates, and for the Chair to write the letter making the case for tenure/promotion.

Another committee on which I serve every year is the Graduate Selection Committee. My Indian background is useful to the committee to evaluate various graduate applicants of Indian origin. It is not only the assessment of the applicants from India that I give which is useful, but also my opinion on the strength of the universities and colleges in India from where they apply for admission to our Department.

Finally, I have served on the Visitors and Conferences Committee after my period as Chair ended in 2008, because the Department realized my commitment to maintain a vibrant program of visitors and conferences.

Important as these committees are, the time involved in serving on these committees is nothing in comparison to what is involved in being Chair. I accepted the position as Chair in 1998 knowing fully well what I was getting into both in terms of responsibility and the time it would consume. I accepted the job with eagerness because I wanted to make a difference and elevate the performance, productivity, and visibility of the department. And I did succeed in achieving this by introducing a number of new programs with the support of my colleagues and the administration.

Weekdays were fully occupied, but I would come to the office on weekend mornings as well (as all committed administrators would do), but keep weekend afternoons and evenings to be with my family. Joseph Glover, my predecessor as Chair, who subsequently steadily rose in the administrative ladder to become Provost at the University of Florida, told me that the chairmanship of a large department (like mathematics) is one of the toughest positions on campus. One of the most time consuming tasks as Chair is to write the Annual Letters of Evaluation of the (more than 60) faculty, by studying their Annual Activities Reports. Each faculty member takes this letter very seriously, and so it is crucial that in writing these letters, all important aspects of their activities are covered. As Chair, I also spent considerable time talking to faculty, and addressing their needs and concerns, because only then can one manage the department to their satisfaction. The Chair is also in-between the faculty and the Dean. In addition to keeping faculty happy, the Chair has to meet the Dean's expectations. George Andrews, who served very successfully as Chair for two two-year terms at Penn State University, once told me humorously, that every faculty member needs to be in the position as department head at least once to understand how tough it is; if they did hold that position, they would stop complaining afterwards. But I must say, that during my ten year term as Chair, I never had problems with my colleagues — except for one faculty member. The only difficulties I had was with the upper administration towards the end of my term as Chair with regard to the implementation of the recommendations of a positive External Review. I will discuss that in detail in a separate section. But in summary, I was pleased that I served a ten year term as Chair because in that time frame, I was able to accomplish a great deal because of support from both the faculty and the administration.

7.4 The visits of Erdős, Gordon, and Andrews to Florida — inflicted with the q-disease (1988–91)

Richard Askey has humorously remarked that the theory of q-hypergeometric functions is so attractive, that anyone who comes into contact with that field will be smitten by the q-disease — that is be drawn to that subject and never recover from this marvellous disease! I like to explain q-disease as follows: A fruitful way to generalize fundamental mathematical expressions or identities is to obtain their q-analogues, namely to replace the expression with a function of the variable q, so that it reduces to the original expression when $q = 1$ or $q \to 1$. These q-analogues are often very elegant and have significant analytical and arithmetical consequences. Once a person has tasted the pleasure of working with q-analogues, one is drawn into its fold. That is what Askey meant when he used the phrase "smitten with q-disease". And that happened to me at the Ramanujan Centennial when after listening to the lectures of

Andrews, Askey and Berndt, I wanted to pursue research in partitions and q-hypergeometric series (q-series for short), an area where Ramanujan had made spectacular contributions. But I was in awe of the subject and somewhat scared of the bewildering multi-parameter identities that dominated it. When I joined the University of Florida, there was no program in number theory in the mathematics department, and it was expected of me to build that program. But the Department had a strong combinatorics program with senior researchers David Drake, Neil White and Andy Vince, and so I took an active part in the weekly combinatorics seminar giving talks on partitions and Ramanujan's work, emphasizing the combinatorial aspects. Thus I was making a systematic study of partitions and q-hypergeometric series in the hope that eventually I could do some research in the area. And that happened miraculously through the visits of Basil Gordon in 1989 and George Andrews in Fall 1990. Before I describe those visits and how my first research in partitions and q series came about, I will discuss major changes at MATSCIENCE — indeed a turmoil there, and a similar but less turbulent situation in the Florida mathematics department, both in 1988.

As emphasized earlier, George Sudarshan who accepted the post as Director of MATSCIENCE in early 1984, did not resign his position at the University of Texas, but decided to spend at most six months each year away from Austin. So he had G. Rajasekharan appointed as Joint Director of MATSCIENCE to act in his place when he would be away. To develop the mathematics program, he arranged for Seshadri to join MATSCIENCE.

Soon after Sudarshan was appointed Director of MATSCIENCE, he received the 1985 Third World Academy of Science (TWAS) Prize given by the ICTP. Even though Sudarshan's absences caused concern, no one rebelled against him since his stature had risen further after the TWAS Prize. He was given greater support by the science policy makers of India. Thus he made several new appointments at MATSCIENCE. However, in 1988, some MATSCIENCE faculty who were appointed by Sudarshan, demanded that Sudarshan should stay full time in Madras for the efficient management of the Institute. Sudarshan's appointment was for a five year term from 1984. He was coming up for renewal of his contract in 1989 but was weakened by this revolt in 1988 in the last year of his five-year contract. A search for a new Director was announced. In the end, R. Ramachandran, a physicist at the IIT Kanpur, was appointed as the new Director effective 1989. Both Sudarshan and Seshadri left MATSCIENCE. Sudarshan was back in Texas. Seshadri was appointed as Head of a newly formed mathematics unit of the SPIC Science Foundation in Madras. SPIC stands for Southern Petroleum

Industries Corporation. Rajasekharan continued at MATSCIENCE, but not as Joint Director, because that position was not needed after Sudarshan left. Certain members who were close to Seshadri left MATSCIENCE and joined the SPIC Science Foundation. Subsequently, Seshadri along with his colleagues left SPIC to create the Chennai Mathematics Institute (CMI) in the outskirts of Madras, which in a span of two decades became a fine research institute and one of the best in India for undergraduate mathematics education. As for MATSCIENCE, the wounds of the 1988 revolt healed after a few years, and the institute grew significantly with tremendous influx of government funds; it is now considered one of the top research institutes in India. Since its inception, the CMI has had an excellent relationship with MATSCIENCE.

Here at the University of Florida there was a turmoil in the mathematics department in Spring 1988. The College which had promised Gerard Emch a total of 25 new positions at the rate of about five per year, went back on the promise after two years — that is after about ten appointments had been made. Also there were issues relating to counter offers when one of the new appointments made in the first year on Emch's tenure received a much more attractive offer elsewhere the very next year. There were also issues regarding start-up funds for the new appointments. Meanwhile, John Thompson, for other reasons, relinquished his half-time position. Emch tried to get commitment from the College for various positions, and when that failed, he resigned in Spring 1988. Al Bednarek who had been Chair for 17 years, stepped in as Interim Chair until a new chair was chosen. David Drake was appointed as the Chair to start from Fall 1988 for a five-year term, and the healing began.

In 1988–89, I came up for tenure and for promotion to the rank of Professor in Florida. I was informed in Spring 1989 that my term as Professor with tenure will begin in August. When I left MATSCIENCE in December 1986, I was given leave for one year, but as the controversy there was brewing, my leave was extended by Sudarshan. But after I received tenure and got promoted to professorship in Florida in 1989, I resigned my position at MATSCIENCE.

The many new appointments Emch made in Florida, did not all remain. Some of them left after a year or two: Michelle Wachs and Greg Gallaway went back to Miami, Peter Baxendale accepted a position at the University of Southern California, and Mike Fried returned to Irvine after Thompson departed. So the Department enjoyed only a mild increase in size, but was assured by the new Dean Willard Harrison of CLAS, who joined in

Fall 1988, that appointments would be made, at the rate of about two per year. With this prospect, I was asked by the Department to help build the number theory program.

In Fall 1988, I attended the Conference on Analytic Number Theory at Oberwolfach in response to the invitation from Professors H.-E. Richert, Wolfgang Schwartz, and Eduard Wirsing. I found out at that meeting that Jeffrey Vaaler of the University of Texas might be interested in moving out of Austin. Vaaler was well established in analytic number theory, but after spending the year 1982–83 at the Institute for Advanced Study collaborating with Enrico Bombieri in the area of irrationality and transcendence, his research area had significantly broadened. So I felt that recruiting Vaaler would be a major step in strengthening the number theory program at Florida and so I talked to the Drake and the Search Committee about this. Vaaler at that time was planning to get married to Leslie Federer (also in Texas), an algebraic number theorist who had received her PhD from Harvard under the direction of Benedict Gross. So both Vaaler and Federer were invited for colloquia in Spring 1989 and both were made offers. The University of Florida also agreed to appoint as TAs, Vaaler's students who were writing their dissertations under his direction. So it was an attractive package that was offered to Vaaler and Federer, but in the end they decided to stay in Austin because the University of Texas made a counter offer.

Even though my effort to get Vaaler to Florida did not succeed, I continued to be on the lookout for possible recruits in number theory for the following year. A name that came to my mind was that of Frank Garvan who had done such wonderful work on Dyson's conjecture concerning partition congruences. So I wrote to George Andrews in 1988 soon after the Ramanujan Centennial asking for a copy of Garvan's PhD thesis, which he promptly sent me. It is an interesting story as to where I got to read Garvan's thesis, and I describe that now.

In the summer of 1988, I was on a visit to India and I took Garvan's thesis with me to read during my stay there. On most of my trips to India with my family, we visited the Mudumalai Wildlife Sanctuary at the foothills of the Nilgiri mountains, about 350 miles from Madras by car. In Mudumalai, we always stayed at the secluded Kargudi Guest House which is right in the heart of the dense jungles and commands an awesome view of the surrounding forests with the Nilgiri range rising majestically in the background. Kargudi Guest House is reserved for the use of forest officials and VIPS who want such seclusion; my father always wrote to the Forest Department and secured the reservation at the Kargudi Guest

House. There is a lovely mango tree in the front lawn of the Guest House and often we relaxed in the shade of this mango tree after a hectic morning of jeep rides in the jungle to spot wildlife. On this trip to Mudumalai, I took with me Garvan's thesis to read under that mango tree! And the thesis was just as exciting to read as the plethora of wildlife we saw!

After completing his PhD in 1986 at Penn State, Garvan was on two post-doctoral fellowships, first at the University of Wisconsin in 1986–87, and then at the IMA in Minneapolis in 1987–88, and following that he returned to Australia, his native country. In 1989, I wrote to Garvan in Australia suggesting that he could apply to Florida for a tenure-track assistant professorship, and he responded positively to this suggestion. He said that he was also thinking of applying to a few other universities in the USA. When I heard that, I contacted George Andrews and requested him to endorse the University of Florida because I knew that Garvan would consult his thesis advisor. And Andrews assured me that he would most happily provide such an endorsement. Indeed, Florida did make Garvan an offer in the Spring of 1990 and Garvan accepted the offer. But then he had also been offered a prestigious NSERC Fellowship to work with Jon and Peter Borwein in Dalhousie, Canada. Since it would be very beneficial for Garvan to work with the Borweins, it was arranged that Garvan would join the University of Florida in Fall 1990, and after spending one semester in Gainesville, he would take up the NSERC Fellowship for the entire Calendar Year 1991, and return permanently to Florida in January 1992. I feel that the appointment of Frank Garvan to the University of Florida is one of my best contributions to the mathematics department.

Soon after Vaaler's visit to Florida in Spring 1989, I attended a conference in Urbana, Illinois in April, in honor of Paul Bateman for his 70th birthday. I had gotten to know Bateman quite well in 1980–81 when he spent a year at the University of Michigan, and he also gave me the opportunity to organize the Special Session on Number Theory at the AMS Meeting in Notre Dame in March 1981. So I enthusiastically accepted the invitation to attend the conference, which was held at lovely Allerton Park outside Urbana. Most participants like me stayed at Allerton Park. Bateman had such a wide circle of contacts, that the conference attracted many of the top researchers in number theory from around the world.

In the Spring of 1989, Gian-Carlo Rota of MIT visited Florida for a colloquium. Rota not only had done pioneering research in combinatorics, but was one of the most influential figures in the mathematical world. I had heard a lot about him, and it was a sheer delight to listen to his colloquium talk in which he discussed his new idea of logarithmic power series. I requested some time with him, and he graciously came to my office for a discussion. Since Rota had initiated the study of general Moebius functions in combinatorics, I showed him my Duality identity connecting the largest and smallest prime factors using the Moebius function and my monotinicity principle for the sieve using the Moebius function. He said he liked my observations, pulled out a small notepad from his jacket, and wrote down my results. I found him to be extremely pleasant and very encouraging.

In the summer of 1989, I went to India with my family. On the way back, I attended the Second Conference of the Canadian Number Theory Association in Vancouver (August 21–25). I was invited by David Boyd of the University of British Columbia on the Organizing Committee which had assembled a fine group of leading number theorists such as Enrico Bombieri, Ken Ribet, Peter Sarnak, Wolfgang Schmidt, Robert Tijdeman, Dale Brownawell, Andrew Odlyzko, and Carl Pomerance, to give hour lectures, augmented by nearly 100 twenty-minute talks in two parallel sessions each day. One of the things I remember about the meeting is the talk of Bombieri for which we had all assembled in the auditorium with eagerness and expectation. There was just five minutes left for the talk to begin, but Bombieri was nowhere to be found. There was great anxiety among the organizers. Then suddenly Bombieri entered the lecture hall to a thunderous applause; he said he was coming straight from Victoria where he had gone to see the Butchart Gardens, whose loveliness has to be seen to be believed! We did not see the Butchart Gardens on that trip, and indeed it was only in 2017 that we made a trip to Victoria.

Vancouver is a beautiful city with the Pacific ocean on one side and snow capped mountains in the backdrop. Mathura, my two daughters and I enjoyed the sightseeing we did in and around Vancouver, especially the cable car ride to the top of Grouse Mountain from where we had a spectacular view of Vancouver. As we drove around Vancouver, we were surprised to see cricket being played in Stanley Park.

While in Vancouver, I contacted Professor T. K. Menon, who had hosted my parents and me so graciously in Honolulu in the sixties. I was a teenager then. He had settled down in Vancouver as a professor at the University of British Columbia. I called on him at his home in Vancouver. He and his wife Rema were pleased to see me — now with my wife and daughters — after a gap of twenty years.

Even though I was inflicted with the q-disease during the Ramanujan Centennial in December 1987, and started lecturing on certain aspects of the theory of partitions and q-series in the Combinatorics Seminar at the University of Florida, up until the end of 1989, I continued to work in analytic number theory. Indeed my talks at the various conferences I attended — the Ramanujan Centennial in India in 1987, the Oberwolfach Conference in Fall 1988, the Bateman-70 Conference in Urbana in April 1989, and the Canadian Number Theory Association Conference in August 1989, — were all in analytic number theory. But then in Fall 1989, an event happened, namely the visit of Basil Gordon of UCLA to Florida, and that shifted the focus of my research to the theory of partitions and q-series.

Sometime in early June 1989, just before I left for India, Basil Gordon contacted me and said that he would be on a fully paid sabbatical in the academic year 1989–90, and he would like to spend one month at the University of Florida. This was like a gift from Heaven to me! Gordon was a master in the field of partitions and q-series, and one of my former teachers. Since 1987 I had been studying the theory of partitions on my own, and made some observations, but I was not yet ready to do research in the area. I realized that Gordon's visit would help me make the transition. Gordon was a leader in the world of combinatorics, and the Managing Editor (along with Bruce Rothschild of UCLA) of the Journal of Combinatorial Theory (Ser A). Thus the combinatorialists at Florida including the Chairman David Drake, were delighted at the prospect of Gordon's visit. So it was arranged that Gordon's room and board for one month would be fully paid, as well as his round-trip travel between Los Angeles and Florida. In Fall 1989, Gordon stayed at the Holiday Inn University Center, a stone's throw from the mathematics department, so that he could come easily to the Department and the Library. We spent many hours discussing various aspects of the theory of partitions, and the net result was that I was able to collaborate with him on two papers — my first two on partitions. During his one month visit, he gave three talks — a departmental colloquium talk on modular forms, a $\pi\mu\epsilon$ talk on the Dirichlet Pigeon Hole Principle, and a Combinatorics Seminar on plane partitions.

The theory of partitions and q-series could be approached from many directions — (i) combinatorially, (ii) by the use of q-hypergeometric techniques, and (iii) via modular forms. Basil Gordon was one of the very few mathematicians who had expertise BOTH in combinatorics AND in modular forms. In the sixties he worked on partitions from a combinatorial angle, but from the seventies, began using modular form techniques. Thus his colloquium on modular forms appealed to the whole department. His talk to the Undergraduate Mathematics Club $\pi\mu\epsilon$ on the Dirichlet Pigeon Hole Principle was a real gem.

To describe a certain aspect of Gordon's research and my joint work with him, I need to first say something briefly about the celebrated Rogers-Ramanujan (R-R) identities. These are two q-hypergeometric series $R(q)$ and $S(q)$ which have product representations:

$$R(q) := \sum_{n=0}^{\infty} \frac{q^{n^2}}{(1-q)(1-q^2)\cdots(1-q^n)} = \prod_{m=0}^{\infty} \frac{1}{(1-q^{5m+1})(1-q^{5m+4})},$$

and

$$S(q) := \sum_{n=0}^{\infty} \frac{q^{n^2+n}}{(1-q)(1-q^2)\cdots(1-q^n)} = \prod_{m=0}^{\infty} \frac{1}{(1-q^{5m+2})(1-q^{5m+3})}.$$

Indeed in the entire theory of partitions and q-series, these two identities are unmatched in simplicity of form, elegance and depth.

The identities have a remarkable history. Ramanujan discovered them in India around 1910 and noticed that the ratio $R(q)/S(q)$ admits a continued fraction expansion with a representation as a ratio of products:

$$\rho(q) := \frac{R(q)}{S(q)} = 1 + \cfrac{q}{1 + \cfrac{q^2}{1 + \cfrac{q^3}{1+\cdots}}} = \prod_{m=0}^{\infty} \frac{(1-q^{5m+2})(1-q^{5m+3})}{(1-q^{5m+1})(1-q^{5m+4})}.$$

This continued fraction plays a fundamental role in the theory of modular forms. Ramanujan made some striking evaluations of the continued fraction. In one of his letters to Hardy, he communicated one such evaluation.

Ramanujan did not have a proof of these identities. We know this because after his arrival in England, when Hardy asked Ramanujan whether he had a proof, he said he did not. Interestingly, in 1917, while Ramanujan was going through some old issues of the Journal of the London Mathematical Society in Cambridge, he came across three papers of the British mathematician L. J. Rogers dating back to 1896–97 where these two identities and many related ones were proved. Rogers was a mathematician whose work was largely ignored by his British peers, but after Ramanujan's rediscovery of his work, he received some recognition. In particular, he was elected Fellow of the Royal Society in 1924.

The Rogers-Ramanujan identities have nice combinatorial interpretation. The combinatorial version of the first identity is: *The number of partitions of a positive integer into parts that differ by at least 2 equals the number of partitions of that integer into parts which are of the form 5m+1 or 5m+4.* But neither Rogers nor Ramanujan emphasized the partition theorem; this combinatorial interpretation is due independently to the famous combinatorialist Major P. A. McMahon in England and the great

German mathematician Issai Schur — both contemporaries of Hardy and Ramanujan. In view of this interpretation, results that relate partitions whose parts satisfy gap conditions with partitions whose parts satisfy congruence conditions are called "Rogers-Ramanujan type partition theorems" and the q-hypergeometric identities which would the analytic counterpart of such partition theorems are called Rogers-Ramanujan (R-R) type identities. In an R-R type identity, the series would be the generating function of the partitions defined by gap conditions, and the product side would be the generating function for partitions given by congruence conditions. In the 1960s, Basil Gordon found a beautiful and important generalization of the Rogers-Ramanujan (MacMahon-Schur) partition theorem to all odd moduli $2k + 1$, for $k \geq 2$.

After the Ramanujan Centennial, I started studying the theory of partitions systematically and made some interesting observations on my own especially relating to Rogers-Ramanujan type identities and partition theorems, but I was not sure of their novelty. When Gordon arrived in Florida, I discussed my observations with him. He then suggested that we ought to systematically study the partition identities underlying a general continued fraction, namely

$$R(a,b) = 1 + \cfrac{bq}{1 + aq + \cfrac{bq^2}{1+aq^2+\frac{bq^3}{\cdots}}},$$

that Ramanujan had communicated in his first letter to Hardy in 1913. Observe that R(0,1) is the famous Ramanujan continued fraction $\rho(q)$. The advantage of having free parameters a and b is that by replacing q by an integral power of q, say q^m, and by replacing a and b by aq^α and b by bq^β, one could get integers in congruence classes modulo m. Thus by a detailed combinatorial analysis of $R(a,b)$ Gordon and I got new partition results as well as fresh insight into some classical partition identities. This was my first paper (with Gordon) in the theory of partitions and q-series and it appeared in the Journal of Combinatorial Theory (Ser A) in 1993.

By interpreting the Rogers-Ramanujan identities in partition form, Schur was led to the next level partition theorem, by replacing the modulus 5 by the modulus 6, and gap ≥ 2 between parts by gap ≥ 3 between parts. However, in doing so, he needed an extra condition, namely that consecutive multiples of three should not occur as parts. More precisely, Schur's celebrated partition theorem of 1926 is: *The number of partitions of an integer n into parts of the form $6k \pm 1$ is equal to the number of partitions of n into distinct parts of the form $3k \pm 1$ and also equal to the number* *of partitions of n into parts that differ by at least 3 with no consecutive multiples of 3 as parts.*

I asked Gordon the following question: The R-R identities come as a pair, and their ratio gives a continued fraction. But Schur's partition theorem does not have a companion. Can we still get a continued fraction for Schur's theorem?

Gordon said that this is an interesting question, and we started thinking about this. What we got was a continued fraction $S(a, b)$ in two free parameters a and b whose "numerator" had a product representation. The combinatorial interpretation of this gave a generalization of Schur's partition theorem. The presence of free parameters a and b gave us a refinement of Schur's theorem. By "refinement" we mean that we can keep track of the number of parts in the various residue classes in the partitions counted.

The ideas underlying the continued fraction for Schur's theorem led Gordon and me to formulate a new technique we called *the method of weighted words* which led us to generalize and refine many fundamental partition theorems of the Rogers-Ramanujan type. This method is now used widely in the theory of partitions. An important consequence of the weighted words method was that we could cast the generalized Schur theorem in form of a q-hypergeometric *series = product* RR-type identity in two free parameters. We called it a *key identity*.

The one month visit of Gordon was extremely fruitful for me. After collaborating with him, I was totally infected with the q-disease. So from the end of 1989, I moved away from classical analytic number theory and the focus of my research became the theory of partitions and q-series.

I submitted my joint paper with Gordon on Schur's theorem to Peter Roquette, Editor-in-Chief of The Journal für die Reine und Angewandte Mathematik (Crelle's Journal). Roquette responded by saying that the Crelle's Journal was dealing with a backlog problem, and so he would like to accept the paper for Manuscripta Mathematica for which he was also the chief editor. So my paper with Gordon on generalizations of Schur's theorem appeared in 1993 in Manuscripta Mathematica.

In December 1989, I went to Madras with my family. Inspired by the Ramanujan Centennial, Anna University had launched a "Ramanujan Endowment Lecture" to be delivered each year around Ramanujan's birthday, December 22. In 1989, the First Ramanujan Endowment Lecture was delivered by Professor R. P. Bambah, a distinguished number theorist who at that time was Vice-Chancellor of Punjab University in Chandigarh. I was invited to attend that Endowment Lecture, and on that occasion the refereed proceedings of the Ramanujan Centenary Number Theory Symposium that I had conducted at Anna University in December 1987, and published in Springer Lecture Notes Series as Volume 1395 that I had edited, was released, and the first copy was presented to Bambah. In the next few days, I gave a series of lectures at Anna University, attended by mathematics faculty and graduate students. Thus my association with Anna University continued strongly even after the Ramanujan Centennial.

After return from India in January 1990, I felt that I needed to go to UCLA to work with Gordon to solidify the work we had done and to prepare our two papers for publication. I wanted at least two weeks with Gordon, and the only time I could find was in the summer. The University of Florida summer term of three months (mid-May to mid-August) is split into two half terms — Summer A from mid-May to the end of June, and Summer B in July and first half of August. I was scheduled to teach a course on the History of Mathematics in Summer A, and so I visited Los Angeles in early August, during Summer B. Not only did we bring our two papers to completion, my stay in Los Angeles led us to a new approach to a deep theorem of Göllnitz, which turned out to be very significant.

Before my visit to Los Angeles in August 1990, I did make one more trip earlier in May to Kansas State University to attend the CBMS-NSF Lectures of Hugh Montgomery on "The interface between analytic number theory and harmonic analysis". Montgomery is the ultimate expert on topics at this interface and it was a perfect choice of topic for his lectures. The conference was organized by Todd Cochrane on the faculty of Kansas State University. He was a former PhD student of Don Lewis at the University of Michigan, and in 1979–80 when both Lewis and Montgomery were away from Ann Arbor on their sabbaticals, Cochrane took my graduate course in analytic number theory.

Gordon and I worked both at UCLA and at his stately home in Santa Monica, a charming suburb of Los Angeles. We used to go out to lunches at Westwood where UCLA is located, and dinners in Santa Monica after working at his home. Our mathematical discussions continued during these lunches and dinners. Gordon was very much like Erdős in that he was single, and he thought of mathematics all the time. One day, as we were walking to lunch in Westwood he told me: "Krishna, we should apply our method of weighted words to a deep theorem of Göllnitz, and use our two parameter generalization of Schur's theorem to get a three parameter generalization and refinement of Göllnitz's theorem. But this will be quite difficult." I was new to partition theory, and I did not know what Göllnitz' theorem was until Gordon explained its significance.

Theorem G (Göllnitz):

The number of partitions of an integer n into parts $\equiv 2, 5,$ or 11 (mod 12) is equal to the number of partitions into distinct parts $\equiv 2, 4,$ or 5 (mod 6), and is also equal to the number of parts that differ by ≥ 6, where the inequality is strict, if a part is $\equiv 0, 1,$ or 3 (mod 6), and with 1 and 3 not occurring as parts.

Göllnitz viewed his theorem as an extension of Schur's theorem to one higher dimension in the sense that the two distinct residue classes mod 3 are replaced by three distinct residue classes mod 6. Gordon and I had viewed the generalization of Schur's theorem as involving partitions (words) on integers in two primary colors a and b, and one secondary color ab. For Göllnitz' theorem, we studied the combinatorics of partitions (words) in three primary colors a, b, c and three secondary colors ab, ac, bc. We found a refined three parameter theorem and its marvellous *key identity* that extends the key identity we had for Schur's theorem. The method of weighted of weighted words approach not only showed why the gap and congruence conditions in Göllnitz' theorem are natural, but also why it is an extension of Schur's theorem; this is because our generalized Schur theorem follows from our generalized Göllnitz theorem when $c = 0$.

Although we had found a significant generalization of Göllnitz' theorem, we could not prove the key identity! So we were stuck, and when I left Los Angeles after a two week stay, Gordon and I agreed to think about the proof. This proof came about in a stunning way during a visit of George Andrews to Florida later in the year that I will describe below. In any case, the stay in Los Angeles for two weeks was immensely productive for me.

One more thing happened in Los Angeles. Gordon taught me how to type manuscripts in tex! Every day we went to UCLA, and Gordon would make me sit in front of the computer and type our joint papers in tex; he would sit next to me and guide me through each step. Thus I owe to Gordon what little of know of tex. This has been of immense value to me. Indeed I now type all my manuscripts, even non-mathematical ones, in tex.

Back in Gainesville from Los Angeles, I was getting prepared for the start of the Fall Term of 1990. Early in the Fall, it was a pleasant surprise when I received a message from the physics department that my proposal to them to invite Andrews for a Frontiers of Science Lecture was approved and they were ready to invite him. The Frontiers of Science lecture series, organized by the physics department, met once a week during the regular academic year and featured very eminent speakers from all science disciplines. Earlier I had attended two talks in that lecture series, one on human ancestors by Johanson (the discoverer of the famous "Lucy" skeleton), and prior to that, a talk by the eminent group theorist John Conway. So I was quite familiar with the quality of the lecture series, and therefore suggested the name of Andrews in 1987 during the Ramanujan Centennial Year. The Frontiers of Science was offered as a one credit course to the university undergraduates, who basically had to attend the lectures and answer one simple question to mark their attendance. Thus the lectures attracted a large audience. I knew that Andrews was a charismatic speaker who could charm audiences at all levels especially with the story of the discovery of Ramanujan's Lost Notebook, and so I confidently suggested his name. I was surprised and disappointed that there was no acknowledgement or response to my letter, and so I had no hope that my proposal would be picked up. Thus I was delighted to hear three years later (!) from the physics department about the decision to invite Andrews. I immediately contacted him and made arrangements for him to visit the University of Florida for a few days in November to deliver the Frontiers of Science Lecture, and to give a colloquium and a number theory seminar in the mathematics department.

As I was getting prepared with arrangements for Andrews' visit, we heard that Mathura's father had a major setback in his health. So Mathura and my daughters rushed to India in October, and I was to join them in December after my Fall classes were over. Fortunately, Mathura's father recovered by November to the relief of all of us.

When I met Andrews at the airport on his arrival, I told him about my discussions with Gordon on the Göllnitz theorem using the method of weighted words, and handed him a note containing the three parameter "key identity" Gordon and I had found. He was totally taken in by the identity and started working on its proof. The Frontiers of Science lecture of Andrews on November 6 had an audience of 750 which he kept in rapt attention as he described in his inimitable style the incredible story of the discovery of Ramanujan's Lost Notebook. But during the four days he was in Florida, he thought of nothing else but the key identity. On his last day, on the way to the airport, he handed me an eight page proof! That is how the Alladi-Andrews-Gordon paper came about and how my collaboration with him started. The proof was a clear demonstration of Andrews' mastery over q-hypergeometric series, and his unwavering concentration to conquer a difficult problem.

In early Fall 1990, I was pleasantly surprised to receive a handsome invitation from Dr. M. Anandakrishnan, the new Vice-Chancellor of Anna University, to deliver the Second Ramanujan Endowment Lecture on

December 22, Ramanujan's birthday. Dr. Anandakrishnan had returned to India that year to take the position as Vice-Chancellor of Anna University after a successful career in the United Nations. In December, when I went to India to join Mathura and our family, I delivered the Ramanujan Endowment Lecture on the theme "Ramanujan and continued fractions". Thus began a long, fruitful and warm relationship with Dr. Anandakrishnan with whom I had many more interactions while he was Vice-Chancellor at Anna University and later after his retirement from that position.

In the Spring Term of 1991, there were several visitors to the Department whom Mathura and I had the pleasure of hosting at our home for dinners and parties. Paul Erdős was of course there on his annual visit in March. Another visitor in the Spring was Ed Bertram from the University of Hawaii and we had the pleasure of reciprocating in some small measure his magnificent hospitality in Honolulu. He gave a colloquium in the Department on March 25. Then in early April, Jean-Marc Deshouillers from the University of Bordeaux, who had so kindly hosted me in 1986, gave an inspiring talk on "The Goldbach and Waring problems". Deshouillers was accompanied by his wife; they were visiting Florida to see their son who was living in Miami at that time.

During the three years 1988–91, as I was delving deeper into the theory of partitions, Paul Erdős visited the University of Florida each year in the Spring for two weeks either before or after the combinatorics conference, held annually in Boca Raton on the South Eastern Coast of Florida. These two week annual visits began during the time when Stan Ulam was a Graduate Research Professor in Mathematics at the University of Florida (1975–84) and continued until Erdős died in 1996. Although we often got together to discuss mathematics, no joint paper came out of these discussions primarily because my interest had shifted to the theory of partitions and q-series, and Erdős was not interested in such things. With a tinge of disappointment that I had moved away from classical number theory, he said, "I was never good at identities. I always preferred inequalities." I then told him: "Behind every inequality lies an identity waiting to be discovered" to which he simply shrugged!

1991 was one of the years when my parents did not come to the United States. So Mathura, my daughters and I spent our entire summer vacation in India. Enroute to India we vacationed in Malaysia and enjoyed the sylvan beauty of the island of Penang, a pearl of the Orient.

Professor K. Ramachandra had very graciously invited me to the Tata Institute, and so I visited him in Bombay for a week in July. It was my first visit to the Tata Institute and I gave five lectures there, one each week day, on a variety of topics in number theory. Naturally I spoke on partitions, my new area of research at that time. Ramachandra wanted me to speak on my work on irrationality even though it was about a decade old, since he was quite excited about irrationality measures and its applications to the finiteness of solutions of certain Diophantine equations.

Mr. C. Subramaniam was the Governor of the State of Maharashtra at that time, and he invited me to the Raj Bhavan (Governor's Residence) in Bombay for breakfast. I did not want to be late, so I left early taking into account possible delays due to heavy traffic congestion in Bombay. I reached the Raj Bhavan early and Mr. Subramaniam was doing his morning Yoga exercises before breakfast. He was man of great discipline, and consequently had excellent health until his nineties.

After I returned to Florida from India, Basil Gordon visited me for two weeks in August/September 1991, and it was just as productive as his visit in 1989. One of the questions Gordon and I investigated was the expansion of "Rogers-Ramanujan products"

$$\prod_{m=1}^{\infty} \frac{(1 - q^{km-j})(1 - q^{km-(k-j)})}{(1 - q^{km-\ell})(1 - q^{km-(k-\ell)})} = \sum_{n \geq 0} a_n q^n.$$

We determined the conditions on k, j, ℓ that would ensure that the coefficients a_n would be zero in certain residue classes modulo k. The above product with $k = 5, j = 2, \ell = 1$ is the famous product associated with Ramanujan's continued fraction $\rho(q)$, for which Richmond and Szekeres had established a vanishing coefficient theorem in the seventies; my result with Gordon was for general moduli k.

Invigorated by my discussions with Gordon, I started an intense study of the theory of partitions and q-series on my own. In particular, I looked closely at a specific set of six identities due to L. J. Rogers which are companions to the celebrated Rogers-Ramanujan identities. Of these six, the first two were especially beautiful and bore a striking similarity to the Rogers-Ramanujan identities. They are combinatorially equivalent to the Rogers-Ramanujan identities as was shown by David Bressoud in 1978. I was spellbound by the beauty of these two identities of Rogers. They are just like the celebrated Rogers-Ramanujan identities, but each has a extra common term as a factor in the denominator of the product. So I asked myself the question: What happens if one considers the ratio $r(q)/s(q)$, where $r(q)$ and $s(q)$ are the series in these two identities, just as Ramanujan considered $R(q)/S(q)$? What continued fraction would this give? As to the ratio of the products, the common factor would cancel, and so we would get the ratio of products as in Ramanujan's continued fraction identity. I was able to show that

$$\rho^*(q) := \frac{r(q)}{s(q)} = q + \cfrac{1}{q^3 + \cfrac{1}{q^5 + \cfrac{1}{q^7 + \cdots}}} = \prod_{m=1}^{\infty} \frac{(1 - q^{5m-2})(1 - q^{5m-3})}{(1 - q^{5m-1})(1 - q^{5m-4})}, \quad \text{(mc)}.$$

The (mc) in the above identity means *modified convergence* which means the approximations have to be truncations at the numerators $1, 1, 1, \ldots$ in succession instead of the standard truncations at the denominators $q, q^3, q^5, \ldots$. What is more, the other four Rogers series appear when considering the odd and even convergents to $\rho^*(q)$ in the standard fashion. Thus the continued fraction $\rho^*(q)$ I had stumbled upon embodied the six companions to the Rogers-Ramanujan identities due to Rogers that I was studying! I found a few more examples of continued fractions that do not converge on the ordinary sense, but do converge in the modified sense. A similar study of their modified truncations and standard truncations led to some new connections between certain fundamental identities in the theory of partitions and q-series. I was really excited and wanted to write a paper on these results without delay. The year 1991 was unusual for me. My parents did not come to America in the summer, and I did not go to India in December! Instead, I spent the entire Christmas break of 1991 typing the paper "Modified convergence of continued fractions of Rogers-Ramanujan type" with my newly acquired knowledge of AMStex given to me by Gordon! In January 1992 I submitted the paper to the Journal of Combinatorial Theory (A) and it appeared there in 1993.

Since I considered myself a new entrant to the theory of partitions and q-series, I sent preprints of my papers to about a dozen or more experts in the field to keep them informed of my progress. On seeing my work, Bruce Berndt invited me to speak at the Illinois Number Theory Conference in Urbana in early April 1992, and also to deliver a Colloquium and a Number Theory Seminar there. My number theory seminar talk in Urbana was on modified convergence. The Berndts also invited me to stay at their house, and I had the pleasure of enjoying the warm hospitality of Bruce and his wife Helen. One evening, Professor Ranga Rao, a well known mathematician on the faculty of the University of Illinois and an old friend of my father, invited me for dinner at his house.

In mid-March 1992, I was invited by *The American Scientist* to write a Lead Review of Robert Kanigel's classic biography of Ramanujan entitled "The Man Who Knew Infinity". A review of this book by Raghavan

Narasimhan had appeared in the American Mathematical Monthly which I had read. As I had said earlier, Raghavan Narasimhan was quite outspoken in his views, and in his review he expressed the opinion that Kanigel did not fully appreciate or understand various Hindu habits and practices. He also said that Kanigel's treatment of Ramanujan's mathematics was not satisfactory. But this was a book on Ramanujan's life, and not about his mathematics, about which there are several excellent expositions due to Hardy, Andrews and Berndt, for instance. On my part, I felt that Kanigel had done a marvelous job in describing Ramanujan's life in full detail including the Hindu beliefs and practices of his family, and brought out the excitement of various developments in Ramanujan's life admirably. Kanigel's book is in a sense a dual biography — of Ramanujan and his mentor Hardy. Indeed there are about 40 pages on Hardy's life as well. My only criticism of Kanigel's book was that it was perhaps not necessary to have spent so much time discussing whether or not Hardy was a homosexual when there was, and still is, no evidence. In the end of the discussion, Kanigel does say that there is no evidence at all! My review of Kanigel's book appeared in the American Scientist [A1] in 1992.

In the four years 1988–92, I had successfully changed the focus of my research from analytic number theory to the theory of partitions and q-series. Basil Gordon was instrumental in helping me make the transition. What I needed next was to have a close interaction with George Andrews, the leader in the field. This happened in an unexpected and pleasant manner.

In Fall 1991, David Drake, the Department Chair, suggested that I apply for a sabbatical. I started work at the University of Florida in January 1987, and so 1991–92 was my fifth full academic year and not the sixth. So I was not thinking of applying for a sabbatical. But Drake told me that there were several sabbaticals that were not claimed, and he would strongly support my application even if it was one year early. Since I was keen to work with Andrews, I applied for a full-year half-pay sabbatical hoping to get Penn State to pay for the remaining half of my salary. Andrews reacted very kindly, and in May 1992 he called me and made the offer. Since Penn State was picking up the other half of my salary, I had to teach one course per semester there during 1992–93, which I did not mind at all. So I all was set to go to Penn State with my family in August 1992. It was going to be another turning point in my career.

7.5   Sabbatical at Penn State (1992–93)

Since I did not visit India in December 1991, I decided to go to India in the summer of 1992 with my family to attend to matters there, and to proceed to Penn State directly from India. We rented out our house in Gainesville for one year starting from July 1992. My parents had arrived in Florida in the early Spring for a three-month visit. So after seeing them off to India, and vacating our house, I left for India with Mathura and my daughters in mid-June. Just before our departure, we heard that Mathura's father again had a heart attack, and that he miraculously survived. So when we arrived in Madras, we were relieved to see him recovering.

In Madras, throughout July, I gave a series of lectures on modified convergence at MATSCIENCE and at the Ramanujan Institute and followed this with talks on this topic at other venues like the Tamil Nadu Academy of Sciences and Anna University. On the way from Madras to Penn State we halted in Singapore where I lectured at the National University of Singapore (NUS) on modified convergence. Professor Peng, the Chair of mathematics at NUS, was very pleased that Jean-Pierre Serre had visited their department and had made several suggestions which they were following. In view of my contact with Peng, I visited NUS often in the next few years. We also broke journey in Los Angeles to visit friends and relatives, but I used the occasion to see Basil Gordon at UCLA to put the finishing touches on our joint paper "Vanishing coefficients in the expansion of Rogers-Ramanujan type products". Even though our house was rented out, we had to come back to Gainesville to pick up our car which we definitely needed during our one-year stay at State College, Pennsylvania.

On our way north from Florida to Pennsylvania, we stayed overnight in Knoxville at the home of Professor Balram Rajput of the mathematics department of the University of Tennessee, a friend of my father. We also spent a night at the apartment of our friends the Morrises in Chapel Hill. Paul and Ruth Morris were in Madras in the sixties when Paul was Director of the USIS. They were close friends of my parents and had visited Ekamra Nivas on many occasions for parties that my parents would host for the international visitors of MATSCIENCE. So I knew the Morrises since by boyhood days, and it was a pleasure to reconnect a quarter century later. After Paul retired from diplomatic service, he and Ruth settled down in Chapel Hill because that region provided them the "four seasons" — all mild — and an invigorating intellectual and cultural atmosphere as well.

We arrived in State College on Monday, August 24, just two days before the start of classes. We had arranged a nice two bedroom furnished apartment at Parkway Plaza, about ten minutes away from campus. There was a shuttle bus to take the tenants to and from campus that operated with good frequency. I took this shuttle bus on most of the days to permit Mathura to use the car. My older daughter Lalitha went to Easterly Parkway Elementary School, a few blocks away from Parkway Plaza. Following the suggestion of George Andrews, we arranged for my younger daughter Amritha to go a private pre-school called the Red Satchel — also close by — where the teacher Ms. Dudley was one of most pleasant persons we had ever met. Andrews' two daughters Amy and Katy, and his son Derek had all attended the Red Satchel, and Andrews had formed a very favorable impression of that school from their experience. So all arrangements for a one-year stay at State College were totally satisfactory.

The campus of Penn State University in the summer is one of the greenest in the United States, with expansive lawns, stately trees, and exquisitely beautiful flower beds. And so when we arrived in August, the campus was in its loveliest form. The mathematics department is housed in Macalester Building, one of the historic buildings of State College. Leading up to Macalester Building is a promenade bordered by stately Dutch elm trees, the pride of Penn State campus. On the ground floor of Macalester was the Mathematics Library. Having the library in the same building as your office is very conducive for research. And in the basement of Macalester Building was, and still is, the University Post Office — a great convenience especially in those days when we wrote and received letters instead of email! Mike the Mailman at the Post Office was almost a Penn State icon, because of his friendly disposition. The students and faculty loved him, and so did I.

The Chair of the Mathematics Department at that time was Jerry Bona, an eminent applied mathematician known for work in fluid mechanics and PDEs. Penn State had an internationally known program in fluid mechanics. There was a Fluid Mechanics Lab located in the basement of Macalester, a lab that was founded due to efforts of William Pritchard. Sadly Pritchard died in 1994 and the lab is now named after him.

Andrews' office was in the top most floor and one had a feeling of being in the attic! I was given a shared office one floor below. My officemate was a very friendly chap — a Russian assistant professor in applied mathematics by name Leonid Berlyand. He is now a full professor there.

The course I was assigned to teach in Fall 1992 was for senior undergraduates on the topic of primality testing and factoring, and the text was the lovely book by David Bressoud on the Penn State faculty. In addition to being a fine research mathematician, Bressoud is a very successful author of textbooks. In view of its 'applied' nature, this course was taken by students from outside mathematics — such as computer science. Teaching this course gave me an opportunity to learn the basics of this important subject. In the Spring/Winter Semester of 1993, I was asked to teach an undergraduate course in Discrete Mathematics from the book "Concrete Mathematics" by Ron Graham, Donald Knuth, and Oren Pataishnik — a classic in every sense. Concrete mathematics has two interpretations — (i) continuous plus discrete because of the use of generating functions to understand discrete objects, and (ii) concrete means practically useful. This course too was attended by students from the computer science department.

George Andrews was giving a year-long graduate course on the theory of partitions and it was my luck that he was doing this during the year of my visit. I sat in on that course and it was a great opportunity to learn from the master of the field. As my father would say, it was like learning The Ramayana from Sage Valmiki himself! It was like sitting in on Montgomery's course in analytic number theory in 1981–82 at the University of Michigan. I have detailed notes of Andrews' course just as I have of Montgomery's course, and I have benefited from both. Andrew Sills and Dennis Eichhorn were graduate students taking Andrews' course and both became active researchers in the area of partitions and q-series.

There were two organizers of the weekly colloquium that year — George Andrews and Anatole Katok. Katok (recently deceased) was a powerhouse in the area of Dynamical Systems, and so with two mathematical giants running the colloquium, there were many stellar speakers that year. I remember the talks of Barry Mazur (Harvard) and of Serge Lang (Yale). Andrews and Katok chaired the talks on alternate weeks. When Andrews was Chair, his choice for dinner with the colloquium speaker usually was "The Tavern" on College Ave, serving good food in a rustic setting at moderate prices. Katok always chose exotic and expensive restaurants; for the colloquium dinner in honor of Mazur, I remember joining the group at a fancy restaurant called the Gambler's Mill Inn in Bellefonte near State College, having an exotic menu that included rabbit!

During the summer of 1992, the Rademacher Centenary Conference was held at Penn State University. Andrews was a former student of Rademacher and so it was natural for him to organize this conference (along with Bressoud) in memory of his teacher. He invited me to the conference, but I could not attend it because I was in India for the summer before arriving in Penn State. Since I was an invitee for the conference, Andrews and Bressoud asked me when I arrived in State College if I would be interested in submitting a paper to the refereed proceedings that they were editing. I had just finished writing up my joint paper with Gordon on "Vanishing coefficients in the expansion of Rogers-Ramanujan products", and so I submitted this to the Rademacher Centenary Proceedings. Andrews and Bressoud accepted this for the proceedings which appeared in the Contemporary Mathematics Series of the AMS as V.166.

Even though I did not attend the Rademacher conference, by a stroke of luck I ended up writing a joint paper with Andrews and Gordon while at Penn State generalizing a partition theorem of Capparelli which was stated at the conference. The way the conjecture in its original form was proved and how we generalized it is an exciting story which I will relate here.

On the opening day of the Rademacher conference, Jim Lepowsky of Rutgers University, who had famously obtained Lie theoretic proofs of the Rogers-Ramanujan identities (in collaboration with Robert Wilson), gave a talk on Rogers-Ramanujan type identities and Lie algebras. During his lecture, he stated two partition conjectures of his PhD student Stefano Capparelli which came up in a study of Vertex Operators in Lie Algebras.

Upon seeing these conjectures, Andrews started working intensely. Even though he was one of the organizers of the conference, he went into hiding during the breaks to work undisturbed. By the end of the conference, he had proved the conjectures, and so on the last day he gave a talk outlining a q-hypergeometric proof! Andrews' proof of the Cappareli conjectures appeared in the Proceedings of the Rademacher conference.

I heard this story from Basil Gordon with whom I was talking regularly by phone from Penn State. Gordon had attended the lectures of Lepowsky and Andrews and felt that the Capparelli Conjectures could be generalized and proved by our Method of Weighted Words. So when I heard this from Gordon, I went about working on the conjectures.

The first thing I did was to reformulate the Capparelli conjectures in terms of partitions into distinct parts, so that I could give a combinatorial proof of a refinement of them. On seeing this, Andrews made an improvement of the refinement by connecting it with his q-hypergeometric proof. Once we had the refinement, Gordon and I, using the method of weighted words, were able to construct and prove the q-hypergeometric *key identity* that yielded a generalization of the Capparelli partition theorems.

Capparelli's paper in which his partition conjectures are stated, had appeared in the Journal of Algebra, and so I suggested to Andrews and Gordon that we submit our paper also to the Journal of Algebra, to which they agreed. It took several months for me to write the paper and so I submitted only in 1993 to my former UCLA professor Robert Steinberg who was an editor of the Journal of Algebra. It was a pleasure for me to receive a letter in the spring of 1994 in his beautiful handwriting accepting our paper for the Journal of Algebra.

In addition to drawing my attention to the Capparelli conjectures, George Andrews showed me a companion to the famous partition theorem of Schur that he had obtained by a computer search. Andrews was a pioneer in the use of computers to discover and prove partition identities by q series techniques. He had a project funded by the IBM which he humorously referred to as a "Mining Project" because he used the computer to search for (= dig up) partition theorems as in say, a gold mine! Such a search led him to a companion to Schur's theorem.

Andrews had proved this the companion theorem using generating functions and he asked me if the method of weighted words would lend insight into the combinatorial structure underlying this theorem.

In the method of weighted words approach to Schur's theorem, Gordon and I used integers occurring three colors — primary colors a, b and secondary color ab with the ordering $ab < a < b$. I noticed that the Andrews companion could be obtained by changing the order of the colors to $a < ab < b$. Since there are six permutations of ab, a, b, I found a constellation of six companion theorems of which one was Schur's theorem, another was the Andrews companion, and four new companions. Thus the method of weighted words yielded four new partition theorems which the computer missed! I then corresponded with Gordon and we connected these six companions to the six fundamental recurrences for the q-multinomial coefficients of order 3. So I wrote a paper with Gordon entitled "Schur's partition theorem, companions, refinements and generalizations". It appeared in the Transactions of the AMS in 1995.

During my year at Penn State, I also wrote up my long paper with Andrews and Gordon on generalizations and refinements of the partition theorem of Göllnitz via the method of weighted words. As I had mentioned earlier, this paper germinated in the Fall of 1990 when Andrews visited Florida, but it took some time to bring the work to completion. In a sense this was good because we could incorporate the idea of changing the ordering of the colored integers in out treatment of Göllnitz' theorem. For Göllnitz' theorem, our method of weighted words approached utilized integers in six colors — three primary colors a, b, c, and three secondary colors ab, ac, bc. Thus we have $6! = 720$ orderings of these six colors and these yield a constellation of 720 partition companions all equivalent to Göllnitz' partition function. Göllnitz' original paper had appeared in the Journal für die Reine und Angewandte Mathematik (= Crelle's Journal), and so I submitted our joint paper to Peter Roquette, Chief Editor of Crelle's Journal. Our paper appeared in Crelle's Journal in 1995.

Fall 1992 was extremely productive for me at Penn State in my new area of research. So for the weekly Colloquium I gave a talk in mid-November on the method of weighted words describing how this technique yields generalizations and refinements of a wide class of partition theorems. I also gave talks in the number theory seminar on my recent work on q-series and on my earlier work on multiplicative functions and Brun's sieve. The number theory seminar met weekly and was well attended by an active group of number theorists and those with a keen interest in number theory. Its participants besides Andrews and Bressoud included Dale Brownawell (Transcendental Number Theory), Winnie Li (Algebraic Number Theory), Leonid Vaserstein (Algebra and Dynamical Systems), and Gary Mullen (Finite Fields), all of whom I got to know very well. So with the participants having expertise in different areas, the number theory seminar featured talks on a wide range of topics.

As far as discussions relating to my research, I had long sessions every week with both George Andrews and David Bressoud. Both of them were extremely generous with their time and were excellent sources for valuable references. In addition, I had frequent telephone conversations with Basil Gordon. His home number was unlisted, but he kindly gave it to me. I made sure that every time I called him, I had a interesting mathematical idea to discuss so that the phone call would be worthwhile and not a disturbance of his privacy. Los Angeles was three hours behind State College, and so my phone calls were around early mornings his time. Sometimes my phone calls would wake him up, but he would not mind that at all; he said he enjoyed waking up to a wonderful mathematical discussion. In Hindu temples, there is the *Suprabhatham*, a sonorous vedic chanting with which the Lord is woken up in the morning as the temple opens for worship by devotees. In Sanskrit, Suprabhatham means Good Morning. I used to tell Gordon that the mathematical ideas that I would offer him is like the Suprabhatham! He said that he was flattered being compared to the Lord!!

But it was not all work at Penn State. We had a very active and enjoyable social life as well.

George Andrews and his wife Joy were very gracious and caring. Within a week of our arrival, they took us on an excursion to Penn's Cave, about half an hour away from State College. There is a stream inside the cave, and one has to take a boat to see the impressive limestone formations. On another weekend, they took us on a picnic to a nearby park, and Joy had thoughtfully prepared excellent vegetarian food to suit our palate.

We also made contact with the Indian community in State College and attended many social gatherings. Out first and primary contact were Professor and Mrs. Vedam, who are best described as "goodness personified". They were our God Parents just as Prof. and Mrs. M. S. Ramanujan were to us in Ann Arbor. Through the Vedams we got to know of various events among the Indian community and made several friends there. Of special note was a "Story Hour" that was held every other Sunday morning, when a gentleman by name Gandhi told stories from the Indian epic *The Mahabharatha* to the children of the Indian community. Gandhi, as one would expect from such a name (!), was very genial, and the children loved him as much as his story telling. The Story Hour met at the homes of different members of the community, and Mr. Gandhi always made small talk about Penn State Football before starting the story, to gain the attention of the kids. My daughters were eleven and five years old, and so they attended the Story Hour with great interest. Of course at the end of the story telling, an excellent pot lunch was served by the parents, which was equally delightful! Mrs. Vedam was a cook par-excellence, and everyone looked forward to the items she would bring to various gatherings.

Professor Vedam, Emeritus of the Material Sciences Department, had known my father, and so we bonded instantly. Mrs. Vedam took a special liking for Mathura who attended some of her cooking classes at their home. The Vedams were passionate about their early morning walks, and since Mathura and I were used to getting up very early, we joined them on walks in the Gym before sunrise. Walks are good at any age, but since I was in my thirties, I needed more rigorous exercise, and to me it was tennis. Since it was not possible to play outdoors in the Fall and in the Winter, I joined the Penn State Indoor Tennis Club on campus. It was not inexpensive, but it was worth it. I was re-living my days at the University of Michigan, when I used to play tennis in the fast indoor courts of the Men's Gym. At Penn State, I joined "the ladder" and moved up steadily to reach the top. Based on my performance, I was selected for a tournament in the indoor club in which the Penn State tennis team players could not participate. I made it to the final and my opponent was a member of the local High School Tennis Team. He was much stronger and faster than me, and had a fiery temper. But I had more experience, and could mix hard hitting

shots with touch play — things I had learnt from seniors at the Mylapore
Club in Madras during my school days. I knew that if I could gain an early
advantage, he would simply disintegrate in frustration and anger. That is
what happened because he underestimated me as we started our game. So
I won easily against a stronger, faster opponent. When I went up to the net
to shake his hands, I said that he was the better player, but that made him
so angry that he threw his racquet into the stands! My daughters observed
this entire match as did some of my club friends.

From time to time, I had friendly informal tennis games with Mike the
Mailman. I even had tennis mates in the mathematics department. I used
to play occasionally with my office mate Leonid Berlyand and also with
Edward Formanek, who is actually a Grand Master in Chess! Formanek is
a reputed mathematician specialising in ring theory. He used to attend the
number theory seminar quite often.

To our pleasant surprise, State College also offered us a high quality
Indian classical music experience throughout the year. There was a group
called *Raaga* which met once a week, typically on Friday evenings after dinner, by rotation at the residence of one of its members. There were vocalists
and instrumentalists in that group and there would be individual singing for
about three hours by the members in both Carnatic and Hindustani music
with instrumental accompaniments. Mathura and I participated in these
musical meetings regularly. The organizer of the Raaga meetings that year
was a young Indian undergraduate student in computer science by name
Arijit Mahalanobis who is a very talented singer of Hindustani music, and
with a deep interest in Carnatic music. Arijit was a student in my class
on primality testing, and he was told by the Vedams that Mathura and I
were passionately interested in Indian classical music. So he approached
me early in the Fall term and invited us to the Raaga meetings.

We also took part in functions associated with Indian festivals such
as Deepavali (or Diwali). Indeed for the Deepavali function, Mathura and
Amritha sang, while Lalitha danced. There were many Indian faculty members of Penn State University with whom we interacted socially at these
get togethers. Prof. Ram Kanwal a senior analyst in the mathematics department and his wife were very kind towards us and invited to their home
for dinner. Another Indian mathematics faculty member whom we used to
meet regularly on a social basis at various Indian gatherings was Paromita
(Pashka) Chowla. Pashka was the daughter of the famous Indian number theorist Sarvadaman Chowla whom Mathura and I had the pleasure
of meeting in Princeton in 1981–82; my parents knew S. Chowla back in

1958–59 when they visited Princeton. S. Chowla was at Penn State from 1963 to 76, and Pashka moved to State College in 1963 with him. Mathura got to know Pashka very well from the "Indian Ladies Circle" meetings that both attended regularly.

Professor C. R. Rao, one of the most eminent statisticians in the world was Eberly Professor at Penn State. He and his wife, graciously invited us to their apartment for dinner. Rao and my father had a long association in India when my father was Director of MATSCIENCE and Rao was Director of the Indian Statistical Institute in Calcutta. Rao had an outstanding career at the Indian Statistical Institute, and after retirement from there, he moved to the United States, where his remarkable productivity continued. After spending time in Pittsburgh, he settled down at State College. He was kind enough to attend my colloquium talk at Penn State.

Another famous Indian faculty member at Penn State was Rustum Roy of the material science department. As a leader in the area of materials research, he had founded a research institute at Penn State. Dozens of Indian researchers were employed there and so he was considered as a "God father" in the Indian community. He was so busy that he never attended any of the social gatherings, and so Mathura and I never met him or his wife. But I read his columns in the university newspaper *The Collegian* in which he expressed his strong preference to applied and inter-disciplinary research, and criticized research in the pure sciences (especially mathematics) as unnecessary and wasteful. As a rebuttal, George Andrews strongly made the case for not just pure mathematics, but more generally for theoretical research. So the two Evan Pugh Professors clashed regularly both in committee meetings and in the newspaper columns. My own feeling was that Andrews was correct and that Rustum Roy did not have the faintest idea of the "usefulness of useless knowledge" — which is the basis (see [B8]) for the creation of the Institute for Advanced Study in Princeton!

During Fall 1992, we had friends and relatives visit us at State College. My paternal cousin V. S. Ravi from India (about ten years older to me) came to spend two nights in September. He was in the Philadelphia area visiting relatives, and it was a quick train journey from there to State College. Ravi in his high school and college days in Madras lived in Ekamra Nivas in the fifties, and had mingled with Nobel Laureates P. A. M. Dirac and C. F. Powell whom my father had as his personal guests. So he admired eminent academicians. At Penn State he wanted to meet George Andrews — the Ramanujan authority. We therefore invited Andrews to our apartment for tea so that Ravi and Andrews could have a leisurely conversation.

Another guest of ours was Mathura's youngest uncle Srinivasan from Bangalore in the month of October. With him we went to Pittsburgh to see the famous temple for Lord Venkateswara — indeed the first Venkateswara temple in the United States. His visit also coincided with that of our friends Bhavani Sankar and Mira from Gainesville; Sankar is a Professor in Aerospace Engineering at the University of Florida.

Whenever we had guests with us, we always went to the Penn State Creamery on campus, reputed for its special ice creams. If one orders one scoop at the Creamery, one gets two large scoops, and ice cream lovers do not complain about this generosity. In later years, our guests have told us that the Creamery was one of the highlights of their visit to State College.

So as we were having the time of our lives at State College, tragedy struck our family. We received a phone call on December 7 night from Madras, that Mathura's father's health had worsened considerably and there was no chance of recovery. She was asked to leave for Madras immediately. Unfortunately the Snow Storm of the Century blasted all of Northeastern USA, and Penn State was in its grip. So she could not leave rightaway and could not be with her father who died on December 11. She did reach Madras by December 13 to be with her mother and brothers for various ceremonies in her father's memory. He was one of India's most successful CPAs and a man committed to public service as well. About 2000 persons came to pay their last respects, a testimony to the high regard he earned from society in general. To our family it was a great personal loss, and to field of accountancy in India, the end of an era!

Mathura left for India by herself and so our daughters stayed with me at Penn State. To relieve them from the rigors of the Penn State winter, and for them to enjoy the company of their friends in Florida, we decided to drive down to Florida for the Christmas-New Year break. During the Christmas vacation, a wonderful mathematical development happened for me pertaining to the famous Göllnitz-Gordon identities.

The Göllnitz-Gordon identities are to the modulus 8 what the Rogers-Ramanujan identities are to the modulus 5. The ratio of the two Göllnitz-Gordon series, yields a continued fraction as beautiful as the Ramanujan's continued fraction, thereby yielding a ratio of products as lovely as in Ramanujan's continued fraction identity. Like the Rogers-Ramanujan identities, the Göllnitz-Gordon identities have elegant partition interpretations. Göllnitz discovered the identities from a study of partitions, whereas Gordon found them by investigating the underlying continued fraction.

One of things I wanted to study was the odd and even parts of various q

series identities — that is, their bisections. I started by bisecting the series and the product in the celebrated Euler's Pentagonal Numbers Theorem, and evaluating the odd and even parts in two different ways. This led me to two identities equivalent to the Göllnitz-Gordon identities from which I was able to easily derive the Göllnitz-Gordon identities. In college, when we are told that every function can be decomposed uniquely into an odd function and an even function, we are given the example of $\sinh(x)$ and $\cosh(x)$ as the odd and even parts of the exponential function e^x. But beyond this, nothing is done with regard to bisecting functions. So here I was like a kid bisecting the well known Euler's Pentagonal Numbers Identity, and to my surprise, it yielded a new and simple proof of the Göllitz-Gordon identities! What is more, when I bisected the famous Gauss Triangular series identity, I got an even simpler proof of the Göllnitz-Gordon identities. I was thrilled to say the least, and I called Andrews from Florida. He confirmed the novelty of my proof. That was the best Christmas/New Year gift I could ask for! Spurred by the success of this approach, I bisected the celebrated Ramanujan continued fraction identity, and obtained lovely products modulo 80. So on return to Penn State from Florida in January, I started writing a paper "Some new observations on the Göllnitz-Gordon and the Rogers-Ramanujan identities". I submitted this to the Transactions of the AMS where it was accepted.

After spending a month with her family, Mathura returned to State College in mid-January. Naturally the loss of her father was especially painful for her. My parents decided to come to State College for a two-month stay starting in mid-February, and that relieved some of the loneliness for she had company at home during the time I was in campus.

My parents joined us for all the Indian cultural gatherings that Mathura and I attended. In the Raaga weekly music meetings, my mother and Mathura were asked to play a major role in vocal renderings. In early April, Arijit Mahalanobis arranged a special demonstration of vocal Carnatic music by my mother and Mathura for his music class at Penn State University attended by two hundred students.

Many of our Indian friends — the Vedams, the Kanwals, Professor and Mrs. C. R. Rao, Pashka Chowla, and others, had either known or heard of my father. So when my parents were there, we were invited to their homes for dinner. In return we had them over at our place as well. George Andrews and Joy hosted a magnificent dinner at their lovely home when my parents were in State College.

We made several trips with my parents. Of special note was the visit to the Corning Glass Works to re-live my father's memories of his visit there in 1963. As a memento, we bought a beautiful stained glass peacock which adorns the window of the breakfast nook at our Gainesville home.

Another memorable trip with my parents was to Cleveland during the Easter weekend in April to attend the Thyagaraja Aradhana there. Saint Thyagaraja is the greatest among the Carnatic music composers, and in India an annual festival — the Thyagaraja Aradhana — is held in his birth place. The impact of Indian classical music is so great in America that annually in Cleveland, this Aradhana is held during the Easter weekend featuring several top professional singers and instrumentalists from India. Thyagaraja composed his songs in the Telugu language, and my father being proficient in Telugu, made a deep study of his compositions. So it was particularly appealing to my father that America, the land of technological marvels, was paying tribute to a Saint-Composer from India! The event attracted more than a thousand carnatic music aficionados.

My parents left State College for India in mid-April. Mathura and I had six more weeks to wind up our establishment in State College.

In view of the many new results that I was able to obtain in the Fall of 1992, I spent much of the Winter/Spring semester at Penn State preparing various papers for submission. But the second semester at Penn State was also fruitful for my research exploration.

In the course of my study of the odd and even parts of Euler's celebrated Pentagonal Numbers Identity, I had to use the famous Quintuple Product Identity, originally due to G. N. Watson in 1929, but rediscovered by Bailey in 1951 and Basil Gordon in 1961. I wanted to understand the quintuple product identity in my own way, and as I looked at it closely, I came up with my own proof of it. This proof of mine led me to theorems of the type

$$P_T(n - k) = P_S(n),$$

where $P_T(n)$ and $P_S(n)$) represent the number of partitions of n whose parts come from the set T and S respectively. We call this a *shifted partition identity*. Andrews was the first to provide an example of a shifted partition identity in the form $P_T(n - 1) = P_S(n)$, with S and T given as integers in certain residue classes (mod 32). Indeed this was his contribution to the American Mathematical Monthly for the Ramanujan Centennial Year 1987. Subsequently Bressoud, Sinai Robins and others found similar shifted partition with moduli other than 32. My work showed that the quintuple product identity and some of its variations provide a natural setting for general shifted partition identities. So this led to my paper "The quintuple product identity and shifted partition functions" which appeared in 1995 in a special issue of The Journal of Computational and Applied Mathematics devoted to q-series. Thus it was a remarkably productive year for me at Penn State, and now I was completely immersed in the theory of partitions and q-series, or should I say, I totally succumbed to the q disease. For the next quarter century, I worked only in that area.

A few weeks before I departed from Penn State, in May 1993, I attended an American Mathematical Society Regional Meeting at the University of Northern Illinois in DeKalb. I was invited by Andrew Granville to speak there at a Special Session in Number Theory in honor of Paul Erdős for his 80th birthday. The connection between DeKalb and Erdős was that his great friend and collaborator John Selfridge was at Northern Illinois University. DeKalb is 700 miles away from State College, yet I decided to drive to DeKalb. I rented a car and drove the 700 miles distance in each direction in one day — twelve hours of driving per day! That sounds crazy, but I was much younger then. It was worth the drive because the Erdős 80th birthday conference was attended by leading number theorists. My talk at this conference was not in analytic number theory, but on my recent work on Schur's partition theorem and its companions.

On Friday, May 28, I took leave of George Andrews in whose company I was inspired to work at my best. He very graciously presented me the Collected Works of Major P. A. MacMahon that he had edited, and a lovely enlarged framed photograph of the Avenue of Dutch Elm Trees leading to the Macalester Building! Both are in the study of my Gainesville home.

On Saturday, May 29, there was a send-off party for us by our Indian friends at the home of the Vedams. There was even a farewell party for my daughter Lalitha at Easterly Parkway Elementary School! Thus State College had touched our hearts and so we wanted to be back there in the future. Indeed I have returned to State College many times since then to attend conferences or to work with Andrews, and have enjoyed the gracious hospitality offered by Penn State University. With all the new results I had obtained in 1992–93, I needed to be back again soon to complete certain research projects I had begun at Penn State. So I returned to Penn State in Fall 1994 on a research leave. But then a lot happened between Summer 1993 and Fall 1994 which is what I describe next.

7.6 Exciting aftermath of the Penn State visit (1993–94)

We returned to Gainesville from Penn State on June 2, 1993, and immediately Mathura and I departed for Europe on June 7, leaving our daughters in the care of our close friends Bhavani Sankar and Mira. I had received many invitations to lecture on my new research which was different from what I had presented during my trips to Europe in the eighties. I also felt that the European trip would lift Mathura's spirits after the loss of her father. So we combined academic assignments with enjoyable sightseeing.

Our first stop was Trieste, Italy, for my lecture at the International Centre for Theoretical Physics (ICTP). Enroute, we had a long layover in Milan, and we utilized the time to take a quick ride into the city to see the magnificent Duomo cathedral which was radiant in the late afternoon sunlight. In Trieste, we were accommodated at the very comfortable Adriatico Guest House, a walking distance from the ICTP and next to the lovely Castle Miramare. This Guest House used to be the Adriatico Palace Hotel where I had stayed with my parents in the sixties when my father visited ICTP at Salam's invitation. The hotel was purchased by the ICTP and converted to Guest House to accommodate guests and visitors whose numbers had significantly increased since the sixties.

The Summer Institute on Complex Analysis of 1975, which I had the privilege of attending as the sole undergraduate participant, was organized with the idea to create a mathematics division at ICTP. By now the mathematics division was flourishing and its head was M. S. Narasimhan who had moved over to Trieste in 1992 after his retirement from the Tata Institute. Although it was Abdus Salam, the Founding Director, who invited me to the ICTP, he referred the matter to M. S. Narasimham who made arrangements for my seminar and our stay. Since it was a mixed audience of mathematicians and some physicists, I gave a talk on "The combinatorics of words with applications to partitions".

It was heavenly weather in Trieste, and so I took Mathura to the nearby Castle Miramare which commands a breathtaking view of the azure waters of the Adriatic. As we strolled through the exquisitely manicured gardens of the Castle, I told her that I wanted to relax in these sylvan surroundings with her now — a spot where as a boy I had spent many hours in the company of my parents! We also made a day trip to Venice, world renowned for its history, its architectural beauty, and its amazing labyrinth of canals.

From Trieste we took the Simplon Express to Paris, an overnight journey that was perhaps the most spectacular and memorable train ride we have had. But it started with a frightening experience! Mathura and I had purchased Eurail tickets but we did not realize that these have to stamped at the first station of use (which in our case was Trieste). So on the first leg of the train journey from Trieste to Venice, the conductor said that our unstamped ticket was invalid. After a long conversation, he took pity on our distress, and kindly offered to go into the station during the brief stop there to get it stamped. In Europe, trains do not stop for very long, and so in those brief ten minutes, he had to come back with the stamped tickets. What a relief when he returned within seconds of the train's departure; the tension and relief can only be compared to what we see in James Bond movies, where Bond would diffuse a bomb just a few seconds before it was set to explode! With the stamped tickets in our hands, we had a restful night as the train sped through Verona, Milan and Domodosalla, before entering the famous Simplon Tunnel connecting Italy and Switzerland. What an awesome sight it was at the break of dawn of the snow capped Alps with clouds hanging on to the mountain tops. We changed trains at Lausanne, Switzerland, and before boarding the famous and elegant TGV to Paris, we had a most delightful breakfast at the station in Lausanne — so enjoyable after that nerve wracking experience at the Venice station.

We arrived in Paris around noon on Saturday, June 12, and so we had almost the full weekend for sightseeing in that charming city before my lectures on Monday. Weather was heavenly, and so the sights of Paris looked even lovelier in the bright sunshine — the Arc de Triomphe, the Eiffel Tower, the Notre Dame Cathedral,... We spent half a day at the Louvre — barely enough to see a few of its highlights like Leonardo da Vinci's Mona Lisa. The sightseeing concluded with a Siene River Cruise at night.

On Monday, June 14, I gave two talks, the first on Modified Convergence at the Group de Travaix in Number Theory organized by Hedi Daboussi. After an enjoyable lunch at an Indian restaurant, we returned to the Institut Henri Poincaré for my second talk in the afternoon in the Paris Number Theory Seminar arranged by Michel Waldschmidt. My talk on the Combinatorics of Words and Partitions was well attended and it was a pleasure to see Richard Askey walk in as I started my lecture. At the end of my lecture, during the questions period, Askey said: "You gave a very nice talk, but having said that, let me tear you apart!" In the next five minutes he systematically told me how I should have presented certain details pertaining to q-hypergeometric series, and also gave some historical background which I did not know. I was relatively new to the subject and so I benefited from Askey's constructive suggestions.

Following my Paris Number Theory Seminar lecture, there was a magnificent dinner at the home of the Waldschmidts to which the Daboussi, his wife, and Catriona Byrne of Springer were invited. I had corresponded with Catriona in connection with the proceedings of the number theory conferences in India which I had edited and published in Springer's Lecture Notes Series, but it was nice to meet her in person. The annual proceedings of the Paris Number Theory Seminar were published by Birkhauser, and Catriona was in discussion with Waldschmidt about its future publication. My paper on the combinatorics of words and partitions appeared in the proceedings of the 1993–94 Paris Number Theory Seminar published by the London Mathematical Society.

From Paris, Mathura and I took a train to Nancy where we were received by Gerald Tenenbaum. The Tenenbaums are gracious hosts, and they had an impressive vegetarian spread for dinner at their home. The next day, after a morning game of tennis with Tenenbaum at his tennis club on the slow red clay, we went to the University of Nancy for my colloquium in the afternoon on Schur's partition theorem and its companions.

Our next academic destination was Lyon where my host was Jean-Louis Nicolas of the University Claude-Bernard, Lyon. I also had another host — Albert Fathi, well known in the field of dynamical systems. He was my colleague in Florida from 1988 to 92, and moved to Lyon as the Director of the recently formed Ecole Normale Superiere in Lyon. Nicolas and his wife hosted a lunch on campus before my talk, but in the evening, the dinner was hosted by Fathi at a fancy restaurant serving famed Lyonnaise cuisine. Wine flowed freely and the bill — absorbed by the Ecole, I guess — was sky high. Mathura and I had a vegetarian dinner and no alcohol, and so our share was small compared to the total cost. Dinner the next night in Lyon at an Indian restaurant was just as tasty, but the cost was much less!

We spent three days in Lyon, and so on Saturday we made a day trip to Geneva by train. The boat trip on the famous Lac Leman of Geneva was delightful as was the cable car ride to the top of the surrounding mountains from where we got a spectacular view of the Alpine panorama.

Our final academic stop in France was Nice where our host was Professor Fredric Pham who had visited MATSCIENCE in 1966. He had deep interest in Carnatic music and had learnt the flute from the maestro N. Ramani in Madras. So we discussed both mathematics and music. It was my father who put me in touch with Pham.

It was a Sunday when we arrived in Nice, and it was a gorgeous cloudless day. So after a drive along the famous Cote D'Azur, Pham took us to neighboring Monte Carlo in Monaco. The views during the drive along the winding highway were breathtaking, and in Monaco we relaxed on the exquisitely manicured grounds of Prince Rainier's Palace.

On Monday, my colloquium in Nice was on Schur's partition theorem. The mathematics department of the University of Nice was housed in a majestic old castle on campus. Jean Dieudonne, one of France's most well known and outspoken mathematicians, had convinced the administration into giving this lovely castle to the mathematics department. Dieudonne had founded the mathematics department at the University of Nice in 1964. In France, both mathematics and its practitioners — the mathematicians, are highly revered and so mathematicians get what they ask for!

Within a few weeks of our return to Florida, we left on a two-week trip to Honolulu, Vancouver and Seattle in early August. It was a professional trip for both Mathura and me: Mathura was invited to give a Bharathanatyam dance program at the East-West Center on the campus of the University of Hawaii owing to the suggestion of our good friend Ramanathan. I was invited to speak at the Straus Memorial Session at the AMS Meeting in Vancouver. The charms of Honolulu and Vancouver are irresistible (but very different), and so we took Lalitha and Amritha with us.

The mission of the East-West Center is to understand the relations between Asia, the Pacific, and the USA. So Honolulu at the cross-roads of the Pacific, is an ideal location for it. For Mathura's program, the Jefferson Auditorium of the University of Hawaii was packed to capacity. Since she explained each dance piece with an introductory demonstration, it was well appreciated by the international audience of about 350 at the event. Our older daughter Lalitha joined Mathura in a couple of items; this was her first experience of dancing before an international audience. I had the pleasure of being the MC for the program. I also visited the mathematics department for discussions with Ed Bertram and Ken Rogers.

We spent a few days in Honolulu after the program enjoying the idyllic environs of the island of Oahu in the company of our friends the Jayaramans and the Ramanathans. During our tour around the island, we visited Hanauma Bay, world renowned for snorkeling. Hanauma Bay is a natural wonder because it was created when one side of a volcanic crater collapsed and fell into the sea. The ocean water rushed into the crater to form the bay. Now the floor of Hanauma Bay is a coral bed, and a breeding ground for fishes. The water is very shallow and so it is ideal for snorkeling. Hanauma Bay is encircled by steep cliffs, and at the top where you enter to take a descending road to the bay, there was a member of the local Hare Krishna group selling soft drinks. We struck up a conversation, and he told us that this simple job is quite remunerative. I was full of envy because, here we are spending thousands of dollars to fly to this island paradise for a short holiday, whereas he is in one of the most scenic spots of that paradise making money! He invited us to his house to join the puja for Lord Krishna's birthday followed by dinner. We were impressed by the beauty of his house and the devotion of the Hare Krishna community.

On this trip to Honululu, we spent a whole afternoon at the Polynesian Cultural Center on the northern coast of the island of Oahu. The center is owned and operated by the Mormon Temple and the Brigham Young University (of the Pacific) adjacent to it. The University offers free education to students of the Pacific islands, with all classes being in the forenoon. In return, the students are required to work at the Polynesian Cultural Center in the afternoons in various capacities — as performers, guides, and so on. The Polynesian Cultural Center is a cash cow since there is hefty entrance fee and the tourists surge into it everyday. But it is worth visiting (once) because you get a quick introduction to the Pacific islands. So it is a win-win situation for all — the tourists, the students, the Mormon Church, and the Brigham Young University.

After a week in Honolulu, we departed for Seattle where we picked up a rental car and drove to Vancouver. The International Joint Mathematics Meeting of the AMS and the MAA (August 15–19) was held on the beautiful campus of the University of British Columbia. A highlight of the conference was the brilliant series of Hedrick lectures by 1966 Fields Medalist Michael Atiyah on "Recent developments in Geometry and Physics". I had heard so much about Atiyah as a mathematician and a speaker. His talks lived up to my expectations for they were a treat both for their content and delivery. Equally enjoyable was the $\pi - \mu - \epsilon$ Frame Lecture by George Andrews entitled "Ramanujan for Students".

The Straus Memorial Session (for the 10th anniversary of his passing) in which I lectured, had several speakers from the world of q such as M. V. Subbarao, Basil Gordon, Jon and Peter Borwein, George Andrews, and Bruce Berndt. My talk was on the Quintuple Product Identity and Shifted Partition Functions — work that I had done towards the end of my sabbatical year at Penn State. We had time to do sightseeing in Vancouver. Since the weather was so good, we took the ski lift to the top of Grouse Mountain not only to enjoy the mountain landscape made attractive by the forests of pine and spruce, but also the spectacular views of Vancouver.

After the AMS Conference, we drove to Seattle to stay with my niece Radhika and her husband Arun who both worked for Microsoft. But instead of going straight to their house, we visited Mount Rainier National Park in the morning. It was a cloudless day, and we enjoyed breathtaking views of that majestic mountain from afar and up close, especially from an area called "Paradise" on the slopes of Mount Rainier — and paradise it was in every respect! When Radhika and Arun heard our tales about a cloudless day at Mount Rainier National Park, they were envious, because in their many years in Seattle, they never had such luck with Mount Rainier. It was just one of those rare days of the Pacific Northwest, the like of which you only see pictures or in posters and calendars.

While we were in Seattle, I took my family to the Boeing factory so that they could see the assembly line of the Boeing 747 that I enjoyed seeing as a boy in 1968. Whereas I saw the prototype 747 in 1968, now a quarter century later, we were seeing the 1000th 747 in production, and this one was for Singapore Airlines by special arrangement. The great Pan Am which launched the 747 no longer existed, and TWA which closely followed Pan Am, was facing a slow death as well. Although my family and I regularly travelled on Singapore Airlines across the Pacific and have enjoyed their superb on board service, it was sad for me to see that the aviation world had changed dramatically in 25 years: legacy carriers like Pan Am which opened up new routes were gone, and the new powers were those like Singapore Airlines which made their way to the top by emphasizing inflight service and attractive uniforms for their stewardesses!

We returned to Florida from Seattle in time for the Fall term to begin in the third week of August.

In Fall 93, one of the distinguished visitors to the Institute of Fundamental Theory (IFT) of the University of Florida was I. M. Singer of MIT who had done groundbreaking work in differential geometry, a highlight of which is the Atiyah-Singer Index Theorem which has applications in physics. Since the IFT connects the Mathematics and Physics departments, Singer addressed the mathematics colloquium. He is an avid tennis player and so my colleague Gerard Emch who was Singer's host, told him that I would give him a challenging game of tennis. So Singer and I had a competitive game during that visit. I found him to be extremely pleasant and easy going for someone so eminent.

During the Christmas-New Year break of 1993–94, I was in India with my family. Mathura left with our daughters during the Thanksgiving break since she was scheduled to give dance performances in December during the music and dance season in Madras. I followed them two weeks later after my classes were over. I had a busy schedule of lectures at Anna University, the Ramanujan Institute, and at MATSCIENCE, where I was a visiting professor for one month. My talks covered the research papers that I had completed as a result of my visit to Penn State in 1992–93.

During the first week of January, just before my departure to Florida, Professor K. Ramachandra of the Tata Institute called on me at Ekamra Nivas. He said that with his retirement coming up in a few months, he wanted me to take over the editorship of the Hardy-Ramanujan Journal that he had founded. Ramachandra was running this journal with his private funds, and with the help of his former student R. Balasubramanian as a co-editor. I told Ramachandra that I was honored that he thought of me in this capacity, but that I would like the journal to have an Editorial Board and have an international publisher for it. He then said that he had founded the journal to preserve and nourish tradition in classical number theory, and he would not want an editorial board to change the tradition of the journal. One of the reasons he founded the journal was that he felt his colleagues at the Tata Institute did not give classical number theory the respect it deserved. I assured him that the persons I would contact would respect his wishes for emphasis on classical number theory.

Back in the United States in early January, I first contacted George Andrews and Bruce Berndt about Ramachandra's proposal, because they are leaders in the Hardy-Ramanujan world, and also because of their classical leanings. After discussions with them, it became clear that it would preferable to start a separate journal in Ramanujan's name emphasising ALL areas of mathematics influenced by Ramanujan — both classical and modern. Papers relating to, or stemming from, Ramanujan's work were scattered in the literature, and such a journal would bring such papers together. In view of this idea, I wrote to Professor Ramachandra that I would not be able to take over the editorship of The Hardy-Ramanujan Journal. Meanwhile, at the suggestion of Bruce Berndt, I was contacted by John Martindale of Kluwer Academic Publishers in the second half of January 1994 expressing interest in a journal of the type I was planning. The suggestion was to send out a proposal to a number of leading mathematicians to get their reactions about the creation of such a journal. I therefore prepared a proposal for a new journal which I called *The Ramanujan Quarterly*. I had quarterly in the title because I planned it to appear four times a year, and also in view of the famous quarterly reports that Ramanujan had submitted to the University of Madras when he had a scholarship there. My proposal was sent to 100 top researchers worldwide and almost everyone endorsed the creation of the new journal enthusiastically. There were a few who said that a print journal would not be good, because the world is moving towards electronic publication. And there were a few who suggested that instead of *The Ramanujan Quarterly*, it should simply be called *The Ramanujan Journal* because in the future we might publish more than four issues a year. In the end, when the journal was launched in 1997, it was called *The Ramanujan Journal* and it was a print journal. Although we started with four issues a year of 100 pages per issue, we have steadily increased in size; Springer, after merger with Kluwer, is the current publisher, and indeed we now publish 12 issues a year of about 225 pages per issue, an increase due the significant worldwide interest in mathematics influenced by Ramanujan. The journal is now available in electronic and print form.

In February 1994, we had the visit of Mourad Ismail to give a talk on orthogonal polynomials, his field of expertise. He said that he would be happy to serve on the editorial board of The Ramanujan Journal, if/when that would be launched. Ismail's visit was followed by that of Bruce Berndt in March. Berndt gave a colloquium on the unorganized portions in Ramanujan's Notebooks, and a $\pi - \mu - \epsilon$ talk for undergraduate students. I had long discussions with him about the new journal being proposed.

Besides my effort to reach out to mathematicians in the Spring term of 1994, the main research developments then were the acceptance of my papers that were outcomes of my visit to Penn State. I mention one such acceptance now because it had an interesting and immediate positive effect: After returning from Penn State in the summer of 1993, I had applied for an NSF grant in Fall 1993 for funding to begin from 1994. This was my first research proposal in the theory of partitions and q-hypergeometric series. When I called the NSF in March 1994 to enquire about the status of my proposal, I was told that decision would soon be made. In early April, I received a letter from Robert Steinberg of UCLA, in his beautiful handwriting, informing me that my paper with Andrews and Gordon on generalizations and refinements of the Capparelli partition theorem, was accepted for publication in the Journal of Algebra. I immediately informed the NSF about this acceptance. In early May I received a call from the NSF saying that my proposal will be funded; one of the reasons (given over the phone) was that the Journal of Algebra acceptance indicated that my work had impact beyond just number theory! I was delighted to get this positive news from the NSF, but I wondered whether the Journal of Algebra acceptance had any effect on the quality of the problems I had proposed! What if the Journal of Algebra had accepted my paper later in July or the paper was submitted to a number theory journal?

With the academic year 1993–94 having concluded on a positive note, I was ready for a number of international academic assignments in 1994–95.

I was going to India in the summer of 94 and enroute, following the Mathura's suggestion, we vacationed in the enchanting island of Bali for a week. Indonesia is a muslim country, but Bali is Hindu. As you exit the airport in Bali, you see a magnificent sculpture from the Hindu epic *The Mahabharata* — of Lord Krishna driving the chariot for the warrior prince Arjuna. You do not see this even in India! In addition to enjoying the sylvan surroundings of Bali and its sun drenched beaches, and being pampered at the Intercontinental Bali as one their first guests (we were there one week after the hotel opened!), we also enjoyed our visit to historic Jogjakarta on the island of Java, which has in its vicinity the temples of Prambanan and Borobudur, built in the 8th century. Of the three principal Hindu male Gods, Brahma (the Creator), Vishnu (the Protector), and Shiva (the Destroyer), only Vishnu and Shiva have temples in India. The only Brahma temple in the world is at Prambanan, and it was wonderful see that. Very close to Prambanan is Borobudur, the largest Buddha temple in the world. It was Sir Stamford Raffles, the founder of Singapore and British Malaya, who oversaw the excavation of Borobudur (when he was Lieutenant Governor of South Java) which was buried under volcanic ash for centuries, and whose existence was forgotten when Indonesia came under the influence of Islam. From our room at the Ambarukkumo Palace Hotel in Yogjakarta, we could see in the distance the fuming volcano Merapi that destroyed Borobudur. We bought models of the Prambanan Brahma temple and of the Borobudur Buddha temple as souvenirs for our home in Florida.

Usually on our transpacific journey, we go via Singapore, which is what we did in 1994 as well. Between Singapore and Indonesia, we travelled on Garuda Indonesia Airlines. I mention this because even though Islam is the main religion of Indonesia, the airline is named after Garuda, the eagle, which is the vehicle for the Hindu god Vishnu!

Back in Madras after a thrilling visit to Indonesia, I had several lecture engagements in the city at various academic institutions. Since I was on leave during Fall 1994 with the primary goal of returning to Penn State to continue my work, we were not returning to Florida in August. Thus Mathura and my daughters stayed back in Madras until late October, while I returned to the USA in August. Since the International Congress of Mathematicians (ICM) in Zurich was taking place during August 3–11, I decided to attend it on my way to Penn State from India in August.

It was thrilling to attend the Opening Ceremony in the grand auditorium of ETH, that great institution where Einstein was a faculty member for a few years. Fields Medals were awarded to Jean Bourgain (IHES and

Illinois), Pierre-Louis Lions (Univ. Paris, Dauphine), Jean-Christophe Yoc-
coz (Univ. Paris - Sud - Orsay), and Efim Zelmanov (Chicago). I had met
Yuccoz a few years before at the University of Florida when he came to
deliver a colloquium at the invitation of Albert Fathi. Zelmanov visited
Florida in 2008 to deliver the Erdős Colloquium during my last year as
Chair when I had the pleasure of hosting him.

Paul Erdős was there at the Zurich ICM and we had discussions over
lunch and dinner about my work on partitions. He said that with regard to
partitions, his main interest was in asymptotics, not in identities. He had
skillfully used an identity for partitions to asymptotically evaluate $p(n)$, the
number of partitions of n.

In describing his work on the asymptotic estimation of $p(n)$, Erdős told
me about *perfect partitions*. A perfect partition of an integer n is a partition
such that the sub-sums will give all the integers from 1 to n. He then said
that he proved that "almost all" partitions are perfect because almost all
partitions have lots of ones. This got me thinking. Back in my hotel room,
I pursued the idea of perfect partitions and found beautiful q analogs of
the famous identities and inequalities of Viggo Brun in sieve theory. I have
not published this, because I have not found any applications of these q
analogs, but they are quite elegant. So a conversation with Erdős always
has beneficial mathematical effects!

After an exciting few days in Zurich at the International Congress of
Mathematicians, I left for State College, Pennsylvania. The two month
visit to The Pennsylvania State University (Aug 6 to Oct 3) was most
fruitful in many ways. The focus of my research was to develop a new
theory of weighted partition identities. The theory of partitions is full of
beautiful identities expressing the equality of partition functions defined
in very different ways. In my theory of weighted partition identities, I
consider the number $p_S(n)$ of partitions of an integer n from a set S of
partitions, and the number $p_T(n)$ of partitions of n from a subset T of S.
Obviously $p_T(n) < p_S(n)$, but the idea is to find positive integral weights
such that when the partitions of n from T are counted with these weights,
they add up to give the value $p_S(n)$. I found several elegant weighted partition identities connecting many famous (but unequal) partition functions.
In many instances these weights were powers of 2, and they gave combinatorial explanations as to why certain well known partition functions are
almost always multiples of powers of 2. My first paper on weighted partition identities appeared in 1997 in the Transactions of the AMS. While
at Penn State, I was also able to obtain finite (polynomial) analogues to

the many fundamental partition identities that I had worked on such as those of Lebesgue, Rogers and Capparelli. Being at Penn State was so useful because I could regularly check with Andrews (the ultimate expert on partitions and q-series) as to which aspect of my work was new.

During the period I was visiting Penn State, the MathFest of the Mathematical Association of America (MAA) took place on the campus of the University of Minnesota in Minneapolis in mid-August, and I attended that because I was invited to speak at a Special Session on q-series. My talk was on new observations on the Göllnitz-Gordon and the Rogers-Ramanujan identities — work I had done at Penn State in 1992–93.

Minneapolis, Minnesota, which is dreadfully cold in the winters, was like paradise in the summer, and so it was a perfect choice for the location of the MathFest. The highlights of the MathFest included the Hedrick Lectures delivered by Ron Graham, and the Progress in Mathematics Lecture given by Ken Ribet of UC Berkeley. George Andrews delivered the History of Mathematics Lecture on "The Well-Poised Thread - Some amazing sums of Gauss, Kummer and Ramanujan", explaining how the idea of "well-poisedness" was a connecting thread binding the works of many stalwarts in the theory of q-hypergeometric series. The lecture was so good that I requested Andrews to give it to me to be published as the Opening Article in the very first issue of The Ramanujan Journal. With his characteristic grace, Andrews agreed to my request, and his paper made a superb opening to *The Ramanujan Journal* when the first issue appeared in January 1997.

With regard to The Ramanujan Journal, several positive developments took place while I was at Penn State. Richard Askey agreed to serve on the Editorial Board and that convinced George Gasper to follow suit. John Martindale of Kluwer came to Penn State to hold discussions with me and Andrews in preparation for the launch of The Ramanujan Journal.

After an eventful two-month stay at Penn State University, I returned to Madras on October 3 where Mathura and my daughters were staying during that period. It was a lovely flight from Amsterdam to Delhi on board the KLM Boeing 747 named "Albert Plesman', after the famous founder of KLM, The Royal Dutch Airlines. It was nice to be back in Madras just in time for my daughter Amritha's birthday on Oct 4, and for the Navarathri Festival so elaborately organized by my mother at our family home Ekamra Nivas. At the end of the festival, in mid-October, I left Madras with Mathura and my daughters to be back in Florida.

Since I had two months left in the Fall 1994 term without any teaching, I made a one-week trip to UCLA to work with Basil Gordon. Visiting

Gordon is literally like being with a Guru as in the days of ancient India. The pupil lives with the Guru throughout the day and learns from the Guru as the two go about the day's activities together. Even though I was staying at the Brentwood mansion of my UCLA college mate Kumar (alias Sampath), I would go to Gordon's home soon after breakfast and work with him there and at UCLA until after dinner. The result of this immersion in work was that we found the weighted words generalization of the famous two hierarchies of partition theorems due to Andrews in the sixties. In 1966 and 67, Andrews had obtained two infinite families of partition theorems to the moduli $2^k - 1$, for $k = 2, 3, 4, \ldots$, and these two families coincided in the starting case $k = 2$, which was Schur's theorem. The method of weighted words generalization that Gordon and I found, was a single infinite family with general parameters, which yielded the two Andrews families as the two extreme cases by specialization of the parameters.

After returning to Florida from UCLA, one morning in mid-December, when I was in the shower, it occurred to me that I could get a bijective correspondence that would explain the deep three parameter generalization of Göllnitz' partition theorem that Andrews, Gordon and I had obtained earlier. I was so excited by this, that even though I did not run out of the shower naked like Archimedes, I called and talked to both Andrews and Gordon about this soon after I got dressed! Both confirmed the novelty and importance of the result, and so in the next few months I wrote a paper entitled "A combinatorial correspondence for Göllnitz' (Big) partition theorem". It appeared in 1997 in the Transactions of the AMS.

The last two months of Fall 1994 in Florida were eventful in other ways as well. John Martindale (from the Boston area) and David Larner of the Kluwer head office in The Netherlands, came to Gainesville in late October to finalize plans for the launch of *The Ramanujan Journal* with me as the Editor-in-Chief. In December, I was informed by the President of the University of Florida that I would receive the TIP Award for distinction in Teaching. Also in Fall 1994, I received an invitation from the Fields Institute in Toronto (at the suggestion of George Andrews), to deliver one of the Plenary Lectures at a major conference on q-series to be conducted there in the summer of 1995. So the year ahead looked bright and promising.

7.7   The Oberwolfach and Fields Institute Conferences and other events (1995)

During January 15–21, 1995, there was a conference at the Mathematisches Forchunginstitut, Oberwolfach, on the theme "Combinatorics of the

Symmetric Group". Partitions play a central role in the study of the symmetric group because the conjugacy classes of the symmetric group S_n are labeled by partitions; so there was emphasis on the theory of partitions at this conference where I was invited to give a one-hour lecture. Just as the Oberwolfach Conference in May 1979 put me in touch with the leaders in Diophantine Approximations and Transcendence this conference put me in contact with some of the major researchers in Algebraic Combinatorics.

I arrived in Frankfurt Airport on Sun, Jan 15 morning by Northwest Airlines DC-10, and there I joined some of the conference participants to take the train to Oberwolfach. It was winter, but the *Schwarzwald* (Black Forest) in which Oberwolfach was nestled, looked lovely with the mountain slopes studded with snow laden trees. It was a pleasant surprise to meet A. O. Morris from Wales at dinner on the first night of the Oberwolfach conference! He had visited MATSCIENCE in the seventies because his research on Clifford algebras intersected with that of my father, and now he was attending this conference because he had done significant work on projective representations of the symmetric group.

Mine was the Opening Lecture of the second day of the conference, and those of us who spoke that morning, were honored with the presence of Mathias Kreck, the new Director of the Institute. Kreck took over as Director in 1994 from long time Director Martin Barner who served in that capacity from 1963 to 94. I had prepared a talk on my joint work with Andrews and Gordon on the Capparelli partition theorems, because of its connections with vertex operators on Lie Algebras. But on the first day of the conference, when I was discussing with David Bressoud and Christine Bessenrodt (Univ. Magdeburg, Germany), I mentioned the new combinatorial correspondence I had obtained for Göllnitz' theorem. Bressoud and Bessenrodt are leading authorities on partition bijections and have given beautiful bijective proofs of some fundamental partition theorems. They got quite excited by the combinatorial correspondence I showed them, and urged me to include it in my lecture the next day. I could do that since I had one hour for my lecture, and I was employing the "method of weighted words" which applied both to the Capparelli and Göllnitz theorems. Thus the Oberwolfach conference was a rewarding experience for me.

One of the most enjoyable lectures at the conference was by Viennot. Those were the days of overhead transparencies, and Viennot had many smaller size transparencies attached to the standard size ones. It was amazing to witness his sleight of hand as he flipped these smaller size transparencies with speed and ease for dramatic effect!

Back in Florida from Oberwolfach, the main development was that on February 9, I received the official contract from Kluwer, appointing me as Editor-in-Chief of The Ramanujan Journal. So I sent out invitations to 26 leading mathematicians to join the Editorial Board. I had spoken to, or corresponded with, all of them, and knew that they were positively inclined to the launch of the new journal. Over the next few months I received letters from all 26 accepting to serve on the Editorial Board.

With 1995 academically off to a good start, I was keen to enjoy my Spring Break in the first week of March. My elder daughter Lalitha was in her final year of Middle School, and even though her Spring Break and mine did not coincide, I decided to go to Hawaii with the entire family, because once she would start high school, she could not take a full week off outside of her Spring Break. Of course going to Hawaii means mixing work with pleasure for me — pleasure because of the sylvan beauty of the islands, and work because of my strong contacts with the University of Hawaii.

At the invitation of my host Edward Bertram, I gave a colloquium at the University of Hawaii on "Some new observations on the Göllnitz-Gordon and the Rogers-Ramanujan identities". The night before the Colloquium, the former Chairman Bill Lampe and his wife entertained us at dinner at an Indian restaurant in Honolulu. In the evening following my Colloquium, Ed Bertram and his gracious wife Alice, hosted a dinner at their lovely home, to which they invited their colleague Kenny Rogers and his wife, and Marcel Herzog from Tel Aviv who was visiting Hawaii to work with Ed. The new home of the Bertrams was nestled in the rain drenched gulches on the "northern" side of the mountain spine of the island of Oahu, and so the home was surrounded by immensely beautiful and dense foliage, with waterfalls in the backdrop on rainy days.

My parents were making their usual three month visit to the USA and stopped in Hawaii enroute to meet us there. So the extended family went from Honolulu to Hilo on the Big Island of Hawaii. It was nostalgic to stay at the Naniloa Surf Hotel in Hilo, where Mathura and I had our honeymoon in 1978, and it was a pleasure for Mathura and me, 17 years later, to have our daughters and my parents stay with us at that charming hotel overlooking the tranquil waters of Hilo Bay. The highlight of the trip was the visit to the Hawaii Volcanoes National Park which is not far from Hilo. We were incredibly lucky to see from close proximity, red hot lava ooze out of the ground, cool down and solidify a few feet beyond. What was even more exciting was to see at dusk, lava falling into the sea and the huge plumes of smoke emerge as the molten lava made contact with the ocean water. Even though this happened at a distance, we could see this clearly, because at dusk, the red hot lava can be spotted from afar. We watched with wonder and amazement new land being created!

It is always sad to leave Hawaii, a paradise on earth, but after Spring Break, work had to resume in Florida. In late March, Paul Erdős was on his usual annual sojourn for a week, but we also had the visit of the famed algebraic geometer Shreeram Abhyankar from Purdue at the same time. He came to Florida at the invitation of our distinguished colleague, the group theorist John Thompson. My father and Abhyankar were old friends, and so we had the Abhyankars and Erdős along with the Thompsons and other colleagues for a dinner at home, as well as a party in their honor so graciously hosted by Mathura for which all our department colleagues and their spouses were invited.

During the third week of May there was a conference at the University of Illinois in honor of Heini Halbertstam who was to retire the next year at the age of 70. Although I spoke about my work on weighted partition identities and published a paper on this topic in the conference proceedings, it was good to meet many of the active researchers in analytic number theory, some of whom I had known for a long time. One night we were entertained after dinner with a piano recital by Peter Elliott from the University of Colorado, and by the comedy of Halberstam's son Michael which made the audience roar in laughter.

I got an interesting idea during the conference that it would be beneficial to apply a redistribution technique of Bressoud to the weighted version of Göllnitz' theorem that I had proved combinatorially in Fall 94. I found time to work out the details of this on the flight back from Urbana to Gainesville. The result was a *New Key Identity* for Göllnitz' (Big) partition theorem, which although very intricate, was simpler in structure to the original key identity that Gordon and I had found. I have found many interesting partition theorems while waiting at airports or while flying, and this was one of them. Andrews jocularly refers to such discoveries of mine as "airport theorems". Like the original key identity which Gordon and I constructed combinatorially, I found this identity by combinatorial reasoning. And like the original key identity that I could not prove, I did not have a q-hypergeometric proof of this. So the drama of 1990 was repeating, and I called Andrews and gave him the new key identity by phone. He reacted saying that it was an important simplification, and the next day he faxed me a six page proof using Jackson's q-analogue of Dougall's summation. Thus we wrote a joint paper on the new key identity describing my combinatorial construction and Andrews' q-hypergeometric proof.

During the third week of June, there was a Conference on "Special functions, *q*-series, and related topics" at the Fields Institute in Toronto to which I was invited to give an hour lecture. Actually this was a two week meeting, the first week being an Instructional Workshop and the second week being a Research Conference, and I attended Week II.

The Research Conference was to start on Monday, June 19 morning with the lecture by the great I. M. Gelfand of Rutgers University. Besides being a mathematical luminary who had made seminal contributions to several branches of mathematics, he was also world famous for the "Gelfand Seminar" he conducted in Moscow for many years, the discussions of which he dominated — often intimidating the speakers regardless of their high reputation. Gelfand emigrated to the United States when he was around 75 years of age, and at Rutgers University he continued his seminar, but in the USA, perhaps out of respect for the "local culture", or because he had mellowed with age, he was not like the ferocious Siberian tiger he used to be in Moscow! I was looking forward to meeting Gelfand and was prepared to face the Tiger during my talk, but unfortunately, he could not come to the conference due to ill health (he was 82 then). Richard Askey on the Organizing Committee was the one who invited Gelfand. So Askey gave the Opening Lecture on June 19 saying that if the person you invite is unable to be present, then you need to take the responsibility to give the talk in the place of the invitee. And of course, Askey gave a superb talk on special functions in his inimitable style.

My talk on the third day was entitled "Refinements of Rogers-Ramanujan type identities" and I gave an overview of how the method of weighted words could be used effectively to generalize and refine a variety of Rogers-Ramanujan type identities. My paper on this topic appeared in the Fields Institute Proceedings two years later. In introducing me, George Andrews said that I am the Editor-in-Chief of the soon to be launched Ramanujan Journal, and so the conference turned out to be an effective scouting ground for good papers for the journal.

I returned to Florida from the Fields Institute Conference on June 22 afternoon, and had barely a few hours to pack before leaving with my family to India the next day via Europe where I had several lecture engagements. My parents who were visiting us in Florida also left on June 23 for India, but via the Pacific. I had such a wonderful trip to Europe in 1993 with Mathura, followed by my visit to Zurich in 1994 for the ICM, that I decided this year that my daughters should accompany us and enjoy many of the wonderful sights of Europe.

In Europe, our first stop was Paris where my charming host Michel Waldschmidt had arranged my talk in the Paris Number Theory Seminar. Since we were on our way to India, I had arranged our air journey via Amsterdam on KLM, and decided to travel in Europe by train using the Eurail Pass. Having suffered a frightening experience in 1993 at the Venice railway station by not having our Eurial passes validated, I made sure all our passes were validated at the Amsterdam airport train station before boarding the train to Paris. But we had to change trains in Brussels.

The Europeans are very proud, and justifiable so, that their trains run on time, but the trains there often stop for barely a few minutes enroute. This could be a problem if you have luggage, because you need to load the luggage within the few minutes that the train would stop there. We had two suitcases besides one hand piece for each of the four of us since we were going to India. When loading the luggage at Brussels on to the train to Paris, my family boarded first, and I gave the luggage to them one by one from the platform. Before I could complete this process, the train door closed and I almost had a heart attack because I thought the train would leave without me! There were no cell phones those days and so the thought of being separated from my family flashed through my mind. I let out a cry so loud that it could be heard throughout the station!! Fortunately, a guard appeared, and opened the door so that I could get in when the last piece of luggage was loaded. Although the European trip started out so scary, the rest of the trip was quite delightful.

Our first full day in Paris was a Sunday and the city was breathtakingly beautiful under bright sunny skies. We showed our daughters many of the great attractions of Paris. We were quite hungry at dinner time since we had walked a lot, and we enjoyed the warm hospitality and the lavish dinner at the home of the Waldschmidts that night.

The next day (Monday), the Paris Number Theory Seminar was at the Institut Henri Poincaré at 2:00 pm. My talk was entitled "Some new observations on the Göllnitz-Gordon and the Rogers-Ramanujan identities". I was told that Jean-Pierre Serre usually would attend the Paris Number Theory Seminar, and he was there for my talk. I was also told that often he would ask probing questions. Having Serre in the audience can be intimidating because of his stature, and his emphasis on thoroughness.

I started the talk by saying: "In the entire theory of partitions and q-hypergeometric series, the Rogers-Ramanujan identities are unmatched in simplicity of form, elegance and depth. These are the three qualities that we as mathematicians seek when we discover a theorem — we want it to be simple to state, for it to be elegant AND deep." As soon as I finished saying this, Serre whispered something into the ears of Waldschmidt, who smiled and nodded. Serre asked several questions during the talk, and even corrected me with regard to a terminology. This correction occurred within the first few minutes. After stating the Rogers-Ramanujan identities, I mentioned their combinatorial interpretation involving partitions with difference between parts at least two. So the minimal *permissible* difference is two, but it is standard to simply say "minimal difference two" and drop the word *permissible*, which is what I did. Serre immediately pointed out that this was misleading unless this convention is stated at the beginning. I said he was right, and indeed he was. So he heard the lecture keenly, and I was glad that he took my presentation seriously. It was indeed an honor for me that he attended my talk.

While I was at the Institut Henri Poincaré, my family was at the Louvre. So we picked them up in the evening from there, and Waldschmidt and other number theorists took us to dinner at the Indian restaurant "The Maharaja" on Boulevard St. Germain. Serre did not come to dinner. So I asked Waldschmidt what he whispered to him at the start of my talk. Waldschmidt said that Serre told him that my opening lines were very good! So I came out of the lecture unscathed.

After one more day of sightseeing in Paris, we left by train for Nancy. On the night of our arrival, the Tenenbaums hosted a magnificent dinner at a fancy French restaurant atop a hill which commanded a spectacular view of Nancy. The Tenenbaums had us seated on the lawns of this hotel so that we could enjoy the view. They had ordered excellent vegetarian food for us. I needed a bit of salt to be added, but there was no salt or pepper on the table. So I asked Tenenbaum if he could request for some salt and pepper. In a low voice he said that French chefs are very proud of their creation and so the chef at this restaurant might feel insulted if we ask for salt. He said the chef feels HE KNOWS exactly the amount of salt to be put in the food, and so asking for extra salt might upset him. What surprised me was that from the chef's point of view, our taste preferences were not at all important; he knew what was right, and we had to accept it. I told Tenenbaum to request salt and to let the chef know that an uncivilized person is asking for extra salt, and so this is no reflection on the chef's culinary abilities, but rather a reflection on my "poor taste"! I do not know what Tenenbaum told the chef, but salt and pepper were brought to the table, and we enjoyed the dinner immensely.

The next day, June 29, was our Wedding Anniversary. While Mathura and my daughters roamed about Nancy, I gave a colloquium at the University, following which the Tenenbaums hosted a dinner for us at an Indian restaurant, where I had no reason to ask for either salt or pepper!

My next academic visit was in Lyon, but we decided to spend the weekend in Zurich enroute. I had such an enjoyable time in Zurich in August 1994 when I attended the ICM, that this time I felt that I should be there with my family and enjoy the sights around Zurich with them. We stayed at the lovely Atlantis Sheraton Hotel in a park like setting outside Zurich where I stayed in 1994. We took an all day tour on Saturday, July 1, to Mt. Titlis in the Swiss Alps and enjoyed the breathtaking Alpine scenery. Enroute we stopped in the exquisitely beautiful town of Lucerne where there was a profusion of flowers.

Our host in Lyon was Jean-Louis Nicolas, a reputed number theorist with the most friendly disposition. He is an expert on arithmetical functions, and was a close friend and collaborator of Paul Erdős. He had accepted my invitation to serve on the Editorial Board of The Ramanujan Journal, and he contributed a fine article on an unpublished manuscript of Ramanujan on highly composite numbers for the first volume of the Journal. He and his wife invited us to dinner at their home after my colloquium, and it was indeed a multi-course French feast. Lyon is famous for its cuisine, and Mrs. Nicolas made very interesting vegetarian variations of some local specialities. Of course, one course was devoted to cheeses, and we were given such a wide array of fine cheeses that it was difficult to choose from them. The true connoisseurs prefer aged strong cheeses with wine, but my family and I preferred the milder cheeses with grape juice — like children!

From Lyon we had to make our way back to Amsterdam to catch our flight to India, but just as we stopped in Zurich on the way to Lyon, we broke journey in Geneva on the way back from Lyon. One of the pleasures of Geneva is the boat ride on the Lac de Geneve (= Lac Leman), and I enjoyed it with Mathura and my daughters as much as I did with my parents in the sixties. But this time we also went to Montreaux via Lausanne and saw the magnificent Chateau de Chillon.

We had two nights in Amsterdam, and so one day was for enjoying the sights of this Venice of the North. Of note was our visit to the House of Anne Frank. My younger daughter Amritha (who was just seven years old) was so touched by the story of Anne Frank, that three years later, at the suggestion of my older daughter Lalitha, she wrote a fine article on her impressions about this. My father sent the article to The Hindu, India's

National Newspaper based in Madras, and the Editor N. Ravi published it! Thus Amritha revealed her writing skills early. Little did we imagine that a decade later, she would get an undergraduate degree in Journalism, work as an intern in The Hindu for a few months, and receive the endorsement of N. Ram, the Editor of The Hindu, for her reporting skills!

I had a busy schedule of lectures in Madras at various institutions. I spoke about different aspects of partitions in these talks, so that it would not be repetitive to those attending more than one of my lectures.

Back in Florida for the Fall Term, the main events were the visits of George Andrews in mid-October and Richard Askey in early December. They both gave two talks — a Colloquium and a Teaching Seminar since both had been involved in a major way in addressing the changes in mathematics education that was sweeping the United States. So for their Teaching Seminars we invited the faculty from the College of Education, but they boycotted both talks because Andrews and Askey had been critical that all that the Colleges of Education do is to address teaching methods and completely ignore the mathematical content of the courses. What is more, the influence of the Colleges of Education is negative because their contribution is to water down the course content. Both Andrews and Askey said that they were not surprised at the lack of interest from the College of Education to their seminars. Of course the mathematical community has a high respect for the views of Andrews and Askey in mathematical education. Indeed, Andrews was invited to give a talk on Mathematics Education at the International Congress of Mathematicians in Berlin in 1998.

In December–January, I made a trip to India because I was organizing an International Symposium on Number Theory at Anna University in January 1996, and I had other academic assignments as well. Mathura and my daughters stayed back in Florida because they had a hectic summer travelling in Europe and to India. Since my family was not with me, there was no sightseeing on this trip; I just focused on academic assignments.

Enroute to India via the Pacific, I broke journey in San Francisco and attended the West Coast Number Theory Conference in Asilomar on the Monterrey Peninsula. I was returning to this conference after a gap of more than a decade, but it was good to see that the conference format had not changed. After Asilomar, I stopped for a day in Singapore to give a talk at the National University of Singapore before reaching Madras.

Even though it was the Christmas-New Year break, I had a full schedule of lectures at various academic institutions in Madras. I spoke at a meeting of the Association of Mathematics Teachers of India (AMTI) on the theme

"Ramanujan for students". It was nice to attend this meeting because as a college student I had participated in the mathematics competitions organized by the AMTI. It was also a pleasure for me to deliver the K. Subramaniam Endowment Lecture at Vivekananda College on the theme "A theory of weighted partition identities". Prof. Subramaniam had admirably served as the Head of the Mathematics Department of Vivekananda College for a number of years, including the years when I was a student there in the BSc class. He encouraged my research efforts as an undergraduate by arranging seminars for me, and generously provided me leave to attend the Number Theory Institute at the University of Michigan and to visit the Australian National University in Canberra to work under Prof. Mahler, both in 1973. In December 1995, I also spoke at the Academy of Sciences, the M. S. Swaminathan Research Foundation, the IIT, and the Ramanujan Institute. Since December 22 is Ramanujan's birthday, many institutions in Madras celebrate that with pride by arranging lectures in the latter part of December.

Some invitees to the International Symposium in Number Theory at Anna University that I was organizing in January, arrived in late December — such as Wolfgang Schmidt (University of Colorado, Boulder) and Dale Brownawell (Penn State University), and so I took them to the Madras Music Academy for some fine Carnatic Music Concerts. My friend GP showed them around Madras and took them on an excursion to Mahabalipuram with its magnificent rock temple ruins, some submerged in the ocean waters off the beach, but with their tops visible. Basil Gordon (UCLA) was also supposed to arrive on Dec 22 (Ramanujan's birthday), and so I had arranged a public lecture by him on Dec 23 on Ramanujan's mock theta functions in the seminar room of our home under the auspices of the Alladi Foundation. But on Dec 22 night when GP and I went to airport to receive him, he did not show up! We waited several hours, came home, and made a phone call to Gordon. He said that he went to the Los Angeles airport to board the flight, but forgot to take his passport — an excellent example of absent mindedness! So he could not make the flight. He was profusely apologetic and said that he would come in the summer of 1996 at his own expense and give the lecture (which he did, see next section). Just as Askey gave Gelfand's talk at the Fields Institute Conference when his invitee I. M. Gelfand could not attend, I gave a talk on mock theta functions because a sizeable audience had turned up at our home on December 23 morning to hear Basil Gordon!

7.8   The launch of The Ramanujan Journal and prelude to Chairmanship (1996 and 97)

The first issue of The Ramanujan Journal was in January, 1997, but all work for that was done in 1996, and the issue came out in print by the end of December 1996. So I will describe the launch of the Journal against the backdrop of many significant academic events of 1996 and 97 that were a prelude to my taking up the position as Chair of the Mathematics Department at the University of Florida in 1998.

The year 1996 opened with an International Conference on Number Theory that I organized at Anna University on Jan 1 and 2. The conference had an impressive list of lectures by internationally renowned researchers — Wolfgang Schmidt (University of Colorado), Dale Brownawell (Penn State University), Michel Waldschmidt, Daniel Bertrand, and Sinnou David (University of Paris), Maurice Mignotte (University Strasbourg, France), Francesco Amoroso (University of Pisa, Italy), and Shigeru Kanemitsu (University of Fukuoka, Japan), as foreign participants, along with Professors K. Ramachandra and S. Raghavan who had retired from the Tata Institute and settled down in Bangalore and Madras, respectively. There were more than 150 attendees, and on the evening of January 1, all the speakers were invited to a Reception on the lawns of the lovely home of Mr. C. Subramaniam fronting the Adyar River.

The conference was especially strong in the areas of Diophantine Approximation and Transcendence owing the presence of Schmidt, Brownawell, Amoroso, and the four French mathematicians. Analytic number theory was represented by Kanemitsu and Ramachandra. I spoke on partitions and Raghavan on quadratic forms. So a broad spectrum of topics were covered in those two days. The refereed proceedings of the conference appeared two years later as Issue 4 of Volume 1 of The Ramanujan Journal in December 1997. Basil Gordon was missed at the conference, but he kindly sent a paper on mock theta functions for the proceedings.

Sinnou David hails from Pondichery, a former French outpost in India. His first name Sinnou is a French variation of the Indian name "Cheenu" which is a short form for Srinivasan. My contact with Mignotte was due my work with Robinson on irrationality measures; Mignotte at that time had the best irrationality measure for π. Walsdschmidt and Kanemitsu love India and everything it has to offer. So they visit India regularly. My association with Kanemitsu strengthened from this conference onwards and he invited me often to conferences he organized in Japan.

On January 2 evening, after the sessions were over, the foreign participants left for Trichy by air to attend the Conference for the Tenth Anniversary of The Ramanujan Mathematical Society that was starting the
next day. I too was invited to speak at the Trichy conference, but I took
an overnight train to on January 2 to arrive in Trichy the next morning.
My friend GP accompanied me on the trip.

Why was Trichy chosen as the venue of the conference? To understand
that, we need to know the origins of The Ramanujan Mathematical Society.

During the 50th Annual Conference of the Indian Mathematical
Society in December 1984 in the state of Gujarat, Prof. K. S. Padmanabhan of the Ramanujan Institute in Madras, proposed the creation of a
new journal, separate from the journals published by the Indian Mathematical Society. In response to this idea, Prof. Sampathkumar suggested
the creation of a new mathematical society and said that the new journal
could be what is published by that society. So a meeting was held in April
1985 at the National College in Trichy organized by Prof. R. Balakrishnan
to discuss the formation a new mathematical society. Then on July 15,
1985, The Ramanujan Mathematical Society was officially registered, and
the creation of The Journal of the Ramanujan Mathematical Society was
endorsed. The Journal and the Society bore Ramanujan's name because
he was the greatest mathematician India had produced, but both the Society and the Journal were devoted to all aspects of mathematics, unlike
the Ramanujan Journal which I was to create in 1997 devoted to areas of
mathematics influenced by Ramanujan.

Both the Society and its journal did not take off as anticipated. So in
1995, ten years after the Society was created, Michel Waldschmidt offered
to help raise the visibility of the Society and give a boost to the Journal.
The decision was to hold an International Conference in Trichy where the
Society was founded, and to get many leading researchers from around the
world for the conference. It turned out that conference could be organized
only in early 1996, but it did attract about 200 participants of which several
were from overseas, including an impressive delegation from France led by
Waldschmidt. To help raise funds for the Society, a Registration Fee of
$250 was asked (only) from the Foreign participants (including me), with
participants from India paying a much lesser amount. As for the journal,
Michel Waldschmidt and Kumar Murty (Toronto) became the Co-Editors
in Chief. Even though the Journal enjoyed a boost with a new editorial
board, it was decided that the refereed proceedings of the conference would
be published in the Contemporary Mathematics series of the AMS. My

contribution to the Proceedings was my joint paper with Andrews on a new *cubic key identity* for Göllnitz' (Big) partition theorem that I had discovered combinatorially in June 1995 on the flight from Urbana back to Florida, which Andrews proved by q-hypergeometric techniques.

I left Trichy by train on the night of January 3 after delivering my lecture at the conference, to depart Madras for the USA. After the Trichy conference, Dale Brownawell, Michel Waldschmidt and Wolfgang Schmidt spent a few days in Madras before leaving India. So my father arranged public lectures by all three on three consecutive days under the auspices of The Alladi Foundation, in the Seminar Room of our family home Ekamra Nivas. On January 9, Brownawell spoke about the work of the late Sarvadaman Chowla, his former distinguished colleague at Penn State University. This was very appropriate for an Indian audience and also because my parents and Chowla interacted closely at the Institute for Advanced Study in Princeton when my father visited there in 1957–58. The next day (Jan 10), Waldschmidt gave a lecture on transcendence results for Ramanujan's τ function. Finally, on January 11, Wolfgang Schmidt, the world's greatest authority on Diophantine Approximations, provided a magnificent survey of this active field of research. The lectures were very well attended by faculty and students at various academic institutions in Madras. I missed these lectures, but back in Florida, I was involved in other academic events.

I returned to Gainesville on Monday, January 8 for the start of classes for the Spring 1996 term, and Bruce Berndt arrived the same evening. We arranged his colloquium the next day and he spoke on the theme "Ramanujan was a radical!". This had a double meaning. Ramanujan had established some amazing results involving radicals (successive k-th roots), and he was a radical (unconventional) in the way he thought!

From January 10–13, 1996, the Annual Meeting of the AMS took place in Orlando, which is not a favourite choice for the AMS. Even though Orlando has gorgeous weather in January, it is viewed by most academics as a recreational city, and is associated with Disney World and Mickey Mouse; so that may be off-putting for research mathematicians.

The Annual Meetings, wherever they may be held, are an attraction for the scientific publishing companies who have book exhibits that attract huge numbers of the participants. So John Martindale of Kluwer was there, and I had a discussion with him and some of the editors of the Ramanujan Journal regarding the first issue to be launched at the end of the year. After the Annual Meeting, Martindale came to Gainesville to continue discussions with me and my colleague Frank Garvan who, along with Bruce Berndt, was to be a Co-ordinating Editor of the journal. One of things I wanted was to have on the cover of each issue of the journal, a certain page from the last letter Ramanujan wrote to Hardy in which he said that he had discovered the mock-theta functions. That page of the letter also contains Ramanujan's signature. Martindale liked my suggestion, and so we wrote to Narosa and were granted permission to use this letter on the front cover of each issue. It was a pleasure for me to design the cover of the journal, which has not changed since the first issue.

In early February, I was invited to give a talk at a number theory conference at DIMACS on the campus of Rutgers University. DIMACS, an acronym for Discrete Mathematics and Computer Science, was founded in 1989 as an NSF Funded Institute with Daniel Gorenstein, a celebrated group theorist, as its Founding Director. Gorenstein led the effort to write the proposal for DIMACS. Such proposals are very time consuming and difficult to prepare because they involve getting support from the client disciplines and the Administration. Gorenstein told us that preparing the proposal sapped all his energy. It took a toll on his health and he died in 1992 at the age of 69 while he was still the Director of DIMACS. Gorenstein also had played a major role in the classification problem of finite simple groups. That type of work, even though long and intense, does not take a toll on one's health.

At the DIMACS conference, I gave a talk entitled "A theorem of Göllnitz and its place in the theory of partitions". It was an honor that Fields medallist Enrico Bombieri attended my talk. Paul Erdős, who was visiting Bell Labs in nearby Murray Hill, was at the conference. At DIMACS, it was nice to reconnect with Professor Richard Bumby on the Rutgers faculty, who had been so kind to me back in 1973 at the Summer Institute in Ann Arbor when I was an undergraduate participant.

After the DIMACS conference I went to Boston to see the Kluwer office in Norwell (on the outskirts of Boston) to meet Martindale and his staff who would be working on The Ramanujan Journal. It is always nice to meet people you correspond with, and to put a face to a name. Kluwer agreed to send me very nice stationery for the journal. Those were days when people sent out letters, and the stationery I had was lovely. In a few years, everything changed to email, and so about 50% of the stationery is still unused.

Paul Erdős made his annual visit to Florida in early March. This time he was accompanied by his long time friend and collaborator John Selfridge. I had known Selfridge since my UCLA graduate student days and it was therefore a pleasure to host him in Gainesville. My parents had arrived in late February on their annual trip to the USA, and my father who regularly attended lectures in the mathematics department, enjoyed listening to the talks of Erdős and Selfridge.

In mid-April, I received a very warm letter from Dean Willard Harrison of the College of Liberal Arts and Sciences (CLAS) of the University of Florida congratulating me on my appointment as Editor-in-Chief of the soon to be launched Ramanujan Journal, and informing me that he will be recognizing me at the CLAS Baccalaureate ceremony in early May. I was happy that Mathura and my parents attended the Baccalaureate and met Dean Harrison at the Tea that followed the ceremony.

My parents left for India in early June, and Mathura, my daughters and I followed them via the Pacific two weeks later. But we broke journey in Hawaii enroute to spend a few days in the enchanting island of Kauai.

In Kauai we stayed at the fabulous Sheraton Princeville Resort which has a breathtaking location on Hanalei Bay that has to be seen to be believed! Kauai, the oldest among the large Hawaiian islands, is the most picturesque, since both wind and water have over long time sculpted the island with fantastic formations. Waimea Canyon, the Grand Canyon of the Pacific, although much smaller, is very impressive, and even more colorful. Kalalau Lookout at the end of the canyon road is from where you get an awesome view of a lush valley surrounded by steep cliffs plunging into the ocean that has the most brilliant turquoise blue water. The view will take your breath away. The northern part of the island where we were staying, is lined with gorgeous beaches of which the most spectacular is Lumahai Beach, voted by Conde Nast Traveller as the most beautiful in the world, and was the location of the movie *South Pacific*. Kauai which has the wettest spot on earth, is so lush that it has been chosen as the location for many movies such as Jurassic Park.

After this vacation in heavenly Hawaii, we went to India for two months. While in Madras, in the leisured comfort of our family home, I found some new weighted partition identities by using quartic transformations on the Göllnitz key identity. But the highlight of the trip to India for me was the visit of Basil Gordon to Madras, a visit he promised he would make after missing his flight to India in December the previous year. Gordon gave two lovely lectures in Madras. The first was at the Academy of Sciences on July 22 where he spoke about the di-logarithm function, q-hypergeometric series and modular forms — joint work with his former PhD student Richard McIntosh. Toward the end of the lecture, he sketched a proof of a conjecture of mine relating to certain partition functions I had introduced using the quartic transformations. Following Gordon, I gave a talk at the Academy the same day on weighted partition identities, and having Gordon confirm my conjectures was indeed helpful. Actually, while in Madras, I finished work on a paper entitled "On a partition theorem of Göllnitz and quartic transformations" for which Gordon wrote an Appendix. This paper appeared in the Journal of Number Theory two years later.

Two days after his talk at the Academy of Sciences, Gordon gave a lecture under the auspices of the Alladi Foundation at our home on the theme "The final problem - an analysis of the mock theta functions" in which he announced his discovery along with McIntosh of some new 8th order mock theta functions. A large audience of academicians, college teachers, and students, who attended his lecture, were much impressed by his presentation. There was a prominent report of his lecture in The Hindu the next day. The Hindu always reports significant events relating to Ramanujan.

The day after his Alladi Foundation lecture, Gordon left on a sightseeing tour of South India by car with my good friend GP as his companion and guide. Gordon returned to Madras in time to attend the party on July 30 in honor of my parents for their 50th Wedding Anniversary. He had become very close to my parents and hosted them magnificently in Los Angeles every year as they stopped there enroute to Florida from India.

After a most productive and enjoyable stay in India, Mathura, my daughters and I departed for America on August 13 via the Pacific. This time, Mathura's mother, Mrs. Gomathi Krishnan, accompanied us for a four-month stay in Florida. We halted in Tokyo for a four-night stay with our good friends Ganesh Kumar and his wife Prema who had moved to Tokyo a few months earlier from Jacksonville because Ganesh had accepted a major assignment with Johnson and Johnson in Japan. Tokyo's Narita Airport where our flight arrived, is about 40 miles from the eastern boundary of Tokyo. So we took a fast train to TCAT (Tokyo City Air Terminal), where Ganesh met us and took us to his elegant home.

Ganesh and Prema had arranged a Bharathanatyam dance program for Mathura on August 16 for the start of the 50th Anniversary Year Celebrations of India's independence. (India received its independence from Britain on August 15, 1947, and so the 50th year of Indian independence was August 16, 1996 to August 15, 1997.) The dance program organized by the India Society of Japan was at the Yamaha Hall in the busy Ginza district of central Tokyo. My elder daughter Lalitha joined Mathura in two of the dances, and I was the MC for the program. It was a great experience for Mathura and my daughter to perform before a large appreciative audience of 500 Indians and Japanese. It was also nice that Mrs. Krishnan saw this program in Tokyo because she had encouraged Mathura in her school days to learn Bharathanatyam.

Ganesh and Prema, like Mathura and me, are deeply interested in Carnatic (South Indian classical) music. A common passion for us is the music of Semmangudi Srinivasier, one of the greatest Carnatic vocalists of the 20th century. In Jacksonville, Prema and Ganesh play a leading role in organizing both music and dance programs. It came as no surprise to us, that during their stay in Japan, they were much involved in arranging such programs in Tokyo.

Since this was the first visit to Japan for my daughters, we showed them some of the great sights of Tokyo such as the Grand Palace, the Tokyo Tower, the Asakusa temple, and the Meiji Shrine. We even made a side trip to Kamakura to see the Giant Buddha atop a hill, from where we got a splendid view of the city, its curved beach, and the Pacific ocean.

One of the great conveniences in Tokyo is that you can check your bags for the international flights out at TCAT itself, and we availed this facility. We were back in Gainesville on August 17. Classes for the Fall Term of 1996 were starting a week later, and so I had just enough time to squeeze in a trip to Los Angeles to work with Basil Gordon!

Back in Florida for the start of the 1996 Fall Term, I heard the sad news that our dear Erdős passed away while attending a conference in Warsaw, Poland. It was not only the end of a great era in mathematics, but the end of a major chapter in the lives of hundreds of mathematicians he influenced. He made an indelible mark in the mathematical world with his numerous research contributions, and influenced the careers of more mathematicians than any other person in history. There was, and will be, no one like him, and his passing left a void that cannot be filled in time. There were several ways in which over the next several years that I was able to pay tribute to him. The first thing I did was, as part my annual series of articles on Ramanujan for "The Hindu", India's National Newspaper, I wrote an article "Erdős and Ramanujan - Legends of Twentieth Century Mathematics" which appeared in December 1996. Next I sent a message to the Editorial Board of The Ramanujan Journal and proposed a Special Issue in memory of Erdős, which the Board readily approved. This Special Issue appeared in Volume II of the Journal.

In late October, I visited the University of Colorado at Boulder, at the invitation of Wolfgang Schmidt to give a Colloquium and a Number Theory

Seminar. I arrived in Boulder on a Saturday night, and so I had a full day of relaxation before my talks. On Sunday night, Schmidt took me to dinner at "The Himalaya", a nice Indian restaurant in Boulder.

Heini Halberstam who had just finished his term as Chair at the University of Illinois, was spending a good part of his research leave (following his Chairmanship) at Boulder, and so I spent the morning of Monday in discussion with him. My colloquium at the University of Colorado was on "The (Big) Theorem of Göllnitz and its place in the theory of partitions". Following my Colloquium, the Schmidts hosted a dinner at their home and they invited the Halberstams, and the Elliotts, to join.

My Number Theory Seminar on the theme "Some new observations on the Göllnitz-Gordon and the Rogers-Ramanujan identities" was on Tuesday at 4:00 pm, following which, Halberstam, Elliott and Schmidt took me to dinner. Thus I enjoyed the warm and gracious hospitality of Wolfgang Schmidt and the number theorists at the University of Colorado.

Back in Florida, it was a pleasure to receive in November, the galley proofs of the Opening Issue of The Ramanujan Journal (January 1997). The Founding Editorial Board Members besides me were: George Andrews (Penn State), Richard Askey (Univ. Wisconsin), Bruce Berndt (Univ. Illinois), Frits Beukers (Utrecht), Jon Borwein (Simon Fraser Univ.), Peter Borwein (Simon Fraser Univ.), David Bressoud (Macalester College), Peter Elliott (Univ. Colorado), Paul Erdős (Hungarian Academy of Sciences), George Gasper (Northwestern Univ.), Dorian Goldfeld (Columbia Univ.), Basil Gordon (UCLA), Andrew Granville (Univ. Georgia), Adolf Hildebrand (Univ. Illinois), Mourad Ismail (Univ. South Florida), Marvin Knopp (Temple Univ.), James Lepowsky (Rutgers Univ.), Lisa Lorentzen (Norwegian Inst. Technology), Jean-Louis Nicolas (Univ. Lyon), Alfred van der Poorten (Macquarie University), Robert Rankin (Univ. Glasgow), Gerald Tenenbaum (Univ. Nancy), Michel Waldschmidt (Univ. Paris), Don Zagier (Univ. Bonn), and Doron Zeilberger (Temple University). I was sad though that Paul Erdős was not there to see the first issue in print.

The first article of the opening issue was by George Andrews on "The Well-Poised Thread - an organized chronicle of some amazing summations and their implications", based on his talk at the MAA MathFest in Minnesota in August 1994. His article definitely set the standard for the journal. It was a pleasure and pride for me to write the Editorial (Preface) to the very first issue. In concluding the Editorial, I said: "The very mention of Ramanujan's name reminds us of the thrill of mathematical discovery. By naming the journal after Ramanujan, we have set ourselves a very high standard. We hope that the mathematics that will fill the pages of this journal will be a source of excitement to experts and non-experts alike."

Mathura, her mother, and my younger daughter Amritha left for India in late November. My older daughter Lalitha waited for her Fall term school classes to end and so she accompanied me to India only in late December. Just before departing to India, I made a short trip to Las Vegas for the West Coast Number Theory Conference. Well, Las Vegas in not on the west coast! In order for the conference to be held in places like Arizona and Nevada, the name of the conference was changed to "The Western Number Theory Conference" starting from the 80s. My talk was the Opening Lecture of the Conference and I used the opportunity to announce the first issue of The Ramanujan Journal and show the galley proofs to a large gathering of number theorists. The formal launch of The Ramanujan Journal would be in Madras later in December: On December 30, at the M. S. Swaminathan Research Foundation, the First Issue of the Ramanujan Journal was "released". The Hindu, India's National Newspaper, gave a prominent report of the launch. The Hindu also published my article "The Ramanujan Journal - its conception, need, and place" on January 17, 1997.

Professor M. S. Swaminathan is arguably India's foremost scientist. His field of research is agriculture, and he is responsible for India's green revolution. Among all Fellows of the Royal Society today, he is the one who got it the earliest, and so he is currently the longest standing FRS! Prof. Swaminathan is a close friend of the Alladi family, and when he heard of the launch of the new Ramanujan Journal, he spontaneously and graciously said that the release function should be at his Research Foundation.

Within one week of our return to Gainesville, my dear friend and colleague Prof. Kermit Sigmon passed away due to skin cancer. He was a gentleman to the core and a fine tennis player. He was my singles opponent for a decade, and always invited me to play on the clay courts of the 300 Club where he was a member and I was not. Never once did he collect from me the guest fee that he had to pay for me to play there, even though I offered it to him numerous times. The only way I could repay his generosity was to help him with arrangements in Madras in December 1992 when he and his wife visited India. In Madras, my parents and my friend GP took care of them, and for this he was most appreciative. In later years, I played singles on weekdays, often at the 300 Club, with Professors Alex Stephan (Slavic Studies) and James Walker (Physics). Walker and his wife Isobelle, became close friends of ours. On weekends, I used to regularly play tennis (singles and doubles) with my Indian friends Atul Gokhale, Arun Someshwar, and Siddharth Thakur.

In late February 1997, John Martindale visited Gainesville and we discussed the promotion of The Ramanujan Journal. I suggested that at least in the first few years, along with the 50 free reprints that each author would receive, a copy of that issue of the journal in which the paper appeared be also sent to the author. Several authors informed me in writing that they appreciated this "gift". Sadly, nowadays, journals do not even offer the 50 reprints gratis and have replaced them with an author's pdf from which one can make cheap looking copies. I am part of a very small minority still ordering reprints of papers!

I hosted a party for John Martindale to mark the launch of The Ramanujan Journal, and invited my mathematics colleagues and some faculty from other departments. Martindale and I also discussed the launch of a new book series *Developments in Mathematics* (DEVM) — name suggested by me — that would publish research monographs, refereed proceedings of conferences, and edited volumes of special interest. This book series was launched in 1998. Both the Ramanujan Journal and the book series are now published by Springer, after Kluwer and Springer merged in 2004.

In early March, my parents arrived in Florida for a three-month stay, travelling as usual via the Pacific. After celebrating Father's Day in Gainesville, my parents left for India in mid-June. I also left for India the same morning with Mathura and my daughters and halted in Hawaii for a one-week vacation in Kauai and Maui.

The Kauai visit was a repeat of what we did the previous year. The Hawaiian islands are so enchanting, that repetition does not make the visit less enjoyable. What was different this time was the visit to Maui after a ten-year gap. Maui is favourite of tourists and is known for its endless stretches of beaches. We stayed at Sheraton Maui which began the legend of Kaanapali beach, and from whose Black Rock Promontory, Hawaiians jump into the ocean. One of the highlights of Maui is the 10,000 ft Haleakala Volcano, from the summit of which you get to see the colorful cider cones of a gaping gigantic crater that descends into the ocean below. We also made a trip to Hana at the other end of Maui and at the foot of Haleakala. The drive along the winding Hana Highway skirting the coast is somewhat dangerous, but worth every bit, because of the breathtaking scenery of the jagged coast line with its black sand beaches and plunging cliffs.

After a memorable week in Kauai and Maui, we left Honolulu for Singapore via Tokyo. We reached Madras just in time to celebrate our Wedding Anniversary on June 29 with a lunch at the Taj Fisherman's Cove Hotel, 20 miles south of Madras on the beach along the Coromandel coast —

India's version of Hawaii! During our two-month stay in India, we also had a Sweet 16 party for my older daughter Lalitha at Ekamra Nivas on July 4 attended by most of our relatives and friends.

In the second half of July 1997 our entire family made a two-week South Indian trip combining academic assignments for me with sightseeing for the family. I gave a talk at the Raman Research Institute (RRI) and was pleasantly surprised to see my Florida colleague Peter Sin in the audience! He was visiting the Indian Institute of Science in Bangalore during the summer. After my talk, Professor Ramachandra who had settled in Bangalore after his retirement from the Tata Institute, invited me to lunch at the National Institute for Advanced Studies (NIAS), where he had an office.

Sightseeing on this trip included visits to the famous temples of Belur, Halebid and Somnathpur from Mysore. These temples are world renowned for their exquisite carvings on stone. In addition we visited Sravanbelgola, where there is a gigantic monolithic statue of Gomateswara atop a hill. During the trip to Somnathpur, we made visit to the roaring Sivasamudram waterfalls, and on our way back in the evening, saw a banded Krait, one of the most venomous snakes, moving across the road! The trip also included visits to the wildlife sanctuaries of Bandipur and Mudumalai, known not just for their fauna — tigers, elephants, wild dogs, spotted deer, the Indian Gaur (the world's largest wild oxen), and peacocks, — but for the incredibly beautiful scenery owing to the location at the foothills of the Nilgiri mountain range. We also enjoyed seeing the Ranganathittu Bird Sanctuary just outside Mysore, which offered a fantastic array of bird species nesting on tops of trees that were on the Cauvery river. On our boat ride in Ranganathittu, we observed nearly a dozen marsh crocodiles on the rocks, waiting to swallow any hapless bird that would fall into the water!

After the South Indian trip, my family stayed for three more weeks, but I had to make a ten-day trip (July 29–Aug 8, 1997) to the USA to attend two conferences, the first at Penn State University and the second, the MAA Mathfest in Atlanta.

I arrived in State College on July 30 and stayed at the home of our good friends, the Vedams. The conference was to start the next day, but on the night of my arrival, the Vedams had invited Prof. C. R. Rao and his wife for dinner. C. R. Rao, arguably the most eminent statistician in the world today, was Eberly Chair at Penn State after his long innings as Director at the Indian Statistical Institute in Calcutta. Defying the passage of time, he continued to be productive, and was the inspiring leader in the Statistics department at Penn State.

The conference began on July 31, and it was a nice gesture on the part of the organizers to include one copy of the first issue of The Ramanujan Journal in the Registration Packet of each participant. I had arranged for Kluwer to send several copies of the first issue of the journal to Penn State University. So it was very good publicity for the journal.

The lead conference organizer was Ken Ono who had joined Penn State the previous year. The conference was organized in a modern conference center call Scanticon just outside of the town of State College, and the arrangements and facilities there were excellent. Basil Gordon who was Ken Ono's thesis advisor was honored at the conference for his 65th birthday year during a banquet held at the stately Nittany Lion Inn on campus on August 2. Since I had a long association with Gordon, and had collaborated with him, I was asked to speak at the banquet. Gordon was an expert pianist, and so at the request of the participants, he played the piano to everyone's enjoyment. The great Indian mathematician Sarvadaman Chowla who had been on the faculty of Penn State University, was remembered at the conference; the Department had launched the Chowla Assistant Professorship the previous year and this was brought to the attention of the conference participants at the banquet.

On the morning of August 2, I was asked to chair a session in which the brilliant Doron Zeilberger was speaking. Just before the talk, Dennis Stanton asked me: "Krishna, how will you handle Doron Zeilberger?" Actually, it was a pleasure to be Chair for Zeilberger's lecture which was electrifying, inspiring, and entertaining!

My own talk was on the previous day. I spoke on an invariant in the theory of partitions which had not been exploited, namely the number of different parts. This statistic is invariant under conjugation, and I used this invariance to obtain new identities including a symmetric six-parameter extension of Heine's fundamental transformation.

I even had a good game of tennis at the indoor club at Penn State University. My tough opponent was Jim Haglund, and we split sets in a duel that lasted one and a half hours. Anticipating that I would get to play tennis on this trip, I had my tennis attire and racquet in my suitcase!

From State College, I flew to Atlanta where I was invited by Andrew Granville to speak at the MAA Mathfest in a Special Session on August 4 in memory of Paul Erdős. I stayed at the spacious home of our good friends Sham and Aruna Navathe. Sham is a world leader on Data Base Systems and is a Distinguished Professor of Computer Science at Georgia Tech. Sham was on the faculty of the University of Florida before moving to Georgia Tech, and so our acquaintance began in 1987 in Gainesville.

One of the featured lectures at the MAA Mathfest was by Ron Graham on the work of Paul Erdős on the morning of August 4. That afternoon, for the Special Session on Erdős, there were four speakers: Carl Pomerance, myself, Neil Calkin, and Andrew Granville, in that order.

I left Atlanta the next day to return to India via Detroit. It was interesting that on the long segment from Detroit to Tokyo, I was on a Northwest Boeing 747-400 called the *World Plane* which was painted in various colors to represent different parts of the globe. I was back in Madras on August 7, just in time for my father's birthday (Aug 9).

I had just ten days in Madras before leaving for Hong Kong for the International Congress on Algebra and Combinatorics. Hong Kong, which for decades was a British colony, was returned to China on July 1, 1997. Thus this conference was just six weeks after China took over Hong Kong, and so we did not know what to expect under the Chinese regime. But Hong Kong was just as exotic and enjoyable (to us visitors) as before.

One evening, to relive old times, I strolled down Nathan Road which is the main artery of the throbbing business district of Kowloon. It ends in front of what was at that time the Regent Hotel (now the Intercontinental), from whose lobby, I enjoyed a stunning view of the harbor and the glittering lights of the island of Hong Kong across the causeway as I had dinner there.

The Congress on Algebra and Combinatorics opened on August 19 morning at the magnificent HITEC Centre in the busy Kowloon section of Hong Kong with the inspiring address of 1994 Fields Medallist Efim Zelmanov. This was followed by the plenary talk of Anders Björner from Sweden. The oldest mathematician at the Congress was Professor B. H. Neumann, and he was honored at the Opening Ceremony. It was a pleasure to reconnect with him 24 years after my visit to the Australian National University in 1973 at his invitation to work with Kurt Mahler.

From the afternoon of August 19, the talks were on the campus of Hong Kong University. Mine was a 40 minute talk the next day on the theme "The combinatorics of words in the theory of partitions". It was chaired by Professor M. Ito of Tokyo. That night there was a banquet at the Chun-Chi Club hosted by the Vice-Chancellor of the University of Hong Kong. All through, the organizers had made very good arrangements for me as a vegetarian; for the banquet, they had a special table for vegetarians, and the service there was impeccable because B. H. Neumann, as a vegetarian, was seated at that table!

I was scheduled to leave Hong Kong the next day (Aug 21) at 4:00 pm by United Airlines for Singapore where I would unite with Mathura and my daughters to leave for the United States together on Northwest Airlines. That morning we were told that a typhoon was approaching Hong Kong and so the afternoon flights would be cancelled. Luckily for me, my United flight was among the last to leave Hong Kong before the typhoon hit the island. Although it left on time, the United Boeing 747 took a circuitous route to avoid the typhoon clouds, and so by the time I reached Singapore, it was past midnight. I took a taxi from Singapore Changi Airport to the Westin Plaza Hotel, where Mathura and my daughters were in deep sleep in our spacious room. We spent a relaxed day in Singapore, shopping and enjoying the wonderful cuisine in that gastronomic capital of the world. We were back in Florida for the start of classes for me and my daughters.

The main academic event for me in Fall 1997 was the International Conference on Number Theory in the first half of November at the Research Institute of Mathematical Sciences (RIMS) in Kyoto, Japan. Kyoto represents the quintessential Japanese culture, and so Mathura accompanied me on the trip. Hence it was academic work for me combined with some fine sightseeing around Tokyo, Kyoto, and Nara, with a stop in Honolulu on the way back for a talk at the University of Hawaii.

We arrived in Tokyo on Friday, Nov 7 evening, and spent the weekend at the home of our friends Ganesh and Prema. On Saturday, Prema took us to Hakone National Park. Weather was glorious and so we had fabulous views of Mt. Fuji. After a heavenly boat ride on Lake Aashi, we saw the hot springs at Hakone, which looked lovely against the backdrop of a forest that glowed crimson and yellow with Fall colors at its best.

On Sunday, we took the *Shinkansen* (Bullet Train) from the Shin Yokohama station of Tokyo to Kyoto. We admired the way in which the Japanese stood, each one very quiet, in a perfect line in front of a mark that signified the train door opening of a compartment, and how the Shinkansan, speeding at about 200 mph, came to a smooth halt at the station in such a way that the compartment door opening was at the position indicated by the mark on the platform! This was a good demonstration of the perfection the Japanese achieve in every thing they do, and their discipline and integrity. It is due to these qualities, that even though they were totally crushed at the end of World War II, they were able to come back with a few decades to become such a dominant economic force in the world.

On arrival in Kyoto, we were warmly received at the station by Shigeru Kanemitsu, the main organizer of the conference, and Bruce Berndt who was one of the main speakers. Berndt loves Indian food, and so after we checked into the comfortable Hotel Kyodokaikan, he joined us at dinner at Ashoka, an Indian restaurant.

Kyoto has the greatest concentration of temples in all of Japan, and there are several which are so beautiful, that it is difficult to see them all in one short visit, especially with a conference to attend. But we did the best we could and managed to see the finest temples of Kyoto. The conference opened only in the afternoon of Monday, November 10, and so in the morning Mathura and I visited the Sampansangado temple with its 1000 Buddhas (!), and the Kiyomizo temple which has a breathtaking hillside location and commands a splendid view of Kyoto.

The conference opened with a warm welcome address by Kanemitsu. My hour lecture was the later that afternoon, and I presented my work on invariants under partition conjugation and q-hypergeometric series identities. Volume I of The Ramanujan Journal was on display at the conference and it attracted the attention of several Japanese participants.

The next day, Mathura and I walked from RIMS to the nearby *Ginkakuji Temple* (the Silver Temple) where the Fall colors took our breath away. Just as we entered the temple grounds, we saw an elderly Japanese artist work on a painting of the temple. It took about two hours for us to finish touring the gardens of the Ginkakuji temple, and by the time we came out, the Japanese artist had finished his painting. Without any hesitation we bought the painting, because in it, the temple and its surrounding trees looked EXACTLY as we saw them! That night, the Japanese participants very graciously hosted a dinner at a Japanese restaurant in central Kyoto, where we were treated to lavish hospitality by Kimono clad hostesses.

The third day afternoon was kept free for sightseeing. The entire program was meticulously planned by Kanemitsu. He led us on a tour, the highlight of which was the *Kinkakuji Temple* (the Golden Temple). Both the temple and its sprawling gardens looked radiant under the afternoon sun. Mathura and I bought a cloth wall hanging of the Kinkakuji temple. That night, during a dinner at an Indian restaurant, Kanemitusu proposed that he would edit a series of five proceedings of the Japan-China Number Theory Conferences in alternate years over a ten year period for my newly proposed book series Developments in Mathematics (DEVM).

On Thursday, November 13, Mathura and I made a day trip to Nara. Kanemitsu very thoughtfully sent a student of his as a guide. After seeing the Shofikuzu temple, we visited the *Todaiji Temple* that boasted the largest bronze Buddha in the world — the *Daibatsu* — housed in the largest wooden construction on earth! And how charming it was to see deer on the sprawling grounds of the Todaiji temple!

The next day it was rainy, but it did not bother us because we were departing Kyoto which had offered us glorious weather throughout. We took the Bullet Train from Kyoto to Osaka Kansai Airport, which is located several miles outside the city limits and is built on land reclaimed from the sea. It is both an engineering and architectural marvel. Our Northwest DC-10 flight from Osaka to Honolulu was delayed, but the airline kindly put us on the earlier DC-10 flight which we could board because, as is our practice, we had arrived at the airport well ahead of our flight! After crossing the International Date Line, we arrived in Honolulu on the same day (Fri) in the morning. That afternoon I went to the University of Hawaii where I gave a colloquium on invariants under conjugation at the kind invitation of Edward Bertram. That night, Ed and Alice Bertram, and Kenny Rogers and his wife Choleng, hosted a dinner for me and Mathura at the Sheraton Waikiki, where Mathura and I had our honeymoon in 1978! On return to Florida, I informed Martindale about Kanemitsu's idea for publishing five conference proceedings in DEVM over a ten year period, and he heartily endorsed this proposal. Thus the trip to Japan in Fall 1997 was enjoyable and fruitful in many ways.

In view of our international travels in 1997, we did not go to India in December. We enjoyed Christmas in Florida and looked forward to 1998 which quite unexpectedly changed my academic life owing to my appointment as Chair of the mathematics department which I discuss next.

The mathematicians responsible for the Capparelli partition conjecture, its proof and generalization in 1992. Standing — Stefano Capparelli, George Andrews and Krishna Alladi. Seated — James Lepowsky and Basil Gordon. At Alladi House, Gainesville, during a party for the participants of the Additive Number Theory Conference, November 20, 2004.

International speakers of Krishna Alladi's session at Anna University for the Ramanujan Centennial, at Ekamra Nivas. Left to right: David Bressoud, Bruce Berndt, Mrs. Askey, Alladi Ramakrishnan, George Andrews, Krishna Alladi, Richard Askey, Basil Gordon, and M. V. Subbarao, December 23, 1987.

Paul Erdős visited the University of Florida annually in the Spring. Here he is explaining a mathematical idea to Krishna at the Alladi House, Gainesville (Feb 26, 1989).

Part II
Chairmanship and the Ongoing Ramanujan Mission

Krishna in the Mathematics Department Chair's Office, at the start of his term as Chair, Fall 1998.

Prof. John Thompson wearing the 2000 National Medal of Science, after the Medal Ceremony in Washington D.C., Dec 1, 2000. Dr. (Mrs) Diane Thompson is next to him.

Left to right: 1970 Fields Medallist John Thompson, 1978 Fields Medallist Daniel Quillen, and 1970 Fields Medallist Sergei Novikov, in the University of Florida Mathematics Department after Novikov's Erdős Colloquium, Apr 12, 2002.

Chapter 8

Some major programs and initiatives (1998–2008)

8.1 Appointment as Chairman (1998)

The year 1998 was the beginning of a change in my academic life because I was appointed as Chair of the Mathematics Department at the University of Florida. Although I initially accepted the position only for three years, I ended up serving for ten years as the Chair.

The year started with the receipt of the December 1997 issue of The Ramanujan Journal which was the refereed proceedings of the International Symposium in Number Theory that I had conducted during January 1–2, 1996. A few days later, in mid-January, Dr. David Larner and John Martindale of Kluwer Academic Publishers approved the launch of a book series called *Developments in Mathematics* (DEVM) with me as the Series Editor. The book series was to publish both research monographs as well as refereed proceedings of conferences. My proposal also was to link the Ramanujan Journal with the book series in the sense that if there are volumes or issues of the journal of exceptional interest, then these could be also printed as spin-off volumes in the book series. The benefit of this would be that an individual who is interested in the special issue/volume of the journal, but who does not subscribe to the journal, could purchase the book version of it from the book series. It turned out that the first volume of DEVM (that appeared in 1998) was the book version of the Erdős Memorial Issues (Volume 2, Issues 1–2, 1998) of The Ramanujan Journal, and of course several mathematicians purchased this book. The second volume in DEVM that also appeared in 1998 had nothing to do with the journal. It was the refereed proceedings of a China-Japan Number Theory Conference, and indeed the first of five such proceedings that Shigeru Kanemitsu edited (along with various co-organizers) for DEVM over a ten year period.

In late January, Joseph Glover, who had been Chair of Mathematics since 1992–93, announced that he was stepping down at the end of the Academic Year 1997–98 after serving for five years. There was an opening for the position of Associate Dean in the College of Liberal Arts and Sciences (CLAS) and he intended to apply for that. He did, and was chosen for the post to begin on July 1. In early February, a Chair Selection Committee of mathematics department faculty was chosen. The Dean then appointed one external member to the Committee. Some colleagues, including members of the Chair Search Committee, urged me to accept the nomination as a Chair candidate, which I did. I had some ideas to improve the department's performance and enhance its international visibility, and felt that by being Chair, I could accomplish these goals by introducing some new programs.

The Chair selection process was very rigorous. All candidates were asked to prepare vision statements. These were first circulated to the faculty and later the candidates faced the faculty to answer questions on their vision. This process is very different from India where candidates are questioned only by the Search Committee. My vision was to improve the Department's performance and productivity by building on its strengths, and to enhance its visibility and gain recognition both within the University and internationally. In particular, to achieve these, I proposed:

(i) the formulation of a Hiring Plan to do targeted hiring each year,

(ii) a budget to conduct conferences annually in various areas of faculty strength so as to bring leading researchers worldwide to interact with faculty and graduate students,

(iii) create a pair of annual Distinguished Colloquia, one each in pure and applied mathematics, that would bring luminaries to the Department,

(iv) and create a named assistant professorship that would be offered to fresh PhDs without being field specific.

After hearing our vision, based on the faculty vote, and after interviewing the candidates, the Search Committee endorsed and sent the names of all candidates to the Dean with comments. Dean Willard Harrison invited all of us for interviews. I was invited for a second interview, following which he offered me the job on May 7. Thus the whole process took three months even though it was an internal search.

Over the ten year period I was Chair, I accomplished all four goals above (and more), and can say now with some satisfaction that the Department's performance and stature grew significantly. In the next few sections, I will highlight some prominent programs during my term as Chair, but before that I will describe in this section some special events that took place before my appointment started on July 1.

As the Chair Search was beginning, simultaneously in early February, I was appointed Chair of the Cultural Council of the India Cultural and Education Center (ICEC) which was to get its own new building in March. This position entailed overseeing all the cultural activities of the ICEC. For the inauguration of the new building, a fantastic two day program was conducted by the ICEC. Mathura was invited to perform Bharathanatyam at the Banquet, which she did admirably. Several students from Gainesville and surrounding areas performed earlier in the day, one such featured dance being that of Mathura's students. In the second interview with Dean Harrison, he referred to my role as Chair of the Cultural Committee of the ICEC, as Editor-in-Chief of The Ramanujan Journal, and as Series Editor for DEVM, as testifying to my organizational and leadership capacities. During the interview, I stressed that, I am, and will always be, a researcher. To ensure that I had time for my research on top of my administrative responsibilities, I requested a zero teaching load. I also requested three administrative positions — an Associate Chair, an Undergraduate Coordinator, and a Graduate Coordinator — with reduced teaching loads, to assist me in the smooth running of the Department's teaching portfolio. The three positions were given to Professors Neil White, Rick Smith, and Paul Robinson, respectively, and they performed their tasks with great efficiency. The Dean said he had noted my research productivity, and would agree to my requests provided I could show each year that my research was progressing well. In fact, as we were discussing my research program, I mentioned two upcoming conferences where I was invited to give talks: (i) the AMS-IMS-SIAM Joint Summer Research Conference on q-series, Combinatorics and Computer Algebra, at Mt. Holyoke College (June 21–25) in honor of Richard Askey for his 65th birthday, and (ii) a conference in Maratea, Italy (Aug 30–Sept 5) for the 60th birthday of George Andrews. To my pleasant surprise, the Dean spontaneously offered to support my trips to these conferences.

Richard Askey was a leader in the field of special functions and a mentor to dozens of researchers in the area. Thus the conference in his honor was attended by an impressive group of senior and junior researchers. My talk at the conference was on "Reformulations of the (Big) partition theorem of Göllnitz and q-series identities". My paper on this topic appeared in the Contemporary Mathematics series of the AMS dedicated to Askey. I will describe the Maratea conference in §11.4 where I will discuss all the George Andrews Milestone Birthday Conferences.

In early June, before the conference in Mt. Holyoke, there was an

Orientation for New Chairs at a resort in Howie in the Hills, about an hour south of Gainesville, to which I was invited to participate. This Orientation was for new department chairs from all over the state of Florida. I found this generally useful but what I DID NOT GET was information as to how university budgets are set up, how the funds come down from the President, to the Provost, to the Dean, and finally, to the department chairs. I had to learn some of this on my own as I started performing the role as Chair. I sensed that there are a number of "secrets" in budgeting, and that no one wants to share the secrets of the trade with others, because each administrator wants to somehow get a good piece of the pie.

After the meetings at Howie-in-the Hills and at Mt. Holyoke, and before starting on my term as Chair, I decided to make a trip to India with Mathura and my daughters as a relaxation. We vacationed in Bangkok and Phuket enroute, and both were charming in their own ways. Bangkok boasts a fantastic collection of Buddha temples and we enjoyed seeing the exquisite Emerald Buddha inside the fabulous Grand Palace, and the gigantic Reclining Buddha at Wat Po (*Wat* means temple). We stayed at the magnificent Royal Orchid Sheraton that is right on the banks of the Chao Phraya River. From Bangkok we took a day trip to historic Ayuthya going out by a motor bus and making stops at the lovely Summer Palace, after which we saw the ancient ruins of Ayuthya which was the capital of ancient Siam. The Buddhist temple architecture there is what has influenced the temples all over Thailand. We returned to Bangkok by boat from Ayuthya, and the skyline of Bangkok with its myriad temple spires looked dazzling under bright sunny skies. The boat dropped us right at the jetty of the Royal Orchid Sheraton on the Chao Phraya River.

The trip to Phuket was a dream come true. Ever since we had seen enchanting scenes of that region in the James Bond movie "The Man with the Golden Gun", we wanted to see James Bond Island and the limestone outcroppings that jut out from the sea like gigantic pillars. And we did that on June 29, 1998 — the 20th Wedding Anniversary for me and Mathura! The visit to Phuket was made even more delightful by our stay at the Sheraton Grand Laguna, where we had to approach our cottage fronting the sea by taking a motor boat from the main lobby.

After a memorable vacation in Thailand, we spent six weeks in India where my academic assignments included lectures at various educational institutions in Madras. I also had an opportunity to give talks in Bangalore at the Raman Research Institute and the Tata Institute center there. Between my two talks, Prof. K. Ramachandra, who was retired from the

Tata Institute in Bombay and living in Bangalore, took me lunch at the National Institute for Advanced Studies (NIAS) which is adjacent to the sprawling campus of the Indian Institute of Science. There at lunch I met Dr. Raja Ramanna, who had by then retired from his post as Director of the Department of Atomic Energy and settled in his native town of Bangalore. In 1983, when my father retired as Director of MATSCIENCE, Ramanna "advised" me never to aspire for an administrative position, and in particular the Directorship of MATSCIENCE, even though he himself was an administrator! So it was interesting for me to inform him that I was going to take up the position as Chair of Mathematics upon return to Florida from India, contrary to his "wise advice"!

8.2   The Erdős and Ulam Colloquia

When scholars of world repute visit a university or a department, they inspire both faculty and students, and the topics they expound often influence the faculty to enhance their own research programs. Keeping this beneficial effect in mind, I proposed to Dean Harrison the creation of two annual Distinguished Colloquia — The Erdős Colloquium in pure mathematics and the Ulam Colloquium in applied mathematics. I did not want to show preference to either pure or applied mathematics, and therefore proposed two such colloquia, and the Dean readily agreed to this suggestion.

Stan Ulam was a Graduate Research Professor in the mathematics department for about a decade, from 1975 until his death in 1984. He made significant contributions to several branches of mathematics, and was particularly known for his role in the Manhattan Project at Los Alamos during WW II, and for his seminal work on the Monte Carlo method of computation. At the University of Florida, he created and headed the Center for Applied Mathematics, and through the activities of the Center, brought in a number of eminent researchers as visitors; three of the regular visitors were Lamberto Cesari, an analyst from the University of Michigan, Gian-Carlo Rota, a world famous combinatorialist from MIT, and Paul Erdős, a legend of 20th century mathematics. Another visitor (though not annually) was the great probabilist Marc Kac of Rockefeller University, who along with Erdős had created the subject of Probabilistic Number Theory. So it seemed to me appropriate to name the distinguished colloquium in applied mathematics as the Ulam Colloquium.

Paul Erdős, whose visits to the University of Florida began during the time of Ulam's term as Graduate Research Professor, continued to visit Gainesville two weeks each year in the Spring either before or after the annual combinatorics conference at Boca Raton on the Atlantic coast of Florida. As I once said: "Like migrating birds, Erdős hovered around the isotherm 70 degrees Farenheit. He could be found in Florida in the Spring, Boulder and Calgary in the summer, and at UCLA in December." Erdős influenced the development of the strong program in Combinatorics at the University of Florida and interacted closely with several faculty members. David Drake, Neil White and Andy Vince in Combinatorics, Jean Larson in Set Theory, and Arun Varma in approximation theory. He was also instrumental in my appointment at Florida. In view of his strong influence on the department, I proposed that the distinguished colloquium in pure mathematics be named as the Erdős Colloquium.

Both Colloquia were launched in Spring 2009. The first Erdős Colloquium speaker was Ron Graham, a world famous mathematician, and a great ambassador for mathematics. Owing to his eminence and close connections with Erdős, it was ideal that the Erdős colloquium was launched with his talk.

In 1998–99, as I started out my term as Chair, I discussed the creation of a biomath program in the Department. So we felt that it would be beneficial to invite a leading researcher in biomath for the First Ulam Colloquium. So we had James Keener of the University of Utah as the speaker and he gave a fine lecture on mathematical models to study heart arhrythmias.

In the last two decades, we have had some of the most illustrious mathematicians address these two colloquia, including several fields medalists. The list of speakers in the two distinguished colloquia are:

Erdős Colloquium Speakers: 1998–99: *Ronald L. Graham* (UCSD); 1999–00: *Daniel Quillen* (Oxford); 2000–01: *Hyman Bass* (Michigan); 2001–02: *Sergei Novikov* (Moscow Univ. and Univ. Maryland); 2002–03: *Benedict Gross* (Harvard); 2003–04: *Stephen Smale* (Berkeley); 2004–05: *Richard Borcherds* (Berkeley); 2005–06: *William Arveson* (Berkeley); 2006–07: *Alexander Kechris* (Caltech); 2007–08: *Efim Zelmanov* (UCSD); 2008–09: *Akshay Venkatesh* (Stanford); 2009–10: *George Andrews* (Penn State); 2012–13: *Michael Barnsley* (Australian Nat'l Univ.); 2013–14: *Fan Chung* (UCSD); 2014–15: *Janos Pintz* (Renyi Inst., Budapest); 2015–16: *Hugh Montgomery* (Michigan); 2016–17: *Jeffrey Remmel* (UCSD).

Ulam Colloquium Speakers: 1998–99: *James Keener* (Utah); 1999–00: *Gilbert Strang* (MIT); 2000–01: *Persi Diaconis* (Stanford); 2001–02: *John Milnor* (SUNY Stony Brook); 2002–03: *Gerhard Frey* (Univ. Essen); 2003–04: *Stan Osher* (UCLA); 2004–05: *Richard Stanley* (MIT); 2005–06:

Louis Nirenberg (Courant Institute); 2006–07: *Anil Nerode* (Cornell); 2007–08: *Dominique Foata* (Univ. Strasbourg); 2008–09: *Daniel Mauldin* (North Texas); 2009–10: *James Keener* (Utah); 2010–11: *Sebastian Schreiber* (UC Davis); 2011–12: *Rudolf Kalman* (UF and ETH, Zurich); 2012–13: *Robert Devaney* (Boston Univ.); 2013–14: *Gunnar Carlsson* (Stanford); 2014–15: *Jun Cheng Wei* (Univ. British Columbia); 2018–19: *Yuri Tschinkel* (Courant Institute and the Simons Foundation).

Besides the Erdős and Ulam Colloquia, I also launched the (annual) *Ramanujan Colloquium* in 2007, so generously sponsored by George Andrews and The Pennsylania State University (§8.7). In the sequel, I will discuss the personalities and the work of many of the speakers of these distinguished colloquia as I describe the various events year by year.

8.3 A hiring plan

Within a few weeks of my appointment as Chair, we heard that the university president had a "biology initiative" — that is, he had set apart funds for appointments in, or in areas that intersect with, the biological sciences. In particular, this meant biomathematics or mathematical biology for our department. Even though we had an applied mathematics program, biomath was new to us. So the first order of business was to convince the department that including biomath would be beneficial. So I formed a small committee and gave it the charge to get data about various leading universities investing in biomath. With this information, I convened a faculty meeting in which the Committee and I made a presentation about the benefits of having a biomath program. The faculty overwhelmingly supported the creation of a biomath program, and so I could confidently approach the Dean with a proposal as to how to go about it. Dean Willard Harrison agreed to an appointment to begin in Fall 1999, with two more in biomath to follow in due course. Combinatorics was in need of young blood, and so the Dean approved a position in this area. In the end we got three positions in my first year as Chair, and we succeeded in making three fine appointments — Miklos Bona (combinatorics), Sergei Pilyugin (biomath) and Jindrich Zapletal (mathematical logic). All three remain in our department today as full professors and have contributed to the growth of the programs in their areas in significant ways.

When approaching the Dean with a request for new faculty positions, it is prudent to have compelling reasons to back up your request. It is also good to have a long range vision for departmental hiring, in line with the vision of the College and the University, and with this vision, make hiring requests to the Dean. Thus in my second year as Chair, I proposed to the Department to come up with a comprehensive Hiring Plan. To this end, I formed a Committee and charged it with gathering data about the composition of a broad spectrum of major mathematics departments, and how they have expanded. This data helped up formulate a comprehensive Hiring Plan which was well balanced and not polarised. The plan provided a systematic way of enhancing existing strengths in various areas and also make appointments in new areas.

In formulating the Hiring Plan, we took into account the mathematical needs of other departments as well. James Brooks (Associate Chair from 2000–01) and I met with Chairs and Graduate Coordinators of certain science departments to find out what areas in mathematics would be of special relevance for them. We included in our Hiring Plan a proposal to hire in areas like applied combinatorics as a result of such discussions, and we could make a compelling case to the Dean for such hires.

Even though we always approached the administration with well artic-ulated reasons for hiring, it was always an effort to convince the university that it is beneficial to invest in mathematics. But I never gave up. I was constantly writing letters to, and speaking up at meetings with, the ad-ministration, making a case for investment in mathematics. For example, I used to point out that when hiring an assistant professor in the experimen-tal sciences or in engineering, the university was gladly providing start-up funds of the order of magnitude of $200,000 to $300,000 to set up labs. The new appointees in those disciplines have to bring in grants exceeding a million dollars, so that the overhead part of the grant would pay back the start-up money that was given out. This process may take at least five years. If the young appointee is a star, then he/she might leave the university for a more lucrative position elsewhere, in which case, the sig-nificant investment would not benefit the university. In contrast, I would point out that start-up funds in mathematics are minimal, and in the range of $20,000 to $30,000 for seminars, visitors and travel, with no labs to be supported. In addition, mathematics is a "cash cow" (to use the phrase of George Andrews) because the Department generates significant revenue for the university by the number of students it teaches. So I used to advocate that investing mathematics was fair and prudent.

While the Hiring Plan focused on getting young assistant professors, it also emphasized senior appointments from time to time if new areas are to be developed. Finally the Hiring Plan struck a balance between hiring in applied and pure mathematics. I have always told the Dean that pure and applied mathematics are like our two arms. We need both arms and should not develop one at the cost of the other. At this point I wish to stress that even though there is talk in the United States of investing in the STEM disciples (STEM = Science, Technology, Engineering and Mathematics), the M in the STEM usually means Applied Mathematics to the upper level administrators and generally does not include pure mathematics. What administrators and members of other disciplines miss is that applied mathematics thrives on the foundation laid by pure mathematics.

During my ten years as Chair, we made several excellent appointments that strengthened combinatorics, number theory, analysis, topology, mathematical logic, and of course applied mathematics. And we did initiate and build a strong biomath program. Due to budgetary reasons, we did not succeed in getting any senior positions. All positions authorized were at the assistant professor level, but we got a good number of these. When I started as Chair in 1998, the Department had 56 faculty positions including 5 lecturers. In 2005, during my seventh year as Chair, the department size had increased to 67 faculty positions including lecturers. And simultaneously, from 1998 to 2005, the number of graduate students had grown from about 70 to slightly more than 100. So I was fortunate that during most of my term as Chair, the Department grew considerably in size. Its productivity grew similarly as well, as will be seen in the sequel.

8.4   The Special Years Program (2001–07)

To stimulate the research atmosphere in the Department and to enhance its productivity and visibility, I proposed to conduct several Special Years. Each Special Year was to have an area (or two areas) of focus, and in these areas have a vibrant program of high level research conferences, training workshops for graduate students, and featured lectures by eminent visitors. I emphasized that there ought to be conferences on both the pure and applied aspects so as to appeal to client disciplines. Similarly, the lectures by the eminent visitors were to be a good mixture of talks on cutting edge research and expository lectures providing historical surveys.

The choice of topics for the Special Years was influenced by departmental strengths. The six Special years that we organized were:

1) Special Year in Topology and Dynamical Systems (2001–02)

2) Special Year in Algebra (2002–03)

3) Special Year in Applied Mathematics (2003–04)

4) Special Year in Number Theory and Combinatorics (2004–05)

5) Special Year in Analysis and Probability (2005–06)

6) Special Year in Mathematical Logic and Set Theory (2006–07)

Algebra was chosen as the topic of the second Special Year because our distinguished colleague John Thompson turned 70 in Fall 2002.

In conducting these Special Years, every research group in the Department was involved, and as a result, all the research programs in the Department were invigorated. Every faculty member, in one way or the other, participated in the organization of one of these Special Years. There was a steady flow of eminent visitors who got to know about the work of our graduate students. As a result, some of our graduate students upon finishing their PhDs, were offered prestigious post-doctoral positions at UCLA, Minnesota, and Ohio State, to name a few. Without any hesitation I would say that the distinguished colloquia and the Special Years brought international visibility to the Mathematics Department.

The Special Years were very well funded by the National Science Foundation (NSF), The National Security Agency (NSA), and in some cases by the Number Theory Foundation (NTF). CLAS Dean Neil Sullivan and Vice-President of Research Win Phillips were very generous in providing seed money — so important to get the program going each year.

In the next chapter, in narrating the events of the individual academic years, I shall describe in some detail each of these Special Years, and the fascinating personalities and work of many of the mathematical luminaries who visited the Department.

8.5   The John Thompson Research Assistant Professorship

Mathematics departments in major universities have named instructorships or assistant professorships. There is the Benjamin Pierce Instructorship at Harvard, the Ritt Assistant Professorship at Columbia, the Gibbs Instructorship at Yale, the Hedrick Assistant Professorship at UCLA, the Hildebrandt Assistant Professorship at Michigan, to name some very well known ones. These are all two or three year positions with reduced teaching loads offered to fresh PhDs. When advertising for such positions, it is "field open", as opposed to tenure-track or tenured positions which almost always are targeted searches. The influx of young talent stimulates the research atmosphere in the department. So with that in view, I had proposed in my *Vision Statement* in 1998 when I applied for the position of Chair, the creation of a named assistant professorship. But unlike the Distinguished Colloquia which is a small budget item, or the Special Years where the bulk of the support is from funding agencies, a named assistant professorship would involve significant funding from the University, or from a donor. So this proposal of mine had to wait until the time was right.

Dean Sullivan of the College of Liberal Arts and Sciences, a reputed physicist in his own right, knew of the eminence of 1970 Fields Medallist John Thompson in the mathematics department. He told me that he wanted something to be done to honor Thompson, but was waiting for the right moment. And it happened in a dramatic manner.

In December 2000, John Thompson was awarded the National Medal of Science by President Clinton. I had the pleasure of attending the award ceremony in Washington DC. Upon hearing the news of the award, I approached the Dean and said the now would be the time for the University to recognize Thompson, and the best way to do that would be to create the John Thompson Research Assistant Professorship. Actually, I proposed this to the Dean one night at a party in the University President's House after a football game that the university won! I am no fan of football, and have never attended a game, although most citizens of Gainesville are madly in love with them. I knew that the Dean would be at the party, and it would be beneficial to propose this when he and all administrators would be in a good mood. And as I expected, the idea worked. The Dean agreed, but the question remained as to how to fund this.

My proposal was for the John Thompson Assistant Professorship to be a three year position, offered every year, so that in a steady state, there would be three such assistant professors in the Department. This meant that we needed money equivalent to three junior faculty lines. The College did not have funds for this. So I proposed that one line be given by the College, one line by the Provost, and the equivalent of one line be given by the office of the Vice-President of Research. After all, Thompson was without a doubt the most eminent faculty member not just in the university, but in the entire state university system of Florida, like the legendary physicist Dirac was when he was at Florida State University! Also Thompson was bringing such great recognition for the University with the award of the National Medal of Science, and so I felt that the upper administration in the university would be glad to support this assistant professorship. And indeed Provost David Colburn and Vice-President of Research Win Phillips both gave their nod for the position to be launched in Fall 2001. The question now was who was going to give the money in the very first year? The answer came is a very surprising way.

One of the big money makers for the University of Florida is *Gatorade*, a drink which supplies all the essential salts to athletes. Gatorade is sold worldwide, and the revenue for the university from Gatorade is significant. To my pleasant surprise, Vice-President Win Phillips said without any hesitation, that the support in the first year would come from the Gatorade funds! It has often been said that the money amassed by the football program could be helpful to fund academic programs. This is just a dream! But here was an instance when Gatorade money was used to support a mathematics research assistant professorship, at least for the first year. And it did start the program as desired. But the administration as a whole — the College Dean's office, the Office of Research, and the Provost's office, told me that positions such as these ought to be funded with private money.

When philanthropists give donations to the university or the college it is either given with (i) a specific target in mind, or (ii) given without any conditions attached. In the case of (i), it is very rare to find a donor who will target the mathematics department for the benefit, and that too a research assistant professorship. We did have one donor (by name Chandler) who gave funds directly to the Department for the Thompson Assistant Professorship, and supported one assistant professor for three years; this pleasant outcome was due to the efforts of my enterprising colleague Tim Olson who cultivated a relationship with the donor. Chandler wanted his name associated with Thompson and the professorship and Thompson graciously agreed to this. So it was called the Thompson-Chandler Assistant Professorship when Chandler supported it.

In the case of (ii), even though the donation could be used for any purpose, seldom will the university or the college make the mathematics department as the beneficiary because, sadly, mathematics is generally not on their list of priorities!

The John Thompson Assistant Professorship was offered every year until my term as Chair ended. In Spring 2006, we had an External Review of the Mathematics Department. The stellar review team was much impressed with all the new initiatives in the mathematics department. In particular, the review team recommended that not only the Thompson Assistant Professorship be continued, but that more of these should be offered each year. Contrary to the recommendations of the eminent panel of external reviewers, the John Thompson Assistant Professorship was terminated in 2008 when I stepped down as Chair, even though Thompson won the Abel Prize in May 2008! The administration said that the decision was made on budgetary reasons, but when the budget situation improved a few years later, there was no interest on the part of the administration to revive the Thompson Assistant Professorship.

8.6 The AMS Chairs Workshops

The American Mathematical Society has for many years conducted a one day Workshop for Chairs at their Annual Meetings. I have attended these workshops every year during the ten years I was Chair. Even though nowadays one can get information from websites about any university, it cannot compare to the insight one gets by direct conversation with the administrators at other universities. Thus I went to these Workshops for Chairs mainly to talk to chairs of other mathematics departments, to find out what issues they are dealing with, and how they tackle them. The AMS puts together a very informative program for the day with presentations both by chairs as well as higher administrators who were chairs previously. Those presentations were also immensely beneficial to me.

The first Chairs Workshop that I attended in January 1999 in San Antonio was organized by John Conway, Jim Lewis, and Douglas Lind, all of them extremely successful chairs.

I had used Conway's fine book on Complex Analysis when I taught the graduate course in that subject in Florida and so was aware of his accomplishments as a researcher and writer. I also knew that he was a successful Chair at the University of Tennessee at that time. He shared with us his approach in dealing with thorny faculty issues, and how he builds the research programs. His book "On being a department head - a personal view" (AMS 1991), was enjoyable to read and useful.

Douglas Lind was Chair at the University of Washington in Seattle. He has a cheerful and genial personality. He told us how important it is to understand the Dean's mind, and make proposals in ways that would appeal to the Dean. He also said that in order to be successful in recruiting faculty, he would give each candidate being considered for a job offer, a tape or CD of the beautiful landscape of the State of Washington!

Jim Lewis was Chair Extraordinaire! He revolutionized the atmosphere in the mathematics department of the University of Nebraska at Lincoln by introducing a number of successful programs encompassing research, teaching, and outreach, as well as some special programs for women and minorities, such as the Nebraska Conference for your women researchers. He gave us very helpful suggestions as to how to convince the adminstration to support programs in mathematics.

In 1991, the AMS had brought out a book "Towards excellence: leading a doctoral mathematics program in the 21st century", which was written by a Committee of leading mathematicians and successful chairs. Conway, Lewis and Lind have contributed valuable articles to that book.

During the decade that I attended the Chairs Workshops, John Ewing was the Executive Director of the AMS and Sam Rankin III was the Assistant Director who was at the AMS Washington Office. Rankin often testified before the Congress making a case for greater support for mathematics. He therefore played a very crucial role. It was Rankin who coordinated the Chairs Workshops and I got to know him very well. I was an active participant at these Workshops, and so Rankin asked to me to be one of the three main organizers of the workshops for a period of three years (2003–06), an experience I found both enjoyable and fruitful. Rankin used to take the workshop organizers to dinners at high class Zagat rated restaurants, and I enjoyed his warm hospitality very much.

Similar in spirit to the AMS Chairs Workshop was the BMS Chairs Colloquium, a two day meeting (Nov 8–9, 2002) at the National Academy of Sciences in Washington DC organized by the Board of Mathematical Sciences of the NSF. There were talks at this meeting about various programs of the NSF and it was an opportunity for Chairs to learn about them so that they could identify which might suit their departments. There were a few invited talks by Chairs, such as by Peter March who told us how Ohio State University developed a biomath program and eventually received funding from the NSF for the Mathematical Biosciences Institute. I was invited to speak alongside Peter March. My talk was entitled "Enhancing visibility and strengthening links with other disciplines", in which I gave a summary of various programs at Florida and how it gained international visibility for us and strengthened ties with different science departments on campus.

8.7   George Andrews in Florida and the Ramanujan Colloquia

George Andrews' close association with the University of Florida in three decades has had an immense effect on the Mathematics Department in general, and the Number Theory Program in particular. I relied on his help and advice to recruit Frank Garvan and Alexander Berkovich in my effort to build the number theory program. With his regular presence, and with the work of the three of us, the Department became a major center for research in the theory of partitions, q-hypergeometric series, modular forms and the work of Ramanujan. It is also due to his sponsorship that the annual Ramanujan Colloquia are held. Andrews is not only a world class researcher, but also an outstanding teacher and a successful administrator, having served two terms as Chair of Mathematics at The Pennsylvania State University. In addition, he has contributed significantly to issues on mathematics education and served the profession admirably in leadership roles such as being President of the American Mathematical Society. All this had an impact on our department in ways more than one. Here I shall describe briefly various aspects of Andrews' association with the Department and the significant positive effects this has had. Details of events that I relate here may be found in the book in various chapters.

Andrews' first visit to the University of Florida was in 1990 to deliver the *Frontiers of Science Lecture*. He spoke about the discovery of Ramanujan's Lost Notebook and charmed the audience of more than 700 that had assembled at the University Auditorium to hear him. But this first visit of his also had a major effect on my research because in the three days that he was in Gainesville, he gave an incredible proof of a remarkable *key identity* that Basil Gordon and I had found for a refinement and generalization of a deep partition theorem of Göllnitz. This resulted in a major paper that the three of us wrote and that started my collaboration with him on several related projects.

When I arrived at the University of Florida in December 1986, I was the only researcher in number theory. I was asked to help the Department build a program in number theory. After the Ramanujan Centennial, the focus of my research shifted to the theory of partitions and q-series, and so I was on the lookout for someone who would work in this area but have expertise in modular forms which I did not have. It was with Andrews' help that I was able to recruit Frank Garvan,, his former PhD student. Details as to how this appointment took place are given in Chapter 7.

In 1998, when the opportunity arose for me to become the Chair of the Mathematics Department, Andrews encouraged me to accept the position and congratulated me when the appointment was made. He knew of my ideas to stimulate the productivity of the Department and increase its visibility internationally. Andrews had served successfully as a Chair, and so unlike many academicians who view Chairs as "operators" with hidden agendas, he knew that a good chair can raise the department's performance. Of course he cautioned me that the responsibility would be heavy, and so I should make sure that I have enough time for research. During my ten years as Chair, I have consulted him many times on both pleasant and thorny issues.

In 1998, as soon as I became Chair, Andrews drew my attention to Alexander Berkovich who was finishing a post-doctoral position at Penn State that calendar year. Berkovich is a mathematical physicist who had turned his attention to q-hypergeometric series due to his collaboration with the eminent physicist Barry McCoy and his group. I was able to offer Berkovich a visiting position to start in Spring 1999. We immediately started collaborating on the difficult problem of Andrews whether there exists a partition theorem lying beyond Göllnitz's theorem, just as Göllnitz's theorem is the next level theorem beyond Schur's. A correspondence with Andrews followed, and in Spring 2000 this difficult problem was resolved in the affirmative. It resulted in a triple paper that was accepted by *Inventiones Mathematicae*. In view of this breakthrough, it was possible to convert Berkovich's position into a permanent appointment. Thus Andrews played a crucial role in the appointment of Berkovich.

After Garvan arrived at the University of Florida, he and I organized conferences regularly in the areas of partitions, q-series, and modular forms. At all these conferences, Andrews was a principal speaker. He also visited Florida often to deliver colloquia and special history lectures.

Andrews has spent considerable time in addressing issues on mathematics education. He has written articles on the negative effects of calculus reform. His ideas on mathematics education are so highly regarded that he was invited to deliver an address on this topic at the International Congress of Mathematicians in Berlin in 1998. Taking all this into account, he was invited more than once by our department to address the Teaching Seminar and also to address the College Dean's Seminar on Education.

In view of his eminence in the world of mathematics and his strong ties with the mathematics department, the University of Florida awarded him an Honorary Doctorate in 2002.

Starting from 2005, an arrangement was made so that he could spend the entire Spring Term in residence in Florida by doubling his duties in the Fall Term of each year at Penn State. Andrews and his wife stay in Belleview which is an hour south of Gainesville and an hour north of Tampa where his second daughter Katy lives with her family. Thus it is a one hour drive in each direction for him to come to Gainesville or to visit his daughter. When he is residence in Florida, he comes to campus every Tuesday to attend the Number Theory Seminar without fail, and on many Mondays to attend the weekly department colloquium. When he comes to campus, he spends time in discussions with Berkovich, Garvan and me.

What is most impressive about Andrews is that his research productivity has remained unabated defying the passage of time. Two of his recent most significant discoveries happened during his springtime visits to Florida. The first was his work on *Durfee symbols* and the second was on the remarkable *spt function*, where $spt(n)$ is the number of occurrences of the smallest part among all partitions on the integer n. Both works have had a huge impact and I am proud to say that his first announcements of these two discoveries were in the Florida Number Theory Seminar.

In January 2006, at the start of his second full spring term in Florida, Andrews told me that he was very much impressed with many new programs I had introduced in the department, and that he would like to make an annual contribution to launch another program. He asked my suggestion on this. I responded immediately that in view of his regular presence each year in the Spring Term, we should launch *The Ramanujan Colloquium* to complement the highly successful Erdős and Ulam Colloquia that we have. The Ramanujan Colloquium is to be on areas of mathematics influenced by Ramanujan in the broad sense. I said that the Ramanujan colloquium speaker should start with a Public Lecture of wide appeal following which the speaker should give two or three technical seminars. I suggested a handsome honorarium which I felt would be appropriate for high profile speakers to deliver these lectures. Without the slightest hesitation Andrews said YES to my suggestion and asked me to start the Ramanujan Colloquium the very next academic year in Spring 2007. And indeed we have been able to attract the most eminent mathematicians year after year as the following list shows:

List of Ramanujan Colloquium Speakers: 2007 - *Manjul Bhargava* (Princeton University); 2008 - *Peter Sarnak* (Princeton University and the Institute for Advanced Study); 2009 - *Dorian Goldfeld* (Columbia University); 2010 - *Kannan Soundararajan* (Stanford University); 2011 - *John Thompson* (Cambridge University); 2012 - *Ken Ono* (Emory University); 2013 - *Freeman Dyson* (The Institute for Advanced Study, Princeton); 2014 - *Peter Paule* (Johannes Kepler University, Linz, Austria); 2015 - *Robert Vaughan* (The Pennsylvania State University); 2016 - *James Maynard* (Oxford University); 2017 - *Peter Elliott* (University of Colorado); 2018 - *Ken Ribet* (University of California, Berkeley); 2019 - *Alex Lubotzky* (Hebrew University, Jerusalem).

Since the speakers in this colloquium series are all world class mathematicians, we have had a few active researchers outside our University, even some from overseas, come to Gainesville to attend these lectures. To reciprocate the generosity of Andrews and The Pennsylvania State University in sponsoring the Ramanujan Colloquium and seminars, the mathematics department has annually provided a matching amount to support some of these visitors. In particular, each year a few students of the Penn State MASS Program, or graduate students at Penn State University, have attended this colloquium with such support. In summary, the research atmosphere in the Department has been significantly stimulated owing to the inspiring presence and support of George Andrews.

8.8   John Thompson in Florida

The most significant appointment at the Mathematics Department of the University of Florida was that of John Thompson as Graduate Research Professor. Thompson's presence helped the Department make several fine appointments in group theory and thereby made it one of the premier centers in the world in group theory. Also, the National Medal of Science that he received in 2000 and the Abel Prize that he won in 2008 brought worldwide attention to the Department and the University. Thompson's tenure in Florida as Graduate Research Professor was in two separate periods: It was a half-time position (at his request) only in the Fall, starting in Fall 1986. He relinquished this position in 1988, but five years later, David Drake, in his final year as Chair, was able to get Thompson to accept the position full-time in 1993 — that is spend the Fall and Spring terms in Florida, returning to Cambridge in the summers. Here I describe briefly some of the major developments in the department that took place owing to his presence.

My colleague Chat Ho (a former PhD student of Thompson) was appointed on the Search Committee in 1986, and three very active group theorists were recruited in the next two years — Geoffrey Robinson from England, Helmut Voelklein from Germany, and Alex Turull from the University of Miami. Thompson's association with the department was definitely an attraction for these three.

When Thompson accepted the appointment in Florida, his research focus was on the Inverse Galois Problem, namely the important problem of deciding which groups arise as Galois groups of field extensions. So at Thompson's suggestion, Mike Fried of the University of California, Irvine, was appointed since Fried and Thompson had begun a collaboration on the inverse Galois problem. Fried's appointment began in Fall 1986 coinciding with Thompson's first term in Florida.

The members of the combinatorics group, led by David Drake, decided that the combinatorics seminar, the longest running seminar in the Department, would focus on theta functions and the modular group in view of Thompson's involvement in the inverse Galois problem. As a new faculty member, I took an active part in the combinatorics seminar, and I gave a few talks on Jacobi's theta functions and the construction of elliptic functions from ratios of theta functions. I was pleased that Thompson appreciated my lectures and said that the style of my delivery was like my father's! Helmut Voelklein who had just joined as an assistant professor, turned the focus of his research to the inverse Galois problem, collaborated with Thompson and Fried, and in a few years became known as an authority on this topic. That is an example of how the presence of Thompson molded the research programs of those who interacted closely with him.

Hiring in algebra continued: Peter Sin (group theory) joined the Department as assistant professors in 1989, as did his wife Jan Cheah (algebraic geometry), thanks again to the effort of Chat Ho on the Search Committee. Thus when Thompson returned, there was a vibrant program in group theory, a program that was launched in 1986 owing to his inspiring presence.

Thompson and his wife Diane stayed in a furnished apartment a few blocks south of campus. Diane was in Gainesville only part of the time each year because she had a teaching position in Cambridge. So Thompson was by himself most of the time. He did have a car, but from time to time, Chat Ho and David Drake assisted him when he needed help.

Whenever Thompson lectured, he never used any notes. His writing on the board was exceptionally clear, and he never made corrections on anything as he wrote. This indicates his clarity of thought and the depth of his understanding. I heard that the great S. S. Chern was like that when he lectured on differential geometry.

Thompson attended the Algebra Seminar regularly and spoke from time to time in that seminar. I remember a beautiful lecture he gave on Eisenstein series. He did not attend the Number Theory Seminar regularly, but he was present in most my lectures there. In 1989, I had shifted the focus of my research to the theory of partitions and q-series, and so in the early nineties I was putting forth a number of new ideas in that area. After one of my talks on the modified convergence of continued fractions of Rogers-Ramanujan type, Thompson suggested that I send my paper to a certain "bright young" mathematician at Cambridge University. That mathematician whom Thompson referred to as bright and young, was Richard Borcherds, who a few years later won the Fields Medal!

One thing I noticed about Thompson was that whenever he attended a colloquium, seminar, or a faculty meeting, he would be in the room ten minutes before the start of the event, often by himself until others slowly came in one by one. He was never late. These little things all indicate his sense of commitment and how meticulous he is.

My interaction with Thompson was closer after I accepted the chairmanship in 1998. Thompson was on the Chair Search Committee in 1998, and his endorsement of me both in the Search Committee and to the Dean was crucial in the Chairmanship being offered to me.

In the Spring of 98, based on the suggestion of Thompson and Chat Ho, the Department made an offer to a bright young algebraist by name Pham Tiep, who joined us that Fall as an assistant professor. So when I began my term as Chair, the Department had a stellar group in algebra, and it was a pleasure for me to work with the algebraists and Thompson to conduct several outstanding programs. I shall now describe my interactions with Thompson during my term as Chair from 1998 to 2008.

As Chair, I initiated a number of new programs aimed at stimulating the research environment and productivity, and to bring increased visibility for the Department both on campus and internationally. John Thompson endorsed every one of these programs. The first program was a pair of Distinguished Colloquia — The Erdős Colloquium in pure mathematics and the Ulam Colloquium in applied mathematics. The idea was to bring mathematical luminaries to the Department to deliver public lectures of wide appeal. Great as this idea was, since these eminent mathematicians are very busy, I felt that not all would accept our invitation. Chat Ho suggested that I request Thompson to co-sign the letter of invitation from me, and I felt it was an excellent idea. So I approached Thompson with this request and he gladly agreed. With Thompson's signature, the invitation would be hard to decline; indeed every mathematician we invited to address these colloquia, accepted our invitation.

One of the greatest recognitions for the Department was the award of the 2000 National Medal of Science to Thompson. It was thrilling to get a phone call in November 2000 from the News and Public Affairs office of the University of Florida, that Thompson will be awarded the Medal in Washington DC in December that year. It was my pleasure and honor to provide News and Public Affairs with all background information about Thompson and his accomplishments, for a Press Release. This was a big recognition for the Department, the College and the University, but it was simply one more in the list of recognitions for Thompson. (For details on the award of the National Medal of Science to Thompson, see §9.3.)

Thompson's 70th birthday was in Fall 2002 and so we declared the year 2002–03 as the Special Year in Algebra in his honor. The Department had launched the Special Years Program in 2001–02, and it was natural to devote the second year of this program to algebra. For details of the activities of the Special Year in Algebra, see §9.5.

Just before my term as Chair concluded, Thompson was awarded the 2008 Abel Prize jointly with Jacques Tits. I had the privilege of attending the Abel Prize events in Oslo, and to serve as one of the Press Contacts. The Abel Prize to Thompson was the greatest recognition for the University of Florida in its entire history. For a description of the Abel Prize Ceremony and related events in Oslo, see §11.2. For an account of the festivities at the University of Florida for Thompson's Abel Prize, see §9.11.

In 2010 Thompson retired from the University of Florida and returned to live in Cambridge, England. Without a doubt, his tenure in Florida was the greatest chapter in the history of the University. It was privilege and honor for me to have had close association with John Thompson, one of the most eminent mathematicians of our time.

Benedict Gross (Harvard University) delivering the Erdős Colloquium at the University of Florida on February 24, 2003, during the Special Year in Algebra (2002–03) in honor of John Thompson's 70th birthday.

George Andrews in cap and gown after receiving an Honorary Doctorate at the University of Florida, Dec 21, 2002. Mrs. Joy Andrews is next to him.

Krishna and Mathura with Fields Medallist Alan Baker, at Gonville and Caius College, Cambridge University, after the banquet in honor of John Thompson's 70th birthday, September 28, 2002. Baker and Thompson both received Fields Medals in 1970.

Chapter 9

Riding the wave as Chair: Ups and downs

9.1 My honeymoon year as Chair (1998–99)

The first year as Chair is called a Honeymoon Year because the administration is very supportive and is willing to excuse you for mishaps as long as they are not major. It was a very successful year except for one matter — I exceeded the allotted budget! It was not by a large amount, and so I was excused by Dean Harrison because the excess expenditure was for good reasons. Actually, the year was filled with exciting events.

Within a week of the start of the Fall term, I had to depart for Maratea, Italy, for the George Andrews 60th Birthday Conference. This fantastic conference (see §11.4 for details about this conference and ALL the Andrews milestone birthday conferences) at an enchanting location on the Amalfi coast of Italy, was organized by Dominique Foata, one of the leaders of combinatorics in France. For now I will say that it was at this conference that I first met Alexander Berkovich, a mathematical physicist who had turned his attention to the theory of partitions and q-hypergeometric series. Berkovich was completing his visiting position with George Andrews at Penn State and so I invited him to come to the University of Florida in Spring 1999 as a visitor. This was the beginning of his association with Florida where he is now at the rank of Professor. There was another conference for Andrews' 60th birthday at Penn State University in October 1998 that I attended (see §11.4).

Since it was my first year as Chair, there was a lot for me to learn. Also I was slowly getting used to the demands of time that the chairmanship entailed. So I did not go to India in December 1998. However, my senior colleague and good friend James Keesling went to India in December–January. In Madras, my father played host to him. My father knew Keesling very well because he and I used to play tennis doubles with Keesling and my colleagues Louis Block and Doug Cenzer quite often in Gainesville. In Madras, Keesling attended the Indian Science Congress which is always inaugurated by the Prime Minister, a tradition begun by India's first Prime Minister Jawaharlal Nehru, a great patron of science. My father arranged for Keesling to deliver the Ramanujan Endowment Lecture at Anna University in Madras, and also introduced him to South Indian classical music and dance by taking him to concerts at the Madras Music Academy. Keesling, who is very friendly and outgoing, is always keen to learn about other cultures and traditions. So he thoroughly enjoyed his visit to India.

For the Academic Year 1998–99, the most significant events for the Department all happened in the Spring 1999 term.

James Keener of the University of Utah delivered the First Ulam Colloquium on Monday, January 11 on "The mathematics of sudden cardiac arrest". It was a resounding success because it was attended by faculty and graduate students from several science and engineering departments, as well as from the medical school. Dean Harrison made the Opening Remarks and stayed for the entire duration of the talk. The next day, we sat with Keener to go over various applications for the position in biomath we had advertised. It was with his advice that we appointed Sergei Pilyugin to start our biomath program in Fall 1999. Pilyugin, who helped us with subsequent hiring in biomath, is now a full professor in the department.

One of the research centers we have at the University is the Institute of Fundamental Theory (IFT), which links the physics and mathematics departments. It was founded, and is headed by, physics professor Pierre Ramond, who happens to be a grandstudent of my father! From time to time, the IFT holds a Colloquium in the mathematics department. On February 23, Shlomo Sternberg of Harvard University delivered a superb lecture on various aspects of mathematics that were influencing current research in physics. In March, Bert Kostant of MIT, another IFT visitor, addressed the mathematics colloquium. I was much impressed with the lucid presentations of Sternberg and Kostant. Their visits were co-sponsored by the mathematics department and the IFT. Sternberg and Kostant had just completed writing a major paper on the Weyl character formula with Benedict Gross of Harvard University and Pierre Ramond, and it was that collaboration which led to their visit to Florida in Spring 1999.

An unusual event took place two days after Sternberg's IFT Colloquium. At the Florida — South Carolina basketball game at the O'Connell Center, Associate Chair Neil White (now deceased) was presented a cheque for $30,000 by GTE in front of an audience of 10,000 during the half-time festivities! This cheque was for a grant for a summer program for minority students that our department had applied for in Fall 1998 due to the initiative of Neil White. Dean Harrison was especially pleased that we applied for this grant and were awarded in such a dramatic fashion!

Starting in late February, we had the one month visit of the famed algebraic geometer Shreeram Abhyankar from Purdue University. He was a guest of John Thompson. Abhyankar believed in a nuts and bolts approach to algebraic geometry, and criticized the use of very heavy machinery or abstraction. In particular, he was not a fan of the Grothendieck program in algebraic geometry, and therefore did not agree with the viewpoint of the mathematicians of the Tata Institute who were followers of the Grothendieck approach. Abhyankar had expounded his views on algebraic geometry in a wonderful article entitled "Historical ramblings in algebraic geometry" that was published in the American Mathematical Monthly in 1976, for which he received the Chauvenet Prize in 1978. I was well aware of all this, and so I asked him to give a talk to our Undergraduate Mathematics Club $\pi - \mu - \epsilon$. He gave a marvelous talk in which he stressed that algebraic geometry is really analytic geometry in deeper and more general terms. He lamented that not enough time is spent nowadays in teaching analytic geometry properly to high school students. My father who was a close friend of Abhyankar, enjoyed his $\pi - \mu - \epsilon$ talk very much.

A few days after the talk, my family (including my parents), John Thompson and his wife Diane, and Abhyankar and his wife Yvonne, had dinner together. During the dinner, we were talking about Abhyankar's "elementary" treatment of algebraic geometry. In that connection, I said that in number theory, the champion of the elementary approach is Paul Erdős. To this Abhyankar immediately shot back: "I do not consider Erdős to be a great mathematician." I was shocked to hear this, and asked Abhyankar why he had formed that opinion of Erdős. He replied that Erdős had published some very simple papers. I then said that a mathematician must be judged by his greatest contribution or his total contribution, and not by his least significant paper (see my discussion on Selberg in §5.3). Abhyankar then asked me to give some examples of great theorems proved by Erdős. I therefore provided a few examples. At this point Abhyankar asked how I knew so much about Erdős' work. I responded that Erdős was like a mentor to me, that I had collaborated with him, and studied many of his papers. On hearing this, Abhyankar exclaimed: "Oh! Erdős is your *Guru*! Therefore I withdraw everything I have said because I should not criticize your Guru in your presence. For instance, I would defend my Guru (Oscar) Zariski against anyone."

Abhyankar withdrew not because he felt I was correct, but out of great respect for the Guru. In Indian culture, the Guru is venerated, and almost worshipped, and Abhyankar had a great regard for the culture of India.

During March 12–13, 1999, the Southeastern Sectional Meeting of the American Mathematical Society was held on campus, and it attracted about 400 participants. Planning for this meeting began long before; AMS Secretary Bob Daverman visited the University in Fall 1998 to discuss the organization of the meeting. It was a major recognition for the Department that two of the four one hour addresses were delivered by our colleagues — 1970 Fields Medallist and Graduate Research Professor John Thompson, and Alexander Dranishnikov, a world authority in low dimensional topology, who in August 1998 had delivered an invited address at the International Congress of Mathematicians in Berlin. Dranishnikov spoke on "Dimension theory and the Novikov conjecture." Dean Harrison made the Opening Remarks for Thompson's lecture on the theme "Double cosets of the upper triangular group", and I had the privilege of introducing Thompson.

Many faculty members of the Department conducted Special Sessions at this meeting. In fact, out of the sixteen special sessions at this meeting, fourteen were organized by our department faculty, some in collaboration with mathematicians from outside. It is customary to have a special session associated with each of the hour lectures. Thus Thompson and Helmut Voelklein organized a special session on Galois theory; there was a second special session organized by Chat Ho and Peter Sin on groups and geometries. Similarly, Dranishnikov and Keesling organized a special session in geometric topology. It would be twenty years before the next AMS Sectional Meeting would be held at the University of Florida.

Since Paul Erdős had passed away in 1996 and he visited Gainesville annually, Jean Larson and I organized a special session on the legacy of Erdős. Jean gave the first talk in the session giving an account of Erdős' annual visits to Florida and the influence it had on the Department. This session featured talks in number theory, combinatorics, and set theory, which are areas in which Erdős had made fundamental contributions and had collaborated with the faculty in Florida.

Immediately after the AMS Meeting, on Monday, March 15, we had the First Erdős Colloquium by Ronald Graham of AT&T Bell Labs and the University of California San Diego (UCSD). Graham, who was a combinatorialist of world repute, was one of the leaders in the field of Ramsey Theory. Graham was a close friend and collaborator of Paul Erdős, and indeed it was due him that Erdős used to visit AT&T Bell Labs so often.

Graham handled all of the finances of Erdős, including the cheques that needed to be given every time an Erdős problem was solved. Graham also was an outstanding speaker, full of sparkling wit that sent audiences roaring in laughter. Thus it was appropriate to invite Graham to deliver the First Erdős Colloquium.

Some of the speakers of the special session that Larson and I conducted extended their stay to attend Graham's lecture, such as Vera Sös, Jean-Louis Nicolas and John Selfridge. Vera Sös of the Hungarian Academy of Sciences, a leading figure in combinatorics and graph theory, was visiting Bell Labs to work with Graham. She came down to Florida with Graham, and on the Tuesday after Graham's colloquium, and spoke in our combinatorics seminar. Vera Sös, Nicolas, and Selfridge were all close friends and collaborators of Erdős. I had met Graham and Selfridge first at UCLA in 1975 when I was a graduate student there; they had come to visit Erdős who would spend every December at UCLA. On this visit to Florida, Selfridge was accompanied by Ethel Rathbun, the former Chair's Secretary at the University of Michigan. Mathura and I knew Ethel from our Michigan days. After she retired as the Chair's Secretary, Ethel took care of Selfridge and his finances; she managed the accounts of the Number Theory Foundation that was founded by Selfridge. Mathura and I hosted a party at our home in honor of Ron Graham and the visitors in number theory, a party that was attended by many of our colleagues and their spouses. My parents were in Gainesville for their annual three month stay, and it was a pleasure for them to meet our distinguished visitors. My father attended several lectures of the AMS Sectional Meeting, including the hour talks by Thompson and Dranishnikov, and some of the talks in the Special Sessions.

In the end of March, I made a short trip to Penn State University, to speak at the Seaway Number Theory Conference (March 27–28) organized by Scott Ahlgren and Ken Ono. I understand that the name "Seaway" for the conference derives from the St. Lawrence river seaway, and the conference had a tradition of being organized at places around the seaway. It was an informal conference. There were no plenary lectures; all talks were for thirty minutes or less. My thirty minute talk at this conference was on my joint work with Basil Gordon on generalizations and refinements of the two Andrews hierarchies of partition theorems.

Upon my return from Penn State University, I had the pleasure of receiving Prof. Ingrid Daubechies of Princeton University, a world authority on wavelets. She gave the "Women in Science Lecture", and the Opening Remarks for her talk were made by Provost Betti Capaldi.

A few days after this great wave of significant events, my father was hospitalized due to a heart condition. He had to undergo an emergency bypass surgery in mid-April which caused considerable anxiety for our family. The presence of Dr. Ravindra in Lake City, a highly regarded cardiologist, and a relative of ours, gave courage to my father to have the surgery here in Florida instead of in India. The bypass surgery was done by the highly competent Dr. Wesley. Prior to the surgery, my father asked Dr. Wesley what the chances of success are. The doctor said that it was about 95%, and he felt that my father would fall in that 95% range and not the 5% failure range! Thankfully, the surgery was successful, and my father returned home two weeks later to recover. Indeed, he recovered quite rapidly and was fit enough to travel back to India with my mother, Mathura and my daughters, on June 7. Since he did recover well, we were in a good mood to celebrate my older daughter Lalitha's high school graduation with a party in June before my parents departure. My father gave a fine speech at this event which was appreciated by all the guests.

Shortly before my departure to India in mid-June, I had a meeting with Dean Harrison to go over the budget request for the next academic year. It turned out that I had exceeded the 1998–99 budget allotted by him, but he realized that all the expenditures were for worthwhile causes. Since this was my first year as Chair, and I had not mastered the art of managing the budget, he excused me for this lapse, but cautioned me not to go beyond the budget; he said that if I needed more funds, I should approach him early, and he would consider providing the additional amount, if appropriate.

I had a full schedule of talks in India. Once again, Professor Ramachandra arranged my lecture at the TIFR Centre in Bangalore. In Madras, I gave several lectures at Anna University, the Ramanujan Institute, the Swaminathan Research Foundation, the IIT, and Vivekananda College. The talk at Vivekananda College was special for me since I was inaugurating the activities of the Mathematics Club for the year 1999–2000, a quarter century after being a student there!

In between my stay in India, I made a trip to Budapest (July 4–11) to attend and speak at the Erdős Memorial conference. For a discussion of this conference and the Erdős Centennial Conference, see §11.5.

On the social side, the main event in India was the *Arangetram* (full two hour graduation dance performance) on July 24 of my daughters Lalitha and Amritha conducted under the guidance of Mathura and her dance teacher. The Arangetram is extremely important in *Bharathanatyam* (South Indian classical dance). In traditional Indian society, a teacher would not let the student perform in public, until the Arangetram is completed. The Arangetram was at the Madras Music Academy and it was a great opportunity for Lalitha and Amritha to perform before a very discerning audience.

A few days after the Arangetram, I returned to Florida with my family to begin work for the 1999–2000 academic year that proved to be just as exciting as my "Honeymoon Year" as Chair.

9.2   Into the new millennium (1999–2000)

Within a few days of my return from India, and before the start of the Fall term, during August 6–7, I attended a Leadership Conference organized by the AMS on the campus of Indiana University in Bloomington. It was nice to meet Don Lewis, who as the Director of the Mathematics Division at the NSF, was attending the meeting and involved in its organization. Lewis always had innovative ideas, and it was during his term as Director, that the mathematics division of the NSF launched programs like VIGRE and IGERT. At this Leadership Conference, on the opening day, I was asked to lead a session entitled "Development - a new frontier" which emphasized fund raising as crucial and beneficial for departments. Colleges in universities do their own fund raising and have Development Officers who work closely with the College. In this session departments were urged to do their own fund raising. I was asked to summarize the discussions of the session.

Back at the University of Florida, we had a new President. It was none other than Charles E. Young, who was Chancellor at UCLA when I was a graduate student there! He had admirably served as UCLA's top administrator for 21 years. On November 2, I had the pleasure of meeting him at the University Auditorium, and four days later, Mathura and I were at a party at the President's home. I still cherish the 1975 letter from him offering me the Chancellor's Fellowship at UCLA for my graduate studies.

The mathematics department of the University of Illinois, Urbana, was celebrating the academic year 1999–2000 as the Millennial Year with several conferences and lectures. I was invited to give an hour lecture in September at the Illinois Number Theory Conference which started the activities of the Millennial year. One day before the conference, I addressed the mathematics colloquium on the theme "A fundamental invariant in the theory of partitions". My host Bruce Berndt hosted lunches and dinners for me at Indian restaurants in Urbana and Champaign throughout my three day stay there. After my colloquium, it was gracious of the Halberstams to host a party at their lovely home.

In Urbana it was a pleasure to meet Professor Walter Hayman from

Imperial College. He addressed the Analysis Seminar an hour before my colloquium. As a school boy, I had met Hayman in the mid-sixties when he visited MATSCIENCE, and he recalled the good game of table tennis we had played at our home in Madras when he came for dinner.

Here in Florida, in mid-November, we had a highly successful conference on "Symbolic computation, number theory, special functions, physics, and combinatorics" owing to efforts of Frank Garvan. The conference was well funded by the NSF and the NSA since it emphasized work at the interface of these fields. Dean Harrison inaugurated the conference following which George Andrews gave an outstanding opening lecture. Some of the other one hour speakers were Dennis Stanton (Univ. Minnesota), Sergei Suslov (Univ. Arizona), and Doron Zeilberger (then at Temple University) who gave a brilliant and hilarious lecture. Barry McCoy from Stonybrook, a world famous mathematical physicist who had just been awarded the Heineman Prize, spoke about his prize winning work: "Rogers-Ramanujan identities and conformal field theory in physics". Both Dean Harrison and Associate Dean of Research Neil Sullivan, were much impressed with the conference, and so the College Newsletter had a prominent report of it. The refereed proceedings of this conference edited by Frank Garvan and Mourad Ismail appeared two years later as Volume 4 in my book series Developments in Mathematics (published by Kluwer at that time).

Some conference participants stayed in Gainesville for a few days after the conference. One of them, Richard Lewis (University of Sussex), gave a colloquium on the Monday after the conference. He was an expert in the theory of modular forms and had close interaction with Frank Garvan.

During November 20–27, I made a trip to China, the first of many for me to that enchanting country steeped in history. It was an academic visit because I was invited to give lectures at several institutions in Beijing and Shanghai, but the country has so much to offer by way of sightseeing, that I decided to take Mathura and my second daughter Amritha with me. My older daughter Lalitha had just joined the University of Florida and so she could not come; an opportunity for her to visit China with us was to come in the summer of 2002. On hearing about our plans to visit China, Mathura's aunt Gigi and uncle Dr. Ramakrishnan of the Harvard Institute of Development, decided to join us.

We arrived on a Saturday, and so we had all of Sunday free before my talks on Monday. So we took a bus tour to the Great Wall of China, which has to be seen to be believed. The height of the massive wall of stone and its length spanning more than half of China as it meanders over mountainous terrain, is staggering! We all had our share of exercise by walking on the wall up a mountain slope, but it was cool, and so we were not exhausted since we were not perspiring. Upon return to Beijing, we were treated to a magnificent dinner by our host Professor Jia, who took us also to a lovely Chinese Opera.

On Monday, I gave a lecture in the morning at the Mathematics Institute of the Academia Sinica arranged by Professor Jia. From there I was taken to Peking University where my host was Professor Chengdong Pan. Both the Academia Sinica and Peking University have a great tradition in analytic number theory going back to L. K. Hua, who had studied under G. H. Hardy at Cambridge University. In fact Pan was a PhD student of Hua. While at Peking University, our hosts took us to a specially arranged banquet on campus where we were served a vast array of vegetarian dishes so well prepared and elegantly presented. Following that, we were shown the exquisitely beautiful summer palace which has an enchanting location on the shores of a lake. In the evening we were taken to the historic Lao She Tea House where we witnessed a Chinese play as we sipped Chinese Tea.

No visit to Beijing is complete without seeing the Forbidden City (the great palace). From the expansive Tiannenmen Square, as we entered the Forbidden City, we saw a gigantic portrait of Chairman Mao adorning the walls, reminding us of the stern communist regime under Mao as a stark contrast to all the opulence we were to witness inside the palace. Jia spent the entire day with us at the Forbidden City and at night took us to a Chinese Acrobatic Show which took our breath away.

My contact with Jia was from the China-Japan Number Theory Conference arranged by Kanemitsu. But I had another host in Beijing, Dr. Miao, who was introduced to me by my colleague Yunmei Chen. Dr. Miao was working in the Defence Ministry and he arranged my colloquium there on Wednesday morning. We flew out of Beijing that afternoon for Shanghai. Even on that short sector, China Airlines provided us the massive Boeing 747-400, and the aircraft was quite full. Sofitel Hotel where we stayed was located next to Nanjing Road, one of the main arteries of Shanghai.

My two talks in Shanghai were on Friday and so we had all of Thursday for sightseeing. A short distance from the hotel is the *Bund*, a promenade on the Yangtze River, the third longest river in the world. The Yangtze joins the sea just after Shanghai. We walked to the Bund to enjoy spectacular views of Shanghai on both sides of the river. We also visited the Shanghai Art Museum which has an outstanding collection that revealed the skill of

Chinese craftsmen over the millennia. What was surprising was that the museum had an impressive Rockefeller collection of Indian art!

My colleague Yunmei Chen hails from Shanghai and she introduced me to my hosts there. In the morning, my host Prof. Hu arranged my lecture at Shanghai Jaiotong University following which he hosted a banquet at a vegetarian restaurant attached to the Longhua Buddha Temple. In the afternoon, after my talk at Tongji University, our hosts took us to dinner at the world famous Gundelin vegetarian Chinese restaurant in central Shanghai. Gundelin has several private dining rooms, each one decorated in a different way. It was a memorable conclusion to our stay in China.

This first visit to China was an eye opener in every sense. We were impressed by the excellence of their mathematical tradition, the lavishness of our hosts, and the magnificence of the country as a whole.

In December, I made my usual visit to India, but my family did not accompany me because on Dec 31 we were scheduled to visit Maui to greet the New Millennium in that island paradise. So I went alone to India on a short academic visit in mid-December.

Enroute to India via the Pacific, I halted in Singapore for a day to give a colloquium at the National University of Singapore at the invitation of Heng-Huat Chan. In Madras, I had a full schedule of talks, but this time two were unusual. On December 22, Ramanujan's birthday, I was invited to deliver the Millennium Lecture at M. S. Swaminathan Research Foundation. The title of my talk was "Ramanujan for the new millennium". The next day I addressed a group of mathematics college teachers at the Tamil Nadu State Council for Higher Education (TANSCHE) at the invitation of its Director Dr. M. Anandakrishnan, who after a successful term as Vice-Chancellor of Anna University, was heading TANSCHE. In my talk "Mathematics education in the new millennium", I stressed how important a solid rigorous training in basic mathematics would be in the new millennium which is going to be dominated by information technology and computer science. I pointed out that even though mathematics is the most rigorous of all disciplines, sadly, in schools and in the undergraduate classes, mathematics is taught nowadays without emphasis on rigour thereby not doing justice to the subject.

The next day (Dec 24) was memorable because my father's autobiography "The Alladi Diary - Part I", published by East-West Press, was released at a function in the lecture hall of *Ekamra Nivas*. Mr. N. Ravi, Editor-in-Chief of *The Hindu*, received the first copy. Professors M. S. Swaminathan and M. Anandakrishnan paid glowing tributes to my father. My father gave a magnificent speech recalling how his Theoretical Physics Seminar conducted in that lecture hall, led to the creation of MATSCIENCE. I was glad that I came to Madras to participate in the function and be with my father on that momentous occasion. A prominent report of that function appeared in *The Hindu*. After spending Christmas in Madras, I left the next day to join my family in Florida for our trip Maui.

To greet the New Millennium, we wanted to do something special. So we decided to spend a week in the enchanting island of Maui from December 31 to January 7. The specific choice of Maui for us instead of the other Hawaiian islands was because Mathura's aunt and uncle Gigi and Ramakrishnan who had joined us on the trip to China, were also going to Maui with their daughter Raji, son-in-law Jay and their entire family during the same period. We heard reports in the papers that at the very instant when the new millennium would be born, all computers would collapse, and so flights would not operate on December 31 or January 1. As it turned out, this catastrophe did not happen, and all flights were operating as scheduled. But many passengers out of fear avoided travelling on December 31, which meant that seats were wide open. So I was easily able to get seats on a Northwest Airlines flight to Honolulu. We arrived in Honolulu around 4:00 pm in gorgeous weather and were to take a Hawaiian Airlines flight to Maui. As we were waiting for our flight to Maui, I noticed a tall handsome man at the gate, and he looked familiar. I then realized he was David Hasselhof, the famous star of the popular TV series Baywatch! He was also going to Maui (with his girlfriend) to greet the new Millennium. When I informed my daughters Lalitha (age 18) and Amritha (age 12) that Hasselhoff is on our flight, they could not believe what I said. But once they saw him, they wanted a photograph with him, and he graciously agreed to this request.

In Maui we stayed at the fabulous Westin Maui on the famous Kaanapali Beach. After dinner, we all went to the beach to watch the fireworks at midnight. It was a perfect setting for us at the start of the year 2000, because among all the places in the world, Hawaii is my favourite; I fell in love with Hawaii on my first trip outside India in 1962 as a boy of seven.

It was a dream vacation in Maui, where for the next six days we enjoyed many things that paradise island had to offer. December to April is the season when whales migrate to the waters between the islands of Maui and Lanai. So we took a whale watching boat tour from the town of Lahaina and saw several whales up close. Thas was the highlight of the trip.

After the dizzying vacation in Maui, we returned to reality — back in

Florida to resume work. The calendar year 2000 started with Gilbert Strang of MIT delivering a delightful Ulam Colloquium on the theme "Small world networks and partly random graphs". Quite often in striking up conversation with a stranger, we end up discovering that we are actually connected through mutual acquaintances, and then we exclaim that it is a small world! It has been found experimentally, that on average, two individuals can be linked by a chain of length six. Professor Strang beautifully explained this six degrees of separation using graph theory. The next day Strang gave a talk in the Teaching Seminar on Linear Algebra. He had published a very popular and influential textbook on Linear Algebra and so we invited him to speak in the Teaching Seminar on this topic.

After hearing Strang's talk on Linear Algebra, I left for Tampa to catch a flight for Newark at 8:30 pm to attend the DIMACS Workshop on "Unusual applications of number theory", at Rutgers University. I arrived at Newark Airport at 00:30 hrs, and took a shuttle to the Holiday Inn near Rutgers University. The reason for rushing like this was because my lecture was the first in the morning at 9:00 am. I spoke on the (Big) partition theorem of Göllnitz, its generalizations, refinements and reformulations, which included my joint work with George Andrews and Basil Gordon. After my lecture, Berkovich presented our joint work with Andrews on the four dimensional extension of the Göllnitz theorem. My joint paper with Andrews and Berkovich describing the construction of the four dimensional extension of the Göllnitz theorem appeared in the workshop proceedings edited by Nathanson, who was one of the conference organizers.

A few days after returning to Gainesville from the DIMACS Conference, I departed for Washington DC to attend a Workshop for Chairs organized by the NSF (Jan 18–19). There I had good discussions regarding plans for the Department with Don Lewis who had just finished his term as the Director of the Mathematics Division of the NSF, and Philippe Tondeur of the University of Illinois who had succeeded Lewis as the Director.

Although I was immersed in administration as Chair, I made sure to have time for research. This was helped by the zero teaching load I was given. The most significant development in my research was the resolution (jointly with Andrews and Berkovich) of a thirty year old problem of Andrews on whether there exists a four dimensional extension of the Göllnitz theorem. Even though the construction of the four dimensional partition identity was sketched at the DIMACS Conference in January, the q-hypergeometric proof was completed only in March. Dean Harrison wanted appropriate publicity for this in the College Newsletter, but what was most surprising was that the local newspaper, The Gainesville Sun, carried a front page announcement with a photograph! The real academic development of this research progress was that my three year NSF grant proposal was approved for funding to begin in the summer of 2000. Dean Harrison had said that my zero teaching load would continue if I would show progress in research, and this resolution of the Andrews problem convinced him to extend the zero teaching load for me. I would like to stress that the chairmanship of any mathematics department is one of the toughest jobs on campus because (i) mathematics departments are large, and (ii) mathematics departments have a huge service component which leads to thorny and complex issues in dealing with the demands of client disciplines. Since this is a very challenging and time consuming job, I feel strongly that relief from teaching is necessary to do the job well AND to conduct research.

On April 10, Wanflette Professor Daniel Quillen of Oxford University gave the Second Erdős Colloquium on the theme "Module theory for non-unital rings". Quillen was awarded the Fields Medal in 1978 for his work in algebraic K-theory. It was a fine gesture on the part of John Thompson, himself a Fields Medallist, to have flown to Gainesville from England (where he was on sabbatical), to give the Opening Remarks for Quillen's lecture!

My parents had arrived in Gainesville in March for their customary three-month stay, and so my father was able to attend Quillen's talk. Mathura and I arranged a reception at our home on March 7 in honor of my father and to mark the publication of his autobiography *The Alladi Diary - Part I*. He gave a fine speech to a large group of mostly Indian friends, about the great contributions in the field of law of my grandfather Sir Alladi, of which he was a direct witness, and of his own efforts to create MATSCIENCE in the face of seemingly insurmountable obstacles.

In May there was the Millennial Number Theory Conference at the University of Illinois, where like at DIMACS, Berkovich and I gave back to back talks, on the Göllnitz theorem and its four dimensional extension (joint work with Andrews). When the Gainesville Sun announced this in April, Andrews was not there in the picture that accompanied the article. But then, Win Phillips, the Vice-President of Research, wanted a photograph of the three of us together to appear in the University of Florida research magazine *Explore*. By special arrangement, such a photograph was taken during the Millennial Conference and sent to Win Phillips.

Dean Harrison decided to step down on June 30, and Neil Sullivan, the Associate Dean for Research, was appointed as the Interim Dean. So after the Millennial Conference, I met with both Harrison and Sullivan to go over the budget and hiring requests for the year 2000–01 before departing for India in the end of June with my family. Enroute I visited Austria and Hungary in response to invitations from Vienna and Linz in Austria, and to speak at a number theory conference in Debrecen, Hungary.

The start of our journey to Europe was a bit scary. We were on Continental Airlines and scheduled to fly via Newark to Amsterdam, but my family and I were on two different flights from Orlando to Newark separated by an hour. We had a long connecting time in Newark and so were not worried about this. But as luck would have it, there was a heavy thunderstorm in Orlando, and after my family had boarded, their flight was not allowed to take off for two hours! The aircraft in which I was to fly had not yet landed, and was circling above Orlando for more than an hour before landing. My family reached Newark in time to make it to the connecting gate, but I was not there yet. So even though a boarding announcement was made, they REFUSED to board unless I would join them at the gate. That is how strong family love is! Luckily, I reached Newark about 15 minutes after they had arrived there, but since I had very little time, I dashed across the airport to reach the boarding gate just before it closed. I was much younger and more agile then! Yes, we were all on board the flight to Amsterdam, but we were exhausted by the tension. Miraculously, all our bags reached Amsterdam where we took a KLM flight to Vienna.

We arrived in Vienna at 3:30 pm on June 27 and checked in at the Holiday Inn Crowne Plaza which had a fantastic location right on the banks of the Danube. We were given a spacious room which commanded a fine view of that great river. Since we had the whole evening to spare, we went, at the recommendation of the Hotel Concierge, to a Mozart-Strauss Opera at the Hofburg Palace after dinner at the hotel. It was really very enjoyable. The next morning my host Christian Krattenthaler of the University of Vienna picked us up after breakfast to take us on a walking tour around Vienna. When we told him how much we enjoyed the concert at the Hofburg Palace, he said that such concerts are for tourists and do not represent the high class concerts that Vienna has to offer! Krattenthaler is nearly a professional level pianist and has exceptionally high standards.

The walking tour lasted all morning and Krattenthaler showed us (from outside), the Rathaus (City Hall), the Parliament Building, the Museums, the National Library, and the Stephen Cathedral. They all looked magnificent under bright sunny skies. He then took us to one of the most famous Viennese cafes which had a fantastic selection of cakes, and he said they were all excellent. So with his nod of approval, my family and I tried different cakes, and they were all delicious.

In the afternoon I gave a Colloquium at the University of Vienna on the theme "Going beyond the theorem of Göllnitz - an exciting voyage". The great mathematician Victor Kac was in the audience. While I was at the University of Vienna, Mathura and my daughters were out shopping. After my talk, we got together, and Krattenthaler took us to dinner at a very traditional restaurant called Heurigen which was on the outskirts of Vienna in a rural setting.

The next day, June 29, was the Wedding Anniversary for me and Mathura (our 32nd). We celebrated it by going to the magnificent Schonbrunn Palace in the morning. We then checked out of the hotel and took the train to Linz where we were warmly received by our host Peter Paule. He put us up in a suite at the lovely Wolfinger Hotel and took us out to dinner that night. Peter has a genial personality and is a charming conversationalist, and my family took a liking to him instantly.

On the following day, Peter arranged my lecture in the Combinatorics Seminar of the Research Institute of Symbolic Computation (RISC) where he was the head of a very active research group. I spoke about the Göllnitz theorem and its four dimensional extension. It was well received, and indeed one of the members of the group, Axel Riese, using software packages, had made some improvements on our work. RISC has a lovely countryside location about half an hour from Linz, and is housed in a castle which had been renovated. Its interior is ultra-modern, and it has advanced computing facilities. After lunch at RISC following my lecture, we returned to Vienna.

July 1 was a Saturday and so we made a day trip to Salzburg on the "Sound of Music Tour" from Vienna. The bus ride to Salzburg took us past countryside of exquisite beauty. Salzburg is a lovely town on a river and surrounded by hills. All of us had enjoyed seeing the movie "The sound of music" in our school days, and so this tour of Salzburg was particularly delightful and nostalgic.

On Sunday we departed Vienna by the morning train to Budapest, leaving most of luggage at our hotel in Vienna since we were to return in two days. We checked in at the magnificent Hyatt Regency, Budapest, which was on the banks of the great Danube, and right across from the Hungarian Academy of Sciences. I had stayed at the Hyatt when I was here the previous year for the Erdős Memorial Conference at the Hungarian Academy (see §11.5), and so I decided to stay here now with my family. After lunch at the Hyatt, I took the afternoon train to Debrecen for a two-day stay leaving my family in Budapest for sightseeing.

The number theory conference (July 3–7) at the University of Debrecen was in honor of Kalman Gyory and Andras Sarkozy for their 60th birthdays. Sarkozy enjoys the reputation as the mathematician with the most number of joint papers with Erdős! Debrecen has a long mathematical tradition, second in Hungary only to Budapest. The University had a venerable journal "Publicationes Mathematicae Debrecen" that was founded in 1950. I had heard about this journal from Erdős who had published many papers in it. I spoke about my new observations on the Rogers-Ramanujan identities on the afternoon of the opening day (Mon, July 3). There was a fine banquet that night honoring Gyory and Sarkozy when many Hungarian mathematicians spoke. The next day, after attending Alan Baker's magnificent Plenary Lecture on Transcendental Number Theory, I left Debrecen by train to join my family in Budapest.

For our return journey from Budapest to Vienna, we took the hydrofoil fast boat, whose pier was right next to the Hyatt. It was a six and half hour journey as the hydrofoil sped along the Danube past stately historic buildings and lovely towns. It made one brief stop at Bratislava in the Czech Republic, before docking in Vienna right next to the Holiday Inn Crowne Plaza where we were staying. After spending one night in Vienna, we left for India by KLM Royal Dutch Airlines via Amsterdam.

Our stay in India was both a vacation and an academic visit for me. As a family we went on a trip to Bangalore, Mysore and the wild life sanctuary of Mudumalai. In the forward direction, I lectured at the Raman Research Institute in Bangalore one morning, where several post-docs from Utah and Duke University were in the audience. It is always inspiring to speak in the main lecture hall of the Raman Institute where there is fine bust of the great Nobel Laureate Sir C. V. Raman. That same afternoon, I spoke at the Tata Institute Center located in the campus of the Indian Institute of Science. On the return journey from the wildlife sanctuary, I gave talks at the University of Mysore where there was (and still is) an active group of number theorists. After a one month stay in India, we were back in Florida in early August to begin preparations for the next academic year which turned out to be significant and exciting.

9.3   The National Medal of Science to John Thompson and other events (2000–01)

The crowning success for the Department in 2000 was the award of the National Medal of Science to Graduate Research Professor John Thompson by President Clinton in early December. It brought tremendous visibility and recognition for the Department on campus and across the nation, surpassed only by the award of the Abel Prize to Thompson in 2008. The year was filled with many significant events which I shall describe in the course of providing details of the National Medal of Science to Thompson.

In September, I had to make a short trip to India in connection with two conferences. I arrived in Madras on September 22 morning after a long trans-atlantic journey, and the very same afternoon, I gave a talk at a conference on q-hypergeomteric series that was organized by Professor Srinivasa Rao at MATSCIENCE. He is a former PhD student of my father, and did his doctoral work in nuclear physics. Subsequently he became interested in q-hypergeometric series. Even though I was invited to lecture at various academic institutions in Madras every year during my visits in the past, MATSCIENCE was not on that list of institutions. So I was returning to MATSCIENCE after a seven-year gap. I spoke about how to extend the Göllnitz theorem to four dimensions. Christian Krattenthaler from the University of Vienna was attending the conference and it was nice to meet him in Madras. It was a nostalgic experience for me to return to my high school Vidya Mandir where I was invited to lecture. My talk to a group of computer savvy high school students was on the theme "From Ramanujan's slate to the modern world of the computer".

On September 30, Bruce Berndt who was visiting Madras, gave a lecture on Ramanujan at Ekamra Nivas under the auspices of the Alladi Foundation. The next day, Berndt, Krattenthaler and I departed Madras for Chandigarh to attend the Ramanujan Millennium Conference at Punjab University. It was a pleasure to meet Professor Inder Singh Luthar, a senior member of the mathematics faculty there about whom my parents had spoken so much; Luthar was a visiting member at the Institute for Advanced Study (IAS) in Princeton in 1957–58 when my father visited IAS. Indeed Luthar became close friends of my parents. He recalled with pleasure the hospitality of my parents who had him over for dinner at their apartment regularly.

It was also nice to meet R. P. Bambah in Chandigarh, his home turf. He is a highly reputed number theorist who had retired from Punjab University a few years before. Bambah had been instrumental in building up the number theory program at Punjab University. In the sixties he also held a position at the Ohio State University and served on the Editorial Board of the Journal of Number Theory which was managed by Ohio State.

The conference opened on October 2, but my talk was on the second day with Bruce Berndt in the Chair. I rushed back to Madras that night because I was returning to Florida the next day.

Back in Florida, I had meetings with Neil Sullivan who had taken over the Deanship of the College of Liberal Arts and Sciences from Willard Harrison. Sullivan wanted to begin a discussion of the renewal of my contract as Chair. When Dean Harrison offered my the Chairmanship in 1998, he gave me a five year contract. But I was nervous to accept the position for that length of time, and so I accepted it only for three years. The academic year 2000–01 was my third year and so Sullivan wanted to begin discussing my renewal. It was during one of these meetings with him in Fall 2000 that I proposed conducting Five Special Years which he endorsed with enthusiasm. Indeed this program started the very next year (2001–02).

In late October I made a trip to Urbana and Washington DC for very different reasons. I arrived in Urbana to attend a q-series conference organized by Bruce Berndt. After attending the magnificent Opening Lecture of the conference by George Andrews, I departed for Washington DC to attend the AMS Meeting of the Committee on Education. As Chair it was important for me to keep track of developments both in research and education, and so I attended a variety of conferences organized by the AMS and the NSF. In Washington I had a very fruitful discussion with Philippe Tondeur, Director of the Mathematics Division of the NSF, about the Special Years Program we were planning to commence in Florida the next year. After a day in Washington DC, I returned to Urbana and gave my talk at the q-series conference on my theory of weighted partition identities.

On November 2, I received a call from the Office of Public Affairs of the University of Florida, conveying the sensational news that our distinguished colleague John Thompson will be awarded the National Medal of Science by President Clinton in December. Thompson was a recipient of the Fields Medal in 1970, The Sylvester Medal of the Royal Society in 1985, The Wolf Prize of Israel in 1992, and the Poincaré Medal of France in 1992; so this was one more in the list of awards for him, but it was the great honor and recognition for the Department, the College, and the University. This news had to be kept confidential until a press release by the White House and this happened on November 13. The news made headlines and we were inundated with congratulatory phone calls and messages.

It was very gracious of both Dean Sullivan and Vice-President of Research Win Phillips to say that I should go to Washington DC as the representative of the University for the National Medal Presentation Ceremony. So I departed for Washington DC on November 30. The Medal Ceremony was on December 1 starting at 6:00 pm at the National Museum Building. That morning there was a private meeting in the Oval Office for all the

Medal Winners and their spouses to meet President Clinton. The medals were presented to the winners that evening by Neil Lane, Scientific Advisor to President Clinton. After the medals were presented, there was a fine dinner served in the same hall. I had the pleasure to be seated at the table with John and Diane Thompson. Joining us at the table were two great mathematicians — 1982 Fields Medalist and 1997 National Medal winner S. T. Yau of Harvard University, and Felix Browder of the University of Chicago who was President of the AMS that year and had received the National Medal the previous year. Also at our table was Arthur Jaffe of Harvard University who conceived the idea of the Clay Mathematics Institute and the Millennium Prizes; he was at that time the Director of the Clay Institute. So it was a very memorable evening in very high company.

There are several winners of the National Medal of Science each year since the award is given in various disciplines. In the Mathematical, Statistical and Computer Sciences category, there were two winners in 2000. Besides Thompson, the medal was given to mathematician Karen Uhlenbeck of the University of Texas, Austin, a world authority on partial differential equations and related areas. Like Thompson, she subsequently won the Abel Prize (2019), and was the first female Abel Laureate.

Back in Florida, there was a Christmas Party on December 7 at the Keene Faculty Center hosted by the College of Liberal Arts and Sciences. Everyone was talking about the National Medal of Science to Thompson, and so it was an opportune moment for me to again make a case for the creation of the John Thompson Research Assistant Professorship. I had spoken to Diane Thompson the previous day requesting her to talk to John and get his nod of approval to have his name associated with the professorship. In that ecstatic mood at the Christmas party, the Dean, the Provost, and the Vice-President for research, all agreed to the creation of this named assistant professorship. Thus the calendar year 2000 ended on a very high note and we looked forward to 2001 with great expectation.

Thompson's presence in Florida helped us build a world class algebra program. In January 2001, the College newsletter flashed the news of the National Medal of Science to Thompson, and also highlighted the work of our algebra group comprising professors Chat Ho, Alex Turull, Helmut Voelklein, Peter Sin, Pham Tiep, and Jorge Martinez.

On January 31, Mathura and I had the pleasure of attending a magnificent dinner hosted by President Charles Young of the University of Florida at the President's Mansion in recognition of Thompson being awarded the National Medal of Science. Dean Sullivan who was at this dinner, hosted a

Reception at the Keene Faculty Center in early April on behalf of the College of Liberal Arts and Sciences. At that Reception, he proudly announced the creation of the John Thompson Research Assistant Professorship. During the Reception, we had the pleasure of listening to a fine rendering of a Mozart piece by mathematics professor Steve Saxon (Baritone) accompanied by his wife Brenda on the piano.

In mid-February we had the visit of Professor Hyman Bass, the President of the AMS. After being on the faculty of Columbia University for many years, Bass moved to the University of Michigan where his appointment is both in the department of mathematics and in the College of Education. On this trip to Florida, Bass was accompanied by his wife Deborah Ball who was well known for her contributions in the area of Mathematical Education. On February 16, both Bass and Ball addressed our Teaching Seminar at noon on the theme "Making a change in mathematics education". Later that afternoon, Bass delivered the Erdős Colloquium. Mathura and I hosted a party at our home in honor of Bass and Ball that was well attended by our colleagues and their spouses.

The Ulam Colloquium for the academic year was delivered by Persi Diaconis of Stanford University. In his talk "On coincidences", he showed by quantitative reasoning that certain coincidences which startle us are not so amazing after all, such as when we find in gatherings of 30 or more persons, that some two members have the same birthday! In view of the popular theme of the talk, we arranged this Ulam Colloquium in the University Auditorium which holds about a thousand people. We publicized the colloquium widely, and even had a report of it in the Gainesville Sun that morning. All this helped garner an audience of about 700 for the Ulam Colloquium. Vice-President of Research Win Phillips who made the Opening Remarks, viewed this event as a huge success.

The visit of Diaconis coincided with the Seminar on Stochastic Processes conducted by our department. This is a well known conference which has been held annually but in different locations around the nation. Diaconis gave one of the main lectures at this conference in Gainesville.

My first contact with Diaconis was at UCLA during 1975–78 when I was a graduate student. He had come to see Erdős who was visiting UCLA, and I met Diaconis in the office of Bruce Rothschild. Subsequently, in 1980, in a paper he published in the Journal of Algorithms on the Fast Fourier Transform (FFT), he used a result of mine with Erdős on the function $A(n)$, the sum of the prime factors of an integer n. The main theme of his paper was that the average running time of the FFT on a positive integer n is approximately $2nA(n)$.

Diaconis is a man of many talents. He is an accomplished violinist and a magician. When he tosses a coin and catches it on its way down, he can tell you whether it will be heads or tails because he notes the parity of number of flips! He demonstrated this to the amazement of the Gainesville Sun reporter who interviewed him before his Ulam Colloquium.

Since the Ulam Colloquium was at 4:00 pm, I had time after the Colloquium to attend the E. T. York Lecture that night by Prof. M. S. Swaminathan (FRS), the world renowned agricultural scientist from Madras. E. T. York had founded the Institute of Food and Agricultural Sciences (IFAS) on the campus of the University of Florida. He was a major force in the sphere of higher education in the state of Florida and rose to the position of Chancellor for the Board of Regents.

It was not surprising that Prof. Swaminathan, one of the most influential agricultural scientists in the world, was invited to deliver the E. T. York lecture. The night before this lecture, I had arranged a special Reception in honor of Prof. Swaminathan at the India Cultural and Education Center (ICEC) in Gainesville, so that members and friends of the Indian community in an around town could meet him.

The very next day after the Ulam Colloquium and the E. T. York lecture, I left for Japan to attend the Japan-China Number Theory Conference on the campus of the University of Fukuoka in Iizuka near Fukuoka. It was organized by Shigeru Kanemitsu (of Fukuoka) and Chao Hua Jia (of Beijing). Mathura accompanied me on this trip to Japan. Our transpacific flight by Northwest Airlines arrived at Osaka's Kansai Airport and there we took a connecting flight by Japan Air System (JAS) to Fukuoka, where we were met by a student of Kanemitsu and taken to Iizuka.

On the opening day of the conference, after the sessions were over, Kanemitsu arranged my talk to high school students who had shown a precocity for mathematics. I spoke on "The Indian mathematician Ramanujan - a glimpse of his remarkable life and the magical world of his identities". My conference lecture was the next morning on the theme "Going beyond the (Big) theorem of Göllnitz". The proceedings of this conference appeared as Volume 8 in my book series Developments in Mathematics (DEVM), edited by Jia and Kanemitsu.

After the conference concluded, we departed on Saturday morning to Fukuoka, where we checked in at the Grand Hyatt and left our bags there. Kanemitsu then took us by train to Nagasaki on a day trip. Nagasaki is where the second atom bomb was dropped on August 9, 1945, and it destroyed the city totally. We went to the Peace Park that has been developed around the site where the bomb fell, and saw the Atom Bomb Museum. It was chilling to see pictures of the devastation due to the bomb. After the visit to the museum, we visited the Avalokateswara temple to spend a few moments in solitude and contemplation.

One of the sights at the Nagasaki Peace Park that attracted us was the statue called "Constellation Earth", sculpted by Paul Granlund, who is known in the mathematical world as the sculptor who produced the bronze busts of Ramanujan for the Ramanujan Centenary in 1987. The next morning (Sunday), we went up the Fukuoka tower to get spectacular views of the city in bright sunshine. We also visited the Tochoji temple that housed a gigantic wooden Buddha. Finally, to remember our visit to Fukuoka, we bought a lovely Hakata doll.

Our flight back to the America was only at 7:00 pm out of Osaka Kansai Airport the next day. So that morning we had time to visit the magnificent Osaka castle. The castle and its expansive gardens full of flowers looked lovely under sunny skies. We roamed about the castle and its grounds for three hours and proceeded to Kansai airport to board our flight to Honolulu. There we were joined by our parents from India. After a four-day stay in Honolulu with our friends the Ramanathans at their lovely home in Kalama Valley, we were back in Florida.

A week after our return, on March 30, John Thompson gave a public lecture entitled "Some remarks on the history of group theory" which was attended by faculty and students of other departments on campus as well. We were very eager to have Thompson give a public lecture in Gainesville after his award of the National Medal of Science.

Three days after Thompson's lecture, James Lepowsky of Rutgers University addressed the department Colloquium. Lepowsky is a world authority in Lie algebras which are of great interest to physicists. So his talk was attended by several members of the physics department. One exciting aspect of his work is the use of Lie algebras to discover and to prove Rogers-Ramanujan type identities. I therefore invited him in 1997 to join the Editorial Board of The Ramanujan Journal where he has played, and continues to play, a helpful and crucial role. Following his lecture, Mathura and I hosted a party for both Lepowsky and Thompson at our home.

On June 15, Dean Sullivan confirmed my appointment as Chair for a second term of three years starting July 1. I had big plans for my second term which included the launch of the Thompson Assistant Professorship and the Special Years Program. But before that, I had to make a summer trip in response to invitations for lectures in Madras as well as a conference in France. So I left for India on June 30 with Mathura and my daughters. My parents left a day earlier, but we all had a few days in Singapore together before reaching Madras. In Singapore, I visited the National University to give a talk on my work on weighted partition identities. In Madras, the very next day of my arrival, I gave a talk at the Academy of Sciences on finite identities related to the Schur and Göllnitz theorems; this was joint work with Alexander Berkovich.

Within a fortnight of my arrival in Madras, I had to leave on a trip to France to attend and speak at a joint conference of the American and French mathematical societies in Lyon. But before my departure to France, at the insistence of my older daughter Lalitha, we did a sightseeing trip in North India. She rightfully argued that we had travelled around the world visiting exotic destinations, had explored every nook and corner of South India, but had not seen any of the great sights North India had to offer. So we made a five day "golden triangle" tour of Delhi, Agra, and Jaipur.

The Taj Mahal in Agra is truly one of the wonders of the world. Like what I said about the Great Wall of China, the Taj has to be seen to be believed. Imagine the best you can of the Taj Mahal in beauty and grandeur. It will excel your highest expectations! The Taj Mahal has a special appeal to people around the world because it is the towering symbol of a man's love for a woman; the moghul emperor Shah Jehan built it as the grandest tomb for his wife Mumtaz Mahal whom he loved dearly. Some years later, Aurangazeb, the son of Shah Jehan, imprisoned his father in the Agra Fort, but put him in a room from where he would have a view of the Taj. Everyone who visits the Agra Fort (we did) gets to see this room and admire the view of the Taj in the distance. Enroute to Agra by car, we visited Mathura, the birth place of Lord Krishna. After Agra, we spent two delightful days in Jaipur, known as the "pink city" because of a profusion of historic monuments and buildings in pink color owing to the type of brick, clay, and stone that was used. The most famous of these monuments is *Hawa Mahal* or the "Palace of the Winds" made of pink and red sandstone. Finally, we visited the Sariska game sanctuary where we saw plenty of wildlife, but did not get a glimpse of the elusive Lord of the Jungle — the Bengal tiger! The trip was made enjoyable also because we stayed at fabulous hotels — the Mughal Sheraton in Agra, and the Rambagh Palace in Jaipur. Coming back to New Delhi, we had time to enjoy a fabulous buffet lunch at the Maurya Sheraton before I boarded the flight to Europe while my family returned to Madras.

An amusing thing happened at the Maurya Sheraton. At lunch, the waiter suggested that we should try "The Clinton Plate" and/or "The Chelsea Plate" from the a-la-carte menu. These were combinations of items ordered by Bill Clinton and his daughter Chelsea. And because they had ordered these dishes, they were very expensive. We just opted for the buffet which offered many more items at less than half the price!

I flew from Delhi to Lyon via Amsterdam on KLM Royal Dutch Airlines on July 16. The three-day Joint Meeting of the American Mathematical Society (AMS) and the Societe Mathematique de France (SMF) started the next day. The AMS regularly conducts joint meetings in other countries in collaboration with the mathematical societies of the host countries. This was the first such meeting in which I participated and I enjoyed it immensely. I was invited to speak at a Special Session on Additive Number Theory co-organized by Mel Nathanson (CUNY, USA) and Jean-Marc Deshouillers (Bordeaux, France). I spoke on the opening day on the theme "New weighted Rogers-Ramanujan partitions and their implications" which was based on my joint work with Berkovich that appeared in the Transactions of the AMS [P35]. Even though partitions represent an additive process, mine was the only talk on partitions in this session.

The conference was held at the Ecole Normale Superiere (ENS), the second ENS to be formed in France after the big one in Paris. There was a magnificent conference reception in the evening of the opening day at the Hotel de Ville, which is actually the City Hall, and one of the largest historic buildings of Lyon. On the second day, I was asked to chair the morning session. There were no talks that afternoon, and so we all went on a boat tour on the River Rhone, followed by a walking tour of Lyon. That night, for the participants of the Additive Number Theory Session, there was a party graciously hosted by Nicolas and his wife at their home.

Michel Waldschmidt was the Secretary of the SMF at that time. The job as Secretary of any professional society entails a lot of work. So I asked Waldschmidt if he received any reduction in his teaching duties in Paris while he served as the Secretary of the SMF. He said NO. In France, such activities are treated as service that is expected of you. In America, one would receive a reduction in teaching load as a recognition of the work involved in such an undertaking.

On July 20, I checked out of my hotel and flew back to India via Amsterdam and Delhi. The flight from Lyon to Amsterdam by KLM was by a Fokker F-28 aircraft. At check-in, KLM told me that I could not take my carry-on bag on board since the overhead bin space was very small. There was no facility in Lyon to check the bag at the gate and collect it as you get out the aircraft in Amsterdam. So grudgingly I checked my bag, the only one I had, and told the agent that I am concerned it might not make it to the final destination. She assured me it would, and it was easy for her to say this. But just as I had feared, the bag did not make it to Delhi! I complained to KLM and they apologized profusely. But what is the use of the apology after the mistake has been made and the inconvenience has been caused? The bag arrived in Madras two days later and I had to go to the airport to clear it in customs.

On July 23, I departed Madras for Bangalore by train early in the morning to reach there by 11:00 am. I had lunch with Professor Ramachandra at the Tata Institute Bangalore Center, following which I gave a colloquium there on the finite versions of the Schur and Göllnitz theorems. S. Srinivasan, a former PhD student of Ramachandra at TIFR, and who, like Ramachandra, had settled in Bangalore after retirement, joined us for lunch. He was especially interested in my old problem on bounds for sums involving small multiplicative functions, and wanted to discuss his recent work on that problem with me. I had proposed the problem at the Straus Memorial Conference in Asilomar in 1983; the problem remains unsolved, but progress in various directions has been made.

Mathura has several relatives in Bangalore, and I had dinner with them even though I was staying at the Raman Research Institute (RRI) Guest House. On the last day, I had breakfast at the home of Chandru and Hema following which, I gave a Colloquium at RRI. Chandru, a successful CPA, is the brother of my dear friend GP and we were classmates at Vivekananda College during (1972–75). Hema, a very distinguished scientist, was on the faculty at RRI. I departed Bangalore in the afternoon by train and returned to Madras that night.

On August 7 we left Madras for USA, but had a vacation in beautiful Malaysia for four nights enroute. Mathura's mother accompanied us on our return journey to spend a few months with us in Florida. We enjoyed visiting the idyllic island of Langkawi, off the western coast of Malaysia. Our cottage at the fabulous Sheraton Langkawi resort was right at the water's edge. In addition to an all-day tour by van around the island, we also took an "island hopping" boat tour. The ocean around Langkawi is studded with hundreds of exquisitely beautiful islands, some jutting out of the sea like pyramids. The spectacular scenery was similar to what we saw in Phuket in 1998. After Langkawi, we spent two nights in Kuala Lumpur to enjoy the sights that city had to offer, before returning to Florida. It was a great summer trip for me, just before the start of the Academic Year 2001–02, which had plenty of significant events coming up in succession.

9.4   Special Year in Topology and Dynamical Systems (2001–02)

The highlight of 2001–02 was the Special Year in Topology and Dynamical Systems which was ably conducted by our topologists James Keesling, Alexander Dranishnikov, Beverly Brechner, and Yuli Rudyak, and by our Dynamical Systems experts Louis Block, Phil Boyland and Jonathan King. The Special Year featured four conferences, two each in topology and dynamical systems, and out of two in each category, one was on applications and the other was on the pure aspects. The four conferences were on *Low Dimensional Dynamical Systems, Applications of Dynamical Systems, Geometric Topology and Geometric Group Theory*, and *Topology in Biology*. There were other significant events of the Special Year as well, such as the Erdős and Ulam Colloquia by Fields Medallists Sergei Novikov and John Milnor. Here I shall emphasize the programs of the Special Year as I describe several significant and interesting events that took place in the academic year as a whole.

On September 4, I received a letter from Robert Daverman, Secretary of the AMS, inviting me to serve on a Committee to select hour speakers for the AMS Regional Meetings. In that letter, Daverman said that he is writing to me at the suggestion of Hyman Bass who was the AMS President. I gladly accepted the assignment.

Exactly one week later (September 11) came the horrible news that the twin towers of the World Trade Center were destroyed by terrorists who hijacked airplanes and flew them right into the two buildings. The terrorists had planned to carry out this attack on September 11, because in American practice, the 11th of September is written as 9-11, and 911 is the emergency police number. The World Trade Center towers were chosen for the attack because New York City and these twin towers represented America's dominance and supremacy in the financial world. The planes hijacked were those of America's largest airlines — American and United. The aircraft chosen were the Boeing 757 and 767, because American manufacturer Boeing was, and is, the world leader in the commercial aviation. It was the second day in infamy just as the attack on Pearl Harbor was a day in infamy. We were stunned in disbelief. I was sitting in my office preparing to attend the number theory seminar when this news came. Life came to a standstill. These attacks plunged not only the USA but the entire world in grief and fear. Indeed it completely changed the face of air travel worldwide forever.

In late October, Eriko Hironaka (Florida State University), daughter of the Fields Medallist Hironaka, addressed the number theory seminar. She is a topologist but her talk in the number theory seminar was on Lehmer's Problem. She now is a Book Acquisitions Editor for the AMS.

The first major event of the Special Year was the conference on Low-Dimensional Dynamical Systems during November 8–13. This conference featured a series of talks by John Franks (Northwestern University), Michael Misiurewicz (IUPUI), and Lai Sai Young (NYU). Franks is a highly reputed researcher, who at that time was Treasurer of the AMS.

The dynamical systems conference was followed by the Ulam Colloquium of Fields Medallist John Milnor (SUNY, Stonybrook) on November 19. Milnor had won the Fields Medal in 1962 for revolutionary work in topology, and was at Princeton University. He then moved to the Institute for Advanced Study, Princeton, in 1970 where he spent twenty years before moving to SUNY at Stonybrook in 1991 as the Director of the Institute of Mathematical Sciences. In the latter part of his career, Milnor became interested in complexity questions in biology, and that was theme of his talk for the Ulam Colloquium. It was a pleasant surprise that during his lecture he made a reference to the famous work of my paternal cousin V. S. Ramachandran on phantom limbs!

Soon after Milnor's Ulam Colloquium, we had the conference on Topological Fluid Dynamics (Nov 28–Dec 5). Konstantin Mischaikow (Rutgers) gave a series of talks at this meeting. Thus Fall 2001 was the period of heightened activity in Dynamical Systems for the Special Year. The principal events in Topology were scheduled for Spring 2002.

Mathura and her mother Mrs. Gomathi Krishnan left for India in late November. During the four-month stay with us in Gainesville, Mathura had a packed schedule of cultural events in which her mother could partake. I departed for India a month later for just a ten-day stay in Madras. But during that brief visit, I gave talks at the Academy of Sciences, at Anna University, at the IIT, at the M. S. Swaminathan Research Foundation, and even at my former high school.

As the New Year moved in, Mathura and I departed Madras at 00:30 on January 1, for the United States via Singapore. We rested in Los Angeles for one night at the spacious home of my close friend Sampath, before moving to San Diego for the AMS Annual Meeting. While in Los Angeles, I managed to find time to visit Basil Gordon for a discussion on partitions. The only item on my agenda at the AMS Annual Meeting was the one day workshop for Chairs which I found quite useful. After this workshop, we returned to Florida because I had to attend to various organizational items before my departure to France in mid-January.

In mid-January, I had to go to Marseille, France, to attend a conference in honor of Jean-Louis Nicolas for his 60th birthday. It was a week-long conference (Mon, Jan 14–Fri, Jan 19) at the *Centre International de Rencontres Mathematiques* (CIRM) in Luminy, near Marseille. The CIRM is like Oberwolfach in the sense it has housing for the participants, and the conferences are typically one week long. And just as Oberwolfach has a lovely location in the mountains of the Black Forest, the CIRM has an enviable location fronting the Mediterranean sea. It was a very enjoyable and fruitful trip, but I had some anxious moments before departure.

I was still an Indian Citizen at that time, and so I had to apply for a Shengen Visa to enter France. Since I had to send my Indian passport to the French Consulate in Miami, I could do that only after my return from India in early January. The French Consulate required visa applicants to appear in person, but they realized that I was very busy with administrative work; so they agreed to process the visa my mail. The issuance of the visa was routine, but somehow a mistake happened in returning my passport by US Express Mail. The passport made its way from Miami to Tallahassee and to Ocala, but not to Gainesville! After speaking to officials of the US Postal Service, I was able to get the passport to reach the Sorting Center in Gainesville on the morning of my departure (January 12)! So I picked up my passport from the Sorting Center that morning on my way to Jacksonville, where I caught my flight to Detroit. From Detroit it was by KLM to Amsterdam, and then Air France to Marseille via Paris. I arrived in Marseille at 8:00 pm on January 13, and took a taxi to the CIRM. It was pleasure to meet many mathematical friends at dinner at the CIRM that night. Among them, Andras Sarkozy (Budapest), and Christian Mauduit (Marseille) proposed that I should have a Special Volume of The Ramanujan Journal in honor of Nicolas to which I readily agreed. Nicolas is an authority on properties of arithmetical functions, and in particular on highly composite numbers whose study was initiated by Ramanujan.

During the Conference Opening the next morning, I was asked to announce the Nicolas Special Volume, and the announcement was received with enthusiasm. This Special Volume of The Ramanujan Journal appeared in March 2005 edited by my and Gerald Tenenbaum on the journal editorial board, with the help of Christian Mauduit, Carl Pomerance, and Andras Sarkozy as Guest Editors.

Like Oberwolfach, the CIRM has a fine library where I worked in between the sessions and in the evenings. The library was useful for me in also resolving a problem at the University of Florida! While I was in Marseille, I received a message from the Dean, that the Provost's committee had some questions about a colleague going up for promotion to Full Professor, since that person did not have the "expected number" of 30 papers, but had only 28. I was asked to respond to this. This colleague is an accomplished mathematician and these 28 papers were all in first rate journals, and substantial papers as well. So there was no doubt in my mind, or in the minds of the eminent mathematicians who wrote letters, about the quality and productivity of this individual. By chance at the CIRM Library, I came across the Collected/Selected Papers of the legendary scientist Freeman Dyson. Dyson began his career as a mathematician and later moved to physics. In this book he emphasizes that in mathematics, a theorem proved is a theorem forever, and so it is customary to publish the Collected Papers of a mathematician. He points out that physics is speculative, and so only years later would one know which papers of a physicist are really correct and significant. So it is customary to publish only the Selected Papers of a physicist. I sent this to the Dean pointing out that 28 papers by my colleague are probably equal to 50 by a physicist. The Dean readily communicated my response to the Provost's committee. The promotion to full professorship went through without further questions.

On Friday, I was asked to chair the concluding session. I flew out of Marseille the next morning for Florida.

In early February there was a conference at the University of Florida in honor of our colleague John Klauder for his 70th birthday. Klauder is an internationally reputed physicist who had a long and illustrious innings at Bell Labs in Murray Hill, New Jersey. He then moved to the University of Florida in 1988 where he held a joint appointment in the physics and mathematics departments (50–50) but with tenure in physics. He has done fundamental work in many areas of physics such as general relativity, quantum mechanics and quantum optics, and is the author of several influential textbooks. The opening lecture of the conference in his honor was delivered by Ingrid Daubechies of Princeton University, whom he had mentored. When Klauder was at Bell Labs, he had invited my father for a colloquium, and so my father knew him well. My father sent a message to Klauder from Madras which was read at the party in Klauder's honor that was held at the Keene Faculty Center when many colleagues spoke.

A few days after the Klauder birthday conference, we had the visit of Heini Halberstam and his wife Doreen. He had retired from the University of Illinois, but was still living in Urbana-Champaign. He had a vacation home in the Florida panhandle and so would come to Florida annually.

When I heard from him that he would be in Florida, I invited him to give a History Lecture. He was a magnificent speaker, oratorical in style, and he gave a splendid lecture on "The Riemann Hypothesis". The day before his History Lecture, he addressed our Number Theory Seminar on the theme "The Brun-Hooley Sieve". He was a world authority on sieves and a master in exposition; so it was a treat to hear him on this topic. That night Mathura and I hosted a party for the Halberstams at our home.

During Halberstam's visit, I asked him whether he would consider moving to Gainesville and taking up a courtesy appointment at the University of Florida. He was willing to consider that option, but on returning to Illinois and discussing with his family, he said that he and Doreen decided not to move out of Champaign-Urbana.

During our Spring Break, we had the first of two conferences in topology focussing on Geometric Topology and Geometric Group Theory (March 2–10, 2002). The conference featured talks by very eminent topologists such as Steve Ferry (SUNY at Binghampton), Mladen Bestvina (Utah), Schmuel Weinberger (University of Chicago), Paul Baum (Penn State), and Nigel Higson (Penn State). Both Ferry and Weinberger later served on the External Review Committee of our department in 2006. Paul Baum and Nigel Higson, like George Andrews, are Evan Pugh Professors, and I got to know them well during my visits to Penn State University. I was pleased that Robert Edwards, a senior and highly reputed topologist from UCLA, attended this conference. When I was a graduate student at UCLA, I was told by my classmates that Edwards is a leading topologist, but I did not have occasion to interact with him then. I was also glad that Robert Daverman (University of Tennessee), an eminent topologist, was at this meeting; Daverman was Secretary of the AMS then and I got to know him well since I used to attend AMS conferences and their special events quite frequently. A selection of papers presented at this conference appeared as a Special Issue of the journal *Topology and its Applications* (Elsevier) in May 2004 edited by Dranishnikov and Keesling, who at that time was on the journal's Editorial Board.

In the second week of March, we had an international conference on "Topology in Biology", conducted in collaboration with Florida State University because DeWitt Sumners, the Chair of Mathematics at FSU, was leading a major program on the applications of topology in biology. The conference opened with the lecture of James Keener, a world authority in biomath, who had addressed the First Ulam Colloquium in our department in March 1999. Sumners delivered the mathematics colloquium on the opening day of the conference and spoke about his project: "Topology of the DNA". It was a hugely successful conference since it brought together experts from different areas of science making crucial use of topology.

In late March, David Bressoud from Macalester College gave a lovely colloquium talk on "The alternating sign matrix conjecture". Bressoud is one of the most effective communicators and his talk was appreciated by students and faculty alike. The next day Bressoud addressed the Number Theory Seminar on the theme "Sign variations of MacDonald identities", which was as impressive as his colloquium. Since Bressoud also has been active in mathematics education, we invited him to address the Teaching Seminar as well, and it appealed especially to our lecturers and TAs.

On April 1, my father's long time friend M. M. Rao from the University of California, Riverside, gave a colloquium talk. He was invited by my colleague Nick Dinculaeanu. My parents were in Gainesville at that time, and it was a pleasure for them to reconnect with the Raos. We attended a dinner in honor of Prof. and Mrs. Rao so graciously hosted by Nannette and Nick Dinculeanu at their home.

One week later, Fields Medallist Daniel Quillen (Oxford) addressed the department colloquium on the theme "The geometry of the Poisson summation formula". Quillen had delivered the Second Erdős Colloquium in 2000. My colleague Alex Turull informed me that Quillen's son David was on the faculty of the University of Florida College of Medicine. So we knew that Daniel Quillen would be interested in visiting Gainesville. Thus we invited him to visit the Department for two weeks in April 2002, which he did. Some years later, Quillen suffered from Alzheimer's disease, and spent his last days with his son in Gainesville. Quillen's son was Turull's neighbor, and Turull recalls seeing Quillen walking in the garden of his son's house residence where he was staying.

The visit of Quillen coincided with that of 1970 Fields Medallist Sergei Novikov, a world renowned topologist. He came to Florida to deliver the Erdős Colloquium on Friday, April 12. The title of his talk was "Topological phenomena in metals". Since our colleague John Thompson had won the Fields Medal in 1970 along with Novikov, we asked him to introduce Novikov. Owing to the presence of Thompson, Quillen and Novikov in our department, I advertised the second week of April as the "Fields Medallist week in the Mathematics Department", and this attracted the attention of the University administration. My father enjoyed meeting and discussing with the Fields Medallists. In fact, that week, my father gave a talk for the undergraduate mathematics club $\pi - \mu - \epsilon$ on the theme "Elementary algebra and geometry as sources of creative ideas".

With Novikov's talk, the Special Year in Topology and Dynamical Systems came to a grand conclusion. We looked forward to the Special Year in Algebra (2002–03), but before that, the summer had a packed schedule of events for me and my family.

The first major event for the summer was on Mother's Day (May 12), when Mathura conducted the *Arangetram* (full two hour graduation dance performance in Bharathanatyam) of two her students at the Moroccan Temple Auditorium in Jacksonville. For a dance teacher, conducting an Arangetram is like having a PhD student defend the thesis for us academicians. This was Mathura's first Arangetram in Florida and the event was attended by about five hundred guests. Thus it was a special Mother's Day for her. Since then, she has conducted more than a dozen Arangetrams.

My older daughter Lalitha graduated from the University of Florida and decided to go to Stetson Law School in St. Petersburg — the first law school in Florida. We had a graduation party for her on June 23, the highlight of which was Lalitha's magnificent one and a half hour Bharathanatyam dance recital. Mathura as her teacher provided live vocal music and was ably assisted by my second daughter Amritha. Graduate student Hariprasad gave fine percussion support on the mridangam.

We did not do our customary trip to India in the summer. Instead we visited Japan and China for two weeks, primarily to take Lalitha to China since she could not join us in November 1999. It was also an academic visit for me since I had lecture engagements in Beijing, Xian and Honolulu. Since my parents were in Florida while we were planning the visit to China, we decided to stop in Hawaii enroute to have a one week vacation with them in the island paradise.

We reached Honolulu on June 27, and spent a few days with our friends the Ramanathans. I gave a colloquium at the University of Hawaii the next day on the theme "New weighted Rogers-Ramanujan partitions and their implications", following which the Bertrams, our hosts, took us to dinner at a Thai restaurant in Hawaii Kai that had a lovely location on the marina. June 29 was the wedding anniversary for me and Mathura, and we celebrated that with a dinner at the Hyatt Regency on Waikiki Beach.

From Honolulu, we moved to Kauai, which is the favourite island for us. We spent a whole week in Princeville on the northern part of Kauai, the very definition of heaven. Our condominium at Pahio at Shearwater was perched on the edge of a cliff and it commanded a breathtaking view of the northern coast of Kauai. The spectacular scenery of Kauai has to be seen to be believed — Lumahai Beach, the Kilauea Lighthouse, Poipu

Beach, the Kalalau Lookout, and the Waimea Canyon. But Kauai has one more thing to offer — the Hindu Temple set in a rain drenched valley. This temple, located on the slopes of Mount Waialeale, is surrounded by verdant gardens. The peaceful atmosphere of the temple and its surroundings is just like the *ashrams* of the sages as described in the Hindu epics.

After a delightful stay in Kauai, my daughters and I departed from Honolulu for China on July 10, whereas my parents returned to India. On arrival the next day in Beijing, we checked in at the Great Wall Sheraton whose imposing size befitted its name. At the invitation of Jia, I gave a talk at the Academia Sinica the next day on the theme "New weighted Rogers-Ramanujan identities modulo 6 and 7". My daughter Lalitha got a taste of the lavish Chinese hospitality. It was the first visit for Lalitha to the Forbidden City and the Great Wall and she was taken in by their awesome grandeur. After a three-night stay in Beijing, we left for Xian on the morning of July 12 by an Air China Boeing 777, and checked in at the elegant Sheraton Xian Hotel. My host Wenpeng Zhang picked me up after lunch and took me to Northwest University where I gave a colloquium. Zhang had several PhD students and I had a fruitful discussion with them. That night Zhang hosted a banquet on the campus of Northwest University for which my family joined. He then took us on a walking tour at night to the city center which was surrounded by a protective wall that was several centuries old. We bought a lovely Jade Buddha which we display proudly in our home during the Navarathri celebrations each year.

The next day, Zhang's students took us on an all day excursion to the Terra Cota Soldiers Monument. It is incredible that this mausoleum for a Chinese emperor containing thousands of terra cota figurines remained undiscovered for centuries, and it was unearthed only in the 20th century; its location was realized after a poor farmer accidently pulled out a centuries old artifact from a well! We actually saw this farmer at the Souvenir Store of the monument and he even signed the picture book of the monument that we bought! I asked whether the Government paid a sizeable sum to the farmer for having discovered the monument which has now become a major tourist attraction. Our hosts said NO; he was given the honor to sit in the souvenir store and sign the books bought by the tourists!! When we returned to Xian, Zhang presented us with a small replica of a chariot with horses that is a prime artifact in the mausoleum in Xian.

We had to return from Xian to Beijing before boarding our flight to Japan. Thus we spent a night in Beijing so that we could show Lalitha the lovely Summer Palace. That was fitting finale for our stay in China.

On July 16 we departed Beijing for Osaka and checked in at the luxurious Osaka Hilton. The next day, Professor Hata, a mathematician friend of Kanemitsu, picked us up at the Hilton and took us to Kyoto by train for an all day tour of that beautiful historic city. We enjoyed seeing the marvellous temples of Kyoto — The Ginka Kuji (Silver Temple), The Kinka Kuji (Golden Temple), and The Sanjensangado Temple, and the Kiyomizu Temple. After a delicious dinner at the Indian restaurant Ashoka in Kyoto, we returned to Osaka by train. The next day, we roamed about Osaka, but made sure to spend several hours at the majestic Osaka Castle.

From Osaka we moved to Fukuoka. All Nippon Airways provided the gigantic Boeing 747 on this short segment and the aircraft was nearly full! Our friendly host Shigeru Kanemitsu met us at the airport and took us to the Grand Hyatt where we stayed. It was glorious sunshine the next morning and so we went up the Fukuoka Tower to enjoy fabulous views of the city. That afternoon Kanemitsu took me by train to nearby Iizuka while my family stayed back in Fukuoka. I delivered a talk to advanced high school students at the University in Iizuka. Kanemitsu received favorable comments when he previously had arranged my talk to high school students, and so he wanted me to address the high school students during this visit as well. My talk this time was on "Irrational numbers".

The following day, Kanemitsu took us on an excursion to Mt. Aso, which is actually a volcano. The cable car ride to top of Mt. Aso was spectacular, and views of the crater from the rim was simply breathtaking.

On our final day in Japan, Kanemitsu took us to Nagasaki. We visited the location where Atom Bomb was dropped and this is now surrounded by Peace Park through which we strolled. Adjacent to Peace Park is the Atom Bomb Museum where the images of the devastation moved us to tears. It was the first time my daughters were visiting Osaka, Kyoto, Fukuoka, and Nagasaki, and it made a big impression on them. After the eventful week in Japan, we returned to Beijing for a night before flying back to Florida.

9.5   Special Year in Algebra (2002–03)

The academic year 2002–03 was one of the most significant in the department's history because it was John Thompson's 70th birthday year and it was highlighted by the Special Year in Algebra in his honor. Leading mathematicians from all over the world visited our department to take part in the year-long celebration. It is no exaggeration to say that in 2002–03, the world's action in group theory (the area of mathematics in which Thompson is the undisputed leader) took place in our department. During this period of intense activity in group theory, the Department welcomed the first John Thompson Research Assistant Professor: Larry Wilson who had just received his PhD in the field of algebra from the University of Chicago.

The Special Year in Algebra featured two parallel programs — one in Galois Theory organized by Helmut Voelklein and one in Group Theory organized by Chat Ho, Alex Turull, Peter Sin, and Pham Tiep. The Galois Theory Program was highlighted by an International Conference in November 2002 and a Workshop on Algebraic Curves and Cryptography in early March 2003. The main feature of the Group Theory Program was a week long International Conference in mid-March 2003. Cambridge University also held an international conference in Thompson's honor in September 2002. In 1970 when Thompson won the Fields Medal, he had just accepted the position as Rouse Ball Professor at Cambridge University. He was at Cambridge for 23 years before moving to Florida on 1993, and so it was natural for Cambridge University to have this conference honoring one of its most eminent professors. The Cambridge conference preceded all the events in Florida for the Special Year in Algebra 2002–03, and I have given a detailed account of my trip to England and that conference in §11.1. So here I shall describe the events in Florida in honor of Thompson.

The activities of the Special Year in Algebra 2002–03 in our department started with the Galois Theory Conference (Nov 4–8). Among the distinguished speakers at this conference were Professors Shreeram Abhyankar (Purdue), Mike Fried (UC Irvine), Robert Guralnick (University of Southern California), and Gerhard Frey (University of Essen). Guralnick gave a history lecture was on the theme "Links between Riemann surfaces and group theory". Guralnick and I were classmates at UCLA in graduate school. He was one year my senior and his advisor was Basil Gordon. The refereed proceedings of this conference edited by Helmut Voelklein and his former student Tony Shashka appeared as Volume 12 in my book series Developments in Mathematics, published at that time by Kluwer.

In the third week of November, I made a two-day trip to New York to deliver the Weissman Public Lecture at Baruch College of the City University of New York, CUNY. The lecture was arranged by Jerzy Kakol from Poland who was visiting CUNY. The title of my talk was "The Indian mathematical genius Ramanujan - a glimpse of his magical mathematical formulae" so as to appeal to a wide audience. My long time friend Mark Sheingorn was on the faculty at Baruch, and so I spent one night at his lovely apartment before the talk. I used to meet Sheingorn regularly when I was working in analytic number theory, and have stayed at his apartment a few times; I was meeting him after more than a decade, and so there was a lot of catching up to do.

The main event in December was the award of an Honorary Doctorate to Professor George Andrews of The Pennsylvania State University during the Fall Commencement on December 21 by President Charles Young of the University of Florida. Thus I did not depart for India until this important event was over, but Mathura left early in December.

On December 20, the day before the Commencement, there was a Lunch Reception in honor of Andrews at the University Centre Hotel, during which Vice-President of Research Win Phillips, and Dean Sullivan were present and spoke in glowing terms about Andrews. I gave an account of the long and fruitful association between Andrews and the number theory group at Florida, and how he was, and continues to be, an inspiring influence. That evening, President Young hosted a dinner at his home in honor of George and Joy Andrews which I had the pleasure of attending. There were about 15 select guests for this dinner.

The next day, at the Fall Commencement, Andrews was recognized with the Honorary Doctorate for being the world leader in the theory of partitions and the work of Ramanujan combined, and for his strong ties with the University of Florida through collaborations with all members of our number theory group. Normally when one is given an honorary doctorate, it is the custom to invite the honoree to give a public lecture. Since classes had ended for the holidays in late December, we decided to schedule the lecture by Andrews in Spring 2003 when he would be visiting.

The very next day following the Commencement, I left for India with my daughters. We arrived in Madras on December 24 morning after a long trans-pacific journey, but that afternoon I gave a talk at Anna University on a quadruple product extension of the Jacobi Triple Product identity (joint work with Alexander Berkovich). This lecture was scheduled just before the university closed for the Christmas-New Year break. My stay in Madras was just for two weeks, but in early January, I gave talks at the Academy of Sciences, the IIT, and the M. S. Swaminathan Research Foundation. I returned to Florida on January 5, and had to get ready for spring semester which was packed with numerous significant events.

The year started with a letter from Fields Medalist Gerd Faltings, editor of *Inventiones Mathematicae*, that my joint paper with George Andrews and Alex Berkovich on the four dimensional extension of Göllnitz' theorem is accepted for publication in that prestigious journal.

We had a spate of eminent visitors in the spring semester. In January,

Hershel Farkas of the Hebrew University, Jerusalem, addressed the Number Theory Seminar, while Mike Fried from UC Irvine gave the Colloquium talk. Farkas got interested in partitions through his seminal work on theta functions, and so visited the University of Florida for one to two weeks regularly for a few years. Fried was visiting for the entire spring semester as the guest of Helmut Voelklein who was running the Galois Theory Program. Another visitor in January was John Wright from the University of Aberdeen. He is the son of E. M. Wright, the well-known collaborator of G. H. Hardy for their book *Introduction to the Theory of Numbers*, a classic in every sense. John Wright is an eminent analyst and was visiting the Department as the guest of Jim Brooks. I had a very pleasant conversation with him about his work and that of his father.

During February 1–2, there was a conference in honor of Pierre Ramond of the physics department for his 60th birthday. I had ties with Ramond for two reasons — (i) he is a grand-student of my father, and (ii) he is the Director of the Institute of Fundamental Theory (IFT), to which the mathematics department is a partner. The day before the conference, there was reception in Ramond's honor at the Keene Faculty Center. I attended the reception and handed over a congratulatory message to him from my father which he was pleased to receive.

The Pierre-Fest conference opened with the magnificent lecture of Nobel Laureate Murray Gell-Mann. It was a pleasure for me to meet Gell-Mann after nearly four decades! When I told him that I remember visiting his home in Altadena in the Fall of 1962, he responded that he had pleasant memories of his stay in Ekamra Nivas in 1961, and remembered the lovely sarees that my mother wore! As the conference was about to begin, there was a devastating announcement of the explosion of Space Shuttle Columbia. This put a pall of gloom over the audience, but Gell-Mann's electrifying lecture lifted our spirits. That evening Mathura and I attended the banquet in honor of Ramond, and there I introduced her to Gell-Mann.

In the third week of February, Cole Prize winner Benedict Gross of Harvard University delivered the Erdős Colloquium on "Automorphisms of unimodular lattices". Gross has such a deep understanding of mathematics, and lectures with such clarity, that he makes difficult ideas appear easy and palatable. We invited Pierre Ramond to make the Opening Remarks for Gross' lecture since Ramond had collaborated with Gross.

Our Spring Break was during the first week of March, and towards the end of that break we had scheduled the big conference in Group Theory. So before that conference, I decided to go to Japan to speak at a conference on "Zeta functions, topology and quantum physics" organized by Shigeru Kanemitsu in Osaka. On the long transpacific flight from Detroit to Osaka, I went through the CBMS Lectures of George Andrews on q-series and their manifold uses including applications to physics. The conference opened on March 3 with the lecture of Michel Waldschmidt (Paris). My talk was the next and it was entitled "Insights into the structure of Rogers-Ramanujan type identities, some from physics". In the talk I discussed a set of identities of Rogers that arise as solutions to the Hard-Hexagon Model in statistical mechanics as demonstrated by Rodney Baxter, and my own work on some novel relations involving these identities of Rogers. The next day after the conference sessions were over, both Waldschmidt and I gave lectures to high school students. Mine was on the theme "Prime numbers and primality testing". There were about 150 students and teachers in attendance. Interestingly, Kanemitsu translated every one of our sentences into Japanese! So after every sentence we spoke (in English), we paused so that Kanemitsu could give the Japanese translation. It was a unique and enjoyable experience. That night there was a reception on the campus of Kinki University hosted by the Rector of the university. I returned to Florida the next day owing to the Group Theory Conference.

Provost David Colburn (now deceased) inaugurated the Group Theory Conference on March 6 following which Michel Aschbacher of Caltech gave the Opening Lecture. Aschbacher had played a major role in the completion of the monumental task of classifying the finite simple groups, and Thompson's seminal work was crucial in this grand program. The last talk on the opening day was the magnificent Ulam Colloquium by Gerhard Frey (Essen) on the theme "Fermat's Last Theorem and data security". Frey had made an important observation linking Fermat's Last Theorem with the theory of elliptic curves, and it was that observation that opened the door for the ultimate resolution of the problem. When the news of the resolution of Fermat's Last Theorem made headlines throughout the world, some of my friends on campus from the science and engineering departments asked, "What is so great in showing that certain equations have no solutions?" So it was good that Frey discussed applications to data security in his lecture. On the night of March 8, there was the conference banquet. Michael Broue, Editor-in-Chief of the Journal of Algebra, presented a copy of the special volume of that journal in honor of Thompson to me and Dean Sullivan. Several group theorists such as Walter Feit, Thompson's great collaborator, spoke at the banquet. The next day, there was a fine party at the home of Chat and Virginia Ho. Chat was Thompson's former

PhD student and had organized this great conference so well along with his colleagues Alex Turull, Peter Sin and Pham Tiep. The refereed conference proceedings edited by them appeared as a book entitled *Finite Groups 2003* published by Walter de Gruyter.

In the latter half of March, Ronald Solomon (Ohio State) gave a History Lecture in which he provided a wonderful account of the classification of finite simple groups. Thompson, whose groundbreaking work was crucial in the classification, made the Opening Remarks for the lecture.

My parents arrived from India during the third week of March for their annual three-month stay. There were several significant events in the next few weeks that they could attend.

On March 21, the International Conference on Number Theory and Combinatorics in Physics began with the brilliant talk by Heineman Prize Winner Barry McCoy of SUNY, Stonybrook, The final talk that day was the History Lecture by George Andrews on the theme "Reflections on the Rogers-Ramanujan identities and statistical mechanics". Vice-President of Research Win Phillips gave the Opening Remarks for Andrews' lecture. Earlier that day, McCoy met Win Phillips and expressed his appreciation of the many programs being conducted in the mathematics department.

During the first week of April, Sergei Tabachnikov from Penn State University visited the Department for a few days. Besides being an active research mathematician, he also directed the highly successful MASS Program at Penn State. MASS is an acronym for Mathematics Advanced Study Semesters. Under this program, some of the brightest undergraduates from around the United States gather at Penn State each Fall to be given a rigorous mathematical training. In addition to delivering a colloquium, Tabachnikov addressed the Teaching Seminar in which he discussed how the MASS Program is planned and conducted. George Andrews told me that the MASS Program was the brainchild of the Russian faculty members at Penn State, most notably Anatole Katok (now deceased).

As the Spring Semester ended, we honored four of our retiring faculty members in a ceremony where talks were given about their accomplishments. The work of Beverly Brechner was described by her accomplished former PhD student John Meyer. Zoran Pop-Stojanovic had during his long term in our department utilized his contacts in Yugoslavia (his home country) to recruit bright PhD students, and so it was appropriate that one such former student Horve Sikic who was visiting our department, spoke about Zoran's contributions. Murali Rao, our senior probabilist, summarized the work of Nicolae Dinculeanu in the field of analysis. Finally, Jerzy

Kakol from Poland who was visiting the Department to work with Steve Saxon, spoke about Saxon's work on nuclear spaces.

During May 3–4, there was the AMS Regional Meeting at San Francisco State University which I attended. My talk in a Special Session on q-series organized by Neville Robbins was on the first day. I spoke about new weighted Rogers-Ramanujan partition theorems. On May 3 morning, as I was getting registered for the conference, George Andrews told me that in the evening there was a play on Ramanujan called *Partition* at the Aurora Theater in Berkeley, and that we should try to see it. It was a Saturday evening, and so we felt there would little chance that we could get seats. But to our surprise, when Andrews called the theater, he was told that there were precisely two seats available in the front row due to a cancellation! So we took those seats immediately. It was an incredible stroke of luck, but I would like to believe that the Goddess of Namakkal who came in Ramanujan's dreams to give him remarkable mathematical formulas, had preserved these seats for us!! Upon return to Florida from the conference, I wrote a review of the play for *The Hindu*, India's National Newspaper, where it appeared in mid-May. The article can also be found in my book [B1] containing a collection of my essays relating to Ramanujan.

June 29, 2003, was the 25th Wedding Anniversary for me and Mathura, and in connection with that my daughters Lalitha and Amritha arranged a magnificent surprise party with 250 guests at the India Cultural and Education Center (ICEC) on May 17. Lalitha was at that time in law school at Stetson University in St. Petersburg, Florida, and she arranged everything from afar by contacting all our friends and relatives in Gainesville and outside. Amritha was in high school and staying with us at home; she helped with the preparation of a slide presentation by selecting photographs right under our nose without our knowledge. Many of our relatives attended the celebration from across the USA, and my daughters arranged accommodation for them at the homes of our friends in Gainesville. 2003 was also the 80th birthday year for my father, and since my parents were visiting us at that time, the event was also in their honor. I was amazed how my daughters pulled this off so successfully, and realized they had the organizational skills to succeed in life.

The question that was being asked at the surprise party was, what was I going to do for Mathura for our 25th Wedding Anniversary? It was something very special, but since I am not good at surprises, I planned the event with her — a holiday in paradise (Hawaii) at one of the grandest resorts in the world, the Hilton Waikaloa on the Big Island. Grand as this was, Mathura wanted to see a new destination, and so I added Guam, the pearl of the Pacific, to our holiday after Hawaii.

On June 28, the day prior to our wedding anniversary, Mathura conducted the Arangetram (Graduation Dance Performance) of one of her students Trina Chakravarty in Jacksonville. This was a major two hour program for which the Chakravarty family had invited about 500 guests. But we had to be in Hawaii on June 29. So what Mathura and I did was to depart Jacksonville early in the morning of June 29, and reach Kona on the Big Island by 5:00 pm taking advantage of the 6 hour time difference between Hawaii and Florida. So Mathura and I were in Hawaii on time on June 29 to enjoy a memorable dinner at the Hilton Waikaloa Resort on our 25th Wedding Anniversary. My parents and my daughters stayed back in Jacksonville at the home of our dear friends Venkat Kunisi and his wife Saraswathi, and joined us in Hawaii two days later to allow Mathura and me to have time for ourselves for a couple of days!

Prof. Venkat Kunisi (now retired) was with the Chemistry department at the University of North Florida, Jacksonville, where he successfully served as Chair of Natural Sciences for many years. At his gracious invitation, my father and I have given talks to the science faculty at the University of North Florida. Venkat and I share a passion for Carnatic music and tennis! My father and I have had several enjoyable games of tennis with him and our other Jacksonville friend Ganesh Kumar.

The Hawaiian stay was a fantasy come true. Mathura and I went to Hilo on the eastern side of the Big Island to have lunch at the Naniloa Surf Hotel where we had our honeymoon in July 1978. With our family we roamed about the island enjoying the breathtaking scenery of the Volcanoes National Park, the Waipio Cliffs, the Polulu Valley Lookout, and Humakoa Coastline. After the Big Island, we all spent three days in Honolulu enjoying visits to our favorite haunts.

After ten days in Hawaii, we all flew to Singapore, where after a two-night stay, my parents left for Madras, whereas Mathura, my daughters and I backtracked to Guam from Singapore via Tokyo. For all of us, Guam was a new destination, and for me, it brought back memories of my transit through Guam in 1967 with my parents on a Pan American flight from Honolulu to Singapore.

Guam met our high expectations. We stayed at Westin Guam and from the balcony of our room we had a spectacular view of Tumon Bay. Tumon Bay is like Hanauma Bay on the outskirts of Honolulu in that it is also of a horseshoe shape, and the water is very shallow so that it is ideal for snorkeling. But Tumon Bay is much wider, and unlike Hanauma Bay, it has become the main resort area of Guam with five star hotels lining its shores, but on a smaller scale than Waikiki in Honolulu.

We rented a Ford Crown Victoria and went around the island enjoying the scenic beauty which largely has been unspoilt by tourist exploitation. Indeed, in the interior of Guam, near the Tolofofo waterfalls, there is a small trench which we saw, where a Japanese World War II soldier was hiding for several decades unaware that the war had ended! This gives you an idea of the remoteness of Guam, because this soldier remained unidentified for decades even though there was build-up in many sections of Guam.

Guam has both a US Naval Base and an Air Force Base, and we drove past both of them. Indeed it was from the Andersen Air Force Base in Guam that the Boeing B-29 Superfortresses flew out to drop the atomic bombs on Hiroshima and Nagasaki. Close to Andersen Air Force Base, just a few miles off the shore, lies Mariana's Trench, the deepest point in the oceans at about 35,000 below sea level. We noticed that the water was of a darker blue in the section where Mariana's Trench is. The surf on this side of Guam is very rough, and swimming is prohibited for good reason. Little did we realize that before the end of the decade, Amritha would be taking up an assignment in Guam as a journalist for the Gannett News Network whose flagship is the newspaper USA Today.

From Guam we journeyed to Madras where the first main event was the release of Part II of my father's autobiography *The Alladi Diary* at Ekamra Nivas, on July 27, his 80th birthday in the Tamil (Lunar) Calendar. It was published by East-West Press, Madras, and Mr. N. Ravi, Editor of *The Hindu*, received the first copy. I was glad that Mathura and my daughters were present for the occasion. A few days later, they returned to Florida, but I stayed back in India until mid-August owing to certain programs.

During the first week of August, I went to Bangalore by train accompanied by my dear friend GP in response to several lecture invitations. Prof. Ramachandra had arranged my Colloquium at the TIFR Bangalore Centre, where I spoke on a quadruple product extension of the famous triple product identity of Jacobi. Prior to my talk, he hosted, as always, a lunch at the National Institute for Advanced Studies (NIAS), and this time I had the company of B. V. Sreekantan who had retired as the Director of TIFR. Sreekantan and my father were old associates at TIFR in 1948, and so we talked about the early days of TIFR over lunch. The next morning, I gave a talk at the Raman Research Institute (RRI) on the theme "Ideas from physics leading to Rogers-Ramanujan type identities". That afternoon I delivered another lecture at the Indian Statistical Institute (ISI), Bangalore, on "Applications of the sieve to probabilistic number theory". The ISI campus is located a good distance outside Bangalore. The ISI, which has its headquarters in Calcutta, had at that time, two more campuses — in New Delhi, and Bangalore.

I also utilized my visit to Bangalore to meet up with friends and relatives. Mathura's cousin Dr. Sundar, a highly successful nephrologist, graciously hosted a dinner at one of Bangalore's finest restaurants.

I stayed back in Madras for two weeks after my family left because of an International Conference in Probability and Stochastic Processes organized in honor of my father during August 8–9, to coincide with his 80th birthday (August 9). The conference was organized by two of his former students Profs. S. K. Srinivasan and A. Vijayakumar at Anna University, where Vijayakumar was Head of the mathematics department. Srinivasan was the second of my father's 21 PhD students, and had received his doctorate in the field of stochastic processes in the fifties. My father gave a magnificent talk on the opening session tracing how probability as a research discipline was introduced in Madras by him in the fifties. It was a pleasure and honor for me to speak at this conference and my talk was entitled "Connections between number theory and the solution of the Chandrasekhar-Munch equation by Ramakrishnan and Srinivasan". For the refereed proceedings of the conference published by Springer-Narosa in two volumes, I contributed the opening article to Volume I entitled "Contributions of Alladi Ramakrishnan to the mathematical sciences". I was back in Florida on August 12 to get ready for the events of the Academic Year 2003–04.

9.6 Fall 2003 and the First SASTRA Ramanujan Conference

While 2003–04 was billed as the Special Year in Applied Mathematics, all events of that Special Year (Conferences, History Lectures, etc.), were held in Spring 2004. Since there were many exciting events in Fall 2003 highlighted by the First SASTRA Ramanujan Conference in December, I describe here the events of Fall 2003 separately.

The academic year opened with a beautiful lecture on August 29 by Professor Zaleski on the History of Polish Mathematics. Poland has a great mathematical tradition and its mathematical luminaries include Copernicus, Benoit Mandelbrot, Alfred Tarski, and Stan Ulam who was a Graduate Research Professor in our Department.

A month later, Eberly Professor C. R. Rao (FRS) of The Pennsylvania

State University, arguably the most eminent statistician alive today, gave the Joint Math-Stat colloquium. As Mathematics Chair, I felt it was important to strengthen ties with other departments on campus, and this was part of such an effort. Actually the statistics department had for many years been trying to get C. R. Rao for a talk, and so they were delighted that I could arrange this visit; one reason to have named (or distinguished) colloquia is that under such a banner one is more successful in getting eminent speakers to accept invitations. Professor Rao's talk on "Anti-eigen values and anti-singular values of a matrix and applications to statistics" drew a large audience that included faculty and students even outside of mathematics and statistics since he is so well known.

C. R. Rao and my father were long time friends and had connections related to their research as well. My father's term as Director of MATSCIENCE intersected with Rao's position as Director of the Indian Statistical Institute (ISI) in Calcutta. After a long and highly successful term as Director at ISI, Rao moved to the University of Pittsburgh and then to Penn State University, where he continued his research and the mentoring of students and post-docs with vigor. Since Rao is such a reputed scientist and was so highly regarded, I arranged a reception and dinner at the India Cultural and Education Center (ICEC) for members of the Indian community in Gainesville to meet him.

In early October, the colloquium by Rao was followed that of Jean Bellisard, a professor of mathematics and physics at Georgia Tech, and a senior member of the Institut Universitaire de France. His visit was co-sponsored by France-Florida Research Institute (FFRI), the Institute of Fundamental Theory (IFT), and the Department of Mathematics. So it was a joint effort much like Rao's colloquium. Belissard mathematics colloquium was on "Tilings, aperiodic media and their non-commutative geometry". He gave talks at FFRI and the IFT as well. While we had dinner, the conversation led naturally to irrational numbers, and he evinced much interest in my work on irrationality measures dating back to 1978–79.

During the first week on November, I made a short trip to Penn State University where I delivered the department colloquium on the theme "New weighted Rogers-Ramanujan partition theorems", as well as the MASS Colloquium to bright undergraduates on the topic "Irrationality estimates using integration by parts". Mathura was with me on the trip. My two hosts were Sergei Tabachnikov, the Director of the MASS Program and George Andrews. C. R. Rao and his wife graciously hosted a dinner at their home where many of our Indian friends from State College were present. At

Rao's home we saw a fine collection of souvenirs and art work that they had acquired during their worldwide travels.

Visit to India for the First SASTRA Ramanujan Conference

The Fall term concluded on a high note with the SASTRA Ramanujan Conference in Ramanujan's hometown Kumbakonam that I was invited to organize, and which was inaugurated by the President of India.

SASTRA University located in Tanjore in the state of Tamil Nadu in South India, is a new private university that was founded in 1984. SASTRA is an acronym for Shanmugha Arts, Science, Technology, Research Academy. SASTRA in Sanskrit means the codes as laid down in the Hindu scriptures. Thus at SASTRA University there is a strong adherence to Hindu culture and traditions.

SASTRA also opened a branch campus in Kumbakonam (Ramanujan's hometown) called the Srinivasa Ramanujan Centre (SRC). In 2003 SASTRA purchased home of Ramanujan in Kumbakonam to maintain it as a museum. That home was in a dilapidated condition. So its purchase and maintenance by an academic institution is one of the most significant developments in the preservation of Ramanujan's legacy for posterity.

In the summer of 2003, I received a call from S. Swaminathan who was a PhD student at the University of Virginia. He introduced himself as the son of Vice-Chancellor R. Sethuraman of SASTRA University. He told me all about SASTRA, and said that to mark the purchase of Ramanujan's home, SASTRA would be conducting an International Conference on Number Theory and Secure Communications at SRC during Dec 20–22 to coincide with Ramanujan's birthday, Dec 22. He also said that the President of India, Dr. Abdul Kalam, would inaugurate the conference and declare Ramanujan's home as a national treasure. Finally, he said that SASTRA would like me to help organize the conference and that I would be provided funds to invite about ten speakers for whom international airfare would be given. I found this last statement quite surprising because seldom is international airfare paid to foreign participants of (mathematics) conferences in India. Swaminathan then clarified and said that the conference would be funded by the Indo-US Forum through a grant administered by the Department of Science and Technology (DST) of India, that approval had been obtained, but to formalize it, a PI from America was needed for the proposal being prepared; SASTRA wanted me to be that PI to which I readily agreed. After Swaminathan's phone call, I spoke to George Andrews, told him that something major was happening in India pertaining to Ramanujan, and that I would want him to attend the conference. To this he replied: "In view of your assurance, I will definitely go to India."

As expected, the full funding came through in early November and so I was all set to leave on December 9. I flew by United Airlines and Lufthansa, and reached Madras on December 10 via Washington DC and Frankfurt.

One week prior to the SASTRA Conference was a symposium in Bangalore at the Indian Institute of Science in honor of Professor Ramachandra for his 70th birthday. He was the father of analytic number theory in India, and had encouraged me all through my career. So two days after my arrival in Madras, I left for Bangalore to attend the conference. My talk on the theme "New polynomial analogues to the Jacobi and Lebegue identities" was the Opening Lecture of the conference on December 13. Imre Katai from Budapest and Shigeru Kanemitsu from Fukuoka, spoke after me. I had to return to Madras that night because I had to receive George Andrews who was arriving late night by the Lufthansa flight.

On Sunday, December 14, Andrews gave a magnificent lecture at our home Ekamra Nivas, arranged by my father under the auspices of the Alladi Foundation. It was a pleasure for us that Professor C. R. Rao who was in Madras then, attended the lecture. Following Andrews' lecture, we all had dinner at a friend's house next door. I mention this because as we were having dinner, we heard the sensational news on TV that Saddam Hussein, who had been in hiding for months, was finally captured!

The next day (December 15), I arranged a lecture of Andrews at Infosys, a very highly successful software company that had become a giant in the corporate world. He then left for a two-day trip to Bangalore on a lecture assignment before returning to Madras to join me on the trip to Kumbakonam for the SASTRA conference.

On December 19, a SASTRA van took all the international speakers including Andrews and me, by road to Kumbakonam. Mathura accompanied me on the trip. A very interesting thing happened on the drive to Kumbakonam which I shall narrate now: The van was going at breakneck speed dodging bicycles, motorcycles, bullock carts, stray dogs, and cattle. The foreign participants in the van, unused to this kind of driving, were too scared keep their eyes on the road. But Andrews who was fully experienced with India in all its forms, was cool as a cucumber and even laughing every time the van swerved to avoid running over a dog or a chicken! When I asked him about his opinion on the drive, he said he admired the skill of the driver, and that it all seemed like a ride in Disneyland!! Then suddenly we heard a loud noise from underneath the van. There was a mechanical problem and the van had to be repaired. Luckily, there was an auto mechanic shop right on the roadside and we stopped there for the van to be repaired. Right next to that workshop was a huge pit with garbage, and pigs were foraging in it. I did not want my foreign guests to see the ugly parts of India, but alas, now they had a full view of this pit next to us!

While we were waiting in the van as it was being repaired, there was a phone call to the SASTRA representative in the van from *Rashtrapathi Bhavan*, the mansion of the President of India in New Delhi, saying that the President's personal secretary wanted to talk to Andrews! The secretary had first called SASTRA University and was told that Andrews could be contacted in the van. The President of India wanted to know something about Ramanujan's mathematics and so Andrews patiently explained to his secretary some highlights of Ramanujan's work including the Rogers-Ramanujan identities!! Some of the foreign participants were amused that Andrews was talking something very important to the President's secretary from a rural roadside location with foraging pigs next to our van!

The repair of the van was quickly completed, and so we resumed our drive to Kumbakonam. After being shown the impressive main campus of SASTRA University in Tanjore, we were taken to the exquisite Sterling Resorts, where five star comfort was provided in a traditional Hindu village setting. Sterling Resorts (now called Indeco Resorts), is an amazing place. Upon arrival, we were all given a welcome foot massage. In ancient times, people travelled by foot, and so such a foot massage was especially soothing. We enjoyed it even though we had travelled by van. To provide a Ramanujan touch, the resort had no notepads or pens in the beautifully furnished rooms; instead in each room beside the bed, there was a slate and a piece of chalk, like the one Ramanujan used to write down the incredible formulae that the Goddess of Namakkal gave him in his dreams!

The International Conference was inaugurated on December 20 morning at a grand public function by Dr. Abdul Kalam, President of India. There were about a thousand guests for the inauguration. After the inauguration, the invited speakers were introduced to the President and a group picture was taken in front of the Ramanujan bust at SRC. The conference opened with an outstanding lecture by George Andrews entitled "Ramanujan for the twenty-first century", that set the tone for the academic program. I followed Andrews with a talk on "A new companion to Ramanujan's continued fraction". After me there were talks by Noam Elkies (Harvard University), and Anatol Balog (Hungarian Academy of Sciences, Budapest).

There was much sightseeing we did in and around Kumbakonam in between the sessions and in the evenings after the sessions concluded. We saw the Town High School in Kumbakonam where Ramanujan had studied. It was inspiring to see Ramanujan's humble home from where a thousand theorems had emerged! We offered prayers at the Sarangapani Temple, down the street from Ramanujan's home. We were impressed with the Ramanujan Museum at SASTRA University that contains a comprehensive collection of letters, documents, and photographs. We also went to Tanjore to see the gigantic Brihadeeswara Temple, whose grand Gopuram housing the sanctum sanctorum of Lord Siva was carved out a single stone! This temple is a UNESCO World Heritage sight. Mathura was especially pleased that our trip included visits to several sacred temples, for which the Tanjore district is so well known. While in Tanjore, we were taken to a shop of a local artisan, and there we bought a beautiful Tanjore painting, which adorns the wall of the *puja* (= worship) room of our house in Gainesville to remind us of that wonderful first visit to Kumbakonam and Tanjore.

Vice-Chancellor Sethuraman wanted to have a Ramanujan Commemoration Lecture as a public lecture on December 22, Ramanujan's birthday. Andrews gave this commemoration lecture entitled "The meaning of Ramanujan" as the concluding talk of the conference, and it was a fitting finale. Following Andrews' concluding lecture, all speakers met with the Vice-Chancellor who asked us if we had any suggestions for the future. All speakers said that SASTRA should conduct annual conferences like this every year around Ramanujan's birthday. The Vice-Chancellor heartily agreed to this suggestion. He then turned to me and said, "Professor, I request you to help SASTRA organize these conferences each year, like this one." That is how I became involved with SASTRA in organizing these annual conferences. Thus since 2003, I have spent December 22, Ramanujan's birthday, each year (until 2019) in Ramanujan's hometown Kumbakonam, as an annual pilgrimage!

9.7 Special Year in Applied Mathematics and other events (2004)

I first describe a few events of January 2004 before providing an account of the Special Year in Applied Mathematics.

After the SASTRA Conference, I stayed in India for two weeks with my family because I had been invited to the Harish-Chandra Institute (HRI) in Allahabad in early January. It was a very fruitful academic visit, but the journey in both directions was full of tensions because there were flight delays caused by the notorious fog in New Delhi.

HRI is in Allahabad which is not directly connected by air from Madras. So I purchased a Jet Airways ticket Madras to Varanasi (also known as Benares) via New Delhi. Varanasi, situated right on the river Ganges, is the holiest city in India, and is only a couple of hours away by car from Allahabad. In December–January 2003–04, ALL domestics flights in India on ALL domestic airlines were delayed because of the worst fog in decades in Delhi, and the delays in flights in an out of Delhi caused by the fog had a ripple effect throughout India.

I arrived in Varanasi on January 1 at 4:30 pm, about three and a half hours late. The chauffeur of the HRI car who was anxiously waiting for several hours, was relieved to see me. It was a bumpy two hour ride from Varanasi Airport to HRI which is on the outskirts of Allahabad. Upon arrival at HRI, I was greeted by my gracious host Professor Adhikari, whom I had known since the time he was a PhD student at MATSCIENCE in the 80s. He had arranged a spacious room for me in the very comfortable HRI Guest House where I had a relaxed conversation with him over dinner.

HRI was originally called the Mehta Research Institute when it was founded in 1965 as an institution devoted to research in mathematics and physics, much like MATSCIENCE. In 1990, the Institute moved to the spacious campus I was now visiting, and in 2000 it was renamed the Harish-Chandra Research Institute since Harish-Chandra who hailed from Allahabad, is the greatest Indian mathematician since Ramanujan. The campus is big enough that many faculty members are given houses on campus.

My two talks at HRI were both scheduled in the morning of Friday, January 2. In the first talk at 10:00 am, I spoke about "Schur's partition theorem, companions, generalizations and refinements". This talk provided the background for my next lecture at 11:30 am on the theme "A partition theorem of Göllnitz and its place in the theory of partitions", which included my joint work with George Andrews and Basil Gordon in Crelle's Journal (1995), and my later work.

After a lunch at the HRI Guest House with Adhikari, I departed by car for Varanasi airport but the return journey to Madras was an incredible drama owing to delated flights, and missed connections. With a great deal of effort on my part negotiating with the airlines to accommodate me on alternate flights, I managed to reach Madras on January 3 night, one day later than originally scheduled. I had just enough time to pack and leave for the USA the next night.

My Lufthansa flight to the USA via Frankfurt was at 10:00 pm on

January 4, but that evening Professor Abhyankar called on us at Ekamra Nivas. He was visiting MATSCIENCE and on every such visit, he made it a point to call on my father. He was a true friend and a man of principle.

The reason for my rushing back to the USA was that I had to attend the Annual Meeting of the AMS in Phoenix, and be at the Chairs Workshop all day of January 6. Mathura stayed back in Madras for a few more days. SASTRA very kindly bought my ticket so that I could be in Phoenix on the return journey. A useful item for me at the meeting was the Graduate Programs Focus Group Discussion led by Hyman Bass.

I departed Phoenix on January 8 to return to Florida via Minneapolis. At Minneapolis airport, Mathura met me; she was returning from India via the Pacific on Northwest Airlines, and so we travelled on the final leg from Minneapolis to Florida together. Back in Florida, I had to get ready for a spring semester packed with events, several of which were connected with the Special Year in Applied Mathematics.

On January 16, we had the visit of Fields Medallist S. T. Yau of Harvard University. He was visiting the University of Florida to deliver the Barr Systems Distinguished Lecture, but my colleague David Groisser who was the Colloquium Chair, arranged for Yau to address the mathematics colloquium. Yau's talk was entitled "Local mass in general relativity". It was an honor and pleasure for me to welcome Yau to our department, especially because I had known him since my graduate student days at UCLA.

In view of the Special Year in Applied Mathematics, the Department launched the Center for Applied Mathematics (CAM) Colloquium in Spring 2004. The CAM was created when Stan Ulam was a Graduate Research Professor. If a department has a Center, and if a research grant is run through the Center, then in addition to a part of the overhead money coming to the researcher, a certain percentage of the overhead is also given to the Center by the University Office of Research. Thus by having a Center, a department gets more funds to support various programs.

Quite fittingly, the First CAM Colloquium was delivered on January 24 by the eminent applied mathematician Tony Chan, Dean at UCLA, on the theme "Variational PDE models and algorithms for image processing". As Chair, Chan had been instrumental in creating the Institute for Pure and Applied Mathematics (IPAM) at UCLA, as an NSF funded institute. Since Chan was Dean now, we had Dean Neil Sullivan to deliver the Opening Remarks for the CAM Colloquium. Chan's CAM Colloquium was the first featured public lecture of the *International Conference on Mathematical Methods in Imaging and Vision* (January 24–27) organized by Yunmei

Chen, Tim Olson, and David Wilson. The second featured lectured of the conference was the Ulam Colloquium on January 26 by Stan Osher (also from UCLA) on the theme "Mathematics in the real world and in the fake world". By fake world he meant the world of movies. I had known Osher when I was a graduate student at UCLA in the seventies and he had just joined the department then. He has a wry sense of humor and it came through in the lecture. I felt pleased that during the Conference Banquet, both Dean Sullivan and Tony Chan in their after dinner speeches, praised the Department for various programs it was conducting. One direct consequence of both Tony Chan and Stan Osher visiting the Department and forming a positive opinion after interacting with the students, was that the very next year, one of Yunmei Chen's graduate students received the Hedrick Assistant Professorship at UCLA on completing the PhD.

In mid-February, I made a one week trip to China to attend the Third China-Japan Number Theory Conference at Northwest University in Xian. I departed on February 9 by Northwest Airlines, and after spending the night of February 10 in Beijing, I flew to Xian the next day by Air China. Professor Wenpeng Zhang, my host, and Professor Jia of Beijing received me at the airport and took me to the university guest house. The room was spacious and generally comfortable, but it was cold, so much so, I slept every night with a sweater on! Professor Zhang had very thoughtfully made sure that excellent vegetarian dishes were made available during the buffets at the Guest House — breakfast, lunch and dinner.

The conference was inaugurated by the Vice-President of the University on February 12 morning. Following this, I gave the Opening Lecture entitled "Sums of the Moebius function". My good friend Shigeru Kanemitsu (Fukuoka), the overall organizer of the China-Japan conference series, chaired the session when I spoke. During the break after my talk, Kanemitsu and Professors Akiyama, Matsumoto and Tanigawa spoke to me about the refereed conference proceedings which they wanted to be published in my book series Developments in Mathematics (DEVM). These proceedings edited by Tanigawa and Zhang appeared as Volume 15 in DEVM two years later. On the second day of the conference, in the evening after dinner, I had a one and half hour discussion with Zhang's graduate students on "Sieve dimension".

On February 14 morning, I checked out of the Guest House, and after a visit to the celebrated Terra Cota Soldiers Monument outside of Xian, I took a late evening Air China flight to Beijing where I spent the night before returning to Florida the following day on Northwest Airlines.

During late February–early March there were three conferences back-to-back at the University of Florida. First was the *Conference on Multiscale Optimization Methods and Applications* (Feb 26–28) organized by my colleague Bill Hager along with Professors Tim Davis (Computer Science) and Panos Pardalos (Industrial Engineering). Dean Tony Chan returned to Florida to deliver a featured lecture at this conference on "Multi-level optimization for circuit placement".

The next conference was on *Computational methods in multi-scale analysis and applications* (Feb 29–Mar 2), organized by colleagues William Hager, Shari Moskow, and Jay Gopalakrishnan along with faculty in Mechanical and Aerospace Engineering, which I was asked to inaugurate. The featured lecture of the conference was the Erdős Colloquium on March 1 by 1966 Fields Medalist Stephen Smale of UC Berkeley and the Toyota Technical Institute, Chicago. He spoke on "Shannon sampling and reconstruction of functions from point values". Smale was awarded the Fields Medal for his work on topology and dynamical systems. During lunch with Professor Smale, when talking about dynamical systems, our conversation led to irrational numbers, and in connection with a question he had, I made an observation. I was flattered that a few months later, in one of his papers, he acknowledged my comment. I was also impressed with his honesty that he acknowledged even a simple comment I made.

These two conferences were followed by a Student Workshop (March 3–4). In connection with these two conferences, we also had three History Lectures of appeal to graduate and undergraduate students. Thus the programs of the Special Year in Applied Mathematics were comprehensive and of interest to students and faculty of various departments on campus.

On March 10, we moved into our new home in the community of Richmond, a home in which we are now living, a home which has since then been the venue of so many parties and get-togethers (both academic and social) well attended by our Indian friends in Gainesville and surrounding areas, and by several colleagues at the University.

My parents arrived in mid-April, a few weeks later than usual, to spend a good part of the summer with us. On the day of their arrival was the colloquium of the eminent mathematical physicist Alexander Varchenko from the University of North Carolina. Since he had also done significant work in Combinatorics, we arranged his talk in the Number Theory Seminar the next day.

With all major academic activities in the department having ended for the Spring Term, I arranged the first get-together at our new home on

April 23, it was for the release of Volume II of The Alladi Diary. Volume II was released in Madras in December 2003, but since my parents were here, I wanted a release function in Gainesville as well; it was attended by a number of our Indian friends and some of my colleagues. It was a fine gesture on the part of one of my father's old students Dr. V. K. Viswanathan (VKV) and his wife Selvi, to come all the way from Los Alamos to spend a few days with us and attend this function. VKV and Selvi are among our closest friends, and so we were delighted that VKV spoke on the occasion describing how my father conducted the Theoretical Physics Seminar in our family home Ekamra Nivas between 1959 and 61, and how it inspired him and other students of the seminar.

A hectic summer of travel to Hawaii, Italy and India (2004)

On June 7, just before our summer travels, Dean Sullivan renewed my contract as Chair for another four years, that is until June 30, 2008.

I departed for Honolulu with Mathura on June 8 to attend and speak at the Hawaii International Statistics and Mathematics Conference (June 9–11). Our outbound journey had one of the most terrifying flying experiences. About two hours after our Northwest DC-10 departed Minneapolis for Honolulu, while we were enjoying lunch as we were flying over the Rockies, the Captain announced that we were to make an emergency landing because the smoke alarm in the cargo hold had come on. Fortunately the airport at Billings in Montana nearby could receive such a large aircraft, and so he initiated a rapid descent, ejecting much of the fuel as we descended. The reason for dumping the fuel is that if this is not done, the aircraft would be overweight for landing; also having less fuel minimizes fire danger. The 20 minutes it took to descend were the most nerve-wracking moments of our lives! The aircraft touched down smoothly and halted far away from the terminal. There were fire trucks at various points on the runway in case there would be a fire at touchdown. The aircraft was thoroughly inspected, and we were told that it was a false alarm, that there was no smoke in the cargo hold. But the faulty fire alarm sensor had to be replaced. By the time this was brought, more than eight hours had lapsed since our original departure from Minneapolis, and so a new crew had to be flown in to operate the aircraft. All this caused a delay of about 12 hours and Billings airport where we waited had meagre facilities. But we were glad that we were alive, and informed my parents and daughters in Florida about the situation. After the smoke alarm was replaced, the Northwest DC-10 took off and the flight to Honolulu was smooth and enjoyable.

The conference venue was the fabulous Sheraton Waikiki Hotel where Mathura and I had our Honeymoon in 1978, but we stayed at the magnificent Hilton Hawaiian Village nearby where we had a spacious condominium on a high floor that commanded a splendid view. The conference opened on June 9 morning with the keynote address of Andre Agresti of the statistics department of the University of Florida. While I attended the conference lectures, Mathura roamed about the International Market Place in the throbbing heart of the Waikiki beach area.

I always use any trip to Hawaii to visit the mathematics department at the University of Hawaii. It was kind of Edward Bertram to arrange my colloquium there on June 10. I spoke on "A monotonicity principle for the sieve and applications to probabilistic number theory". My talk on "Sieve methods in Probabilistic Number Theory" was on the afternoon of the final day of the conference, with Maurice Mignotte from Strasbourg, France, following me. He is an authority on Diophantine Approximations. I was asked to Chair the concluding session on number theory.

For relaxation during the breaks or in the evenings, Mathura and I got together with our long time friends the Ramanathans at their lovely home, or enjoyed the wide range of facilities at Waikiki beach. We spent three relaxed days in Honolulu after the conference before returning to Florida.

On July 31, Mathura conducted the Arangetram of one of her students. following which, my parents left for India the next day.

Mathura, Amritha and I departed on August 3 on a twelve-day trip to Italy. The academic part of the trip was the visit to the International Center for Theoretical Physics (ICTP) in Trieste.

Trieste is a lovely town on the Adriatic coast bordering Yugoslavia. We flew to Rome and took a train there to Trieste via Venice.

The ICTP is located in Grignano, a lovely seaside resort area outside Trieste. We were accommodated at the Adriatico Guest House of the ICTP, and from the balcony of our room, we had a beautiful view of the Adriatic coastline. This guest house used to be the Adriatico Palace Hotel where as a boy I had stayed with my parents in the sixties. As a guest house it was very comfortable, but gone was the refined luxury as a palace hotel.

The new Director of the ICTP was Katepalli R. Srinivasan, a physicist, who during our meeting on August 6, said he wanted to do something related to Ramanujan and requested some information which I was able to provide after returning to Florida. At that time I did not realize that he was thinking of launching the ICTP Ramanujan Prize (see next section for my idea to launch the SASTRA Ramanujan Prize).

My lecture at the ICTP was that afternoon on "Sieve methods and probabilistic number theory", which was attended by the head of the mathematics division Professor Trang and the former head of mathematics M. S. Narasimhan, who was visiting ICTP. Trang was educated in France and belongs to the Grothendieck school of algebraic geometry.

I have wonderful memories of my visits to Trieste in the sixties in my boyhood days. I was glad Mathura and Amritha were with me so that I could show them my favorite haunts in and around Trieste. The ICTP is located right next to the elegant Miramare Castle which has one of the finest locations fronting the sea. The castle is surrounded by the sprawling and beautiful Miramare gardens. So on the afternoon of the first day, after checking in at the ICTP, we went to see the Miramare Castle and gardens. We also took a boat from the Grignano pier to Trieste, where we went to Piazza Oberdan to see the first premises of the ICTP from 1964 to 67.

After this nostalgic and fruitful visit to the ICTP, we had a splendid vacation in Italy visiting Venice, Florence, Pisa, and Rome, in that order. We saw many of the major attractions in these historic cities, but here I report only on one item of sightseeing due to its mathematical connection.

Every visitor to Florence sees Michelangelo's David, but relatively few see the Sagrestia Nuova in the Basilica di San Lorenzo where four fantastic male statues representing Day, Night, Dawn and Dusk sculpted by Michelangelo sit over the tombs of Giuliano Medici and Lorenzo Medici. According to G. N. Watson who investigated Ramanujan's 3rd order mock theta functions, the grandeur of Ramanujan's identities can be compared to the austere beauty of these four statues. Basil Gordon, my great Guru, stressed that if at all I visit Florence, I must see these four statues for otherwise, the trip would be incomplete. So following his advice, we went to the Medici Chapel. We felt that these statues were incredibly beautiful and well worth seeing.

After a memorable visit to Italy, we flew back from Rome on the new Airbus A-330 that Northwest had introduced to begin the phasing out their DC-10 workhorse.

Two days after returning from Italy, I departed for Madras by myself in response to an invitation from INFOSYS, the giant software company in India. My contact at INFOSYS was a young energetic fellow by name Sriram Gopalan who, prior to joining INFOSYS, had received some mathematical training under my father. Sriram had suggested to Mr. N. R. Narayana-murthy (NRN), the visionary Founder and Chairman of INFOSYS, that INFOSYS should conduct a Mathematics Olympiad for high school students. NRN liked this idea, and following Sriram's suggestion, invited me to set the question paper for the Olympiad, which I did over the summer. Along with the questions, I sent all the solutions. The actual grading of the answer sheets was done by a group of professors from academic institutions in Madras. The prize ceremony was on August 26 in Madras, but Mr. Narayanamurthy not only wanted me to speak at the Prize Ceremony, but also to give lectures both in Madras and in Bangalore, the head-quarters of INFOSYS, to the students who took the Olympiad; these lectures were be on mathematics related to the Olympiad problems. I accepted this invitation and went to Madras via the Pacific on Malaysia Airlines. Enroute, I stopped in Los Angeles for a day to work with Basil Gordon.

My four lectures in Madras, spread over two days were before the prize ceremony. The themes of my talks were: (i) The pigeon-hole principle and Diophantine approximation, (ii) Arithmetical functions and the Ramanujan number, (iii) Partitions, q-series and Ramanujan, and (iv) Gems from geometry, the charm of Fibonacci numbers, and set theory.

I was taken aback by the beauty and cleanliness of the INFOSYS campus in Madras — what a contrast to the dirt and confusion in the city surrounding it! Sriram told me that Mr. Narayanamurthy inspects everything and has the highest standards of efficiency and perfection.

Mr. Narayanamurthy was there for the Prize Ceremony. The Chief Guest was M. S. Ananth, the Director of the IIT. I was asked to give a one hour speech since I had set the Olympiad question paper. Mr. Narayanamurthy liked my speech on "The importance of spotting and grooming young mathematicians". He invited me to the Bangalore campus of INFOSYS, requested me to repeat the Madras lectures and said he would attend them in Bangalore, which he did.

NRN very graciously put me up at the magnificent Taj Hotel in Bangalore and sent a car each day to take me to the INFOSYS campus. Sriram who accompanied me to Bangalore told me that the Bangalore campus is an order of magnitude bigger and more beautiful than the Madras campus! And indeed it was. I was surprised and impressed that even though his day was packed with appointments and meetings, Mr. Narayanamurthy would come down from his office to attend every one of my lectures to the Olympiad students. When I went to his office to take leave of him and thank him for his gracious hospitality, he said that the explosion in information technology in India snatched away the brightest minds from fundamental sciences, especially mathematics. He felt that for the kind of work INFOSYS and other software companies do, bright people are needed, but not the most brilliant minds who should really be doing research in the fundamental sciences. Since the IT sector has benefitted immensely from mathematics graduates, he said that he would like to do something in return to encourage talented youngsters to take to mathematics. So when Sriram suggested the Olympiad, he heartily endorsed it. A few years later he instituted the Infosys Prize in mathematics.

I returned to Madras from Bangalore on August 29. Two days later I left for Florida, accompanied by Mathura's mother who agreed to spend a few months with us in our new home. Back in Florida, I looked forward to an event-packed Special Year in Number Theory and Combinatorics.

9.8 Special Year in Combinatorics and Number Theory and the idea to launch the SASTRA Ramanujan Prize (2004–05)

Although each of the Special Years were filled with significant events, the Special Year in Combinatorics and Number Theory was especially dear to me since it pertained to my area of research and my professional activities in the field. The Department received significant external funding for this, with a sizeable NSF grant being well supplemented by grants from the NSA and the Number Theory Foundation (NTF).

Events of Fall 2004: The Special Year got off to a stormy start. The *Fifth International Workshop on Automated Deduction in Geometry* organized by Neil White of our Combinatorics group, took place during September 16 and 18, between two hurricanes Frances and Jean! We were spared the wrath of the hurricanes for the two days of the conference, and all but one of the participants were present. This was part of a well established series of workshops that take place every two years in different locations across the globe. In conducting the Special Years, we felt that in addition to having our own conferences, it was equally important to host certain well established conferences. One of the two plenary speakers was Doron Zeilberger, Board of Regents Professor at Rutgers University, who delivered an electrifying lecture. Several participants spent extra time in Gainesville to give talks in our seminars. The longest stay was by Don Ming Wang from Paris and Beihang University, China, who gave a History Lecture on September 13 on the theme "Automated Geometric Reasoning".

The Seventh Ulam Colloquium was delivered on October 11 by Richard Stanley (Norman Levinson Professor of Applied Mathematics at MIT), arguably the most influential combinatorialist of our generation. His talk was entitled "A survey of lattice points on polytopes". The previous night there was a party in Stanley's honor at the home of Neil White who knew Stanley from the seventies because both were PhD students under Gian-Carlo Rota at that time. Stanley's book on *Enumerative Combinatorics* is a Bible in the field. Stanley also has been the PhD advisor to dozens of brilliant students many of whom have had very successful careers as research mathematicians; Miklos Bona who was appointed at the University of Florida during my first year as Chair, is a former PhD student of Stanley.

The month of October ended with the magnificent History Lecture (in Number Theory) by David Bressoud (Macalester College) on "The alternating sign matrix conjecture", the second time he was lecturing on this topic in our Department (at our request), but this time since it was a featured History Lecture, there were several students in the audience. Bressoud's lecture was followed by the History Colloquium (in combinatorics) on November 1 by Gyula Katona (Director, Alfred Renyi Institute, Budapest) who spoke on the "History of the extremal problems for set systems".

In between the events of the Special Year, I made a two-day trip to Ohio State University in the second week of November with a two-fold purpose. First on Sunday, November 7, I participated in a one day Meeting for Chairs and Partners at the Mathematical Biosciences Institute (MBI) at the kind invitation of its Founder and Director Avner Friedman. The MBI is one of the NSF Funded Institutes, and it was useful to learn about their comprehensive program. Prior to coming to Ohio State, Friedman was at the University of Minnesota, where he served as Director of the Institute for Mathematics and its Applications from 1987 to 97. I was told that Ohio State approached Friedman for help with the proposal for the MBI with the understanding that if it is funded, then he would move to Ohio State as the Director of the MBI. This happened in 2002.

Next on Monday, at the invitation of Steve Milne, I addressed the Algebra and Number Theory Seminar on the theme "New weighted Rogers-Ramanujan identities modulo 6 and 7 and their implications". I had lunch with the Mathematics Chair Peter March and we had a fruitful discussion comparing the programs in our two departments. March was so successful as an administrator, that he later moved up to hold deanships at several universities, and was also Director of the Mathematical Sciences Division at the NSF.

The next major event in the Department was the *Additive Number Theory Conference*, November 17–20, attended by more than sixty active researchers from around the world. By additive number theory was meant the areas of classical number theory dealing with additive questions like the

Goldbach and Waring problems, as well the theory of partitions. Thus it was the first conference to bring together researchers in classical analytic number theory and the theory of partitions. The NSF liked the cross-fertilization that this would bring about, and thus funded our proposal.

The conference opened with the splendid lecture of George Andrews on "Partitions with short sequences and Ramanujan's mock theta functions". Associate Dean Jack Sabin who inaugurated the conference, stayed to attend Andrews' talk and was charmed by his charismatic lecturing style.

One of the highlights of the conference was Basil Gordon's talk on the second day on the exciting theme "The return of the mock theta functions". Gordon spoke on joint work with his former student Richard McIntosh (University of Regina) on the new 8th order mock theta functions they had discovered. The title was especially appropriate and catchy because G. N. Watson in delivering the very first ever lecture on mock theta functions titled it as "The Final Problem", after the famous concluding Sherlock Holmes episode, because the mock theta functions were the last great discovery of Ramanujan from his deathbed. Following Gordon, McIntosh spoke about mock-theta functions giving details as to their asymptotics. Gordon had actually arrived a few days before the conference to work with me and so we asked him to address the Number Theory Seminar the week prior to the conference. The sessions for the second day concluded with Bruce Berndt's History Lecture entitled "Ramanujan: his life, friends, notebooks, and identities for the Rogers-Ramanujan functions". That night there was a conference party at our home so graciously arranged by Mathura. There were about 80 guests because in addition to the conference participants, several colleagues of mine attended the party with their spouses. These colleagues helped by giving rides to several participants to come to the party, but we also had arranged for a bus to bring many of the participants to our house.

The third day of the conference opened with the plenary talk of Mel Nathanson who spoke about bases and asymptotic bases, a topic on which he is a world authority. The afternoon plenary talk was by James Lepowsky (Rutgers) on partitions and vertex operator algebras. In the 80s, Lepowsky and Robert L. Wilson had given proofs of Rogers-Ramanujan type identities by a study of vertex operators in Lie algebras. Following that seminal work, the theory of vertex operators has become a fertile ground for the discovery of several new Rogers-Ramanujan type partition theorems. A good example is the pair of new partition theorems discovered by Stefano Capparelli (a PhD student of Lepowsky) in the early nineties, for which George Andrews,

Basil Gordon and I provided proofs, generalizations, and refinements. In fact, Lepowsky's talk was immediately followed by that of Capparelli.

The refereed proceedings of this conference edited by me and titled "Surveys in number theory" appeared in 2008 as Volume 17 in the book series Developments in Mathematics (Springer).

In early December, Mathura, her mother, and my older daughter Lalitha, departed for India via the Pacific. My second daughter Amritha and I followed them a week later.

The second SASTRA Ramanujan Conference and the idea to launch the SASTRA Ramanujan Prize, December 2004

For this second conference at SASTRA University in Ramanujan's hometown Kumbakonam, Vice-Chancellor R. Sethuraman very kindly invited me to inaugurate the conference, and to deliver the Ramanujan Commemoration Lecture on December 22, Ramanujan's birthday. The conference turned out be extremely significant because it was there that I proposed the creation of the SASTRA Ramanujan Prize. I was pleased that this time my second daughter Amritha who flew with me to India, accompanied me to Kumbakonam. I was also pleased that my colleague Frank Garvan attended the conference. Finally, Andrews' elder daughter Amy and her husband Mark Alznauer who were visiting India in December, were at this conference in Kumbakonam as well.

Amritha and I flew to India on Northwest Airlines from Orlando via Minneapolis and Amsterdam to Bombay. Before boarding our Jet Airways flight to Madras, we enjoyed a fine Indian breakfast in the Jet Airways lounge in Bombay. It is sad that Jet Airways is now gone owing to passengers preferring the no frills low cost carriers in spite of their poor service on the ground and in the air. Frank Garvan arrived in Madras by Air India a day later but his luggage was not there! So GP and I took him to a clothing store near my house where he bought some essential items before boarding the van to Kumbakonam with us.

On December 20 morning, just as we were ascending the dais for the conference inauguration, Vice-Chancellor Sethuraman whispered into my ear that he would like to set apart $10,000 every year for a worthy cause in the name of Ramanujan. He requested me to decide how it should be used and announce this in my inaugural speech to follow in a few minutes. I was surprised by this offer from the Vice-Chancellor and asked him whether he wanted me to make the decision rightaway without consulting any committee? He said committees will cause unnecessary delays, and there will be lack of agreement as well.

Inaugural ceremonies are elaborate events in India starting with a prayer song, then the auspicious lamp lighting ceremony (the lamp representing enlightenment or knowledge), followed by a fairly long welcome of the audience and the chief guest (in this case, me), then a presentation of a memento to the chief guest, etc. Thus I had a full 15 minutes to think about how to use the $10,000 as all this was going on. When I was asked to inaugurate the conference and deliver my speech, I said that the Vice-Chancellor had just informed me that he would set apart $10,000 annually for something significant to be done in the name of Ramanujan. I also said that the Vice-Chancellor asked me how these funds ought to be used and requested me to announce it at this inauguration. So I suggested the creation of the SASTRA Ramanujan Prize with the $10,000 to be given annually to a mathematician not exceeding the age of 32 for path-breaking contributions to areas influenced by Ramanujan. I pointed out that the Fields Medal is given to mathematicians not exceeding the age of 40, and that with 32 as the age limit, we will be recognizing mathematicians much earlier, some of whom might go on to win the Fields Medal (this has actually happened!). I said that the age limit should be 32 because Ramanujan lived only for 32 years and so this is a challenge to the candidates as to what they can accomplish in that time frame. The Vice-Chancellor liked this proposal of mine and said in his response to my speech that the prize will be given at the conference next year in December 2005. He then turned towards me and said, "Professor, I would like you to chair the Prize Committee". Everything happened in Kumbakonam within the span of an hour and that is how I got involved with the Prize. I have chaired the Prize Committee since the inception of the prize.

Frank Garvan gave the Opening Lecture of the conference on ranks and cranks of partitions. This was very appropriate since it was he, in collaboration with Andrews, who solved the famous conjecture of Dyson and found the crank for partitions that combinatorially proved all three of Ramanujan's partition congruences modulo the primes 5, 7, and 11. Following Garvan, I gave my regular conference lecture entitled "How many prime factors does a number have?"

During the afternoon of Dec 20, we did plenty of sightseeing in Kumbakonam and Tanjore, visiting either famous temples, or places especially connected with Ramanujan. Inspired by what she saw in Kumbakonam, Amy decided that she would write a children's book on Ramanujan, which just got published in January 2020. The conference talks all ended on December 21. The only event on December 22, Ramanujan's birthday, was

my Ramanujan Commemoration Lecture "Srinivasa Ramanujan and probabilistic number theory". Frank, Amy, Mark and I returned to Madras
on December 22 by the beautiful coastal highway. Some participants were
flying out of Madras that night to be with their families for Christmas.

Amy, Mark, and Frank stayed back for a few days in India. At my
suggestion they rented a chauffeured car and went on a one week excursion
trip of South India. My dear friend GP accompanied them as a wonderful
companion and guide. Since GP was taking care of them, I stayed back in
Madras to spend time with family and friends.

On the morning of January 26, when I was playing tennis with my
good friend V. K. Venkatesan at a club near his house in the Kotturpuram
section of Madras, we felt a slight tremor, as if due an earthquake! Unlike
the Himalayan region where the mountains are being pushed up due to
collision of the Indian subcontinent with the continent of Asia, South India
does not experience earthquakes. So this tremor was very unusual. The
conversation at the club turned to earthquakes. I said that in the far east,
earthquakes trigger what are called Tsunamis, but that no such thing has
ever affected India as far as we know. I returned home around 8:30 am after
tennis. Within a few minutes, we heard people screaming and running along
our street. They said they were running away from a gigantic tidal wave
that had engulfed the beach due to an upsurge of the ocean water. Indeed
it was a TSUNAMI that had hit the Madras coast, and beach was just a
two miles away from our house!

The effect of the Tsunami was horrendous. Several villages on the eastern coast of India were destroyed, some even wiped out. Hundreds of people
relaxing on the famous Madras Marina Beach were washed away. We later
heard that the Tsunami affected the eastern coast of Malaysia and even
Phukhet in Thailand. It was caused by a massive earthquake in the sea
bed of the Bay of Bengal.

On the day of the Tsunami, Frank, Amy, Mark, and GP, were in the
hill station Ootacamund (Ooty), about 7000 ft above sea level and in the
interior of South India. So they were unharmed. News of the Tsunami
spread across the world, and George Andrews who heard this, called me
anxiously to talk to his daughter and son-in-law. I assured him that they
were safe in the hills of Ooty. It was ironic that just a few days before the
Tsunami, we had travelled along the coastal road and everything looked
lovely. Now it was a scene of disaster.

Amy, Mark, Frank and GP returned to Madras on December 28. The
next day, GP took them and my older daughter Lalitha to see the Madras

High Court, where my grandfather had reigned as one of its most eminent lawyers for four decades. It was especially important for Lalitha to see the High Court since she has taken to a career in law. There at the High Court, they happened to meet Mr. Parasaran, the Attorney General of India. When GP introduced Lalitha to her, he waxed eloquent about my grandfather Sir Alladi! Frank, Amy and Mark departed for the USA that night after a most memorable visit to India.

December 26, 2004, will be remembered as the day the Tsunami hit the coastline around the Bay of Bengal with devastating effect. But something nice also happened that day. Motivated by the Infosys Mathematics Olympiad that took place earlier in August, SASTRA held a Mathematics Olympiad on December 26 and several thousand school students across India took part in it. SASTRA asked me to set the question paper for their Olympiad, which I did. The winner of the Olympiad was given a scholarship by SASTRA to spend one month with me in Florida in spring 2006 for mathematical training.

Finally I should say that 2004–05 was the Centenary of Special Relativity. To mark that occasion, my father brought out a book which was a compilation of all his research papers and popular articles on Special Relativity. This book was released at a well attended function at our home in Madras on December 30, and I was glad that my family and I were present for that memorable occasion.

Events of the Spring and Summer of 2005: On New Year's Day morning, my daughters and I departed Madras for Florida. Mathura stayed back for two weeks to help her mother who took ill after returning to India from Florida in December.

Upon return to Florida, I departed for Atlanta the same day to take part in the AMS Chairs Workshop held on January 4, one day before the start of the Annual Meeting of the AMS. I had taken a very active part in Chairs Workshops since my first year as Chair, and so I was invited to be one of three Workshop Leaders and to serve a three year term starting with 2005. I led a discussion on "Strategic Planning and Development" which had become so important for mathematics departments in view of unreliable funding from university administration for special projects. That night, Sam Rankin III, the Assistant Executive Director of the AMS who manages these workshops, took the three of us workshop leaders for a high class dinner at a Zagat rated restaurant at the Four Seasons Hotel. Rankin is a superb host, and we enjoyed his lavish hospitality. During the dinner,

Rankin told us how aggressive the program directors in physics and chemistry are; they would ask for huge chunks of money and their allocation would consume most of the NSF budget for the sciences, leaving a paltry amount for mathematics. I stayed on in Atlanta for two more days to attend the AMS meeting and to have discussions with John Martindale about the Ramanujan Journal and my book series. Since I am actively involved in publishing, the Annual Meeting provides an excellent opportunity for me to discuss with various publishers.

In connection with the activities of the Special Year, we had a spate of visitors during the Spring Term. After the AMS Annual Meeting, George Andrews arrived to spend the entire spring term with us and to give a series of six lectures on Ramanujan's Lost Notebook between January and March. The first of these lectures entitled "Ramanujan's Lost Notebook, An Overview: What did Ramanujan have up his sleeve?" was on January 24, and it attracted a huge audience of faculty and students. Dean Sullivan who made the Opening Remarks was very pleased at this lecture series and its wide appeal. In addition, Andrews gave two more lectures on "The joy of collaboration - why pure mathematicians should not mind their own business" (February 14) and on "Research mathematical scientists and mathematics education" (February 22) arranged by the College Dean's Mathematical Sciences Committee. These talks were at the Keene Faculty Center so as to attract faculty from various departments in the College.

During the previous year, I had discussions with Andrews about his long term association with Florida, following which he made arrangements with Penn State University to double his duties there in the Fall so that he could free all of the Spring term to visit Florida. Spring 2005 was the first of his annual full semester visits to the University of Florida.

Another visitor to spend a substantial part of the Spring Term with us was Hershel Farkas of the Hebrew University, Jerusalem. He gave two talks on "The theta function in combinatorial analysis and number theory", the first of which was on January 25. Farkas has a long standing collaboration with Irwin Kra on the study of Riemann surfaces. In 2001, Frakas and Kra published a major book with the AMS on "Theta constants, Riemann surfaces, and the modular group". Toward the end of the book, they provide several partition identities that emerge from their study of theta constants. Since Florida has a strong research program on partitions, he chose to speak about this aspect of his work during his visit to Gainesville.

Our next visitor was Michel Waldschmidt (University of Paris) who gave a marvellous History Lecture on February 21 on "Elliptic functions and transcendence". This lecture was co-sponsored by the France-Florida Research Institute (FFRI). After the revolutionary work of Alan Baker in the sixties, the theory of transcendental numbers experienced a worldwide resurgence in activity. Waldschmidt was one of those who led the development of transcendence theory in France since the seventies.

Even our Spring Break was full of academic activity. On February 28, at the start of Spring Break, Dean Sullivan inaugurated the five day *Conference on Arithmetic-Algebraic Geometry* organized by my colleagues Richard Crew, Kevin Keating and Norm Levin. This was immediately followed by the *Conference on Pattern Avoiding Permutations* (March 7–11) due to effort of our colleague Miklos Bona, a world authority on this topic. Like the Workshop on Automated Reasoning that we had in September 2004, this conference was part of a well known series which was brought to Florida on account of our Special Year.

On March 7 we had a special Reception in honor of John Thompson and George Andrews at the Keene Faculty Center in the morning. At that reception, the two-volumes we had brought out as the Proceedings of the Special Year Algebra 2002–03 in honor of Thompson were released by UF President Bernie Machen who had just taken office in Fall 2004. All the participants of the Permutations Conference, our colleagues, and members of the Dean's office attended this reception and so it was a grand affair. That afternoon, Fields Medallist Richard Borcherds of UC Berkeley delivered the Erdős Colloquium on the topic "Feynman path integrals and the Bernstein polynomial". It was fitting that Dean Sullivan, who is a physicist, gave the Opening Remarks. My father who had done important work on Feynman graphs in the sixties, enjoyed the lecture of Borcherds.

I had gone to the airport that morning to receive Borcherds who had taken an overnight red-eye flight from California. I noticed that he did not have any hand piece, nor did he have any checked-in luggage. When I expressed surprise at this, he reached to his back, and from inside his undershirt, he pulled out a few sheets of paper which were the notes of his lecture. That is all he needed, and he carried them right on his body!

On March 9, Doron Zeilberger, the brilliant mathematician from Rutgers University gave a talk on "The W-Z Method". It was full of mathematical ideas, but hilarious as well — a typical Zeilberger lecture! Here W and Z stand for Herb Wilf (University of Pennsylvania) and Zeilberger who jointly came up with a technique to verify and prove large classes of combinatorial and q-hypergeometric identities. Quite appropriately, Herb

Wilf gave the Opening Remarks for Zeilberger's talk. Both had come here in connection with the conference on pattern avoiding permutations.

The next day, Carl Pomerance (Dartmouth) gave the Center for Applied Mathematics (CAM) Colloquium on primality testing, which is of great relevance to the study of secret codes. Pomerance had made significant contributions to primality testing and the related area of factorization of integers. Vice-President of Research Win Phillips who made the Opening Remarks, complimented the mathematics department for conducting so many high profile programs in rapid succession.

March 14 was Albert Einstein's birthday, and 2004–05 was the Centenary of Einstein's theory of special relativity. Even though we had a function in Madras at our home for the Relativity Centenary and released my father's book on the topic there in December, since our parents were visiting us, we also had a celebration at our home in Gainesville on March 14 night and had the book released in Gainesville as well. Interestingly, Professor Jim Howland who had communicated some of my father's papers on Relativity to the Journal of Mathematical Analysis and Applications (Academic Press), was visiting the University of Florida as the guest of John Klauder, and we were pleased that both of them attended this celebration!

Since George Andrews unearthed Ramanujan's Lost Notebook, and since Ramanujan is a demi-God to every Indian, it is not surprising that Andrews is greatly admired by the Indian community. So I arranged a special reception and dinner on March 18 night at the India Cultural and Education Center (ICEC) in Gainesville for our Indian friends in this community to meet him and hear him speak. This event was a great success and Andrews gave an excellent talk on "Ramanujan as an inspiration and what India means to me".

Richard Askey who could not attend our Additive Number Theory Conference in November 2004, made a separate trip to Gainesville in Spring 2005 and delivered a fine History Lecture on March 21 on "Orthogonal polynomials", a topic in which he is a world authority. The next day he gave a Mathematics Education lecture "What can we learn from Math Education in other countries?" We had announced his lecture widely throughout campus. Interestingly, even though there was a large audience for his Math Education lecture, only one member from the College of Education attending his talk! Faculty in the colleges of education in the United States think of Askey as their enemy, because he had been openly critical that their Math Education programs have no substance, and are even harmful to the cause of good education. The same evening after the Math Education lecture, Askey addressed our $\pi - \mu - \epsilon$ undergraduate mathematics club on the topic "Fibonacci numbers and hyperbolic functions". In view of his informal, yet engaging lecturing style, he was a hit with students and faculty alike. With Askey's talks, the program of the Special Year in Number Theory and Combinatorics came to a close.

The Spring Term concluded with John Thompson being recognized by the Dean and the President at the Graduation Ceremony at the O'Connell Center in front of 14,000 people. I was pleased that my tribute to Thompson was published prominently in the Brochure for the Graduation Ceremony that everyone was given. That night, at the Performing Arts Center, Thompson was given the CLAS Distinguished Scholar Award. I would say that the College honored itself by recognizing Thompson in this manner.

The highs and lows of Summer 2005: Just as we were feeling elated on having conducted and completed such a successful Special Year program, we received the distressing news that Mathura's mother was terminally ill. So Mathura left for Madras on March 31 to spend time with her mother. My daughters and I were to join her only a few months later because of various important events that we could not miss.

On May 14, Lalitha graduated from Stetson Law School in St. Petersburg. It is the oldest law school in Florida and has a beautiful campus with Spanish style architecture and manicured gardens. During a well conducted Award Ceremony the previous day, Lalitha won six awards, the most for any student that year. Of note was her award for Distinguished Performance and Leadership. It was a proud moment for us and we missed Mathura at the graduation and the award ceremony. Of course it was nice that my parents were there to see their grand daughter graduate with a law degree. Some of our close friends from Gainesville attended the graduation.

Since 2005, Lalitha has had considerable experience in different aspects of law — first in criminal law when she worked as a prosecuting attorney for the State of Florida in Tampa, then as a defense attorney in Orlando working on bankruptcy, wills and estates, and now as an immigration attorney in Atlanta. She also served the profession admirably as President of both the Florida and Georgia chapters of the South Asian Bar Association, and in an organization called SHAKTHI for the empowerment of women.

Next on May 28, Amritha graduated from High School. She was chosen by her school to be the student speaker at the graduation, and she spoke very well before an audience of several thousand at the O'Connell Center of the University of Florida. It was sad that Mathura could not be present for this function as well, but she was doing her duty in India as a daughter assisting her mother during her final weeks.

In between the two graduation ceremonies, I went to New York to attend the Combinatorial and Additive Number Theory (CANT) Conference in New York during May 19–21, organized by Mel Nathanson. The CANT is an annual conference that Nathanson has been organizing for many years. I attended the beautiful lecture by George Andrews on partitions at 5 pm on the opening day. The CANT typically has had speakers in traditional additive number theory and that was Andrews' first talk at a CANT conference. I think our November 2004 Additive Number Theory Conference that brought together researchers in traditional additive number theory and the theory of partitions had a good effect on CANT to invite partition theorists as well. Following that talk, we all went to a Chinese restaurant for a dinner in honor of Nathanson for his 60th birthday.

My talk was in the forenoon of the second day. I spoke on "New polynomial analogues to Jacobi's triple product and other fundamental q-series identities", and chaired the afternoon session. The third and last day featured talks by Andras Sarkozy from Budapest, and the brilliant Gregory Chudnovsky (Columbia University). The conference activities concluded with an enjoyable party at the home of the Nathansons in New Jersey.

Before leaving for India, I made a one week trip to Japan in June to attend the International Conference on Probability and Number Theory organized by Shigeru Kanemitsu in Kanazawa. My parents who had been in Florida for four months, left for India also via the Pacific on the same day as I departed for Japan. I flew into Osaka's Kansai airport but my arrival there was one hour behind schedule. Since I did not have any check-in piece, I was able to jump into the Haruka Express train to Shin Osaka station, where I took the Thunderbird Express to Kanazawa. I had to dash across the train station to make the connection in the nick of time. During the registration at the conference the next morning (June 20), Kanemitsu very thoughtfully presented me a sweater made by the brand *Erdős*! This had no connection with Paul Erdős, but is a well-established Chinese clothing line!

I gave the Opening Lecture of the conference on "Moebius function identities and inequalities, and their q-analogues" blending classical analytic number theory with partitions and q-series. In a sense this talk represented my previous life (!) as an analytic number theorist, and my current life in the theory of partitions and q-series. After lunch, Peter Elliott (University of Colorado), gave his plenary talk on the behavior of multiplicative functions over shifted primes. For me the conference was a nice opportunity to reconnect with researchers in probabilistic and analytic number theory whom I had known for many years. One of the highlights of the conference was the plenary talk Hillel Furstenberg (Hebrew University, Jerusalem), who among other things, was world famous for his proof of Szemeredi's theorem by methods of ergodic theory.

I was asked to chair the session on the morning of June 22. That afternoon, Kanemitsu took us on a tour of the Kanazawa Castle and its sprawling gardens. The next day, Wolfgang Schwartz gave a survey talk on the History of Probabilistic Number Theory. I departed Kanazawa the following morning by train for Osaka where I boarded a Northwest flight to return to Florida the same evening.

One day after my return to Florida, Amritha and I departed for Madras, India, to see Mathura's mother Mrs. Gomathi Krishnan who was rapidly declining in health. She passed away on July 12 at her home surrounded by all members of her family. She never ever heard uttered a harsh word, and was always calm and composed even in the most difficult of circumstances. She had deep faith in the Hindu religion and traditions, and Mathura imbibed this faith from her. It was due to her encouragement that Mathura pursued Bharathanatyam (Indian classical dance) as a lifelong pursuit. The passing away of Mrs. Krishnan was a blow to the whole family; to Mathura it was a loss of her philosopher and guide as well.

During this very difficult time for our family, while I was in Madras, I received a letter from the Indian Ambassador to the United States inviting me and Mathura to attend a reception in Washington D.C. on July 19 to meet the Prime Minister of India, Dr. Manmohan Singh. Mathura could not attend this reception because she wanted to be with her family for at least a month after her mother passed away. So Amritha and I returned to the United States on July 18. The very next day, in the morning, I departed Florida to reach Washington D.C. in time for the Reception at the Marriott Wardman Hotel where I also stayed.

Dr. Manmohan Singh is a highly accomplished economist, having received his PhD degree from Oxford University. Prior to being Prime Minister, he was Finance Minister, and during that time, he was instrumental in getting several important economic treaties between the USA and India signed. There were familiar faces at the Prime Minister's Reception such as Eberly Professor C. R. Rao of Penn State University, the most eminent statistician in the world. Also a the reception was N. Ravi, Editor of The Hindu, India's National newspaper, who had come from Madras, and

M. Prakash who had worked with the Hindu in the eighties, but was now in New York as the Editor of the news magazine The Urban Indian. There were about 600 persons at the reception. I also met Dr. V. S. Ramamurthy of the Department of Science and Technology (DST) of India, and had a conversation with him about the DST funding for the upcoming SASTRA Conference in December when the first SASTRA Ramanujan Prize would be given. Thus is was a fruitful visit to Washington D.C. in many ways.

Mathura returned from India on July 28, and in the third week of August, we had a grand celebration for Amritha's graduation when Amritha herself gave a fine Bharathanaytam performance before dinner, just as Lalitha did in 2002. Mathura had missed the graduation of Lalitha from law school, and Amritha from high school. So we were glad that Mathura was present for this event and sang for the dances as well. The Bharathanatyam program was dedicated to the memory of Mathura's mother.

The Academic Year 2005–06 was to start in a few days, with the highlight being the Special Year in Analysis and Probability.

9.9 Special Year in Analysis and Probability (2005–06) and the First SASTRA Ramanujan Prizes

The year 2005–06 was especially significant for the Department not only because it was the Special Year in Analysis and Probability, but also because it was year that the Department underwent an External Review. Although the External Review was positive, the way the upper administration acted following the review was disappointing and shocking. I have discussed the External Review and its aftermath in separately in §9.12. Here I shall focus on the Special Year and other events of 2005–06.

When discussing with Dean Sullivan regarding renewal of my contract as Chair for a third term of four years starting Fall 2004, I asked for a year's paid leave in between to which he agreed. This leave was in addition to the research leave I would get on completing my tenure as Chair. So I decided to take leave during the Academic Year 2005–06, and had my colleague Scott McCullough as Interim Chair. This was also convenient because McCullough's field is analysis and so he could manage the activities of the Special Year more effectively as Chair during my period of absence.

Events of Fall 2005: The Fall term started with a letter to me from the Indian Ambassador to the United States inviting me and Mathura to attend a Reception at the New York Hilton on September 15 to meet the Prime Minister of India, Dr. Manmohan Singh, who was to come on a second visit to the United States. Since Mathura had missed the first reception in Washington D.C. in July, I was pleased that this second reception would give her an opportunity to meet the Prime Minister.

Mathura and I departed for New York on the morning of September 15 and we checked in at the Sheraton La Guardia Hotel next to the airport. The Reception at the New York Hilton started with a buffet dinner for 700 guests following which we assembled in the grand ballroom to hear the Prime Minister. He spoke about Indo-American cooperation and the new treaties that had been signed. Manjul Bhargava, the brilliant young mathematician from Princeton University who was to receive the First SASTRA Ramanujan Prize in December (see subsection below on this), was there at the Reception along with some other eminent academics of Indian origin in the United States. I found Bhargava in the company of the tabla wizard Zakir Hussein; Bhargava has a deep interest in Indian classical music and had taken lessons on the tabla from Zakir Hussein himself!

New York City is a gourmet's paradise, and it is not surprising that the city has excellent Indian restaurants. The day after the Reception, our friends Prakash and Uma, who are New Yorkers and who also attended the Prime Minister's Reception, took us to lunch at *Saravanas*, a very popular South Indian restaurant, before our flight to Florida in the evening.

During the leave from chairmanship, I spent some time in Fall 2005 at Penn State University to interact with George Andrews and to conduct research. Soon after I returned from New York, I departed for State College, Pennsylvania, on September 21, first on just a two-day trip. It turned out that Don Zagier was visiting Penn State University and that our talks were back to back the next day. Zagier actually gave two talks, first in the Number Theory Seminar before lunch on modular forms, and then the MASS Colloquium at 2:30 pm. My Colloquium to the Department was at 4:00 pm on the theme "A theorem of Göllnitz and its place in the theory of partitions". That night, there was dinner at the India Pavilion for me and Zagier which was attended by several mathematics faculty.

While I was in State College, I received the sad news that my dear colleague Chat Ho passed away due to complications following an open heart surgery. Chat was a former PhD student of John Thompson, and in addition to conducting first rate research, he was instrumental in building our world class program in group theory, by working with the Chair in recruiting ALL our algebra faculty. He had a god-like worship of John Thompson, his Guru. When I saw him in his hospital room before I departed for Penn State, he was recuperating from surgery; he looked his cheerful self and he proudly showed me a memorial article by Thompson on Walter Feit that he was reading. Quite unexpectedly, things turned for the worse, and Chat passed away a few days later to the shock of everyone. Since this trip to Penn State was so short, I was back in Gainesville to attend his funeral and offer my condolences personally to his bereaved family.

I was back in State College on Monday, October 17, for a one month visit. So I stayed at the Days Inn close to campus where I received a discounted rate for the long term stay. The only condition was that I had to vacate the room on football weekends! So on such weekends, I either stayed with my good friends the Vedams at State College, or came on a fleeting visit to Gainesville!

Penn State University always inspires me to be productive. I think it is due to the presence of George Andrews and my close interaction with him. During this visit I made considerable progress on my research and gave a half a dozen seminars as well.

The day after my arrival in State College, I spoke in Andrews' Partitions Seminar, on the theme "Some new observations on partitions into parts $\not\equiv \pm 1 \pmod 4$". I was glad that my talk evinced the interest of some of the graduate students working with Andrews. Two days later, I spoke in the Number Theory Seminar on "New finite versions of fundamental q-series identities", following which the number theorists hosted a lunch at the Faculty Club in the Nittany Lion Inn.

During the first week of my stay in State College, I had several discussions with Andrews on the series representations I had obtained for weighted partition identities modulo 6 that Berkovich and I had found which were related to the Rogers-Ramanujan identities modulo 5. My series were similar but different from series that David Bressoud had obtained previously for the modulus 6, but I could transform my series into his by a duality idea. Andrews confirmed the novelty of the approach and so I wrote a joint paper [P36] with Berkovich that appeared in the Proceedings of the 2005 Kanazawa Conference (see previous section) in 2007.

On Tuesday of my second week, I addressed the Partition Seminar again on "Göllnitz-Gordon partitions with weights and parity conditions". This too was joint work with Berkovich and had just appeared in the Proceedings of the 2003 Number Theory Conference held in Osaka. I also spoke in the MASS Colloquium on the number of prime factors of integers. "How many prime factors does a number have?"

I had to return to Gainesville for the weekend, but on the outbound and return flights, I found a nice two dimensional version of a famous identity of Sylvester that extended the celebrated Pentagonal numbers theorem of Euler. I derived this identity combinatorially and when I showed it to Andrews upon return from Gainesville, he joked that it was another of my airport theorems, and that the Days Inn should throw me out more often! He gave a beautiful proof of my identity in the spirit of Cayley's proof of Sylvester's identity. Inspired by his proof, I was able to obtain a multi-dimensional generalization that week, but I wrote up the results only about a decade later when I was back at Penn State on a sabbatical!

On Monday, November 7, I addressed the Penn State (Undergraduate) Math Club on the theme "A pilgrimage to Ramanujan's hometown" describing Ramanujan's life story, and my annual visits there for the SASTRA conferences. Mathura arrived in State College the next day. Since she was there, we had get togethers all of that week with our Indian friends in State College. Also George Andrews and his wife Joy graciously took us to dinner at the classy Terragon Restaurant at the Atherton Hotel. That weekend, after visiting some of our friends and relatives in New Jersey, Mathura and I returned to Gainesville so that I could attend the Ulam Colloquium of Louis Nirenberg on Monday, November 14.

The first major event of the Special Year was the *International Conference on Martingales and Stochastic Processes* (November 10–13), organized by colleagues Murali Rao and Liqing Yan with the support of Joe Glover (Associate Provost) and Zoran Pop-Stojanovic (Emeritus). One of the main speakers was Daniel Strook (MIT). I remember my father speaking very highly of Strook after hearing about his work at a Summer Institute in Probability and Stochastic Processes at the University of Colorado in the eighties. I was glad that Prof. Vittal, my former teacher at Vivekananda College, came from India to attend this conference and present a paper.

The conference on martingales was immediately followed by the *International Conference on Partial Differential Equations* (November 13–15) organized by Yunmei Chen and Lei Zhang, and for this conference one of the main speakers was Louis Nirenberg of the Courant Institute. On November 13 night, there was a banquet for the two conferences at which both Nirenberg and Strook were present. I had arrived in time to attend the banquet, and the conference organizers kindly asked me to speak after dinner. In my speech I recalled meeting Nirenberg in 1973 at MATSCIENCE when he participated in the International Conference on Functional Analysis; I was an undergraduate then, and I heard his beautiful lecture on the Laplacian. I also met Nirenberg later in 1973 when I accompanied my father to the Courant Institute where he gave a colloquium at Nirenberg's invitation.

Nirenberg was one of the most eminent mathematicians in the world. He was very soft spoken. For his monumental contributions, he was awarded the Abel Prize later in 2015. The 2005 Abel Prize had just been awarded to his colleague Peter Lax. Since we had just invited Lax to visit us in Spring 2006, I requested Nirenberg to put in a good word to Lax about the University of Florida so that Lax would accept the invitation in the midst of his heavy academic schedule especially after winning the Abel Prize. And my strategy worked!

On November 14, Nirenberg delivered the Ulam Colloquium on "The maximum principle". I had the honor of making the Opening Remarks for Nirenberg's lecture. The very next day, I returned to Penn State University for my final week there.

I had no lectures at Penn State University during the last week of my visit, but I had fruitful discussions with George Andrews on my work that I was writing up there for publication. It was indeed a most productive one month at Penn State. But it was not all work and no play in State College. I had taken by tennis racquet with me and I played regularly at the Indoor club (both singles and doubles). The tennis had to be indoors because it was too cold in the late Fall to play outdoors.

On December 2, John Thompson gave a magnificent lecture in our department entitled "A tribute to Saunders Maclane". Maclane was a giant in the field of mathematics, and was a co-founder of Category Theory along with Samuel Eilenberg. Thompson was Maclane's PhD student at the University of Chicago. Maclane had passed away in April 2005, and so we requested Thompson to speak about Maclane.

In mid-December I had to go to India for the SASTRA Ramanujan Conference when the first SASTRA Prizes were to be given. But before that, I made a two-week trip to Germany in response to a handsome invitation from Helmut Voelklein to give lectures at the University of Essen.

Although Helmut was very happy with many things happening in the Department during my term as Chair, he was not pleased with certain decisions of the University. So he accepted a position as Professor and Head of a group in Algebra at the University of Essen and told me that I should NOT try for a counter offer from Florida. Voelklein's departure was definitely a big loss to our algebra program. He invited me to spend two weeks in December 2005 at the Institute for Experimental Mathematics (IEM) of the University of Essen and give a few talks. So I departed on December 5 for Paris by Delta Airlines and connected at Charles DeGaulle Airport, Paris, to an Air France flight to Dusseldorf, from where I took a train to Essen. Voelklein had arranged my accommodation at Hotel Europa in Essen where I checked in on December 6 evening.

The next morning, Voelklein picked me up from the hotel and took me by train to show me how I should travel to the Institute. My colloquium at the IEM was on that afternoon at 2:15 pm. I lectured on "The theorem of Göllnitz and its place in the theory of partitions", which was well attended by all members of two separate research groups under Voelklein and Gerhard Frey, and by several faculty as well. That night, Frey, his wife, and Voelklein, took me to dinner at a very classy restaurant in the Coal Museum and Casino of the World Cultural Heritage District of Essen.

The IEM is located in a wooded area on the outskirts of Essen. Thus in-between my periods of work, when I took breaks, it was pleasant to walk in the vicinity of the Institute. Voelklein would accompany me to lunch every day when we would discuss mathematics and issues facing the mathematical communities in USA and Germany. I was by myself on the weekend and so one day, I went to see Gruga Park in Essen, which is the largest metropolitan park in Germany. On Sunday, since all city shops were closed, I took a train to Dusseldorf to shop in one of best aviation memorabilia stores at Dusseldorf Airport!

During my second week of stay in Essen, I spoke in the Arithmetic-Algebraic Geometry Seminar on "New observations on the Göllnitz-Gordon and the Rogers-Ramanujan identities". While in Essen, I had discussions with Voelklein and Frey about cooperation between the IEM and the UF Mathematics Department, to jointly conduct conferences, and to have an exchange program. Unfortunately, in 2006, in spite a very positive External Review report, the University imposed severe budget cuts, and so I could not proceed with these plans for cooperation with the IEM.

During my stay in Essen, I was receiving messages from India — from SASTRA University and Department of Science and Technology (DST) — regarding arrangements for the SASTRA conference. Everything well going well, and so I was looking forward to my trip to India.

Trip to India and the award of the First SASTRA Ramanujan Prizes

I departed Dusseldorf on December 14 for Paris where Mathura joined me. We spent two nights in Paris and reached India on December 16 night. My colleague Alexander Berkovich (who was to speak at the SASTRA conference) arrived earlier that day in Madras with his wife Larissa.

The next morning, I gave a talk at the Presidency College on

"Probabilistic Number Theory" for the conference on probability organized in connection with the 150th Anniversary of the Mathematics Department. My talk was arranged in the auditorium style lecture hall of the physics department where several decades earlier, my father, and Nobel Laureates Sir C. V. Raman and Subrahmanyam Chandrasekhar, had their classes. So it was inspiring for me to give a talk in that same lecture hall. That night, several participants of the SASTRA conference arrived in Madras in order to depart for Kumbakonam two days later.

With regard to the SASTRA Ramanujan Prize, I should say that there was wholehearted support from the international mathematical community. We received outstanding nominations from some of the most eminent mathematicians in the world. The Prize Committee for the first year was Manindra Agarwal (IIT, Kanpur), George Andrews (Penn State Univ.), Jean-Marc Deshouillers (Univ. Bordeaux), Tom Koornwinder (Univ. Amsterdam), Jim Lepowsky (Rutgers Univ.), and Don Zagier (Univ. Bonn), with me serving as the Chair. The six world renowned mathematicians on the Committee represented a broad spectrum of fields, and I have maintained this breadth over the years. I am a non-voting chair, and I vote only to break a tie. It turned out that the Committee chose two candidates as the top contenders — Manjul Bhargava (Princeton University) and Kannan Soundararajan (then at the University of Michigan). They were the strongest young mathematicians in algebraic and analytic number theory respectively. They were tied at number 1, and I did not want to vote to break the tie the very first year. The Committee also felt that both deserved to win. So I wrote a letter to Vice-Chancellor R. Sethuraman of SASTRA that this very first year, two FULL prizes of $10,000 each should be awarded to Bhargava and Soundararajan, and that with this decision, the SASTRA Ramanujan Prize will be on the world map. It was very generous of the Vice-Chancellor to agree to this suggestion, and thus the SASTRA Ramanujan Prize got off to a splendid start. Everyone on the prize committee was happy with the outcome.

Manjul Bhargava had made phenomenal contributions to number theory, most notably by his discovery of higher order composition laws. This was his PhD thesis, written under the direction of Professor Andrew Wiles of Princeton University (of Fermat's Last Theorem fame) and published as a series of papers in the Annals of Mathematics. Gauss, the Prince of Mathematicians, constructed a law of composition for binary quadratic forms. Bhargava introduced entirely new and unexpected ideas that led to his discovery of such composition laws for forms of higher degree. Bhargava then applied these composition laws to solve a new case of one of the fundamental questions of number theory, that of the asymptotic enumeration of number fields of a given degree d. Bhargava's work had created a new area of research in a classical topic that has seen very little activity since the time of Gauss. Bhargava was already a Full Professor at Princeton University, the youngest at that rank in that prestigious institution.

Kannan Soundararajan had made brilliant contributions to several areas in analytic number theory that included multiplicative number theory, the Riemann zeta function and Dirichlet L-functions, the analytic theory of automorphic forms and the Katz-Sarnak theory of symmetric groups associated with families of L-functions. Even as an undergraduate at the University of Michigan, Soundararajan had made outstanding contributions, and was awarded the First Bennie Morgan Prize of the AMS for undergraduate research. Subsequently, in his PhD thesis, written under the direction of Professor Peter Sarnak of Princeton University, he proved spectacular results on zeros of L-functions. A part of his PhD thesis was published in the Annals of Mathematics. Soundararajan is also a leading expert on random matrix theory and its implications in analytic number theory. Soundararajan was considered to be one of the most creative young minds to emerge in the last decade, was at that time a Full Professor at the University of Michigan, Ann Arbor.

On a personal note, I would add that Soundararajan's very first paper, written when he was in the 11th grade of Padma Seshadri High School in Madras, and published in the Journal of Number Theory in 1992, was on an improvement of an inequality on multiplicative functions due to me, Erdős, and Vaaler (see Chapter 6). I first met Soundararajan when he attended one of my lectures at the Ramanujan Institute in 1990. Just before the start of my lecture, he was introduced to me by Mr. P. K. Srinivasan as a brilliant young boy interested in research in number theory.

Bhargava was accompanied by his mother Mira, herself a mathematician. Soundararajan was accompanied by his parents. There were thirteen of us, — Mathura and me, Bhargava and his mother, Soundararajan and his parents, Berkovich and Larissa, Shigeru Kanemitsu (Fukuoka), Eugene Mukhin (IUPUI, Indianapolis), Ken Ono (Univ Wisconsin), and B. Ramakrishnan (Harish Chandra Institute) — and so we departed for Kumbakonam on the morning of December 19 from Madras in two vans reaching Sterling Resorts around 8:00 pm. Enroute, we stopped at a rest area, where Bhargava and his mother relaxed on a swing at the rest area's park! Two other participants — Karl Mahlburg and Michel Waldschmidt, did not come with us in the van but arrived in Kumbakonam separately. Everyone appreciated the traditional welcome foot massage on arrival. The next morning, we had the grand inauguration of the International Conference on Number Theory and Mathematical Physics. Dr. Arabinda Mitra, Executive Director of the Indo-US Forum of Science and Technology, as the Chief Guest, gave the SASTRA Prizes to Bhargava and Soundararajan after I read their citations. Ken Ono was so much impressed with what he saw, that he wrote a fine article [A4] for the Notices of the AMS describing the prize ceremony, the conference lectures, and the wonderful sightseeing we did in and around Kumbakonam.

After the inauguration, Soundararajan gave the opening lecture of the conference. Bhargava delivered the concluding Ramanujan Commemoration Lecture on December 22 morning, Ramanujan's birthday. He spoke about his recent solution with Jonathan Hanke of the problem of determining all universal quadratic forms — a problem that has its origins in Ramanujan's work. I sent a report of this lecture to *The Hindu* where it appeared the next day. In March 2006, Ivars Peterson, referred to my article in his report in *Science News*, of which he was the Editor.

Sightseeing in an around Kumbakonam included visits to Ramanujan's home, the Town High School where he studied, the Sarangapani Temple, the Swamimalai Temple and the gigantic Brihadeeswara Temple in Tanjore. We also visited the Government College in Kumbakonam from where Ramanujan had borrowed *Carr's Synopsis*. But to our shock, we were told that the copy of Carr's Synopsis that Ramanujan had borrowed and returned, was now missing from the library of the college!

After Bhargava's concluding Ramanujan Commemoration Lecture on December 22 morning, we departed Kumbakonam for Madras. On that return journey, at the request of Bhargava and his mother Mira, we stopped to observe a farmer prepare fresh sugarcane juice on the roadside using a machine to crush the sugarcanes. January is the *Pongal* or harvest festival in the state of Tamil Nadu, and a month before that, sugarcanes appear in the market. Even though the juice was freshly prepared by the farmer, fearing infection, Bhargava and Mira did not want to drink it; I also have the same fear of infection. Instead, they bought a whole sugarcane from the farmer and enjoyed it in their hotel room in Madras.

We returned to Madras on December 22 night, but some of the participants stayed for a few more days and enjoyed wonderful Carnatic music concerts because it was the music season in the city.

Berkovich and Larissa went on a one week trip by car to Mysore and

Ooty with GP as their companion and guide. Bhargava and his mother Mira came to Ekamra Nivas for a leisurely evening and my father showed them my grandfather's chambers and the lecture hall upstairs where he conducted his Theoretical Physics Seminar. Another evening, Mathura took Mira to some sari stores for which Madras is so famous.

It was during the last week of December that my high school Vidya Mandir celebrated its 50th Anniversary. I was one of four old students invited as a chief guest — the others being tennis champion Ramesh Krishnan, test cricket batsman Srikkanth, and a well known painter. It was a wonderful occasion for me to meet several schoolmates and my teachers after 35 years. It was nostalgic when we were all asked to assemble in various classrooms and recite the prayer which we used to do as schoolboys every morning! (In those days, Vidya Mandir was a boys-only school from sixth grade upwards; now it is completely co-educational.)

Spring and Summer 2006: After an event packed December in India, Mathura and I went to Coimbatore for a short holiday to spend time with Mathura's brother Sriram and his family, and visit a wildlife sanctuary, before returning to Florida on January 7. However, one day before my departure from Madras, I did give a lecture at the Academy of Sciences on "Series and polynomial representations for weighted Rogers-Ramanujan partitions" based on the paper I wrote at Penn State University two months before. Upon return to Florida, I departed for San Antonio to attend the Annual Meeting on the AMS, where I was one of the leaders for the Chair's Workshop. The emphasis at this workshop attended by more than thirty chairs was Long Range Planning for mathematics departments.

George Andrews arrived in Florida after the AMS Meeting to spend the entire Spring semester with us. He told me that he was much impressed with the success of the Special Years program and the distinguished lecture series I had launched. So he expressed his desire to make an annual contribution to the Department from his Evan Pugh fund at Penn State to conduct a high quality program at the University of Florida, and wanted my suggestion. I told him that we have two very successful distinguished colloquia — the Erdős and Ulam Colloquia, but now that he will be here every Spring, we should launch the Ramanujan Colloquia, because with his regular presence in Gainesville, and with the work of our number theory group, Florida is one of the world's major centers for work related to Ramanujan. I said that the Ramanujan Colloquium should be for areas influenced by Ramanujan and that the eminent speakers who are invited to deliver the Colloquium as a Public Lecture of wide appeal should also give two specialized research talks in seminars the next day so that it would benefit our research program. Andrews heartily endorsed this proposal and said that I should prepare to launch the Ramanujan Colloquium in the Spring of next year. So with this wonderful new development confirmed, I was in a great mood to depart on an academic trip to Spain in the end of January.

Mathura decided to accompany me on the trip to Spain. She had a US Passport, but I had only an Indian passport and so I needed a visa and had to appear in person at the Spanish consulate in Miami to be interviewed before I could get the visa.

I was invited to deliver a public lecture at the Royal Spanish Academy of Sciences in Madrid on Wednesday, February 1, and at the University of Valencia on Friday, February 3. But we decided to arrive in Madrid three days earlier to do some sightseeing there.

Mathura and I departed Jacksonville on Saturday, January 28, and reached Madrid in the early afternoon the next day via Detroit and Amsterdam. We stayed at the very comfortable Holiday Inn in Madrid for four nights. Our sightseeing in Madrid started with the Prado Museum on Sunday and it was outstanding. This was followed by high tea (tapas) at the Westin Palace Hotel right across from the museum. On Monday we spent a leisurely four hours enjoying the regal beauty of The Royal Palace, which in opulence and luxury surpassed the Versailles Palace. The chandeliers there were fantastic. In front of the palace is an impressive bronze statue of Philipe IV of Spain on a horse. The statue is of scientific significance: The horse is standing on its hindlegs with its front legs up in the air. Before beginning his work, the sculptor was grappling with the problem of balancing the statue because the center of gravity will fall outside the base of support. The genius Galileo came up with the solution. He suggested that the rear of the statue be made of solid bronze and the front of hollow bronze so that the center of gravity will be in the rear! Galileo made precise calculations which helped the sculptor finish the statue.

On Tuesday, we roamed the city, visited the Thyssen Museum, the botanical gardens, and bought souvenirs at the famous Plaza Mayor.

I was introduced to the Royal Spanish Academy by Professor Jerzy Kakol who visited Florida regularly to work with my colleague Steve Saxon. Both had ties with the Spanish mathematical community due to their research. My host at the Spanish Academy was Prof. Manuel Lopez-Pellicer, who is an authority on problems at the interface of topology and functional analysis. His work had connections with the research of Saxon and Kakol, and he had visited our department once when Kakol was spending an extended time in Florida to work with Saxon. Pellicer picked us up at the Holiday Inn on Wednesday afternoon and took us to the Spanish Academy for my talk at 7:00 pm. I spoke on "The Indian genius Ramanujan - a glimpse of his magical mathematical formulae". My talk in the main auditorium was attended by several members of the Academy, many from other disciplines. Following my lecture, Professor Pellicer took Mathura and me to *Artemis*, a Middle Eastern restaurant where we could get excellent vegetarian fare. That night, at Mathura's insistence, we attended a Flamenco dance performance at one of the traditional theaters in Madrid, and it was fabulous. The Flamenco performance started at 10:30 pm, which is very late for me. But it was riveting and so I did not fall asleep!

On February 2 morning, we took a train to Valencia, where our host Professor Gabriel Navarro and his wife Isabelle received us at the railway station and took us to our hotel, the lovely Astoria Palace. Navarro had previously visited Florida with his family for an extended stay to work with our colleague Alexandre Turull and that is when I got to know him well. Navarro is a reputed group theorist who among other things had done important work on p-groups, and thus had ties with Turull not only because both hailed from Spain, but also because Turull's speciality is p-groups. Navarro is a gentleman to the core and his wife is equally gracious. In the evening, the Navarros took us around the beautiful city of Valencia in their car before having us at their lovely home for dinner with their family. At dinner were Isabelle's two sisters and their husbands. Isabelle told us that she had bought a sari, but then Mathura and I were surprised to hear that she and her sisters wanted to try the same sari! Mathura showed them how the sari is worn, and the three sisters took turns wearing that sari. It was a very enjoyable evening at the Navarro home.

The next day I addressed the Algebra Seminar at noon at the University of Valencia on some aspects of Ramanujan's work as related to my own research. Following this, Navarro took us to an enchanting seaside restaurant where he specially ordered vegetarian paella. Valencia is where paella originated as a vegetarian dish, and it was only later that seafood and meat were added to paella as it is now commonly done. Paella is one of most well known dishes of Spain, and one of the tastiest. So what we enjoyed at that beachside restaurant, thanks to Navarro, was the paella as it was originally done (vegetarian), and it was excellent — more so with the splendid view of the Mediterranean sea!

From Valencia we proceeded by train to Barcelona for two days of sightseeing. We stayed at the magnificent Hilton Diagonal Mar on the outskirts of Barcelona, and took the metro to the city. One of the things to see in Barcelona is the Columbus statue fronting the sea at the end of the famous boulevard called Las Ramblas, and we did this all by walk. Two other attractions are the Picasso Museum and the Sagrada Familia both of which we enjoyed seeing. On February 7, we departed Barcelona by KLM for Amsterdam. It was a clear day, and so the view of the Pyrennees and the Alps was breathtaking. We were back in Florida the same night.

In second half of February, we had the visits of John Martindale, Hershel Farkas and Sinai Robins. Martindale who was originally with Kluwer, had been inducted into Springer after Kluwer and Springer merged in 2004. When Martindale was in Florida, he had discussions with me and Andrews about the Ramanujan Journal, which from 2004 was a Springer publication. We had a party at our home for Martindale attended by members of the algebra, number theory, and combinatorics groups.

On February 14, George Andrews gave his first talk in the joint Combinatorics and Number Theory Seminar on the serendipity of partitions. This was followed the next week by Farkas' talk in the seminar on "Theta functions and combinatorial number theory." Farkas always visited Florida with his wife and so Mathura and I had a party for them that night which was well attended by my colleagues. Farkas and his wife strictly adhere Kosher requirements, and so during parties at our house, even though everything is vegetarian, they just have the fresh fruit and vegetables (uncooked). The next day, Farkas addressed the department colloquium. Following that, Mathura and I had dinner with him at the new Hillel Center just outside campus where the food was Kosher.

The last visitor in number theory in February was Sinai Robins from Temple University. I had known him since the early nineties when he was a graduate student at UCLA working under Basil Gordon. His research interests had since then broadened considerably. He addressed the Colloquium on February 27 and the Number Theory Seminar the next day. We took him, at his request, to the restaurant Merlion which served Singaporean cuisine. A few years later, Sinai resigned his position at Temple and took up a job at the Nanyang Technological University in Singapore!

In early March, we had the Southeastern Analysis Meeting (SEAM) organized by colleagues Scott McCullough, Jim Brooks and Steve Summers as one of the conferences of our Special Year. The featured lecture of the conference was the Erdős Colloquium by William Arveson of UC Berkeley on "Operator theory and the K-homology of algebraic varieties". It was the third successive year that the Erdős Colloquium was delivered by a faculty member from Berkeley!

On Monday, March 6, we had the Colloquium of Kathrin Bringmann, who at that time was a Van-Vleck Assistant Professor at the University of Wisconsin, working with Ken Ono. She and Ono had obtained very significant results on mock theta functions and she presented these in her Colloquium talk and in the number theory seminar the next day.

2005 Abel Prize Winner Peter Lax of the Courant Institute arrived on the morning of March 7, and that same afternoon, he delivered the Center for Applied Mathematics (CAM) Colloquium entitled "A Phragmen-Lindelof principle in harmonic analysis". My parents had arrived in late February for their usual four-month stay, and so my father attended Lax's lecture. My father and Lax had first met in 1956 at the Courant Institute in New York, and it was nostalgic for both of them to meet again exactly after 50 years! That night we had a party at our home in honor of Lax and it was attended by my colleagues as well as faculty from physics, statistics and engineering. Actually Lax's lecture was originally scheduled for February 13, but due a major snowstorm in the northeast of USA, he could not come then. So his lecture was rescheduled on March 7, and the person who was most happy with this was my father who could not have met Lax if his lecture had been in mid-February!

Prior to the CAM Colloquium, Lax and I had a discussion in my office. On seeing pictures of me and Erdős, Lax said that he too was encouraged by Erdős in the early forties when he first arrived in New York as a teenager escaping from Hungary due to the terrors of WW II. Erdős who was visiting the Institute for Advanced Study at that time, invited Lax to the Institute and introduced him to Albert Einstein. Lax repeated this story about his meeting with Erdős at the start of his colloquium talk.

The next day, George Andrews delivered the IFT Colloquium in the physics department. I used the occasion to release the special issue of the Ramanujan Journal which had just appeared. It was the proceedings of the Number Theory and Mathematical Physics Conference conducted in 2004 jointly by the Mathematics Department and the IFT. That night we had a dinner in honor of Lax and Andrews at the Hilton.

A major family event during the summer was the *Nischayathartham* (= Engagement Ceremony) of my older daughter Lalitha to Aditya Srinivasan. The Ceremony conducted in traditional South Indian style followed by dinner for about a hundred friends was at our home on May 28 during the Memorial Day Weekend. I was glad that my parents were there to bless the couple. Following that event, our whole family had a one week vacation in early June in lovely Honolulu. Mathura and I also had another vacation in the summer — four days in the US Virgin Islands (July 30–Aug 3). We stayed at the fabulous Marriott Frenchman's Reef Hotel which commanded a breathtaking view of the bay at Charlotte Amalie in St. Thomas. We also visited the neighboring idyllic island of St. John.

Between August 17 and September 11, I had to make an academic trip to India and China. Mathura accompanied me on the trip which was transatlantic between USA and India, and a separate trip to China from India.

We arrived in Madras on August 18, and the next morning we departed for Hyderabad because I was scheduled to deliver the First Musili Memorial Lecture at the Central University of Hyderabad on August 19 at 3:00 pm. My contact with Prof. Musili goes back to my graduate student days at UCLA when he visited there for two years (1975–77) as the guest of Professor Steinberg. Musili and his wife were very kind to me. They had me over at their apartment for dinner several times. I also used to have discussions with him regularly on subjects like differential geometry and algebraic geometry that I was learning for the first time. Musili was an algebraic geometer and he did PhD under Seshadri at the Tata Institute. In view of my close association with Musili, Professor Kannan, the Head of the Mathematics Department at the Central University, invited me to deliver this memorial lecture. I spoke on "New views on Schur's partition theorem". I had met Kannan first at the University of Madurai in the seventies when I was there for a lecture at the invitation of his guru Prof. M. Venkataraman, who was a contemporary of my father; Venkataraman and my father started the departments of mathematics and physics at Madurai University in 1959. It was nice to reconnect with Kannan after three decades.

After returning from Hyderabad, Mathura and I rested in Madras for a couple of days before departing for China on Malaysia Airlines via Kuala Lumpur. The flight from Kuala Lumpur to Shanghai on August 24 morning was by a Malaysia Airlines Boeing 777. It was a wonderful flight and we enjoyed the on board service immensely. I mention this because it was a Malaysian 777 from Kuala Lumpur to Beijing that vanished without a trace in 2014! The mystery of its disappearance is still unsolved.

Our host Professor Kong received us at Pudong Airport, Shanghai, and took us to the magnificent Hilton Shanghai Hotel where we were given a room on a high floor that provided a fine view of the city. The next morning we left for Guilin — a place we wanted to visit ever since we saw a movie at the Chinese pavilion at Epcot Center in Orlando in which the fantastic scenery around Guilin is highlighted.

On Sunday, August 27, we did an all day excursion to Hong Shao, the highlight of the trip being a four hour boat cruise on the Li River. On both sides of the river there are limestone outcroppings jutting into the sky like pyramids or skyscrapers. Indeed this area has hundreds of these formations and the whole region looked like a city filled with skyscrapers, except the skyscrapers were limestone hills with near vertical sides! The landscape was both breathtaking and eerie. In Hong Shao we saw a spectacular sound an light show and returned to Guilin by car instead of by boat.

From Guilin, we were back in Shanghai by China Southern Airlines on August 28. On August 29 morning, Kong had arranged my lecture at the new campus of Shanghai Jiao Tong University (SJTU). This was my second visit to SJTU, the previous one having been in 1999 at their old campus. This time I spoke on "A new approach to Jacobi's triple product identity leading to a quadruple product extension". After lunch at SJTU, Mathura and I departed by China Eastern Airlines for Weihai to attend the Fourth China-Japan Number Theory Conference. One of the conference organizers on the Japan side, Professor Kanemitsu, was our co-passenger. The conference venue was the Shandong University Academic Center where all participants were accommodated. This was a gigantic building with an enviable location on the beach fronting the Yellow Sea, with Korea (S and N) on the opposite shore of the sea.

The conference inauguration was the next morning with welcome and introduction by Shigeru Kanemitsu (Fukuoka) and Jianya Liu (Weihai). I was asked to speak at the inauguration in view of my association with the China-Japan Number Theory Conferences organized by Kanemitsu at various locations. Following the inauguration, the opening lecture was by Professor Shinzel of Poland, who for many years was the Chief Editor of *Acta Arithmetica*. The second plenary talk in the morning was by the world famous mathematician Shou-Wu Zhang of Columbia University who spoke on period integrals and special values of *L*-series.

I was the chair of the afternoon session when there were several shorter talks by Chinese and Japanese number theorists. My talk was the first on the morning of the next day. I spoke on the same topic as in the Colloquium in Shanghai. I was followed by Trevor Wooley (University of Michigan), a world authority in additive number theory and the Circle Method. He spoke on Waring's problem in function fields. The proceedings of the Weihei Conference edited by Kanemitsu and Liu was published by World Scientific as Volume 2 in the book series Number Theory and its Applications for which Kanemitsu is the Founder and Series Editor.

We flew from Weihai to Shanghai on September 2, spent a night there before returning to Madras. After one week in India, Mathura and I returned to Gainesville. I had a busy academic year 2006–07 to work on.

9.10   Special Year in Logic and Set Theory and the launch of the Ramanujan Colloquia (2006–07)

The year 2006–07 was the most difficult for me as Chair because of the issues I had to face with the Provost's office in connection with an External Review, even though it was positive. Since I will discuss that separately, I will focus on the Special Year in Mathematical Logic and Set Theory and other academic events in this section.

One of the successes of 2006–07 for the Department was a grant from the NSF providing complete support for the Special Year in Logic and Set Theory — airfare and local expenses for ALL conference participants, as well as for ALL speakers during the year. The NSF launched a new program enabling such support, and the organizers of the Special Year — Douglas Cenzer, Jean Larson, Bill Mitchell, Rick Smith and Jindrich Zapletal — submitted a well planned and comprehensive proposal which was recommended for full funding. There were six conferences during the academic year, the first of which was on *Combinatorial Set Theory* during September 15–17, for which I was back from my trip to China to attend it. Jean Larson led the effort for this conference. One of the principal speakers was Menachem Magidor who was the President of Hebrew University, Jerusalem. Another principal speaker was Neil Hindman of Howard University who I knew from my graduate student days when he used to come to UCLA to discuss with Bruce Rothschild; he was at California State University at that time. In connection with this conference, Peter Komjath of the Eotvos Lorand University in Budapest visited us for two weeks. He is a leader in combinatorial set theory and as a young mathematician (not exceeding the age of 40) had won the Paul Erdős prize in 1991.

On September 21, I made a day trip to Daytona Beach to give a talk there. Tulsian Gajendra, a former graduate student of ours, had accepted a permanent faculty position at the Daytona Beach Community College (DBCC), and had secured a grant from the Provost to conduct a series of public lectures in mathematics. At his invitation, I gave a talk on "Paul Erdős - his wonderful life and mathematics" which was attended by a large number of faculty and students.

During November 1–5, there was the Computational Complexity in Analysis (CCA) 2006 Conference jointly organized by my colleague Doug

Cenzer along with Klaus Weihrauch of Hagen University, Germany. As the concluding talk of the first day, we had the History Lecture on computability theory by Robert Soare of the University of Chicago. The conference organizers kindly invited me to make the opening remarks. The second featured lecture of the conference was the Ulam Colloquium by Anil Nerode, Goldwin Smith Professor at Cornell University, for which I was again given the opportunity to make the opening remarks. On the Monday following the conference, we had the Colloquium of Jeff Remmel of UCSD who was briefly my colleague in Florida during 1987–88, and who had an active collaboration with Doug Cenzer. Remmel was an authority in combinatorics, but his research encompassed computability theory as well.

The 2006 SASTRA Ramanujan Prize to Terence Tao

After the department Christmas Party on December 8, Mathura and I departed for India the next day on Northwest Airlines to Singapore where we connected to a Jet Airways flight to Madras. The main event during our three-week stay was the award of the 2006 SASTRA Ramanujan Prize to Terence Tao of UCLA. Tao had done revolutionary work in many different areas of mathematics and was at the young age of 31 considered the most influential mathematician in the world. He had just won the Fields Medal in August 2006, and the SASTRA Prize came right after that. He was nominated for the SASTRA Prize only in 2006 and was the unanimous choice by the prize committee.

One of Tao's most notable contributions was to the famous Kakeya Problem in higher dimensions, which has major applications in Fourier analysis and partial differential equations. One important aspect of the problem is to determine the fractal dimension of the set generated by rotating a needle in n-dimensional space. In joint work with various mathematicians, Tao significantly improved all previously known estimates for the fractal dimension using new and surprisingly simple combinatorial ideas in an ingenious way. Another of Tao's outstanding contributions was his joint work with Ben Green on the existence of arbitrarily long arithmetic progressions of prime numbers. Yet another fundamental contribution of Tao concerns the sum-product problem which is due to Erdős and Szemeredi. Roughly speaking, this problem states that either the sumset or the product set of any set of N numbers must be large. Tao was the first to recognize the significance of this problem in combinatorial number theory and harmonic analysis. Tao's contributions have also had impact in physics, especially in general relativity.

Tao was born in Adelaide, Australia in 1975 and lived there until 1992. He then went to Princeton University in 1992 for his PhD, which he completed in 1996 under the direction of Elias Stein.

Among the speakers for the 2006 SASTRA conference were Cilanne Boulet, James Haglund, and my colleague Miklos Bona. Boulet had received her PhD from MIT in 2005 under the supervision of Richard Stanley and had a visiting position at Cornell University thereafter. Haglund had graduated from the University of Georgia in 1993 with Rodney Canfield as his PhD advisor. He came to the conference from the University of Pennsylvania where he had just taken up a permanent position. Bona was also a former PhD student of Stanley at MIT; he was one the three appointments I made at Florida during my first year as Chair. Also attending the conference was my father's long time friend Professor J. Sethuraman, a noted statistician from Florida State University. He was accompanied by his wife Brinda. All guests were put up at Hotel Savera near my house.

As scheduled, I departed for Kumbakonam on the morning of December 18 at 10 am in a van accompanied by the Sethuramans, Boulet and Haglund. We stopped at the Madras airport enroute to pick up Bona who had just arrived. Bona said that his suitcase was not there and so we waited till he filed a report. Tao and his wife were scheduled to arrive only that night and so Mathura stayed back to come with them by car to Kumbakonam the next day. John Martindale of Springer arrived with his wife only on the morning of December 19 and he was received by GP and brought to Kumbakonam separately.

The International Conference on Number Theory and Combinatorics was inaugurated on the morning of December 19. The opening lecture of the conference was by Professor Sethuraman who discussed some combinatorial questions in statistics. There were some French participants who had come to India to speak at a combinatorics conference in Bangalore the previous week, and so we invited them to speak at the SASTRA conference. Their talks were arranged soon after Sethuraman's lecture. Bona, a leading expert on permutations, spoke in the afternoon.

The French participants had a guide book of India with them and they were keen to see the Darasuram Temple, a UNESCO World Heritage site. Since this temple was close to Kumbakonam, I took our guests there on the evening of December 19. The Darasuram temple is for Lord Shiva, and is a marvel in art and architecture. It is not so well known as some of the other temples in the Tanjore district, but it ought to be. I was glad that the French participants brought it to my attention.

My talk was the first the next day and it was on the theme "Two extensions of an idea of Sylvester and their implication in the theory of partitions". Cilanne Boulet followed me and gave an excellent lecture. She had just finished her PhD and I was impressed with the maturity she showed in her lecture. That afternoon, Michel Waldschmidt who was in India visiting Pondichery, a former French outpost in the Madras state, came to Kumbakonam from there. He gave a fine lecture on the work of the famous Indian mathematician S. S. Pillai. Earlier that morning, I interviewed Tao to send a write up to The Hindu, India's National Newspaper.

On the morning of December 22, Ramanujan's birthday, the SASTRA Ramanujan Prize was awarded to Terence Tao. He then gave the Ramanujan Commemoration Lecture on "Long arithmetic progressions among the primes". With that the conference concluded on a high note and we all returned to Madras. On the way back we dropped of Bona at the airport for his flight back to Florida. His suitcase had arrived the previous day and was kept for him at the airport. So he collected it and checked it back for his return flights to the United States!

The next day, our guests who stayed back in Madras — Terence Tao, his wife Laura, John Martindale, his wife Liz, and James Haglund — came over to Ekamra Nivas for High Tea! They enjoyed seeing our old historic home. After that I took them all to the Madras Music Academy to attend the concert of Mandolin wizard U. Srinivas (now deceased). Our guests were impressed how the Mandolin, a western instrument, was beautifully adopted in Carnatic music. Following the concert, the Taos left for the airport for their return to Los Angeles.

The following morning, John Martindale and Liz departed Madras in a chauffeured car for a one week tour of South India before returning to the United States. GP accompanied them and acted as a companion and guide. Being of British descent, John and Liz especially enjoyed the visits to the hill station Ootacamund, the summer capital of the British in the Madras Presidency, and Coonoor, the British cantonment near Ootacamund.

After the visitors had all left, Mathura, my daughters and I spent the week between Christmas and New Year attending to things that had to be done in Madras in connection with Lalitha's wedding scheduled for February 2007 in Florida. We were all back in the USA on January 3.

Spring 2007 and the first Ramanujan Colloquium

Whereas my family flew back to Florida from India, I actually went to New Orleans from Madras because I had to attend the Annual Meeting of the AMS, from January 4–7. This was the third and final year for me as Leader of the Chairs Workshop which took place on January 4. The theme of the Chair's Workshop was "External Perception", which was a very crucial topic. Within a university, the administration and other departments view the mathematics department as a service department because of the large teaching component, and not as a research department because in terms of grant size, mathematics departments are low compared to other science departments. I worked hard at the University of Florida to change this perception, and succeeded to a large extent.

One of the highlights of the AMS Annual Meeting was the Willard Gibbs Lecture delivered by Peter Lax. Even though the lecture was at night, the hall was overflowing because many participants wanted to hear Lax who had won the Abel Prize in 2005.

In late January I went on a one week trip to the United Arab Emirates (UAE) in connection with the accreditation of the new BS Mathematics Programs at the American University of Sharjah and the University of Sharjah. This was my first visit to the Middle East, and I was invited as Chair of a three person committee from the USA. In subsequent years, I made several visits to the Middle East to evaluate mathematics programs there; I have discussed all these visits together in §9.13.

Within a week of my return from the UAE was the wedding of my elder daughter Lalitha to Aditya Srinivasan on February 10 in Tampa. It was an elaborate affair and my parents arrived from Madras on February 4 for the event. The wedding was in two parts: a South Indian style traditional wedding done in the proper Hindu vedic style in the morning from 6:00 am to noon followed by lunch, and a western style reception that evening with dinner. There were about 350 guests for each of these two events. George Andrews and Roger Baker (Brigham Young University) were visiting the University of Florida and so they attended the wedding. It was an interesting experience for them and for some of my department colleagues and their spouses to see an elaborate six-hour traditional South Indian style Hindu wedding, complete with the vedic ceremonies. Mathura worked tirelessly to ensure that the morning Hindu wedding ceremony was conducted as it is traditionally done in India. The wedding was in Tampa instead of Gainesville for two reasons: (i) Lalitha was working as a prosecuting attorney in Tampa, and she wanted to be hands on with regard to every arrangement, and it was easier for her to accomplish this in Tampa, and (ii) there was no Hindu temple in Gainesville, and there is one in Tampa, and this was crucial for the morning Hindu wedding ceremony.

My father gave a moving speech at the reception when he said it was such a happy moment for him and my mother to witness their granddaughter's wedding in the USA, and how blessed he was that his health was good enough for him to travel to the USA for the wedding. My friends who attended the reception told me how much they appreciated my father's speech. He was 84 at that time, and his comment on being in good health was very pertinent, because the very next year, in June 2008, he passed away at our home in Gainesville (see Section 9.11).

Our close friends Ramanathan and Sakuntala who came from San Diego to attend the wedding, stayed a few days at our home in Gainesville after the event. It was an opportunity for us to reciprocate in small measure the gracious hospitality of the Ramanathans during our numerous visits to Honolulu. Mathura's younger brother Sriram came all the way from India for the wedding representing Mathura's side of the family.

During February 2007, there was a *Conference on Model Theory and Computable Model Theory* organized by my colleague Doug Cenzer as part of the Special Year Program, but I could not attend any of the events of this conference because upon return from the UAE I was busy with my daughter's wedding arrangements.

On February 15, we had two talks back to back. Roger Baker addressed the Joint Combinatorics — Number Theory Seminar, while Andrews gave the second of his series of six lectures. Baker had an arrangement at Brigham Young University that his two teaching terms would be the Fall and the Summer(!) each year so that he would be free in the Winter term to visit other universities. Owing to this arrangement, Baker came to Florida regularly for a month to two months each year during a five-year period.

On February 22, I received a phone call from John Martindale, informing me that he is retiring from the world of publishing. He had a long and successful term at Kluwer Academic Publishers which is when he approached me for the launch of The Ramanujan Journal. After Kluwer merged with Springer in 2004, he continued to oversee the Ramanujan Journal at Springer. I was glad that he came to India with his wife in December 2006 before his retirement and I was able to play host to him.

The first half of March was intense with two conferences of the Special Year in Logic and Set Theory: The first was the *Conference on singular cardinal combinatorics and inner models* (March 5–9) with Bill Mitchell as the lead organizer. Next was the *Annual Meeting of the Association of Symbolic Logic* (March 10–13) for which my colleagues Jean Larson, Bill Mitchell, Jindrich Zapletal and Doug Cenzer were the main organizers.

The Ramanujan Colloquium sponsored by George Andrews and the Pennsylvania State University was launched on March 19 with the inspiring lecture of Manjul Bhargava of Princeton University who spoke about his joint work with Jonathan Hanke on the complete resolution of a problem stemming from Ramanujan's work, namely the determination of all integer valued universal quaternary quadratic forms. Many colleagues who attended the talk wrote to me saying that it was the best colloquium they had heard. Bhargava followed this with three number theory seminars over the next two days in which he gave the details of the proof of this definitive result. Lisa Lorentzen and her husband came all the way from Norway to attend Bhargava's lectures as did some graduate students from Penn State University and some students of Ken Ono.

I had sent a report to *The Hindu*, India's National Newspaper, about the launch of the Ramanujan Colloquium, and it was published on March 20 in their National Page. I also sent a report to India Abroad (a weekly), perhaps the leading Indian newspapers in the USA, which published it prominently two weeks later.

On Tuesday, March 20 night, Mathura and I hosted a party at our home in honor of Bhargava. It was attended by more than 60 guests which included not only faculty and spouses from the mathematics and science departments, but by some members of the Indian community in Gainesville who were keen to meet this rising star of Indian origin. In fact our friend Dr. Durai came all the way from Tampa to meet Bhargava at our party! In view of the presence of members of the Indian community at the party, I requested Bhargava to speak briefly and he kindly obliged; he spoke about how he entered the domain of mathematics, and why he was fascinated by the subject. Bhargava was totally relaxed at the party, having conversation with faculty, students, and friends. After all the guests left, Bhargava spent an hour at home with my family chatting on topics ranging from Indian classical music to Indian food in the USA.

On Wednesday, March 21, after Bhargava's final lecture, Lisa Lorentzen gave a talk on Ramanujan's ideas on continued fractions. Thus it was an intense and fulfilling week in number theory.

As these wonderful events were going on, we heard the sad news that Al Bednarek, who had been Chair of the Department for 17 years (1969–86), passed away after a long battle with cancer. It was Bednarek who offered me the position in Florida in March 1985, and indeed I was his last appointment. He was extremely patient and understanding while I waited in India for more than a year for my immigrant visa before joining the

Department in December 1986. He never once got annoyed at my frequent and anxious phone calls from India pertaining to various procedures relating to my green card application. Some of these phone calls were early in the morning (for him). I felt his loss dearly.

The Spring Term concluded with an Appreciation Tea on April 26. At the Tea, we expressed our appreciation to David Drake who was retiring at the end of the term. It was with Drake's appointment that the combinatorics program at the University of Florida started in the seventies. The seminar continues actively today, and it is the longest running seminar in the Department. Before the Tea, Neil White of the combinatorics group gave a thirty minute presentation of Drake's research accomplishments. Drake was also the chair of the search committee during the year when the offer was made to me, and was chair of the mathematics department from 1988 to 93. It was he who worked to get John Thompson appointed as Graduate Research Professor in 1993. So, on behalf of the Department, I expressed my appreciation of Drake's stewardship of the Department.

At the Appreciation Tea, we also presented a plaque to William Chandler and his wife Cynthia for supporting The John Thompson Research Assistant Professorship as a three year position starting in 2008. The position was called the Thompson-Chandler Research Assistant Professorship during the years when it was supported by the Chandlers. It was my colleague Tim Olson who convinced the Chandlers to make a donation to support this position.

Events of Summer 2007: Two days after the Appreciation Tea, I was off to New York to serve on a panel to hear the reading of a play called "The First Class Man" at the Tribeca Film Festival. Back in mid-March, I was contacted by the Tribeca film group saying that they are approaching me at the suggestion of Terence Tao. I was told that George Andrews had agreed to serve on the panel, and so I too accepted the invitation.

I flew to New York from Gainesville via Atlanta on Saturday, April 28, but the departure from Gainesville was delayed. While waiting at the airport, and during the flight, I constructed an elegant variation of the famous Rogers-Fine identity combinatorially and noted a beautiful partial theta series that it yielded. Upon arrival at New York's JFK Airport at 7:45 pm, I was picked up by a Tribeca limousine and taken to the Exchange Hotel in Manhattan where Andrews was also staying. When I had dinner with Andrews that night, I showed this "Airport Theorem" (to use his phrase) that I had proved and he confirmed that the results were novel.

The next morning Andrews and I were taken to the Tribeca Film Festival venue, where we were introduced to David Freeman, an award winning writer, who was the author of this script "The First Class Man" which was the story of the interaction between Ramanujan and Hardy in England. We heard the reading of the script for about an hour starting at 2:00 pm following which there was a panel discussion at 3:00 pm. The script had received a Guggenheim Award (grant) and was being considered to be made into a film. There were two things about the script that bothered Andrews and me and we mentioned our concerns during the panel discussion:

(i) As in the case with all stage productions on Ramanujan, in this script also there was focus on the remarkable Hardy-Ramanujan series representation for the partition function $p(n)$. The script described Major MacMahon's help in verifying the formula computationally and said that MacMahon had collected all his students and assigned to each one of them the task of writing down a set of partitions of 200, and that collectively these students would have listed ALL partitions of 200 from which the value of $p(200)$ could be obtained. This is totally wrong and misleading for several reasons. Firstly, the number of partitions of 200 is of the order of magnitude 3 trillion (!), and so even if MacMahon had assembled a hundred students, they could not list all partitions on 200 in their lifetime! Secondly, MacMahon computed $p(200)$ from a celebrated recurrence of Euler, and showed that the Hardy-Ramanujan series gave the same value. The point here is that to compute the value of $p(200)$, one NEED NOT write down ALL partitions of 200. So Freeman missed this point completely. Since he wanted something about student assignments in the script, Andrews and I said that he could alter the script as follows — "Imagine a hundred students working non-stop and each writes down $1/100$ of the partitions of $p(200)$. Even then it would take a million years or more...." After hearing our arguments, Freeman agreed to this suggestion.

(ii) Our second objection pertained to Ramanujan's behavior as portrayed by Freeman. The script said that when Ramanujan had difficulty getting vegetarian food in England, a certain Indian student invited Ramanujan to his residence for dinner. This Indian student had an English girlfriend who took a liking to Ramanujan and admired his genius. I was not bothered by this. The script went on to say that Ramanujan enjoyed the interest and admiration that the English girl showed, and he accepted her offer to bring vegetarian food cooked by her to his chambers the next time. I was bothered when I heard this, because not only did this not happen, it projected an incorrect version of Ramanujan's character; he had great affection for his wife Janaki and never showed interest in any other woman even when he was alone in England. As per the script, the English girl arrives in Ramanujan's chambers and delivers the food. Ramanujan and girl then engage in a conversation which is when the girl says that she is sculptor and would like to make a statue of Ramanujan in the nude! I was shocked when I heard this and literally jumped out of my bar stool type seat on the stage! During the panel discussion I strongly objected to portraying Ramanujan as being unfaithful to his wife. Also, this story is totally false and completely unnecessary to be brought into the script when describing Ramanujan's life in England. The author Freeman said in his response, he knew this was his fantasy, but insisted that he will continue to have it in the script.

During the panel discussion, the bright stage lights were on, but the auditorium was not lit. So I could not see any part of the audience that numbered about 300 to note their reaction to our objections. When the panel discussion had finished and we were leaving the auditorium, Freeman's wife told Andrews and me that she too felt uncomfortable with the episode of the British girl showing interest in Ramanujan, but that her husband refused to change it. In any case, I was no longer a panelist after I voiced my objections so strongly, but Andrews continued as a panelist. Andrews told me that at the next reading of the script a few months later, he noted that some changes were made in the script. But with regard to the episode of the English girl, the revised script said that Ramanujan went to bed with her!! At this point Andrews objected strongly and said that Freeman had portrayed Ramanujan as an adulterer; so he resigned from the panel. In the end, The First Class Man was never made into a movie!

In any event, after the panel discussion, Andrews and I got together for dinner that night at an Indian restaurant in Manhattan called the Tandoori Palace. There Andrews suggested that it would be worthwhile to provide a purely combinatorial proof of the partial theta identity I had found as a special case of my variant of the Rogers-Fine identity. I found this in the wee hours of the next morning and communicated it to Andrews at daybreak when he was in a taxi on his way to Newark airport. So even though the "First Class Man" never made it to the silver screen as a result of our comments, my trip to New York was mathematically rewarding. My paper on the Rogers-Fine identity was published two years later in the International Journal of Number Theory (IJNT).

I checked out of the Exchange Hotel in the morning, went to La Guardia where I took the Delta shuttle to Boston Logan Airport. There I took a taxi to the Springer office in Cambridge, Massachusetts, where I had a fruitful discussion about the Ramanujan Journal with Ann Kostant who had taken over the management of the journal from John Martindale. Ann and her colleague Kyle Cavenaugh took me to lunch at the India Pavilion where we discussed the Ramanujan Journal and my book series Developments in Mathematics. Both agreed to increase the number of issues of the journal per year to clear the backlog, and in view of the increase in the number of submissions annually — a measure of success of the journal. My discussions with Ann continued after lunch. She presented me Gian-Carlo Rota's lovely book "Indiscrete Thoughts" (Birkhauser), in which Rota shares his thoughts on mathematicians and mathematical topics. I have read this book many times, and never felt tired of its contents. That night Joachim Heinze of the Springer office in Heidelberg and Ann took me to dinner at a fancy restaurant to put a nice finishing touch to my visit to Springer.

Back in Florida, we had the Ninth Erdős Colloquium by Alexander Kechris of Caltech as the final event of the Special Year in Mathematical Logic and Set Theory. Kechris is a noted logician and a former PhD student of Yiannis Moschovakis at UCLA. When I was a graduate student at UCLA, I had met Kechris in the mathematics department because he and Moschovakis were the organizers of the Caltech-UCLA Logic Seminar which met alternately at Caltech and UCLA. I was asked to make the Opening Remarks for the Erdős Colloquium by Kechris, and in doing so I alluded to meeting him at UCLA, and he appreciated this very much.

One week later, I departed for Urbana via Indianapolis to participate in the Illinois Number Theory Fest (May 16–20). This meeting was in honor of the 80th birthdays of Heini Halberstam and John Selfridge. Halberstam was Chair of the mathematics department at the University of Illinois from 1980 to 88, and Selfridge had a long association with the department.

My talk was on the first day in the afternoon and I spoke on a multidimensional extension of Sylvester's identity. That evening, after the sessions were over, a movie was shown about how children of Jewish families were taken to England from mainland Europe during WW II to escape the Nazi regime. Halberstam was a child who arrived in England in a such a manner and grew up there.

I could not stay for the conference in full because I had to finish some work in Gainesville before leaving for the next conference trip on May 21. It was kind of the organizers of the Illinois conference to accommodate my request to speak on the first day. I was back in Gainesville on May 18.

I departed for New York on May 21 to attend two conferences: (i) Dorian

Goldfeld's 60th birthday conference at Columbia University, and (ii) the Combinatorial and Additive Number Theory Conference (CANT 2007) at the City University of New York (CUNY) organized by Mel Nathanson.

I arrived at Newark Airport on May 21 morning and preferred to stay at the Hilton Newark instead of in Manhattan where the hotel prices were sky high. The Newark Hilton is conveniently connected by a covered walkway to Penn Station Newark from where I took a train to Manhattan. I attended the afternoon sessions of the Goldfeld-60 conference that day. It was exciting to hear the talk of Fields Medalist S. T. Yau because he mentioned mock theta functions in connection with certain problems in differential geometry. I attended the banquet in honor of Goldfeld at the Faculty Club of Columbia University, which was on a high floor and commanded a fine view of the campus that looked splendid under bright summer skies. Goldfeld is one of the leading figures in number theory and so it was not surprising that the conference and the banquet were attended by dozens of mathematical luminaries some of whom spoke at the banquet. The following night, there was a nice party at the home of Dorian Goldfeld in New Jersey.

The next morning I was at the CUNY Graduate Center for the first day of the CANT 2007 Conference. My 45 minute talk was at 11:30 am, and I spoke about my multi-dimensional version of Sylvester's identity. In the afternoon I was back at Columbia University to attend the talk of Akshay Venkatesh. Goldfeld had spoken to me very highly of Venkatesh during the party the previous night, and so I did not want to miss Venkatesh's talk. That night Peter Sarnak, Manjul Bhargava and I went to dinner. There I talked to Sarnak about our newly launched Ramanujan Colloquium and he kindly agreed to address this colloquium the following year. So the trip to New York was quite fruitful.

After attending to a variety of departmental matters, I departed on July 7 with Mathura on a one month summer trip to India via South East Asia. We flew by Northwest Airlines to Singapore from where we took Malaysia Airlines to reach Siem Reap in Cambodia via Kuala Lumpur. When one mentions Cambodia, the first thing that comes to mind are the horrors of the *Khmer Rouge* regime between 1975 and 78. But Cambodia is the home of the remarkable *Angkor Wat* Hindu temples built over a thousand years ago. During the Khmer Rouge rule, a vast majority of the Buddhist temples and historic monuments were destroyed, but fortunately Angkor Wat survived the regime. Now that Cambodia was back to normalcy, we decided to go to Siem Reap to see the great ancient temples of Angkor Wat.

We stayed at the magnificent Le Meridien Angkor from where we took a chauffeured car to see the temples of Angkor Wat. What we saw was simply breathtaking. The main temple of Angkor Wat has to be seen in the morning because it looks radiant under the rising sun. At the Bayeon area, trees have grown over some on the ancient temple ruins. It was an awesome but eerie sight. One day we took a boat ride to a floating village where we bought souvenirs to remind us of our fantastic visit.

9.11 My final year as Chair (2007–08): The 2008 Abel Prize to John Thompson and the Program in ANTC

My final year as Chair had its share of important developments, crowned by the award of the Abel Prize to John Thompson in May 2008. This brought worldwide recognition to the Department. Since the six special years had ended, I began a Program in Algebra, Number Theory and Combinatorics (ANTC) in 2007–08 and it ran for three years. This program, although not as elaborate as the special years, still had a good number of conferences and featured lectures. The program was funded by the NSF, the NSA, and the Number Theory Foundation (NTF).

The first event of the Program in ANTC was the Conference on Group Representations and Combinatorics (September 10–13) organized by Pham Tiep and Peter Sin and inaugurated by our distinguished colleague John Thompson. One of the principal speakers for the conference was our colleague Alexandre Turull who gave a series of lectures on Schur indices. Among the main speakers was Robert Guralnick (University of Southern California), who was one year my senior in graduate school at UCLA. I was also pleased that Bhama Srinivasan (University of Illinois, Chicago) was a main speaker at this conference; she hails from Madras, India, and did her undergraduate studies at Madras University in the fifties. Bhama and I are acquainted not only through mathematics, but through our family connections as well. As part of this conference program, Michel Broue (Institut Henri Poincaré, Paris) gave a History Lecture on local group theory.

In May 2007, Professor Srinivasa Varadhan of the Courant Institute, NYU, was awarded the Abel Prize for fundamental contributions to the theory of probability. In the midst of his especially busy schedule after winning the Abel Prize, he accepted our invitation to deliver the Center for Applied Mathematics Colloquium on October 26.

Varadhan received his undergraduate degree at the Presidency College of the University of Madras in 1959. He then joined the Indian Statistical Institute (ISI) in Calcutta to do his PhD under the direction of Professor C. R. Rao, the Director of the ISI. He moved to the Courant Institute in 1963 after receiving his PhD and has been there ever since.

Varadhan delivered the CAM Colloquium on Large Deviations, a topic in which he had made seminal contributions. We invited Provost Joseph Glover to make the Opening Remarks because Glover's area of expertise is the theory of probability. That night Mathura and I hosted a party at our home for Varadhan and his wife Vasundhara, which was well attended by faculty in the mathematics and statistics departments with their spouses.

On October 31, I left for State College, Pennsylvania, on a short two-day trip to have discussions with James Sellers and Ae Ja Yee of the Penn State Mathematics Department about the George Andrews 70th Birthday Conference to take place in Fall 2008. Sellers and Yee had very kindly invited me to be one of the organizers of this conference. While at Penn State University, I addressed the Algebra and Number Theory Seminar on the theme "Partitions with non-repeating odd parts and q-hypergeometric identities".

On November 7, I arranged the Joint Mathematics-English Colloquium by Professor David Leavitt of the English Department. Earlier in the year, he had published a book called "The Indian Clerk" which is a fictitive novel based on Ramanujan's life. I thought it would be of general interest to students and faculty if Leavitt would give a talk about his book. As expected, his lecture attracted a sizeable audience from several departments on campus. There was excellent audience participation in the question portion after the lecture. Leavitt's book was subsequently reviewed by Heini Halberstam in the Notices of the American Mathematical Society.

==

December 2007 trip to India and the SASTRA Ramanujan Prize to Ben Green

On December 8, Mathura, Amritha and I departed on a three-week trip to India. Enroute, we vacationed for a few days in the Phillipines, staying in Manila at the Grand Hyatt. There were two unique sightseeing experiences for us. First was a day trip to Corrigedor Island which is right at the entrance to Manila Bay. Corrigedor was a crucial outpost for the Allies during World War II. It was captured by the Japanese and taken back by the Allies a few years later. The island is picturesque, but the main attraction there are the many ruins of war time fortifications which are very impressive. The second trip we made was to Lake Taal and the Taal Volcano. Both the lake and the volcano are natural wonders because we have a volcano on the Taal Island in the sea, and inside this volcanic crater is a lake, at the center of which is a smaller volcano! We took a boat to the Taal Island and there we were on mules which took us to the top of the volcano to view the crater. When we arrived in Madras, my father was keen to hear about Corrigedor Island because he had followed the events of World War II very closely.

The main academic event during my visit to India was the SASTRA Conference (December 21–22) and the award of the SASTRA Ramanujan Prize to Ben Green, Hershel Smith Professor at Cambridge University. Green had made phenomenal contributions to several important problems in combinatorial additive number theory, by himself and in collaboration with Terence Tao of UCLA, who won the SASTRA Prize the previous year. Green's PhD thesis of 2002, written under the direction of Fields Medalist Tim Gowers of Cambridge University, is a collection of several outstanding papers. In one of them he solved the Cameron-Erdős conjecture which is a bound for the number of sum free subsets among the positive integers up to a given number N. Over the years several top mathematicians had worked on this problem which was finally solved by Green.

Green's most spectacular contribution is the study of long arithmetic progressions of primes, starting with his seminal paper of 2005 in the Annals of Mathematics. This paper contained very fundamental ideas which he and Terence Tao could subsequently develop to settle the long-standing conjecture that the primes contain arbitrarily long arithmetic progressions.

Green was born in Bristol, England in 1977. He went to Cambridge University to do his BA (1995–98) and continued there to do his PhD (1999–2002) during which time he was awarded the Smith's Prize (2001). He was quickly elevated as Professor at the University of Bristol in 2005. He received the prestigious Fellowship at the Clay Institute in 2005 and was appointed Hershel Smith Professor at Cambridge University in 2006 at the young age of 29.

Several noted number theorists from around the world spoke at the SASTRA Conference — Roger Baker (Brigham Young University, Utah), Michel Waldschmidt (University of Paris), Tom Koornwinder (University of Amsterdam), Ole Warnaar (University of Melbourne), Christian Krattenthaler (University of Vienna), and Ken Ono (University of Wisconsin). Amanda Folsom, who was a post-doc at Wisconsin with Ken Ono, was also a speaker. Koornwinder served on the Prize Committee that year, and it was good that he could attend the conference.

Roger Baker arrived in Madras on December 17 and so I took him to a couple of concerts at the Madras Music Academy which he enjoyed very much. Ben Green arrived in Madras on the early morning of December 18. I took him and Baker on a tour around Madras showing them the famed

Marina Beach and the historic buildings fronting the sea, including some of the Madras University. My father was happy to show them around our family home Ekamra Nivas before I took them out to dinner.

Sightseeing in Kumbakonam on December 20 and 21 evenings included all the important sights which we now show every participant on the SASTRA conferences. In December 2007, there was unseasonal torrential rains around Kumbakonam. The main auditorium at SASTRA was flooded and so all lectures were in a large lecture hall on one of the upper floors. On December 22, Ben Green gave the Ramanujan Commemoration Lecture entitled "Going beyond the circle method of Hardy-Littlewood-Ramanujan".

The participants all departed Madras by December 23. The next day, my older daughter Lalitha and son-in-law Aditya arrived in Madras for a one-week stay. On Christmas Day December 25, we had a grand party at Ekamra Nivas for all our friends and relatives to meet the newlyweds. Thus 2007 ended on a high note.

===

Lalitha and Aditya returned to America in the end of December via Europe. Mathura, Amritha and I departed Madras for the USA on January 2 via the Pacific. My parents departed Madras one day later. We enjoyed a short stay with them in Singapore before departing for Minneapolis. From there my family left for Florida while I went to San Diego to attend the Annual Meeting of the American Mathematical Society.

One of the highlights of the meeting for me was the award of the AMS Cole Prize to Manjul Bhargava who in 2005 had won the First SASTRA Ramanujan Prize. For the SASTRA Prize, we select very young mathematicians and a confirmation of the quality of our selection is when these young mathematicians, in the next few years, go on to win prizes with a hallowed tradition.

At the AMS Conference, I had fruitful discussions with Joachim Heinze, Ann Kostant and Elizabeth Loew of Springer about the expansion of The Ramanujan Journal (both in the number of issues per year and the number of pages per issue) and about additional editors for my book series Developments in Mathematics. The expansion of the journal is a measure of its success because the expansion was needed to accommodate the steady rise in the submission of excellent papers.

Back in Florida, the first main event of 2008 was the Erdős Colloquium on January 25 by Fields Medallist Efim Zelmanov (University of California at San Diego), who spoke on "Asymptotic properties of infinite families of finite groups". I was in Zurich at the International Congress of

Mathematicians in 1994 when Zelmanov received the Fields Medal, and I felt that since our Department had such a strong program in group theory, we should invite Zelmanov to Florida. So it was a pleasure to have him in our midst. Mathura and I hosted a party for Zelmanov at our home, and my father enjoyed talking to him in leisure during the party. I found Zelmanov to be extremely pleasant and easy to converse with. George Andrews had arrived in Florida right after the AMS Annual Meeting in San Diego, and so he was there to attend Zelmanov's colloquium.

That same day, after Zelmanov's colloquium, I received a message from the NSF that the proposal that my colleague Tiep and I had submitted for special conferences on Quadratic and Higher Degree Forms, to be organized in consultation with Manjul Bhargava in Spring 2009, with be fully funded — that is complete funding for every participant. Such complete funding was crucial because at the University of Florida there were significant budget cuts and so there was little hope of getting any partial support for such conferences. This was wonderful news coming from the NSF right after the close of our six Special Year Programs in Spring 2007.

In early February, Andrews gave a magnificent History Lecture on the theme "Euler's contributions to partition numerorum". He had just published a long paper on this topic in the October 2007 Issue of the Bulletin of the AMS. Andrews is well known to many on campus as a fine speaker and so his lecture attracted both faculty and students from several departments.

In February we received Roger Baker from Brigham Young University and Dominique Foata from the University of Strassbourg, France, both for visits of several weeks. In connection with the ANTC Program, Foata gave a one month course of lectures on "q-series in Combinatorics - permutation statistics". This theme resonated well with the research interests of both the number theorists and combinatorialists in the Department.

On February 12 we had back to back lectures by Foata and Baker — Foata in the Combinatorics Seminar and Baker in the Number Theory Seminar. Doron Zeilberger, a close friend and collaborator of Foata, was also in our department in connection with Foata's visit. Zeilberger gave a lovely lecture on "q-Foata" on February 18, following which Foata delivered the Ulam Colloquium on the theme "Eulerian polynomials".

On February 21, I was asked to inaugurate the "Conference on Combinatorics and Groups" in memory of our late colleague Chat Ho. Dominique Foata gave the Opening Lecture of the conference. Other eminent speakers at the conference included John Thompson, Chat Ho's former thesis advisor, and George Andrews. It was gracious of Viginia, Chat's wife, to host the conference party at their lovely home.

==

Announcement of the 2008 Abel Prize to John Thompson

The most sensational news of my ten-year term as Chair was the phone call I received on March 7 from the Norwegian Academy of Sciences that the 2008 Abel Prize would be awarded to John Thompson and Jacques Tits! I was told to keep this news secret until March 27 morning when the Norwegian Academy would call John Thompson to give him the news. The reason the Norwegian Academy called me with the news twenty days earlier was to allow me time to prepare for a Reception in honor of Thompson on March 27 afternoon, and to prepare a news release on behalf of the University of Florida to be sent out on March 27. Since the top administrators of the University of Florida were to be invited to the Reception, such as the President, the Provost, the Dean, and the Vice-President of Research, they all had to be informed of this great development and asked to keep it secret for three weeks.

On March 11, I received another call from the Norwegian Academy, this one from Anna Marie Astad, who said that the Academy would like me to be one of the Press Contacts for the 2008 Abel Prize. I felt honored, and so I readily accepted to be a press contact.

It was a pleasure for me to plan the Reception for Thompson for which the College very kindly made available the Keene Faculty Center. Of course, some key members of the Department, like our secretarial staff, and my colleagues in the algebra group, were informed of the prize, because they had roles to play in the reception, but they were sworn to secrecy. I informed Thompson and Diane that there would be a Reception at the Keene Faculty Center on March 27 afternoon, when the President of the University would be present, and I wanted both of them at that Reception. I did not divulge to Thompson that the reception was in his honor.

I came very early to campus — at 5:30 am on March 27. At 7:00 am, Thompson received a call from the Norwegian Academy informing him that he and Tits will share the 2008 Abel Prize. Minutes later, Anna Marie Astad called and told me that Thompson had been informed. So I called Thompson and Diane, congratulated them on the wonderful news, and said that the Reception was to honor of Thompson for the Abel Prize!

Within a few minutes of talking to Thompson, I sent an email to the President's office, requesting that my email announcing the Abel Prize to Thompson be sent throughout campus. The news spread like wildfire, and more than a hundred faculty from around campus showed up for the Reception in a few hours time. Dr. Berit Johne from the Norwegian Embassy in

Washington D.C., came to attend the Reception and presented Thompson with a special plaque. University President Bernie Machen was in Tallahassee that morning for budget discussions; he returned to Gainesville in his private airplane to attend the Reception.

I opened the Reception program with a welcome and tribute to Thompson. Next, our colleague Steve Saxon, an almost professional opera singer (bass - baritone), sang with inspiration: "Quia fecit mihi magna" from the 'Magnificat' in D Major, BMV 243 (Bach). President Machen in his speech said that this recognition to Thompson and the University is bigger than the winning of National Championships in football and basketball! The President also noted that the Mathematics Department has to be credited for providing a conducive environment for Professor Thompson. After the President's speech, the Provost and the Dean spoke briefly, following which we played a tape that I had prepared in which there were tributes to Thompson paid by the members of the algebra group, and by Bert Kostant (MIT), Peter Sarnak (Princeton) and George Andrews (President Elect of the AMS), who are all members of the National Academy of Sciences. In the end, Thompson responded with his characteristic humility.

My parents and Mathura attended this Reception. I was pleased that they were there to witness this event of great significance during the last few weeks of my term as Chair. That night, Mathura and I had dinner with Thompson, Diane, and Dr. Berit Johne at the Hilton. We all looked forward to the Abel Prize Ceremony in May in Oslo (this is described in detail separately in §11.2).

===

While preparations were going on during March 7–27 for the Abel Prize Reception, those three weeks were packed with other significant events.

During March 8–15, we had the Conference on Partitions, q-series, and Modular Forms that Frank Garvan and I organized as one of the main events of the Program in ANTC. For the first three days, there was a Student Workshop to provide a background for the fifteen undergraduate students attending the conference. I gave a talk to the students on the first day on the theme "The Indian genius Ramanujan - a glimpse of his magical mathematical formulae". Garvan gave a talk on the second day of the workshop entitled "A page from Ramanujan's Lost Notebook" — the page that contained a formula which inspired Garvan to discover his Vector Crank to explain all three of Ramanujan's congruences modulo 5, 7, and 11. The research conference began on March 12 with the hour lecture by Herb Wilf (University of Pennsylvania) entitled "Towards a global theory of integer partitions". Wilf's lectures were always marked with elegance and clarity. Ken Ono gave three lectures on mock theta functions and harmonic Maass forms, and in doing so discussed links with several major problems in number theory, such as the Birch–Swinnerton-Dyer Conjecture. Thus the main lectures of the Student Workshop were delivered by Ono's student Sharon Garthwaite with the aim to provide them the background needed to follow Ono's lectures. Other main speakers for the conference were George Andrews, Bruce Berndt, Peter Paule, Ole Warnaar, and Sander Zwegers. Indeed it was Zwegers's PhD thesis work linking mock theta functions with Maass forms that started this new wave of activity on mock theta functions. The conference attracted 75 leading researchers — senior and junior — and was one of the best we had organized. The refereed proceedings edited by Garvan and me appeared as Volume 23 in the book series Developments in Mathematics (Springer) in 2012.

Two days after the conference, we had the visit of Bert and Ann Kostant. Bert's work is of interest to both mathematicians and physicists, and so he addressed the IFT Colloquium on March 17, jointly arranged by the mathematics and physics departments. He gave a magnificent lecture on the structure of the exceptional Lie group E_8. The next day, Hershel Farkas of the Hebrew University in Jerusalem, who was visiting the Department for a few weeks, gave a talk in the Number Theory Seminar on theta constants.

On March 19 we had the Second Ramanujan Colloquium by Peter Sarnak of Princeton University, who in 2007 had also been appointed as a Professor at the Institute for Advanced Study. Sarnak is one of the most influential mathematicians of our generation, and a brilliant and charismatic speaker. My father who attended Sarnak's colloquium, compared his lecturing style to that of Nobel Laureate physicist Murray Gell-Mann! Sarnak spoke on "Sieves, the Ramanujan Conjectures, and expander graphs". That night, Mathura and I hosted a party at our home in honor of Sarnak, Kostant, Farkas, and Andrews. During the next two days, Sarnak gave talks in the Number Theory Seminar. We also had him at our home for a private dinner, when he and my father had a leisurely conversation about the development of mathematical sciences in India.

During the months of April and May, I was preparing to wind up my term as Chair, and simultaneously preparing for my trip to Oslo for the Abel Prize Ceremony and events (May 19–21). Mathura agreed to accompany me on the trip. During April–May, I received several official communications and invitations regarding various events related to the Abel Prize Ceremony. For instance, there was an invitation from the King and Queen of Norway to attend a dinner at the Akershus Castle in honor of Thompson and Tits. So Mathura and I were in high spirits to embark on this momentous trip. Details of our trip to Norway, the Abel Prize Ceremony and related events are given in separately in §11.2. Here I will just say that it was perhaps the most thrilling academic trip in my life.

Back from Norway in the last week of May, I had just one month left before handing over the charge to my colleague and good friend James (Jed) Keesling.

==

My father passes away

As we were in an exhilarated mood brought about by the Abel Prize to Thompson and the resulting recognition for the Department, tragedy struck our family. My dear father passed away at our home within two weeks of my return from Oslo. It happened in a most unexpected manner:

On June 7, Mathura had the Annual Recital for her dance school. About twenty five of her students performed in a fantastic two-hour program attended by about one hundred and fifty members of the Indian community in Gainesville. The final item of the program was a dance on Lord Panduranga (= Lord Krishna) by Mathura and my daughters Lalitha and Amritha. My father was eagerly looking forward to this dance which was a fitting finale to the program. After the program ended with this dance, many members of the audience went up to the stage to take pictures and have pictures taken with the performers. We asked my father to come up to the stage, but he said he was feeling weak, and could not climb the few steps to ascend the stage. So I took him home immediately for him to rest, while my family wound up and came home a bit later. While at home, he complained of breathing difficulty, and so we called our family friend Dr. Neeta Pohani, who came immediately. As Dr. Pohani was examining my father, his pulse went down, and so we called an ambulance. He died before the ambulance arrived, surrounded by all members of his immediate family. It was a shock to all of us, but we took solace in the fact that he led a full life, did not suffer much, and was taken to heaven by Lord Panduranga himself! The Lord beckoned him, and he answered His call.

The loss of my father was a terrible blow to all members of our family who looked up to him for guidance and support. It was especially hard on my mother to whom he was married for 62 years, and who had been with him through thick and thin. It was as much of a blow to me because he was not just a father, but my academic mentor and my greatest academic supporter. He was an internationally reputed scientist, who in the face of obstacles, had created quite miraculously, the Institute of Mathematical Sciences (MATSCIENCE). I was a direct witness to all his efforts that led to the creation of MATSCIENCE. His oratorical lecturing style, his passion for scientific research, and his desire to rub shoulders with the leaders of the profession, made an indelible mark on me as a young boy. All through life, I have tried to follow his example for he was my guiding spirit.

In view of his annual visits to Gainesville from 1987 onwards, my father got to know many members of the Indian community in Gainesville, and several academicians at the University, including many from the Department. And everyone who got to know him admired him for his many accomplishments, and his boundless optimism. So his funeral on June 10 was attended by more than 150 members of the Gainesville community. Meanwhile, my dear friend GP in Madras contacted Mr. N. Ravi of The Hindu, and other members of the press, to have an obituary column of my father published in the Madras newspapers.

My father was disappointed that I left MATSCIENCE and settled down in the United States, but he supported my decision for the sake of my happiness. A few years before he passed away, he told me that I had made the right decision to move to the United States. He also said that I should not wait further to become a US Citizen. So I applied for US Citizenship. My citizenship swearing-in ceremony was scheduled for June 17, 2020; he knew about the event, but passed away ten days before that.

June 30 was my last day as Chair. Jed Keesling took over the Chairmanship on July 1 and on that day, he sent a message thanking me for my services, and pointed out that the 2000 National Medal of Science to Thompson, and the 2008 Abel Prize to Thompson, were like two fine book ends to my term as Chair.

My family and my parents were scheduled to leave for India in early July, which we did, except that my father was not with us. Before we left, we performed all the Hindu ceremonial rites for my father, almost all of which were at our home in Gainesville. One of the ceremonies required immersing the ashes of the departed soul in the ocean. Our dear friend Pradeep Rawal graciously offered the use of his house on St. Augustine beach where we performed a specific Hindu ceremony before immersing my father's ashes in the waters of the Atlantic ocean.

We arrived in Madras on July 6. Our family home Ekamra Nivas, a home of a thousand memories, looked empty and desolate without my father's presence. The next day, our family priests performed a religious ceremony in memory of my father at our home. Professor M. S. Swaminathan

(FRS), arguably India's most distinguished scientist, arranged a Condolence Meeting at his Foundation on July 18. Besides me and Professor Swaminathan, some members of MATSCIENCE spoke on that occasion. Jean-Marc Deshouillers from Bordeaux, who knew my father well, and who happened to be visiting MATSCIENCE then, also paid his tribute at that function. On my part, I arranged a Memorial Function on July 20 in the upstairs seminar hall of Ekamra Nivas. It was attended by academicians, family, friends, and admirers of my father.

A few days before my father passed away in Gainesville, he selected photographs of him in conversation with eminent mathematicians at our home in Florida — John Thompson, George Andrews, Peter Sarnak, Bert Kostant, Efim Zelmanov, and Dominique Foata. He wanted these enlarged and displayed at Ekamra Nivas. So I put them on display there during the Memorial Function. In addition, I made an enlargement of a lovely photograph of my father with Niels Bohr at Bohr's house in Copenhagen, and had this portrait unveiled in the seminar room of Ekamra Nivas by Professor Anandakrishnan. Mr. N. Ravi, Editor of *The Hindu*, India's National Newspaper, was one of those who spoke on that occasion; he published a prominent report of the function in The Hindu the next day.

R. Balasubramanian, the Director of MATSCIENCE, called by phone to inform me that a new lecture hall at the Institute was soon to be opened, and that it would be named the *Alladi Ramakrishnan Hall*. He invited me to cut the ribbon to open the hall on July 23 and give the first lecture there. I felt honored to do this and gave a talk on "Some new observations on the Göllnitz-Gordon and Rogers-Ramanujan identities".

Mathura and I left Madras on July 27 to return to the United States. My mother stayed back in Ekamra Nivas. For the next six years, she spent six months in Madras and six months in Gainesville each year, until she moved permanently to Florida in 2015.

===

On our return trans-pacific journey, Mathura and I halted in Singapore for two nights. My cousin Geetha, and her son Rajesh (who was working in Singapore), called on us at our hotel to offer their condolences personally. Mathura and I travelled together up to Detroit, from where she proceeded to Florida, but I went elsewhere due to a couple of academic assignments.

I stayed overnight at the Westin Hotel attached to the terminal at Detroit Airport, and left the next day for Madison, Wisconsin, to attend the MAA MathFest at the University of Wisconsin. James Sellers of The Pennsylvania State University organized a session on "The Legacy of

Ramanujan" at the MathFest, and had invited me to speak in that session. The other speakers were George Andrews, Bruce Berndt, and Ken Ono. I spoke on "Hardy-Ramanujan and the Probabilistic Number Theory".

The next day (Aug 2), I left Madison early for Boston where I had discussions with Ann Kostant of Springer about The Ramanujan Journal and my book series Developments in Mathematics. After I checked in at the Marriott in Cambridge, Massachusetts, Ann picked me up and took me to her home for lunch. There Ann and Bert Kostant, and Ann's daughter Elizabeth, suggested that I should edit a volume entitled "The Legacy of Alladi Ramakrishnan in the Mathematical Sciences". Ann said that Springer would be pleased to publish this. I was touched by this warm gesture of the Kostants and Elizabeth, and I readily agreed.

The next morning, John Martindale of Kluwer, now retired, joined me for breakfast before my departure. Hariprasad Babu and his wife Shubha, picked me up at the Marriott and took me to Boston Logan airport for my flight back to Florida. Hariprasaad, who was initially trained by my father, had joined the University of Florida graduate program in mathematics. After getting a Masters Degree, he had taken up a position at a software company in Boston. He offered to take me to the airport because he and Shubha wanted to meet me and personally convey their condolences. Thus as the academic year 2007–08 ended, I looked forward to my year of research leave after my Chairmanship.

9.12 The External Review of the Mathematics Department (2005–06)

The External Review of the Mathematics Department conducted in Spring 2006 was an exercise in futility. It was a review commissioned by the Administration, and the Department took it very seriously. The review was positive, but recommendations of the review report were not implemented. Instead, major cuts to the department budget and size were carried out from 2008 onwards after my term as Chair ended, due to budget cuts across the university. Here I describe how the External Review happened, what the recommendations were, and what the aftermath was.

In Fall 2005, during my year of leave in the middle of my third term as Chair, while I was visiting Penn State University, I got the message that the newly appointed Provost would be commissioning an External Review of the department to be conducted in Spring 2006. We were informed that more than a dozen departments will undergo external reviews over the next three years, and three or four departments including Mathematics and English were to be reviewed first. The decision to have the review of the mathematics department in Spring 2006 disturbed me during my leave when I had so much academic travel planned, but I managed to deal with the situation with the help of my colleagues. I felt that if the Administration was genuinely interested in moving the department forward, the Provost's office could have discussed the matter with me, and scheduled the review in Spring 2007 after my return in Fall 2006 to full time duty as Chair.

The last External Review of the Department was in 1997 as part of the Board of Regents (of the State of Florida) review of all mathematics departments in the State. My term as Chair began in 1998, and so this review amounted to an evaluation of the Department's performance during my term as Chair. We welcomed this as an opportunity to analyse ourselves and suggest ways to improve. I was back in Gainesville for a few days from Penn State University in mid-November to attend the Ulam Colloquium of Louis Nirenberg, and during that visit, I met with the Steering Committee to discuss how we would prepare for the External Review. Scott McCullough was Acting Chair in 2005–06, and after discussions with him and the Steering Committee, I requested my senior colleague Alex Turull to prepare a draft of the Self Study Report. I had a lot more travel scheduled during the academic year, but I would be finished with that by February 2006, when I would take over from Turull to finish the Self Study Report.

One of the first things to be done was to choose the external reviewers. Our choices were Professors Peter March (Ohio State) — Chair, Ron Douglas (Texas A& M), Jon Hall (Michigan State), and Steve Ferry (Rutgers). Their fields of expertise were Probability and Applied Math, Analysis, Algebra, and Topology, respectively, and it had the balance of areas of speciality that we desired. Peter March was Chair at Ohio State and had been intimately connected with the NSF-funded Mathematical Biosciences Institute (MBI) there. Ron Douglas had served as Provost at Texas A& M. Steve Ferry and Jon Hall were Distinguished Professors. Thus the Committee combined academic accomplishment with administrative experience. The Dean approved our choices and so we wrote to them and were pleased that all four agreed to serve on the External Review Team.

The Self Study Report is the most important document for the External Review. Our Self Study Committee consisted besides me and Scott McCullough (Acting Chair during Oct 1, 2005–March 31, 2006), the members of the Steering Committee, all Administrative Faculty — The Associate Chair Jim Brooks, the Graduate Coordinator Paul Robinson, and the Undergraduate Coordinator David Groisser — and one other faculty member — Alex Turull — who chaired the Committee and acted as the scribe.

In view of the six Special Year Programs, the tenure track faculty in the Department were classified into six major research groups: (i) topology and dynamical systems, (ii) algebra, (iii) applied mathematics, (iv) number theory and combinatorics, (v) analysis and probability, and (vi) logic and set theory. We also had the group of Lecturers who teach the beginning level undergraduate classes. We asked each group to prepare a presentation to the External Review Team when they would visit the Department in April.

Throughout January, February and March, there were several meetings of the Self Study Committee, the six research groups, and the faculty as a whole, not only to prepare the Self Study Report, but also plan the presentations to be made in April to the Review Team. George Andrews who was visiting the Department in Spring 2006, as he does every Spring, told me that he was impressed with the level of commitment of the faculty and staff towards the External Review process.

The Self Study Report was delivered to the External Review Team, the Dean, and the Provost, in March 2006, one month before the site visit in April to provide enough time to study the report. The Department also assembled substantial data regarding its undergraduate and graduate programs to be given to the Review Team during the site visit.

The site visit of the Review Team was on Monday, April 10 and Tuesday, April 11. The entire first day was for meetings with the Department members. After a brief welcome at the Dean's office, the morning program for the Review Team comprised meetings with the Department Administration, hearing presentations on the undergraduate and graduate programs, discussions with the post-docs including the Thompson Assistant Professors, meetings with the graduate students, the staff, and representatives of the client disciplines. The afternoon of the first day was devoted to presentations by the six research groups and a meeting with the lecturers. After these meetings on the first day, the Review Team had one-on-one meetings on the morning of second day with select persons including George Andrews, the Associate Chair, the Directors of the Center for Applied Mathematics, and some faculty from other departments and other colleges. Associate Chair Jim Brooks and I had lunch with the Team who told us they were impressed with the performance of the Department and the positive comments they heard from those they met. They said that the staff — especially Sharon Easter (Administrative Assistant), Sandy Gagnon (Office Manager), and Margaret Somers (Chair's Secretary) — had very positive comments about the faculty and the department administration, adding, that if there were any problems, one would find out by talking to the staff.

In the afternoon the Review Team had meetings with the Dean and the Provost and that concluded the site visit. We were told that they conveyed their positive impressions to the Dean and the Provost.

Peter March, the Chair of the Review Team, said that it would take about a month for him to send us the External Review Report. And indeed we received it in mid-May. It was a positive report. The highlights of their observations and recommendations were the following:

I) The Department's reputation and international visibility have risen significantly in the past few years owing to several new programs of high quality, most notably the Distinguished Colloquia, the Special Years, and the Thompson Assistant Professorship.

II) The Department has done extremely well in spite of its limited budget by being very creative in the ways in which it ran its new programs. Thus the Department deserves more resources and it is well worthwhile investing in the Department.

III) The atmosphere in the Department is very collegial and the faculty, staff, and students, all seem enthusiastic to work towards taking it to a higher level of productivity.

IV) By forming liaison committees, the Department could create more effective channels of communication and enhance interaction with other departments on campus.

V) The Department deserves additional resources including more faculty positions, more teaching assistants, and space.

The External Review Committee did identify certain specific needs and programs that it recommended for immediate consideration. One of these was the Thompson Assistant Professorship. The Committee recommended that the number of such positions offered each year should be increased.

Upon receiving the report, I called the Dean who said he was very pleased to receive such a positive report. Of course, any decisions based on the report were to be made by the Provost and so the Dean told me that he and I would meet the Provost in a few weeks.

The meeting with the Provost was on July 20 to provide ample time to read the report. The Provost however wanted to focus on the graduate program and see ways to trim expenditure. I stressed that the External Review was commissioned by the Provost's office, that the Department took this very seriously and spent considerable time in the review process, and that the meeting was to discuss the External Review Report. So I said that we need to consider that first before taking up other matters. But the Provost wanted to focus on streamlining the budget and identify programs that either needed to be trimmed or eliminated. I assured the Provost that the Department would cooperate with the Administration in addressing the budget shortfall, but as the Review Report pointed out, the Department was already strained due to limited funds; thus trimming the budget would be like cutting the Department to the bone since there was no flesh (excess funds) left to be cut! I came away from the meeting disappointed that the external review process which consumed our energy for the entire academic year, was a waste of time! The mathematics department was not the only one identified for cuts. Indeed the budgets of all colleges were slated for cuts, but the College of Liberal Arts and Sciences (CLAS) suffered the most. One reason for this was that over the previous few years, CLAS went over its budget by about 3% to 5% each year. The previous Provost, realizing that CLAS had spent its funds for worthwhile causes, made up for the shortfall at the end of each year. However, between the time the previous Provost stepped down and the new Provost took office, there was a time-lag and the deficit had accumulated to about 10% to 15% in the CLAS annual budget. The new Provost insisted that the College come up with a plan to trim its programs and offerings.

I have never really understood how the university allocates budgets to the various colleges, and administrators do not share their knowledge of this process. You are considered a successful and astute administrator if you can work with the system and get significant funds. But there is something I learnt about budgeting which made me uncomfortable and I will share it with you now, since it relates to the budget woes of CLAS.

One way money is allocated is by measuring the number of student credit hours that a college generates. But the important point to be noted is that the University allocates a certain number of dollars for each such credit hour but this amount varies from college to college! For example, the College of Engineering would be given E dollars per student credit hour, and CLAS would be given C dollars per credit hour, with E much larger than C. The reason given for E to be much larger was that engineering courses require uses of labs and so such courses require more funds for their operation. CLAS also has courses that use labs, but the courses in the humanities do not use any labs, and that lowers the value of C. While I understood and accepted that E has to be larger than C, I could not understand how exactly the values of E and C were determined. I could not get an answer to this question. The reason this question is important is because, if there is a degree of arbitrariness in choosing the values of E and C, then by lowering E slightly and increasing C a little bit, there would have been no budget shortfall of 3% to 5% each year for CLAS. And the College of Engineering would have continued to be in the black even if there would have been a slight reduction in the value of E. But the University did not want to touch the values of E and C. So Dean Sullivan of CLAS, who I consider as an excellent administrator with a genuine interest to raise the performance of various units in his College, was forced to come up with a drastic Five Year Plan of deep cuts to various programs. This Plan never really materialized because of opposition from the various units in the College, but it consumed the energy of all department administrators in CLAS. Indeed, during much of Fall 2006 and even later, I spent considerable time writing letters and memos both to the Dean and to the Provost showing how various programs in mathematics were doing well, how vital they were to the University, and why they should not be trimmed. Dean Sullivan had always argued correctly, that the College of Liberal Arts and Sciences is the backbone of any university. You could have a great university without a medical school, an engineering school, or a law school, but you cannot have a great university without a leading college of liberal arts and sciences. Quoting Dean Sullivan, I argued that just as CLAS is vital to the University, the mathematics department with its research and service components is vital to CLAS and the University. Indeed, there is no university without a mathematics department, and no university in the top ten without a comparably ranked mathematics department.

Another source of revenue, and indeed the most significant, are the faculty lines assigned to a college. From what I understand, a faculty line is a permanently flowing river of money (my phrase), with different faculty lines having different annual amounts allocated to them. If a college is authorized N faculty lines, the Dean may use only M of those lines for faculty appointments (salaries), obviously with $M < N$. The annual funds in the remaining $N - M$ faculty lines are left to the Dean's discretion to use as he pleases. A college is rich if $N - M$ is large, not just if N is large, because N could be large but M may be close to M to meet teaching demand. I heard that to deal with the budget crunch, the Administration wanted to take back a good number of unused faculty lines from various colleges and so the college deans were not happy with this idea.

As we were grappling with the problem of budget shortfall, news came in March 2008 towards the end of my term as Chair, that John Thompson would be awarded the Abel Prize two months later. Without a doubt, this was the greatest recognition to the Department, the College, and the University. President Machen of the University of Florida acknowledged at the Reception for Professor Thompson in March 2008, the day when the Abel Prize news came, that it was a recognition for the University exceeding the championship titles in football and basketball! Although the Abel Prize brought unparalleled recognition to the University, it did not save the Department from budget cuts. Indeed, in the decade after my term as Chair, the mathematics department was also a victim of the budget cuts sweeping the university. In particular, the Thompson Assistant Professorship program was discontinued contrary to the Review Committee recommendation to increase the number of such assistant professorships. The faculty strength which had grown from about 55 to 67 from 1998 to 2005 during the first seven years of my term as Chair, shrunk drastically in ten years after my term ended. The number of graduate students which increased from about 70 to 105 during my term, also shrank significantly in ten years. But the number of undergraduate students enrolled in the University and in the mathematics courses continued to be as high as before. However, after a decade, when the budget situation improved, the university suddenly gave 16 faculty positions (!) to the Department in a two year period (2018–2020) to make up for all the lost positions in the previous ten years. The department was quick to take advantage of this windfall. Several excellent appointments were made and the Department is doing extremely well now. The three Chairs who followed me — James Keesling (2008–13), Doug Cenzer (2013–18), and Kevin Knudson (from 2018) — all very competent, steered the department efficiently during the period of budget cuts, and made fine appointments when positions were made available. Indeed five of these new faculty have won prestigious NSF Career Awards.

In summary, I received full support from the Administration all along and thus enjoyed my work as Chair until I encountered this unpleasant experience with the Administration unwilling to implement the recommendations of the External Review. It was this that made my term as Chair a bittersweet experience. The irony is that starting from 2007, I have been invited regularly to the Middle East to be part of teams — often in the capacity as Chair — to do external reviews and accreditations of mathematics programs (see the next section). The universities in the Middle East keep American universities as the model for their academic aspirations and for the review process. Thus my experience working at several American universities was helpful in these reviews.

9.13 Evaluation of mathematics programs in the Middle East

Among my special and rather unique academic experiences are my trips to the Middle East to evaluate mathematics programs in the United Arab Emirates and in Saudi Arabia. I would also say that these have been eye-openers, because I was pleasantly surprised at the level of development in the academic setting and at the steps taken by both the administration at the universities, and by the governments, to ensure the progress of their academic institutions. Mathematics is viewed as indispensable for the sciences, engineering, and medicine, and I found a high regard for the subject among the university administrators I had discussions with at every university I visited. But research in pure mathematics, in subjects like topology or algebraic geometry, is viewed as important only within the mathematics departments. I should say that the university administration does not interfere with such pursuits in pure mathematics as long as the mathematics departments and their faculty address the service to the university by way of teaching in a satisfactory manner. I have made nine separate trips to the Middle East since 2007, and here I shall give an account of my visits and make observations on the universities I have visited, and the state of mathematics there. I also did one review virtually in 2022.

The way universities have been built in Europe and the USA, is to first start with the departments of Mathematics and English (or Classics), and then to form other departments. In the Middle East, usually the universities start with the College of Engineering (or a program in engineering) and then expand by adding departments in the sciences and mathematics. This is because, in the Middle East, owing to their economic priorities, training in engineering is considered most vital, and indeed, most students take to engineering as their major. Since lucrative careers in the Middle East with a mathematics degree are rare, very few male students take to mathematics as a major. Thus the number of mathematics majors was uniformly low in all universities I visited, and most of the students who majored in mathematics in the UAE were female; this is because, female students are not pressured to pursue lucrative careers, and also local culture deters female students from working in labs, especially in biology and medicine. However both the administration and the faculty in universities in the Middle East consider Level 1 research universities in the USA as their models to emulate as far as how their departments should function, and the instruction is to be imparted. A good percentage, and in some cases, almost all, mathematics faculty are expatriates because the locals take to careers in industry which are more lucrative. In all the universities I visited, faculty are not given tenure, but renewal of the contract is almost certain if good work is done. But lack of tenure is made up by attractive salaries and benefits such as housing on campus and an international travel allowance.

With regard to funding of various units in the University, it was not surprising to find that the engineering colleges were funded very well, but the university administration was also keen to provide adequate support to the College of Sciences and to the mathematics department.

The middle eastern countries that I visited, namely the UAE and Saudi Arabia, have become very prosperous due to oil money. They now want to invest in higher education, and money is not an issue. The universities who want external reviews to be done, treat the review teams extremely well. The members of the review team are flown in business class and put up at five star hotels. The idea is not to pamper them so as to get a positive review. The fact is that the work during the four or five day site visit is intense, and one has to start work right away on the very first morning after arrival. So there is no time to recover from jet lag. If you fly in business class, you can sleep well on the long flight and hence will be ready to start work the next morning. On the long flights I took, I would sleep half the time and spend a good portion of the time I am awake working on the review report. In fact, all members of the review team start working on the report a few weeks before the trip.

The External Review Team (ERT) usually consists of three members and is assisted by a Commissioner from the Commission of Academic Accreditation (CAA) of the Ministry of Higher Education. The Chair of the ERT assigns each member (including himself) the various sections of the report for which he is responsible. These External Review Reports are very elaborate, and so it is a lot of work before and during the visit. The CAA is very particular that the Course Learning Outcomes (CLO) and the Program Learning Outcomes (PLO) are satisfactorily met and that the report by the ERT must address the CLO and the PLO. This is very time consuming. We the ERT members start working on the report before the trip by going through the Self Study Report, and later complete our report on site after talking to various persons. The report is to be completed and handed over to the CAA on the last day of the site visit. On rare occasions, the report can be finalized a week or two later if there are issues that cause the delay. The ERT Members are given a handsome honorarium as well as a per-diem amount. The honorarium for the Chair is slightly higher. Of the nine visits I have made, I have served as Chair half a dozen times.

The CAA takes this review process very seriously. A few weeks before the site visit, the CAA sends the ERT a set of guidelines to be strictly followed, one of which is that the ERT members are to be dressed formally — not just when on site (campus), but also during discussions at a conference room in the hotel where the ERT members would be staying. A full suit is preferred, but a jacket and tie are also accepted.

I shall now describe my visits to the universities in the Middle East, whose programs I have evaluated.

The American University of Sharjah and The University of Sharjah

My very first visit to the Middle East was in Jan–Feb 2007, when I was invited to be Chair of a three-person committee to evaluate the proposals of the American University of Sharjah (AUS) and the University of Sharjah (UoS) to start BS programs (Majors) in mathematics. The UAE consists of several states (Emirates), namely, Abu Dhabi, Ajman, Dubai, Fujairah, Ras Al Khaimah, Sharjah, and Umm Al Quwain. All Emirates have a Sheikh (a ruler). Abu Dhabi, the largest of the Emirates, and with the most oil reserves, is the capital. Among the Emirates, Sharjah has invested the most in higher education and AUS and UoS are among the best universities in the UAE. So I accepted the invitation to Chair the ERT enthusiastically. The other two ERT members were David Mandersheid of the University of Iowa and Ted Hill of Georgia Tech.

I departed for Dubai on January 26. I flew Delta Airlines to Amsterdam where I connected to a KLM flight to Dubai. Upon arrival in Dubai at 11:30 pm, I was met by a representative of the Merhaba (Meet and Greet) Service, so thoughtfully arranged by the CAA for all ERT members. This was especially convenient, because there was a mile-long queue but owing to the Merhaba service, I was whisked past this queue in a golf cart and taken straight to the immigration counter. Without the Merhaba, it would have taken hours for me. Sharjah is the next Emirate to Dubai, and it was just a half hour drive to the Millennium Hotel in Sharjah, where I checked into my comfortable rooms just past midnight. The two other ERT team members also arrived late and I met them the next morning at breakfast after which we were taken to AUS for the start of our work on site.

The American University of Sharjah and the University of Sharjah are next to each other forming one mega sprawling campus. The main buildings of AUS and UoS are gigantic constructions in moghul style architecture with white marble and they rise majestically above the surrounding landscape like the Taj Mahal! The UAE has no trees except for a few date palms, and so you see these magnificent buildings of the AUS and UoS from miles away.

Both AUS and UoS were founded in 1997, and although they are sister campuses, their missions are very different. AUS is very cosmopolitan, as cosmopolitan as can be within the boundaries of the culture and practices of the Emirates. AUS has several foreign students and all students, both local and foreign, are permitted to wear modern western style dresses on campus. In contrast, there is a strong emphasis on Islamic culture and traditions at UoS, where there is even a major in Shariah (laws) that is offered. In UoS there are two separate colleges for men and women. The male and female students do take some classes together, but when they do, they sit on opposite sides of the aisle that runs through the middle of the lecture hall — boys on one side of the aisle and girls on the other. This is not surprising, and is common in ancient societies. In India for instance, even in a co-educational school, the boys all sit together, and girls sit separately from them in the classroom.

When AUS and UoS started, so did their mathematics departments. But the mathematics faculty on both campuses were teaching courses that were requirements for majors in other areas such as engineering. Only in 2007, based on the growth that took place in the first decade, did both universities come out with proposals to create a mathematics major.

Of the five days we were on site, we split the time evenly, visiting AUS first and then UoS.

The faculty at AUS who are at the assistant professor level and higher, have their doctorates from reputed universities in Europe and the USA. Indeed, one faculty member, Thomas Wunderli, in the area of PDEs, received his doctorate from the University of Florida. Even the regular faculty who are not from Europe or the USA, are from outside the UAE from countries like Syria, Lebanon, Egypt and Iran. Since the faculty are expatriates, they are provided housing on campus. We were shown some of the homes of the faculty, and they were very modern and elegant.

Our two main tasks were (i) to identify gaps in the undergraduate mathematics instruction, and suggest modifications in the course offerings to make it comprehensive and complete, and (ii) to suggest areas of hiring for the next five years to ensure that faculty expertise will cover a broad arena of mathematics. We were pleased that the upper administration of the University was committed to provide adequate support for the mathematics department.

Our impressions on UoS regarding the quality of the faculty, the scope of the BS Program, and level of commitment by the administration to mathematics, was similar to what we had formed of AUS. So, we were pleased to have been invited to assess the proposals from AUS and UoS.

My visit to AUS and UoS ended on the morning of Fri, Feb 2. It had been all work for five days with no time for sightseeing. My flight out of Dubai was not until late that night, and so Mathura's cousin Mala and her husband Gurumurthy, who were living in Dubai, showed me around the city, and took me shopping. My daughter Lalitha was getting married on February 10, one week after my return from the UAE and so I utilized the last day in the UAE to get some gifts for the family from Dubai.

I have returned to both AUS and UoS a few years later (2011, 2014, 2018) but on separate visits to these universities after their BS programs had been implemented to ascertain whether the programs had been formed as per our recommendations. I was pleased to see that the programs had taken shape and that faculty appointments in certain areas we had suggested had been made. Based on the success of the BS Program, and on the strength of the faculty, AUS also proposed to launch a Masters degree program in mathematics. So I made a third visit to AUS to evaluate that proposal. I am glad to have had the opportunity to evaluate and endorse the BS and MS mathematics programs in these two leading universities in the UAE. In all subsequent visits, we were put up in Dubai at the fabulous Intercontinental, and not in Sharjah, but the commute was only half an hour. My first visit for evaluation made a positive impression on the process and the sincerity of the effort on the part of the universities and the Commission of Accreditation. Thus I accepted subsequent invitations to do such evaluations. I shall next describe visits to other universities.

The King Fahd University of Petroleum and Minerals

The King Fahd University of Petroleum and Minerals (KFUPM) in Dhahran (now Dammam), Saudi Arabia, invited me as Chair of a three-person committee to evaluate a proposal of the mathematics department to start a BS major in mathematics. The other two members of the ERT were Christian Houdre (Georgia Tech) and Maarten d'Hoop (Purdue). Christian is a probabilist, and Maarten an applied mathematician, and with my expertise being in number theory, we three formed a well-balanced ERT. One cannot go to Saudi Arabia as a tourist. One has to be invited in order to enter the country. I accepted the invitation since it gave me an opportunity to observe the higher educational system in Saudi Arabia.

I departed for Saudi Arabia via New York on November 12, 2009, on a one week trip. At New York's JFK Airport, at the Korean Airways lounge before boarding the Saudi Arabian Airlines flight, I met Christian and Maarten. It was good that all three of us were on the same flight because we could discuss various aspects of the KFUPM proposal before arrival in Dhahran.

The trans-atlantic portion of the journey from New York to Jeddah on the Saudi Boeing 777 was very comfortable. I was able to work and sleep on this 13 hour flight. Some of the male members of the crew were Saudi, but the female cabin crew were not Saudi — they were from the Philippines, Malaysia, and the Middle East outside of Saudi Arabia, such as Iran. On a screen in front of the cabin was an arrow that always pointed towards Mecca. We arrived at Jeddah at 10:30 am the next day, and had a long layover there because our flight to Dhahran was to depart only at 7 pm. So we three had time to discuss the proposal in some detail at the Saudi Arabian Airlines lounge at Jeddah. Upon arrival at the ultramodern airport in Dhahran at night, we were taken to the fabulous Hotel Le Meridien Al-Khobar and given rooms with a breathtaking view of the Persian Gulf. The next morning, after breakfast, we were taken to KFUPM.

The campus of KFUPM is next door to the gigantic Saudi Arab-American Oil Company (ARAMCO) headquarters. Thus the flagship program at KFUPM is engineering, and a majority of the engineering graduates of KFUPM are recruited by Saudi ARAMCO. The mathematics department is large, and its faculty are actively involved in teaching mathematics courses that are requirements for engineering and the sciences. As in the case of AUS, the mathematics department of KFUPM was proposing the launch of a BS Mathematics Major. Our task was to make sure that the course offering was comprehensive and that there was the required faculty expertise to teach all the courses of BS Major.

A majority of the faculty members were expatriates. There were a good number of faculty from Pakistan, some from Iran, Syria and Egypt. There were some faculty who were Saudi. Only those faculty who are Saudi are given administrative positions. At the time of our visit, Prof. Al Sabah, the Dean of the College of Sciences, a charming man, was a mathematics department faculty member. The faculty are active in research, and indeed the department publishes its own research journal. We were impressed with the enthusiasm of the faculty in carrying out their mission of research and teaching, and in starting the new BS Major. The real question was whether there would be enough number of students enrolling in the BS Mathematics

Major. In view of the opportunities for jobs afforded to graduates in engineering of the sciences, and the lack of similar job opportunities for mathematics graduates, it was expected that the number of BS Mathematics Majors would be very low. This problem is not unique to KFUPM, but is faced by universities even in the UAE. In the UAE, the BS Mathematics Major is primarily populated by female students, for reasons explained above, but here in Saudi Arabia, KFUPM is an all male campus.

To help the enrollment of students in the mathematics major, we suggested that they be offered attractive scholarships — for example with some travel money to attend workshops in the Middle East and may be even in Europe and the USA. The Dean of Sciences was receptive to this idea.

My two colleagues and I attended a few classes taught by the faculty. Owing to my expertise, I attended a class in number theory that was taught by Professor Aslam Chaudhry, who was of Indian origin, and a muslim by religion. He was lecturing on prime numbers. I was surprised when after 45 minutes of the one hour class, he requested me to give a short talk to the students about some unsolved problems on prime numbers — which I did! I found the students to be very attentive and interested.

During the week that we were there, Benedict Gross of Harvard University was visiting the department. I had the pleasure of attending his colloquium talk on elliptic curves at KFUPM. He was visiting KFUPM to work with his former PhD student who was on the faculty there. I had lunch with Gross one day and found that he had a favourable impression of the work of the mathematics faculty. I should also add that the King Fahd library of KFUPM is excellent.

While we worked collectively on campus on our report, we worked individually in our hotel rooms in the evenings upon return from campus. At night, I used to take walks along the beautiful corniche and the promenade along the Persian Gulf next to the hotel. Since I get up very early in the morning, I would watch the lovely sunrise over the Persian Gulf, and see Bahrain in the distance.

On November 16, which was our last full day on campus, Dean Al Sabah hosted a magnificent sit down dinner at a traditional Arabian restaurant in Dhahran, and took special care to ensure that I was given vegetarian food. Yes, we experienced true Arabian hospitality that night. The next morning, as Chair of the ERT, I presented our report to the Rector Al Sultan in the company of the Deans of Engineering, the Sciences, and the Chairs of Mathematics, Physics, and Chemistry, and handed over the report to Dean Al Sabah of the College of Sciences. Our flight out of Dhahran was only the next morning and so we had the whole afternoon and evening that day for some sightseeing. It was kind of Professor Bokhari, a Pakistani, to take us in his car on a drive along the causeway that connects Dhahran to Bahrain. At the center of this causeway highway on the sea, is an island where there are immigration counters to screen persons travelling to Bahrain from Saudi Arabia, or entering Saudi Arabia from Bahrain. On this island there are restaurants on both the Saudi and Bahrain sides. We had refreshments at one of the restaurants on the Saudi side watching a gorgeous sunset from the middle of the ocean. It was an awesome sight with the lights of Bahrain on one side and of Saudi Arabia on the other!

The next morning, we flew out of Dhahran to Jeddah. Out flight from Jeddah to New York was only at 2:45 am (after midnight). So we had time to go around the city of Jeddah. One of the faculty members of KFUPM whom we met in Dhahran, and who was in Jeddah briefly on his return to Egypt for vacation, offered to show us around Jeddah in a chauffeured rental car that he hired for us. Jeddah not only has a major international airport, it is also a vital port for Saudi Arabia on the Red Sea. Jeddah is also close to the holy city of Mecca and so is a major entry point for international pilgrims going to Mecca. After pointing to us the road that proceeds to Mecca, our faculty friend showed us some beachside residences, a floating mosque, and the old city, before taking us to the gigantic and opulent Red Sea Mall for us to buy some souvenirs. While we were completing our shopping at the Red Sea Mall, the siren sounded indicating that soon it would the time for the final evening prayer of the day. The shops would not sell anything during this time of the prayer and so we finished our shopping quickly. We were told that after this final prayer, the mall was open only for ladies. These ladies could be accompanied by their husbands, male children or their brothers. But men like us who were not with our wives or sisters, had to leave the mall. Since we had to get back to the airport, and we had finished our shopping, it was time for us to leave anyway. I had bought a lovely turquoise bracelet for Mathura as a gift/souvenir since she cannot visit Saudi Arabia.

Our return trans-atlantic flight on Saudi Arabian Airlines from Jeddah to New York JFK was just as comfortable as the outbound flight. As we parted company at New York JFK Airport to proceed to our respective destinations, Christian Houdre said that he would invite me to Georgia Tech for a colloquium. And this he did a few months later.

I was back at KFUPM in May 2012 as part of a team to review the progress of the BS programs in the sciences that had been started. I was the lone person assessing the mathematics program along with Dennis Leavens, from the Commission of Accreditation who was also working with the other members of the ERT. Yes, the mathematics department had launched the program as was agreed in 2009, but the main problem of low enrollment in the BS Mathematics Major remained. The mathematics department was concerned that the BS Mathematics Major may be terminated. In our report we stressed that it is worthwhile to continue to offer the mathematics major and it survived.

In 2012, Mathura and I were visiting Turkey during the first week of May on our way to India. Since she could not visit Saudi Arabia, she departed for India from Istanbul. I returned to Florida, and two days later left for Dhahran. It was kind of KFUPM to convert my round-trip ticket Florida-Dhahran-Florida to Florida-Dhahran-Madras (India). So after my work in Dhahran at KFUPM, I joined Mathura in Madras (see §10.4).

New York University, Abu Dhabi

Certain leading universities in the USA have branch (affiliate) campuses, or manage certain programs, overseas. Cornell University manages a medical school in Doha, Qatar, whereas Texas A&M has an affiliate campus in Doha. New York University (NYU) has an affiliate campus in Abu Dhabi (NYUAD) and in Shanghai, China. I was invited to be on the team to review the College of Science of NYUAD in November 2015, and I accepted the invitation. My good friend Sartaj Sahni of the Computer Science Department of the University of Florida served as Chair of the ERT. He does at least half a dozen such reviews in the Middle East each year. I was the only one on the ERT to assess the mathematics program at NYUAD. NYUAD was launched in 2010 and so we were conducting the review of the programs after five years of operation of the university.

I flew by Delta on October 30 out of Orlando to Amsterdam where I connected to a KLM flight to Abu Dhabi. Pablo Laguna (Georgia Tech), who was to do the assessment of the physics program, was a co-passenger and so we had an opportunity to compare our notes on NYUAD during the trip. All ERT members were accommodated at the luxurious Rotana Beach Hotel at Sahni's request. The next morning we were all taken to NYUAD to begin our work.

All programs of NYUAD are managed from the USA by New York University. Thus the quality of all the programs is high. We were told that the exams questions were set by NYU and even the grading was done in New York to ensure high standards in evaluating student performance. What we as ERT members had to look into, among other things, was the effectiveness of the delivery of the instruction and the progress of the students. The permanent faculty were all expatriates, and there were eminent academicians from NYU visiting NYUAD on a regular basis to work with the permanent faculty on the courses or to conduct research with them. When I was there, my friend Andy Majda, a world famous applied mathematician from NYU, was visiting NYUAD delivering a series of lectures. Majda was on the UCLA faculty when I was a graduate student there, and we played tennis regularly (see §3.5). In Abu Dhabi, I was on a fleeting visit with no time for relaxation. So Majda and I did not play tennis in Abu Dhabi, though I would have loved to.

NYUAD has a lovely seaside campus with a cluster of modern buildings that included also the residences for students and faculty. The university is well funded and well managed by NYU. So I saw no real problems with the mathematics BS program except for some small gaps in the course offerings. I made some suggestions regarding faculty hiring in the next few years to provide more balance in the expertise. I also suggested some alternate textbooks for certain courses. The impressions from other team members regarding the programs in physics, chemistry, biology, and computer science, were also positive. So it was an easy report to write in terms of recommendations, but the Commission of Accreditation which manages all these external reviews, requires us to fill in details on a wide variety of items, and so the actual completion of the report is quite time consuming for each of the ERT members. Sahni as Chair combined the reports on the various science departments given by the ERT members and made the presentation to the Vice-Chancellor, the Provost, and Dean of the College of Science at the Exit Meeting on November 4. I was back in Florida the next day and felt pleased that I had the opportunity to observe how an affiliate campus of a top university functioned in the Middle-East.

The United Arab Emirates University

The United Arab Emirates University (UAEU), the very first university of the UAE, was founded in 1976. It is located in Al Ain which does not have a coastal location; it is in the interior of the UAE and borders Saudi Arabia. It is the largest university and has a current enrollment of 13,000 students, with a good mix of Emirati and international students.

I received an invitation to serve on an External Review Team for the College of Sciences. These external reviews for universities in the UAE were commissioned only after 2000 and so, since its inception, UAEU never had an external review or accreditation in the strict sense as is being done by the COE nowadays. And UAEU was not keen on participating in such reviews initially even though other newer universities in the UAE did participate. Finally, UAEU agreed to undergo the review process and in 2017 all the colleges of UAEU underwent reviews for the first time. The College of Sciences review was scheduled in the first half of April.

I departed Orlando for Dubai on April 7 morning, but my departure was plagued by delays in succession. Three lanes of the Beechline Route 528 to Orlando airport were closed due to an accident and so near the airport, traffic was moving in a trickle. Next there was a huge crowd at the Sky Priority check-in counters of Delta and that increased the delay. Finally our flight was delayed by about an hour because the pilots were caught in a traffic jam coming to the airport! This last delay was good because otherwise, I may have missed my flight to Atlanta. Anticipating delays, I usually ask my hosts in the Middle East to book me on flights with plenty of transit time at airports. They usually issue tickets with very tight connections unless they are told otherwise. I do not mind long transit times at airports because I can work comfortably in the airline lounges.

This time I was travelling on Air France from Atlanta to Paris and on to Dubai. Both Air France flights were extremely comfortable, but I need to point out a key difference. On both flight segments, the aircraft was a Boeing 777-300. On the Atlanta-Paris segment, the seating in Business Class was 2-3-2 in each row, compared to 3-3-3 in Economy Class. However, on the Paris-Dubai segment on the same type of aircraft, the seating in Business Class was 1-2-1 in each row, and they were all booths which offered total privacy. I think this superior arrangement on the flight to Dubai is because the sheiks demand greater comfort and Air France is competing on this segment with Emirates Airlines. Charles DeGaulle Airport, Paris, is not the most convenient for making connections, owing to its size, and the complications of navigating through the airport. Fortunately my Dubai flight was leaving from the same terminal and so I had time to freshen up in the Air France Lounge and have a light meal.

I arrived at 10:30 pm on April 8 at Dubai Airport where the Merhaba agent quickly took me through immigration and customs. I was put in a Audi taxi and whisked away to Al Ain. After an hour and a half drive, I checked in just past midnight at the luxurious Al Rotana Hotel in Al Ain where all ERT members were accommodated.

Work began in earnest the next morning. Compared to the other universities in the Middle East that I had visited, here at UAEU, it was less of a problem with the number of students in the mathematics major because since its creation in 1976, UAEU had held its high position quite well, and had over the years worked on improving its course offerings both in quality and range of topics. However with AUS and UoS having started BS Mathematics Majors and a Masters program, there was competition, and UAEU was well aware of that. The ERT felt that the programs were run satisfactorily, and so we as a team had only a few suggestions for improvements. On my part, I suggested some changes in the course offerings in mathematics, and the possibility of introducing internship in industry in lieu of a few upper division courses, so as to attract students who might be tempted by AUS or UoS.

The Al Ain area, even though it is in the interior, has quite a few oases, and so there is greenery. I was told that Emirati come to Al Ain for relaxation owing to this. One evening, there was an Emirati wedding at our hotel, and I was stunned at the opulence of the wedding and the number of luxury cars parked at the hotel.

Watching news was my only relaxation, for I was working late into the night in my room finishing up my portion of the report, which was long because we were reviewing both the Bachelors and Masters programs. As in the case with my other visits, here too in Al Ain, when I had to work all evening on completing the report, I would simply order room service and not step out for dinner. One advantage of staying in five star hotels is that the room service menu is excellent as are the facilities in the room to work. The Exit Meeting was on April 14 morning and that night I departed Dubai, flying KLM as usual to Amsterdam and Delta to Atlanta from The Netherlands. On my visits to AUS and UoS, I heard a lot about UAEU from the commissioners, and so I was glad that I had an opportunity to visit UAEU and review its mathematics program.

University of Balamand, Dubai

Although the middle east nations like Syria, Lebanon, Iraq, and Iran, are not as prosperous as Saudi Arabia, the UAE, Bahrain, Kuwait, and Qatar, they have a longer and stronger academic tradition, in spite of all their political turmoils. Indeed there are reputable universities in these countries. One such is the University of Balamand (UOB) in Lebanon. This private university, founded in 1988, is secular in its policies. Its main campus is located in the north eastern hilly part of Lebanon, next to the Balamand Monastery, but it has two smaller campuses in Beirut. Spurred by its success, UOB proposed in 2017 to start a branch campus in Dubai —

The University of Balamand Dubai (UOBD). I was invited to be part of a team to evaluate this proposal and make recommendations. It was an interesting proposal for a middle east university to launch a branch campus in another middle east country. So I accepted the assignment.

I departed Orlando for Dubai on October 6, and my outbound route on Air France was identical to my trip earlier in the year. And like the previous trip, the seating on the Paris-Dubai segment was much more comfortable reaffirming my opinion as to why this was the case.

UOB had just bought the land in Dubai and had begun construction of essential buildings on this campus, but there were not facilities yet on this new UOBD campus. So we worked all the time in the conference rooms of the Al Rotana Bustan Hotel near Dubai airport where all ERT members were accommodated. It was nice to once again work with Ian Cumbus as our Commissioner — Ian was the commissioner I worked with on my very first visit to the UAE in 2007. Badr Abulela, the Chief Commissioner at the CAA, was also with us from time to time to help us in our discussions.

It was a new kind of proposal for us to examine because we had to assess whether the programs with which UOBD was to start could be effectively implemented with the funds available. We had long discussions with faculty in the sciences of UOBD who had come to Dubai, and some of whom were planning on helping the launch of various programs by being in residence in Dubai in the initial years until permanent new faculty for UOBD are appointed. We also had discussions with the Dean of UOB because he and the Rector were deciding on the allocation of resources. We recommended that the proposal be revised in certain ways and found the faculty of UOB very receptive to our suggestions. The ERT members were all taken to the new campus and shown the buildings that were being constructed. The Exit Meeting was on October 10 morning, and at lunch that day hosted by UOB, one of its faculty members brought us home made baklava, claiming proudly that the baklava from Lebanon is the best in the Middle East!

Work was over but my flight was only the next day. So that night I met up with Mathura's cousin Chitkala and her husband Rajaram who took me to dinner at a fine Rajasthani restaurant. Dubai has a fantastic collection of restaurants serving different types of Indian cuisine. The next day, after purchasing some gifts for my family, I departed Dubai by car for Abu Dhabi airport where I caught an Alitalia flight to Rome. As an aviation enthusiast, I read a lot of airline news, and I knew that Alitalia was in deep financial difficulty. Yet, the flight was on time, and the service on board the Alitalia Airbus 330 was superb. I was back in Florida the next day, feeling pleased that I had an opportunity to help in the launch of a new university in Dubai by a critical assessment of the proposal.

Khalifa University, Abu Dhabi

During 2020 and 2021, I received invitations to do reviews of programs in the UAE, but they were through virtual meetings. My strong feeling is that for such reviews, live visits are essential, and so I declined the invitations. Then in November 2021, I received an invitation to Chair a team on a live visit to Abu Dhabi to evaluate a proposal by Khalifa University to launch an MSc Program in Applied Mathematics. Since it was a live visit, I accepted the invitation. The CAA kindly agreed to my request to hold the visit during the University of Florida Spring Break in March 2022, so that I would not miss classes. In January 2022, due the Omicron scare, the live visit was canceled and replaced by a virtual visit. Although I was not pleased with this change, I agreed, since I had accepted the assignment.

My team mate was Frank Neubrander of Lousiana State University (LSU) who had teamed up with me twice before. He is a Distinguished Professor of Mathematics at LSU and the Director of the Cain Center for STEM Research. He is quite experienced in such reviews having done ten of them. In 2021, he chaired a team to do a review of a combined PhD Program at Khalifa University for four science disciplines of which mathematics was one. That team recommended that instead a combining the PhD degree programs, these programs ought to be separated. Khalifa University followed this recommendation and so a Mathematics PhD at Khalifa University was launched in 2021. Also in that review, it was suggested that Khalifa start a Masters in Applied Mathematics Program to act as a bridge between their BSc Program and the new PhD program. It was for the launch of this MSc Program in 2022 that I invited to do a review.

Khalifa University was formed in 2017 by combining four scientific institutes in Abu Dhabi. From the start, it was envisaged to be a high performance university. Indeed, in five years, it has become one of the top universities in the UAE and has secured a good international ranking as well. So Khalifa University was well positioned to start this MSc Applied Mathematics Program. However, in order to be able to write a thorough report, I felt that it is best that Neubrander and I are together during the virtual meetings scheduled during Mar 7 and 8 so that we can discuss and complete the report as is the case with live meetings. So I went to LSU. The Commissioner from the CAA, Prof. Perumal, visited the Khalifa Campus and provided us information based on what she saw there.

Not only did Prof. Neubrander offer gracious hospitality, he contacted Professors Ling Long and Fang-Ting Tu of the mathematics department who arranged my talk in the Algebra and Number Theory Seminar on March 8 afternoon. Long is a former PhD student of Winnie Li at Penn State, and Tu, who hails from Taiwan, had Winnie Li as her post-doctoral mentor. I spoke on my joint work with my former PhD student Todd Molnar in analytic number theory. Thus the visit served a two-fold purpose.

In summary, although each review involved hard work before and during the site visit, I enjoyed evaluating various mathematics programs in the Middle East. I learnt a lot how universities in that part of the world functioned, and was impressed with the sincerity with which the universities, their administration, and the government, conducted these evaluations.

Krishna and his father Alladi Ramakrishnan with 2005 Abel Laureate Peter Lax (Courant Institute), after Lax's Center for Applied Mathematics Colloquium at the University of Florida, March 6, 2006. Krishna's father and Peter Lax were meeting after exactly 50 years!

Krishna Alladi being introduced to Dr. Abdul Kalam, President of India, on the opening day of the First SASTRA Ramanujan Conference, December 20, 2003. To Krishna's left are George Andrews (Penn State) and Noam Elkies (Harvard).

Manjul Bhargava (Princeton) and Ken Ono (then at Wisconsin) relax on a swing at a rest area (Dec 19, 2005) on the road from Madras to Kumbakonam to attend the SASTRA Ramanujan Conference. Krishna and Mathura are standing in the rear.

Manjul Bhargava (Princeton) and Kannan Soundararajan (then at Michigan), winners of the First SASTRA Ramanujan Prizes, in front of the home of Srinivasa Ramanujan in Kumbakonam, December 20, 2005. "Sound", as he is called, went on to win the Ostrowski and Infosys Prizes in 2011. Bhargava won the Fields Medal in 2014.

Terence Tao (UCLA), Fields Medal and SASTRA Ramanujan Prize Winner in 2006, at the Ramanujan Statue at SASTRA University in Kumbakonam, Dec 22, 2006. Along with Krishna and Tao is Krishna's dear friend G. P. Krishnamurthy (alias GP), who helped with arrangements for all foreign visitors for the SASTRA Conferences.

2008 SASTRA Ramanujan Prize Winner Akshay Venkatesh (then at Stanford University) garlanding the statue of Ramanujan at SASTRA University, Kumbakonam, December 21, 2008. Venkatesh went on to win the Fields Medal in 2018.

2013 SASTRA Ramanujan Prize Winner Peter Scholze (University of Bonn), delivering the Ramanujan Commemoration Lecture at SASTRA University, December 22, 2013. Scholze went on to win the Fields Medal in 2018.

Left to right: Krishna Alladi, John Thompson, George Andrews, Richard Askey, and
Alladi Ramakrishnan, at Alladi House, Gainesville, during a party, Mar 21, 2005.

Krishna in conversation with Peter Sarnak (Princeton University and Institute for Advanced Study) during a party at Alladi House after Sarnak's Ramanujan Colloquium,
March 19, 2008. John Thompson and Bert Kostant (MIT) are in the background.

Chapter 10

Post chairmanship

10.1 An eventful year of research leave (2008–09)

After ten years as Chair, I was given a full year of research leave in 2008–09. That academic year turned out to be as eventful as any.

The year started with a visit in September to the University of Illinois, Urbana, for one week in response to an invitation from Bruce Berndt to give several lectures. On September 8, the day of my arrival, Bruce and his wife Helen had me at their home for dinner, and I enjoyed the vegetarian Italian dishes so thoughtfully prepared by Helen. The next day, I gave a talk in the q-Series Seminar on "A new combinatorial study of the Rogers-Fine identity and related theta series". Berndt always had several PhD students and post-docs working with him and I met them all at this seminar. In particular, I had fruitful discussions with his PhD student Atul Dixit. Byungchan Kim, one of Berndt's Korean students, approached me with a question on a partial theta series of Ramanujan. In answering his question, I was subsequently led to a new proof of Ramanujan's partial theta identity and some partition theorems.

Right after the q-series seminar, I attended a talk by Rajmohan Gandhi at the University Bookstore, on the theme "Lincoln and Gandhi" based on his recent book. Rajmohan Gandhi has the distinction of being the paternal grandson of Mahatma Gandhi and the maternal grandson of C. Rajagopalachari, a great statesman in India who was a close friend of my grandfather Sir Alladi. Rajmohan Gandhi is a research professor at the Center for South Asian and Middle Eastern Studies at the University, and the author of several acclaimed books. When I introduced myself to him after his talk, he referred to my grandfather Sir Alladi in very high terms not only as a brilliant jurist, but also for his role on the Drafting Committee of the Constitution of India.

On Thursday, September 11, I gave two talks in Urbana. The first was in the Number Theory Seminar on "Partitions with non-repeating odd parts and q-hypergeometric identities". This was followed by a Colloquium on "New weighted Rogers-Ramanujan partitions and products modulo 6". I was pleased that Paul Bateman (former Chair of the mathematics department, who was retired by then), and his wife Felice, attended my talks. That night, the Berndts graciously hosted a party at their home for me, and it was attended by several faculty members and their spouses. I was back in Florida the next day.

On September 22, I departed for Salt Lake City in response to an invitation from Roger Baker for lectures at the Brigham Young University in Provo, Utah. The view of the great Salt Lake from the air on a bright sunny afternoon as we descended into Salt Lake City Airport was spectacular. I took the Xpress Shuttle to Provo where I checked in at the comfortable Courtyard by Marriott. Roger Baker picked me up from the hotel and took me to the home of his colleague Jasbir Chahal for an Indian dinner. Members of the number theory group joined us at Chahal's home that evening. Chahal is an algebraic number theorist whose focus is elliptic curves.

The next morning, Roger picked me up at the hotel and took me to the campus Brigham Young University. The campus, set in the lovely Utah Valley, looked gorgeous with well manicured gardens surrounding various buildings. After a lunch at the Faculty Club hosted by the number theory group, Jasbir took me Sundance Park, owned by the famous filmstar Robert Redford. The granite hillside walls of Sundance Park reminded me of Zion National Park. Back from Sundance Park, I gave a Colloquium talk on "A theorem of Göllnitz and its place in the theory of partitions".

The following day, Roger walked me through campus to an upscale restaurant called Chef's Table, where his wife Lynette joined us for lunch. That afternoon, I addressed the Number Theory Seminar on "New observations on the Göllnitz-Gordon and the Rogers-Ramanujan identities". This concluded my visit to Brigham Young University made memorable by the gracious hospitality of Roger Baker and his colleagues.

In mid-October, my mother had a cataract surgery in Madras and so I made a three-week trip to India (Sept 30–Oct 20) to be of assistance to her. On the long flight from Detroit to Tokyo on board a Northwest Boeing 747-400, I worked out a new proof of the Ramanujan partial theta identity that Byungchan Kim had shown me during my visit to Urbana a few weeks earlier. This new proof gave a dual identity as well. I enjoy, and am inspired by, airports and air journeys!

To take a break from the packed academic programs, I took a family vacation for four days (Nov 8–12) in the US Virgin Islands, something we had always wanted to do. We stayed at the Marriott Frenchman's Reef Resort on the island of St. Thomas. The Marriott is located at one end of the magnificent horse-shoe shaped bay of Charlotte Amalie, the capital of St. Thomas, and from our room we enjoyed a spectacular view of the bay, and the cruise ships that visited the island. We also made a day trip to the neighboring island St. John, and enjoyed breathtaking views of about half a dozen beaches on our tour around the island.

During the first week of December, I participated in the George Andrews 70th Birthday Conference on the campus of The Pennsylvania State University. That visit to Penn State University is described in §11.4 along with the other Andrews milestone birthday conferences.

Upon return from Penn State, I received a letter from Professor Peter Goddard, Director of the Institute for Advanced Study, inviting me to write an obituary article on my father to be published in the Institute Letter. Goddard is an accomplished astrophysicist, and he and my father had extensive correspondence over the years. I felt it was a special honor — posthumous for my father — that the Institute wanted to publish an obituary article on him. Such obituaries are typically confined to permanent members of the Institute, but Goddard said he wanted the obituary piece on my father because he knew that it was my father's visit to Institute for Advanced Study in 1957–58 that inspired him to create MATSCIENCE. Goddard also said in his letter, that he is approaching me to write the obituary since I had followed my father's academic life very closely. I felt honored being asked to write this article.

==

Visit to India (Dec–Jan) and the 2008 SASTRA Ramanujan Prize to Akshay Venkatesh

Mathura and I departed for India on December 13. The highlight of the trip was the SASTRA Conference, and the award of the 2008 SASTRA Ramanujan Prize to Akshay Venkatesh of Stanford University.

Akshay Venkatesh had made far reaching contributions to a wide variety of areas in mathematics including number theory, automorphic forms, representation theory, locally symmetric spaces and ergodic theory, by himself and in collaboration with several mathematicians. His paper with H. Helfgott containing a number of very striking and original ideas, gave the first non-trivial upper bound for the 3-torsion in class groups of quadratic fields. His fundamental work with Jordan Ellenberg on representing integral quadratic forms by quadratic forms has its roots in the work of Ramanujan.

An important and difficult topic in number theory is the problem of asymptotically counting number fields of a given degree according to their discriminants. In the case of degree up to 5, the problem was solved by Manjul Bhargava. In another paper, Ellenberg and Venkatesh provided the first major improvements (when the degree is large) over earlier bounds of Wolfgang Schmidt, thereby breaking an impasse of several years. Also of great importance was his own individual work on subconvexity of automorphic L-functions and his joint work with Elon Lindenstrauss which settled a famous conjecture of Sarnak on locally symmetric spaces.

Venkatesh, who is of South Indian descent, was born in New Delhi in 1981 but was raised in Perth, Australia. He showed his brilliance in mathematics very early. His entry into research began as a PhD student at Princeton in 1998 under Peter Sarnak. After completing his PhD in 2002, he was C. L. E. Moore Instructor at MIT for two years before being selected as Clay Research Fellow in 2004. He was then appointed as Associate Professor at the Courant Institute, NYU. In 2008, at the young age of 27, he was elevated to the rank of Full Professor at Stanford University. After winning the SASTRA Prize, he went on to win the Fields Medal in 2018 and was appointed Professor at the Institute for Advanced Study that year.

Venkatesh came to India accompanied by his wife Sarah. His parents flew from Perth, Australia, to be present at the conference as well.

On December 19 morning, I departed from Madras for Kumbakonam in the SASTRA van with Akshay, Sarah, Shaun Cooper and his wife. Shaun, a former student of Richard Askey, is an expert on special functions. He is at Massey University in New Zealand. Venkatesh's parents were not in the van with us. They were touring South India and reached Kumbakonam separately. Two other international participants — Michel Waldschmidt (University of Paris) and Jean-Marc Deshouillers (University of Bordeaux) — also travelled separately to Kumbakonam.

Akshay Venkatesh was presented the SASTRA Ramanujan Prize on December 20 morning at the inauguration of the SASTRA Conference. Immediately after the ceremony, he gave the Prize Lecture on his work on sub-convexity. In the afternoon we had the talks by Cooper, Deshouillers, and me. I spoke on my recent work: "A partial theta identity of Ramanujan and its number theoretic interpretation". That evening we were treated to a fine carnatic music concert at SASTRA.

The next day, there was only a morning session which featured the talk of Michel Waldschmidt. In the afternoon, we visited Ramanujan's home and the Sarangapani Temple. That night, I found a pretty connection between

Ramanujan's partial theta identity, and an identity of George Andrews. I was glad I did this work in Kumbakonam, Ramanujan's hometown!

On December 22, the conference concluded with the Ramanujan Commemoration Lecture by Venkatesh after which we returned to Madras. Enroute all of us including Venkatesh and Sarah enjoyed fresh coconut water that was given on the roadside by a farmer who had a pile of tender coconuts and was opening them up for passers-by.

==

After the SASTRA conference was over, my second daughter Amritha arrived in Madras on December 25 for a six-month stay. She had just graduated with a bachelors degree in journalism from the University of Florida one semester early because she had received college credit for some courses while she was in high school. She decided to do her internship with the news media in Madras, most notably The Hindu, India's National Newspaper. By spending six months in Madras, she would also give company to my mother who was by herself in Ekamra Nivas until June 2009. Amritha started her internship with CNN India which she did for two months before transferring to The Hindu in March 2009.

Mathura and I departed Madras on January 9 for Kuala Lumpur from where we flew to Taipei next morning, all on Malaysia Airlines. We had always wanted to see Taiwan, but in previous years we could not go because we were citizens of India. Diplomatic ties between India and Taiwan had been severed after WW-II because India sided China in not recognizing Taiwan. So it is difficult for Indian citizens to visit Taiwan. Now that both Mathura and I were US Citizens, we decided to visit Taipei for a few days, and we enjoyed our visit immensely. Especially impressive in Taipei were the National Museum and the Chiang Kai Shek Memorial. We also enjoyed seeing some Buddhist temples in Taipei as well as a temple for the philosopher Confucius. From Taipei, we returned to Singapore via Kuala Lumpur, and boarded a Northwest flight in Singapore for our return to the United States on January 14.

Back in Florida, I received a letter from Steve Krantz, the new Chief Editor of the Notices of the AMS, inviting me to serve a three-year term as an Associate Editor of the Notices. I always enjoy reading the Notices with its interesting blend of articles on mathematics and mathematicians, and various news items about the profession. So I readily accepted the invitation. Also, I had known Krantz since my graduate student days at UCLA (1975–78) when he was an assistant professor there. So I looked forward to working with him.

In late January, George Andrews arrived for his four-month stay in Florida. He addressed the department Colloquium on February 16 on "Lessons from Ramanujan's Lost Notebook", which attracted even faculty and students from other departments. The next day he gave a lovely talk in the Number Theory Seminar on Catalan numbers. During the first month of his stay, I had several discussions with him on my work pertaining to Ramanujan's partial theta identity, and on his related partial theta identity.

On February 23, Dorian Goldfeld of Columbia University delivered the Ramanujan Colloquium on "Multiple Dirichlet Series", a topic to which he had made major contributions. He gave two more talks in the next two days — one in the Combinatorics Seminar of his generalization of Ramanujan sums, and another in the Number Theory Seminar on automorphic representations. Mathura and I hosted a party for Goldfeld and Andrews on February 24 when Goldfeld gave brief speech and told us how he got into mathematics research and how Bombieri encouraged him.

During the Spring Break (March 7–14), we had the highly anticipated *Conference on Quadratic Forms* that my colleague Tiep and I were organizing with the help of Manjul Bhargava. During the first four days (March 7–10) we had a Student Workshop to provide suitable background for many undergraduate students attending the conference. The main speaker for the Workshop was Jonathan Hanke, who along with Bhargava had solved the problem of determining all universal quaternary integer valued quadratic forms — a problem going back to Ramanujan. Roger Baker (Brigham Young) and Hershel Farkas (Hebrew University) also were Workshop speakers. They gave two lectures each. We took all the Workshop participants, the speakers, and Bhargava to Paines Prairie for sightseeing; they were all amazed to see several large alligators basking on the shores of the lake on a bright sunny afternoon, some only a few feet away.

The Conference on Quadratic Forms began on March 11 morning with the Hour Lecture of Noam Elkies (Harvard). At the end of the first day, we had the Erdős Colloquium of Akshay Venkatesh (Stanford) entitled "The Geometry of Numbers - old and new".

The next day we heard the main lectures of George Andrews (Penn State), and Robert Greiss (Michigan) and Parimala who had retired from the Tata Institute and taken up a permanent position at Emory. Mathura did a fantastic job in arranging the conference party at our home that night attended by all the participants and some of my colleagues.

The third day of the conference featured the lectures of Rainer Schulze-Pillot (Saarbrucken, Germany), Ken Ono (Wisconsin), and John

Thompson. I had met Rainer when he visited the University of Michigan in 1980 at the invitation of Don Lewis. It was after this conference that I invited him to join the Editorial Board of The Ramanujan Journal mainly to deal with papers pertaining to quadratic forms, and he accepted the invitation.

On the night of the third day, we celebrated the 70th birthday of Bruce Berndt with a dinner at an Indian restaurant. Berndt loves Indian food, and so this was a natural choice. Since Berndt's birthday falls in March, and since we typically hold our Florida conferences in March, Berndt often celebrated his birthday with us in Florida.

On March 14, the final day of the conference, we had the lectures of Bruce Berndt and Stephen Milne (Ohio State). Tiep and I were happy with the success of the conference, but more importantly, Bhargava felt that the conference lived up to expectations.

The final academic event of the Spring semester was the Center for Applied Math (CAM) Colloquium by Bert Kostant (MIT) on "Subgroups of E_8 and cyclotomic polynomials". Kostant's lectures are smooth as silk, and his talk attracted a sizeable crowd from the physics department. Prior to Kostant's talk, at a tea meeting, George Andrews, as President of the AMS, led a discussion on the theme "The view from Washington". That night, we hosted a party at our house on honor of Kostant and Andrews. Kostant enjoyed Mathura's eggless Mango Mousse so much that he said she should patent it!

On May 1, Amritha departed Madras for the USA after completing her internship with The Hindu. Mr. N. Ram, The Editor-in-Chief of The Hindu, was much impressed with her and so he wrote a strong letter about her excellent performance. Amritha received a job offer from Pacific Daily News of the Gannett group of newspapers (whose flagship is USA Today) to work as a lead reporter in far off Guam. This gave her an opportunity to deliver news reports about Guam and the Pacific rim, and so she accepted the job in spite of the far flung location of Guam. Amritha arrived in Florida from India on May 2 and had one month time to prepare for her departure to Guam. But before that, on May 7, we had the Annual Recital of Mathura's dance school JATHISWARA, this one dedicated to my father's memory. So Amritha joined my older daughter Lalitha to perform a very special dance on Lord Shiva in that recital.

One week after the JATHISWARA Recital, I was off to Providence, Rhode Island, to attend the meeting of the Committee on Committees of the AMS. Providence is the headquarters of the AMS. The Committee on

Committees is advisory to the President of the AMS and its role is to help
the President appoint members to various AMS committees. As President
of the AMS, George Andrews appointed me as Chair of the Committee on
Committees for a two-year term.

As soon as I returned from Providence, Mathura left for Madras on May
19. I followed her a few days later, because I had to stay in Gainesville for
the Higher Degree Forms Conference (May 20–23) which I was organizing
along with Tiep and Manjul Bhargava. This conference was a natural
successor to the Quadratic Forms Conference we conducted in March.

The Higher Degree Forms Conference opened with Bhargava's lecture.
This was very appropriate because his work on higher reciprocity laws was
key to the recent major advances in higher degree forms. One of the senior speakers at the conference was Benedict Gross of Harvard University.
Gross has such a smooth delivery that when you hear him, you feel that
you understand everything about the subject! That is the hallmark of a
great speaker. The Higher Degree Forms conference, although on a smaller
scale compared to the quadratic forms conference, was just as successful.
The refereed proceedings of the two conferences, edited my Bhargava, Tiep
and me, appeared as a single volume in the book series Developments in
Mathematics (Springer) three years later.

One day after the conference, I departed Florida for Madras. The main
academic event for me in India was the "Ramanujan Rediscovered" Conference (June 1–5) organized by Bruce Berndt and held on the beautiful
campus of INFOSYS in Bangalore. Mathura accompanied me to Bangalore. While I stayed at the INFOSYS Guest House along with other
participants, she spent a few days with our in-laws, Captain Srinivasan
and Gayathri at their home. The conference was dedicated to Venkatachaliengar for the centenary of his birth. He was a mathematician from
Bangalore who had written a book entitled "The development of elliptic
functions according to Ramanujan" and this was published by Madurai
University in 1987 during the Centenary of Ramanujan's birth. Although
Venkatachaliengar was President of the Indian Mathematical Society during
1979–80, his work did not get the recognition it deserved. It was George
Andrews and Bruce Berndt who recognized Venkatachaliengar's work on
partitions and elliptic functions respectively. So it was appropriate that
Bruce Berndt delivered the opening lecture of the conference in which he
highlighted Venkatachaliengar's work. My talk was the first on the second
day and it was entitled "Investigations of representations due to Andrews,
Ramanujan and Fine of a certain partial theta series". On the last day,

Professor Ramachandra, in spite of his feeble health at the age of 76, gave a talk on some problems in analytic number theory.

After the conference in Bangalore, Bruce Berndt spent a few days in Madras and so I asked him to deliver the First Alladi Ramakrishnan Memorial Lecture at Ekamra Nivas on June 12, to commemorate the first anniversary of my father's passing. He spoke on "Ramanujan's Lost Notebook: recent discoveries, continued fractions, transformation formulas, and the Riemann zeta function".

The next major event was a memorial function for my father organized by the Presidency College (of the University of Madras) on June 26. My father did his BSc Hons at Presidency College. After MATSCIENCE was founded in 1962, its offices during the first two years were at Presidency College. At my request, the function was arranged in the Old English Lecture Hall of the Presidency College where MATSCIENCE was inaugurated on January 3, 1962. I had a large portrait of my father prepared by the famous G. K. Vale photo studio in Madras, and presented it to the Presidency College on that occasion. This portrait was subsequently displayed in Hall of Fame of the College right next to similar portraits of two other distinguished alumni — Nobel Laureates Sir C. V. Raman and Subrahmanyam Chandrasekhar. In my speech, I recalled the magnificent address of my father on January 3, 1962 titled "The miracle has happened" that he delivered in that very hall. Justice Mohan who was the Chief Guest said that English diction in that speech of my father is so wonderful that it should be made a standard reading assignment for high school classes in English! Others who felicitated my father on that occasion were Mr. N. Ravi, Editor of The Hindu, and Prof. Anandakrishnan, who unveiled my father's portrait. A prominent report of the function appeared in The Hindu the next morning.

On June 29, my 31st Wedding Anniversary, Mathura, my mother and I departed Madras for America. We halted in Singapore for two days and arrived in Hawaii on July 1 for a one-week vacation in Honolulu and Kauai. Amritha joined us in Honolulu from Florida to enjoy the Hawaiian vacation before taking up her assignment in Guam.

In connection with Amritha's and our flights to Guam, I must mention an amazing offer we received from Northwest Airlines.

The merger of Northwest with Delta was approved in October 2008, but the merger was completed only in 2010 with the combined airline being called Delta. In 2009, to celebrate this merger, Delta offered a fantastic deal on transpacific fares in the summer: For a payment of $1 (a mere formality), one could upgrade from a mid-level economy class fare to Business Class!! We took this deal and while we were in Hawaii, we made reservations for our flights to Guam. The transpacific flights were still operated under the Northwest name with Northwest aircraft. In particular we all enjoyed the superb comfort and inflight service on the upper deck of the Northwest Boeing 747-400 on the Detroit to Tokyo sector enroute to Guam.

Amritha departed Florida for Guam on July 16. Mathura and I followed her two days later. Our goal in visiting Guam was to help her get settled there. We bought Amritha a car and arranged an apartment for her. Within two days of her joining Pacific Daily News, was Guam Liberation Day — (July 21, 1944 was when American forces liberated Guam from the Japanese during World War II) — and Amritha's first assignment was to cover that event. Her full page report appeared in the newspaper on July 22 while we were in Guam! So with Amritha well set in Guam, Mathura and I departed Guam on July 26 with a tinge of sadness to be back in Florida.

10.2   Back to teaching and the Program in ANTC (2009–10)

During the Academic Year 2009–10, I returned to teaching full time, which meant two courses per semester. Except for one year since then, I have always taught one graduate course and one undergraduate course each semester. For the graduate course, I started doing a three year rotation of the following subjects: (i) analytic number theory, (ii) irrationality and transcendence, and (iii) partitions and q-hypergeometric series. Throughout 2009–10, I taught analytic number theory using as a base the notes that I had prepared at the University of Michigan in the 80s, but enhancing these notes considerably. With regard to research activities, I was able to run the Program in ANTC for one more year, that is for 2009–10, because the NSF had provided significant funding for the two conferences on quadratic and higher degree forms, and there was a good amount of funds left over which I could use by getting a no cost extension of that grant approved. So I conducted two Focused Weeks in 2009–10 — one on quadratic forms and one on integral lattices. Unlike conferences which have a large number of participants, these Focused weeks brought together fewer experts but enabled close interaction over a one-week period.

During the weekend of October 24–25, there was an AMS Sectional Meeting on the campus of The Pennsylvania State University. George Andrews was not at this meeting since he was in Providence attending to work as the AMS President. I had a strange feeling being at Penn State without the inspiring presence of Andrews. Since I arrived in State College one day prior to the conference, I addressed the undergraduate students of the MASS Program and informed them about the Focused Weeks we were planning in Florida. At the AMS Conference, I spoke in the Special Session on q-Series organized by Penn State number theorists David Little, James Sellers and Ae Ja Yee.

I departed State College for Atlanta to attend the wedding of the daughter of our friends Drs. P. K. and Vimala Nair (University of Florida), and to speak in the Algebra Seminar at Emory University at Parimala's invitation. Mathura and my mother joined me in Atlanta due to the wedding.

My talk in the seminar was on Göllnitz' theorem and its place in the theory of partitions. Prior to the talk, Parimala and her colleague Skip Garibaldi took me to lunch at the Faculty Club. There by chance I met Ken Ono's older brother Santa Ono who was a Dean at Emory. Within a year, Ken Ono joined the faculty at Emory and Santa Ono took up the position as Provost at Case Western University! So they never were together at Emory!

In mid-November, I made a five-day trip to Dhahran (= Dammam), Saudi Arabia, as Chair of a team to do an assessment of the mathematics program at the King Fahd University of Petroleum and Minerals (KFUPM). It was a unique experience which I narrate in §9.13.

====================================

Trip to India and the 2009 SASTRA Prize to Kathrin Bringmann

Mathura, my mother and I departed Orlando for India via the Pacific on December 10. The main event in India in December for me was the SASTRA conference where Kathrin Bringmann was awarded the 2009 SASTRA Ramanujan Prize for her joint work with Ken Ono (as his post-doc) on mock-theta functions, Ken, who attended the conference, arrived in Madras with a group of students and post-docs: Riad Masri, Matt Boylan, Christelle Vincent, and Marie Jameson.

On December 20, there was a Computer Science and Mathematics Day celebrated at the IIT Madras. I was invited by the IIT to speak on that occasion and to suggest an expert speaker. So I suggested Ken Ono, knowing fully that he would enthrall a crowd of bright IIT students. And that he did as he gave a talk on Ramanujan's legacy to a large gathering of 400 students. Following Ono, at the request of the IIT professors who organized this meeting with the Rotary Club of Madras, I spoke about the creation of MATSCIENCE by my father in 1962.

Kathrin Bringmann, who was to receive the 2009 SASTRA Ramanujan Prize, arrived in Madras on the morning December 21 along with Ben

Kane her post-doc. Bringman was to have arrived the previous day, but she was held up at Brussels Airport due a snowstorm. She called me from Brussels airport whether she should cancel her trip or depart the next day. I responded without hesitation that she should NOT cancel her trip but take the Jet Airways flight the next day, which she did, much to my relief.

Bringmann arrived at 9 am on December 21 and had only two hours to freshen up at the hotel before we all departed for Kumbakonam. But she is a health fanatic, and said she needed to do her exercise in the hotel gym before departure. We said YES because we wanted her in good health and in good spirits for the prize ceremony the next day!

Bringmann had done important work on modular forms and mock theta functions by herself and in collaboration with several mathematicians, most notably Ken Ono. The power series coefficients of Ramanujan's mock theta functions can be evaluated very accurately like those functions which can be expressed in terms of the theta functions. Yet, the exact connections between mock theta functions and theta functions had remained a mystery all these years until recently. The first breakthrough came in the 2003 PhD thesis of Sander Zwegers written under the direction of Professor Don Zagier in Bonn, in which he showed how the mock theta functions of Ramanujan fit into the theory of real analytic modular forms. From here, Bringmann in collaboration with Ono and others obtained far reaching results: these connections with modular forms are made explicit, further questions concerning asymptotics and congruences are addressed, and a comprehensive theory relating holomorphic cusp forms to Maass forms is developed. I should add that if Ono had been younger, he would have been awarded the SASTRA Ramanujan Prize along with Bringmann.

Kathrin Bringmann was born in 1977 in Muenster, Germany. She received her PhD in 2004 at the University of Heidelberg under the direction of Winfried Kohnen. During 2004–07, she was Van Vleck Assistant Professor at the University of Wisconsin where she began her great collaboration with Ken Ono. Subsequently she joined the University of Cologne, Germany, as Professor. Earlier in 2009, she was awarded the prestigious Krupp Prize — a 1 million Euro research grant for a five-year period.

Although the departure from Madras by the SASTRA van was slightly delayed, we were able to reach Kumbakonam in time to see Ramanujan's home and the Sarangapani Temple that evening. The next morning, December 22, Ramanujan's birthday, Bringmann was awarded the SASTRA Ramanujan Prize following which she delivered the Ramanujan Commemoration Lecture. Her lecture was followed by Ono's one hour talk and the half hour talks by his students and post-docs, describing the modern theory of mock theta functions. At 4 pm, the conference concluded and all the participants left Kumbakonam to catch their return flights at Madras late that night. Mathura and I stayed back for one more day to visit a few temples that the region around Kumbakonam is so famous for.

==

Mathura and I departed Madras on January 2. After a two-day stay in Singapore, we flew to Guam via Tokyo on Northwest Airlines. Upon arrival there on January 4, we checked in at the Hilton Guam.

Amritha was well settled in Guam and it was a pleasure to visit her office at Pacific Daily News. There we met the Chief Editor who said that he was very pleased to have Amritha on his editorial staff. We then saw the Visitor Center and Museum attached to the US Naval Base, and watched the award winning movie "Guam 1941–44" dealing with the fall of Guam to the Japanese forces in 1941 and the liberation of Guam by the allied forces in 1944. The next day we visited a Hindu Temple and were impressed that even in this remote island in the Pacific, there are enough Hindus residing to motivate the construction of this temple!

I departed Guam on January 7 because of my Spring Term classes, but Mathura stayed back for three more days with Amritha. After I left, she moved to Amritha's comfortable apartment at Agana Heights.

One week after my return from Guam, I was off to San Francisco to attend the Annual Meeting of the AMS. My main work there was the Editorial Board Meeting of the Notices of the American Mathematical Society, and my appointments with the Springer editors to discuss the Ramanujan Journal, my book series Developments in Mathematics and the volume in memory of my father. Back in Florida from the AMS Meeting, I looked forward to a spring semester packed with academic events.

On February 1, James Keener of the University of Utah delivered the 12th Ulam Colloquium. He had delivered the First Ulam Colloquium in Spring 1999, and I had invited him then at the suggestion of my colleague James Keesling. And now Keesling, my successor as Chair, invited him back for the Ulam Colloquium.

Two weeks later, on February 15, George Andrews delivered the 12th Erdős Colloquium "The world of orthogonal polynomials and Bailey chains". The next day, the Focused Week on Integral Lattices started with a lecture of Eiichi Bannai of Kyushu University on Lehmer's Conjecture that the Ramanujan function $\tau(n)$ is never zero. This important problem, still unsolved, has links with the theory of Euclidean lattices and spherical *t*-designs, which was the emphasis in Bannai's lecture. Two other main speakers of the Focused Week were Wai Kiu Chan of Wesleyan University and Robert Greiss of the University of Michigan.

On February 18, David Bressoud, who was then President of the MAA, addressed the Teaching Seminar and discussed the transition for students from high school mathematics to college level mathematics. Bressoud is a powerful and motivated speaker, and has taken a deep interest in mathematics education. George Andrews who was President of the AMS at that time, introduced Bressoud for his talk. Bressoud was Andrews' colleague at Penn State for many years before moving to Macalester College in St. Paul, Minnesota. Andrews and Bressoud have very different views on mathematics education, but this difference did not affect their close friendship and the regard they have for each other. Interestingly, these two former colleagues at Penn State were Presidents of the AMS and MAA simultaneously!

The Second Focused Week was on Quadratic Forms and Theta Functions. It began on March 22 with the lecture of Bruce Berndt on the famous circle and divisor problems and certain Bessel function expansions in Ramanujan's Lost Notebook. He was followed by Michael Schlosser of the University of Vienna whose talk in the Focused week was also the Department Colloquium that week. Several students of Berndt participated in this Focused Week and gave talks on theta functions. The final day featured a talk in the morning by George Andrews on false theta functions, and an afternoon lecture by Roger Baker (Brigham Young) on asymptotic formulas related to quadratic forms.

On March 24, during the Focused Week, Kannan Soundararajan (Stanford), who had won the First SASTRA Ramanujan Prize in 2005, delivered the Fourth Ramanujan Colloquium on the remarkable solution of the Quantum Unique Ergodicity (QUE) Conjecture for modular domains due to him and Roman Holowinsky (Ohio State). Holowinsky subsequently received the SASTRA Ramanujan Prize in December 2011, for this joint work. After Soundararajan's Ramanujan Colloquium, Mathura and I hosted a party that night in his honor and for the Focused Week participants at our home. At the party, Soundararajan gave a brief speech as to how his research career in mathematics began in Madras.

During Soundararajan's stay in Florida, he told me about some interesting work he was doing on the Dickman function, and its connections with Ramanujan's work on this function. I told him that I would be delighted if he would publish this in the Ramanujan Journal. His paper on the Dickman function appeared in December 2012 in a special volume of the Ramanujan Journal that I edited for the Ramanujan 125 celebrations.

On April 9 evening, there was a party at the home of the President of the University of Florida in honor of John Thompson for his retirement from Florida. The Department invited Richard Lyons of Rutgers University, one of Thompson's former PhD students, to give a Colloquium on the work of Thompson in group theory. During the party at the President's House, I was asked to speak. I recalled how Thompson brought unparalleled recognition to the University and the Department and how his presence helped in building a stellar program in algebra. Thompson's retirement marked the end of a great era for the Department.

On April 22, as the Spring Term ended, we recognized two other colleagues on their retirement — Jorge Martinez who maintained an active program in ordered algebraic structures, and John Klauder who was a distinguished professor with a joint appointment in the physics and mathematics departments. There were talks in the morning about their work followed by a banquet in their honor that night at the Hilton.

Mathura and I departed on May 10 for India on Northwest Airlines via the Pacific. I had no academic assignments in India, but my time was all focused on the sale of the rear portion of our family home Ekamra Nivas, which coincidentally was completed on June 7, the day my father passed away two years before! However, during my stay in Madras, I was able to make progress in the study of variants of q-hypergeometric series that I obtained by combinatorial arguments. Also, the proofs of the book in memory of my father were sent to me in Madras and I was able to correct them while I was there. The first quarter of the book has several articles about my father's life and contributions, all prepared by me. This is followed by a collection of contributions by eminent physicists and mathematicians who knew my father well. These papers are grouped in three sections on (i) mathematics, (ii) probability and statistics, and (iii) mathematical physics — scientific areas to which my father had contributed. I edited section (i). Professor C. R. Rao, a great friend of my father, and arguably the greatest statistician in the world, edited (ii). Finally, my distinguished colleague John Klauder edited (iii). Of note is that the volume contains the massive paper of John Thompson and my colleague Peter Sin on the Divisor Matrix. Thompson's talk in Oslo after he received the Abel Prize in 2008 was on this topic. I told Thompson that I was honored to receive this paper for the volume in memory of my father. Thompson responded by saying "Your father deserves this". Another massive paper in the volume is by Shreeram Abhyankar dedicated to my father and his father. Abhyankar told me that he was planning on submitting the paper to the Journal of Algebra but then decided to publish it in the volume dedicated to my father.

Mathura and I left Madras June 22 night for Singapore where we stayed at the lovely Pan Pacific Hotel. Interestingly, Mathura's younger brother Sriram was visiting Singapore with his family at that time and it was nice to spend time with them there. We left Singapore very early on June 25 and after a long transit at Tokyo Narita airport, reached Guam that night. Since my mother was with us, we arranged a comfortable two bedroom apartment at The Ohana Oceanside, a condominium complex that commanded a fine view of the Pacific ocean. We spent five memorable days in Guam. Amritha would come to our apartment from work in the afternoons to enjoy the delicious South Indian lunch prepared by Mathura. It was a delight for Mathura to cook in Guam since the vegetables were so plentiful and fresh on this lush tropical island. Some vegetables and fruits were brought in daily from the Philippines. We drove around Guam to show my mother some very scenic spots such as the Two Lovers Point from where one gets the most gorgeous view of Tumon Bay. Legend is that a pair of young lovers, who were prevented from getting married, jumped to death off the cliffs at Two Lovers Point. We also visited Magellan point where the great explorer Ferdinand Magellan landed, and went to Merizo Beach in the southern tip of Guam where one can get a fabulous view of Cocos Island.

One of the most awesome experiences in Guam for us was the "view of Mariana's Trench". Five miles off the east coast of Guam lies Mariana's Trench, which at 35,000 feet below sea level, is the deepest point on the surface of the earth! Amritha had some good contacts in Guam, and so she arranged for us to enter an estate of a friend on the east side fronting the ocean from where we could get spectacular view of the Pacific. Just five miles away was Mariana's Trench, and we could see that the water color was a much darker blue, indicating the great depth of the ocean! In contrast, on the west coast of Guam at Tumon Bay and Agana Bay, the water is so shallow and calm that one can see the coral bed below. On the eastern side of Guam the water depth increases significantly just a few meters off the coast. In addition, the surf is very rough, and so swimming is prohibited on the east coast in most places.

After this view of Mariana's Trench, we had dinner at the home of Professor Narayana Balakrishnan who is on the Chemistry faculty of the University of Guam. Since Amritha's reports were appearing regularly in the newspaper, and because she would attend Indian cultural activities, several members of the Indian community such as Professor Balakrishnan, knew her. His residence is near the university and on the east coast of Guam. We drove around the lovely campus of the university before reaching his home. His wife Shuzuko is Japanese, but she had prepared a delicious South Indian dinner for us. Jennifer Balakrishnan, the famous mathematician now at Boston University, is their daughter. Jennifer was a graduate student at MIT at that time, and so we did not see her in Guam.

June 29 was the 32nd Wedding Anniversary for me and Mathura and we celebrated it with a lunch at the Italian restaurant Al Dente at the Hyatt Regency, Guam. Amritha presented us with a book "Portrait of Guam" to remind us of the special visit to this pearl of the Pacific.

We were back in Gainesville on July 1, and the next morning I received a letter from Professor Peter Goddard, Director of the Institute for Advanced Study, inviting me as his guest to attend the 80th Anniversary of the Institute in September. I readily accepted his gracious invitation and informed him that Mathura will be accompanying me.

Throughout July, Mathura was busy with the Arangetram of one of her students that took place on July 31. Amritha flew to Gainesville all the way from Guam to provide vocal support to Mathura for the program — a gesture that was appreciated by all who attended the event.

In mid-August, I made a one week trip to China for an International Conference on the Renaissance of Combinatorics in honor of Doron Zeilberger for this 60th birthday at the Center for Combinatorics at Nankai University in Tianjin. This conference was the final event of 2009–10 and is described in §11.3.

10.3 The 80th Anniversary of the Institute for Advanced Study and other events (2010–11)

At the start of the academic year, I received the first copy of the book "The legacy of Alladi Ramakrishnan in the mathematical sciences" (Springer) that I had edited along with Professors C. R. Rao and John Klauder. The main event at the start of the academic year was the 80th Anniversary of the Institute for Advanced Study in Princeton, during September 24–25, an event that Mathura and I had the privilege of attending as the personal guests of Professor Peter Goddard, Director of the Institute. Since the volume in memory of my father had appeared, I decided to hand a copy of the volume to Professor Goddard as my personal gift to him.

During the Nankai conference, Zeilberger had invited me to Rutgers University at any time suitable for me. When I mentioned to him in China that I would be in Princeton for the 80th Anniversary of the Institute, he immediately said that I should give a talk at Rutgers before this conference.

Mathura and I departed on a five-day trip to New York-New Jersey on September 23. My mother also came with us because she wanted to use the time to visit her elder sister Padma who was living in Westchester County north of New York where two of her sons were working. So we flew into Westchester County airport, where my cousin Ganesh picked up my mother to take her to his house where my aunt was staying. Mathura and I proceeded to New Jersey to the home of our long time Gainesville friends, Ravi and Bhavani Varadarajan for lunch. I then went to Rutgers University where my talk in Zeilberger's Experimental Mathematics Seminar was at 5:00 pm. The brilliant Zeilberger is very idiosyncratic, and I saw this first hand during this visit.

Zeilberger met me at the mathematics building entrance. His office is on a high floor. He said I could take the elevator if I wanted to, but he would walk up the flight of stairs. He hated elevators. He ran up the stairs and greeted me as I stepped out of the elevator!

When Zeilberger invited me to speak in his seminar, his letter of invitation said that my talk was to be for 48 minutes! Zeilberger is a stickler to time, and so, before the start of the lecture, I requested him to give me an indication at the 45th minute so that I can close the lecture in the next three minutes. My talk was on "New companions to Euler's Pentagonal Numbers Theorem", essentially the same as my talk at the Zeilberger 60 conference in August. After my seminar, Zeilberger graciously hosted a dinner at the Indian restaurant Spice Paradise in neighboring New Brunswick. I had a rental car and so I told Zeilberger that we could go to the restaurant in my car. He said he avoided automobiles as much as he could, and so he preferred to walk to the restaurant. By the time I walked to the parking lot at the Rutgers campus, drove through traffic to New Brunswick, and found a parking spot there, and reached the restaurant, Zeilberger was already there! Mathura and Zeilberger's wife joined us for dinner. It was a very pleasant evening. I must say, that in spite of his many idiosyncracies, he is a most kind hearted person and a gracious host.

After dinner, Mathura and I were going to Princeton to spend three nights there in connection with the 80th Anniversary Conference of the Institute for Advanced Study (IAS). Zeilberger lives in Princeton, and since his wife was with us, he agreed to get a ride to their home. So Mathura and I dropped the Zeilbergers at their home before checking in at the guest house of the Institute for Advanced Study.

===

Conference for the 80th Anniversary of the IAS, Sept. 24–25

As guests of Professor Peter Goddard, Director of the Institute for Advanced Study in Princeton, Mathura and I were accommodated at the Marquand Guest House for three nights (Sept. 23, 24 and 25). The 80th Anniversary Conference was during September 24–25, and it was magnificent.

The Institute which is abundant with world class faculty, was teeming with Nobel Laureates and Fields Medalists for this event. Some prominent participants were: *John Milnor* (1962 Fields Medalist) — SUNY Stonybrook, formerly at IAS; *John Nash* (1994 Nobel Laureate in Economics) — Princeton University; *Enrico Bombieri* (1974 Fields Medalist) — von Neumann Prof. at IAS; *Jean Bourgain* (1994 Fields Medalist) — IAS; *Vladimir Voevodsky* (2002 Fields Medalist) — IAS; *Frank Wilcek* (2004 Nobel Laureate in Physics) — MIT; *Pierre Deligne* (1978 Fields Medalist) — Emeritus at IAS; *John Thompson* (1970 Fields Medalist and 2008 Abel Laureate) — Emeritus at University of Florida and Cambridge University.

(NOTE: I should mention that in subsequent years, the following prizes were awarded: 2011 Abel Prize to Milnor, 2013 Abel Prize to Deligne, and 2015 Abel Prize to Nash. I also note with sadness that Nash and his wife died in a tragic automobile accident as they were returning to Princeton after the Abel Prize Ceremony. Also Bourgain passed away in 2018, and Voevodsky died in 2015.)

I felt that the concentration of intellect in the Institute lounge was so great that the earth might cave in under its sheer weight! It was a most exhilarating feeling to be in the midst of so many great minds and hear their thought-provoking lectures. It brought back memories for me and Mathura of our stay at the Institute in 1981–82 when I was a Visiting Member.

Mathura and I met Professor Goddard on the morning of September 24 before the sessions began. I presented him with a copy of the book *The legacy of Alladi Ramakrishnan in the mathematical sciences* (Springer) that had just appeared, and he reciprocated by giving me two books, one on the early years of the Institute, and another on Dirac. I was then taken to the archives section and shown the file on my father that contained among other things the correspondence between my father and Robert Oppenheimer who was the Director of the Institute when my father was a Visiting Member. I requested copies of some letters and received these a few weeks later.

It was nostalgic for me to see the building where I had my office in 1981–82 (the same building where Enrico Bombieri, Harish-Chandra,

Robert Langlands, and Armand Borel had their offices), and for Mathura and me to see 56 Einstein Drive, which was our apartment then.

I attended a good number of talks, not all, because there were parallel sessions. Milnor gave a commentary on the work of D'Arcy Thompson and discussed the role of conformal (and other) maps in understanding how bone structure may have evolved. Milnor had won the Fields Medal in 1962 for groundbreaking work in topology, but in recent years has been studying applications of mathematics to biology. Voevodsky, who was an algebraic geometer/topologist, spoke about the consistency of the foundations of mathematics. Bourgain's talk covered several areas of mathematics — number theory, representation theory, linear groups, analysis, ... Thus these giants presented a global view of the current state of mathematics and related areas. They were introduced by Peter Sarnak, Pierre Deligne, and Enrico Bombieri, respectively.

During the afternoon Tea on the first day, Bombieri asked Thompson where he could find his paper on the divisor matrix. Thompson replied that it was in the volume on my father's legacy. I told Bombieri that this volume could be found in the Institute Library since I gave Professor Goddard two copies, one for him, and one for the Library.

Mathura and I attended the Conference Banquet and were honored to be seated next to Professor Goddard, the Director. The speeches by Freeman Dyson and Robert Langlands at the banquet were most impressive. In giving a fantastic account of 80 years of the Institute, Dyson was critical that Oppenheimer concentrated too much on particle physics. Dyson pointed out that it was at his insistence that a program on astrophysics was started at the Institute with the appointment of Bengt Stromgren. Dyson had initially suggested the great astrophysicist Subrahmanyam Chandrasekhar of the University of Chicago for a permanent appointment at the Institute, but Chandrasekhar was not interested in the offer, perhaps because he wanted to be close to Yerkes Observatory near Chicago for night sky observations. In retrospect, Dyson said that he felt it was better for Stromgren to be a Permanent Member at the Institute because Chandrasekhar was a "lone wolf" who preferred to work alone and so may not have blended with the culture of the Institute where Permanent Members spend considerable time interacting with the visitors.

Langlands in his speech noted that in the public eye the Institute is identified with Einstein (and consequently with physics) but really the Institute was built on the foundations laid by the School of Mathematics; indeed the School of Mathematics was the only school when the Institute was founded in 1930 and some of its early members (during 1930–60) were Einstein, Hermann Weyl, von Neumann, Kurt Goedel, Oswald Veblen, Marston Morse, Carl Ludwig Seigel, Andre Weil, Atle Selberg, Freeman Dyson, and others. Like Dyson, Langlands in his magnificent account of the history of the school of mathematics analyzed both the contributions and the personalities of some of these eminent mathematicians. Bye the bye, Langlands was subsequently awarded the 2018 Abel Prize for the *Langlands Program* he initiated which has revolutionized mathematics.

A visit to such a great seat of learning, and the lectures by the leaders of our profession, inspire us to excel in our work. Thus I came away from this visit with a satisfaction matched only by the stimulating experience I had during my stay at the Institute in 1981–82.

$$====================================$$

Mathura and I checked out of the Marquand Guest House early on September 26 and went to White Plains New York where my mother was staying with her older sister Padma who was living at the home of her youngest son (my cousin) Ganesh. There my aunt Padma's eldest son and his wife Drs. Dharmarajan and Lakshmi joined us for lunch. We were seeing all of them after 30 years! After lunch, we departed from Westchester County Airport and returned to Florida via Detroit.

On October 19, Mathura and I had a party at our house for the release of the Springer volume on the legacy of my father. The guests who numbered sixty included some university faculty and their spouses, plus some members of the Indian community. It was nice that Elizabeth Loew of Springer, who was the editor in charge of the publication of the volume, was there as well.

During the end of October, I made a day trip to Atlanta to give a colloquium at Georgia Tech at the invitation of Christian Houdre. I spoke on "Euler's pentagonal numbers theorem - refinements, companions, and variations". Christian was part of a three member team with me when we visited the King Fahd University of Petroleum and Minerals (KFUPM) in Dhahran, Saudi Arabia, in November 2009 for an assessment of their mathematics program. For an account of the visit to KFUPM, see §9.13. As we left Dhahran upon completion of that assignment, Christian told me that he would like to invite me to Georgia Tech.

Ken Ono who had just joined Emory University in Atlanta as a Distinguished Professor, attended my colloquium at Georgia Tech along with some his PhD students. In the discussion that followed my talk, he urged his students to study these elegant variations of Euler's Pentagonal Numbers Theorem that I discussed. While at Georgia Tech, it was nice to reconnect with Prasad Tetali whom I had first met in Madras in 1986 when he was a Masters student at the Indian Institute of Science, Bangalore.

In mid-November, I made a trip by car to Statesboro, Georgia, to give a colloquium at Georgia Southern University at the invitation of Andrew (Drew) Sills. He and I attended the year along graduate course at Penn State University in 1992–93 when I spent my sabbatical there. After completing his Masters at Penn State in 1994, he moved to the University of Kentucky where he received his PhD in 2002 with Andrews as one of his two advisors. Later he was Hill Assistant Professor at Rutgers during 2003–06 where he interacted closely with Doron Zeilberger. He joined Georgia Southern University in 2007 where he has been since then.

On the night of my arrival, Sills hosted a dinner his colleagues Hua Wang and Steve Damelin joined us. Wang had visited the University of Florida in connection with our combinatorics conferences and special years. Damelin and I had served on the Committee on Committees in 2009 when Andrews was President of the AMS.

My colloquium the next afternoon was on Göllnitz' theorem and its place in the theory of partitions, following which I departed Statesboro to be back in Gainesville afer a five-hour drive.

In mid-December, my mother received her US Immigrant Visa during her interview in Jacksonville. My father was never interested in applying for a green card, but told me that I should become a US Citizen so that after his time, I could sponsor my mother for a green card. My dear friend Nadadur Sampath Kumar, one of the most successful immigration attorneys practicing in Los Angeles, helped with the green card application for my mother. It was thoughtful of him to have come from Los Angeles to Jacksonville to be with my mother during her green card interview.

===

Trip to India and the 2010 SASTRA Prize to Wei Zhang

With my mother's immigrant visa having been processed, she and I departed Florida the next day (Dec 15) for India via the Pacific. The trip had a two-fold purpose: (i) I had to attend the SASTRA Ramanujan Conference (Dec 21–22), and (ii) as a family, we had to perform the *Valaikappu* (= bangles ceremony) at our home in Madras for my elder daughter Lalitha who was six months pregnant with her first child. Lalitha could only get leave from her work for about ten days, and so she could not travel with me. She arrived in Madras a few days later accompanied by Mathura.

On December 20, one day before my departure to SASTRA, I was at Vivekananda College to deliver a lecture in memory of my boyhood friend Parthasarathy (alias Raju). He was the son of Mr. Nambi Iyengar, my father's personal assistant. Nambi Iyengar's family lived in our house behind the main building, and so Raju and I were friends from childhood. He graduated with a Bachelors Degree in Chemistry from Vivekananda College in 1974 and had a large circle of friends there. Even after he moved to the United States where he received his Phd and settled down, he maintained contact with me and all his college friends. In addition, he was a Ramanujan enthusiast and would send messages about Ramanujan under the pen name "Pacha Nambi". After he passed away, at the urging of all his friends, Vivekandanda College agreed to hold the memorial lecture, and in view of my long association with him, I was asked to deliver this talk.

The main academic event of December 2010, was the award of the SAS-TRA Ramanujan Prize to Wei Zhang (Harvard University) who had made far reaching contributions by himself and in collaboration with others to a broad range of areas in mathematics including number theory, automorphic forms, L-functions, trace formulas, representation theory and algebraic geometry. In his PhD thesis at Columbia University, written under the direction of Professor Shou-Wu Zhang, he essentially settled the famous Kudla Conjecture pertaining to Shimura varieties. His thesis opened up major lines of research and led to collaboration with Xinyi Yuan and Shou-Wu Zhang. In addition, Wei Zhang by himself had done outstanding work on trace formulas and Shimura varieties.

Wei Zhang who hails from the People's Republic of China, was born in 1981. After obtaining a Bachelor's degree from Beijing University in 2004, he joined Columbia University to do his PhD. After completing his PhD in 2009, he went to Harvard University where he held the prestigious Benjamin Pierce Lectureship. He is now a professor at MIT.

It was very supportive of Professor Shouwu Zhang, the former thesis advisor of Wei Zhang, to come to India for just two days to be present at the prize ceremony of his student. Showu Zhang was attending a conference in China; he flew into Madras on December 20 night. A few hours later, at 2:30 am on December 21, Wei Zhang arrived at Madras airport from the USA via Brussels on Jet Airways. They barely had a few hours rest at Hotel Savera in Madras. We departed Madras at 10:30 am by the Savera van and reached Kumbakonam by 5 pm giving us enough time to see The Town High School where Ramanujan studied, Ramanujan's home, and the Sarangapani Temple where Ramanujan and his family worshipped. As usual, we stayed at the lovely Sterling Resorts. The next morning, December 22, Ramanujan birthday, after receiving the SASTRA Prize, Wei Zhang delivered the Ramanujan Commemoration Lecture. He was followed by Shouwu Zhang. I noticed the close bond between the Master (Shou-Wu) and Student (Wei), like the bond between the *Guru* and *Sishya* in our Hindu tradition. This time we had three members of the Tata Institute give talks — Dipendra Prasad, Ritabrata Munshi and Arvind Nair. We all departed SASTRA around 4 pm for Madras and stopped at Madras airport on the way for Wei Zhang and Shouwu Zhang to catch their international flights.

December 23 was the *Valaikappu* for Lalitha. The Valaikappu, or bangle ceremony, is like the baby shower in the United States, in the sense that it is a ladies function in honor of the expectant mother and is a lot of fun. But there is an important difference. For the valaikappu, there is a Hindu cultural and even religious aspect to it. In our culture, bangles signify happiness and prosperity. The ladies who attend the function, deck the mother-to-be with colorful bangles, and purchase bangles for themselves from the woman who sells bangles at the ceremony. For Lalitha's Valaikappu, we also arranged for a Sanskrit and cultural scholar to give a short discourse on the significance of having children. In such discourses, stories of Lord Krishna and other Gods as children are told, and the expectant mother is blessed by the scholar, who in this instance happened to be the grandson of the late Ananta Rama Dikshitar, who was hailed as one of the greatest in the art of *Harikatha* (delivering religious discourses). As a boy of five or six, I had attended along with my mother, Ananta Rama Dikshitar's Harikatha on the Hindu epic *The Ramayana* at my school Vidya Mandir. And for Lalitha's Valaikappu, Dikshitar's grandson blessed her for the birth of her first child. We had about one hundred guests and they enjoyed a sumptuous South Indian lunch after the discourse.

On December 30, I arranged the Second Alladi Ramakrishnan Memorial Lecture by Professor C. R. Rao at Science City in Madras. On that occasion, the book on the legacy of my father was released, and copies handed over to Dr. Anandakrishnan, Director of Science City, and Mr. N. Ravi, Editor of The Hindu newspaper. I was pleased that 2007 Abel Laureate S. R. S. Varadhan of the Courant Institute, a former PhD student of C. R. Rao, attended this function. There was a prominent report in the Hindu the next morning about the book release and Professor Rao's lecture. Thus 2010 was very eventful and ended on a high note.

On January 1 night, Mathura, Lalitha, and I departed Madras by Jet Airways reaching Singapore the next morning. We had an enjoyable two day holiday in Singapore and stayed at the Conrad Centennial Hotel. Especially delightful was the English Afternoon Tea served in the Lounge on the 35th floor which commanded a spectacular view of the city and the marina. We flew out of Singapore back to the USA on January 4.

==

After just a day of rest, I departed Gainesville for New Orleans on January 6 for the Joint Annual Meetings (JMM). New Orleans is an interesting city, and JMM is held there once every few years. I was happy to see the book on my father's legacy on display at the Springer booth. On January 7 morning, I was the first of four speakers of the MAA session on "The Beauty and Power of Number Theory" organized by Thomas Koshy. My talk was on "Euler's Pentagonal Number Theorem - companions and variations". The next day I enjoyed the invited addresses of Kannan Soundararajan and Akshay Venkatesh (both of Stanford University) who had won the SASTRA Ramanujan Prizes in 2005 and 2008 respectively. Following that, there was the board meeting of the Notices of the AMS.

The day after my return from the JMM, Jon Borwein and his wife arrived by car from New Orleans, because Atlanta airport was closed due to bad weather. It rarely snows in Atlanta, but when it does, it paralyses the city and the airport because they do not have equipment to deal with snow. I think now the city and the airport are better prepared.

Jon Borwein was a highly accomplished mathematician and a great speaker, and so we arranged for him to give several talks. On January 11, he spoke in the Number Theory Seminar on "Ramanujan's theory of alternate elliptic integrals". That same afternoon, he addressed the Colloquium. The next day he gave a talk in the Graduate Number Theory Seminar on "The life of π" (history of the number π). I too have given lectures on the history of π, and in such lectures I emphasize that π is so fundamental, that a discussion of the history of π amounts to a treatment of the history of mathematics! Jon Borwein and his younger brother Peter Borwein had written a fantastic book entitled "π and the AGM" (1988). AGM stands for the Arithmetic-Geometric Mean of Gauss. The Borwein brothers are experts in computation and had used a formula of Ramanujan to compute several millon digits of π.

Roger Baker arrived on January 12 to spend a good part of the Spring semester with us. George Andrews had also arrived in Gainesville that morning for his semester-long stay. The next week, Roger Baker gave two talks, one in the Graduate Number Theory Seminar on "Weyl sums" and another in the Number Theory Seminar.

During the weekend of January 22–23, there was a conference on partitions at Emory University. Almost immediately after joining Emory in Fall 2010, Ken Ono announced an important result on partitions that he, Amanda Folsom and Zach Kent had obtained — a new p-adic formula for the partition function. This conference was to celebrate Ono's arrival at Emory as well as this new result on partitions. Several active researchers in the partition world were invited to speak at this conference, including Frank Garvan, Alex Berkovich and me from the University of Florida, and (of course!) George Andrews (who was visiting Gainesville).

I arrived in Atlanta on Friday, January 21 afternoon, and checked in at the Emory Conference Center where all speakers were accommodated. After dinner, we all attended a Public Lecture at 8 pm by Ken Ono who spoke about this new breakthrough. Following the lecture, there was a Wine and Cheese Reception where I met Ono's parents. Ono's father, Professor Takashi Ono, is a well known algebraic number theorist who was a professor at Johns Hopkins University until his retirement.

After the conference at Emory, George Andrews began his Spring Term in Florida. During each of his annual visits, we invite him to address the department colloquium, and this time he delivered the 13th Erdős Colloquium on February 7. During that week in February we had the visit of Ken Gross from the University of Vermont. He had enormous success in the sphere of mathematics education, especially with his Vermont Mathematics Initiative (VMI). Mathematics Education receives significant funding but most mathematics education programs are pedagogical and lack real subject content to be effective. The VMI has real content and so has been very successful in the training of high school mathematics teachers. Andrews takes an active interest in mathematics education and he drew our attention to Ken Gross and the VMI. During the week, Ken Gross gave a widely publicized talk on Mathematics Education and the VMI.

I am proud to say that during Andrews' annual visits to Florida, he comes up with important new ideas in the theory of partitions, and always, these ideas are first announced in the Florida Number Theory Seminar. This time, Andrews announced his new ideas on "anti-telescoping" in the Number Theory Seminar in mid-February. Over the next few years, there was a flurry of activity involving the method of anti-telescoping.

The weekend before Andrews' talk, we had the *seemantham* for my daughter Lalitha, a semi-religious ceremony performed a few weeks before the birth of the first child, for the well being of both the mother and the child. This is very different from the valaikappu described above and is performed by the groom's family. So the seemanthan was performed at the home of Aditya and Lalitha

During the weekend of March 11–13, there was a conference at Georgia Southern University on partitions and q-hypergeometric series in honor of Dennis Stanton and Mourad Ismail for their 60th and 65th birthdays. I was invited to the conference, but one week before the meeting, I declined the invitation because I was suffering from a kidney stone which had to be surgically removed. This was the only time I could not attend a conference due to health reasons. Even though I could not be there, I was pleased that the refereed proceedings of the conference appeared as a special volume in The Ramanujan Journal two years later.

The Ramanujan Colloquium for the academic year was delivered by John Thompson on March 21, and I had recovered by then. He spoke about "The divisor matrix, Dirichlet series, and $SL(2, \mathcal{Z})$" which is joint work with my colleague Peter Sin — the same talk he gave in Oslo after receiving the Abel Prize. The Ramanujan Colloquium was followed by two technical talks when Thompson went into the details of his work in the number theory seminars during the next two days.

That week, my brilliant cousin V. S. Ramachandran (Ramu as he is known to his relatives and friends in India, and called Rama by his friends outside India), a distinguished professor at UCSD, was in town to deliver a highly publicized lecture on his famous work in neuro science. His book on phantom limbs, based on his fundamental research, is widely acclaimed. This featured talk was arranged by the Student Government. He is a powerful and charismatic speaker. I attended his talk which attracted about 500 students at the Reitz Union Ballroom, many of whom wanted him to autograph his book for them. Prior to the lecture, Ramu asked me to join him for dinner at a downtown restaurant with the leaders of the student government, which I did. But after his talk which finished at 9 pm, Ramu told me that he really wanted some spicy Indian food! Mathura was not in Gainesville when Ramu was visiting because, after the seemantham, she stayed back in Orlando to be of help to Lalitha during the final weeks of pregnancy. However, before she left, she had cooked plenty of Indian food for me, and kept it in the freezer. So I invited Ramu home for a late night dinner and warmed up some Indian dishes prepared by Mathura which he enjoyed immensely. I can say with certainty, that once your tongue is used to the taste of spicy curry, that is what you always crave for!

I should emphasize here that when Ramu was in medical school in India, he was debating between a career as a doctor or an academic career of research in the health sciences. Seeing his brilliant research which was published in prestigious journals like Nature, my father and an uncle of mine

Dr. Hariharan, himself an outstanding physicist, encouraged Ramu to take to a career of research. In his book on phantom limbs, Ramu acknowledges the crucial encouragement given both by my father and uncle Hariharan.

March 31 was one of the happiest days of my life when Lalitha gave birth to a baby girl — Kamakshi Alladi Aditya! Grandchildren are very special, and the feeling has to be experienced to be believed. Kamakshi, my first grandchild, was born in Orlando where my son-in-law Aditya and Lalitha were living, and I rushed to be by their side. *Kama* in Sanskrit means love (both in a sensual sense and in the sense of compassion) and *akshi* refers to eyes. Thus Kamakshi means one whose eyes show love (in the sense of compassion). Kamakshi is one of the names of the Hindu Goddess Parvathi. I was glad that Lalitha and Aditya, although settled in America, chose a traditional Hindu name for their child. A few days later, Lalitha and Aditya came to our home in Gainesville to perform the *namakaranam* (naming of the child) ceremony, which is done on the eleventh day after birth (the day of birth to be counted as the first day).

April 4 was my mother's 80th birthday. Since she was in Madras, all we could was to call her by phone. Mathura and I had plans for a birthday celebration for her in the Fall when she would be in Gainesville.

On May 16, I departed Orlando for India flying transpacific. I had several hours of transit at Singapore's Changi airport, voted as the best in the world. The airport is a shoppers paradise, and I purchased a Mother's Day gift for my mom. I arrived in Madras on May 18 but Mathura arrived only a week later.

Bruce Berndt was in Madras in response to an invitation from the British Society of Scholars and the Ramanujan Museum, and he delivered a talk they had arranged at MATSCIENCE. I knew of Berndt's visit to Madras and so I arranged a public lecture by him at our family home under the auspices of the Alladi Foundation. He spoke on "Ramanujan's Notebooks, his Lost Notebook, and his creativity". Mr. N. Ravi, editor of The Hindu newspaper, presided over the lecture.

On May 30, I inaugurated a Summer School on Physics and Chemistry at the University of Madras organized by Professor Devanathan, a former PhD student of my father. On that occasion, his book on Relativistic Quantum Mechanics dedicated to my father was released.

June 6 was my father's annual ceremony. This ceremony is called *sraddham*. Sraddham in Sanskrit means "that which is done with concentration and devotion". The sraddham is performed each year on the day of passing and the date is determined by the Lunar calendar. It is considered most auspicious to conduct the ceremony at one's family home. Even though it is the son who performs the sraddham for the departed parents, not the daughter, Hindu custom demands that only married sons can perform this ceremony and that the daughter-in-law should be present as well! Mathura had arrived a few days earlier, and so we performed this four-hour ceremony at Ekamra Nivas with our family priests.

After conducting my father's annual ceremony, Mathura, my mother and I departed Madras the next day by Malaysia Airlines for Kuala Lumpur. We stayed three nights at the magnificent Pan Pacific Hotel attached to the new Kuala Lumpur International Airport (KLIA). We rented a chauffeured car to visit the planned town of Putra Jaya (the Federal Administrative Centre of Malaysia) and the historic seaside town of Malacca. After two enjoyable days of sightseeing in Malaysia, we flew to Singapore to catch our flight to Guam via Tokyo on Delta Airlines.

Amritha was finishing her assignment in Guam on July 1 and so we wanted to help her wind up and also enjoy Guam before her departure from that charming pacific island. At Amritha's recommendation, we stayed at the lovely and expansive Leo Palace Resort in the central highlands of Guam. We had a spacious two bedroom apartment there, and so once again Mathura prepared outstanding South Indian meals for all of us.

One of the things Amritha did for exercise was to go on rather strenuous hikes in various parts of Guam with her colleagues. So she invited me to join on one of the "easier" hikes, but even that was strenuous for me as a fifty five year old. We walked through a dense jungle and the path in many places was either rocky or muddy. I had to be careful not to twist my ankle. Fortunately I survived. The reward for the hike was that it ended on a cliff with a spectacular view of the Pacific and the rocky coast. Of course, one had to be careful not to fall off the cliff into the water which was deep because just a few miles away is where Mariana's Trench is located! While I burnt a lot of calories on that hike, every bone in my body was aching, and so we decided to take it easy at the resort the next day, and enjoy the superb fare prepared by Mathura. The following day, I got to play tennis with Amritha on the (not too well maintained!) grass courts of the Leo Palace Resort, and to go snorkeling in the clear blue calm waters of Fujita Beach on Tumon Bay.

On our previous visit to Guam, Prof. Balakrishnan had invited us to dinner at his apartment. So this time, we invited the Balakrishnans to dinner at the Hilton Guam Resort (where we had stayed twice before), and enjoyed a Polynesian show by the poolside at night as we enjoyed dinner. So that wrapped up a memorable stay in Guam.

In mid-July, I made trip to the United Arab Emirates (UAE) for one week in connection with the Accreditation of the mathematics programs at the American University of Sharjah. This and other visits to the Middle East for the evaluation of mathematics programs, are described in §9.13.

On July 30, Mathura conducted the Arangetrams of two of her students in Jacksonville within a span of three weeks. This took a lot of effort on her part. Amritha, who had returned from Guam the previous week after completing her assignment, was there to provide valuable vocal support to Mathura, thereby contributing to the success of both programs.

During the months of July–September, we had two Turkish visitors in the department — Prof. Ali Bulent Ekin of the University of Ankara and a student of his. They mainly interacted with Frank Garvan. But I got to know Ali Bulent quite well and that was the beginning of a strong friendship. As a consequence, in later years, I visited Turkey twice at his invitation, and he visited Florida twice as well.

10.4   A year of milestones (2011–12)

There were several milestones during 2011–12 both academic and family related. My mother turned 80 in 2011, and MATSCIENCE turned 50 on January 3, 2012. I will describe a variety academic events against the background of these and other milestones.

During the academic year, I gave for the first time a graduate course on "Irrationality, Diophantine Approximations and Transcendence". I prepared elaborate Notes for the course starting from Summer 2011. I have given this course once every three years since then, and each time I have improved on my Notes by adding new topics or elaborating the discussion in certain sections.

The first memorable occasion for the year was the party on September 4 for my mother's 80th-birthday. It was an elaborate event attended by more than 150 guests. I presented a slide show covering important events of her life — with her parents in Trivandrum in the 1930s, her involvement in hosting guests for my father's Theoretical Physics Seminar, her world travels with my father for five decades, and her pursuit of Carnatic music as a vocalist. Mathura who had learnt so many songs from her, delighted the audience with a few favourites. In view of my mother's deep interest in carnatic music, we asked our friend Dr. Meera Sitharam, a professor in the computer science department, to play a few songs on the Veena. Meera is highly accomplished Veena player and was a graded artiste of the All India Radio. The *veena* in an ancient Indian classical music instrument and is associated with *Saraswathi*, the Hindu Goddess of Learning.

Dr. Jayaraman and his wife Kamala, who are dear friends of my parents, and octogenarians themselves, were present for the occasion. My cousin Dr. Dharmarajan (alias Raja) and his wife Dr. Lakshmi, came from New York; Raja's gave interesting recollections of my mother in the 1960's.

Having spent two years in Guam with Pacific Daily News, which is part of the Gannett News Network, Amritha's next assignment was for USA Today (the flagship of the Gannett group) in Munroe, Louisiana, starting from early October. Although it is a long drive to Munroe from Gainesville, we were glad she was not as far away as Guam!

The main academic events for 2011–12 were to start from our December trip to India. But before that, our extended family enjoyed a fabulous one week vacation in sunny Honolulu during the Thanksgiving Break with a stay at the Hilton Hawaiian Village. On the Monday before Thanksgiving, I did visit the University of Hawaii to give a colloquium talk so kindly arranged by Pavel Guerzhoy, my new host there after the retirement of Edward Bertram. Captain Srinivasan, the father of my son-in-law Aditya, was with us on this vacation, and he volunteered to attend the talk. He is a retired Air India Boeing 747 captain. He said he thoroughly enjoyed my presentation, but did not understand one word of it!

==

Trip to India and the 2011 SASTRA Ramanujan Prize to Roman Holowinsky

Mathura, my mother and I departed Orlando for India on December 12 via the Pacific. Upon reaching Singapore, we transferred to a Malaysia Airlines flight to the island of Penang, often referred to as the Pearl of the Orient. We stayed three nights at the exquisite Shangri-La Rasa Sayang Resort on Batu Ferringhi Beach, and enjoyed the scenic tours around this idyllic island. Mathura and I had visited Penang in 1991 with my daughters who were 8 and 3 years old then, but we made this second visit as a holiday for my mother for her 80th birthday, enroute to Madras.

The SASTRA Prize Winner Roman Holowinsky arrived in Madras on the night of December 18, a few days before the conference. So during the next two days, I showed him around Madras, walked with him on the famed Marina Beach, and took him to a concert at the Madras Music Academy. I found him to be extremely pleasant and warm.

Roman Holowinsky of Ohio State University was chosen as the awardee of the SASTRA Ramanujan Prize for his very significant contributions to areas which are at the interface of analytic number theory and the theory of modular forms. Along with Kannan Soundararajan of Stanford University (winner of the SASTRA Ramanujan Prize in 2005), he solved an important case of the famous Quantum Unique Ergodicity (QUE) Conjecture in 2008. This is a spectacular achievement.

In 1991, Zeev Rudnick and Peter Sarnak formulated the QUE Conjecture which in its general form concerns the correspondence principle for quantizations of chaotic systems. One aspect of the problem is to understand how waves are influenced by the geometry of their enclosure. Rudnick and Sarnak conjectured that for sufficiently chaotic systems, if the surface has negative curvature, then the high frequency quantum wave functions are uniformly distributed within the domain. The modular domain in number theory is one of the most important examples, and for this case, Holowinsky and Soundararajan solved the holomorphic QUE conjecture.

The manner in which this solution came about is amazing. Since 1991, many mathematicians had attacked this problem and made major advances, but the problem for holomorphic domains remained open. By a study of Hecke eigen-values and an ingenious application of the sieve, Holowinsky obtained critical estimates for shifted convolution sums and this almost settled the QUE conjecture except in certain cases where the corresponding *L*-functions behave abnormally. Simultaneously, Soundararajan who approached the problem from an entirely different direction, was able to confirm the conjecture in several cases; he noticed that the exceptional cases not fitting Holowinsky's approach, were covered by his techniques! Thus by combining the approaches of Holowinsky and Soundararajan, the holomorphic QUE Conjecture was fully resolved in the modular case.

Roman Holowinsky was born in 1979. He obtained a Bachelors in Science Degree from Rutgers University in 2001. He continued at Rutgers to do his doctorate and received his PhD in 2006 under the direction of Henryk Iwaniec. He held post-doctoral visiting positions at the Institute for Advanced Study and the Fields Institute, Toronto before joining the permanent faculty at Ohio State University.

On December 21 morning, I departed Madras in the SASTRA van for Kumbakonam accompanied by Holowinsky and three other main speakers — Eknath Ghate (Tata Institute), Paul Nelson (Ecole Polytechnic, Lausanne), and Harald Helfgott (University of Paris). Ghate had received his PhD from UCLA in 1996 and Nelson his doctorate from Caltech in 2011. Helfgott's PhD advisor was Iwaniec and so he came to India to see his academic brother Holowinsky get the SASTRA Award.

We reached Kumbakonam by 5 pm and had time for our usual sight-seeing there before checking into the lovely Sterling Resorts for a night stay. The conference began with the Ramanujan Commemoration Lecture by Holowinsky, followed by talks by Ghate, me, Nelson and Helfgott. The conference concluded with the prize ceremony in the afternoon. We were back in Madras on the night of December 22.

After the SASTRA conference, Holowinsky spent some time at the Tata Institute. I understand that he donated the entire $10,000 prize money to a worthy cause for mathematics in India while he was at the Tata Institute.

On December 26 morning, there was a major function at the Madras University Auditorium arranged by the National Board for Higher Mathematics (NBHM) when it was declared that December 23, 2011 to December 22, 2012 would be a mathematics year in honor of Srinivasa Ramanujan's 125th birth anniversary (which was to fall on December 22, 2012). At the function on December 26, Robert Kanigel was honored with a special scroll for his acclaimed biography of Ramanujan "The Man Who Knew Infinity", and it was announced that his book is being translated into the Tamil language. That evening, Kanigel gave a popular lecture on his book, following which there was a dinner at the Taj Coromandel Hotel hosted by N. Ram, Editor-in-Chief of The Hindu, to which I was invited. At the dinner, Kanigel autographed his book for me and said he appreciated my review of his book in *The American Scientist* in 1993.

During the dinner at the Taj, I had a leisurely conversation with Kanigel. He told me that while most Indians still associate him with his biography of Ramanujan, he had moved on to writing other biographies and so the focus of his attention had turned to other personalities.

January 3, 2012 was the 50th Anniversary of the birth of MATSCIENCE. I asked some faculty members of MATSCIENCE whom I met at the December 26 function when Kanigel was honored, whether anything was planned for the 50th anniversary. They said that nothing had yet been planned. So before my departure from Madras on January 1, I drew the attention of Mr. V. Sriram (a historian of Madras) to the 50th anniversary, and sent him some material about the inauguration of MATSCIENCE in 1962 as well as some pictures. On January 3, a prominent article by Sriram appeared in The Hindu recalling the inauguration of MATSCIENCE 50 years earlier, and including some of the photographs I had provided. A few weeks later, MATSCIENCE announced that the Institute will be observing 2012 as the 50th Anniversary year which will conclude with a conference during the first week of January 2013.

===

Mathura and I departed Madras on January 1 night for Singapore where we spent two days before boarding the Delta flight to the United States. We arrived at Detroit airport on January 4 morning but there we bifurcated. She flew back home to Florida, whereas I took a connecting flight to Boston to attend the Joint Mathematics Meetings. In Boston, I stayed at the Sheraton which was connected to the Hynes Convention Center where the conference was held. This was convenient because I did not have to step out into the freezing cold of wintry Boston. I reached the hotel by 5:45 pm, and so I had time to attend some of the events that evening including the featured Gibbs Lecture. The next day I attended Andrews' Retiring Presidential Address entitled "Our challenges".

On January 6, at the Notices of the AMS Board Meeting, I proposed that the Notices should bring out a feature article on Ramanujan 125. This was enthusiastically endorsed by Steve Krantz, the Editor, and the Board. I was asked to edit this featured article scheduled to appear in December 2012 to coincide with Ramanujan's 125th birthday.

Upon return to Florida, I heard the sad news that Basil Gordon had passed away on January 12 after a battle with liver cancer. I had spoken to him in November 2011 when he was convalescing at home. He did not want to talk about his health. Instead he asked me to inform him about my research, for he said, mathematics is the most gratifying to the soul! He was an outstanding mathematician and a remarkable person in many ways, and my heart sank when I heard the news. Since Steve Krantz was a former colleague of his, I informed him about Gordon's passing. He spontaneously said the Notices of the AMS should bring out a feature article in Gordon's memory, and asked me to edit that article as well.

During the Martin Luther King long weekend in mid-January, Mathura and I went to Monroe, Louisiana, to be with Amritha who had nicely settled down in her new job there with the Gannett News Network. Monroe is the town where Delta Airlines started out in the summer of 1925 as a crop dusting operation. Owing to my passion for aviation, Amritha took us to the Chanault Aviation Museum in Monroe where the birth of Delta was highlighted. In view of Amritha's job as a reporter, here in Monroe, as in Guam, she came into contact with several important persons. One night, Amritha arranged for us to have dinner with Dwight Vines, the former President of the University of Louisiana, Monroe, and his wife. It was a very enjoyable evening with that elderly couple who were interested in the role my grandfather played in drafting the Constitution of India.

India's Republic Day (that is the day when the Constitution of India was ratified) is January 26, but in the United States, Indian groups in various large cities observe that on the closest Saturday instead. For the Republic Day celebrations in Tampa on January 22, the India Association in Tampa invited me as the Chief Guest with the request that I talk about Ramanujan since this was his 125th birth anniversary. What was especially nice was that Dr. Jayaraman, a close friend of our family, and a contemporary of my father, shared the stage with me, since he was asked to do the Flag Hoisting for the event. Dr. Jayaraman is a former PhD student of Sir C. V. Raman, and my father knew him since his days as a student of Sir C. V. After spending several years at AT&T Bell Labs in Murray Hill, New Jersey, Dr. Jayaraman retired and came to Tampa with his wife Kamala, a close friend of my mother. The Republic Day function attracted more than five hundred persons, and many of them, especially the youngsters, had questions for me on Ramanujan after my talk.

George Andrews arrived in Florida right after the Joint Meetings in Boston to spend the Spring Semester with us. On February 6, he gave a Special Colloquium on "The Data Deluge as the wave of the future", a talk that attracted faculty from several departments, especially computer science. This talk was almost identical to his Retiring Presidential Address given in Boston in January. The next week, on February 14, in the Number Theory Seminar talk by graduate student Keith Grizzell, Andrews asked a question concerning an infinite product of the type one has in the deep theorem of Göllnitz. The very next day, pursuing this question, Andrews formulated and proved a new companion result to the partition theorem of Göllnitz that he communicated to the number theory group by email. On that morning (February 15), I was being operated for a kidney stone, but during the period of convalescence at home the next two days, I identified the q-hypergeometric version of this theorem in terms of the Göllnitz key identity that Gordon, Andrews and I had in our 1995 paper. Thus a new joint paper for me with Andrews evolved. It was entitled "A dual of Göllnitz' partition theorem" and it appeared in The Ramanujan Journal two years later. Thinking of mathematics right after the kidney stone surgery took my mind away from the inconvenience and pain during the few days of convalescence.

Since the year 2012 was Ramanujan's 125th birth anniversary, I was interested in publishing a book on my own to mark that occasion. Starting from the Ramanujan Centennial in 1987, at my father's suggestion, I wrote articles annually for The Hindu newspaper in India, either about some aspect of Ramanujan's mathematics, or about mathematical luminaries in history whose life and work had things in common with Ramanujan. I also had published articles related to Ramanujan's legacy in other venues. I thought it would be good to publish a compilation of all these articles in the form of a book entitled "Ramanujan's place in the world of mathematics". I gave the collection of these articles to Andrews in February to get his opinion. He went through these articles and said wholeheartedly that I ought to publish these as a book. Enthused by his endorsement, I approached Springer with the proposal to publish a book based of my articles on Ramanujan. Springer reacted favorably and said that I would get a contract from Springer India. Indeed things started moving rapidly because Springer and I wanted the book to be released during the Ramanujan 125 celebrations in India in December 2012. Shamim Ahmad and Sanjiv Goswami of the Springer India office in Delhi were most cooperative and I signed a contract for the book [B1] during the Spring 2012 Term.

March 1–9 was our Spring Break, and to get a break from my hectic academic schedule, Mathura and I went on a one week trip to Dubai in the United Arab Emirates (UAE). I had made a couple of trips to the UAE in previous years for the accreditation of mathematics programs, and I got a very positive impression of the country. So I wanted Mathura to see the UAE. Also Mathura had two cousins living there and to see Dubai in their company was an added attraction to her. My previous trips to the UAE were all official, leaving no time for sightseeing. So on this trip to the UAE we did a lot of sightseeing and enjoyed every bit of it.

We left for Dubai on March 1 taking the Delta 15-hour flight from Atlanta on the Boeing 777-200 extended range aircraft. I had arranged the Marhaba (meet and greet) service in Dubai because of its great convenience that I had experienced on my official trips to the UAE. We were whisked away through immigration and customs and the Marhaba representative put us in a taxi that the Hilton hotel had sent us. It was a luxury version of the Audi-A8 with special lighting inside — a series of small bulbs all around the ceiling, like an illumination! The chauffeur was from India. He spoke Tamil and said that the Arab sheiks like these extra fittings in the car! We checked in at the Hilton Dubai Creek and our room commanded a fine view of the city. For the previous few days Dubai had its notorious dust storms, but by the time we arrived, the storms had cleared.

Sightseeing in Dubai included going to the top of the Burj Khalifa, the tallest building in the world, made more famous with the Mission Impossible movie starring Tom Cruise! There is a lot of demand for the ride to the top of the Burj Khalifa, and so I had bought the tickets before leaving for Dubai. And YES, the view from the 124th floor is awesome.

Another unique experience was the Desert Safari, which included a Camel Ride, and a belly dancing show (!) during a sumptuous dinner in the middle of the desert. Dubai is a shoppers paradise, and so Mathura's cousins took us to three gigantic malls whose opulence has to be seen to be believed. Dubai is also famous for gold jewelry and so we went to the area of the Gold Souks — shops mostly operated by Indians — and Mathura bought gold jewelry for our family.

After five days in Dubai, Mathura and I went to Abu Dhabi, the capital of the UAE for a three-day stay. I was so impressed with the Audi A8, that I requested the Hilton Dubai for the same car to take us to Abu Dhabi. It was a two hour drive on a magnificent highway along the coast. The Hilton Abu Dhabi where we stayed has an enviable location fronting the beach, and a fine restaurant on the beach which served Indian food that we enjoyed immensely. One of the most visited sites in Abu Dhabi is the Sultan bin Sayed Mosque, which with its gleaming white marble and its imposing presence reminded us of the Taj Mahal. After a three-night stay in Abu Dhabi, we drove back to Dubai airport to catch our flight back to Atlanta. I think the visit to the UAE lived up to Mathura's expectations.

Ken Ono arrived in Gainesville one day after our return from the UAE. He delivered the Ramanujan Colloquium on Monday, March 12 on Ramanujan and partitions, with the catchy title "Adding and counting" for. Sure enough, it attracted a large audience. He is an electrifying speaker who can hold the attention of a large and mixed audience in a public lecture. The next day, he gave two more technical seminars on this theme.

March 31 was my grand daughter Kamakshi's first birthday but we celebrated it the next day (Sunday) with a large gathering of friends by inaugurating the new outdoor kitchen and terrace in our house. The builder made sure that it would be ready for this special occasion! That terrace has since become the venue of so many parties, including several that Mathura and I have arranged in honor of our visitors in number theory and for our number theory conferences.

The final number theory seminar for the semester was in the third week of April with two talks back-to-back by Robert Lemke-Oliver and Zach Kent both from Emory University. Robert was a PhD student of Ono, and Zach was a post-doc of Ono who had collaborated with Ono and Amanda Folsom on their surprising p-adic formula for the partition function.

During April 30 to May 9, Mathura and I made a trip to Turkey. This was our first visit to that nation steeped in history and it was an eye opener in every sense. Although most of the trip was for sightseeing, it became the basis of an academic trip two years later.

We took the non-stop flight from New York to Istanbul. It was exciting to see the Bosphorus and the continental boundary between Europe and Asia from the air prior to landing. Ali Bulent, our gracious host, received us at the airport and took us to the fabulous Conrad Istanbul Hotel where I had arranged our stay. Ali Bulent joined us for a fantastic Turkish lunch at the Conrad, following which he took us out to see the city. Every inch of Istanbul breathes history for it is a confluence of Arab and European civilizations, and of the religions of Islam, Christianity, and Judaism.

Our first stop was the Dolgamache summer palace which has an enviable location right on the Bosphorus — the narrow straights separating Europe and Asia. The Palace is exquisitely beautiful both inside and out — its manicured gardens with a profusion of flowers, the gleaming white imposing buildings, and the opulent interior. After roaming through the palace and the grounds, we took a ferry right at the docks of the palace to cross the Bosphorus from Europe to Asia (but all in Turkey)! It was a gorgeous day, and boat ride was one of the most memorable experiences. After reaching the Asian side, we took another small boat to an island on the Bosphorus close to the Asian shore, that has a light house in a tower called *Kiz Kulezi* (Maiden's Tower) featured in the James Bond movie "The world is not enough". After the visit to Kiz Kulezi, Ali Bulent took us to the famous steps on the banks of the Bosphorous where we sipped Turkish Tea and gazed in wonder at the monumental buildings of Istanbul on the European side, made more enchanting by the crimson sunset. Our first day in Turkey couldn't have been better.

The next morning, we visited the famous Grand Bazaar in central Istanbul to buy souvenirs, but we had to be careful not to be pick-pocketed by thieves or to be cheated by the vendors! Having Ali Bulent by our side was very helpful. After lunch, Emre Alkan, a former PhD student of Ken Ono, met us at the Conrad and we went in Ali Bulent's car to Koc University where I gave a colloquium on "Euler's pentagonal numbers theorem - variations and companions". Koc is a non-profit private university founded in 1993 and has since become one of the most prestigious in Turkey. Its beautiful wooded hillside campus is north of Istanbul on the shores of the Black Sea. On the way back from Koc, we met Ali Bulent's sister and mother and had refreshments at a restaurant overlooking the Bosphorus.

The next day was the peak of our sightseeing in Istanbul. The entire morning was at the sprawling Topkapi Palace, the pride of the Ottoman Empire. After roaming for three hours through the Topkapi, we bought souvenirs there, and had lunch at the Terrace Restaurant right on the banks of the Bosphorus. It was exciting to see the ships and ferries plying the straights separating Europe and Asia. The afternoon was spent seeing Hagia Sophia and the Blue Mosque, adjacent to each other. The Hagia Sophia started out as a Greek orthodox cathedral, then became a Catholic Church when the Christians overran Constantinople, and finally was taken over by the Ottomans. Inside the Hagia Sophia is a large mural of King Richard the Lion Heart in Constantinople during the crusades.

On our final day of sightseeing in Istanbul, we saw the Panorama 1453 Museum which featured a movie projected on a circular screen depicting the fall of Constantinople to the Ottomans in 1453.

On May 5, we checked out of the Conrad Istanbul and Ali Bulent took us in his car on a five-hour drive to Ankara, where he lived. The highway connecting Istanbul and Ankara goes through very scenic mountainous areas. On reaching Ankara, we checked in at the Ankara Hilton and went immediately to the Ankara Castle set on top of a hill. That night, we had dinner at the lovely apartment of the Yesilyurts on the campus of Bilkent University. After completing three years at the University of Florida as John Thompson Assistant Professor, Hamza Yesilyurt accepted a permanent position at Bilkent University, which after twenty years of its founding, is recognized as one of the top research universities in Turkey. I was to visit Bilkent for a colloquium on my next trip to Turkey two years later.

The next day, Hamza took Mathura, Ali Bulent and me on an all day excursion to Cappadocia, a region of breathtaking beauty with its fantastic rock cathedrals, cave dwellings, and canyons. It is a charming blend of Bryce and Zion canyons, and it has featured in various movies.

On our last full day in Ankara, Ali Bulent took us to the Ataturk Memorial, and then to the campus of the University of Ankara where he worked. I was to also visit the University of Ankara two years later for lectures.

Mathura was to leave for India from Turkey on May 8. So we checked out of the Ankara Hilton early, and drove straight to Istanbul airport where she boarded an Emirates Airlines flight to India via Dubai. On reaching India, she sent me a message that the Emirates service both on the ground and in the air lived up to its high reputation. I could not go with her because I had an assignment at the King Fahd University of Petroleum and Minerals (KFUPM) in Dhahran starting May 13. So I stayed one more night in Istanbul and returned to the Florida before departing for Saudi Arabia two days later.

I did a quick turnaround and departed Gainesville for Saudi Arabia on May 12 flying the segment from New York to Jeddah on a Saudi Arabian Airlines Boeing 777. After three days of work in Dhahran (now Dammam) I departed for Madras again by Saudi Arabian Airlines. My hosts at KFUPM were kind enough to arrange a one way ticket between Florida and India via Dhahran instead of a round-trip between USA and Saudi Arabia. My trip to Saudi Arabia is described in §9.13.

On May 21, I received a letter from Prof. R. Balasubramanian (Balu as I know him), the Director of MATSCIENCE, inviting me to be actively involved in the 50th Anniversary Celebrations of the Institute, to which I consented. As a follow up to this letter, MATSCIENCE sent three of its faculty members on May 30 to my family home Ekamra Nivas, to see the birth place of MATSCIENCE. After I showed them around the house, and especially the Seminar Room, they took me to MATSCIENCE to discuss plans with Balu for the 50th Anniversary. Next on June 3, Mrs. Indira Chowdhury, an archivist with the organization called Shrishti, visited Ekamra Nivas. She was put in charge of the archival work of MATSCIENCE since she had done similar work for the Tata Institute so well. After seeing Ekamra Nivas in full detail, she came back the next day to interview both me and my mother for our recollections regarding the birth of MATSCIENCE, and to take pictures of us and the house. She told me that she would like to take with her several albums and files of my father containing historic pictures and letters pertaining to the birth of MATSCIENCE. I told her that I cannot part with them, but am glad to have them scanned in my presence at Ekamra Nivas. But I was departing for America in two days, and so this had to be done only later, but well before the 50th anniversary conference in early January. Since classes for me at the University of Florida were resuming in the end of August for the Fall term, I told her that the only time I could return to Madras to enable the scanning of these documents and photographs was in mid-August.

Mathura, my mother and I departed Madras on June 6 by Malaysia Airlines for Singapore where we spent two days, following which we departed for Honolulu via Tokyo. Upon arrival in Honolulu, we connected to a flight to Kona on the Big Island of Hawaii where Amritha arrived from Monroe to spend a few days with us. This time we stayed at the Waikaloa Resort of the Hilton, and from the *Lanai* (balcony) of our room, we had spectacular views of twin volcanoes Mauna Loa and Mauna Kea especially at dawn.

The highlight of our four-day stay on the Big Island was the visit to the Volcanoes National Park. This time we took the Ridge Road between the two volcanoes and got a stunning view of Mauna Kea as the clouds cleared to reveal the observatories at the summit at 14,000 ft above sea level. At the Volcanoes National Park, we drove along the Chain of Craters Road all day, and at dusk we gathered along with other tourists at the Jaggar Museum overlook to see the awesome grandeur of the red fumes that emerged from the Halemaumau Fire Pit. It looked especially brilliant at dusk and at night. Amritha returned to Monroe right after her stay on the Big Island, but my mother, Mathura and I spent a few days in Honolulu before returning to Florida.

Amritha was to complete her assignment in Monroe in July before joining the Emory University MBA Program in the Fall, and it was nice that in May, she received Gannett's Watchdog Reporting Award! N. Ram, Editor-in-Chief of The Hindu, under whom Amritha was an intern in 1999, had advised her to get an MBA degree so that she could seek administrative positions in the broadcasting sector. Usually before joining an MBA program, it is good to have some job experience. So following Ram's advice, Amritha joined the Emory MBA program after working three years in Guam and Monroe with the Gannett news network.

In mid-August, I departed for Madras to get the scans done for MATSCIENCE. The Institute paid for my trip but they could not make the payment by just inviting me for the purpose of scanning the valuable documents and photographs. So I was invited to deliver a lecture at MATSCIENCE to which I agreed.

I flew to India via Amsterdam on KLM, Royal Dutch Airlines. On the forward journey, I broke journey in Delhi on August 12 to meet Sanjiv Goswami and Shamim Ahmad of Springer India to discuss the production of my book to be released at the Ramanujan 125 Conference in December. It was a successful meeting during which both of them said that they would be at the SASTRA Conference for the release of the book.

I was in Madras for just one week, and it was the first time in my life that I was in Ekamra Nivas without any of my family members! Our staff who were on leave, had to come back to work because I was there. My good friend GP stayed in Ekamra Nivas the entire week to help me assemble whatever needed to be scanned. Indira Chowdhury came from Bangalore, brought a scanner from MATSCIENCE, and in two full days in Ekamra Nivas, she scanned hundreds of letters, documents and photographs! So the job that needed to be done was completed satisfactorily. While she was in Ekamra Nivas, she did another two hour interview of me, this time asking what I thought about the academic scene in India and why I decided to leave India and immigrate to the United States. I gave a Colloquium at MATSCIENCE on August 17 on "New weighted Rogers-Ramanujan partition theorems and products modulo 6 and 7". Two days before this, I went to the Raj Bhavan (Governor's Residence) for the Independence Day party. My father and mother would attend this party annually, and it was interesting that even after my father had passed away, this invitation was being sent to Ekamra Nivas! So I had to respond to it in his place. GP accompanied me to the Governor's party.

I departed Madras on August 20 afternoon by Jet Airways for Delhi where I boarded a KLM Airbus A-330 to Amsterdam and the United States. I was back in Florida on August 21 just in time for the Fall Term classes which were to start the next day.

10.5 Ramanujan 125 in USA and in India (2012–13)

The highlight of the academic year was the Ramanujan 125 celebrations both in America and in India. During the time of the Ramanujan Centennial in 1987, I had just made an entry into the Ramanujan world, but by the time Ramanujan 125 came, I had a lot of involvement both in terms of research and in terms of the preservation of Ramanujan's legacy. So I was deeply involved in several major events for this occasion. There were other important events during the academic year, and here I will discuss all of these along with the programs of Ramanujan 125.

At the start of the academic year, I received a letter from Eric Friedlander (AMS President) and Robert Daverman (AMS Secretary) informing me that I will be inducted as one of the Inaugural Fellows of the AMS in the Joint Annual Meetings in San Diego in January 2013. The idea to launch the AMS Fellows Program was proposed initially during George Andrews' term as the President. But many mathematicians who are very egalitarian expressed the view that this might create a "caste system" (my phrase) in the mathematical community and so they opposed it. Hence the Fellows Program was not approved in the initial round since it did not get a 2/3 majority vote. Subsequently, it was emphasized again that the program would encourage more mathematicians to become AMS members (since one has to be an AMS Member to be elected as AMS Fellow), and that mathematics departments would have more clout with the administration if they had Fellows on their faculty, like departments in other disciplines do. It was also agreed that the program could be launched with a simple majority vote, and this was achieved during the term of Friedlander who was the next AMS President. The first group of Fellows were called the Inaugural Fellows and I was pleased to be included in this group.

During September and October, I was busy with the proofs of my book *Ramanujan's Place in the World of Mathematics* [B1], planned for release in December, and the proofs of the Feature article on Ramanujan 125, due to appear in the Notices of the AMS also in December.

The first major conference for Ramanujan 125 was at the University of Florida during November 5, 6 and 7. Besides the funding we received from the NSF and NSA for this conference, we also had support from Penn State University. Thus Ae Ja Yee of Penn State University was a co-organizer of this conference with Garvan and me. We had about 70 active researchers as participants. The opening lecture by George Andrews (Penn State) set the tone for the conference. He was followed by Ken Ono (Emory) and so the conference had a lively start. Of course the two other members of the "gang of three" in the Ramanujan world, namely Richard Askey (Wisconsin) and Bruce Berndt (Illinois) gave hour lectures as well. Besides these four, we had plenary talks by Robert Vaughan (Penn State) on "The Hardy-Ramanujan-Littlewood Circle Method", Dorian Goldfeld (Columbia) on "Ramanujan Sums", and Doron Zeilberger (Rutgers) "Ramanujan as the greatest experimental mathematician". Other plenary speakers were Kannan Soundararajan (Stanford), Kathrin Bringmann (Cologne), Christian Krattenthaler (Vienna), and Gerald Tenenbaum (Nancy). Gerald cancelled his trip at the last minute since he suffered an accident and so I gave his talk on "The core of an integer", at his request using his slides. The refereed proceedings of the conference edited by Frank Garvan, Ae Ja Yee and me appeared in the Contemporary Mathematics series of the American Mathematical Society in 2014. Since I had finished correcting the page proofs on my book "Ramanujan's place in the world of mathematics" in September, the book was scheduled to appear in print in mid-November. So Elizabeth Loew of Springer very kindly had an advance copy on display in the Springer book exhibits at this conference. Mathura graciously hosted the conference party at our home on November 5, and since we had the 70 conference participants along with some of our faculty and their spouses as guests, we catered most of the Indian food for the party.

The Notices of the AMS Feature entitled "Srinivasa Ramanujan - going strong at 125" that I edited, appeared in two parts — four articles in each part — covering various aspects of Ramanujan's work, and the impact on current research. Part I appeared in the December 2012 issue and Part II in the January 2013 issue. I wrote the opening article in Part I on "Ramanujan's thriving legacy" and gave a report of the major happenings in the Ramanujan world since the Ramanujan Centennial in 1987 such as journals that have been launched, conferences that have been conducted, stage productions (movies, plays, operas...) on Ramanujan, prizes bearing Ramanujan's name, books on Ramanujan's mathematics and life, etc.

A few days after the Ramanujan 125 conference in Gainesville, Frank Garvan underwent heart surgery. He had borne the brunt of the arrangements for the conference; he did not want the surgery to impede his work for the conference and so he scheduled the procedure right after the conference. Hats off to his dedication! The procedure was successful and he started working from home a week later. But owing to the surgery, he could not make it to India for the Ramanujan 125 conferences. Since I was on the Scientific Committee of the conference in New Delhi, I volunteered to give Garvan's plenary lecture. So in the second half of November, I went to Garvan's house several times to discuss his slides so that I would be able to give his talk satisfactorily.

Just as M. S. Raghunathan led the effort for the NBHM (The National Board for Higher Mathematics) Conference for the Ramanujan Centennial in India, this time also for Ramanujan 125 he was in charge of the main NBHM conference to be held in India's capital New Delhi. In early 2012, he formed various conference committees and put Bruce Berndt as the chair of the Scientific Committee. I was invited to serve on this committee, to which I consented. One of the roles of the members of the Scientific Committee was to suggest speakers for the conference, and I helped with regard to speakers from outside India. Raghunathan and other mathematicians from India on the Scientific Committee decided on the speakers to be invited from within India. Sometime in mid-2012, when the conference program was being drawn up, Raghunathan wrote to me saying that he would like the SASTRA Ramanujan Prize to be given at the NBHM conference to be held from December 17 to 22 in New Delhi. The prize is given annually at SASTRA's Kumbakonam conference each year concluding on December 22, Ramanujan's birthday. So Raghunathan's request meant two things: (i) hold the SASTRA conference before the NBHM conference, and (ii) not give the prize in Kumbakonam, but in New Delhi instead in 2012. Naturally this caused some concern at SASTRA and so I put a call through to Vice-Chancellor Sethuraman to discuss Raghunathan's request and he graciously agreed to co-operate with the NBHM. It was decided that Vice-Chancellor Sethuraman would give the prize in New Delhi on December 22.

My mother and I departed Florida on December 10 on Delta Airlines via the Pacific. At Singapore we transferred to a Jet Airways flight and arrived in Madras on the morning of December 12. Mathura, my daughter Lalitha, and grandchild Kamakshi, arrived two days later.

The Ramanujan Legacy Conference at SASTRA University (Dec 14–15) and the award of honorary doctorates to Andrews, Askey and Berndt

SASTRA University wanted to do something special for Ramanujan 125 and asked for my suggestion. So in discussions with the Vice-Chancellor, I proposed that George Andrews, Richard Askey and Bruce Berndt be awarded Honorary Doctorates by SASTRA University at the Ramanujan Legacy Conference that SASTRA would hold during December 14–15, 2012. The Vice-Chancellor was delighted to hear this proposal, but said that he had to get this passed in the Senate (of SASTRA University); this endorsement came through as expected.

The invited speakers for the SASTRA conference besides Andrews, Askey, and Berndt, were Ken Ono (Emory), Zhiwei Yun (Stanford), S. D. Adhikari (Harish Chandra Institute), Atul Dixit (Tulane), Robert Schneider (Emory), Peter Paule (Linz), Ole Warnaar, (Melbourne) Michael Hirschhorn (New South Wales), and Wadim Zudilin (Newcastle, Australia) and me.

On December 13 morning, I departed Madras for Kumbakonam in the SASTRA van with many foreign delegates. Andrews, Berndt, and Joachim Heinze (Springer) arrived in Madras only that night and so they did not reach SASTRA until the mid-day of December 14. For this reason, the Convocation Ceremony in which the honorary doctorates were to be presented, was scheduled for December 15 morning. Another group of foreign delegates — Richard Askey, Ken Ono, Wadim Zudilin, and Robert Schneider — arrived by train in Kumbakonam very early in the morning of December 14, and so they were present for the first day sessions starting a few hours later. They were coming after attending a Ramanujan 125 conference in Mysore to which I was invited, but could not attend.

The Ramanujan Legacy Conference at SASTRA was inaugurated at 9:30 am on December 14 by Dr. Rajiv Sharma, Secretary of the Department of Science and Technology (DST). At 10 am, Ken Ono delivered the opening lecture of the conference within a few hours of arriving by the overnight train from Mysore. Since all the speakers had assembled in Kumbakonam by mid-day of December 14, I arranged the visits to Ramanujan's home, the Town High School, and the Sarangapani Temple, that evening.

Since Mathura, my daughter Lalitha and grandchild Kamakshi arrived in Madras only on December 14 around noon, GP stayed back in Madras to receive them. He then accompanied Mathura to bring her by car to Kumbakonam that night.

On December 15 morning, the entire front page of The Hindu, India's National Newspaper, was filled with photographs of: the three recipients of the Honorary Doctorates (Andrews, Askey, Berndt), all the speakers of the 2012 SASTRA Conference, as well as photos of all the SASTRA Prize winners including the 2012 prize winner Zhiwei Yun! This cover of The Hindu was arranged by SASTRA and it was fantastic publicity for SASTRA and the conference. The participants had never seen anything like this before, and each wanted a copy of the newspaper as a souvenir!

In a colourful ceremony, Andrews, Askey and Berndt were awarded Honorary Doctorates on the morning of December 15. The ceremony was held in the main auditorium of the Srinivasa Ramanujan Centre (SRC) of SASTRA University in Kumbakonam, where all conference lectures were given. The Trinity, dressed in full regalia, walked into the auditorium as Indian classical music was played on the *Nadaswaram*, a powerful wind instrument. All three enjoyed the ceremony, but Askey felt the music was too loud! The Nadaswaram music is always loud since it is played in Hindu weddings and festivals attended by hundreds to thousands of people! After the Honorary Doctorates were conferred, as part of the same ceremony, my book "Ramanujan's Place in the World of Mathematics" [B1] was released, and the first copies of the book were handed over to Andrews, Askey and Berndt by Thomas Hempfling of Springer-Birkhauser, Basel. My book is published by Springer India, and Hempfling, among his other responsibilities, is the head of Springer India. Sanjiv Goswami and Shamim Ahmad of the Springer office in Delhi, the editors who oversaw the production of this book, were present at this ceremony as well. Following the release of the book, a Special Volume of *The Ramanujan Journal* that I had edited for Ramanujan 125, was released, and Joachim Heinze of Springer, Heidelberg, presented the first copy to SASTRA Vice-Chancellor Sethuraman.

Since the ceremony and the photo session consumed much of the morning, all talks were in the afternoon, including those of Andrews, Askey and Berndt, and of 2012 SASTRA Prize Winner Zhiwei Yun. In the afternoon session, an alarming event happened which I now describe.

I was feeling a bit tired on the evening of December 14, and so I rested at Sterling Resorts where all participants stayed. I am mildly diabetic, and so on December 15 morning, I took my usual pills. However, I ate much less than normal that day. In the afternoon, while listening to Dick Askey's lecture, I felt very weak. I knew that my sugar level was dropping but I had my chocolates in a bag which I had left with Mathura and GP who were seated at the rear of the auditorium. So I got up from my seat in the first row to walk to the rear to get the chocolates. But then, as soon as I got up, I collapsed and fell headlong on the floor in full view of the audience of 500 in that auditorium to their shock and disbelief! Immediately, several of the SASTRA staff rushed to my side, and luckily, one of them thrust some toffees into my mouth when I asked for something sweet. In a few seconds I felt better much to the relief of everyone. Dick Askey said that in his entire career, he had never witnessed a scene such as this! That scene added a bit of drama to a conference which was memorable in many ways.

We all departed Kumbakonam on the morning of December 16. Many of us including me, who were going to Delhi for the NBHM conference to start the next day, were dropped off at Madras Airport to catch our evening flight to Delhi. Mathura was not accompanying me to Delhi, and so she was concerned I might collapse in Delhi as well! I assured her that this would not happen. After that alarming event in Kumbakonam, I make sure not to take my normal dose of diabetic medicines when I am eating less, to ensure that my blood sugar does not drop to dangerously low levels.

The NBHM Conference in Delhi (Dec 17–22), and the award of the 2012 SASTRA Ramanujan Prize to Zhiwei Yun

Upon arrival at Delhi, we were received at the airport by the local organizers and taken to Hotel Park, a comfortable 4-star hotel where all speakers were put up. After checking in, when I called Mathura at night to say that I am comfortably settled, she gave me the alarming news that Amritha, who was supposed to have departed Atlanta for India a few hours ago, had called from the airport to say that she could not board the flight because her Indian visa had expired, and she had not realized that! So I called Amritha from the hotel, asked her to catch the next flight to Houston, go to the Indian Consulate the next day and get her ten year visa. I assured her that I will call Delta the next day and change her flights to India suitably. All this was going on during the night before the Delhi conference was to start and you can imagine my tension. But there was something else going on in Delhi which caused tension to all citizens and visitors — Delhi was in the grip of gang rapes for the past few weeks! The situation was so alarming and reported all over the world that Kathrin Bringmann called me a few days prior to the conference asking whether she should call off her trip! I assured her that there is nothing for the participants to worry about, and so she came.

The Legacy of Ramanujan Conference of the NBHM was inaugurated on the morning of Monday, December 17, at the Viceregal Lodge on the beautiful and sprawling campus of the University of Delhi. The lectures were all held at the nearby Conference Centre. Ken Ono was the opening speaker and he gave an electrifying start for the program. There was a nice book exhibit at the conference, and I was pleased that at the Springer booth, my book on Ramanujan, the book on my father's legacy, and the special volume of The Ramanujan Journal for the 125th year of Ramanujan's birth, were all on display. When I met Shamim Ahmad at the Springer booth, I told him that the next day, before the sessions start, I would like to go to his office to make a phone call to Delta to change Amritha's ticket. He kindly agreed to this request. That night Amritha called me from Houston to say that her visa will be issued the next day. So knowing that the Amritha would get the visa, I called Delta and changed her reservations with December 19 as her new departure date.

With Amritha's travel plans sorted out, I was in a relaxed mood to enjoy the conference lectures on the second day (Dec 18). My plenary talk on behalf of Garvan was on December 19 morning and it came out well because I had sat with Garvan to prepare for it. Many of the participants, senior and junior, congratulated me after the talk. I think they did not have high expectations of my performance in Garvan's place, and were pleasantly surprised that I did well! That afternoon I was asked to be chair at the plenary talk of Robert Vaughan who spoke on "Partitio Numerorum", a term used by Hardy and Littlewood for their series of papers on additive number theory using the circle method. Vaughan is the world's authority on the circle method, and this was confirmed by the thoroughness and quality of his presentation. That night I received a message from Amritha that she was boarding her flight to Tokyo from Atlanta.

Sanjiv Goswami, the head of the Springer India office in New Delhi was very kind to host a dinner for me on the night of December 20 at a fancy restaurant. His colleague Shamim Ahmad in Delhi, and Joachim Heinze from the Heidelberg office of Springer, joined us for dinner.

The first plenary talk on December 21 afternoon was by George Andrews and it had the title "Ramanujan's magic". In his inimitable style, he took us on a tour of some of the most startling identities of Ramanujan that are engaging our attention today. My own plenary talk was in the afternoon and I spoke on "Some remarkable partial theta identities of Ramanujan, Andrews and Rogers-Fine". That night there was a special dinner for all the plenary speakers on the lawns of the home of the Vice-Chancellor Prof. Dinesh Singh of Delhi University.

December 22, Ramanujan's birthday, was the final day of the conference.

There were plenary talks in the morning by Manjul Bhargava, Ole Warnaar, Zhiwei Yun, the 2012 SASTRA Prize Winner and Richard Askey. After this final set of lectures, M. S. Raghunathan, the convener of the Conference, and in his capacity as the President of the Ramanujan Mathematical Society, announced that the refereed proceedings of the conference would be published by the Society. The concluding session of the conference on December 22 afternoon was the Award Ceremony for the SASTRA Ramanujan Prize to Zhiwei Yun of Stanford University.

Zhiwei Yun had made fundamental contributions to several areas that lie at the interface of representation theory, algebraic geometry and number theory. His PhD thesis on global Springer theory at Princeton University, written under the direction of Professor Robert MacPherson of The Institute for Advanced Study, opened up new vistas in the Langlands program. Yun had also done significant work in algebraic geometry. His work on the uniform construction of motives with exceptional Galois groups is considered to be very fundamental.

Zhiwei Yun was born in Changzhou, China in 1982. He showed his flair for mathematics early by winning the Gold Medal in the 41st Mathematical Olympiad in 2000 in Korea. He received his PhD from Princeton University in 2009. He was a Visiting Member at the Institute for Advanced Study in 2009–10, and held the C. L. E. Moore instructorship at MIT during 2010–12. In fall 2012 when he received the SASTRA Prize, he had just joined the mathematics faculty at Stanford University.

As at SASTRA University, I was asked to read the citation of the prize in Delhi. SASTRA Vice-Chancellor R. Sethuraman was on stage to give the $10,000 cheque to Yun. The award plaque to Yun was given by Jitin Prasada, Minister of State for the Government of India. The SASTRA Prize Ceremony was a fitting conclusion to the NBHM Conference. That night I returned to Madras from Delhi.

I needed a vacation after this hectic academic schedule, and so on December 23 morning, I departed Madras for Coimbatore for a brief holiday to be spent with Mathura's younger brother Sriram and his family. As I was finishing up with the conference in Delhi, Amritha had arrived in Madras, and therefore she, Mathura, my elder daughter Lalitha and my granddaughter had left Madras for Coimbatore two days before. My son-in-law Aditya also arrived in Coimbatore from Florida a few hours after I reached there. We all had a fun-filled holiday with the highlight of the sightseeing being a visit to the hill station of Coonoor.

Upon return from Coimbatore, we had a party at Ekamra Nivas on December 26, for our friends and relatives to see our granddaughter Kamakshi.

The main reason that Lalitha, Aditya and Kamakshi came to India was to visit the temple of Thirupathi, 100 miles north of Madras, for Kamakshi to be tonsured — that is have her head shaven! It is Hindu belief that for one's health and well being, one should be tonsured at a temple at least once in life. Especially in the case of girls, and for most boys, tonsuring is done at a very early age. Kamakshi was not yet two years old. The temple in Thirupathi is where most children get tonsured. Lord Venkateswara (Lord Vishnu, the Protector), is the deity in Thirupathi, and our family deity. So it was especially significant that my granddaughter was tonsured at that temple. This was done on December 27. We had arranged the party one day before so that Kamakshi would have a full head of hair when all our friends and relatives would see her for the first time!

The three day Golden Jubilee Conference at MATSCIENCE was scheduled for January 2 to 4, 2013. On New Year's day morning, The Hindu newspaper gave a prominent announcement of the conference. It was inaugurated by Mr. M. R. Srinivasan, a nuclear scientist who was Chairman of the Atomic Energy Commission of India in the late eighties, and was on the Board of MATSCIENCE at that time. He knew my father well and had followed the development of the institute. After his inaugural speech, George Sudarshan who followed my father as MATSCIENCE Director, shared his thoughts on the occasion. After the inaugural ceremony, I gave the first lecture (one hour) on "Alladi Ramakrishnan's Theoretical Physics Seminar and the creation of MATSCIENCE". It was a slide show and a historical tour from 1950 to 62. Mr. Srinivasan stayed back to attend my talk and said he enjoyed it immensely. Many younger members of MATSCIENCE who were not familiar with its origins came up to me and said that they enjoyed hearing about the glorious beginning of the Institute. After lunch, I attended the talks by my father's former PhD students — Professors Mathews, and Devanathan, both retired by then. They recounted their experiences in my father's Theoretical Physics Seminar. I was pleased to note, that outside the Chandrasekhar Auditorium where these lectures were held, in the long corridor, there were historical pictures of my father's Theoretical Physics Seminar and of the Institute printed on the wall, including several that I had given in August 2012. Also, a commemorative booklet of the Institute was released that day, containing some more pictures from my father's collection. I could not attend the conference on January 3 and 4 because I departed with my family for the United States on January 2 night.

Since both my daughters were with us, Mathura and I decided to have a vacation with them in Kota Kinabalu in Malaysia on our return to the USA. We departed Madras late at night on January 2 by Malaysia Airlines and arrived in Kuala Lumpur early in the morning to catch a connecting flight. Prior to landing at Kota Kinabalu we had a spectacular view of coastline of the island of Borneo and of the volcanic peak of Mount Kinabalu towering over the clouds. It was a delightful stay for two nights at the Le Meridien in Kota Kinabalu. Our sightseeing included tours of the city and a visit to the Kinabalu National Park and the rain forests of Borneo. The highlight was the amazing view of the majestic Mount Kinabalu as the clouds cleared over the rain drenched forests surrounding the mountain. As we descended from the mountain slopes, we stopped at the Poring Hot Springs where we all enjoyed a cleansing bath in the natural springs. We departed Malaysia on January 5, and after a night stay at the Transit Hotel at Changi Airport, Singapore, we were back in the United States the next day.

After a day's rest, Mathura and I departed Gainesville early in the morning of January 8 for San Diego via Atlanta for the Annual Meeting of the AMS. We arrived in San Diego by noon, and so we spent some time with our friends the Ramanathans, at their home in the lovely town of Del Mar near San Diego before checking in at the Marriott Marquis Hotel right on the marina and attached to the Convention Center. That night, during a dinner for me and Mathura that the Springer Editors Elizabeth Loew, Marc Strauss and Joachim Heinze hosted, we discussed the enhancement of the Ramanujan Journal and the book series DEVM. The next day, my distinguished cousin V. S. Ramachandran (a professor of Neuroscience at UC San Diego) and his wife Diane, hosted a lunch for us in Del Mar.

During the next two days, I attended the talks in the Special Session on "The influence of Ramanujan on his 125th birthday" organized by George Andrews, Bruce Berndt and Ae Ja Yee. The special session had a fine collection talks on a wide range of topics such as partitions, q-hypergeometric series, congruences, ranks and cranks, modular forms, and mock-theta functions. My talk on the second day of this special session was on a generalized Lebesgue identity in Ramanujan's Lost Notebook.

Our good friends Edward Bertram (retired from the University of Hawaii) and his wife Alice were attending the meeting and so we got together for lunch. That afternoon I attended the Board Meeting of the Notices of the AMS where my proposal to have a feature on Paul Erdős for his Centenary was approved. Steve Krantz the editor of the Notices requested me to edit this feature to which I agreed.

On the night of January 11, there was the Dessert Reception for the Inaugural Fellows of the AMS. Mathura and I attended this along with the Bertrams (Ed Bertram was also inducted as an Inaugural Fellow). It was a very well arranged reception with classical music on the piano being played as the Fellows were in conversation. George Andrews who was there at the reception, hosted a dinner before that for all the students of his 1992–93 partitions class at Penn State. I had attended that course during my sabbatical at Penn State in 1992–93 and so I was invited to that dinner.

George Andrews arrived in Florida in mid-January after the AMS Annual Meeting, and delivered a few lectures in the Number Theory Seminar in February. During the first week of March which was our Spring Break, I went to Auburn University at the kind invitation of Professor Narendra Govil to deliver a Math Club Lecture. Govil had been organizing these lectures of wide appeal to undergraduate students for many years with funding from the Provost's office, and he had invited some high profile speakers before, including George Andrews, who spoke about Ramanujan. So I gave a talk instead on "Paul Erdős at 100" to an audience of about a hundred undergraduates. That night, the Govils had me over at their home for an Indian dinner for which some mathematics faculty were invited. The next morning, I had a meeting with Govil and the Associate Provost of Auburn University, following which I departed Auburn.

Right after Spring Break, we had the visit of Steve Milne (Ohio State), who addressed both the Colloquium and the Number Theory Seminar the next day. Later in March, the great mathematician turned physicist Freeman Dyson (Institute for Advanced Study) came to deliver the Ramanujan Colloquium and seminars. We had invited Dyson for the Ramanujan 125 Conference in Gainesville in November 2012, but he could not come then. So he agreed to come for the Ramanujan Colloquium.

Dyson arrived on Sunday, March 24 afternoon. I picked him up at the airport, and took him to dinner at an Indian restaurant where Frank and Cyndi Garvan joined us. He enjoyed the food, but ate very sparingly, which is the reason he was healthy and vibrant at the advanced age of 89!

Dyson was a legend in the scientific world and so for his Ramanujan Colloquium on Monday, March 25, it was a full house. He spoke about "Playing with partitions" and told us the story of how he discovered the combinatorial explanation for two of Ramanujan's famous congruences by the idea of the rank of a partition which is the largest part minus the number of parts. Dyson was an undergraduate in Cambridge at that time and working under Hardy. He published his remarkable findings in the

Cambridge undergraduate journal *Eureka* in 1944. Dyson by nature was very frank and forthright — brutally frank sometimes. In this lecture, as he was talking about Ramanujan and Hardy, he shocked everyone by saying, "I hold Hardy personally responsible for the death of Ramanujan!" Dyson went on to say in the lecture that Ramanujan needed a warm and considerate friend, but Hardy was cold and aloof. He pointed out, that when he showed Hardy the combinatorial explation of Ramanujan's congruences by means of the rank, Hardy gave him the cold shoulder!

Dean Paul D'Anieri of the College of Liberal Arts and Sciences, who viewed Dyson as a demi-God, was thrilled to hear that Dyson would deliver the Ramanujan Colloquium. So he enthusiastically agreed to give the Opening Remarks for Dyson's lecture. That night, we all had dinner with Dyson at The Hilton. In a speech after dinner, I recalled how my father had written to Dyson in 1972 seeking his opinion on my fledgeling research in number theory as an undergraduate. I turned towards Dyson and said that he responded to my father's letter saying that my work as an undergraduate showed promise, but a bright undergraduate should do physics which is more serious instead of number theory which is recreational!! He smiled and nodded as I said this. I then apologized that I did not follow his advice but continued in number theory instead! I do not think Dyson was correct in judging number theory as recreational, and most mathematicians would not agree with Dyson's assessment of number theory.

The next day Dyson gave a Number Theory Seminar on "The prisoner's dilemma". That night Mathura and I hosted a party in his honor at our house attended by faculty and their spouses from the mathematics, physics and statistics departments. We had several graduate students at the party, and Dyson was in a relaxed mood conversing with them.

Dyson's third talk was on Wednesday, March 27 in the physics department on the theme "Are graviton's in principle detectable?" He started the lecture by saying, "I hate dogmas and I always question them" in a thunderous voice that one would not expect from someone nearly of age 90! The physics auditorium was overflowing with students sitting on the floor in the aisles, and some standing in the rear. The hall had more individuals than the Fire Marshall would permit, but there was no interference from the police. The physics department had for many years tried unsuccessfully to get Dyson, and so the physics chair was delighted when I called him and offered one of Dyson's lectures to be arranged in the physics department.

After Dyson returned to Princeton the next day, I sent him a letter of thanks to which he replied within minutes. My email to him and his response are given below:

==

Alladi, Krishnaswami

Sat 3/30/2013 3:28 pm

Dear Professor Dyson,

What a fantastic visit it was! Your three talks (each on a different topic) were magnificent. The physics professors told me that they do not remember when on a previous occasion that hall was filled and overflowing. My abstract algebra students thanked me for suggesting that they should attend your talks. The students in my graduate number theory course appreciated the time you spent in conversation with them during the party at my house. Dean Paul D'Anieri thanked me for the opportunity he had to meet you. On our part, Mathura and I were delighted to have you at our home.

With Andrews' support we are getting truly outstanding speakers for the Ramanujan Colloquium. We are honored to have had you as a speaker in this colloquium series.

Warm regards,

Krishna

==

Freeman Dyson < *dyson@ias.edu* >

Sat 3/30/2013 4:43 pm

Dear Professor Alladi,

Thankyou for your message, and for your hospitality, and for your efficient organization, and for inviting me to come to celebrate Ramanujan. I enjoyed enormously the meetings with old and new friends, and with students in particular. I had an easy trip home, and today I am back at the Institute dealing with the 214 E-mail messages that arrived during my absence. Thankyou once more for the glorious holiday in Florida.

Yours ever,

Freeman Dyson

==

During the middle of April, we had the visit of Michael Zarky and Phoebe Liebig from Los Angeles. Phoebe (recently deceased) was the daughter of the great mathematician Marshall Stone. She retired as Professor of Classics at the University of Southern California. Michael Zarky produces musical instruments and has a deep interest in Carnatic music. His hobby is bird watching, and so I took him to Paines Prairie in Gainesville where he could take photographs of a wide variety of birds. Michael and Phoebe were in Madras in 1989 soon after the demise of Prof. Stone (a great friend of our family) at the Madras Woodlands Hotel, and during that trip, they met my parents at Ekamra Nivas. That is how our association with Michael and Phoebe started, and we have been in touch every since. We had them at our home for dinner to which we invited some of our friends who were close to our parents as well.

As soon as the Spring term classes ended on April 24, Mathura and I departed for Israel the next day on a one-week visit in response to an invitation from Professor Hershel Farkas of the Hebrew University in Jerusalam. Israel was as much an eye opener for us as Turkey was. The level of mathematical research in Israel is exceptionally high.

We arrived in Tel Aviv in the late afternoon of April 26 in gorgeous weather. We checked in at the Hilton Tel Aviv and were given a room with a balcony on a high floor from where we had a splendid view of the city and the coast. After enjoying dinner in the lounge on the top floor, which commanded a fine view of the beach below, we strolled along the beach to the port of Tel Aviv nearby. The next day was sabbath because it was a Saturday. It was interesting for us to note that one of the elevators was programmed to stop at every floor of this high rise hotel, because during sabbath, the orthodox jewish people would not operate the electrical switches! We happened to get into one of those sabbath elevators and it took a long time to get the floor where our room was located.

Since most places were closed on Saturday due to the sabbath, at the advice of the Concierge, we walked to the nearby old city of Jaffa to see its ancient and architecturally beautiful buildings on the Mediterranean coast. In the afternoon, we visited downtown Tel Aviv, saw the parliament building from the outside, and the plaza where former Prime Minister Yitzhak Rabin was assassinated in 1995. We were surprised to learn that Israel which is supposed to have the tightest security, could not prevent the assassination of its prime minister! The next morning, after seeing the Diaspora Museum on the campus of Tel Aviv University, we took a taxi to Jerusalem.

It was a pleasant drive from Tel Aviv to Jerusalem where we arrived at noon at the Crowne Plaza Hotel, the place of our stay. In the afternoon, we visited the Israel Museum, where the most impressive display was the Dead Sea Scrolls. That night, Hershel and his wife Sara took us to dinner at a lovely vegetarian garden restaurant on the outskirts of Jerusalem.

My talk at the University was only on Thursday, and so on Monday, Hershel and Sara took us out on an all day excursion to Masada, the famous 2000 year old fort built by King Herod which turned out to be the last stand of the Jewish rebels against the Romans. Mathura and I had seen the TV mini-series Masada starring Peter O'Toole as the Roman leader Lucius Flavius Silva, and so we looked forward to this trip. The historic Masada fort is set atop a hill to which one could either walk up, as most energetic younger tourists did, or take a cable car, which is what we did. Not only is the fort spectacular, but so is the view from the top of the valley below, of the Dead Sea in the distance, and of Syria beyond. It was searingly hot at Masada where it was stone all around. So after coming down, we had a refreshing drink at the Visitors Center where we also purchased some Dead Sea cosmetic and skin care products, which are sold these days in American shopping centers. On our way back from Masada to Jerusalem, we stopped at a beach where I had a bath in the Dead Sea which is so salty, that even I, who does not know how to swim, could not sink! That night we had dinner at the Indian Restaurant Kohinoor at the Crowne Plaza, but the food was just average in taste.

The next morning (Tue, April 30) we went to Bethlehem, the birth place of Jesus. Bethlehem is in the West Bank and is under Arab occupation. So our Jewish taxi driver dropped us at the entrance to the walled city saying that he would pick us up about three hours later. We had made arrangements to be met by a Palestinian taxi driver inside the West Bank, but I must say that it was scary to enter the walled city, go past several barbed wire fences, and past the guards with their menacing rifles! What a relief it was when we emerged out of this high security entrance, our cheerful taxi driver shouted, "Hello Mr. and Mrs. Alladi!" He was a very friendly chap who took us straight to the Church of Nativity in Bethlehem. There was a mile long queue to enter the church, and we were concerned how long it would take for us to see the church. But the smiling taxi driver said that he had made arrangements for us to enter the church from the Exit side (!!) and the spot where Jesus was born, is right next to the Exit! After seeing this holy spot where Jesus was born, we strolled in Bethlehem outside the church and bought a nice souvenir of the Nativity scene made of olive wood. As we emerged out of the walled city of West Bank, we were equally relieved to see our Jewish taxi driver waiting to take us back to our hotel. It was indeed a very memorable experience.

Upon return from Bethlehem, I went to the Hebrew University in the afternoon to have discussions with Hershel Farkas on partitions and theta functions. He is a world authority on theta functions and Riemann surfaces and has authored a book with Irwin Kra on Riemann Surfaces, considered a bible in this field. More recently Farkas and Kra have published a book on theta functions in which applications to number theory, especially to the theory of partitions, are emphasized. Even though Farkas had retired, he is active in research, and the Hebrew University allowed him to continue to use his large office. After our discussion, Farkas took Mathura and me to the Israel Museum where we were met by his Sara who worked there. She showed us the Herod Exhibit which was outstanding. At the museum store, I bought a book on Masada, the great fort of King Herod. The day concluded with a party at the home of Hillel Furstenberg.

Furstenberg is a world famous mathematician, who was awarded the 2020 Abel Prize for his monumental contributions to ergodic theory and allied areas. As a number theorist, I am most familiar with his famous ergodic theory proof of the great theorem of Szemeredi on arithmetic progressions. Furstenberg is the PhD advisor to many eminent mathematicians: Alex Lubotzky, Vitaly Bergelson, and Yuval Peres, to name just three. At his house, I met several of his eminent colleagues, including 2010 Fields Medalist Elon Lindenstrauss, and Alex Lubotzky. I found Furstenberg to be very modest and soft spoken. In view of my interest in number theory, Furstenberg told me about his topological proof of the infinitude of primes that he found as a graduate student. This lovely proof is included in "Proofs from The Book" (Springer) which is God's book of the most elegant proofs of the most fundamental theorems.

We spent all of the next day in the Old City of Jerusalem, which has an Arab section, a Jewish section, and a Christian section. We saw the Church that has been built on the spot where Jesus was crucified, the tomb of King David, and the room where Jesus had his Last Supper. Historic as all this was, we were also saddened by what we saw. Our spirits were lifted later that evening when Farkas and Sara hosted a party at their home for us. They had invited a number of their colleagues and their spouses, and had made several vegetarian dishes for us.

The next morning, Mathura and I visited the Holocaust Museum and were touched to see the horrific scenes of World War II. That afternoon (May 2), I gave my colloquium at the Hebrew University on "A theorem of Göllnitz and its place in the theory of partitions". Lindenstrauss was the Colloquium Chair, and so the invitation letter for the colloquium came from him. But as is the custom, my host Farkas introduced me. It was an honor to give the talk at the Albert Einstein Institute of Mathematics in the Hebrew University with luminaries like Furstenberg and Lindenstrauss in the audience.

That night after my colloquium, we took a taxi from Jerusalem to Ben Gurion Airport, Tel Aviv, to board our Delta flight to New York. The security at the airport was unbelievably thorough, and so we were glad that we arrived there three hours before departure. Looking back, it was a most exhilarating and memorable trip, both mathematically and culturally.

One day after our return from Israel was the MBA graduation ceremony of my son-in-law Aditya at Rollins College, Orlando. He was working at Siemens, Orlando, as a mechanical engineer, but decided to study in the evenings for an MBA degree, which helped him rise in the administrative ladder. It was a pleasant surprise to meet at the graduation ceremony, Mr. Ram Mohan Rao, the former Governor of Madras, who was a student of my father when he (Rao) was doing his Masters at the Presidency College in the fifties. My father had taken me with him to meet Ram Mohan Rao when he was a Governor of Madras, and it was interesting to see him now as a common man. Even as a Governor, he was unassuming, and so nothing had changed in his demeanor.

On May 8, Mathura and I departed Florida for India flying transpacific. In India, my main academic assignment was a brief visit to the Tata Institute of Fundamental Research (TIFR) in Bombay (now Mumbai).

During the Ramanujan 125 conference in New Delhi in December 2012, one evening, on the bus ride back to the hotel from the conference site, I was seated next to Dipendra Prasad, Dean of the School of Mathematics at TIFR. Dean here means Chair or Head of the mathematics department (school). Dipendra asked me when I had last visited TIFR. I responded saying that it was in 1989 at the invitation of the late Professor Ramachandra. He immediately said that was too long ago, and so he would like me to visit the TIFR the next time I would come to India. I told him that I would be back in May 2013. On hearing that he said that I should visit the TIFR, and he promptly send me an invitation which I accepted.

I departed Madras for Bombay on the first flight in the morning of May 21 by Air India. The drive from Bombay Airport to TIFR is more than hour since one has to cross the entire width of this teeming city to reach the Tata Institute which literally is at lands end. I was accommodated at the comfortable Ramanujan Guest House of the TIFR, and met with Dipendra Prasad and Sankaranarayanan for lunch. We had good discussions about mathematics at TIFR and in India. I was given a nice office overlooking the sea, and during the few days that I was there, I prepared the slides for the lecture I was to give at the Erdős Centennial Conference in July. As I walked through the corridors of the School of Mathematics, I had a strange feeling because I did not see on the office doors any of the names of senior mathematicians that I have always associated with TIFR, and the reason was clear. Real estate in Bombay is almost non-affordable unless you have ancestral property. All these mathematicians had retired from the TIFR, and when they retired, they could not stay in the TIFR apartments that were allotted to them. So they left Bombay altogether and settled elsewhere. Thus even though they were Emeriti, they were not living in Bombay and coming into the TIFR. Hence they no longer had offices at TIFR. The TIFR School of Mathematics had a new breed of active researchers, and I had discussions with some of them.

On the night of my first day in Bombay, I had dinner at the posh home of my boyhood friend C. S. Narayanan, alias Ravi. He is one year older to me, and was our next door neighbour in Madras. So we have grown up together, played together — cricket at Ekamra Nivas, and shuttle cock at his house. Ravi now is settled in Bombay as a prosperous boss of a company in the shipping industry. He sent me his chauffeured Mercedes to pick me up. It was a very pleasant evening at his house where I enjoyed meeting his charming family. As a school boy, we always admired his boundless optimism. He would never complain about difficult conditions in India. It was not surprising to me that he has been successful in his line of work.

The next day, I had long discussions with Sankaranayanan about classical analytic number theory at the Tata Institute after Professor Ramachandra had retired and passed away. Sankaranarayanan is the lone carrier there of Ramachandra's torch, and he gave me an account of his research on the Riemann zeta function and Dirichlet series. That night, Dipendra Prasad invited me to dinner at his apartment. He prepared a fine Indian meal by himself. He had invited a few colleagues to join us for dinner and the conversation was about the SASTRA Ramanujan Prize, especially because Dipendra had written letters in support of various candidates.

On my last full day at TIFR (May 24), I gave a colloquium on the theorem of Göllnitz and its place in the theory of partitions. There was a sizeable audience, but no one there really worked in the theory of partitions. So I did not give any technical details, although I did discuss the type of tools needed to establish the theorems. I was back in Madras the next morning by the first flight.

On the day before my departure from Madras to the United States, the page proofs of the volume "Quadratic and higher degree forms" that I was editing along with Manjul Bhargava and Pham Tiep was delivered to me at Ekamra Nivas. Although such proofs are nowadays sent by email, I prefer hard copies. The convenience here was that the typesetting of our volume was being done in Madras, and so it was easy for Springer to arrange for the proofs to be delivered to my house.

Mathura, my mother and I departed Madras on June 6 night for Honolulu via Singapore and Tokyo, and on the 7 hour flight from Singapore to Tokyo, I spent time usefully by checking/correcting the page proofs — time better spent than watching movies! After a delightful vacation in sunny Honolulu, we were back in Gainesville on June 12.

The main academic events for the rest of the summer, were (i) The Erdős Centennial Conference in Budapest during June 30–July 6, and (ii) The Andrews 75th Birthday Conference in Tianjin, China, during August 1–5. These conferences are described in detail in §11.5 and §11.4.

10.6   My second sabbatical (2013–14)

During 2013–14, I was on a fully paid sabbatical. On my previous sabbatical in 1992–93, I received only half-pay from the University of Florida. The other half of my salary was paid by The Pennsylvania State University, where I taught one course during each of the two semesters. Half-pay sabbaticals are automatically given after every six full years of service. To get a fully paid sabbatical for the whole year, one needs to have put in several years of service to the University. When I applied for the fully paid sabbatical in 2013–14, I had served the University for more than 25 years. I had no teaching duties in 2013–14 and so I got a lot of writing done.

At the start of the academic year, on August 28, Rochelle Kronzek, a senior editor in the USA for World Scientific Publishing Company (WSPC), came to my office to meet me. We had very fruitful discussions on two possible projects — (i) for me to prepare an edited version of my father's autobiography "The Alladi Diary", and (ii) for me to write my own academic memoirs, namely this book! It was agreed that I would finish (i) first and then take up (ii). It was that visit of Kronzek which started my close and fruitful association with WSPC. She very kindly sent a letter of introduction to Dr. K. K. Phua, the Chairman, and his son Max Phua, both at the WSPC head office in Singapore.

In view of my sabbatical, I spent one month (Oct 22–Nov 21) visiting Penn State University at the kind invitation of George Andrews. As on my previous visit to Penn State in Fall 2005 during my research leave, I again stayed at the Days Inn, close to campus. I am always very productive whenever I visit Penn State University. On this visit, I wrote my paper "Partitions with non-repeating odd parts and combinatorial identities", which was accepted by the *Annals of Combinatorics*. Also, I worked on the proofs of my paper for the Ramanujan 125 Conference Proceedings to be brought out by The Ramanujan Mathematical Society.

My first talk at Penn State was in the Number Theory Seminar on October 24 on "Multiplicative functions and small divisors". The year 2013 was the centenary of Paul Erdős and so I chose to speak about this joint work of mine with Erdős and Jeff Vaaler.

October 26 was the celebration of Deepavali (or Diwali) — the Hindu Festival of Lights — and so my elderly friend Mrs. Vedam (now deceased) took me the dinner organized by the India Association of State College. There I met many of my old Indian friends, and in particular Paromita (Pashka) Chowla. It was sad to see Pashka, retired from the Penn State Mathematics Department, on a wheel chair.

On October 31, I delivered the MASS Colloquium on the theme "Paul Erdős at 100", and it was well received by the dozen bright students of the MASS Program. As on previous extended visits to State College, I had to check out of the Days Inn during a football weekend. So, like before, I returned to Gainesville during the weekend November 2–3, and during the flights on this trip, I worked out some interesting connections between basis partitions and partitions with non-repeating odd parts that I incorporated into the paper I was writing at Penn State. I spoke about this in the Andrews Partitions Seminar upon return to Penn State from Florida.

At Penn State, I completed one more paper "On a multi-dimensional extension of Sylvester's identity". This was work that had been bottled up for nearly a decade, and finally, was written up at Penn State, like opening a bottle of wine after it has aged well! George Andrews provided a q-hypergeometric proof of the two-dimensional version that I had found and proved combinatorially, and this enabled me to provide combinatorial and q-hypergeometric proofs of the multidimensional version. This paper appeared two years later in the International Journal of Number Theory (IJNT) published by World Scientific. Since I was by myself at State College, I was leading a bachelor's life — working long hours in the department and coming back to my hotel at bed time.

In mid-November, I addressed the department colloquium. My talk "Andrews and the Göllnitz theorem" focused on the insights of George Andrews in understanding this deep theorem. At the end of my talk, I presented the book "Combinatory Analysis" to the department Chair Nigel

Higson. This book, dedicated to Andrews, had appeared in my book series *Developments in Mathematics* (Springer). In return, Andrews gave me as a gift a copy of "Vorlezungen uber Zahlentheorie", the classic text of Edmund Landau, that was in the possession of Gollnitz' father!

I had to check out of the Days Inn again for the weekend after my colloquium, and so I returned to Gainesville for just two days. Back in State College for the final week of my visit to Penn State, I was able to complete the two papers I had started writing there in October.

After the assignment at Penn State, I joined my family for a holiday in Key West during the long Thanksgiving weekend. One of the main attractions at Key West is the home of the great novelist Ernest Hemingway which we enjoyed seeing. But to me, the aviation enthusiast I am, Key West is significant because the great Pan American World Airways (Pan Am) started service in 1927 with a flight from Key West to Havana. I was disappointed that the old Pan Am House in Key West is now a seafood restaurant called the Coral Reef, with just a Pan Am board nailed to a tree! So much for preserving the memory of the greatest airline that ever existed. During our stay at Key West, we visited Miami for a day. There we went to an Aviation Model Store where Lalitha's father-in-law, Captain Srinivasan, bought me a beautiful 1:100 scale model of a Pan Am Boeing 707. To match this thoughtful gift, my son-in-law Aditya bought me a lovely scale model of a Northwest Boeing 747-400. The significance of these two models is that Pan Am was the launch customer for the 707, and Northwest the launch customer for the 747-400. Both airlines do not exist any more, but these models, displayed at my home in Gainesville, preserve their memories. In my opinion, the Pan Am House in Key West ought to have been kept as an aviation museum instead of being converted to a seafood restaurant.

Back in Gainesville from the Thanksgiving break, we had the visit of Gerald Tenenbaum of the University of Nancy, France. He could not come to our Ramanujan 125 conference in Florida in November 2012, and so like Dyson, he visited us later. On December 2, Tenenbaum delivered the Erdős Memorial Lecture (since it was the Erdős Centenary) on "Divisors", a topic which was dear to Erdős, and on which Tenenbaum is a world authority. Indeed the book "Divisors" by R. R. Hall and Tenenbaum, published by Cambridge University Press, is the standard reference on this topic.

The next day, Tenenbaum delivered the Number Theory Seminar on "Friable Fourier series", namely Fourier series with non-vanishing coefficients a_n, where n is a "Friable number" — a term coined by Tenenbaum for numbers all of whose prime factors are small. By the bye, Tenenbaum is also the world authority on Friable Numbers.

Tenenbaum and his wife had hosted us so graciously during our visits to Nancy. So on our part, Mathura and I had a party at our home in their honor, and it was attended by many of my colleagues and their spouses. I took Tenenbaum and his wife to Paines Prairie and they were stunned to see so many large alligators basking on the shores of the lake just a few feet from the walk path. During our Thanksgiving visit to South Florida, we did visit the famous Everglades National Park, but there we did not see as many alligators as we did at Paines Prairie!

Trip to India and Singapore and the 2013 SASTRA Ramanujan Prize to Peter Scholze

Two days after the Tenenbaums left, Mathura, my mother and I departed for California on our way to India. Our close friends Ramanathan and Sakuntala had just celebrated their 50th wedding anniversary and so we called on them in San Diego. Any time we visit Southern California, my dear friend Nadadur Sampath Kumar (Sampath) insists that we spend a few days with him. So we stayed in his magnificent mansion in Brentwood. Since he and I were college mates at UCLA in the seventies he and his wife Vatsala invited some of our old friends for dinner at his house while we were there. After this stay in Los Angeles, Mathura returned to Orlando to be of assistance to Lalitha who was in her last few weeks of pregnancy with her second child. My mother and I proceeded to India.

We arrived in Madras on December 11 and a few days later, I delivered the S. S. Pillai Memorial Lecture at The Ramanujan Institute on "Multiplicative functions and small divisors" because Pillai had worked on multiplicative functions. Pillai was one of India's most distinguished mathematicians in the post-Ramanujan era, and so I felt honored to deliver this lecture in his memory.

The main academic event during my three-week stay in India was the award of the 2013 SASTRA Ramanujan Prize to Peter Scholze of the University of Bonn. Scholze had made revolutionary contributions to several areas at the interface of arithmetic algebraic geometry and the theory of automorphic forms. Already in his master's thesis at the University of Bonn, he gave a new proof of the Local Langlands Conjecture for general linear groups. While this work for his master's was groundbreaking, his PhD thesis written under the direction of Professor Michael Rapoport at the University of Bonn was a more marvelous breakthrough and a step up in terms of originality and insight. In his thesis he developed a new p-adic machine called *perfectoid spaces* and used it brilliantly to prove a significant part of the weight monodromy conjecture due to the Fields Medalist Pierre Deligne, thereby breaking an impasse of more than 30 years. Scholze then extended his theory of perfectoid spaces to develop a p-adic Hodge theory for rigid analytic spaces over p-adic ground fields, generalizing a theory due to Fields Medalist Gerd Faltings for algebraic varieties.

Peter Scholze was born in Dresden in December 1987. As a student he was a winner of three gold medals and one silver medal at the International Mathematics Olympiads. He finished his Bachelor's degree in three semesters and his Masters courses in two semesters. He was made Full Professor at Bonn soon after his PhD. The SASTRA Prize was one of the first awards for Scholze and he was the youngest winner of the prize at age 25! He went on to win the Cole Prize of the American Mathematical Society in 2015, and the Fields Medal in 2018.

Peter Scholze and Michael Rapoport arrived in Madras late at night on December 19. Rapoport was a Visiting Member at the Institute for Advanced Study in 1981–82 when I was there and so I was pleased to reconnect with him. Also that night, Larry Rolen (University of Cologne) and Michael Rassias (PhD student at ETH Zurich) arrived.

The next day, we all left in the SASTRA van for Kumbakonam. Sinai Robins (then at the National University of Singapore) who was also to have arrived on December 19 night, sent a message saying that he would arrive a day later. So GP stayed back in Madras to receive him and bring him to Kumbakonam. On the van drive to Kumbakonam, during conversations with the conference delegates, it was interesting to see the close bond between master and pupil (Rapoport and Scholze). Our van suffered a breakdown enroute, and so SASTRA sent us a replacement van. This caused a delay by a few hours. For this reason, and because of Sinai's late arrival, I arranged all sightseeing in Kumbakonam the next day.

The SASTRA conference opened on the morning of Dec 21 with the talk of Michael Rapoport. Except for Scholze's talk, all other talks were delivered that day — by Sinai Robins, Larry Rolen, Michael Rassias and me. Two speakers from within India were Saradha of the Tata Institute and Thangadurai of the Harish Chandra Institute.

On December 22 morning, Ramanujan birthday, the SASTRA Ramanujan Prize was awarded to Peter Scholze. After the prize ceremony, Scholze delivered the Ramanujan Commemoration Lecture on the topic of perfectoid spaces. He spoke with enormous energy, pacing side to side on the stage to write on two boards held on tripods at the ends of the stage. After lunch at SASTRA following Scholze's lecture, we departed by van for

Madras. Scholze, Rapoport and Rassias left that night, but Rolen and Robins stayed back in Madras.

The next morning, I took Sinai Robins and Larry Rolen to the famed Madras Marina beach. We walked along the shores discussing the mathematical culture in India. That evening, I took both of them to a Carnatic music concert at the Madras Music Academy.

While Larry departed Madras for Germany on December 23 night, Sinai was in India for another week. On December 24, he departed in a chauffered taxi with GP for Mysore. After return from Mysore he had a few days in Madras when GP took him to Mahabalipuram. GP told me that Sinai had taken such a liking for Cafe Coffee Day, the Indian analogue of Starbucks, that every time he saw Cafe Coffee Day, he would stop for refreshments. And that definitely meant that less time was spent on sightseeing!

I departed Madras for Singapore on January 1 morning by Malaysia Airlines via Kuala Lumpur. On that visit to Singapore, I stayed at the Hilton on Orchard Road, the main artery of the city. Within minutes of my checking in, I received a phone call from Mathura, that Lalitha was in the hospital for delivery. Indeed, a few hours later, in the late hours of the night in Singapore, I was informed by a phone call from my son-in-law Aditya, that Lalitha gave birth to a baby boy, who was named Keshav. It was January 1 in Florida, and it was the best news for the New Year that I could get! I really could not sleep that night owing to the excitement.

In the few weeks preceding the delivery, Lalitha and Aditya had been watching the Indian epic *The Mahabharatha* on TV. In that story, Lord Krishna is affectionately called Keshava by his warrior friend Arjuna. Lalitha and Aditya liked that name so much, that they named their son Keshav.

I was in the highest of spirits the next morning as I jogged along Orchard Road, enjoyed breakfast at the Checkers Cafe of the Hilton, and sent messages back home from the Executive Centre of the hotel. That afternoon, I had my first meeting with Dr. K. K. Phua, the Founder and Chairman of WSPC, and his son Max Phua, the Executive Director. World Scientific which was founded in 1981 by Dr. K. K. Phua, had in a span of three decades, become one of the leading scientific publishers in the world. I had a very fruitful meeting with the father and son and it was the start of a close relationship with World Scientific. That evening, Sinai Robins came to meet me at the Hilton. He took me on a long walk along the Marina of Singapore and to dinner at a nice Italian restaurant there.

I departed Singapore very early in the morning of January 3 for the USA via Tokyo and was back in Orlando the same evening to enjoy the company of my grandson Keshav! Thus 2014 was off to a wonderful start!!

On January 14, I departed Gainesville to attend the Annual Meeting of the AMS in Baltimore. The Convention Center where the conference was held, is in downtown Baltimore, an area notorious from crimes. So I chose to stay at the Baltimore Hilton which is connected to the Convention Center by an enclosed walkway.

The next morning I met with Thomas Hempfling of Springer/Birkhauser who said he was pleased that my book on Ramanujan was well received by the readers, and that in the future I should plan a second edition with more articles. I also had a meeting with Rochelle Kronzek to whom I reported about my visit to WSPC Singapore, She was very supportive of the idea of me writing my academic autobiography (this book).

The following morning, I met Carol Meade, the Archivist of the AMS, at the book exhibits. She told me that Carl Pomerance had given her all his correspondence with Paul Erdős to be archived by the AMS. I have with me all the letters that Erdős had written to me since 1974 — more than 100 letters — and so this conversation with Carol gave me the idea to give all these letters of Erdős to the AMS in the near future.

From January 31 to February 6, I was in the United Arab Emirates (UAE) in connection with the evaluation of the mathematics program of the University of Sharjah (UoS). See §9.13 for a description of my visits to the UAE for the evaluations of mathematics programs. Within three hours of arrival in Atlanta from the UAE, I departed for India via the Pacific with Mathura to attend the wedding of her niece in Madras. But I also had some editorial work in Madras.

WSPC has a branch in Madras, and the decision by WSPC was for the autobiography of my father *The Alladi Diary* that I was to edit, the Madras office would handle the preparation of this book. So on February 13, Ms. Ranjana Rajan of WSPC Madras came to Ekamra Nivas for discussions, following which she took me to her office to introduce me to her colleagues who would be working with me on the book. I also utilised my stay in Madras to visit the offices of Special Printing Services (SPS) who were handling the submissions and typesetting of the papers for The Ramanujan Journal.

On February 17, Professor Krishna B. Athreya of Iowa State University delivered the Third Alladi Ramakrishnan Memorial Lecture at Ekamra Nivas on the theme "Alladi Ramakrishnan and branching random walks". There was a sizeable audience for this talk and I was pleased that Professor Kohur Gowrisankaran of McGill University, Montreal, was there. Krishna

Athreya's father Mr. Balasundaram, was one of my father's teachers at the P. S. High School in Madras in the 1930s. So besides the theory of probability, Athreya had another connection with my father! Gowrisankaran had received his PhD from the Tata Institute in 1962 in the area of probability theory, and so he too knew my father.

On our return journey from India to the USA, Mathura and I had a four-day vacation in the paradise island of Koh Samui in Thailand. We departed Madras in the wee hours on February 21 by the new Thai Airways flight to Bangkok, and there we took a Bangkok Airways flight to Koh Samui. We stayed at the stunning Conrad Koh Samui located on the hilly slopes at the southern tip of the island. We enjoyed a car trip around the island and admired the exquisitely beautiful beaches. One day, the hotel concierge arranged a picnic for us at a small private island about three miles away. We were taken in a motor boat to that island and left there for a few hours to enjoy the breathtaking ocean scenery in complete privacy. I heard of Koh Samui first in the movie "Meet the parents", and wanted to have a vacation there, and it surpassed our expectations in every sense.

A few days after return from Thailand, I departed for the University of Illinois, Urbana, Illinois, in response to an invitation from Bruce Berndt. I actually flew into Indianapolis airport, and picked up a rental car there for the two hour drive to Urbana. That night, Bruce hosted a welcome dinner at the Sitara Indian restaurant, where we were joined Armin Straub who was spending the year at the University as a post-doc. I gave two talks the next day, first in the Number Theory Seminar in the morning on partitions with non-repeating odd parts, and then the Colloquium in the afternoon on the dual of the Göllnitz theorem. That night, after my colloquium, Bruce took me to another Indian restaurant — he loves Indian food, and he knew that I would have no objection in his choices. His favourite Indian dish is Palak Panneer (spicy spinach curry containing cubes of Indian home made cheese) which he will not share with anyone(!), and his favourite Indian dessert is Gulab Jamun (deep fried balls made of lentil flour in heavy syrup).

A major academic event in March was the Ramanujan Colloquium on March 17 by Peter Paule, Director of the Research Institute of Symbolic Computation (RISC) in Linz, Austria. To attend his talk, we had a few visitors from Penn State, not just the students of the MASS Program, but faculty members Ae Ja Yee and James Sellers. On the morning of March 17, Ae Ja addressed the Number Theory Seminar. That afternoon, Peter Paule delivered the Ramanujan Colloquium on the theme "Andrews, Ramanujan and Computer Algebra". Peter heads a team of experts at RISC who investigate Rogers-Ramanujan type identities, devising computer algebra packages not only to prove known identities but to discover new ones. He is a fine speaker who always provides a nice overview and just the right amount of detail, and so his talk was much appreciated by the diverse audience. The next day, Peter delivered two talks, one in the Number Theory Seminar and another in the Combinatorics Seminar. The hectic program during those two days concluded with the Teaching Seminar talk by James Sellers when he shared his experiences as the Undergraduate Coordinator at Penn State.

Paule's visit was perfect except for the mishap that his baggage failed to arrive with him. He actually arrived at Jacksonville airport from where he took a taxi to Gainesville. He lodged a complaint with Delta about his missing baggage, and was told the next day that by mistake the bag had been sent to Jackson, Missisippi, instead of Jacksonville, Florida! I don't think he ever got his bag because it could not be located in Jackson either. What concerned him were the medications he had in the bag, and not so much his clothes. While he was in Gainesville, he bought for his use some clothes with the University of Florida Gator logo!

Since 2013–14 was the Erdős Centenary, I received several invitations to give lectures to student audiences on the work of Erdős. At the University of Florida, I addressed the Undergraduate Mathematics Society on April 3 on "Paul Erdős at 100 - reflections on his life and work" and I gave the same talk the next day at the Mathematics Club of the University of Central Florida in Orlando.

My final trip during the Spring Term was to the University of Texas at Lubbock for a Sectional Meeting of the AMS (April 11–12). I spoke at a Special Session on Special Functions and Combinatorics. An unusual feature of this meeting was that the AMS Reception for the participants was arranged in the Press Box located at the top of the Football Stadium!

I was able to travel so much during the regular academic year 2013–14, because I was free from teaching due to my sabbatical. There was more academic travel I did in the summer as well, especially a trip to Turkey to participate in a conference and to deliver more than half a dozen lectures at various universities.

Prior to my international travels in the summer, we were in Atlanta for the Graduation Ceremony of Emory University, when Amritha received her MBA degree. A few days before her graduation, she received a job offer from AT&T in Atlanta which she accepted.

Before the trip to Turkey, Mathura and I visited India for three weeks in May to attend to some family matters. On the way back to the USA, my mother joined us. We broke journey in Bangkok for a two-night stay. The highlight of the sightseeing was the magnificent Wat Arun (Temple of the Sun) on the banks of the Chao Phraya river — the sprawling Buddhist temple that we missed seeing on earlier trips to this fascinating city. The Teak Palace of the King, made entirely of teak wood, was also spectacular.

After Thailand, we had a ten-day vacation (June 6–16) in Hawaii, spending one week in the idyllic island of Kauai, and three days in Honolulu. What made this Hawaiian holiday special was that my friend Sampath joined us with his wife and mother-in-law for a one week stay in Princeville, Kauai, the very definition of heaven. As we walked along the Princeville golf course at dawn, with clouds capping the tops of the mountains on one side, and the sun rising over the ocean on the other, Sampath told me that Kauai is the most beautiful place he had seen in all his travels worldwide. Of course, that is my opinion as well. The highlight of the sightseeing was the spectacular helicopter ride we had that revealed the breathtaking scenery of the Napali coast and cliffs.

In early July we made a trip to Austin, Texas, for a wedding on Mathura's side of the family. I utilized the visit to meet with Carol Meade, the AMS Archivist at the University of Texas. She showed me the collections at the Briscoe Center including the letters of Erdős donated by Carl Pomerance and some of the photographs from the Paul Halmos collection. Seeing the Briscoe Center gave me an idea of what I could donate from my family's collection of academic letters and photographs.

Trip to Turkey: The final event for the year 2013–14 was a two-week trip to Turkey for Mathura and me in August. The trip had a three fold purpose: (i) to participate in the International Conference on Algebra in Samsun, August 5–8, (ii) to give lectures at several universities in Ankara, and (iii) to do a 2000 kilometer sightseeing trip by car to various historic sites. The sightseeing trip actually had an unexpected academic visit of significance as will be described below. My trip was sponsored by Tubitak, the Turkish analog of the US National Science Foundation. My host Ali Bulent of the University of Ankara, who was one of the main organizers of the conference, had applied to Tubitak for a grant to support my trip.

We departed Orlando on August 3 for New York Kennedy Airport, where we connected to a Delta flight to Istanbul. We arrived at Ataturk International Airport, Istanbul the next morning, and had a five hour transit there before boarding the Turkish Airlines Airbus A-320 for Samsun. As we were waiting for our flight in the transit lounge, my colleague Alexander Berkovich joined us; he was on the same flight to Samsun to attend the conference. Upon arrival in Samsun, we were met by our friendly hosts who took us to Hotel Konaks where we were given a comfortable room on a high floor overlooking the Black Sea.

Samsun, a city with a population of about 1.5 million, is a provincial capital, and an important port on the Black Sea. There are two major universities in Samsun, and some of their mathematics faculty were on the conference organizing committee. In view of its location on the northern border of Turkey fronting the sea, the weather in the summer in Samsun is celubrious, and so it is a popular site for conferences. This conference was actually held at Hotel Konaks, and so it was convenient for all participants.

The International Conference on Algebra opened on August 5 morning at 10 am with the Rector of the University of Samsun and the mayors of city attending the inauguration. Ali Bulent gave a fine speech about the state of mathematics in Turkey that pleased the participants and the officiating dignitaries. The Opening Plenary Talk was by Prof. Birkenmeier of the University of Louisiana, an algebraist who had visited Turkey often on collaborative projects. I was asked to chair his opening lecture.

One of the plenary talks in the afternoon, was by C. Y. Yildirim, who became a mathematical hero in Turkey in view of his fundamental work on small gaps between primes in collaboration with Dan Goldston and Janos Pintz, for which the three were awarded the Cole Prize of the AMS in January that year. That night, the Mayor of Samsun hosted a magnificent banquet in the city, following which we all walked to see the Ataturk Memorial before retiring to our hotel for the night.

My talk was on August 7, and I gave a colloquium style lecture on the Göllnitz theorem, its generalizations, and implications, since it was a mixed audience of algebraists, number theorists, and some combinatorialists. I was followed by Bruce Berndt. There were no talks that afternoon and so the participants were taken on an excursion to a bird sanctuary. The scenery was beautiful, but we saw very few birds since their migration to the south had already started before the onset of autumn. Instead we saw plenty of water buffaloes! That night we had a banquet at a garden restaurant, where the rich Baklava dessert was outstanding. Many senior participants avoided the baklava which is very high in calories, but some others who loved sweets, were happy to consume any unclaimed baklava!

The conference concluded on the morning of August 8 with the hour lectures of Armin Straub and Emre Alkan. Our flight out of Samsun to

Ankara was late that night, and so Mathura and I joined the participants on an afternoon excursion to Amasya, but Ali Bulent took us in a car since we wanted to be back in Ankara on time to catch our flight.

Amasya is a spectacularly beautiful town in a valley surrounded by mountains and rugged rocky cliffs, containing dwellings of the Hittite Civilization dating back to 6000 BC! The town is made more beautiful due to a lovely river that winds through it. Upon entering Amasya, we stopped at a museum honoring the arrival of Ali Kamal Ataturk in 1919 who gathered local support for his war of independence. The Dean of Amasya University, was kind enough to meet all the participants and to take us on a walking tour of this enchanting town — showing us the Ottoman houses, the mansion of the Sultan, and the mosque. He was gracious to host a magnificent banquet at the University which is located atop a hill and commanding a breathtaking view of the city. Since we were hungry after the long walk, Mathura and I agreed to attend the banquet, but we really should not have in view of the long drive back to Ankara. But at Ali Bulent's assurance — and he is very persuasive — we attended the banquet. There we were seated next to the Dean and during the conversation, Ali Bulent proposed that the next conference might be held at Amasya University. Although it was an interesting proposal, it did not materialize because a few years later, the political situation in Turkey deteriorated.

We left Amasya right after dinner, but the drive at night through meandering roads took longer than anticipated. Mathura and I were visibly nervous that we would miss our flight, in spite of Ali Bulent's assurance that he would persuade the airline officials to hold the flight for us! — Really?!

We arrived at Samsun airport an hour later than we had planned. There we saw Berkovich and his host Yesilyurt sipping Turkish coffee and smoking cigarettes at a cafe just outside the terminal; they said that our flight was delayed by two hours, and so there was nothing to worry about. For the first time in my life, I felt pleased that flights do get delayed! Berkovich did not come on the excursion to Amasya because his buddy Yesilyurt who hails from Turkey told him "There is nothing to see there" (!), when actually there is so much worth seeing that one afternoon was not sufficient.

We departed Samsun at 11:30 pm by an Andalusian Airlines Boeing 737 and arrived in Ankara an hour later. After resting for a few hours at the comfortable apartment of Ali Bulent, we departed in the morning after breakfast on our 2000 kilometer journey.

Ali Bulent rented a car for us and he kindly agreed to drive the entire distance. Our first stop was Gordion, one hundred miles from

Ankara, where the Phyrgians had settled in 3000 BC. We saw the Museum constructed on the spot where Alexander the Great cut the Gordion knot! I had read so much about Alexander the Great in my middle school history classes in India, that he was my hero. I was amazed that someone so young, could conquer so much territory, in such a short span of time. Right next to the museum, is the Tomb of Midas, and the excavation site. At the gift shop, I bought a lovely plaque showing the face of Alexander the Great.

From Gordion we proceeded to the town of Pummakale, which has fantastic trevartine (limestone) terraces and healing hot springs. We gazed in wonder at the ruins of the town of Heirapolis, dating back to 7th century BC. Many visitors to Pummakale take a bath in the hot springs, but due to time constraints, we satisfied ourselves by just cleansing our feet in the healing waters. That night we stayed at a comfortable government guest house in the nearby town of Denizli, where after dinner we bought some Turkey Towels. These are cotton towels for which Turkey is so famous, that the term Turkey Towels is used generically — at least in India.

Checking out of the guest house in Denizli the next morning (August 10), we drove to Aphrodisias, which dates back to 1000 BC, and has both Greek and Roman culture. Most impressive were the ruins of the Temple of Athena (Greek), and the stadium built by the Romans with seating for 12,000! From Aphrodisias, we proceeded to the town of Celzuk, where we saw the ruins of the Temple of Artemis (= Temple of Diana), one of the Seven Wonders of the (Ancient) World! Its first construction dates back to the Bronze Age, but after it was destroyed in a flood in the 7th Century BC, it was reconstructed in 550 BC. I had read about the Seven Wonders of the World as a young boy, and so I gazed in wonder as I stood on the rock stones of the Temple of Artemis.

From Celzuk, we proceeded to the nearby site of Ephesus of Greco-Roman culture, which has the best preserved ruins in Turkey. The highlights there were the Gate of Hadrian, the massive Library, and the Theater. After roaming about the ruins of Ephesus, we checked in at a government guest house in Celzuk.

After this enjoyable but long and tiring day of sightseeing, Ali Bulent suggested at dinner that we should see the Mathematics Village on the outskirts of Celzuk that very night. It was late and so I felt that the Mathematics Village might be closed. But Ali Bulent said that it was only 9 pm and that the night was still young. So we departed for the Mathematics Village without any appointment to see anyone.

The Mathematics Village is nestled in a mountainous region in the village of Sirince, and it was quite scary as Ali Bulent navigated the narrow unpaved roads at night to reach the Village. It was pitch dark as we parked, but in the glare of our headlights we suddenly saw a female student emerge from a path, proceed to her tent, pick up a notebook, and head back along the path from which she came. So we knew we were at the right place and followed that path. A lovely set of buildings loomed very soon, and we were greeted by a cheerful band of students. We were taken to the terrace where Prof. Ali Nesin, the founder of this wonderful institution, and his daughter, Gabriela, were in discussion with students at dinner. They spontaneously invited us to join them at dinner, but we had to decline the kind invitation since we had already dined in Celzuk. Gabriela then showed us around this remarkable place. We were simply overwhelmed by what we saw.

The Mathematics Village is a unique institution. It offers short but intense courses to mathematics students from high school and college. The students are immersed in mathematics for the few weeks they are there, learning from the professors who teach, as well as from discussions among themselves. The students could relax in the evenings by playing ping-pong or cards or by simply reading in the excellent library or in the wooded environs. Most students are housed in stone and clay houses, but as demand almost always surpasses capacity, some have to be put up in tents. At the time we visited, there were about four hundred students in attendance. Teaching is voluntary; while the Mathematics Village does not provide honoria for the teachers, it provides free accommodations and meals. It is a sylvan setting in which to either learn or teach mathematics. As we toured the Village, I was impressed to see some students in discussion in front of a blackboard in a courtyard and some others in the library, even in the late hours. So in the midst of the sightseeing tour, we had a wonderful opportunity to visit this remarkable institution. Ali Bulent deserves special thanks for taking us to the Mathematics Village.

I was so impressed by what I saw, that I asked Ali Nesin to submit an article on the Mathematics Village to the Notices of the AMS. He appreciated my suggestion but said that it would better for me to write this article in collaboration with Gabriela. The feature article authored by the two of us appeared in the June/July 2015 issue of the Notices of the AMS. The article starts with an Introduction by me, followed by a description of the evolution of the Mathematics Village and its programs by Gabriela. I was delighted that a few years later, at the 2018 International Congress of Mathematicians in Rio de Janeiro, Ali Nesin was awarded the prestigious *Leelavati Prize* for public service for creating the Mathematics Village.

The next morning we departed Celzuk for Troy. We drove past the lovely city of Izmir on the Mediterranean coast to the town of Pergamon where we first saw the ruins of Asclepion, a medieval city dating back to 1000 BC. The ruins of the Temple of Zeus, and the Theater here were impressive. Next we saw the Acropolis situated atop a nearby hill. The stadium of the Acropolis is awesome both in its size and in its location on the slopes of the hill; sitting on the top row, the slope of the massive stadium seems to merge into the valley floor more than a thousand feet below.

After a delightful Turkish lunch at Pergamon, we drove to Cannakkale enjoying breathtaking views of the Aegean sea and the coast from the highway skirting the mountains. We stopped for refreshments enroute, and enjoyed pickled olives with Turkish bread and olive oil at a roadside cafe that had a scenic overlook. By nightfall we reached Cannakkale near Troy, where Ali Bulent had made reservations at a lovely guest house.

The next morning, after breakfast, we departed Cannakkale for Troy, the famous site of the Trojan War. We roamed about the ruins of Troy for about two and a half hours in the morning. The settlements of Troy date back to 3000 BC, and the city was developed in several stages: Troy I (3000–2500 BC), Troy II (2500–2000 BC), ..., until Troy IX (100 BC–500 AD). Thus Troy has Greek and Roman influence, and pre-Greek civilization as well. It was thrilling to see the gate through which the Trojan Horse came into the city. Equally exciting was the Archeological Museum of Troy containing statues of Aphrodite and Hadrian among others.

After lunch at Troy, we boarded a ferry at Cannakkale to cross the famous straits of Dardanneles to Escabat on the Gallipoli peninsula. Our car was on board the ferry as well. After checking in at a guest house in Escabat, we went to the Museum at Gallipoli to see a show of the heroic defense of Gallipoli by the Turks against the British forces in World War I. The final sightseeing for the day, and of the trip, was the magnificent War Memorial at the southern tip of the Gallipoli peninsula, with spectacular views of the Aegean Sea and the straits of Dardanelles.

We departed Escabat the next morning (August 13), and had a long drive back to Istanbul. Mathura was returning to Florida the next day, and since we had bought plenty of gifts for the family, and souvenirs on this trip, we stopped a mall upon entering Istanbul to buy a suitcase for Mathura to carry back the souvenirs and gifts. I was to stay back in Turkey for a few more days for my lectures in Ankara.

We spent the night of August 13 at the fabulous Conrad Istanbul. Ali Bulent and I had a refreshing dip in the Conrad pool before we all enjoyed

Turkish hors d'oeuvres in the roof top open air lounge of the Conrad from where we had a breathtaking view of the historic city of Istanbul on the Bosphorus. After dropping Mathura at the airport the next morning, Ali Bulent and I drove back to Ankara.

It was really gracious and generous of Ali Bulent to not only have me as his house guest, but give the Master Bedroom for me to use. While we were relaxing at his apartment on that first evening, I heard the news that Manjul Bhargava had received the Fields Medal at the ICM in Seoul, Korea. He was the First SASTRA Ramanujan Prize Winner (along with Soundararajan) in 2005, and I was delighted to hear that he went on to win the Fields Medal. The next morning as I was writing a note of congratulations to Manjul, I received an email from The Hindu, inviting me to write an article on Bhargava in the next two weeks to which I gladly consented.

On my first full day in Ankara (August 15), Ali Bulent took me to the campus of Ankara University where I had discussions with two of his students. Ankara University has a beautiful campus, with buildings constructed of stone, surrounded by tall trees and gardens. My talk that afternoon was at Haceteppe University where I spoke on "Two remarkable partial theta identities of Ramanujan and Andrews". My talk was also attended by students in engineering and biosciences. That night, the dinner was at the Faculty Club, a glass walled building which was located on a hilltop, and commanding a splendid view of the city.

I had no talks during the next two days (Aug 16, 17) because it was the weekend. But two of Ali Bulent's students came over to his apartment for lunch on Saturday and to discuss their PhD work on recurrence sequences with me. Ali Bulent is an expert chef, and he prepared an outstanding Turkish vegetarian lunch, which received approval from his mother, a graceful lady who spent that afternoon with us.

On Monday, August 18, I went early to Ankara University to continue discussions with Ali Bulent's students. After lunch at the University hosted by the Chair, I gave a colloquium talk "Paul Erdős at 100 - reflections on his life and work". There was a large student audience for this talk, and at Tea after my lecture, several students were engaged in discussions with me and among themselves. In the late afternoon I was shown the ruins of the Roman Temple of Augustus in central Ankara, following which Ali Bulkent and his students hosted a fine dinner at a restaurant on the slopes of a hill at the top of which is the famous Ankara Castle.

The next day (August 19), I gave my final lecture in Ankara at Bilkent University on "Partitions with non-repeating odd parts - q-hypergeometric and combinatorial identities" at the invitation of Hamza Yesilyurt. Bilkent is one of best research universities in Turkey and my talk was attend by an active group of researchers in algebra and number theory. After my lecture, Hamza arranged dinner at a lovely garden restaurant where we were joined by his wife and infant son.

I flew out of Ankara's Essenboga Airport for Istanbul on August 20 morning by a Turkish Airlines Airbus A-320. Turkish Airlines proudly advertises its service on board, and I was pleasantly surprised that even on this short flight, the refreshments were very good. At Istanbul, I boarded my Delta flight to New York, and was back in Florida that night. Thus concluded a most memorable trip to Turkey, and a fine ending to my one year of research leave. Fall classes were to start in a few days.

10.7 A socio-academic melange (2014–15)

The year 2014–15 was hectic because I had a variety of academic programs and for the whole year was also busy working with Mathura on preparations for the wedding of my second daughter Amritha.

August 25 was the first day of classes for the Fall term, and I taught a year long graduate course in Analytic Number Theory, the outcome of which was that Todd Molnar, one of the students in that course, completed his PhD in that area under my direction two years later. I also attended the year long graduate course of Frank Garvan on "Modular forms and mock theta functions" in which he did a thorough discussion of the classical and modern theory of mock theta functions.

Since my article for The Hindu on Bhargava's Fields Medal work was due in the middle of September, I wrote to Bhargava upon return from Turkey that I would like to speak to him by telephone about his recent work. I was familiar with his work on higher degree composition laws, and his resolution of the problem of determining all universal quaternary quadratic forms (jointly with Jonathan Hanke), because that was the work for which he received the SASTRA Ramanujan Prize in 2005. But I was not completely familiar with his subsequent fundamental work on ranks of elliptic curves. In two long telephone calls Bhargava patiently explained the key points of his recent work and its relationship with the celebrated Birch–Swinnerton-Dyer Conjecture, and this was very helpful for me to prepare the article. What appeared in The Hindu was an abridged version of what I had written, but later my original article appeared in full in the Asia Pacific Mathematics Newsletter (APMN) published by World Scientific.

It was a relatively quiet semester with regard to seminars by visitors, but we did have three talks by outside speakers in the Number Theory Seminar. In late October, Michael Griffin and Olivia Beckwith, PhD students of Ken Ono at Emory University, gave back to back talks on their doctoral work. They just drove down from Atlanta. Later in November, Sinai Robins visited for a week and gave a talk on "Cone theta functions". He had taken leave from his position at Nanyang Technical University, Singapore, and was visiting Brown University for the year. One evening I took him to the Bat House on campus. He was amazed to see several thousand bats emerge at twilight. There are hawks that prey on these bats when they emerge, but the bat population is too large for these hawks to have an effect.

My second daughter Amritha's wedding was fixed for July 4, 2015, but Indian weddings are elaborate affairs and their planning begins about a year earlier. Thus during Fall 2014, we made several weekend trips to Tampa to make arrangements for the wedding that was planned to take place there.

==

Trip to India and the 2014 SASTRA Ramanujan Prize to James Maynard

The main academic event in Fall 2014 was the award of the SASTRA Ramanujan Prize to James Maynard in December. Only my mother and I departed for India on December 10 flying the transpacific route via Tokyo and Singapore. Mathura stayed back in Florida, because she was planning to go to India in mid-January with Amritha for wedding related purchases.

While I was in Turkey earlier during the year, I convinced Ali Bulent that it would worthwhile for him to attend the SASTRA conference and see Ramanujan's home in Kumbakonam. He arrived in Madras on December 16, three days before our departure to Kumbakonam, and so GP and I showed him the city and its surroundings. James Maynard arrived on December 18 afternoon as did Ken Ono with two of his PhD students Michael Griffin and Jesse Thorner of Emory University. That evening I took all the participants to the Madras Music Academy where they enjoyed a Carnatic vocal concert. The next day, all of them were taken by GP on an excursion to Mahabalipuram to see the rock temples by the sea dating bank to 7th Century AD. One other participant Kagan Kursungoz from Sabanci University in Istanbul, and a former PhD student of George Andrews, arrived late that night.

James Maynard of Oxford University was selected to receive the 2014 SASTRA Ramanujan Prize for his spectacular contributions to number theory, especially to some of the most famous problems on prime numbers. The theory of primes is an area where questions which are simple to state can be very difficult to answer. A supreme example is the celebrated "prime twins conjecture" which states that there are infinitely many prime pairs that differ by 2. Maynard had obtained the strongest result towards this centuries old conjecture by proving that the gap between consecutive primes is no more than 600 infinitely often. Not only did he significantly improve upon the earlier path-breaking work of Goldston, Pintz, Yildirim, and Zhang, but he achieved this with ingenious methods which are simpler than those used by others.

A generalization of the prime twins conjecture is the prime k-tuples conjecture which states that an admissible collection of k linear functions will simultaneously take k prime values infinitely often in values of the argument. In the last one hundred years, several partial results towards the k tuples conjecture have been obtained either by replacing prime values of some of these linear functions by "almost primes" (which are integers with a bounded number of prime factors) or by bounding the total number of prime factors in the product of these linear functions. Another major achievement of Maynard was to significantly improve on work of earlier researchers on k-tuples of almost primes.

The Prime Number Theorem implies that the average gap between the n-th prime and the next prime is asymptotic to $\log n$. Two questions arise immediately: (i) how small can this gap be infinitely often (the small gap problem), and (ii) how large can this gap be infinitely often (the large gap problem)? The prime twins conjecture says that the gap is 2 infinitely often, and what we have described above relates to his sensational work on the small gaps problem. In mid-2014, Maynard announced a solution of the famous $10,000 problem of Paul Erdős concerning large gaps between primes. The solution was also simultaneously announced by Kevin Ford, Ben Green, Sergei Konyagin, and Terence Tao, but Maynard's methods are different and simpler. In 1938 Robert Rankin established a lower bound for infinitely many large gaps which remained for many years as the best result on the large gap problem. Paul Erdős then asked whether the implicit constant in Rankin's lower bound could be made arbitrarily large, and offered $10,000 to settle this question.

Maynard's work is so significant, that he was subsequently awarded the AMS Cole Prize in 2020, and the Fields Medal in 2022. The SASTRA Ramanujan Prize was the first major international award he received.

I departed for Kumbakonam on December 20 morning by the SAS-TRA van with all the international participants. We stopped at a roadside cafe/restaurant where Ono and his students enjoyed Indian lunch served piping hot. Ono feels at home in India because he visits often. He is fit as a fiddle with his rigorous exercise regimen, and therefore he has built up a good resistance. Michael Griffin had actually lived in India for a few years doing charity work for a church, and so he had been accustomed to India's rural conditions. He was comfortable eating Indian food using his fingers just as we Indians do!

We reached Kumbakonam by 4:30 pm and so had time to see the Town High School, Ramanujan's home and the Sarangapani Temple that evening before checking into Sterling Resorts.

James Maynard was given the 2014 SASTRA Ramanujan Prize at the conference inauguration on December 21 morning. Except for Ono and Maynard who spoke on the morning of December 22, Ramanujan's birthday, all talks were on December 21. After a fantastic Ramanujan Commemoration Lecture by Maynard on "Small Gaps between Primes" on December 22 morning, Ken Ono gave the concluding lecture of the conference on "Ramanujan's legacy". He concluded his lecture by showing a trailer of the movie "The Man Who Knew Infinity" and invited the entire audience to the stage to sing a Happy Birthday song to Ramanujan, with a photo of the movie in the background!

We returned to Madras later in the evening. Ali Bulent and Maynard stayed in Madras for one more night, while the rest of the guests departed on December 22 night.

On December 23 morning, I had arranged for GP to take Ali Bulent and Maynard to the famous Theosophical Society in Madras. That day, earlier in the morning, Maynard had walked all the way to see the Kapaleeswar Temple (for Lord Shiva) near my house. I told Maynard that if I had known that he wanted to see the temple, I would have sent my car, because I would not want him to walk in the streets which could be crowded and dirty. But he said it was nothing to worry about. As a boy he had lived in India, and so he is used to conditions in the country.

After all the visitors had left, I departed Madras on December 24 for USA via Singapore. I usually return after New Year's day, but this time I came back early because we had arranged a big party at our home in Gainesville on January 1 for my grandson Keshav's first birthday.

==

In mid-January, Mathura and Amritha both departed for India via the Pacific on a one week trip. But they travelled separately and on different days. It was a short stay but intense because of various things they had to buy for Amritha's wedding. In particular, they made a trip to

Kanchipuram, 50 miles south of Madras, to buy about two dozen silk sarees to be given as gifts for our relatives and close friends who would be attending the wedding. India is famous for silk sarees, and Kanchipuram is the place where the best South Indian silk sarees are made.

On my part, in mid-January, I went to San Antonio for the Annual Meeting of the AMS. San Antionio is a favourite of the AMS and they hold their annual meeting there once every five years. I stayed at the Hilton Palacio del Rio close to the Convention Center, but it was cold and wet throughout my stay. I spoke at the Special Session on Partitions and q-Series on my multi-dimensional extension of Sylvester's identity. The session was organized by Ae Ja Yee and Tim Huber. When I attended the Prize Session, I was happy to see Peter Scholze to receive the AMS Cole Prize. It is gratifying for me if a SASTRA Prize Winner goes on to win prizes with a hallowed tradition. I congratulated Scholze and said that I hoped he, like Bhargava, would win the Fields Medal soon. And this he did at the ICM in 2018!

In the second half of January, Ali Bulent arrived in Gainesville to spend the Spring semester at the University of Florida. His visit was fully supported by Tubitak, the Turkish version of the NSF. He spent the first few days at our home before moving into an apartment near campus.

In mid-February, our dear friend Professor P. V. Rao of the statistics department passed away. PV as he was known to friends, had invited my father for a colloquium in the statistics department in 1980. Thanks to my father's suggestion, PV and his wife Premila were among the first contacts in Gainesville for Mathura and me in 1986 when we had just arrived, and our friendship had grown over the years. Premila asked me to speak at PV's memorial service, and during that occasion I recalled how our friendship and contact began.

During Spring Break (February 27–March 7), I made a short trip to Madras. One goal was to bring back with me some items for Amritha's wedding that Mathura had ordered in January during her visit. But I also had academic events during the trip. In particular, I arranged the Fourth Alladi Ramakrishnan Memorial Lecture to be delivered on March 5 by my father's long term friend Prof. J. Sethuraman of the statistics department of Florida State University. His talk was titled "Alladi Ramakrishnan and modern statistics". Normally I take the speaker for dinner at Hotel Savera near my house, but Sethuraman and his wife Brinda prefer home cooked meals, and so we had them for dinner at our home after the lecture.

The next day, there was a function at the Tamil Nadu Academy of

Sciences to which I was invited as a Chief Guest to give the First Young Scientist Award of the Academy. I recalled how the Academy was first proposed by my father in November 1971 at the Madras Center for Development Studies (MIDS) headed my Dr. Malcolm Adiseshiah who asked various enlightened citizens of Madras to propose ways in which science could be fostered in the state of Tamil Nadu. I handed over a copy of my father's proposal on the Tamil Nadu Academy of Sciences to the President of the Academy Professor Anandakrishnan. My father served as the First Secretary of the Academy and later as its President, and I felt honored to take part in this function. Thus the short visit to Madras was fruitful.

Soon after my return from India, Janos Pintz of the Alfred Renyi Institute of the Hungarian Academy of Sciences arrived in Gainesville to deliver the Erdős Colloquium on Monday, March 16. Pintz is an authority in analytic number theory and a former student of Paul Turan, who was a close friend and a long time collaborator of Erdős. So he was an appropriate choice as speaker for the Erdős Colloquium. He gave a marvelous survey of the problem of small gaps between primes, and discussed the ideas in his famous joint work with Goldston and Yildirim. Since there was a weekend between his arrival and his colloquium, we showed him around Gainesville. In particular, he enjoyed seeing Devil's Millhopper, one of the largest sink holes in this region with pre-historic type ferns at the base, where the temperature is always warm.

The final academic event of the Spring term was the Ramanujan Colloquium by Robert (Bob) Vaughan of The Pennsylvania State University on April 6. Vaughan is the leading authority on the Hardy-Littlewood-Ramanujan Circle Method, which is the primary tool to deal with a vast array of problems in additive number theory, and so he gave a general lecture about this in the Colloquium, followed by two technical seminars the next day. In problems utilizing the Circle Method, it is not only the ideas used that are important, but also the details in the various estimates being carried out, because the sharper these estimates are, the better is the result that we would get. Vaughan is technically highly skilled in such estimates, and this came out clear in his lectures which were thorough in the details.

In connection with Vaughan's lectures, we had book exhibits by Springer and World Scientific. In view of my association with World Scientific firming up in 2014, Rochelle Kronzek attended the events of the Ramanujan Colloquium starting from Spring 2015. The representatives from Springer (Marc Strauss and Elizabeth Loew) and from World Scientific (Kronzek), always spend time with me on such visits discussing various projects.

In the month of May, Mathura and I made a short trip to India prior to Amritha's wedding, also with the purpose of bringing my mother from Madras to Florida so that she could be present for the wedding. But in the end of April I was troubled by a kidney stone; this was surgically removed without delay by my doctor friend Dr. Perinchery Narayan, one of most accomplished urologists in the United States. I was thankful that the kidney stone problem occurred after the academic events of Spring 2015, and before my trip to India and the academic events of the summer.

Mathura and I departed on our transpacific journey to India on May 7, and halted for two days in Singapore enroute. On the flight from Tokyo to Singapore, I worked out a new combinatorial proof of the Euler-Glaisher theorem and informed Andrews about this upon reaching Singapore. He not only liked my "airport theorem", but in his later work he graciously called my variation of Schur's theorem that came out of this proof as the Alladi-Schur theorem!

While in Singapore, I had fruitful meetings at the World Scientific head quarters with Max Phua (Executive Director), Lai Fun, and Rok Ting, about my book projects. The special sightseeing on this visit to Singapore was the spectacular Tulip Festival that we saw at the Marina Bay Gardens near the Conrad Centennial Hotel where we stayed.

In Madras, I had meetings with Ranjana Rajan and her colleagues at the World Scientific office regarding my father's autobiography that I was editing, because Ranjana and her team were in charge of the publication of this book. The main academic event for me was the colloquium I gave at MATSCIENCE on May 19 at the invitation of Viswanath, a young researcher in number theory who had joined the institute. He had received his PhD at Berkeley under the direction of Fields Medalist Richard Borcherds.

Mathura and I left Madras on May 23 with my mother for the United State. On the return journey to Florida, we had a holiday in the Hawaiian Islands. What made this special was Dr. Perinchery Narayan, his wife Raji, and his two charming and brilliant sons joined us for the holiday. Indeed it was due to Dr. Narayan's help with the kidney stone surgery that I was able to make the trip as scheduled. I arranged for two condominiums for us at Bali Hai Villas in heavenly Princeville in Kauai. The most memorable event in Kauai was the helicopter ride that we took. Kauai's scenery is breathtaking, and view from the helicopter of the Napali Cliffs, of Waimea Canyon, and of the Kalalau Valley, leaves you breathless in admiration. After a memorable and event-filled week in the idyllic island of Kauai, we spent four nights in Honolulu at the famed Hilton Hawaiian Village. Honolulu is a bustling city, but it is exciting and beautiful.

We returned to Florida on June 1, but the very next day I departed for Baltimore to participate in the meeting of the Society of Industrial and Applied Mathematics (SIAM) at the National Institute of Standards (NIST) in Gaithersberg, Maryland. I was invited by Howard Cohl of NIST to speak at a Special Session on "The Legacy of Ramanujan" organized by Bruce Berndt. My talk was about a generalized Lebesgue identity in Ramanujan's Lost Notebook. One of the plenary talks at the conference was delivered by Frits Beukers. The last time I met him was in 1979 at Oberwolfach, and it was a pleasure to see him again after 36 years.

The entire month of June was spent preparing for Amritha's wedding which was scheduled as a grand two-day event in Tampa during the July 4th long weekend. Amritha was getting married to Jis Joseph, a Catholic Christian boy of Kerala heritage. It was a new experience for us, as it was to the Josephs, to understand a different religion from a personal family point of view, and our two families emerged more enlightened as a result of the bond created by this wedding. Since the Josephs live in Tampa, their guests combined with ours totalled more than five hundred. Many friends and relatives from Gainesville, and from across the USA attended the wedding; in particular, Mathura's brothers Chella and Sriram came all the way from Madras with their spouses. Since this wedding was in the middle of the summer, the only mathematics visitor of mine who attended the wedding was Ali Bulent who was spending six months visiting the University of Florida. Of course, Frank and Cyndi Garvan and Alex Berkovich and Larissa, attended the wedding events in entirety.

Mathura and I arranged all the wedding events at the magnificent Marriott Waterside Hotel in Tampa where we and many of the guests stayed. The events began with an evening of music and dance on June 3 coordinated by Mathura and my older daughter Lalitha and featuring middle-eastern food. On July 4, the entire morning was devoted to an elaborate Hindu Wedding Ceremony. After a sumptuous South Indian lunch, we went to a nearby church to attend a Christian Wedding Ceremony in the mid-afternoon arranged by the Josephs. We then returned to the Marriott for a Western style reception for which the dinner fare was North Indian. In India, there is the saying, "Build a house and conduct a wedding to know what work really is". Yes, an Indian wedding is very elaborate, but the effort is worth every bit when you see the joy it brings to the married couple and to the friends and family members who partake in it.

With the wedding behind us, I was ready for three academic trips in the summer before the end of the academic year:

Trip to South Korea: In the last part of July, Mathura and I visited South Korea in response to a kind invitation from Byungchan Kim and Youn-Seo Choi of the Korea Institute for Advanced Study (KIAS) in Seoul which is perhaps the most reputed among the academic research centers in South Korea. I was to give several talks in Seoul. Alex Berkovich and Larissa were also in Seoul due to invitations from Choi and Kim.

Mathura and I departed for Seoul on July 19. Upon arrival at Seoul Incheon Airport on July 20, we took a taxi to the Hilton Grand Hotel where we stayed for a week. My first talk was on the next morning at KIAS and I spoke on partitions with non-repeating odd parts. Korea has several active groups in the theory of partitions and q-series, and my talk and that of Berkovich in the afternoon were attended by researchers at other academic centers in Seoul such as Jaebum Sohn and Soon Yi Kang. Choi, Kim, Kang, and Sohn are all former PhD students of Bruce Berndt with Choi being seniormost among Berndt's Korean students. When I visited Seoul, Choi was the head of the mathematics department at KIAS. Byungchan Kim also presented his work that afternoon, and so with three talks back-to-back, it felt like a mini-conference. That evening, our hosts took us to a famous Buddhist temple in downtown Seoul, and hosted a dinner at a monastic vegetarian restaurant that is managed by the temple trust.

The following day, at KIAS, Berkovich spoke in the morning, and I gave the afternoon lecture on partial theta identities of Ramanujan and Andrews since this was intimately connected with Kim's work. I had no lectures the next day (Thursday, July 23) and so we spent the entire day sightseeing in Seoul. The highlights included the (main) Royal Palace as well as the Cheing Deo Summer Palace with its impressive gardens. As I had said earlier, I always like to go up a tower to get a view of the city. Since the afternoon was bright and sunny, after seeing the palaces, we went up the Seoul Tower located atop a hill to enjoy panoramic views of the city.

For the weekend, Mathura and I went to beautiful Gyeong Ju in the central highlands of South Korea. Alex and Larissa joined us on this trip. The Gyeong Ju region is very scenic and also rich in the cultural heritage of Korea with several lovely palaces and Buddhist temples.

It was a two-hour train journey from Seoul to Gyeon Ju. Mathura and I checked in at the Hilton Gyeong Ju and were given a room on a high floor overlooking a lovely lake. Alex and Larissa stayed at a suites hotel at the other end of that lake.

Soon after check-in, we all rented a private car with a chauffeur and went sightseeing around the town of Gyeong Ju. Particularly impressive were the tombs of Tumuli Park, and the Anapji Palace with a pond surrounding it. It was spectacularly illuminated at night. The next day we rented a chauffeured car again for a full day of sightseeing. Most impressive was a Buddha located in a grotto on the steep sides of a cliff, and the National History Museum. On our final day (Sunday) in Gyeong Ju, we again went around in a car. The most impressive sights were the Buddha of Gonguram carved out of the rocks in a cliff high above the valley, the twin towers of King Minnum, and the Undersea Tomb of King Minnum in a rocky island just off the beach. We were back in Seoul on Monday, and since I had no lecture that day, we spent the afternoon buying souvenirs and gifts.

On the morning of Tuesday July 28, Byungchan Kim arranged my lecture for high school students at Seoul Technological University. My talk entitled "Paul Erdős - one of the most influential mathematicians of our times" had an audience of about 100 students in the gifted program.

No visit to Korea is complete without seeing the Demilitarized Zone (DMZ) and the border between capitalist South Korea and communistic North Korea. We had an all-day guided tour to the DMZ on July 29. Tensions between the South and the North had escalated during the time we were there, and so it made the visit to the DMZ all the more exciting. It was thrilling to enter a small room where there is a line drawn to indicate the border between the South and the North, and to stand on the southern side of that line to have a picture taken. All the time, a North Korean guard was watching us menacingly! We were instructed not to crack jokes while at the DMZ lest it would provoke the North Korean guards to open fire! I think this was a joke, but we followed the instructions since we did not want to risk being shot or arrested. Another exciting aspect of the visit was a walk through one of the tunnels that the North Koreans had dug in order to infiltrate into the South. When we returned from the excursion to the DMZ, Byungchan Kim met us at our hotel and took us to dinner at an Indian restaurant called the Ganga (Hindu name for the river Ganges). Byungchan is a warm and friendly host who made our trip so enjoyable.

July 30 was our final day in Seoul and we spent much of the afternoon on a second round of shopping for souvenirs. We departed Seoul on July 31, and were back in Florida the same day.

In August I attended two conferences back-to-back — the Illinois Number Theory Conference (Aug 13–14), and the Lattice Paths and Combinatorics Conference (Aug 17–20) in California.

I departed Florida for Indianapolis on the morning of August 12, rented a car there and drove to Urbana, Illinois. After checking in at the Illinois

Union Hotel, I had dinner with Bruce Berndt at an Indian restaurant —
where else?! — before attending a welcome reception at Altgeld Hall. The
Illinois number theorists have perfected the art of organizing conferences
with Adolf Hildebrand attending to every detail. The conference had about
100 registered participants. My talk on the multi-dimensional analogue of
Sylvester's theorem was on the afternoon of the first day (Aug 13). After
attending the sessions on the second day in full, I departed Urbana on
August 15 morning not for Gainesville but for Tampa. August 15 is India's
Independence Day, and the Josephs and their Kerala friends had arranged
a big get-together for the Onam Festival at New Port Richey near Tampa,
for which my daughters Lalitha and Amritha were performing a dance.
I reached Tampa on time to attend the Onam program and witness the
beautiful dance of my daughters, much to their satisfaction. After spending
the night near Tampa airport, I departed for Los Angeles the next morning
(Aug 16) to attend the Lattice Paths and Combinatorics Conference being
held on the campus the California State Polytechnic University in Pomona
(Cal Poly, Pomona), once the estate of the W. K. Kellogg whose name is
synonymous with breakfast cereals. I had lunch at the home of my friend
Sampath in Los Angeles before driving to Pomona to check in at the Kellogg
West Hotel on Cal Poly campus where all participants were staying.

Pomona Conference: It was nostalgic for me to return to Pomona after
42 years! I was reminded of my visit in the summer of 1973 when my father
took me there to meet Prof. LeVeque (see §2.3) as I was just starting my
research attempts in number theory as an undergraduate.

The Lattice Paths and Combinatorics Conference Series was launched
in 1984 by Sri Gopal Mohanty of McMaster University in Hamilton, On-
tario, Canada, and is held roughly once every four to six years at various
venues around the world. Mohanty is ably assisted in his effort by a group
of dedicated researchers in the area. George Andrews has been a regular
participant and lead speaker in this conference series. This conference was
the eighth in the series and its main organizer was Alan Krinik who is on
the faculty of mathematics at Cal Poly, Pomona. The interesting aspect of
this conference was that it was dedicated to four mathematicians: George
Andrews and Lajos Takacs who had made significant contributions to lattice path combinatorics, and to the late Shreeram Abhyankar and Philippe
Flajolet who had used lattice path combinatorics in their fundamental
work. The theory of partitions has strong links with lattice paths combinatorics. Since I had significant collaboration with Andrews in the theory of

partitions, Krinik invited me to deliver one of the main lectures of the conference on my joint work with Andrews. I also had known Abhyankar quite well since my boyhood days owing to my father's close contact with him, and so Krinik asked me also to be the banquet speaker to share my reminiscences of Abhyankar and my long association with Andrews.

There were about forty partitions at this conference from around the world. Christian Krattenthaler from Vienna was there and he helped Krinik and Andrews edit the refereed proceedings of the conference which appeared in my book series *Developments in Mathematics* (DEVM) in 2019. Two of Abhyankar's former PhD students at Purdue University of Indian origin were at the conference: Sudhir Ghorpade, a professor at IIT Bombay, and Devadatta Kulkarni who took up a job in industry after obtaining his PhD. Ghorpade was the Head of the Mathematics Department at IIT Bombay, and following this contact we had at the Pomona conference, he invited me to IIT Bombay in January 2019 for a colloquium.

It was a pleasure to see Jeffrey Remmel of the University of California, San Diego, at the conference. He drove up from La Jolla to attend the conference lectures. We talked about his book "Counting with Symmetric Functions" (co-authored with Anthony Mendes) that had just appeared in my book series DEVM (Springer). It was the last time I saw him for he died quite suddenly in September 2017 at the age of 68. He was a colleague of mine at the University of Florida in 1997–88 and so I got to know him well. Remmel was a prolific and deep researcher in the field of combinatorics. He had an eternally cheerful personality. I never saw him agitated or get upset at anything. At UCSD, he had a major collaboration with Adriano Garsia, and here in Florida he collaborated with my colleague Doug Cenzer in the area of mathematical logic. He was a successful administrator as well.

Two of my papers appeared in the refereed conference proceedings. In my long paper "My association and collaboration with George Andrews", I described my research collaboration with him (as I did in my talk during in the regular conference sessions), as well as my association with him in India, Penn State and Florida, which I described in the banquet speech. The text of my speech at the banquet on my recollections about Abhyankar also appeared in the proceedings several photographs of Abhyankar and Andrews, and the editors kindly included some of them in the conference proceedings. The Pomona conference was both intellectually stimulating and full of memories. It was a nice conclusion to the academic year 2014–15.

10.8　My sixtieth birthday year (2015–16)

I turned 60 on October 5, 2015, and so an International Conference in Number Theory was held at the University of Florida in March 2016. The idea to hold such a conference was put forth by George Andrews and Frank Garvan during the Spring Term of 2015 when Andrews was in residence in Florida. Andrews went on to suggest that for this conference, in addition to our usual list of invitees in the areas of partitions, q-series and modular forms, researchers in analytic number theory, irrationality and transcendence, especially those with whom I had contact during my early years, ought to be invited. I was not sure whether there would be a good response from the analytic number theory, irrationality and transcendence community since I had not worked in those areas for twenty five years. I said yes to Andrews' suggestions, and to my pleasant surprise, the response to our invitations from the researchers in all areas was overwhelmingly positive. So the conference which attracted two hundred participants, was one of the largest mathematics conferences organized in Florida.

Much of Fall 2015 was spent preparing the NSF and NSA proposals for the conference and securing basic support from the Department, the College, and the Office of Research. The conference proposals were prepared thoroughly by Frank Garvan with Andrews providing expert advice. The effort was well worthwhile because the conference received ample funding; in fact, Office Manager Margaret Somers also secured from the Alachua County Tourism Board (!) which helped us arrange bus transportation for all the participants staying in different hotels in Gainesville.

Among the academic events in the Fall was my one week visit (Oct 30–Nov 6) to the UAE to do an accreditation of the College of Science for the New York University, Abu Dhabi (NYUAD), a university administered by NYU. I was the representative on the team for mathematics and I came away impressed with NYUAD. For details of my visit, see §9.13.

On November 21, my family (Mathura, daughters Lalitha and Amritha, and sons-in-law Aditya and Jis) arranged a grand 60th birthday dinner party for me at the Paramount Plaza Hotel in Gainesville. There were about 180 guests including our friends in Gainesville, a few of my relatives from out of town, and some of my colleagues and their spouses. Lalitha, Amritha, Adi and Jis had a humorous slide show conducted in the "Who Wants to be a Millionaire" format to get audience participation. The younger generation is full of novel ideas. I could not have thought of that. Mathura gave a lovely speech and beautifully sang one of my favourite Carnatic music songs and she was followed by my four-year old grand daughter Kamakshi who danced a piece on Lord Krishna to the delight of everyone. Our good friend Prof. Meera Sitharam, gave a short South Indian classical music recital on the Veena. There were speeches by three of my close friends Dr. Perinchery Narayan, Prof. Venkat Kunisi, and Mr. Lavakumar. It was a memorable event, beautifully organized, and my mother was there to witness it.

===

Visit to India and the 2015 SASTRA Ramanujan Prize to Jacob Tsimerman

In the second half of November, Madras had torrential monsoon rains, almost non-stop for two weeks, the like of which the city and the region had not experienced ever before. In view of these rains, the reservoirs around the city were full and overflowing, and so it was decided to release water from these reservoirs to prevent their collapse. As a result of both the rains and the opening of the reservoirs, the city of Madras was in floods — yes the entire city was under several feet of water. Some residents moved to the upper floors and some others had to be rescued from their homes by boats. My family home Ekamra Nivas is constructed in such a way that it is about a foot and a half above the ground. Due to the flooding, our home was surrounded by water as in a moat, but luckily the water did not rise high enough to enter the house. It would have been a disaster if it had. But our prayers were answered. In view of this flooding, I was not sure whether we would be able to make our usual trip to India in December. It was remarkable that the flood waters receded and the city recovered in a week's time. So we did make the trip after all.

Mathura, my mother, my daughter Lalitha, and her kids Kamakshi and Keshav, departed for India on December 12. We halted for two nights in Singapore, staying as usual at the Conrad Hotel. A highlight of the Singapore stay was a visit to the zoo where my grandchildren had "breakfast with the Orangutans"! The Singapore zoo has a beautiful park like setting, and this breakfast show is one of its major attractions. On reaching Madras, we were relieved that there was no major damage to our home due to the floods, but there were several minor repairs that we had to do. In view of the concerns we had about the floods, and the time it would take to make repairs at Ekamra Nivas, we arranged for Lalitha and our grandchildren to leave for Coimbatore upon arrival in Madras and spend a few days there with Mathura's brother Sriram and his family.

The main academic event for me in India was the SASTRA Ramanujan Conference and the SASTRA Ramanujan Prize to Jacob Tsimerman.

Jacob Tsimerman had made deep and highly original contributions to diverse parts of number theory, and most notably to the famous André-Oort Conjecture. Much of Tsimerman's research stems from his PhD thesis at Princeton University in 2010 under the direction of Professor Peter Sarnak. The thesis concerns arithmetical questions around the Andre-Oort conjecture and makes substantial progress towards it.

The Andre-Oort Conjecture states that special subsets of Shimura varieties which are obtained as Zariski closures of special points, are finite unions of Shimura varieties which are special algebraic varieties. Yves Andre initially stated this conjecture for one dimensional subvarieties, and subsequently Frans Oort proposed that it should hold more generally. The conjecture lies at the confluence of Diophantine problems and the arithmetic of modular forms. One of the techniques to attack the Andre-Oort conjecture is to obtain suitable bounds for certain Galois orbits of special points. A major achievement of Tsimerman was to obtain unconditional bounds up to dimension 6, thereby significantly improving on earlier work.

Tsimerman had made major contributions to many other fundamental problems. As a graduate student at Princeton, he collaborated with Manjul Bhargava and Arul Shankar to determine the second term in the asymptotic formula for the number of cubic fields with a bounded discriminant. In collaboration with Jonathan Pila, he also established significant results on multiplicative relations among singular moduli — a topic dear to Ramanujan.

Tsimerman was born in Kazan, Russia in 1988. In 1990 his family first moved to Israel and then in 1996 to Canada, where he participated in various mathematical competitions from the age of 9. In 2003 and 2004 he represented Canada in the International Mathematical Olympiad (IMO) and won gold medals both years, with a perfect score in 2004. He was a doctoral student at Princeton under Peter Sarnak and got his degree in 2011. After holding a post-doctoral position at Harvard, he started his term as a regular faculty member at the University of Toronto in 2014.

Tsimerman, his girl friend Ila Varma, and Ila's mother, arrived in Madras from Delhi on December 18 afternoon. They had a full day's rest in Madras before we departed for Kumbakonam on December 20 morning. Ken Ono who was travelling in India, arrived in Kumbakonam that evening accompanied by his PhD student Robert Schneider and Schneider's wife. Tsimerman was awarded the SASTRA Ramanujan Prize on December 21 morning during the inauguration of the SASTRA Conference. That evening, after the sessions concluded, Ken Ono showed me and a few

SASTRA Deans, the movie "The Man Who Knew Infinity" because he had a CD of the movie with him. The movie had not yet been released, and so he had to get special permission from the producers to show the movie. I very much appreciated this gesture of both Ono and the movie producer. Ono said that since I could not go the official opening of the movie at the Toronto Film Festival in September, he arranged this viewing for me. The movie opened in theaters only in April 2016, and so this was a preview he arranged.

On December 22, Ramanujan's birthday, we had originally scheduled just two talks in the morning: Tsimerman's Ramanujan Commemoration Lecture on the André-Oort Conjecture, followed by Ken Ono on certain equivalent versions of the Riemann Hypothesis. We accommodated one more talk that morning — that of Ila Varma who had just finished her PhD at Princeton under the joint supervision of Richard Taylor and Manjul Bhargava. After the talks on December 22 morning, we all left by van for Madras from where our guests flew out to their destinations.

On the six hour van drive to Madras, I had long discussion with Schneider about my early work on duality between prime factors and how one can obtain partition analogues of various identities in multiplicative number theory. Although I had used these ideas to motivate some of my research in partitions, I never published anything on such connections between multiplicative number theory and the theory of partitions. Schneider was keenly interested in pursuing such connections and he has significantly extended my rudimentary ideas in various ways in his PhD thesis and later work.

==

Lalitha and my grandchildren Kamakshi and Keshav arrived in Madras from Coimbatore on December 24 morning and spent the day at Ekamra Nivas (where all repairs had been completed), and we all departed for Hong Kong that night by the new Cathay Pacific Airways Airbus A-330 flight. The highly acclaimed Marco Polo Service of Cathay Pacific lived up to its reputation and our expectations. We had a delicious South Indian breakfast on board as we enjoyed the view of the island studded South China Sea prior to landing at the new Hong Kong International Airport. It was a long and scenic drive from the airport to central Hong Kong where we checked in at the fabulous Conrad Hong Kong Hotel. We were given inter-connecting rooms on the 52 floor with a breathtaking view of the causeway, of Hong Kong Harbor, and of Kowloon on the other side. This visit to Hong Kong was primarily for the enjoyment of Lalitha and my grandchildren.

We visited the Botanical Gardens where the kids enjoyed seeing a variety of birds in the ponds and in the aviary. We took the Star Ferry from Hong Kong to Kowloon, where after shopping in bustling Nathan Road, we enjoyed a dinner at the Intercontinental Hotel from where one gets the most awesome view of Hong Kong across the causeway at night. Since it was Christmas day, the illuminations throughout the city were spectacular.

We did a day trip to see the Giant Buddha on Lantau Island. Although there is a cable car to the top of the mountain where the Buddha is located, one still has to climb 250 steps to reach the Buddha. We were amazed and impressed that four year old Kamakshi climbed these steps cheerfully and with ease! Kamakshi was counting the steps as she climbed and descended. I was at base camp taking care of Keshav while Mathura, Lalitha and Kamakshi went to the top to see the Buddha.

Even though Hong Kong is now part of China, British culture and habits can still be seen. At the Conrad, an English Afternoon Tea is served daily in the rooftop lounge, and we enjoyed this immensely.

We departed Hong Kong on December 30 morning for Tokyo where we boarded a Delta flight back to Atlanta. It was a pleasure to meet our dear friends Ganesh Kumar and Prema at the Delta lounge at Tokyo Narita Airport and to have them as co-passengers on the transpacific flight. Over the years, we have shared our flying experiences with Ganesh and Prema and it was a nice coincidence that we were on the same international flight.

After resting a few days in Gainesville, Mathura and I departed for Seattle on January 5 to attend the Annual Meeting of the AMS starting the next day. We had all of January 5 afternoon free, and since I am aviation enthusiast, we went to see the Boeing Museum of Flight. That night Mathura and I had dinner with Rochelle Kronzek of World Scientific, but I was feeling uneasy physically. It turned out that in the next few hours, I came down with a terrible bout of Gastro-entitis. So I rested throughout the morning in the hotel room and was fasting. Our good friends Edward and Alice Bertram, who now reside in Seattle after they both retired from Hawaii, joined us for lunch at the Hyatt where we were staying, but I did not eat anything. I felt much better in the afternoon to attend the sessions.

The next day, at the Prize Session, it was a pleasure to see my good friend Doron Zeilberger of Rutgers University receive the David Robbins Prize for his joint work with Christian Koutschan and Manuel Kauers formerly of the Research Institute of Symbolic Computation (RISC) in Linz. This was the second major AMS Prize that Zeilberger received, the first being the Steele Prize in 1998 for his revolutionary joint work with Herb Wilf. Koutchan and Kauers are former PhD students of Peter Paule at

RISC, where Paule heads a group that, among other things, specializes in algebraic combinatorics including extensions of the Wilf-Zeilberger (W-Z) method.

The next day, with the Bertrams we went around the lovely Puget Sound and called on my niece Radhika. That evening Mathura and I attended the AMS Fellows Reception along with the Bertrams. We were back in Gainesville the next day.

The academic year 2015–16 was when I shifted the focus of my research back to analytic number theory because I was working closely with, and guiding, my PhD student Todd Molnar on a problem involving the local distribution of the number of small prime factors since it revealed an interesting variation of the classical theme that was previously unnoticed. So I gave three talks in the Number Theory Seminar in February on this theme.

=======================================

Gainesville Int'l Number Theory Conference, March 2016

The International Conference in Number Theory, affectionately called Alladi60 by the organizers, was a five-day event (Thursday March 17–Monday, March 21) in Gainesville, featuring about 200 participants from around the world, of whom nearly 100 gave talks. Frank Garvan spearheaded the scientific organization of the conference: sending the invitations, creating and maintaining the conference website, and arranging the schedule of talks. He was ably assisted by Office Manager Margaret Somers who took care of the hotel accommodations and other local arrangements. Ali Uncu (PhD student of Berkovich), lent much needed technical support and expertise including the video recording of the main lectures; he was assisted by Chris Jennings-Schaffer (Garvan's PhD student) and graduate student Jason Johnson. Cyndi Garvan was in charge of the social program; she efficiently arranged the slide show and talks at the conference banquet. Of course, all along, George Andrews was providing valuable guidance to everyone.

During the budget negotiations with the Department and the College in Fall 2015, it was agreed that the Erdős Colloquium and the Ramanujan Colloquium in 2016 would be two of the featured lectures of the conference — for the latter colloquium, George Andrews, the sponsor, gave his nod of approval. I had always wanted to host Hugh Montgomery of the University of Michigan, and now an opportunity arose: we invited him to deliver the Erdős Colloquium at the conference, to which he consented graciously. Since James Maynard of Oxford University had done some of the most exciting recent work in analytic number theory, we invited him to deliver the Ramanujan Colloquium and the two more specialized talks; he too accepted our invitation. To give the Opening Lecture of the conference, we invited Manjul Bhargava of Princeton University (whose revolutionary work had earned him the Fields Medal in 2014), and he happily agreed to do so. These featured talks by high profile speakers had a positive effect in attracting a large number of active researchers, young and senior.

Bhargava's brilliant opening lecture on "Square-free values of polynomial discriminants" set the tone for the conference. His lectures are crystal clear and it is an inspiration and pleasure to hear him speak.

The veteran Ron Graham followed Bhargava, and to me it was an honor that Graham came because I had known him since my graduate student days. The rest of the day comprised of main lectures in analytic number theory by Aleksander Ivic (Belgrade), Roger Baker (BYU), Dan Goldston (San Jose), Helmut Maier (Ulm), Janos Pintz (Budapest), Gerald Tenenbaum (Nancy), Cameron Stewart (Waterloo, Canada) and Winnie Li (Penn State). In view of the number of eminent mathematicians who were speaking, the main lectures were each of 40 minute duration.

All lectures on the first two days were at the Straughn Center on the fringe of the campus, but on Saturday and Sunday, we had the talks in Little Hall, the home of the mathematics department, since there were no classes during the weekend. It was kind of Rochelle Kronzek of World Scientific, and Elizabeth Loew and Marc Strauss of Springer, not only to be present at the conference, but to have book exhibits as well.

James Maynard gave the Ramanujan Colloquium at 5 pm on the first day. This lecture alone was arranged at Pugh Hall in an auditorium that seats five hundred. The lecture was attended by several faculty and students from various science departments, and so the choice of Pugh Hall was prudent. Dean Richardson of CLAS made the Opening Remarks. Maynard gave an account of recent progress on gaps between primes including his own path-breaking work, one aspect of which involves small gaps between primes. Another of Maynard's major achievements is his resolution of the Erdős $10,000 problem on large gaps between primes, but this was also simultaneously and independently solved in a quadruple paper by Kevin Ford, Ben Green, Sergei Konyagin, and Terence Tao. It was a wonderful gesture on the part of Ron Graham to give Maynard a cheque for $5,000, when he finished delivering his Ramanujan Colloquium. The cheque for the remaining $5,000 was given by Graham to Kevin Ford after Ford's lecture the next morning, with that amount to be shared by the authors of the quadruple paper. The talks earlier in the day by Goldston, Maier, and

Pintz, dealt with gaps between primes and so this conference featured talks reporting the latest advances on celebrated problems on prime numbers.

After Maynard's Colloquium, there was a Wine and Cheese Reception at the Keene Faculty Center during which there were speeches by Richard Askey, Frank Garvan, Alex Berkovich, and Doug Cenzer recalling their long and close association with me. Christian Krattenthaler (University of Vienna), delighted the audience with a piano recital before the speeches. Christian is very meticulous. On the previous day, he went to the Keene Center to test the piano and practice on it.

On the morning of the second day, after Maynard's second talk, we had the two main lectures by Wadim Zudilin and Doron Zeilberger on irrationality. I was pleased that my early work in 1980 at the University of Michigan with Michael Robinson on irrationality measures still drew attention, but of course major advances had been made since then as reported by Zudilin and Zeilberger. I was also pleased that Michael Robinson attended the conference. He now works for the Center for Computing Sciences in Maryland, and it was nice to see him after many years.

The second day afternoon was mainly devoted to partitions, q-series and modular forms with lectures by Ken Ono (Emory), Mourad Ismail (Central Florida), James Lepowsky (Rutgers), and Christian Krattenthaler (Univ. Vienna). Kathrin Bringmann (Cologne) was invited to speak, but could not come; her post-doc Ben Kane gave a fine lecture in her place.

The final lecture of the second day was the featured Erdős Colloquium by Hugh Montgomery (Michigan), for which Vice-President of Research David Norton gave the Opening Remarks. Montgomery gave a thorough account of the history and the state of the art concerning Littlewood Polynomials, mentioning some open problems in the end. Montgomery's lectures are always enlightening, and indeed his talk inspired Zeilberger to work on this topic and contribute a paper to the conference proceedings. That night Mathura and I hosted a dinner party at our home for all the conference participants. The north Indian dinner catered by the restaurant Amrit Palace in Ocala, was appreciated by all, especially with the food being freshly prepared at our house and served piping hot.

On Saturday the lectures moved to Little Hall. After Maynard's third and final lecture, Kannan Soundararajan (Stanford) and his post-doc Robert Lemke-Oliver gave a beautiful lecture on a surprising bias among the primes they had observed recently. This bias was widely reported in the press soon after its discovery, but this lecture in Gainesville was the first conference presentation of this important observation. It was followed by the talk of Dorian Goldfeld (Columbia) on the additive arithmetic function $A(n)$ that I had investigated with Erdős during my undergraduate days in India. One of the results that Erdős and I had established was that $A(n)$ is asymptotically half the time odd and half the time even. This followed easily from the Prime Number Theorem. In his talk, Goldfeld made a significant extension of this by establishing that $A(n)$ is uniformly distributed in all residue classes of integers $k \geq 2$ using more advanced techniques involving Dirichlet L-functions. I was pleased that my early work on $A(n)$ continues to attract attention. The remaining two main lectures in the morning were by Peter Paule (Linz, Austria) and Mel Nathanson (Lehman College, CUNY). Peter spoke about the uses of computer algebra in discovering and proving Rogers-Ramanujan type identities and in that process discussed my work on the Göllnitz-Gordon identities.

During the lunch break that day, we had a preview of the movie "The Man Who Knew Infinity", one month before its screening in theaters. Manjul Bhargava, who along with Ken Ono was a scientific advisor to the movie producer, made a special arrangement for the movie to be shown to the conference participants. Manjul gave a lovely introduction before the showing of the movie. After the showing, some of the participants including George Andrews, Bruce Berndt, Richard Askey, Robert Vaughan and me, were interviewed by a representative of the movie production company, because they were preparing a video CD to publicize the movie.

The entire afternoon of Saturday consisted of three parallel sessions of talks. I tried my best to attend as many talks as possible.

The Conference Banquet was on Saturday night at the Paramount Plaza Hotel. Cyndi Garvan, who very competently organized the evening program, gave a comprehensive slide show of my academic career. This was followed by talks of George Andrews (Penn State), Bruce Berndt (Illinois), Elizabeth Loew (Springer), Edward Bertram (Hawaii), Kannan Soundararajan (Stanford), Mike Hirschhorn (Univ. New South Wales), Dorian Goldfeld (Columbia), Doron Zeilberger (Rutgers) and Sukumar Das Adhikari (Harish Chandra Institute). Ron Graham started his speech by putting up a group photo of the conference "Computers in Number Theory" that took place at Oxford University in August 1969. It was a major conference attended by several leaders in number theory and a good number of younger researchers who went on to become leaders themselves in the decades that followed. Graham said in his speech that this Gainesville conference compares well with that Oxford conference in quality and numbers, and a healthy mix of senior and junior researchers. He hoped, that as was the case with the Oxford conference, many of the young researchers of the Gainesville conference would go on to become leaders in their areas of work.

Since every knows I wear scarves in cold weather — not one, but two — Mike Hirschhorn presented me two scarves from Australia made of fine merino wool — by the bye, cold according to me is when the temperature drops below 70 degrees Fahrenheit! Adhikari, as per the Indian tradition, draped me in a silk shawl. I was deeply touched by the generous remarks made by the speakers about me and in my response I said I could not adequately thank the conference organizers, the speakers and participants for their kind words and support. In my speech I expressed my gratitude to all who had encouraged me in my career.

It is customary in such conferences held in connection with a person's birthday, to have family members speak at the banquet as was the case with the Andrews birthday conferences. Although my entire family was present at the banquet, none of them spoke because they had all taken part in the program back in November 2015 at my 60th birthday party. I should say, Mathura took great effort and pleasure in planning and arranging the big dinner at our house on the night before the conference banquet.

On Sunday, the talks continued in Little Hall, with main lectures in the morning by Robert Vaughan (Penn State), Hershel Farkas (Hebrew University), Richard Stanley (Univ. Miami), Richard Askey (Wisconsin), Alex Berkovich (Univ. Florida) and Ole Warnaar (Univ. Queensland). Vaughan spoke about the recent advances on Waring's problem, a topic in which he is the world authority. Farkas, whose expertise is on Riemann surfaces, discussed the Schottky problem. Richard Stanley, arguably the world's most eminent combinatorialist, had just retired after a long tenure at MIT and moved to the University of Miami. Dick Askey has such an encyclopedic knowledge of special functions, that his lectures always provide new insights; during his talk, I found Hugh Montgomery paying special attention and taking detailed notes. In a talk entitled "Going down memory lane", Berkovich gave an account of our collaboration in the theory of partitions. Sunday afternoon was again devoted to a set of three parallel sessions. In the evening, after the sessions, my colleagues and I took small groups of participants to various restaurants in Gainesville.

On Monday, the last day, the talks moved back to the Straugn Center. In the first main lecture of the day, George Andrews spoke about refinements of what he has generously dubbed as the Alladi-Schur theorem. I confess that my variation of Schur's theorem, although elegant, is not deep, but in the hands of an expert like Andrews, a significant refinement using my version of Schur's theorem was obtained. Peter Elliott (Colorado), the world's premier authority in Probabilistic Number Theory, gave a comprehensive account of multiplicative functions from a group theoretic point of view. The other main lectures in the morning were by Jean-Marc Deshouillers (Bordeaux), Bruce Berndt (Illinois), Michael Schlosser (Univ. Vienna), and Sergei Suslov (Arizona State). Suslov's talk was in mathematical physics, and I was touched that he started his lecture with a picture of my father and referred to his work in elementary particle physics.

After one more round of parallel sessions in the afternoon, the conference concluded with the Department Colloquium of Wadim Zudilin (Univ. Newcastle, Australia). Our department colloquia are held on Mondays at 4 pm, and so we made the last conference lecture on Monday to be department colloquium. Zudilin, is an authority in transcendence, but more recently has established some of the strongest results on irrationality of values of the zeta function, results such as one out of a set of zeta values at certain odd integers larger than 3 is irrational. It was a fitting conclusion to the five day conference that was packed with lectures reporting the most significant advances in several aspects of number theory.

During the week following the conclusion of the conference, I received several warm messages from participants saying that it was one of the best conferences they had attended. The credit goes to all those who helped organize the conference. The refereed conference proceedings edited by George Andrews and Frank Garvan was published by Springer in March 2018. It was considerate of Marc Strauss of Springer to agree to include about forty pictures of the conference participants in the volume. A fine report of the conference written by Rochelle Kronzek appeared in The Asia Pacific Mathematics Newsletter published by World Scientific. The Hindu, India's National Newspaper also carried a report of the conference based on material sent by George Andrews.

In early June, I received a letter from Hugh Montgomery saying how much he enjoyed the Gainesville conference. Along with that letter, he sent three reprints of Ramanujan's papers — yes, actual original reprints! — as a 60th birthday gift for me. These reprints are of the following papers:

(i) S. Ramanujan, "On the expression of a number in the form $ax^2 + by^2 + cz^2 + dw^2$", *Proc. Cambridge Phil. Soc.*, **19** (1917), part I, 11–21.

(ii) S. Ramanujan, "Some properties of $p(n)$, the number of partitions of n", *Proc. Cambridge Phil. Soc.*, **19** (1917), part V, 207–210.

(iii) S. Ramanujan, "Congruence properties of partitions", *Math. Zeitschrift*, **9** (1921), 147–153.

It was in (i) that Ramanujan wrote down 55 examples of quaternary integer coefficient quadratic forms that take all integer values. Such quadratic forms are called "universal". The proof that Ramanujan's had provided the complete list of all such universal quadratic forms was given by Manjul Bhargava and Jonathan Hanke in 2005, and presented by Bhargava in his Ramanujan Commemoration Lecture in December 2005 when he was awarded the First SASTRA Ramanujan Prize. Papers (ii) and (iii) relate to Ramanujan's path-breaking discovery of the partition congruences and his ingenious proofs. Ramanujan died in April 1920 and so (iii) was published posthumously by Hardy based on Ramanujan's notes.

In his letter, Montgomery said that Hardy gave these reprints to Harold Davenport, who in turn passed them down to Montgomery. Davenport was Montgomery's thesis advisor at Cambridge. I could not have received a more valuable 60th birthday gift. It was thoughtful of Montgomery to have handed these reprints to me. Montgomery wrote apologetically that he had written his name on the top of the cover page of each of these reprints, and that might mar their value. In my response thanking him, I said that I am glad that his name is written on the reprints, because he is my link with Trinity College where Hardy and Ramanujan were!

Mike Hirschhorn actually stayed on in Gainesville for three weeks after the conference. He used his extended visit to the University of Florida to make significant progress on his book "The power of q" which was published in the book series *Developments in Mathematics* (Springer) in 2017. Over the years, Hirschhorn has used the Jacobi Triple Product identity in clever ways to establish several appealing results, and this identity of Jacobi dominates the discussion in his book.

===

Visit to Austria and Hungary, April–May, 2016

As soon as classes for the Spring term ended on April 20, Mathura and I departed for Austria on April 22 in response to a handsome invitation from Peter Paule for me to give several lectures at the Research Institute of Symbolic Computation (RISC) in Linz. Christian Krattenthaler arranged my colloquium at the University of Vienna, and Janos Pintz invited me to give a talk at the Hungarian Academy of Sciences in Budapest.

Mathura and I flew via Amsterdam, arrived in Vienna at 7 pm on April 23, and immediately took a train from the airport to Linz, where we checked in at the elegant Wolfinger Hotel. The next day being a Sunday, Peter Paule picked us up in the morning and took us to the lovely village of Poestlingberg, located atop a hill nearby. He treated us to a fine Austrian lunch at a restaurant there which is attached to a castle that commanded a fine view of the city. The great river Danube on which Linz stands actually flows very close to the Wolfinger Hotel, and so Mathura I strolled along the banks of the Danube in the afternoon admiring the trees and plants which had blossomed profusely due to the advent of Spring.

The next morning (Monday), Peter picked us up to take us to RISC which is housed in a castle in the countryside about thirty minutes from Linz. As Director of RISC, Peter heads a very active group of number theorists. Over the years, he has mentored several outstanding PhD students including Axel Riese, Manuel Kauers, Carsten Schneider, and Silviu Radu, all of whom I met in Linz on this visit and my earlier visit. Many of Paule's students are either at RISC or at the Johannes Kepler University in Linz.

My first talk at RISC was a colloquium that day, and I spoke on "Reflections on the life and work of Paul Erdős". Some months earlier, I had sent Peter a list of titles of talks with their abstracts and asked him to choose from among that list.

Peter did not want to burden me with too much work, and so he allowed us free time for sightseeing in Austria. So on Tuesday, Mathura I went by train to lovely Salzburg, where we spent the entire day. Of course we saw the Mirabell Palace and the Mirabell Gardens where scenes of the movie "The Sound of Music" were filmed. Salzburg is steeped in Western classical music, and is the birth place of Mozart. Especially impressive was the Salzburg Museum depicting the history of the city and of Mozart's life. We visited the modest home of Mozart where he composed some of his famous operas as a boy, and I compared it with the equally modest home of Ramanujan in Kumbakonam from where a thousand theorems emerged.

On Wednesday, Bill Chen from Tianjin, China, was at RISC. Peter is Adjunct Professor at Chen's Center for Applied Mathematics the University of Tianjin, so Peter visits Tianjin regularly, and Chen comes to Linz periodically. Since Chen could not be present at the Alladi60 conference, he gave a Colloquium on Wednesday morning and graciously dedicated it for my 60th birthday! In my seminar talk which was right after Chen's colloquium, I spoke about partitions with non-repeating odd parts, since my paper on this topic had just appeared in the Annals of Combinatorics, for which Chen is the Founder and was Editor-in-Chief.

Thursday was another free day, and so Mathura and I strolled around the city, seeing the majestic cathedral and buying souvenirs. In the afternoon, we had refreshments at pastry shop where the Sacher Torte for which Linz is so famous, was actually introduced.

On Friday, Peter took us on an excursion in his car to Salzgammergut, the Lake District. It was a sunny day, and the view of the snow capped Austrian Tyrol (mountain range) from the banks of the Transee Lake was spectacular. He hosted a lunch at a monastic restaurant on the lake shore, where we enjoyed fine Austrian vegetarian dishes. The day concluded with a visit by cable car to the top of mount Gruenberg, from where we had panoramic views of the lakes, the valley, and the Austrian alps.

On Saturday, Mathura and I went to Vienna to spend the entire day there. We were met at the train station in Vienna by Gaurav Bhatnagar who was on an extended visit to the University of Vienna working with both Christian Krattenthaler and Michael Schlosser. Gaurav is a former PhD student of Steve Milne at Ohio State, and after getting his doctorate, went back to Delhi, India, to work in industry. But he has kept active in his mathematical research. It was very kind of Gaurav to accompany us to Schonbrun Palace whose gardens are world famous. These gardens are laid along the slope of a hill, at the top of which is the Glorietta restaurant where we had lunch and enjoyed fine pastries, while admiring the panoramic view of the palace and the gardens. We spent much of the afternoon strolling through central Vienna, and seeing the St. Stephens Cathedral and the Hofburg Palace. In the late afternoon, we were invited to High Tea at the home of Michael Schlosser. Schlosser's wife had made it almost like a dinner because she had taken pains to prepare so many delicious Austrian vegetarian dishes. Christian Krattenthaler joined us for High Tea at Schlosser's home. It was a nice ending to an eventful day in Vienna.

On Sunday, we returned to Salzburg, this time taking a four hour Sound of Music Tour. The movie Sound of Music, which was a box office hit worldwide, was a total flop in Austria, because the locals did not like the changes to the story that the movie producers made. The movie was a hit in India and so this tour is popular with Indian tourists.

After this relaxed weekend of sightseeing, I gave two talks at RISC on Monday and Tuesday. My talk on Monday was the Second Colloquium and it was in analytic number theory on the distribution of the number of prime factors, and tracing the birth and growth of probabilistic number theory. My last lecture at Linz on Tuesday was given under the auspices of the SFB — the joint program in number theory of RISC and the University of Vienna. For that lecture, the number theorists at the University of Vienna including Krattenthaler, Schlosser and Bhatnagar came to Linz by train. For the SFB Colloquium I spoke on the dual of Göllnitz' theorem. That night, Peter and his family hosted a dinner for me and Mathura at Taj Mahal, the Indian restaurant close to our Hotel Wolfinger. Thus concluded a most enjoyable and fruitful stay in Linz.

The next morning we departed Linz for Vienna and checked in at the Hilton Vienna. That afternoon, I gave two lectures at the University of Vienna at the invitation of Christian Krattenthaler. Christian is one Austria's most distinguished mathematicians. He leads an active research program in number theory which was supported at one time by the prestigious Wittgetstein award. Krattenthaler also is an able administrator and has served both as Chair and Dean. As I entered the mathematics department building, I was impressed to see a tablet containing a list of distinguished former faculty such as Kurt Gödel and Franz Mertens. The name of Mertens was especially relevant to my research in number theory.

My first talk in Vienna was a Student Colloquium entitled "How many prime factors does a number have?" and it was attended by several graduate and undergraduate students (and faculty). This was followed by the regular department colloquium. That night, Krattenthaler and the members of his number theory group took Mathura and me to dinner at Heuringen, a traditional Austrian restaurant in the suburbs of Vienna. This is one of Krattenthaler's favourite restaurants and also has an excellent selection of vegetarian dishes to our liking.

On Thursday, Mathura and I departed Vienna for Budapest by train. We took only our roll bags (small) because we were to come back two days later. After a two and a half hour train journey, we arrived at Kaleti station in Budapest where our host Prof. Janos Pintz met us and took us to the Hilton Budapest where we were accommodated for two nights. The Hilton has a fantastic location on the Buda side of Budapest, since it is situated right in the grounds of the Budapest Castle. After about an hour of sightseeing in and around the castle, Pintz took us to his home where his wife had, so thoughtfully, prepare a delicious and elaborate vegetarian dinner. Pintz' home is on the slopes of a hill, and from the balcony of his house we could get a splendid view of Budapest at night!

Janos Pintz had arranged my Colloquium at the Alfred Renyi Institute of The Hungarian Academy of Sciences on Friday morning. I gave a one and a half hour talk on the distribution of the number of prime factors, including the variation of the classical theme on which Todd Molnar was writing his thesis under my direction. Hungary is particularly strong in the study of additive functions, and in particular on the number of prime factors, and I was pleased that Imre Ruzsa and Gabor Halasz, two veterans of this field, attended my talk. They along with Vera Sos took me and Mathura to lunch after which we returned by train to Vienna. When we checked in at the Hilton Vienna, the desk clerk said with a downcast expression that no room was available for us. As I was about to protest, he smiled and said that we were upgraded to the Presidential Suite! I could not believe what I heard. One entire half of the top most floor was the Presidential Suite which was more than 3000 square feet in area, surrounded by a balcony on all three sides. We literally felt like Royalty when we entered to suite. When the bellboy brought us our luggage, he said the last occupant of the suite was an Arab sheik from Dubai. Lest he would think of us as misers, I gave him a huge tip for bringing our luggage!

It was a bright day and so I walked along the balcony and took panoramic pictures of Vienna. We had only one night at this Presidential Suite, and so Mathura and I decided not to go out for dinner but order room service instead which we enjoyed in the spacious dining room of the Suite. Of course, I was also extravagant in tipping the waiter for room service! The stay at the Presidential Suite was a royal ending to a fantastic visit to Austria and Hungary where our hosts Paule, Krattenthaler and Pintz took solicitous care of us.

==

Two days after our return from Austria, my mother and I departed for India on May 9 via the Pacific. It was mainly to attend to family matters in Madras but I gave a few talks during the visit. My friend GP arranged a lecture at the Madras Book Club, where I spoke about my book [B1] "Ramanujan's Place in the World of Mathematics". There were more than one hundred people in the audience including leading members in several walks of life. I also gave a talk under the auspices of The Alladi Foundation on the theme "Stage productions on Ramanujan - a comparison". Ramanujan is a hero to every Indian, and so this talk was attended by persons young and old and from different segments of society.

My mother and I departed Madras on May 26 for Honolulu via Bangkok. We were joined in Honolulu by Mathura, Amritha, and son-in-law Jis. We had a glorious ten-day holiday in the Hawaiian islands splitting time between the islands of Oahu and Hawaii (Big Island). Mathura's aunt Gigi joined us for stay on the Big Island where the highlight was the visit to the Volcanoes National Park, and the view of the crimson glow of lava fumes emerging from the Halemaumau Fire Pit as seen from the Jaggar Museum

Observatory. An interesting coincidence was that our friends Ganesh Kumar and his wife Prema who share our passion for Hawaii were also at the Big Island at the time, and it was nice to meet up with them.

After return from India and Hawaii, I spent much of June and July writing a long review of the movie "The Man Who Knew Infinity" at the invitation of *Inference*, an International Journal for the Review of Science, published electronically from Paris. It is a fantastic movie in which there is never a dull moment. But there a few unnecessary dramatizations which I point out in my review. Movie producers introduce dramatizations for special effects and to create a sense of thrill. But Ramanujan's life story is so incredible, that dramatization is not necessary to create excitement. Inference modified my review to suit their format and published it [A3] in September 2016; subsequently, in December 2016, the original version of my review appeared in the Asia Pacific Mathematics Newsletter [A4] published by World Scientific.

10.9   Return to analytic number theory (2016–17)

After working intensely in the theory of partitions and q-series for a quarter century, I returned to analytic number theory in 2016. The return was prompted by my PhD student Todd Molnar who wanted to work only in analytic number theory. In 2015, I had observed an interesting variation of the classical theme while looking at the local distribution of the number of "small prime factors", and this was intriguing. So I asked Molnar to pursue this problem and this became his PhD thesis which he submitted in 2017. I started working with him on this problem from 2015 onwards, and spent much of 2016–17 completing this work. Since then, I have been mostly working on problems in analytic number theory. I will now describe briefly the problem I gave Molnar.

There is a vast literature on the distribution of $\nu(n)$, the number of prime factors of n, and indeed this forms the core of the subject of Probabilistic Number Theory. In earlier sections, I have described the fundamental work of Erdős, Turán, Kubilius, Sathe and Selberg on the distribution of $\nu(n)$. The results of Sathe and Selberg, concern the local distribution of $\nu(n)$, with uniform asymptotic estimates in k for $N_k(x)$, the number of positive integers $n \leq x$ with k prime factors. In view of my earlier work on integers all of whose prime factors are large, or all of whose prime factors are small, I was interested in discussing the local distribution of $\nu_y(n)$, the number of prime factors on n which are less than y. This is what I mean by "small prime factors". Here we consider positive integers below a given magnitude x, and the parameter y can vary between 2 and x. What I noticed was that for the quantity $N_k(x, y)$, namely the number of positive integers $n \leq x$ for which $\nu_y(n) = k$, the asymptotic behavior different when y is small in comparison with x and this is a *variation of the classical theme*. Molnar and I determined when this change takes place. Although $\nu_y(n)$ figures prominently in the proof of the celebrated Erdős-Kac Theorem, it is the global distribution of $\nu_y(n)$ that is discussed there and not the local distribution, and so this variation of the classical theme had escaped attention! So this was good thesis problem to give Molnar, but before doing so, I wanted to be sure that it was a new problem. I contacted Gerald Tenenbaum in Nancy, France, the leading authority on such questions, and he confirmed the novelty of my observation. In fact, he got interested in the problem, approached it from a different direction, and wrote a short paper supplementing our results. The Alladi-Molnar paper and Tenenbaum's paper on this problem appeared back-to-back in the same issue of *The Ramanujan Journal* in January 2020.

During September 17–18, I was a principal speaker at the PANTS Conference on the lovely campus of University of North Carolina, Greensboro (UNCG), and there I presented this work on $N_k(x, y)$. PANTS, an acronym for Palmetto Analytic Number Theory Series, is an annual conference that started in 2006. PANTS is organized by a consortium of universities in the southeastern USA and so is run in conjunction with SERMON, the South Eastern Regional Meeting on Numbers. The southeast coast of the USA is known for the Palmetto palms, and the hence the choice of the name of the conference series. The PANTS conferences are organized for the benefit of graduate students and therefore is funded by both the NSF and the NSA. This two day conference has four principal (one hour) lectures, two of which are delivered by senior researchers, one by a junior researcher such as a post-doc or an assistant professor, and one by a graduate student who has excelled in research. These principal talks are supplemented by shorter presentations by graduate students. I was invited by Matt Boylan of the University of South Carolina to deliver an hour lecture as a senior researcher. The second hour lecture by a senior researcher was delivered by Roger Baker of Brigham Young University.

The conference was on a weekend, and so Boylan also arranged for me to deliver a Colloquium in the mathematics department of UNGC on the day before the conference. So I arrived in Greensboro on Thursday evening and checked in at the lovely Troy Bumpas Inn (a house converted into an Inn) on the fringe of the campus. The suite I was given was spacious and elegantly furnished in southern American style.

The next afternoon I gave the Mathematics-Statistics Colloquium on the theme "Stage productions on Ramanujan - a comparison", which appealed to the large general audience. The colloquium host Filip Saidak, was happy with the theme of my talk.

The opening lecture of the conference was mine at 10 am on Saturday. I spoke on "The number of small prime factors with restrictions - variation of the classical theme". Roger Baker spoke after me. PANTS is a well organized conference series, and I was glad to have been given an opportunity to participate in it. Brett Tangedall of UNCG was the main organizer of the 2016 PANTS Conference.

The next major academic event was the SASTRA conference in India in December. Before that, during the Thanksgiving break, our entire family had a holiday in Cancun, Mexico, where we attended the magnificent "destination wedding" of one of Mathura's students. Cancun is known the world over as a fabulous beach resort area, which it is, but there is plenty of Mayan history there. So besides attending the wedding, we had a whole day trip to see the Mayan ruins of which the most spectacular was Chichen Itza, one of the wonders of the world.

===

Visit to India and the 2016 SASTRA Prize to Kaisa Matomaki and Maksym Radziwill

Mathura, my mother and I departed for India on December 9, flying via the Atlantic on KLM and Jet Airways. It was first time I was on a long international sector (Amsterdam-Delhi) on Jet Airways, and the service was excellent. I was sorry when Jet Airways ceased operations in 2019.

Before the SASTRA Conference, I attended a conference in Varanasi jointly organized by The Indian Mathematics Consortium (TIMC) and the American Mathematical Society at the Benares Hindu University (BHU) which had celebrated its centenary in 2015–16. The TIMC is a newly formed organization run by a group of mathematical societies in India such as The Ramanujan Mathematical Society. At this joint meeting, there was a session on number theory to which I was invited to speak.

I departed Madras for Varanasi (formerly known as Benares or Kasi) on December 15 and reached there via Bombay in the early afternoon. The drive from the airport to the city was long and tiring because the roads were congested and in terrible shape. Varanasi is the holiest city in the Hindu religion, and its history dates back to 800 BC. The city is supposed to have been created by the Hindu God Shiva. There are more than 20,000 Hindu temples in Varanasi of which the most famous in the Kasi Viswanath

Temple for Lord Shiva. It is believed that Buddha created Buddhism here in 528 BC and gave his first sermon in Varanasi. In spite of the holiness, I found the city to be extremely dirty, and so the drive to the city center was not pleasant. After checking in to my comfortable hotel, I was picked up and taken to the campus of BHU, which was beautiful. I was surprised because, even though the city surrounding it had declined in cleanliness, BHU had maintained its campus superbly. It was a pleasure to see the old colonial style buildings surrounded by spacious lawns and gardens. Since it was close to 5 pm when I reached campus, the sessions that day had ended and so I could only attend the cultural program and a banquet on the lawns that evening. There were several hundred participants at the banquet and it was good to meet many long time acquaintances.

The next day my lecture was at 11:30 am in the Number Theory Session organized by Sukumar Das Adhikari of the Harish-Chandra Institute in neighboring Allahabad. I spoke about the local distribution of number of small prime factors and the variation of the classical theme, and the Indian number theorists were pleasantly surprised that I had returned to analytic number theory! I received a text message that my Jet Airways flight to Bombay was leaving two hours earlier, and so I departed BHU campus right after lunch to catch my flight.

I was glad that I had an opportunity to visit BHU. I had heard much about this university from my father who used to say that Sir S. Radhakrishnan and Sir C. P. Ramaswami Iyer, two great contemporaries and close friends of my grandfather, had served as Vice-Chancellors of BHU.

Soon after my return to Madras, the SASTRA conference delegates started arriving. On December 18, Steven Weintraub of Lehigh University arrived; as Secretary of the AMS, he had attended the AMS-TIMC Conference in Varanasi. Since we were departing for Kumbakonam only on December 20, we could take him to concerts at the Madras Music Academy that night and the next evening. He enjoyed these concerts very much.

Maksym Radziwill, the SASTRA Prize Winner, arrived on December 19, but Kaisa Matomaki (from Turku, Finland) who shared the prize with him, could not come because she could not leave her infant child in Finland, nor could she bring her child to India. While I was in Austria in May, I invited Gaurav Bhatnagar to attend the SASTRA Conference; he arrived in Madras also on December 19.

It was decided to award the 2016 SASTRA Ramanujan Prize jointly to Kaisa Matomaki of the University of Turku, Finland, and Maksym Radziwill of McGill University, Canada, and Rutgers University, USA, because their revolutionary collaborative work on multiplicative functions in short intervals had shocked the mathematical community by going well beyond what could be proved previously even assuming the Riemann Hypothesis. It also opened the door to a series of breakthroughs on some notoriously difficult questions such as the Erdős discrepancy problem and Chowla's conjecture, previously believed to be well beyond reach.

Kaisa Matomaki made a prominent entrance on the world stage during 2007–09 when she established a number of significant results which are contained in about ten excellent research papers as well as in her outstanding PhD thesis of 2009 submitted to the University of London under the direction of Glynn Harman. Her papers span several fields — classical analytic number theory, sieve theory, the theory of modular forms, and Diophantine approximation. Her collaboration with Radziwill began in 2014.

As an undergraduate at McGill University during 2006–09, Maksym Radziwill wrote an incredible undergraduate thesis on large deviations of additive functions under the direction of Andrew Granville. His exceptional PhD thesis at Stanford University submitted in 2013 was written under the direction of Kannan Soundararajan who won the very first SASTRA Ramanujan Prize in 2005. Radziwill had several original papers on a variety of deep problems in probabilistic number theory, the theory of the Riemann zeta function and more generally on L-functions, as well as the theory of modular forms and elliptic curves.

The Matomaki-Radziwill joint work on multiplicative functions in their paper in the Annals of Mathematics in 2016 concerns the behavior of multiplicative functions in short intervals, and especially that of the Liouville lambda function which takes value 1 at an integer with an even number of prime factors (counted with multiplicity) and value -1 at an integer with an odd number of prime factors. It is well known that the statement that the lambda function takes values 1 and -1 with asymptotically equal frequency is equivalent to the famous Prime Number Theorem. More refined statements concerning the relative error in this equal frequency are related to the celebrated Riemann Hypothesis. Such equal distribution results were also known for short intervals, where by short one means intervals of type $[x, x + h]$, with h typically being a fractional power (< 1) of x. One would expect that such equal frequency would hold when h is of size x to an arbitrarily small positive power, but even with the Riemann Hypothesis, the best that is known is that powers larger than $1/2$ will work. Instead of asking for equal frequency in every short interval of length h, if we require that equal frequency should hold for "almost all" short intervals of length h, then one can reduce the size of h considerably; it was known that h can be made of size the 1/6-th power of x, and assuming the Riemann Hypothesis, Gao showed that h can taken as a power of $\log x$. Matomaki and Radziwill shocked the world by showing unconditionally that equal frequency holds almost always as long as h tends to infinity with x.

Kaisa Matomaki was born in Nakkila, Finland, in 1985. She did her Masters at the University of Turku and received the Ernst Lindelöf Award for the best Masters Thesis in Finland in 2005. After completing her PhD at the Royal Holloway College of the University of London in 2009 she returned to Turku where she is an Academy Research Fellow.

Maksym Radziwill was born in Moscow in 1988. In 1991 his family moved to Poland and then in 2006 to Canada. After completing his under-graduate studies at McGill University in 2009 he joined Stanford University to work for his PhD which he completed in 2013. He was a Visiting Member at the Institute for Advanced Study in Princeton in 2013–14, and held the Hill Assistant Professorship at Rutgers University during 2014–17, before joining the permanent faculty at McGill University, Montreal.

As usual, we departed Madras on December 20 morning for Kumbakonam and reached there early enough to show our guests the usual sights. That evening, Henri Darmon (McGill University) and Sujatha Ramdorai (University of British Columbia) arrived in Kumbakonam by train from Bangalore where they had attended a conference.

Henri Darmon inaugurated the SASTRA Conference on December 21 morning, and Steven Weintraub handed the SASTRA Ramanujan Prize to Radziwill. Darmon is a very distinguished number theorist with expertise in the area of elliptic curves. He is a former PhD student of Benedict Gross of Harvard University. Following this inauguration ceremony, Darmon gave the opening lecture. I followed Darmon and spoke about the local distribution of the number of small prime factors. Radziwill who had established some deep results on local distribution of additive functions, made some interesting comments after my lecture.

Radziwill delivered the Ramanujan Commemoration Lecture on the morning of December 22, Ramanujan's birthday, on his joint work with Matomaki. Following Radziwill, Sujatha gave the concluding lecture of the conference emphasizing the importance of good mathematical training in youth. In her talk, she described the programs for children being conducted at the *Mathematics Park* that she had created in India and is managing. We returned to Madras that night and dropped our foreign guests at the airport for their return flights.

On December 19, the day before my departure to Kumbakonam, my daughter Amritha and son-in-law Jis arrived in Madras from Cochin. They had come to India in connection with the 90th birthday of Jis' grandmother in Kerala, and after attending that celebration, came to Madras to spend a few days in Ekamra Nivas. They departed Madras only on December 25 and so I could spend some time with them after my return from the SASTRA Conference. We actually had a party at Ekamra Nivas on December 24 so that many of our friends and relatives could meet Amritha and Jis. That was a festive ending to the year.

Mathura, my mother and I departed Madras for the USA on January 2 morning. We broke journey in Atlanta for a few days before returning to Gainesville because of the AMS Annual Meeting (Jan 4–6) in Atlanta. I attended the Joint Prize Session where Henri Darmon was awarded the AMS Cole Prize for number theory. I spoke in a Special Session on Partitions and Related Topics organized by Dennis Eichhorn, Tim Huber and Amita Malik. After a relaxed weekend with Amritha and Jis in Atlanta, we were back in Gainesville on January 8 for the start of the Spring Term.

====================================

An interesting event at the start of the year was the India Republic Day Celebrations in Jacksonville on Saturday, January 28, for which I was invited to be the Chief Guest. January 26 is India's Republic Day, but in the USA, it is celebrated on the nearest Saturday so that many persons can attend. January 26, 1950 was when the Constitution of India was ratified and India became a Republic. My grandfather Sir Alladi played a pivotal role on the Drafting Committee of the Constitution, and so on January 26, 2107, The Hindu, India's National Newspaper, carried a feature article by V. Sriram (Historian of Madras) on my grandfather's role on the Drafting Committee. The India Society of Jacksonville had invited me specifically to speak on the theme "Sir Alladi and the Indian Constitution". I have with me in Gainesville copies of several important documents relating to my grandfather, and so I was able to give a nice account of how the Constitution of India was drafted. There was an audience of about 400 that day in Jacksonville, and to many young Indian boys and girls growing up in the USA, there were many things I told them they had never heard of before.

====================================

Visit to Mauritius, March 3–10, 2017

The main academic trip in Spring 2017 was to Mauritius where I was invited by the Ramanujan Mathematics Trust to deliver the Ramanujan Memorial Lecture. Ramanujan has made a deep and lasting impact on many branches of mathematics and his name is well-known throughout the world. But it would come as a surprise that he is a major force even in the remote island of Mauritius in the southernmost part of the Indian ocean off the coast of Africa! I was asked to be the Chief Guest at a Ramanujan memorial event at the University of Mauritius in March 2017 when prizes were distributed to the Third Ramanujan Mathematics Contest Winners. Mathura and I enjoyed our visit immensely and I now describe some of the unique aspects of this Ramanujan movement in Mauritius and the attractions (academic and geographic) of this charming island country.

Dr. Sattianathan Sangeelee, now retired, is a distinguished consultant physician from Mauritius who practiced in England. He was captivated by the legacy of Srinivasa Ramanujan and so he started a Ramanujan Trust in Mauritius in 2013 with support of the Mauritius Ministry of Education and the University of Mauritius. There are several prominent members of Mauritius who currently support the Trust and with their approval, Dr. Sangeelee takes the lead in every action of this Trust and attends to every detail. The University of Mauritius, in particular its mathematics department, provide academic expertise, advice, and facilities to the Trust.

Mauritius is a small island about 1200 miles east of the South African coast in the Indian ocean, and east of, but quite close to, Madagascar. The Republic of Mauritius consists of the island of Mauritius and a few smaller islands further east. Very close to Mauritius is the French island of Reunion, which made headlines when pieces of wreckage of the ill fated Malaysia Airlines Boeing 777 that had disappeared into the Indian ocean were found washed ashore on its beaches.

Mauritius was a Dutch colony in the 17th century and a French colony in the 18th century, when sugarcane plantations abounded. The French used African slave labor to clear the forests and brought immigrant laborers from India to work on the plantations. In addition, the French brought in skilled artisans from Pondicherry (a French outpost in India) to construct most of the stone buildings in Port Louis, the capital. The British took over Mauritius in 1810, but slavery was not abolished until 1835, and that too only after a bloody rebelion. Mauritius gained its independence in 1968 and became a Republic. Today more than 60% of the island's population are third generation Indians, like Dr. Sangeelee. French influence is still very strong; indeed both English and French are spoken in Mauritius.

There is much more to the Mauritius economy today than just sugarcane plantations. The verdant landscape is filled with orchards growing different types of tropical fruits like lychees and mangoes, and the breathtaking scenery attracts tourists and honeymooners for a vacation that is far less expensive than in Hawaii or the Cote d'Azur. Also, Mauritius is expanding in the sphere of education and the quality of mathematics instruction at the University is very good.

To identify and encourage the mathematically gifted students of Mauritius and to promote pure mathematics, Dr. Sangeelee had the brilliant idea to hold a Ramanujan Mathematics Contest each year for senior high school students, but undergraduates are also allowed to take part in it. The age limit is set at 32 which was Ramanujan's life span. The First Ramanujan Mathematics Contest was held in 2014; the Talent Exam questions were selected by the local mathematicians and the prizes were awarded by some prominent members of the Mauritian community. For the Second Ramanujan Mathematics Contest in 2016, Dr. Sangeelee wanted an international input and presence. So Dr. Sangeelee approached Bruce Reznick (University of Illinois, Urbana) who had considerable experience in making the exam questions for mathematics olympiads. Reznick not only graciously agreed to select the questions for the Second Contest, but also for future contests. Sattianathan then invited Reznick to Mauritius as the Chief Guest for the function in March 2016 when the prizes were awarded, and asked Reznick to deliver the Ramanujan Memorial Lecture. Thus the Ramanujan Mathematics Contest in Mauritius was off to a great start with international support for its mission. Since it is a Ramanujan Mathematics Contest, questions related to Ramanujan's work are included. The contest is for one and a half hour duration and one hundred applicants are selected to take part in it. The First Place Winner is given a Srinivasa Ramanujan Gold Medal and a cash prize of (Mauritian) Rupees 25,000 (about $700). Two runners-up are awarded a book prize and a cash prize of (Mauritian) Rupees 10,000 each. The book that was given as a gift was "The Man Who Knew Infinity", Robert Kanigel's widely acclaimed biography of Ramanujan.

After having read my book *Ramanujan's Place in the World of Mathematics* (Springer, 2012), Dr. Sangeelee invited me to be the Chief Guest for the function in March 2017 when the prizes for the Third Mathematics Contest would be awarded, and asked me to deliver the Ramanujan Memorial Lecture on the topic given by the title of my book. I had not heard of this Trust and was pleasantly surprised that it functioned in far off Mauritius. I was delighted to hear that Ramanujan had such a strong presence even in a far flung island, and since fostering the legacy of Ramanujan is one of my priorities, I readily accepted the invitation. I did not know what to expect, and so I communicated with Bruce Reznick to get an idea of what had happened the previous year.

It is a long journey to Mauritius from America by any route. Mathura and I travelled to London from where we flew by Air Mauritius on a 12 hour non-stop flight. We had a 12 hour transit in London and so we rested at the Heathrow Airport Hilton before boarding the Air Mauritius Airbus A-340 flight, which although very long, was extremely comfortable. I always admire the beauty of four engined jets, and the Airbus A-340, designed as a long range aircraft with four engines, is a beauty in every sense. The in-flight service on board Air Mauritius was superb. Upon arrival in Mauritius we were pleasantly surprised that such a small remote island had such a magnificent ultra-modern airport which was capable of receiving the mammoth double-decker Airbus A-380. We were well rested when we arrived there on a Sunday at noon, and we spent the rest of the day relaxing by the pool at the gorgeous Hilton Mauritius Resort (where we stayed) which has an enviable beach front location on the west coast of Mauritius. I was ready to begin work the next day.

On Monday morning, Dr. Sangeelee arranged a 40-minute television interview with me on the theme "Ramanujan as an inspiration to students." I was interviewed by four persons — (i) Dr. Sangeelee who was the interview coordinator, (ii) Professor Bhuruth, Chair of the mathematics department at the University of Mauritius, (iii) Dr. Vyapoori, the Vice-President of Mauritius, who is a mathematics enthusiast, and (iv) Dr. Aman Maulloo, Director of the Rajiv Gandhi Science Centre, Mauritius. As I was interviewed, there was a live audience of about 20 students, some of whom asked me questions in the end; that was also part of the program. It was a very enjoyable experience. The entire interview was shown on the Mauritius National Television that night. Mathura who was with me at all other events, could not attend this TV interview because at the same time she was asked to address a group of dance students next door at the Mahatma Gandhi Centre, on how she trains her students in Indian classical dance in Florida.

The Ramanujan Memorial Function was on Tuesday afternoon on the campus of the University of Mauritius. The auditorium was packed and the audience of about 200 comprised students, faculty, and dignitaries from different professions including the Mr. Arjoon Suddhoo, Chairman of Air Mauritius, who had very kindly sponsored my ticket on Air Mauritius (the trans-atlantic ticket was sponsored by the Ramanujan Trust). My one hour talk "Ramanujan's place in the world of mathematics", contained mathematical snapshots interspersed with stories and anecdotes to make it palatable to a general audience. On the dais with me, and attending the lecture, were the Vice-President of Mauritius, the Minister for Education, the Vice-Chancellor (= President) of the University, the Indian High Commissioner (= Consul General of India), the Mathematics Chair, and Dr. Sangeelee. Following my lecture, the prizes were awarded to the winners by the Vice-President of Mauritius. It was a fantastic function.

After the Ramanujan Memorial Function, we walked across campus to see the Bust of Ramanujan that Dr. Sangeelee had installed on the university campus in 2013; Dr. Sangeelee was inspired by Ramanujan's 125th birth anniversary in 2012 to commission this impressive bust which was sculpted in India, and flown to Mauritius in 2013.

On Wednesday morning, I addressed a group of undergraduate and graduate mathematics students at the University on the theme "The remarkable life and mathematics of Paul Erdős". Since I was in Mauritius on a visit in memory of Ramanujan, I emphasized how Erdős' proof of Bertrand's postulate, his very first paper that catapulted him to world fame, was related to Ramanujan's own proof, and that was how Erdős made his first mathematical connection with Ramanujan. I then discussed the Erdős-Kac theorem and the birth of probabilistic number theory, and showed the students how the origins of this subject could be traced back to a paper of Hardy and Ramanujan on round numbers. After my lecture on Erdős, the Vice-President of Mauritius graciously hosted a lunch.

The island of Mauritius has much to offer by way of natural scenery. On Monday afternoon, following my TV interview, Sattianathan's younger brother Ramanathan took us on an excursion to the beaches of the southwestern shore. Many beaches have gorgeous turquoise blue water. Parasailing is a favorite sport on these lovely beaches.

On Wednesday, after my lecture on Erdős, Dr. Sangeelee took us around Mauritius to see the beaches on the northwestern part of the island.

We had all day Thursday for sightseeing before boarding the Air Mauritius flight to London at 10:00 pm. So Mathura and I went to the hilly interior of the island in a chauffeured car we hired at the hotel. It was a very different terrain from the beaches we saw. In the mountainous region, there is the lovely Black River Gorges National Park which among other things has spectacular water falls, and the "seven colored earth" — a breathtakingly beautiful area of multi-coloured soil, just as impressive, although on a much smaller scale, compared to the colored volcanic cinder cones you see in Haleakala in Maui. And throughout the mountainous interior of Mauritius, there are several gigantic granite cliffs comparable in majesty to what one sees at Zion National Park.

Mauritius is the only place in the world where the *Dodo*, a large flightless bird, existed. The Dodo is now extinct, but well preserved remains of the Dodo can be found in the National Museum in Port Louis. Bruce Reznick was fortunate to see the remains of the Dodo (he made that his sightseeing priority!), but we missed it because the National Museum was closed for renovation when we were there. But even though we missed seeing the Dodo, there was a memorable gift that I received when Mathura and I went into Port Louis to buy souvenirs and gifts for our family.

In a conversation with Mr. Suddhoo, Chairman of Air Mauritius, I told him how much Mathura and I enjoyed the flight on the Air Mauritius Airbus A-340. But I said that I could not find a post card of the aircraft or a scale model to purchase on board. On hearing this, Mr. Suddhoo said that when I come into Port Louis, I should stop by his office and he will have a gift for me. That gift was a lovely 1:200 scale model of the Air Mauritius A-340 which is now proudly displayed at our home in Gainesville along with my other aircraft models.

It was a most memorable visit to Mauritius that surpassed our expectations thanks to the untiring efforts of Dr. Sattianathan Sangeelee and the kindness of several prominent citizens of Mauritius. It is also satisfying to see that in Mauritius, the preservation of the legacy of Ramanujan and the encouragement of mathematical talent among youth was being carried out by both academic and non-academic individuals. The world of mathematics will profit from support given by members of different professional backgrounds. It was an eye opener for me that Ramanujan's name has cast a magic spell over this enchanting island, and I am fortunate to have been given an opportunity to take part in the Ramanujan celebrations in a distant part of the globe.

On the night of our arrival in Mauritius, Dr. Sangeelee called me by phone at the hotel and said that a cyclonic storm was threatening the island of Mauritius, and had it come through, the entire event would have been canceled. Fortunately the storm missed Mauritius and went further north, and so a decision was taken that morning to proceed with the event as scheduled. We were not aware of the impending danger as we were flying for nearly 24 hours to get to Mauritius. I think the Goddess of Namakkal played a role in diverting the storm so that the Ramanujan Memorial Function would not be hampered!

==

Upon return from Mauritius, I gave a full report of my trip to George Andrews (who was in Florida for the Spring Term) and told him that he should visit Mauritius next year. And indeed he did at the invitation of the Mauritius Ramanujan Trust and enjoyed his visit as much as I did.

A week after my return from Mauritius, we had the Ramanujan Colloquium and seminars by Peter Elliott of the University of Colorado. Elliott is the greatest expert in Probabilistic Number Theory whose origins can be traced back to a 1917 paper of Hardy and Ramanujan on the number of prime factors. He had written a two volume book which is a Bible in this field. Thus it was an appropriate to invite him as a Ramanujan Colloquium speaker. His visit also gave me an opportunity to discuss with him my recent work with Molnar on the number of small prime factors.

Peter Elliott and his wife Jean arrived in Gainesville late on Saturday, March 18 night. They had a full day Sunday to rest and relax, and so I took them to Devils Millhopper, the huge sink hole and the park surrounding it that Erdős used to visit regularly. That night, we took the Elliotts to dinner at Amrit Palace in Ocala, our favourite Indian restaurant in Florida.

Elliott delivered the Ramanujan Colloquium on Monday afternoon on Probabilistic Number Theory following which he delivered two number theory seminars the next day on additive and multiplicative functions and groups of rationals. On Wednesday, at my request, he addressed my graduate analytic number theory class on some open problems in number theory.

Before the end of the Spring Term, in early April, I made a one-week trip to the UAE in connection with the evaluation of the mathematics program at the United Arab Emirates University (UAEU) in Al Ain. A report of my visit to Al Ain can be found in §9.13.

In May, Mathura, my mother and I were in India for about two weeks. On the way back, we vacationed in Hawaii (where else?!) — splitting time between the islands of Kauai and Oahu. While we were in Kauai, our good friends Sethuram, his wife Uma, and his mother joined us. Uma Sethuram works as the Budget Officer in the College of Liberal Arts and Sciences.

The visit to Hawaii was delightful as always, but the highlight of the summer travel was the Alaskan cruise that the entire family took (August 6–13) at Mathura's urging. The cruise departed from Vancouver and so two days before the cruise, we went to Victoria to see the world famous Butchart Gardens. I had heard a lot about the Butchart gardens, but my desire to see it increased because during a conference in Vancouver in 1994, Bombieri made a special trip to see the gardens, and he has exceptionally high standards. The Butchart gardens were the loveliest botanical gardens I had seen, and met my expectations! As for the Alaskan cruise, it excelled our expectations!! The two highlights of the cruise were the fantastic close-up views of the Hubbard Glacier from the ship, and the flight out of Juneau on a sea plane over the snow capped peaks and glaciers. The view on a cloudless day was simply breathtaking. The Alaskan cruise was a fine relaxed ending to an eventful academic year.

10.10 My mission continues: 2017 and beyond

My involvement in fostering the legacy of Ramanujan in various ways continues as strongly as ever. More academic centers and organizations the world over are conducting programs related to Ramanujan's mathematics, and it is a pleasure to participate in many of them. This enables me to meet new groups of researchers, keep abreast of the latest developments, and widen my academic horizon as well. I shall also report about other academic events of interest not related to Ramanujan.

During October 6–13, I made a trip to Dubai in connection with the accreditation of the College of Sciences programs at the newly formed University of Balamund, Dubai (UOBD). The University of Balamund is a well known university in Lebanon, and they were opening a branch campus in Dubai. For details about this trip to Dubai and UOBD, see §9.13.

The major academic trip in December 2017–January 2018 was to India where after the SASTRA Ramanujan Conference, I visited IIT Gandhinagar for a significant Ramanujan event.

===

Trip to India and the 2017 SASTRA Prize to Maryna Viazovska

Mathura, my mother and I departed for India on December 7, one day after my classes for Fall 2017 ended. This time we flew on China Southern Airlines from Los Angeles to Singapore via Guangzhou (formerly Canton). On the long trans-pacific flight from Los Angeles to Guangzhou, we were on the Super Jumbo Airbus 380. It was a most enjoyable experience on the upper deck of this gigantic aircraft.

Professor M. S. Swaminathan (FRS) always wants me to deliver a popular lecture on Ramanujan at his Foundation on each my visits to Madras in December. So on that trip, before departing for the SASTRA Conference, I gave a lecture on "Stage productions on Ramanujan - a comparison", giving an assessment of productions such an the *Opera Ramanujan*, the play *Partition*, and the movie "The Man Who Knew Infinity".

We had a good spectrum of speakers for the SASTRA conference, and many of them arrived on December 19 — Henry Cohn with his wife, Ken Ono with his PhD students Ian Wagner and Madeline Locus, Ravi Kulkarni, and my PhD student Ankush Goswami. The SASTRA Prize Winner

Maryna Viazovska and her senior colleague Philippe Michel from the Ecole Polytechnique in Lausanne, Switzerland, arrived only on the morning of December 20. They were met by SASTRA representatives at the airport and we picked them up on our way to Kumbakonam. I was worried whether Viazovska would be too tired after the road journey to Kumbakonam that she was taking after a long international overnight flight, but she and Michel seemed to handle it quite well.

Maryna Viazovska of the Swiss Federal Institute of Technology, Lausanne, Switzerland, was awarded the 2017 SASTRA Ramanujan Prize especially for her stunning solution in dimension 8 of the celebrated sphere packing problem, and for her equally impressive joint work with Henry Cohn, Abhinav Kumar, Stephen D. Miller and Danylo Radchenko resolving the sphere packing problem in dimension 24, by building upon her fundamental ideas in dimension 8.

Viazovska has made deep contributions to several fundamental problems in number theory. In her PhD thesis of 2013 written under the direction of Don Zagier at the Max Planck Institute, she resolved the famous Gross-Zagier Conjecture in a substantial number of cases, including the important case pertaining to higher Green's functions that had been open for 30 years.

The sphere packing problem has a long and illustrious history. Johannes Kepler asked for the optimal way to assemble cannon balls (of uniform radius) and conjectured a configuration, but he could not prove it. This is the sphere packing problem in three dimensions, and can be generalized to arbitrary dimensions. The sphere packing problem in three dimensions is known as Kepler's problem or The Kepler Conjecture. This conjecture in three dimensions was finally resolved by Thomas Hales in 1998 who gave a proof which was a tour-de-force that combined ingenious geometric optimization arguments with machine calculations. The sphere packing problem in higher dimensions remained open.

In dimension 8 there is $E8$, an exceptional simple Lie group with a root lattice of rank 8, and in dimension 24, we have the Leech Lattice, and both have remarkable structures. This gave some hope that the sphere packing problem could be resolved in dimensions 8 and 24. Noam Elkies (Harvard) and Henry Cohn (Microsoft Research) conjectured the existence of certain magic auxiliary functions in dimensions 8 and 24, which if determined, would resolve the conjecture in those dimensions. But these magic functions remained elusive. Viazovska produced these functions by an ingenious use of modular forms. Her attack was viewed as audacious, but the mathematical world applauded in disbelief.

Once Viazovska had succeeded in dimension 8, the immediate question was whether her methods could be extended to dimension 24. Indeed, in the span of a week, by working at a furious pace, Viazovska in collaboration with Cohn, Kumar, Miller and Radchenko, successfully resolved the 24 dimensional case by building upon her ideas in dimension 8.

Maryna Viazovska was born in Kiev in the Ukraine in 1984. After completing her high school and BSc education in the Ukraine, she went to Germany to earn a Masters Degree at the University of Kaiserslautern. She then joined the University of Bonn for her PhD which she obtained in 2013. Since her PhD, she has received several awards such as the Salem Prize in 2016 and the Clay Research Award in 2017 prior to the SASTRA Prize. She went on to win the Fields Medal in 2022.

The SASTRA prize to Viazovska was given at the inauguration of the conference on December 21 morning, following which, Phillipe Michel gave the Opening Lecture. He was followed by Ken Ono. Right after these lectures, Michel departed to Madras in a car provided by SASTRA so that he could catch his return flight to Switzerland. I was amazed and impressed that he came all the way for just one day to attend the prize ceremony and give his talk. Michel was on the Prize Committee that year and so I was glad that a representative on the Prize Committee attended the conference.

Ankush Goswami, my student, was working on a problem in analytic number theory for his thesis which was a generalization and extension of some earlier work of mine on the parity on the number of restricted prime factors. So we gave back-to-back talks in the afternoon, in which I first provided the background and he later summarised his thesis work.

On December 22 morning, Ramanujan's birthday, Viazovska delivered the Ramanujan Commemoration Lecture on the resolution of the sphere packing problem in dimension 8. Following this, Henry Cohn gave the concluding conference lecture on the solution of the problem in dimension 24. Cohn is a fantastic speaker and he took pains to explain to the large gathering of students, how one could go about understanding the geometry in higher dimensions. It was fitting that both Viazovska and Cohn spoke on December 22 and in their talks covered the important aspects of the proofs in dimensions 8 and 24 respectively.

In many universities and academic institutions in India, usually there are conferences or lectures arranged around December 22, Ramanujan's birthday. If such an event is arranged when the SASTRA conference takes place, I am unable to attend it. Atul Dixit invited me to be the Chief Guest at a prize distribution ceremony of a Mathematics Talent Test at

IIT Gandhinagar on December 22. When I told him that I am unable to accept his invitation because I need be at SASTRA University on that date, he kindly moved that event to January 4.

I departed Madras for Ahmedabad on January 3 in the early afternoon by the direct flight offered by Indigo Airlines. Dixit sent the IIT car with chauffeur to pick me up and take me to the guest house on campus. The IIT Gandhinagar campus is relatively new, and since a regular guest house had not yet been completed, the Institute used an unoccupied building consisting of a set of large apartments for full professors as the guest rooms. So I had a suite of rooms, because the apartment was big.

The Indian Institutes of Technology (IITs) were started in the 1950s by the Congress government under Jawaharlal Nehru. Initially, there were five campuses in Madras, Bombay, Delhi, Kanpur and Kharagpur, and they attracted the brightest students. The IITs are an Indian success story, and an IIT degree guarantees a lucrative career in the USA. In view of this phenomenal success, and to meet demand, several more IITs were opened in the last two decades, including this one in Gandhinagar, Gujarat, on the outskirts of Ahmedabad, the main city in the state. The campus is spotlessly clean with wide well laid out roads, and so it was a pleasure to take an early morning walk. Dixit picked me up at 9 am on January 4, and took me to the mathematics department. I gave two talks that day. My first talk in the morning was on "Ramanujan's place in the world of mathematics" (based on my book) to a group of 100 high school students who took the Talent Test earlier in the day. There were more than 200 in the audience because my talk was also attended by several faculty and graduate students. Besides doing research in the area of q series, Dixit is taking an active role in conducting such mathematical contests.

Atul Dixit completed his Phd under the supervision of Bruce Berndt at the University of Illinois in 2012. After spending a few years as a Post-Doc at Tulane University, he returned to India and joined IIT Gandhinagar. He is happy in India and doing very well in research, teaching, and service. He is able to get young researchers to spend a year or two at IIT Gandhinagar and work with him. When I was there, Bibekananda Maji from Gauhati was on an extended visit to work with Dixit.

My talk in the afternoon was the Colloquium. I spoke about my recent work in analytic number theory. Following that, we returned to the main auditorium to watch the movie "Hidden figures" that was shown to the students who had taken the Talent Test. The day concluded with the Prize Ceremony and the distribution of certificates. I was pleased to have taken part in this ceremony recognising talented students in mathematics. I departed Ahmedabad at 7 pm and was back in Madras that night.

===================================

Two days after my return from Ahmedabad, Mathura, my mother and I, departed for Singapore on our way back to the USA. In Singapore, I had a very fruitful meeting with Rok Ting and Rajesh Babu of World Scientific, discussing publication details relating to my father's autobiography, The Alladi Diary, which I was editing. The next day, we departed Singapore for Los Angeles on China Southern Airlines via Guangzhou, once again enjoying the superb service on the trans-pacific sector on board the super jumbo A-380. What a pleasant surprise it was during our long transit at Guangzhou airport, to meet Michel Waldschmidt in the China Southern Lounge! Waldschmidt is a world traveler, and he was on his way back to Paris from Sydney. We compared notes on our experiences with China Southern, and of course caught up on news pertaining to developments in the world of mathematics.

Upon arrival in Los Angeles in the afternoon on January 9, we drove straight to San Diego to attend the Annual Meeting of the AMS. After dropping my mother at the home of our long time friends the Ramanathans (formerly of Hawaii), Mathura and I checked in at the Marriott in San Diego. I had fruitful meetings with Elizabeth Loew and Marc Strauss of Springer, and Rochelle Kronzek of World Scientific regarding my journal, book series, and book projects. I attended the Joint Prize Session and it was pleasure to see my former UCLA college mate Robert Guralnick get the Cole Prize, David Bressoud receive the MAA Award for Distinguished Public Service, and Henry Cohn get the AMS Conant Prize for mathematical exposition. I was not surprised that Cohn was awarded the Conant prize for it was clear from his lecture at SASTRA that I had heard just a few weeks earlier, that he is an outstanding expositor.

Soon after my return to Florida from San Diego, I was off to Dubai for the accreditation of the new Masters Program at the University of Sharjah. Both the American University of Sharjah (AUS) and the University of Sharjah (UoS) invited me back after I did their initial accreditation a few years earlier. It was nice to see the progress they had made, and the support they received from the Administration based on our reports. The unusual aspect of this trip was that on the way back, I halted in Rome to attend a Conference on Discrete Mathematics arranged by Stefano Capparelli. The Commission of Accreditation of the UAE was kind enough to purchase my ticket with a stop in Rome on the way back, and this did not raise the price of the ticket.

After completing his PhD at Rutgers University in 1988 under the direction of James Lepowsky, Capparelli returned to Italy to take up a permanent position at the Sapienza University of Rome. He was organizing this International Conference on Discrete Mathematics, Italy — called DICRETALY — and invited me to give an hour lecture, to which I agreed.

After completing my work for the American University of Sharjah, I took a chauffeured taxi to Abu Dhabi Airport on January 31 to catch my Alitalia flight to Rome at 11:20 pm. I had no time for dinner in Dubai and so I enjoyed a light meal at the well appointed Etihad Airlines Lounge at Abu Dhabi Airport before boarding the Alitalia flight to Rome. Even though Alitalia was under bankruptcy protection, I must say that service on board the Airbus A-330 was excellent. We arrived at Rome's Fumicino airport at 6 am on February 1, half an hour earlier than scheduled. When I came out of customs and immigration, I did not see Capparelli there and so I got a bit anxious. But as I was about to call him, he appeared and apologized that he did not realize that my flight had arrived early.

Capparelli took to me the elegant hotel where all participants were staying. I showered and changed quickly to walk with the participants to the conference venue. There were about 80 participants, mostly Italian mathematicians, and it was a pleasure for me to see Jim Lepowsky (Rutgers) and Mirko Primc (Belgrade) there. The talks by the three of us — Lepowsky, Primc and me — were scheduled on the second day.

Even though I had established a generalization of the Capparelli theorems in collaboration with Andrews and Gordon in 1992, I spoke about the new dual of Göllnitz' theorem that George Andrews and I had worked on recently. My talk was in the morning. Lepowsky and Primc gave talks in the afternoon, and both spoke about Lie algebras and Rogers-Ramanujan identities. They both explained connections with the Alladi-Andrews-Gordon work on the Capparelli partition theorems. Primc had submitted his paper to The Ramanujan Journal where it appeared two years later.

It was very gracious of Capparelli to take Lepowsky, Primc and me to dinner on both nights at Italian restaurants in Rome. The polite waiters were proud to say that the pasta and the sauces were all made fresh. When in Rome, dine like a Roman!

On February 3, I departed for the airport very early. The Alitalia transatlantic flight from Rome to New York on the Airbus A-330 was just as good as the flight from Abu Dhabi to Rome. All in all, it was a very fruitful visit to Rome and I was glad to be have been able to squeeze it in along with my trip to the UAE.

Visit of AMS President Ken Ribet

A major academic event in Spring 2018 was the visit of AMS President Ken Ribet (University of California, Berkeley), to deliver the Ramanujan Colloquium on March 26, followed by two seminar talks the next day. Ribet arrived at mid-day on Sunday, March 25, and so I took him to dinner that night. During the relaxed conversation at dinner, he asked me how the SASTRA Ramanujan Prize was launched. So I told him the remarkable story as it happened in 2004. He was amazed at what he heard and said that I should write an article on the SASTRA Ramanujan Prizes for the Notices of the AMS and relate this story. The next morning, I received a message from Erica Flapan, Editor of the AMS Notices, inviting me to write this article. Obviously, Ribet had written to Flapan that night after dinner and Flapan acted immediately. The article appeared in the January 2019 issue of the Notices.

Ken Ribet is a world authority on mathematics related to Fermat's Last theorem, and so he spoke about this in the Ramanujan Colloquium. He made the connection with Ramanujan by talking about Ramanujan's taxi cab equation $x^3 + y^3 = z^3 + w^3$ and stressing how the addition of the variable w produces non-trivial integer solutions, but the Fermat equation for cubes, namely $x^3 + y^3 = z^3$, has no non-trivial integer solutions. Dean Richardson made the Opening Remarks, and in doing so, he also released the refereed Proceedings of the Alladi60 Conference (Springer) [B11] that had just appeared. The first copy was presented to Ribet.

Ribet gave two talks the next day in the Number Theory Seminar on Elliptic Curves. He is a fine speaker and lectures in a way that even non-experts can understand. After his talks, we took him to Paines Prairie where he was thrilled to see the Alligators basking in the sunshine just a few feet away! That night, Mathura and I hosted a party in his honor at our home that was attended by several faculty members, their spouses, and students from my graduate number theory course. I was impressed that at the party, he spent time talking to the students.

During May–June 2018, my mother and I went to Madras, halting in Singapore on the way back. While in Singapore, I had a fruitful meeting with representatives of World Scientific to discuss the completion and publication of my father's autobiography that I was editing. Within a week of my return from India was the George Andrews 80th birthday conference at Penn State University, and report of this is given in §11.4.

Summer Research Institute in China

In July, I was off to China to participate and speak at a Summer Research Institute on q-series in Tianjin, China, for which Mourad Ismail of the University of Central Florida and Ruiming Zhang of Northwest University, China, were the organizers. It was a three week workshop, but I was there only for the first week. Some of the speakers of the Workshop who came from abroad like me, were not present for the entire duration, but most Chinese participants were.

I departed Detroit for Beijing on July 23, 2018, by the new Delta Airbus A-350. This aircraft is technologically the most sophisticated and is the answer of Airbus Industrie to the Boeing 787 Dreamliner, with almost the entire airframe made of composite materials. The 12-hour flight was exceptionally smooth and quiet. We flew by the polar route which took us over the northern coast of Alaska, the closest to the North Pole that I have travelled. I enjoyed the panoramic views of Prudhoe Bay, Alaska, and gazed in wonder at the icy Arctic Ocean. Since it was summer, I could see the sun set and rise again shortly thereafter. Upon arrival at the gigantic Beijing Capital airport, I was met by a PhD student of Bill Chen and taken to Tianjin by car. I was accommodated at the Teda International Guest House and given a palatial room — yes, it was regal. Bill Chen really rolls the red carpet for his guests!

The Workshop was inaugurated on July 25 morning by Bill Chen Director for the Center for Applied Mathematics of Tianjin University. He was formerly Director of the Center for Combinatorics at Nankai University. The two universities, although separate entities, are next to each other, and work in cooperation. This workshop was held at the new Chern Institute for Mathematics. The bust of the great differential geometer S. S. Chern (who hails from Tianjin) is there in the entrance hall of this Center. The Workshop was sponsored by the two Centers and the Chern Institute of Mathematics, Nankai University.

The opening hour lecture of the workshop was by Peter Paule on computer algebra proofs of q-hypergeometric identities. Peter actually was giving a mini-course of four lectures in which he reported the work of his group at the Research Institute of Symbolic Computation (RISC) in Linz. Carsten Schneider of RISC, Paule's former PhD student and colleague now, gave two talks on recent advances in symbolic computation.

There were actually four other mini-courses given by Mourad Ismail on "Introduction to q-series", Michael Mertens of the University of Cologne on

"Classical modular forms", Larry Rolen (then at Georgia Tech.) on "Harmonic Maass forms", and Alexei Zhedanov of Renmin University, China, on the "Askey-Wilson algebra". These mini-courses combined with hour lectures by senior researchers and shorter 30 minute talks, made this a highly successful workshop. Michael Mertens and Larry Rolen had recently completed their PhDs under the direction of Kathrin Bringmann and Ken Ono. Both Ono and Bringmann have not only mentored several excellent doctoral students, but have also trained them to be effective expositors.

Just outside the Center for Differential Geometry, is a spot where the great S. S. Chern is buried. Chern was one of the most influential figures in differential geometry, and has trained generations of students many of whom have made their marks in the field. His most illustrious student is Fields Medalist S. T. Yau with whom I had the pleasure of interacting at UCLA during my graduate school days. Chern was a distinguished professor at UC Berkeley for many years, but he returned to his native city of Tianjin after retirement. He is viewed as a demi-God in China and justifiably so. The home of Chern on the campus of Tianjin University is preserved as a museum, and the participants were taken on a tour of this home on the second day after the sessions ended. What an inspiring place this is! The home has several valuable well preserved documents and a fantastic collection of photographs on display. I noticed a photograph of the 1970 International Congress of Mathematicians in Nice showing Fields Medalists Alan Baker, Heisuke Hironaka and John Thompson seated in the front row with their medals. Chern is right behind them. I immediately took a photograph of this photo and got it enlarged upon return to Florida. It is a picture that I treasure. It also serves as a fine memento of my visit to the historic home of Chern.

After the visit to the home of Chern, Bill Chen hosted a dinner for the speakers at an Indian restaurant in Tianjin, saying that he chose the restaurant due to my presence! I appreciated the gesture very much.

During the weekend of July 28–29, a few of the participants like me were taken to Beijing for sightseeing in an around the capital. The sights included the gigantic fortress castle of the Chinese emperors called the Forbidden City, and the Great Wall of China. I had seen these before in 1999 and 2002, but I did not mind seeing them again. The crowds were much greater than they were a decade and a half earlier, and security was much tighter. Whereas we had comfortable spacing between visitors previously when we climbed the Great Wall, now we had to jostle past people who were packed like sardines. The main difference is that there is a cable car now that takes you to a certain elevation from where it is a relatively short walk up the wall to the top of a certain section. We took the cable car ride which by itself is impressive because you get a panoramic view of the mountains and the Wall that meanders over the hills. The Forbidden City and the Great Wall are doubtlessly impressive, but the thrill I got on seeing Chern's home was greater!

One of the participants with whom I had long discussions was Gaurav Bhatnagar who had come from Vienna where he was on a long visit. Gaurav was interested in some of my earlier work on multiplicative functions and small divisors. I was pleased that this work I did more than thirty years ago was attracting some attention.

My talk at the workshop was on July 31, and it was entitled "Revisiting the Riemann zeta function at the even positive integers, and new q-analogue". There is a beautiful formula of Euler for the values $\zeta(2k)$ of the Riemann zeta function at the even positive integers $2k$. The evaluation of $\zeta(2)$ was a famous problem and Euler shot to fame by brilliantly solving this. He then extended his method to show that for each positive integer k, $\zeta(2k)$ is a rational multiple of π^{2k}, where the rational is given in terms of Bernoulli numbers. In a graduate course on irrational and transcendental numbers that I gave in 2014–15, I showed the class a short proof that $\zeta(2k)$ is a rational multiple of π^{2k}, and posed a problem to the class to determine this rational value by the method I had outlined. Colin Defant, a brilliant undergraduate student who was taking my graduate course, solved this beautifully and in that process obtained a very new recurrence for the Bernoulli numbers. Over the centuries, several proofs of Euler's formula for $\zeta(2k)$ have been given, but ours was considered novel and so Colin and I published it in the International Journal of Number Theory in 2018. The first half of my hour lecture was on my joint work with Colin. So what has this to do with q-series for me to present this at this workshop?

One of the ways classical results are generalized is to establish their q-analogues. More precisely, a q-analogue is an identity involving a free variable q which reduces to the classical identity when $q = 1$. The Chinese mathematician Zhi-Wei Sun, established a lovely q-analogue of Euler's formula for $\zeta(2)$. On seeing this, my PhD student Ankush Goswami perked up, and with a span of two weeks found a lovely q-analogue for Euler's formula for $\zeta(2k)$ for all $k \geq 1$. In the second half of my talk, I presented Goswami's q-analogue of $\zeta(2k)$.

Colin Defant had published a number of papers in combinatorial number theory as an undergraduate. So after completing his undergraduate studies at the University of Florida, he went to Princeton University where is now working for a PhD degree under the direction of the eminent combinatorialist Noga Alon. Goswami completed his PhD thesis in 2019 under my direction, and included this work on q-analogues of $\zeta(2k)$ as a chapter in his thesis. After getting his doctorate degree, he went to the Research Institute of Symbolic Computation in Linz on a two year post-doctoral fellowship to work in Peter Paule's group.

I departed Beijing on August 1 and returned to Florida, flying once again on the Airbus A-350 on the trans-pacific sector Beijing-Detroit. This was just as enjoyable as the outbound flight. The trip to Tianjin was memorable and fruitful, and a fine ending to a hectic academic year 2017–18.

The year 2018–19 opened with a magnificent Triple Arangetram that Mathura conducted for three of her students in Jacksonville in early September. This is like having the PhD defense of your three students all on the same day! She had spent the entire summer training her three senior students for this event and the effort paid off. The girls performed magnificently and in unison to melodious music by Mathura and Amritha. It was indeed Mathura's hour of glory as a dance teacher. There were about 1200 guests for this grand event in Jacksonville and all of them were treated to dinner hosted by the families of the students.

====================================

Royal Society Conference for the Centenary of Ramanujan's election as FRS (Oct 15–16, 2018)

The first major academic event of 2018–19 was the conference held at The Royal Society in London during Oct 15 and 16 to commemorate the Centenary of Ramanujan's election as Fellow of the Royal Society. The conference was held at Carleton House Terrace, London, the current premises of the Society. The organisers of the conference were Ken Ono (then at Emory University), George Andrews (The Pennsylvania State University), Manjul Bhargava (Princeton University), and Robert Vaughan, FRS (The Pennsylvania State University).

Fellows of the Royal Society are citizens of the British Commonwealth of nations; those outside of the Commonwealth can be elected as Foreign Members. To be elected as Fellow or a Foreign Member of the Royal Society is one of the most prestigious recognitions for an academician. Each year, about 50 Fellows and 10 Foreign Members are elected. This number was smaller in the past. In any case, each year is the centenary of some group of eminent academicians elected as Fellows a century earlier. But the Royal

Society does not hold conferences for the centenary of the election of its Fellows. That the Royal Society held such a conference in London in 2018 for the centenary of Ramanujan's election as FRS indicates the high esteem in which he is held by the Society, and how unique he is among the scientific luminaries in history.

In early Fall 2018, Ken Ono invited me to deliver a talk at the Royal Society Conference on the theme "The Legacy of Ramanujan - the work of the SASTRA Prize Winners". I accepted the invitation but I told Ono that on Saturday, October 13, for the Dasara festival, Mathura and I had arranged a grand program at our home in Gainesville featuring a carnatic music professional concert by the noted vocalist Suryaprakash accompanied by violin and percussion professionals all from Madras. Thus I could leave Florida only on Sunday, October 14, and reach London on Monday, October 15 early morning. Ono said that was fine because he was scheduling my talk only at 10 am that day which meant I would have to rush straight from the airport to the Royal Society to deliver the lecture! Since the cause was high, namely to honor Ramanujan at the Royal Society, I agreed to this invitation, even though the program was tight. Over the next few days I worked furiously on the conference lecture with Ankush Goswami helping me put my talk in power point format.

The conference featured talks by 15 leading experts worldwide on Ramanujan's extraordinary legacy across mathematics, computer science, physics, and electrical engineering. They were (in alphabetical order): Krishnaswami Alladi (University of Florida), George Andrews (Penn State University), Bruce Berndt (University of Illinois), Miranda Cheng (University of Amsterdam), Amanda Folsom (Amherst College), Jeff Harvey (University of Chicago), Winnie Li (Penn State University), Alex Lubotzky (Hebrew University), Ken Ono (then at Emory University), Sujatha Ramdorai (University of British Columbia), Peter Sarnak (Princeton University and the Institute for Advanced Study), Kannan Soundararajan (Stanford University), P. P. Vaidhyanathan (Caltech), Maryna Viazovska (Ecole Polytechnique Federale de Lausanne), and Trevor Wooley, FRS (University of Bristol). There were about 100 registered participants who had come to attend the lectures, including several Fellows of the Royal Society.

My travel to London had high drama. I departed Orlando on Sunday, October 14 at noon to catch a Delta Airbus A-330 flight at Detroit at 6 pm to arrive in London the next morning at 6 am. As I was on board sipping a pre-take-off beverage, the Captain announced that one of screws on an engine was missing, and that even though the A-330 was certified to fly with one less screw, he did not want to take any chances. He assured us that it would just take a few minutes, but from all my travel experiences over five decades, I knew it would take longer. It turned out that the screw was not available in the maintenance department at Detroit airport, and so eventually it was obtained by removing the same screw from another Delta A-330 at Detroit that was not scheduled to fly until the next day. All this caused a delay of an hour and forty minutes. I had planned to take this flight to London because of its early arrival; even the delay of an hour and a half would give me time to reach the Royal Society for my lecture. Well, one delay causes another. On arrival at London Heathrow Airport, we were told that the gate originally assigned to us had been given to another aircraft and so we had to wait for more than half an hour on the ground before being allowed to disembark. After quickly changing into fresh clothes at the airport, I took a taxi to The Royal Society. As luck would have it, the main highway from Heathrow Airport to London was clogged as were also several streets in London. The taxi driver cleverly avoided the worst of these roads, but in spite of his presence of mind and skill, when I arrived at the Royal Society, it was just ten minutes before my talk. I rushed in only to find out that George Andrews was introducing Ken Ono who as the main organizer, had volunteered to speak in my slot, and had moved my talk to the next day in the slot originally allotted to him. Nevertheless, as I entered the lecture hall at the Royal Society, George Andrews announced: "There is Krishna Alladi". On hearing this and seeing me enter the hall, the audience broke into a thunderous applause! No exaggeration. The result of my arriving late was that I missed the opening two lectures of George Andrews and Bruce Berndt, but I was able to attend and enjoy all other talks. I will now give a report of the lectures.

Besides the well known fundamental fields such as number theory, analysis, modular forms, and automorphic forms, where Ramanujan has had an enormous influence, the conference covered areas like signal processing and black holes where Ramanujan's ideas are having an impact.

In his last letter to G. H. Hardy of Cambridge University written in January 1920 from Madras, India, a few months before his death, Ramanujan described his latest invention — the remarkable mock theta functions, now considered to be among his deepest contributions. George Andrews discussed how Ramanujan may have discovered the mock theta functions. It is nearly impossible to understand how the mind of Ramanujan worked, but Andrews can probably make the best guess of this! No one knew (until recently) how mock theta functions are connected to the classical theta functions or modular forms. During the Ramanujan Centennial year 1987, the great mathematical physicist Freeman Dyson said that understanding this link is one of the tantalising puzzles of mathematics. Amanda Folsom's talk "Mock theta functions - then and now" began with the classical theory of mock theta functions of Ramanujan, and culminated with the modern theory developed in the last two decades by Zwegers, Ono, Bringmann, and others, which solves this tantalizing puzzle.

Bruce Berndt, who had spent most of his life analysing Ramanujan's Notebooks, gave a talk entitled "Living with Ramanujan for forty years". Jeff Harvey spoke about Ramanujan's influence on string theory, black holes, and a mathematical topic of great importance called moonshine. Miranda Cheng's talk also dealt with black holes, but it is to be noted that these are experimental black holes produced in labs that have very short lives. Alex Lubotzky dealt with the fascinating topic of Ramanujan graphs and Ramanujan complexes, while Vaidhyanathan showed us how Ramanujan sums are used in signal processing.

One of the contributions that Ramanujan is most famous for, is his invention jointly with Hardy in 1917 of the *Circle Method* to obtain an asymptotic series for the partition function. Subsequently this method was developed by Hardy and Littlewood to be applicable to a wide class of problems in Additive Number Theory such as the celebrated Goldbach and Waring problems. Today, a century later, the circle method is applied in analytic number theory, Diophantine approximations, discrete harmonic analysis, and even arithmetic geometry. Trevor Wooley discussed recent progress on some of the deepest problems involving crucial use of the circle method, and indicated formidable challenges that remain.

Ken Ono described his recent work with Michael Griffin, Larry Rolen, and Don Zagier on an equivalent version of the celebrated Riemann Hypothesis involving the Polya-Szego polynomials, and in this process showed how the circle method is utilized.

Ramanujan was always interested in special properties the integers, and in this spirit, Soundararajan talked about the problem of classifying families of integer factorial ratios. A close study of such ratios reveals connections with deep topics such as the Selberg integral, the Macdonald-Morris conjectures, and the Riemann Hypothesis.

The celebrated Ramanujan Conjectures on bounds for the coefficients of certain types of automorphic forms, are at the very heart of mathematics. This central topic was covered by Peter Sarnak and Winnie Li in two separate lectures. In talking about Selmer groups in Iwazawa theory, Sujatha

Ramdorai illustrated Ramanujan's influence within arithmetic geometry. So the lectures at the conference dealt with the latest developments in a wide range of fields stemming from, or influenced by, Ramanujan's ideas. My talk at the conference was on "Ramanujan's legacy: the work of the SASTRA Prize Winners". I described the work of the seventeen SASTRA Ramanujan Prize Winners as of 2018 and the impact this has had on the development of mainstream mathematics.

During the conference, the bust of Ramanujan sculpted by Paul Granlund, that is normally in the Royal Society Library, was placed next to the podium in the auditorium where the talks were given. It was inspiring for all of us to speak with the Ramanujan bust next to us. This bust was sculpted by using the now famous passport photograph of Ramanujan. But back in the 1930s, it was the great astrophysicist Subrahmanyam Chandrasekhar (later FRS and Nobel Laureate) who took this passport photograph from Mrs. Ramanujan and handed it to Hardy. The bust at the Royal Society was actually owned by Chandrasekhar who donated it to the Society. Outside the hall, in the foyer, in a glass case, the letter written by G. H. Hardy and cosigned by several Royal Society Fellows proposing Ramanujan for the Fellowship, was on display in a glass case.

One of the Fellows of the Royal Society who attended this conference was M. Vidyasagar, who is the son of the late number theorist M. V. Subbarao. When I used to meet with Subbarao in Madras as an undergraduate to discuss number theory, he told me about his son Vidyasagar in the area of electrical engineering who had done research in number theory, some in collaboration with him. At that time Vidyasagar was in Canada, but now he is settled in India as a distinguished professor at the IIT in Hyderabad. Vidyasagar was honored with the Fellowship of the Royal Society in 2012 for his important work in control theory in electrical engineering. I was glad to have met Vidyasagar at this conference since I had heard of him from Professor Subbarao even in the seventies.

After the close of sessions on the first day, there was a screening of the movie "The Man Who Knew Infinity" for all registered participants, following which there was a formal dinner hosted by the Society for the speakers and their spouses.

It was fitting that the Royal Society decided to publish the refereed proceedings in *The Philosophical Transactions of the Royal Society A* as a Special Issue. This journal has an illustrious history. It was established in 1665 as the very first journal devoted to science. It is therefore the world's longest running scientific journal. The word *philosophical* refers to natural philosophy, nowadays called science. The Philosophical Transactions was the journal that introduced the peer review process. Isaac Newton's first paper on "A new theory about Light and Colours" appeared in this journal in 1672. It is to be appreciated that the refereed proceedings of the conference appeared in this journal with such a hallowed tradition. Twelve of the invited speakers including me (see [A6]) have published their talks in this Special Issue of the Philosophical Transactions that has been edited by Ken Ono (now at the University of Virginia). The Philosophical Transactions publication based on the Royal Society conference, is yet another recognition of Ramanujan's stature in the mathematical firmament.

====================================

Before our usual trip to India in December, my entire family enjoyed a relaxed Thanksgiving vacation in the islands of Maui and Kauai. We were returning to Maui after two decades. Our previous visit to Maui was in 1999 to greet the new millennium. This time we stayed at the magnificent Hilton Waldorf Astoria Grand Wailea Resort, which surpassed our highest expectations in beauty and luxury. The highlight of the sightseeing in Maui was the visit to the top of the 10,000 ft Haleakala volcano and the awesome view from there of the multi-colored volcanic cinder cones in the open jawed crater that descends into the ocean below.

Kauai is our favourite and we visit this paradise about once every two years. We never get tired of its idyllic beauty, where the water falls descending from the rain drenched mountains, and the spectacular beaches with their lava rocks and turquoise blue waters, are a feast to the eyes.

====================================

Trip to India and the 2018 SASTRA Ramanujan Prize to Yifeng Liu and Jack Thorne

Mathura, my mother and I departed Atlanta on December 11 for India via Tokyo and Singapore. The main academic event in India was the SASTRA Ramanujan Conference in Kumbakonam during December 21–22 when Yifeng Liu of Yale University and Jack Thorne of Cambridge University were jointly awarded the 2018 SASTRA Ramanujan Prize for outstanding work in the areas of algebraic geometry, arithmetic geometry, representation theory and number theory.

Yifeng Liu had made spectacular contributions to arithmetic geometry and number theory on a wide spectrum of topics. His PhD thesis at Columbia University (written under the direction of Shou-Wu Zhang) on arithmetic theta lifts went well beyond the work of earlier notable researchers. By himself and in collaboration with Binyong Sun, he made real progress on the famous Gan-Gross-Prasad conjectures in the representation theory of classical groups. Also, in collaboration with Shou-Wu Zhang and Wei Zhang, he studied the p-adic logarithm of Heegner points in terms of a p-adic L-function and proved p-adic versions of theorems of Waldspurger and of Gross-Zagier. In addition, Liu had established a non-Archimedean analogue of the famous Calabi Conjecture for certain abelian varieties.

Yifeng Liu was born in Shanghai, China, and received his BS Degree from Peking University in 2007 after which he joined Columbia University, where he received his PhD in 2012. He held the C. L. E. Moore Instructorship at MIT (2012–15) and an assistant professorship at Northwestern University (2015–18) before being appointed as an associate professor at Yale University in 2018 when he won the SASTRA Prize.

Jack Thorne had made far reaching contributions to number theory, representation theory, and algebraic geometry. He works in two rather different areas: modularity of Galois representations and arithmetic invariant theory. His 2012 PhD thesis on the arithmetic of simple singularities at Harvard University was jointly supervised by Richard Taylor and Benedict Gross, two of the dominant figures in contemporary number theory. His thesis leads to new results about the sizes of Selmer groups for abelian varieties of small dimension, and bounds for the number of rational and integral points on various types of algebraic curves of genus greater than one. Regarding modularity of Galois representations, Thorne has been a central force in eliminating certain major restrictions on the Taylor-Wiles method. Some of his most striking results have appeared in joint papers with Laurent Clozel. Another crucial aspect of Thorne's ideas that in his joint work with Clozel, was his discovery of a new automorphy lifting theorem. These works of Thorne and of Clozel-Thorne have greatly extended the scope of the Taylor-Wiles method.

Jack Thorne received his BA Mathematics degree from Cambridge University in 2007. He then went to Harvard University where he completed his PhD in 2012. Following that he was appointed Reader at Cambridge University in 2013, but during the period 2012–17, he was also a Clay Research Fellow of the Clay Mathematics Institute. In 2017 he was awarded the Whitehead Prize of the London Mathematical Society and was promoted to full professorship at Cambridge University in 2018, the year he won the SASTRA Prize.

Thorne arrived in Madras on December 19 in the early morning accompanied by his mother. Since we were leaving for Kumbakonam only the next day, my friend GP showed them around Madras. Yifeng Liu arrived in the evening as did some other SASTRA conference partici-pants — Kalyan Chakraborty from the Harish Chandra Institute in Al-lahabad, Sudhir Ghorpade from IIT Bombay, Chris Jennings-Schaffer from Cologne, and Ali Uncu from the Research Institute of Symbolic Compu-tation (RISC) in Linz, Austria. Chris received his Phd degree from the University of Florida in 2017 under the direction of Frank Garvan and was spending two years at the University of Cologne working with Kathrin Bringmann on a post-doctoral fellowship. Ali Uncu also completed his PhD degree in 2017 from the University of Florida (under the direction of Alex Berkovich) following which he took up a post-doctoral fellowship at RISC to work with Peter Paule and his group. Our guests had a restful night at Hotel Savera where the accommodation was attentively arranged by GP. We all departed Madras by the SASTRA van on December 20 morning and by 5 pm we reached Kumbakonam, where our guests enjoyed seeing the Town High School, Ramanujan's home, and the Sarangapani Temple before checking into the lovely Sterling Resorts (now known as Indeco Re-sorts). There Uncu shared a room with GP, and the next morning he told us that they had long a discussion comparing the ancient cultures of India and Turkey.

Liu and Thorne were awarded the SASTRA Ramanujan Prize at the inauguration of the conference on December 21 morning following which Chakraborty, Ghorpade, Jennings-Schaffer, Uncu and I all spoke. Jennings-Schaffer's talk was on mock theta functions, whereas Uncu spoke about the combinatorics underlying certain partition identities. I spoke the Riemann zeta function at the even positive integers and the new q-analogue (as I did in Tianjin), presenting my joint work with Colin Defant and the work of my PhD student Ankush Goswami.

The unusual feature of this conference was that we had two Ramanujan Commemoration Lectures back to back on morning of December 22, Ra-manujan's birthday, by the two winners of the prize. We all left for Madras after lunch, but Mathura and I did not travel with the guests because we had to visit a special temple on a pilgrimage enroute to Madras.

On December 26, at the request of Professor M. S. Swaminathan, I gave a talk at his Foundation in Madras on "Ramanujan's Legacy: the work of the SASTRA Prize Winners" repeating my talk at the Royal Society, but on a more popular level (not technical).

Mathura departed for USA on December 30 via Hong Kong to be back in Florida for my grandson Keshav's birthday on January 1, and also to be of assistance to Amritha during the final weeks of her pregnancy. I stayed

back in India because I had lecture assignments at the Bhaskharacharya
Pratisthana (BP) in Poona and at IIT Bombay on January 2 and 3.

On New Year's Day 2019, I departed Madras in the late afternoon by
the Jet Airways direct flight to Pune (formerly Poona). Professor Ravi
Kulkarni, the Director of BP, had sent a car to pick me up. The taxi driver
stopped in front of the main building of BP, but it was pitch dark. We
called out for the housekeeper or watchman, but no one was there. The
taxi driver said he would stay until someone from BP showed up. After
a fifteen minute wait, the housekeeper who had gone out for a late night
Tea(!) showed up and took me to the next building which was the Guest
House. This too was pitch dark since there were no guests to stay there
that night. There had been a big conference that finished the previous day
and all guests had vacated. The room that was allotted to me had to be
cleaned, and the housekeeper worked hard to finish this job in half an hour.
He told me the Professor Kulkarni resides on the premises of BP in an
adjacent building and will arrive shortly. And yes, Kulkarni did arrive at
around 9:30 pm and asked me if I would like to have dinner. I was famished,
and so I said YES. He took me to a nearby restaurant where there were
several young people having dinner late at night.

My lecture at BP on January 2 was only at 4 pm, and I spent the
morning in discussion with Kulkarni about BP and mathematics in India.

The Bhaskharacharya Prathisthana was founded by the eminent mathematician Shreeram Abhyankar in the seventies. Abhyankar hails from
Poona and belonged to the Chitpawan Brahmin community. He was very
proud of Indian culture and India's heritage in mathematics. Although he
was settled at Purdue University in the USA, he wanted to create a research
institution in India, and in his home town of Poona. So he founded BP
and named it after Bhaskhara, one of India's greatest mathematicians in
history. Acharya means teacher in Sanskrit. Kulkarni also has a passion
for the Indian heritage in mathematics, and is from Poona. So he is a worthy successor to Abhyankar as the Director of BP. Kulkarni had received
his PhD from Harvard University in 1968 under the guidance of Shlomo
Sternberg. Prior to arriving at BP, he was Director of the Harish Chandra
Institute in Allahabad.

Before my colloquium, Kulkarni took me lunch at a nearby restaurant
called Krishna which served fine Maharashtrian food, including the Maharashtrian speciality — the Shreekhand (a dessert). He joked that I might
sleep during my lecture after such a fine meal!

At Kulkarni's request I gave the Colloquium on "Ramanujan's Legacy -

the work of the SASTRA Prize Winners", the same talk I gave at The Royal Society. Kulkarni said that this was very appropriate because one of the missions of BP is to foster the legacy of Indian mathematics.

My Colloquium concluded at 5 pm, and there was a taxi waiting outside to take me to IIT Bombay where I was to speak the next day.

Poona is connected to Bombay by a magnificent highway which goes over hilly country and so is very scenic. But even though the highway is fast, once you reach the outskirts of Bombay, the traffic grinds to a halt. After fighting the traffic jams of Bombay, the taxi driver took me to the Jal Vihar Guest House on the sprawling campus of IIT Bombay where I was accommodated. Bombay is disgustingly crowded, but within Bombay, the IIT campus, spacious and full of trees, is an oasis. Sudhir Ghorpade, the Chair of the mathematics department, met me at the Guest House at 9:30 pm and walked me to his apartment on campus, where his wife had so graciously prepared a Maharashtrian dinner for us.

My talk at IIT Bombay was only at 3:30 pm the next day, and so I spent the morning working in my comfortable room at the guest house. Several senior faculty of the ITT Mathematics Department joined me for lunch at the guest house hosted by Ghorpade — Dipendra Prasad, Anand Vardhan, Ravi Raghunathan, and Balasubramanian, who after retirement from MATSCIENCE, was visiting IIT Bombay for two years. Dipendra had recently moved from the Tata Institute to IIT Bombay. The timing was perfect for IIT Bombay because soon after he arrived at the IIT, he was awarded the Third World Academy of Sciences (TWAS) Prize by the Abdus Salam International Centre for Theoretical Physics (ICTP). So I congratulated Dipendra on his recent recognition.

Retirement at the Tata Institute is at age 60, and once you retire, you lose the apartment given to you on campus. Bombay real estate prices are sky high, and so most TIFR faculty leave the city of Bombay altogether after retirement. But some like Dipendra, move to the IIT Bombay where the retirement age is 65 and housing is provided on campus for faculty.

My talk at IIT Bombay was on my recent work in analytic number theory. Immediately after my talk, I took a taxi to Bombay airport, which although is quite close "as the crow flies", took about an hour to reach owing to the maddening traffic. Upon reaching the Jet Airways counter, I was told that my flight was an hour and half late. So I arrived in Madras only at 2:30 am, but dear GP was there to receive me at the airport. A person who comes cheerfully to receive you at the airport in the wee hours is definitely a true friend!

Before departing Madras, I had a long meeting with Ranjana Rajan and Nisha Rahul of World Scientific in Madras to go over the final details before the publication of The Alladi Diary, my father's autobiography that I was editing, and scheduled to be in print in the next couple of months. My mother and I left Madras on Sunday, January 6 for the USA via the Pacific. We halted in Singapore for two nights because I had an important meeting on Monday with Max Phua, Rok Ting, and Rajesh of World Scientific regarding the printing of The Alladi Diary in Singapore.

======================================

A week after my return from India and Singapore, I was off to Baltimore to attend the Annual Meeting of the AMS. Gaurav Bhatnagar (from the University of Vienna) who was at the Annual Meeting, came to Gainesville to visit the University of Florida the following week. He gave a talk in the number theory seminar on his recent work on arithmetical functions and we had fruitful discussions during his one-week stay.

======================================

On January 31, a new light entered our lives — my daughter Amritha gave birth to baby Sahana! As a dutiful mother, Mathura was in Atlanta during the delivery. I rushed to Atlanta by car on Friday, February 1 after my class. Sahana is the name of a melodious *raga* (= scale) in South Indian classical music, and it was a perfect choice by Amritha and Jis for the name of their daughter. Jis' parents had also rushed to Atlanta from Tampa on February 1. After spending most of the weekend there, I was back in Gainesville because I had a full academic program for the Spring Term.

======================================

George Andrews arrived in Gainesville after the Baltimore meeting. In mid-February, he spoke in the Number Theory seminar on "How Ramanujan may have thought of mock-theta functions", which was the theme of his lecture at the Royal Society. The next day, I departed for Texas in response to an invitation from Baylor University for a colloquium talk.

It is about an hour and a half drive southbound from Dallas, Texas, to Baylor on the highway that goes to Austin, the capital of Texas. So I took a flight to Dallas-Fort Worth Airport (DFW), rented a car there and drove to Baylor. My host Lance Littlejohn, was waiting for me at the modern Indigo Hotel where he had made arrangements for my stay. We had a late night dinner and a leisurely conversation. Lance had been very much influenced by George Andrews in his early research days, and he was present at the Andrews80 conference at Penn State in the summer of 2018, which is where I met him (see §11.4) for a description of this conference). After my lecture at that conference, he came up to me and said that he would like to invite me to Baylor and that is how this visit materialized.

My talk at Baylor was in the afternoon the next day, and so in the morning, Lance took me on a long walk around the lovely campus and to lunch at the Faculty Club. My colloquium at Baylor was to a general audience and so I spoke on "Ramanujan's place in the world of mathematics". This is the title of my book with Springer and I presented Lance a copy of the book. That night, Lance and his colleagues hosted a dinner at the classy Baylor Club atop the Football Stadium at Baylor.

On Monday, March 11, after our Spring break, Alex Lubotzky of the Hebrew University in Jerusalem, delivered the Ramanujan Colloquium. I was so impressed with the work of Lubotzky, Sarnak and others on Ramanujan graphs, and with Lubotzky's lecture at the Royal Society (Oct 2018), that I invited him to deliver the Ramanujan Colloquium, to which he graciously consented. Lubotzky arrived in Gainesville on Sunday afternoon and so we had time to have dinner that night. Lubotzky observes Kosher guidelines, but since vegetarian food is Kosher, I took him the newly opened Indian vegetarian restaurant in Gainesville.

Lubotzky delivered the 13th Ramanujan Colloquium on the theme "From Ramanujan graphs to Ramanujan complexes". Ramanujan graphs, and more generally expander graphs, are remarkable objects with connections to many parts of mathematics and computer science. These are graphs which are highly connected; to separate them into disconnected pieces, one must remove a large number of edges. In the last decade, mathematicians have formulated the notion of expansion to higher-dimensional complexes. Lubotzky is a powerful speaker and lectures with absolute clarity. The next day he delivered two lectures in the number theory seminar on expander graphs. Mathura and I hosted a party for him that night at our home where the food is always vegetarian and so it met his Kosher needs.

My PhD student Ankush Goswami who had his thesis defense in mid-February, submitted his thesis in early May. His thesis was in two parts. Part I concerns his work relating to the parity of the generalized divisor function with restrictions on the size of the prime factors. This was a problem I gave him based on my earlier work on the parity of the number of prime factors with restrictions on the size of the primes. This work involves a variety of analytic techniques, including the study of the behavior of functions satisfying difference-differential equations. Goswami demonstrated his skill in handling these analytical tools. The second part of the thesis was on the lovely q-analogue to Euler's identity for the Riemann zeta function at the positive even integers. This was a problem he identified on his own, and solved it impressively. Peter Paule offered him a two year post-doctoral fellowship at the Research Institute of Symbolic Computation (RISC) starting in May 2019 which he accepted. Before joining his position at RISC, Goswami returned to India to get married, and following that, went to Europe with his new bride. His wife got a position in Lithuania and so he and his wife were commuting between Lithuana and Austria regularly to spend time together.

==

Launch of The Alladi Diary in Madras and Gainesville

My mother and I departed Florida for India on April 24 for a three week stay. The main purpose of the trip was to have a function in Madras for the release of The Alladi Diary published by World Scientific. Mathura could not come to India because she had the separate Arangetrams of two of her students and had to be in Florida for intense training of the students in the final few weeks before their Arangetrams. We flew via the Pacific on Korean Airlines, once again enjoying the supreme comfort of the Airbus A-380 super jumbo on the Los Angeles to Seoul segment. I am glad I flew on the A-380 quite a few times, because this mammoth aircraft is being pulled out by various airlines in preference to smaller, less comfortable, aircraft they are able to fill more easily. The Covid pandemic is hastening the demise of very large aircraft such as the A-380 and the Boeing 747.

The Alladi Diary edited version published by World Scientific was released on May 8 at a grand function held at the M. S. Swaminathan Research Foundation in Madras. It was gracious of Prof. Swaminathan to offer the use of his Foundation for this function. He knew my father well and has great regard for my father's contribution to the development of mathematical sciences in India. For his eminence and worldwide recognition, he is to the field of agricultural sciences what C. R. Rao is to the statistical sciences. Besides him, there were three others who spoke at the function: Mr. Gopalakrishna Gandhi — former Governor of West Bengal, Prof. Anandakrishnan — former Vice-Chancellor of Anna University, and Mr. N. Ravi, Publisher of *The Hindu*, India's National Newspaper. Prior to their speeches, I presented a slide show providing a kaleidoscopic account of my father's remarkable life.

Mr. Gopalakrishna Gandhi had a distinguished career both as a diplomat and as a politician. He rose to the position of Governor of the state of West Bengal in India. He is the paternal grandson of Mahatma Gandhi and the maternal grandson of the great statesman C. Rajagopalachari (CR as he was well-known), a contemporary and close friend of my grandfather. In the Alladi Diary, while describing the life of my grandfather Sir Alladi, my father makes several references to CR, and so it was nice to have Gopalakrishna Gandhi speak on this occasion. Back in 1975, my father and I met Gopalakrishna Gandhi at the Raj Bhavan in Madras when he was working with Governor K. K. Shah.

Dr. Anandakrishnan was one my father's closest friends. He held several leadership positions in administration and in governance in the realm of higher education. He was Vice-Chancellor of Anna University and was the Chairman of the Indian Institute of Technology, Kanpur. He served as the President of the Tamil Nadu Academy of Sciences (TNAS) whose creation was proposed by my father in 1971. He also had been a Chief Guest at several events organized by my father at the Alladi Foundation at our family home Ekamra Nivas. Thus I was pleased that Dr. Anandakrishnan an eminent academician who had held leadership roles in various capacities, and who had a long association with my father, spoke on this occasion.

My father had a long and close association with The Hindu, and in particular with Mr. N. Ravi, its Publisher. It was gracious of Mr. Ravi not only to have spoken at the function, bur also to have arranged for a prominent report in the newspaper the next morning.

The release function was jointly arranged by The Alladi Foundation, The M. S. Swaminathan Research Foundation, and The Academy of Sciences, Chennai (formerly TNAS). My father's old student Prof. V. Devanathan, who had served for many years as the Secretary of TNAS, was instrumental in ensuring the active participation of several members of the Academy. I was pleased that many relatives, and friends of the Alladi family, attended the event. GP helped me get in touch with various persons in Madras as I was organizing this event from Florida.

My mother and I departed Madras on May 10 to the USA via Singapore and Seoul. Three days after our return, Mathura's paternal uncle Gangadharan and his wife Geetha arrived in Florida. We had a short but intense and enjoyable three days with them in Gainesville and in Orlando.

Just as we had the release of the Alladi Diary in Madras, we had a release function at our home in Gainesville on July 13, to inform our friends here about this autobiography of my father. Rochelle Kronzek, the Editor of World Scientific in the USA, was most enthusiastic and supportive of this event. It was owing to my first meeting with Rochelle in my office in Fall 2013, that my association with World Scientific began. She was the one who put me in touch with Dr. K. K. Phua, the founder of World Scientific, and his son Max Phua, the CEO of WSPC. In her fine introductory speech at our home, Rochelle gave an account of how WSPC evolved and in a few decades became a major worldwide publishing house, how my association with her and WSPC evolved as well, and how pleased WSPC is in publishing the edited version of The Alladi Diary. As in Madras, here in Gainesville, I presented a slide show about my father's life which the guests who numbered more than 90 enjoyed very much. For this slide show using the computer, graduate student Jason Johnson, who has superior technological skills, was most helpful. Over the years my father had enjoyed the company of so many friends in Gainesville, and they also had a great liking and admiration for him. So I felt pleased that I organized this event. Also, I was happy that it provided an opportunity for Mathura and my daughter Lalitha, son-in-law Aditya, and my grandkids Kamakshi and Keshav, to partake in this function because they could not be present for the release in Madras. My grandchildren watched the slide show with keen interest and concentration.

==

In between the two release functions of The Alladi Diary was the 80th birthday conference of Bruce Berndt at the University of Illinois in Urbana from June 6 to 9. I could not be present at this conference in full because June 7 was the day my father passed away in 2008, and so on June 7, I had to perform a religious ceremony in his memory. It is very important to perform this ceremony called the *Sraddham* annually, and so I wrote to the organizers of the conference that I will be there only for the last two days. My father's ceremony was conducted at the Hindu temple on Inverness, Florida, on June 7 morning, and after completing that, I flew from Orlando to Indianapolis where I rented a car to drive to Urbana. It was 10:30 pm by the time I reached there.

At the conference, World Scientific had a book exhibit, and Rochelle Kronzek had thoughtfully displayed the Alladi Diary. The banquet in honor of Bruce Berndt was on the night of June 8 and I was able to attend that and give a short speech recalling our long association. Since Berndt knew my father well, had visited Ekamra Nivas, lectured at the Alladi Foundation many times, and delivered the very first Alladi Ramakrishnan Memorial Lecture after my father passed away, I presented him a copy of the edited version of The Alladi Diary that had just appeared. My talk at the conference was the first lecture on the last day (June 9). I spoke about Part I of Goswami's thesis on the parity of the generalized divisor function, because we were going to write a joint paper on that theme.

In the latter part of June, I had the interesting experience of giving a talk to the REU students of Ken Ono at Emory University. I enjoyed speaking about problems involving the number of prime factors to these exceptionally bright students. Ono is an outstanding mentor at all levels and conducts a vibrant NSF funded REU (= Research Experience for Undergraduates) Program each summer. He gathers about twenty talented undergraduates from around the nation and over a period of two months gives them an intense dose of number theory. Some of his current and former PhD students assist him in the REU as instructors.

Just as I was having a busy academic program in the summer, Mathura's schedule was packed in connection with the Arangetrams of two of her students in Jacksonville. The first was on June 29, our wedding anniversary, and the second was on July 27. We both needed a vacation together after these events and so we took a two week cruise with Holland America to Iceland and Norway (August 4–18). The cruise departed out of Amsterdam, and we were on the gigantic cruise boat *The Nieuw Statendam* which has been commissioned just a few months earlier. Mathura always wanted to see Iceland, and I wanted to see the fjords of Norway. This cruise provided what each of us wanted, and the timing was perfect — after Mathura's Arangetram, and before the start of classes in the fall. This cruise was just as amazing as the Alaskan cruise we had taken two years earlier. Iceland is a land of both fire and ice. Iceland has several spectacular waterfalls and large number of fantastic geysers. Indeed the word geyser for the fountains of hot water stems from the name of the place Geyser in Iceland where these hot springs abound. The highlight of the fjord tour of Norway was the drive down the Trollsteigen, a meandering road with about 20 hairpin bends descending steeply into the valley below. As a bonus to the visits to Iceland and Norway, the cruise also took os to Newcastle-upon-Tyne and Edinburgh, where we could get a taste of English history and culture.

The main academic event in Florida in Fall 2019 was the AMS Sectional Meeting in Gainesville during November 2–3. The AMS meeting was returning to UF campus after twenty years, the last having been in March 1999 during the first academic year of my chairmanship. About three hundred mathematicians were on campus for this two-day conference which was on a slightly smaller scale compared to the 1999 meeting. Of the 15 Special Sessions conducted, eight were by my colleagues in collaboration with mathematicians at other universities. Frank Garvan conducted a Special Session on "Partitions and Related Topics" along with Dennis Eichhorn (University of California, Irvine) and Brandt Kronholm (University of Texas, Rio Grande Valley). There was good participation in this session by active young researchers. I spoke on "q-analogues of Brun's identities and inequalities" — work that I had done in August 1994 in Zurich during the International Congress of Mathematicians after a conversation with Erdős! I had never published this nor spoken about this before, and since some young researchers like Robert Schneider and Madeline Locus Dawsey took interest recently in certain ideas of mine in analytic number theory, I decided to speak about this work.

==

Visit to India and the 2019 SASTRA Prize to Adam Harper

The concluding academic event for the calendar year was the SASTRA Ramanujan Conference (Dec 21–22). But there were some anxious moments for me before the trip.

Mathura departed for India on November 24 with Lalitha, Aditya and my grandchildren. They all left earlier than me because they were to attend a wedding in Pune on Aditya's side of the family and go on a pilgrimage to nearby Shirdi. I was to leave with my mother only in December after the end of my classes. They travelled on Qatar Airways via Doha and enjoyed the superb service of the airline both on the ground and in the air. Knowing my passion for aviation, my grandson Keshav thoughtfully bought a model of the Qatar Airways Airbus A-380 as a gift for me.

About two weeks before their departure, I was bothered by a kidney stone, but I felt fortunate that I could have the procedure done before Mathura's departure. And yes, it was completed as scheduled. But as luck would have it, a second procedure had to be done, and this problem occurred AFTER Mathura had left. Realising that I was by myself and that I was scheduled to depart for India on December 4 with my mother, my good friend, the highly accomplished urologist Dr. P. Narayan, quickly arranged the second procedure to be done at the Ocala General Hospital on Wednesday, November 27, one day before Thanksgiving Thursday. This would allow me time to recuperate before my departure. It was most thoughtful and kind of Dr. Narayan to personally take me to Ocala in his car, check me in at the hospital, perform the procedure, and bring me home that night when he was returning from his work! I cannot adequately thank him for his kindness and care. Yes, I recuperated in time to fly to India on December 4 with my mother. We travelled to India via the Pacific on Delta, Korean Air and Air India and arrived in Madras on December 6.

The SASTRA Ramanujan Conference was to start on December 21 and my international guests were not to arrive before December 19. Since

Mathura and Lalitha's family were already in India, we went on a five-day trip (December 8–13) to the charming city of Mysore and the Mudumalai wildlife sanctuary. Mysore and Mudumalai had been our family's favourite haunts since the sixties, and Lalitha wanted to take Aditya and her kids to those very spots she had enjoyed since her childhood. The most wonderful experience was the boat ride on the Cauvery river at the Ranganathittu Bird Sanctuary outside of Mysore, where we saw a thousand birds of all sizes, colors, and shapes, nesting on clumps of trees that are at the banks of the river or in the middle of the river. Hungry marsh crocodiles lay beneath the trees waiting for some hapless bird to fall into the water. Our boatman took us within a few feet of the crocodiles much to the delight of my grandchildren. The trip to Mysore and Mudumalai was also a relaxation for me after the tension caused by the kidney stone procedures.

Back from the trip to Mysore and Mudumalai, I was ready to receive the international speakers of the SASTRA Conference. Adam Harper, the winner of the SASTRA Prize, arrived just past mid-night on December 19 and was received by GP at the airport. That same morning at 7:00 am, Ken Ono arrived from Bangalore with his wife. He checked in at Hotel Savera (where all of our guests for the SASTRA Conference were accommodated), rested for a few hours, and proceeded to MATSCIENCE in the afternoon for a colloquium. Ono was in India for one month as a Jubilee Professor and his entire trip was supported by the Indian Academy of Sciences. Under that scheme, Ono visited about a dozen academic institutions in India giving lectures of wide appeal. He was an ideal choice as Jubilee Professor since not only is he working on cutting edge research, but he is also an outstanding speaker. Ono was corresponding with me right from the start regarding his lecture tour of India and it was decided that the SASTRA conference will be his final assignment on his tour. His colloquium at MATSCIENCE was also part of his assignments as Jubilee Professor. Other international speakers for the SASTRA Conference were Larry Rolen (Vanderbilt), Paul Jenkins (Brigham Young), Nick Anderson (Brigham Young), Josh Males (Cologne), and Brad Rodgers (McGill). Rolen and Jenkins were former PhD students of Ono. Rodgers received his PhD from UCLA under the direction of Terence Tao. Josh Males was finishing his doctorate at the University of Cologne under the direction of Kathrin Bringmann.

From within India we had two speakers — A. Sankaranarayanan of the Tata Institute and Kaneenika Sinha of IISER, Pune.

We all departed Madras on December 20 morning for Kumbakonam. Our guests enjoyed seeing the Town High School, Ramanujan's home, and the Sarangapani Temple before checking in at Sterling Resorts for a two-night stay. Ken Ono inaugurated the SASTRA Conference on December 21 morning and presented the SASTRA Ramanujan Prize to Adam Harper.

Adam Harper has made path-breaking contributions to analytic and probabilistic number theory by establishing a number of deep and surprising results. His fundamental researches, both individually and in collaboration, cover the theory of the Riemann zeta function, random multiplicative functions, S-unit equations, smooth numbers, the large sieve, and the recent highly innovative "pretentious" approach to number theory.

Even as a second year PhD student at Cambridge University, Harper disproved a famous conjecture on sums of random multiplicative functions. It was widely believed that sums of random multiplicative functions ought to have a normal distribution (Gaussian law). In his PhD thesis, Harper demonstrated that sums of random multiplicative functions taken over integers in a large interval $[1; x]$, do not behave like sums of random functions.

Another great advance due to Harper concerns estimates for the $2k$-th moments of the Riemann zeta function on the critical line. This problem going back to Hardy and Littlewood, is of fundamental importance in analytic number theory. There is a conjectured asymptotic formula by Keating for the $2k$-th moment on the critical line. This conjecture is known to be true for $k = 1$ and $k = 2$, but has remained open for larger values of k. In 2009, assuming the Riemann Hypothesis, Soundararajan obtained almost the correct order of magnitude. But Harper, using the Riemann Hypothesis was able to get the correct order of magnitude upper bound for all values of k as per the conjectured formula.

Adam Harper was born in Lowestoft in the United Kingdom. He completed his PhD in 2012 at Cambridge University under the guidance of Professor Ben Green, and won the Smith Essay Prize. He was a Post-Doctoral Fellow with Andrew Granville at CRM Montreal during 2012–13, following which he was a Research Fellow at Jesus College, Cambridge University during 2013–16. In 2018 when he won the SASTRA Prize, he was an Assistant Professor at the University of Warwick.

After the award of the prize, all the speakers, except for Ono and Harper gave talks that day. The next day, December 22, Ramanujan birthday, we just had two talks in the morning — Harper's Ramanujan Commemoration Lecture on his prize winning work, and Ono's talk on the Riemann Hypothesis (RH) in which he presented his recent joint work with Larry Rolen, Michael Griffin, and Don Zagier on an approach to RH via the Polya-Jensen polynomials.

After the SASTRA conference, I had a few relaxed days in Madras with Lalitha and my grand children Kamakshi and Keshav. It was amusing to see them take a fascination to the toy airplanes that I played with as a kid; these are still in good condition at Ekamra Nivas after nearly sixty years!

Mathura, Lalitha and my grandchildren departed for the USA on December 26 early morning by Qatar Airways via Doha. My mother and I followed them three days later by flying trans-pacific.

==

The year 2020 started with a number of fruitful and happy events. New Year's Day was my grandson Keshav's sixth birthday, and we celebrated it by visiting the Kennedy Space Center.

In mid-January, I attended the Annual Meeting of the AMS in Denver. I was delighted and proud to see 2014 SASTRA Ramanujan Prize Winner James Maynard get the prestigious AMS Cole Prize for Number Theory. After the prize ceremony, in congratulating Maynard, I said that I hoped he would next win the Fields Medal like the previous SASTRA Prize Winners Manjul Bhargava and Peter Scholze who won the Cole Prize and later the Fields Medal. At the Denver meeting, I enjoyed listening to Ken Ribet's Retiring Presidential Address entitled "Fermat's Last Theorem in 2020". It was an interesting coincidence that while I was in Denver for the AMS Meeting, Amritha was also in the city in connection with her project for Ernst and Young! So she met me at my hotel for breakfast one morning.

My youngest grandchild (Amritha's daughter) Sahana's first birthday was on January 31, and we had a double celebration of that — a party at our home in Gainesville on January 22 for our Gainesville friends to see the child, and a party in Atlanta on February 1 arranged by Amritha and Jis.

Just as we were savouring all these lovely memories, news came in mid-February that a new virus which originated in China in November 2019 was spreading rapidly across the world. This new Coronavirus, dubbed Covid-19, was different from all the scourges of the past. It is a virus which attacks the lungs, injures the immune system, and spreads rapidly through contact and through the air we breathe. It is also more lethal than viruses that we have endured in the past. The world did not know how to stem the tide of the virus except through self isolation and lockdowns. In spite of this, millions of people have died, with the elderly and those with co-morbidity being the most affected. Our sympathies and prayers to all the families struck by this tragedy. A few long time friends of our family passed away in India due to this pandemic.

Covid-19 brought everything to a near standstill — it had a devastating effect on travel, education, and business, and our lives. However, we should be thankful that we are living in the era of internet communication and so it was possible for most of us to work from home. In the sphere of education, classes at all levels from elementary school to graduate level classes were taught using zoom. This is not as effective as live instruction in class, but we have to manage with the situation at hand. Many things started to open up in the last quarter of 2020, but the long periods of lockdowns were distressing to all of humanity. But human beings have overcome the most formidable circumstances. There was hope that in the near future, a cure for the virus will be found, and things will be back to the way it was. This pandemic reinforced the following thoughts: we should enjoy every possible moment with family and friends, and continue to work with the same dedication in the career path we have chosen. On my part, I will continue to pursue my research problems in number theory with the same passion, serve the profession with the same dedication, and work in fostering the legacy of Ramanujan with the same determination.

10.11   Epilogue — return to normalcy from Covid-19

Although Covid-19 cast a pall of gloom worldwide through much of 2020 due to lockdowns, and caused misery both healthwise and psychologically, yet people across the globe worked from home with determination. Scientists worked tirelessly to rapidly find a vaccine to counter Covid-19, and indeed from the beginning of 2021, vaccinations were systematically administered. Thus there was a slow return to normalcy with the lifting of lockdowns, but at the same time, considerable caution is being exercised to avoid the resurgence of the pandemic. I used the period of isolation in 2020 to complete the Second Edition of my book *Ramanujan's Place in the World of Mathematics* [B1], that was published by Springer in 2021, as well as to make progress on this autobiography! The second edition of my book on Ramanujan has six more articles including an elaborate review of the movie "The Man Who Knew Infinity" (the life of Ramanujan), and a report of the 2018 Royal Society Conference for the Centenary of Ramanujan's Election as FRS that was discussed earlier.

==

2020 SASTRA Ramanujan Prize to Shai Evra

Owing to the pandemic, there was no live conference at SASTRA, but we did have a SASTRA Ramanujan Prize for 2020. The winner of the 2020 Prize was Shai Evra of the Institute for Advanced Study, Princeton, and the Hebrew University, Jerusalem. Evra was a PhD student of Alex Lubotzky of the Hebrew University. Evra's fundamental research concerns locally symmetric spaces of arithmetic groups and their combinatoric, geometric, and topological structure. He employs deep results from representation theory and number theory pertaining to the Ramanujan and Langlands conjectures to establish expander-like properties. He has had two major collaborations, one with Tali Kauffman in the study of higher dimensional Ramanujan complexes, and another with Ori Parzanchevski on three dimensional unitary groups. The Ramanujan Conjectures and their generalizations are a central piece in Evra's work. On December 22, Ramanujan's birthday, Evra gave a talk on zoom arranged by SASTRA University.

Shai Evra was born in Be'er Yaakov, Israel. He received his BSc, MSc, and PhD (2019) degrees from the Hebrew University in Jerusalem. He has been recognized with several prizes, most notably the Hebrew University Dean's Prize in 2010, the Perlman Prize in 2015, and the Nessyahu Prize in 2020. He spent the years 2018–20 as a Visiting Member at the Institute for Advanced Study, Princeton, and 2020–21 at Princeton University, following which he returned to the Hebrew University permanently.

In July 2021, there was the Subbarao Centenary Conference sponsored by his distinguished son Dr. M. Vidyasagar (FRS) and organized by the Indian Institute of Science Education and Research (IISER) in Pune, India. Professor M. V. Subbarao had done significant work in two areas of number theory — (i) arithmetical functions, and (ii) the theory of partitions and q-hypergeometric series. He was on the faculty of the University of Alberta since the 1960s. I first met him in Madras in 1972 on Ramanujan's birthday, December 22, when he came to MATSCIENCE to deliver a lecture at my father's invitation. I was a first year BSC mathematics student then, and had started working on number theory on my own. He kindly asked me to come to his apartment in Madras and informed me about some important results in number theory. I remember him proudly mentioning his son Vidyasagar, then a young graduate in engineering in Canada, who had collaborated with him in number theory. Thus it was fitting that Dr. Vidyasagar, now a Fellow of the Royal Society, arranged this conference for the centenary of his father. Indeed in 2019, when the conference was planned for 2021, it was to be a live conference sponsored by Vidyasagar, but owing to the pandemic, it was converted to a virtual event. My talk at the conference was on certain parity questions associated with the generalized divisor function with restrictions on the prime factors (joint work with my PhD student Ankush Goswami). By the bye, I had the pleasure of meeting Vidyasagar at the Royal Society in London in October 2018, when he came to attend the lectures of the conference to commemorate the centenary of Ramanujan's election as FRS.

==

And as these academic events were taking place, on a personal level, the most happy development was the birth my fourth grandchild on August 1, 2021: Amritha gave birth to a baby boy who was named *Shreyas*, which in Sanskrit means — prosperity, or being virtuous. The birth of a child not only brings boundless joy, but gives us hope for the future.

==

2021 SASTRA Prize to Will Sawin

As for the 2021 SASTRA Ramanujan Conference, even though international travel with some restrictions had resumed with the pandemic seemingly under control, out of caution, SASTRA decided not to have a live conference. The 2021 SASTRA Ramanujan Prize winner was Will Sawin of Columbia University whose revolutionary work (some jointly with Philippe Michel and Tom Browning) includes the resolution of celebrated problems in number theory such the Prime Twins and Goldbach Conjectures in a finite field-function field setting. He presented some aspects of his revolutionary work during a two-day zoom conference (Dec 21–22, 2021) arranged by SASTRA University. Other speakers at this conference were Kaisa Matomaki (winner of the 2017 SASTRA Prize), Philippe Michel (collaborator of Sawin), and Ken Ono.

Sawin is a prodigy who earned both his high school diploma and BSc degree simultaneously at age 17! He is a tremendously talented mathematician who has made pathbreaking contributions at the interface of number theory and algebraic geometry due to his great technical ability and deep understanding of a variety of powerful methods.

Thus the SASTRA Ramanujan Prize continues to recognize pathbreaking work by mathematicians early in their careers, in the spirit of Ramanujan who perhaps is the finest example of exemplary academic achievement in youth. Working on the SASTRA Ramanujan Prize Committee has been one of the most satisfying aspects of my ongoing Ramanujan mission.

Honorary Doctorate: After this book went to Press, I received a message on September 4 that SASTRA University would be awarding me an Honorary Doctorate (Honoris Causa) during their XXXVI Convocation at their main campus in Tanjore on September 18, 2022, in recognition of my research accomplishments, and my service to the profession in various capacities. The Convocation was grandly conducted and I was honored and humbled to receive this recognition.

Doron Zeilberger (Rutgers University) accepting a porcelin vase as a lovely gift from William Y. C. Chen (Vice-President, Nankai University), during the Zeilberger-60 Conference at Nankai University, Tianjin, China, August 16, 2010. Photo - courtesy of William Chen.

Left to right: Krishna Alladi, Frank Garvan, George Andrews and Freeman Dyson (Institute for Advanced Study) in the mathematics department, Mar 25, 2013. In 1987, Andrews and Garvan had solved Dyson's "crank conjecture" for partitions.

Krishna and Mathura with Peter Goddard, Director, Institute for Advanced Study, Princeton, in the Director's office, September 24, 2010. Krishna and Mathura were the guests of the Director for the 80th Anniversary Conference of the Institute.

Krishna Alladi listening intently to a conversation between Fields Medallists Enrico Bombieri (Institute for Advanced Study), and John Milnor (SUNY at Stony Brook) during a conference for the 80th Anniversary of the Institute, September 24, 2010.

Group photo of the participants of the Gainesville International Number Theory Conference in honor of Krishna Alladi's 60th birthday, March 18, 2016. Picture courtesy of Ali Uncu.

Dorian Goldfeld (Columbia) giving a talk on extensions of the Alladi-Erdős results on the additive function A(n), at the Alladi-60 Conference, Gainesville, March 19, 2016.

The $10,0000 problem of Paul Erdős was solved independently by James Maynard, and (jointly) by Kevin Ford, Ben Green, Sergei Konyagin and Terence Tao. Here Ronald L. Graham is presenting a $5,000 cheque to Maynard during Maynard's Ramanujan Colloquium at the Alladi-60 conference, on March 17, 2016. The next day, Graham presented another $5,000 cheque to Ford to be shared with Green, Konyagin, and Tao.

Krishna before the start of his lecture at the Royal Society during a conference to celebrate the Centenary of Ramanujan's election as FRS, Oct 16, 2018. The famous bust of Ramanujan sculpted by Paul Granlund is next to the podium.

Krishna with Prof. M. Vidyasagar (FRS) at the Royal Society Conference to celebrate the Centenary of Ramanujan's Election as FRS, October 16, 2018.

Mathematicians at The Alladi House, Gainesville - A sample

Dominique Foata (University of Strasbourg) with Krishna's mother, Mrs. Foata, and Mathura, during a party at the Alladi House in his honor on February 19, 2008. Foata, a world leader in Combinatorics, had delivered the Ulam Colloquium the previous day.

Gerald Tenenbaum (University of Nancy), Mrs. Tenenbaum, Mathura and Krishna, at a party at the Alladi House in Tenenbaum's honor on December 3, 2013. Tenenbaum had delivered the Erdős Memorial Lecture at the University of Florida for the Centenary of Paul Erdős.

Participants of the Alladi60 Conference, during a dinner at the Alladi House, March 18, 2016.

Ken Ribet (UC Berkeley), President of the AMS, during a party at the Alladi House in his honor, going through a file of letters of Paul Erdős to Krishna Alladi.

Chapter 11

Some special conferences and events

11.1 The Thompson-70 Conference, Cambridge, UK, Sept. 2002

John Thompson's 70th Birthday Conference at Cambridge University was organized by Richard Lyons (Rutgers University), Chat Ho (University of Florida), Gordon James (Imperial College, London) and Jan Saxl (Cambridge University). Thompson was the PhD advisor of Lyons, Ho and James. Although I was not a researcher in group theory or Galois theory, the organizers of the Cambridge conference invited me to speak at the banquet honoring Thompson as the representative of the Department and the University. It was kind of Dean Sullivan to spontaneously agree that the College would support my trip.

Mathura and my younger daughter Amritha also came to England and so we spent a few days in London before going to Cambridge for the conference. Even though we all departed Florida from Tampa on September 21, I was on Northwest Airlines to London via Detroit, whereas Mathura and Amritha were on American Airlines via St. Louis. There were no hiccups on our flight schedules; we arrived in London Gatwick Airport the next morning within minutes of each other and met at the baggage claim. We took a taxi to the stately Park Lane Sheraton hotel facing Hyde Park. On that day (Sept 22), there was a huge demonstration in London that the fox hunts, a time-honored British tradition, should not be banned. Tens of thousands of persons from all around England gathered in London for this demonstration which took place in and around Hyde Park. Thus the direct approach to our hotel was not possible; we were dropped in an alley behind the hotel from where we walked to the hotel with our luggage.

Our elegantly furnished room overlooked Hyde Park and so we could see the demonstration. It was peaceful and ended by mid-day. So we could move about the city in the afternoon without any hindrance. It was a beautiful sunny day and so we roamed the city all afternoon and evening to the Buckingham Palace, the Wellington Arch, James Park, Trafalgar Square, and Picadilly Circus. The next morning we took a tour of Buckingham Palace and it met our high expectations. In the afternoon we saw the Westminster Abbey, Big Ben, The Houses of Parliament, St. Paul's Cathedral, and the Millennium Bridge, all in splendid weather. It was Amritha's first visit to London and we showed her as much as we could of this great city in the few days we were there.

On Tuesday, September 24, I made a day trip to Brighton to give a talk at the University of Sussex at the kind invitation of Prof. Richard Lewis (now deceased). I departed by train from Victoria Station, London. It was a two-hour train journey to Falmer where Lewis picked me up and took me to Brighton. He first showed me the Pariyon Palace of splendid oriental architecture — Indian moghul style on the outside, and Chinese design inside. After this we took a walk to see the coast of Brighton and the English Channel. My talk on "New weighted Rogers-Ramanujan identities" was in the afternoon and it was well attended. I departed Sussex by train to at 4:30 pm to be back in London by 6:00 pm. Mathura had two cousins in London, and so upon return from Sussex, as well as on the previous night, we had dinner with them.

The day after my visit to Sussex, there was a tube strike in London. I understand that such strikes are quite common, but we had no sightseeing plans in London that day. Instead we were booked on a bus tour to Oxford and Stratford-upon-Avon. The ride through the lovely Cotswald countryside was thoroughly enjoyable. We were shown the great campus of Oxford University. Shakespeare's home at Stratford-upon-Avon was equally inspiring. Another highlight was the magnificent Warwick Castle.

After a wonderful stay in London, we were ready to go to Cambridge for the conference in honor of John Thompson. On September 26 morning, we took a train from London's Kings Cross Station and reached Cambridge by 11:00 am. After checking in at the Holiday Inn Crowne Plaza, I proceeded to the Isaac Newton Institute which was the conference venue, and attended the Opening Plenary Lecture of the conference by George Glauberman (University of Chicago). This was followed by the talk of Ian Macdonald after whom the Macdonald polynomials and the Macdonald identities are named. Some of the plenary speakers at the conference were Michael Aschbacher,, John Conway, Walter Feit, Robert Guralnick, Richard Lyons, Geoffrey Robinson, Helmut Voelklein, and Efim Zelmanov.

After attending several inspiring lectures at the conference on the opening day, I met Mathura and Amritha in front of St. Mary's Church from where we went to see King's College which perhaps has the most imposing design among all the Cambridge colleges.

On the second day of the conference, I attended the morning plenary lectures by Robert Guralnick (University of Southern California), and my colleague Helmut Voelklein. Both Voelkein and Guralnick had major collaboration with Thompson on the Inverse Galois Problem. That afternoon, I took time off from the conference to visit Trinity College and the Wren Library along with Mathura and Amritha. Since I am so much involved in the mathematical world of Ramanujan, the visit to Trinity and the Wren was of special significance to me. That night we attended a party at the home of John Thompson so meticulously planned by his wife Diane.

While in Cambridge, I was hoping to meet Prof. D. G. Kendall, one of my father's PhD mentors. But Kendall was in his nineties and so did not attend the conference. So I called him by phone and conveyed my father's greetings to him. He said he had pleasant memories of my parents in Manchester during 1949–51, and that he had keenly followed my father's academic progress.

The final day of the conference was September 28. In the afternoon session, Fields Medallist Efin Zelmanov (UCSD) spoke. One of the organizers who introduced him said that the conference was honored with the presence of a Fields Medallist as a speaker. With characteristic humility, Zelmanov responded that it was he who is honored to speak in the presence of the most eminent group theorist (John Thompson) of our generation. Both the organizer and Zelmanov were correct in what they said!

On the concluding night, there was a banquet in honor of Thompson at Gonville and Caius College, where Thompson was a Fellow. Mathura and I attended the banquet. We had a very nice conversation with Professor Alan Baker of Cambridge University who had won the Fields Medal in 1970 along with Thompson. Baker enquired about my research and said that he had noted the progress of the Ramanujan Journal that I had founded.

The College of Liberal Arts and Sciences of the University of Florida was one of the sponsors of this conference. So Dean Sullivan sent a congratulatory message which I read as I concluded my speech at the banquet in which I described Thompson's association with the University of Florida, and the impact it has had. Thus concluded a most memorable visit to England. We returned to Florida the next day.

11.2   The 2008 Abel Prize Ceremony and events in Oslo

It was the most exhilarating academic experience of my life to be present at
the Abel Prize Ceremony in Oslo on 20 May, 2008 and witness the award
of the Abel Prizes to my distinguished colleague John Griggs Thompson of
the University of Florida and an equally eminent mathematician Jacques
Tits of the College de France. The 2008 Abel Prizes of about \$1.2 million
in total, were awarded to Professors Thompson and Tits by King Harald V
of Norway for their monumental mathematical contributions that shaped
modern group theory. The King and the Queen of Norway graced the
Prize Ceremony which was attended by several Norwegian dignitaries, foreign diplomats, mathematicians and academicians from within and outside
Norway, and members of the press. There were several events held in Oslo
during May 19–21 related to the Abel Awards. I was invited to all these
festivities and this is a report of these events in the order in which they
took place.

Wreath laying at the Abel monument: Niels Henrik Abel (1802–29)
was Norway's most distinguished mathematician, and indeed one of the
greatest mathematicians in history! The Norwegians are justly proud of
Abel and have his statue right inside the gardens of the Royal Palace in
Oslo. This statue was made by Gustav Vigeland, a highly reputed Norwegian sculptor. On Monday, May 19 at 5:00 pm, one day before the Abel
Prize Ceremony, there was a function at the Abel Monument during which
the 2008 Abel Laureates John Thompson and Jacques Tits laid wreaths at
the monument. By 4:30 pm, a crowd had gathered at the monument in
eager expectation to witness this first of a series of events related to the
2008 Abel Awards. Before requesting the laureates to place the wreaths,
Professor Ragnar Winther, Chairman of the Abel Board, pointed out in a
short speech that the work of the 2008 Abel laureates is actually quite well
connected to that of Abel and another great Norwegian mathematician So-
phus Lie who made great efforts to honor Abel during the Abel Centenary
in 1902. Thus Professor Winther said that it was of particular significance
that Thompson and Tits lay these wreaths at the Abel monument.

**Mathematicians dinner at the Norwegian Academy of Science and
Letters:** On May 19 at 7:00 pm, there was a dinner hosted by the Norwegian Academy of Science and Letters. The dinner was only for a select
group of about 40 mathematicians and was held at the beautiful building of
the Academy, a short distance from the Palace. For all other events, spouses
were also invited. The only spouses at this dinner were Dr. (Mrs.) Diane

Thompson and Mrs. Tits. Professor Reidun Sirevag, Secretary General of the Academy, welcomed the gathering and said that the Abel Award honors mathematicians and recognizes the greatness and importance of mathematics. Thus the Academy felt that there ought to be an event in which mathematicians could meet and discuss among themselves and this dinner provided such an occasion.

During the dinner, an event of unusual importance took place, namely the release of a book containing three famous handwritten manuscripts by Abel, and the presentation of these books to the laureates. Professor Ole Didrik Laerum, President of the Norwegian Academy of Science and Letters, explained the incredible saga of these three manuscripts. Abel sent them to Leopold von Crelle, the editor and publisher of the Journal fur die Reine und Angewandte Mathematik, otherwise known as Crelle's journal. Two of them were published in Crelle's Journal. After publication, the original manuscripts were not returned to Abel. Some time later, Leopold von Crelle sold these handwritten manuscripts to an Italian manuscript collector Libri. After Libri's death, the Abel manuscripts ended up in the possession of Prince Boncompagni who put them up for sale in 1898. The great Swedish mathematician Mittag-Leffler purchased these three manuscripts of Abel at an auction in Italy and kept them at his mansion. He published the third manuscript in his journal Acta Mathematica in 1902 for Abel's centenary. A few years ago Arild Stubhaug began working on Mittag-Leffler's biography, and while going through the collections in Mittag-Leffler's mansion, he noticed these three manuscripts of Abel. He then drew the attention of Professors Laerum and Sirevag of the Norwegian Academy of Sciences who then had discussions with Prof. Anders Björner, Director of the Mittag-Leffler Institute. It was a great gesture on the part of the Mittag-Leffler Institute and the Royal Swedish Academy of Sciences to turn over these manuscripts to the Norwegian Academy of Sciences. They have now been brought out in the form of a lovely book edited by Prof. Stubhaug and published by Scatec (Scandinavian Advanced Technology). After proposing a toast to Abel's memory, Professor Laerum presented both Abel laureates with a copy of the book. Upon receiving the book, Professor Thompson said that he would like to say a few words. In an emotional response, Thompson expressed his gratitude to Abel for discovering such wonderful results and providing so much for mathematicians of succeeding generations to work on. Witnessing this important presentation at the Norwegian Academy, my mind went back to the Ramanujan Centenary in Madras, India, in December 1987, when the first printed copy of

Ramanujan's Lost Notebook was presented to Professor George Andrews. In a voice choked with emotion, Andrews thanked Mrs. Janaki Ramanujan for preserving the loose sheets that eventually bore the name "Ramanujan's Lost Notebook" and expressed his great appreciation for Ramanujan for having given us so much to work on. I felt that Thompson's emotional response on Abel compared with that of Andrews on Ramanujan!

The 2008 Abel Prize Ceremony: The 2008 Abel Prize Ceremony was held at the Aula of the University of Oslo. The Aula is a stately building with large stone pillars at the entrance and having an impressive auditorium. The Aula is located close to the Royal Palace and faces a wide avenue that leads directly to the palace.

The ceremony was scheduled to start at 2:00 pm on Tuesday, May 20. All members of the audience occupied their seats by 1:45 pm before the arrival of the King and Queen at 2:00 pm. From 1:00 pm onwards, in front of the Aula, there was music by the Staff Band of the Norwegian Defence Forces to welcome the guests.

Seated on stage at the Aula were the Abel Laureates John Thompson and Jacques Tits, Mrs. Tits to be assistance to her husband, Professor Ole Didrik Laerum, President of the Norwegian Academy of Science and Letters, and Professor Kristian Seip, Chairman of the Abel Prize Committee. The program started promptly at 2:00 pm when Their Majesties King Harald V and Queen Sonja arrived and occupied their seats up front in the central aisle from where they could witness the entire ceremony.

The ceremony began with music by three of Norway's noted performers — Ms. Randi Stene (vocal), Mr. Lars Anders Tomter (viola), and Mr. Ketil Bjornstad (piano). The music was both enchanting and uplifting. Following this, Professor Laerum made the Opening Remarks in which he briefly outlined Abel's life and contributions, emphasized the importance of mathematics in our lives, and the described the objectives of the Abel Prize. Then Professor Seip briefly recalled some landmark results and ideas of group theory starting with the foundational work of Galois and Abel, and with that as background, he highlighted the pathbreaking contributions of the laureates Thompson and Tits. He then requested King Harald V to present the 2008 Abel prizes to Thompson and Tits. It was a glorious moment. In his brief acceptance speech, Professor Thompson graciously acknowledged the crucial contributions of several mathematicians over the decades that helped achieve the classification of finite simple groups. This is a monumental accomplishment and Thompson led the worldwide effort of dozens of mathematicians. Out of modesty Thompson did not emphasize his own epoch making work in group theory. Jacques Tits in his equally modest acceptance speech, thanked his contemporaries and collaborators, and appreciated the great support given to mathematics by the King and Queen of Norway, the Norwegian Academy of Science and Letters, and the Abel Foundation. The Abel Prize Ceremony was brief (40 minutes) and all arrangements were tastefully done and executed perfectly.

Press conference: Following the Abel Prize Ceremony, there was a press conference at 3:00 pm organized by Ms. Tora Aasland, Minister of Research and Higher Education. The press conference was held at the classic Hotel Continental (where the laureates stayed) close to the Aula. The minister said that she understands the importance of sound mathematical education and arranged this press conference to get suggestions from eminent mathematicians about how to address problems facing mathematics education at all levels. The students of the Abel Mathematics Competition for high schools were recognized just the day before and so one of the topics discussed was the importance of spotting and encouraging talented youngsters. Another was the declining interest among students to pursue mathematics in view of lucrative careers in other fields, and how to go about attracting students to mathematics.

The Abel Banquet: At 7:00 pm on May 20, there was a grand banquet at the magnificent Akershus Castle in Oslo. The Akershus has been recently renovated and is now used primarily for state receptions. Since King Harald and Queen Sonja were present at the banquet, dress was formal — black tie or national costume. My wife Mathura wore an Indian silk sari for the occasion. It was a pleasure for us to see the Norwegians (men and women) in their beautiful national costumes — these costumes being distinct for different regions of Norway. There were 160 guests in total, and spouses were separated in seating so that we could meet different members. The banquet was attended by several Norwegian dignitaries, members of the diplomatic service from foreign countries, and eminent mathematicians. Before the banquet, there were brief speeches by Ms. Tora Aasland, Minister for Research and Higher Education, and by Professor Ari Laptev, President of the European Mathematical Society. At the end of the banquet, we were treated to lovely music by members of the Norwegian Soloists Choir. It was a very memorable evening in the company of extremely distinguished persons in a grand setting fit for the occasion.

A very interesting incident happened to me and Mathura as we were getting out of the Akershus Castle to return to our hotel. We wanted to take a taxi to the city center and we stopped an empty taxi at the Akershus which was ready to give a ride. But since the city center was not very far (according to the taxi driver), he was not inclined to give a ride. So Mathura and I decided to walk, even though we felt it was long walk, especially at night. Just then a chauffeur driven Mercedes pulled up and the person in the back seat offered us a ride to the city center. It turned out that he was the Mayor of Oslo! We were touched by his gesture of goodwill.

The Abel Lectures: Wednesday, May 21, the day following the Abel Prize Ceremony, was devoted to the Abel Lectures — four of them delivered at the Georg Sverdrups Hus of the University of Oslo. The first two lectures by the laureates Thompson and Tits were in the morning, and the next two by Michel Broue (Director — Institut Henri Poincaré and Editor in Chief of the Journal of Algebra) and Alex Lubotsky (Maurice and Clara Weil Professor at the Einstein Institute of Mathematics, Hebrew University) were in the afternoon. The audience were welcomed by Kristian Seip (Chair of the 2008 Abel Prize Committee) and Geir Ellingsrud (Rector of the University of Oslo). Sir John Kingman, FRS (University of Bristol, former Director of the Isaac Newton Institute in Cambridge, England and former President of the Royal Statistical Society), Member of the 2008 Abel Prize Committee, introduced Thompson who spoke on "Dirichlet series and $SL(2, Z)$", his recent seminal work at the University of Florida.

For the past decade, Thompson has been systematically investigating upper triangular matrices. This was the theme of his invited one hour address in March 1999 to the American Mathematical Society at the meeting held on the campus of the University of Florida. In the past few years, Thompson has been studying the divisor matrix D from a group theoretic point of view. Here D is the infinite upper triangular matrix $[d_{i,j}]$, where $d_{i,j} = 1$ if i divides j, and 0 otherwise. Thompson discussed crucial connections with Dirichlet series such as the Riemann zeta function, with arithmetical functions like the Möbius function, and with $SL(2, Z)$. Certain aspects of this work are in collaboration with Professor Peter Sin of the University of Florida as pointed out by Thompson in his lecture. The Thompson-Sin paper appeared in the book [B2] in memory of my father.

Jacques Tits spoke on "Algebraic simple groups and buildings", areas to which he has made pioneering contributions. He viewed groups as geometric objects and introduced what is now called Tits building. This encodes in geometric terms the algebraic structure of linear groups. Tits was introduced by Professor Hans Folmer (Humboldt University, Berlin), Member of the 2008 Abel Prize Committee.

The afternoon talks by Professors Michel Broue and Alex Lubotsky were on the great advances in group theory made possible by the fundamental contributions of the laureates. Broue spoke on "Building cathedrals and breaking down reinforced concrete walls". He was introduced by Dusa McDuff (SUNY at StonyBrook), Member of the 2008 Abel Prize Committee. Lubotsky spoke on "Simple groups, buildings, and applications". He was introduced by 1994 Fields Medalist Efim Zelmanov (UCSD), Member of the 2008 Abel Prize Committee.

There were several participants (mathematicians) who were registered for the two day Abel program, and the so Abel lectures which were excellently planned and executed, had a large and well informed audience.

The Abel Party: The concluding event was the Abel Party held at the Norwegian Academy of Sciences at 7:00 pm on May 21. There was a large gathering of mathematicians and members of the Norwegian Academy from other disciplines. It was nice touch to have live music at the party — American jazz in honor of John Thompson and French music on the accordion and viola in honor of Tits. It was a cheerful ending to a glorious three days.

Concluding remarks: It was a pleasure and a privilege for me to represent the University of Florida and the Mathematics Department at the 2008 Abel Prize Ceremony and related events, and to see my distinguished colleague John Thompson being given the greatest honor in mathematics along with the very eminent Jacques Tits. Events like this make us feel proud to be mathematicians and happy that our discipline is held in such high regard by the international community of scholars. The fine finishing touch to my visit was the memento I received at the Abel Party from Professor Laerum, President of the Norwegian Academy of Science and Letters, namely the book "Tracing Abel" of the three Abel manuscripts that were first presented to the Abel laureates during the mathematicians dinner on May 19. The epoch making works of mathematical giants like Abel from earlier times, and of distinguished mathematicians like Thompson and Tits of our day, are an inspiration for all of us engaged in mathematical research.

11.3 The Zeilberger-60 Conference, Tianjin, China (2010)

Doron Zeilberger is a brilliant mathematician known for several fundamental contributions to Combinatorics of which we mention just three here: (i) the Zeilberger-Bressoud proof of the q-Dyson Conjecture, (ii) his proof

of the Alternating Sign Matrix Conjecture, and (iii) the Wilf-Zeilberger
(WZ) method to certify and prove various classes of combinatorial identities involving binomials, factorials, and q-products. I am most familiar
with his bijective proof (established jointly with David Bressoud) of the
celebrated Rogers-Ramanujan identities, which is shorter than the first bijective proof of these identities due to Adriano Garsia and Stephen Milne.
Zeilberger is an electrifying and humorous speaker whose talks appeal to
experts and novices alike. I have heard him many times, and thoroughly
enjoyed every one of his presentations which are full of interesting ideas.
He has been, and continues to be, a great influence in the field, and so
it is hardly surprising that there were more than one conference to celebrate his 60th birthday. I missed the conference in May 2010 at Rutgers
University, his home turf, but I had the opportunity to attend a grand
60th birthday conference entitled "The Renaissance of Combinatorics: Advances, Algorithms, Applications" at Nankai University in Tianjin, China,
August 14–16, 2010.

In early April 2010, I received a warm invitation from Professor William
(Bill) Chen, Director of the Center for Combinatorics at Nankai University,
to give an hour lecture at the DZ-60 Conference in August. I readily accepted the invitation. It was only later that I found out that the position
as Director was only one of many positions that Bill Chen held. He was a
Vice-President at Nankai University and a senior member of the political
party. Thus he has enormous influence which he put to good use in encouraging mathematical research, especially in combinatorics. He had George
Andrews and Peter Paule on the Advisory Board of the Center, and on the
Editorial Board of *The Annals of Combinatorics*, a leading journal published by Birkhauser for which he was the Founder and Editor-in-Chief.
Thus it was not surprising that Andrews and Paule were inducted as co-
organizers of this conference with Bill Chen.

I departed from Florida for China on August 11, flying this time by
American Airlines instead of Delta because AA offered a much less expen-
sive ticket. I arrived at 10:50 pm on August 12 at the fabulous and gigantic
new Terminal 3 at Beijing Capital Airport and checked in at the brand new
Airport Hilton which was less than three weeks old. The next day, I was
picked up in the afternoon by representatives of the Center and taken to
Tianjin. We went straight to a restaurant where we joined Chen, Paule,
Andrews, and Zeilberger, at dinner.

The conference opened on August 14 morning with Bill Chen's
welcome and remarks. He stressed the major role played by Zeilberger

in the development of combinatorics. The first lecture of the conference was by Zeilberger himself and it was titled "Why is it so hard to count?" Combinatorial proofs often involve a counting procedure, but to find effective and appealing ones is not so easy. He was followed by George Andrews on the theme "Partitions with early repetitions and Slater's list". With that the morning session ended on a high note. The lunch each day was on campus, but dinners were banquets at famous restaurants or clubs in Tianjin.

I have often said that the meaning of a banquet becomes clear only if you experience it in China, but even these Chinese banquets pale in comparison with what Bill Chen arranges! At the banquet, Chen approached the guests one by one, and graciously asked each guest to propose a toast to Zeilberger and exhorted everyone to drink a drop of that concentrated wine. I was spared of the wine since I am a tee-totaller, and therefore was given strawberry juice instead!

My talk was first in the morning of the second day with George Andrews in the chair. I spoke on "Variations of themes of Euler and Gauss in the theory of partitions" and emphasized underpinnings behind the theorems I presented. Zeilberger liked my talk very much and said that I should visit Rutgers University in the Fall and speak in his Experimental Mathematics Seminar on the same topic. And yes, I did visit Rutgers and enjoyed his gracious hospitality (see §10.11). Peter Paule followed me as the morning speaker. He spoke on "Symbolic computation: new algorithmic methods in physics and special functions". This was very appropriate for two reasons: (i) he leads a group of researchers at the Research Institute of Symbolic Computation (RISC) in Linz who are doing important work in symbolic computation as pertaining to combinatorics, and (ii) the WZ Method is one of the crucial tools employed by Paule and his group. Indeed, the conference featured two more talks by researchers from RISC — Manuel Kauers and Carsten Schneider.

On the morning of the last day, the first talk was by Bill Chen who presented his work on partitions done jointly with various members of his group. The talk included a brilliant combinatorial proof of a certain partition theorem of mine that came up in my analysis of a partial theta identity of Ramanujan; I had spoken about this in my talk the previous day.

Zeilberger's famous collaborator Herb Wilf (University of Pennsylvania) gave a delightful talk entitled "How to lose as little as possible?" in the concluding session of the conference on August 16 afternoon. He started his talk by putting up a slide that contained the following lines:

Doron Zeilberger is

(i) a fine human being,

(ii) an exceptionally creative and important mathematician,

(iii) a pleasure to work with, and

(iv) a warm and supportive mentor to his students.

This was no exaggeration and a fine tribute to Zeilberger from his eminent collaborator.

Following Wilf's lecture was the Closing Ceremony which began with Bill Chen presenting Zeilberger an exquisite porcelain vase. George Andrews conducted the concluding ceremony when several participants paid tributes to Zeilberger. In fact, I was asked to speak first — perhaps for alphabetical reasons(!) — and so I had an opportunity to express my admiration for Zeilberger's brilliant contributions and my appreciation for his support of all my efforts. Near the end, Doron's wife spoke, after which Doron had the last word — he thanked Bill Chen, the organizers, and the participants for making it such a memorable conference. That night there was a fabulous banquet at an exclusive club in Tianjin.

On November 17 morning, the conference participants were treated to a fantastic magic show by a friend of Bill Chen. After this there was a banquet at the Foreigners Guest House on the campus of Nankai University before the participants departed.

Bill Chen gave his personal car to take Herb Wilf and me to Beijing. At my suggestion, Wilf agreed to spend the night at the Beijing Airport Hilton before departing for America the next day. On the drive from Tianjin to the Hilton, Wilf told me about his hobby — piloting his own airplane! He was a licensed pilot and owned a small Cessna airplane. He and his wife would fly this plane when they were going to conferences within the USA. He said that for one of the Gainesville conferences, he flew his Cessna from Philadelphia to Jacksonville where he left it at the airport, then rented a car and drove to Gainesville. He said that his training in mathematics was immensely helpful in flying an airplane; he pointed out that three dimensional geometry was what he needed. In view of my interest in aviation, he suggested that I too should get a pilot's license and buy a Cessna. But my family would not allow me to do that, and I should respect their feelings.

I was told by a mutual friend, that a few years later, when Wilf was piloting his plane with his wife by his side, the wheels would not come down. So he called the airport and told them that he would do a belly landing! The airport had fire engines lined up along the runway which they smeared with foam to reduce friction. Luckily, there was no fire when the belly landing took place and so Wilf and his wife escaped without any incident. But Wilf stopped flying after this. It was for reasons like this that my family did not want me to get a pilot's licence. They asked me to confine my passion for aviation to buying books and memorabilia on aviation!

After a comfortable stay at the Beijing Airport Hilton, I departed the next day by Japan Air Lines for Tokyo where I connected to an American Airlines Boeing 777 flight to Dallas-Forth Worth. I was back in Florida on August 17 night. Although I had taken photographs of my own, Bill Chen sent me a beautiful set of pictures to keep our memories of that wonderful conference fresh in our minds.

11.4   The Andrews Milestone Birthday Conferences, 1998–

George Andrews, without a doubt, is the greatest expert in the theory of partitions and the work of Ramanujan combined. Defying the passage of time, he continues to be a prolific researcher and a powerful influence in his field as an octogenarian. In addition to conducting world class research, he has served the profession admirably in many ways — most notably as Chair of the mathematics department twice at The Pennsylvania State University, and as President of the American Mathematical Society. In spite of his high eminence, he is a gentleman to the core, and a most caring and helpful person who responds to all his emails with surprising quickness in the midst of his vast commitments. It is rare to find eminence and humanity combined. Owing to this precious combination, mathematicians around the world, and his colleagues at Penn State, have conducted conferences in his honor once every five years starting from his 60th birthday in 1998. I have had the pleasure and honor to attend and speak at every one of his milestone birthday conferences, and I will now share my experiences on these.

The 60th birthday conference, Maratea, Italy, Sept 1–5, 1998

The first of these milestone birthday conferences was held in Maratea, a lovely town on the enchanting Tyrrhenian coast, south of Naples in Italy. In late April 1988, I received a warm invitation from Prof. Dominique Foata to speak at the conference in September. He is a leader in the field of combinatorics in France, and so I felt flattered when he said in his letter: "As one of the great scientific collaborators of George, you would be most welcome if you would accept to participate in this seminar." I had just started my term as Chair of the Mathematics Department at the University of Florida in July, and it was gracious of Dean Harrison to offer to support my trip with funds from the College of Liberal Arts and Sciences.

In those days, I used to be a frequent flier with Northwest Airlines and its partners, but Northwest was on strike. So I was on TWA from Orlando to Paris via JFK Airport, New York. Upon arrival at Charles de Gaulle Airport, Paris, on August 21 morning, I was told that there was a problem: my Alitalia flight to Naples was departing from the Schengen Visa area in another terminal, but I had only an Italian visa and not a Schengen visa. To get to other terminal, I needed to get out of the International Terminal, which meant that I needed a visa to enter France, which only the Schengen visa would provide. I was an Indian citizen back then, and so I required visas to enter various European countries. What annoyed me was that there was no way to reach the other terminal without getting out of the International Terminal. I voiced my concerns and asked how could the airport which prides itself as a great transit airport, not provide a direct access to the Schengen area? I told the officers that I am happily settled in the USA, and have no intention of entering France illegally. On hearing my arguments, an immigration officer agreed to walk with me to the other terminal (so as to be sure that I do not stay in France), and get me into the Schengen area to catch my flight. I appreciated his gesture of help.

I arrived as scheduled at Naples Airport at 2:30 pm, but my suitcase was not there! I was not sure whether this was due to the inefficiency of TWA, the Paris Airport, or Alitalia. In any case I lodged a complaint with Alitalia and they promised to deliver the suitcase in Maratea. The conference organizers had arranged a bus to transport us from Naples to Maratea, and I had time to board this bus at 3:30 pm.

It was a gorgeous day, and the scenery along the coast was spectacular. I checked in at the Villa del Mare Hotel in Maratea which was perched atop a cliff, and was given a room overlooking a stunning bay and beach several hundred feet below. The hotel was the conference venue, so carefully chosen by Dominique Foata, the main organizer. The French are known for high standards of excellence and for emphasis on elegance, and the choice of the conference venue and the quality of the lectures, testified to this. Everything was meticulously planned. There were sessions every morning, and after 5:00 pm; the afternoons were kept free for "discussions and to enjoy the pleasant surroundings".

Dominique Foata is a towering figure who contributed significantly to the development of algebraic combinatorics in France. He was a prominent member of the Lotharingien de Combinatoire group which conducted annual conferences in combinatorics and published a journal. By working on partitions and related topics, George Andrews straddles both combinatorics and number theory. So this 42nd Lotharengien Combinatoire meeting was devoted to his 60th birthday.

After a welcome by the Dean of Sciences of the Universita di Basilicata on morning of Tuesday, September 1, the scientific program opened with the talk of Richard Askey. He was the ideal choice as Opening Speaker in view of his eminence and his close association with Andrews for four decades. His talk was entitled "A tribute to Andrews". This was followed by the lecture of Adriano Garsia (UCSD), on the Macdonald-Kostka polynomials.

Several leading members of the Lotharengien group were present at the conference such as Jacques Desarmenien, Alaine Lascoux, and Volker Strehl. There were also talks by major figures from North America like Herb Wilf, David Bressoud, and Steve Milne, to name a few. It was at this conference that I met Alexander Berkovich, who was on a visiting position at Penn State due the efforts of Andrews. This position was ending in December that year, and so Berkovich was keenly looking for positions to start in January 1999.

On Tuesday afternoon, there was a boat ride arranged for the participants. In order to get to the boat pier, we had to walk down the cliffs from the hotel on a steep flight of stairs to the beach below. The boat ride was breathtaking with the azure blue ocean contrasting the colors of the majestic cliffs surrounding us. And as is common with mathematicians, the participants on the boat were immersed in discussion of important mathematical problems while enjoying the magnificent scenery around us. On returning to the hotel from the boat ride, I found that my suitcase, all intact, had been delivered to my room!

On the evening of the second day, Christian Krattenthaler (University of Vienna) gave a lovely piano recital. He was then one of the young stars of the Lotharingien group, but was also a highly accomplished pianist. In his early years, he was debating whether to take to a professional career in piano or become a mathematician; we are glad he chose the latter.

The area around Maratea is well known for its profusion of famous old churches, and so on the afternoon of the third day, an excursion by bus to the town of Padula was arranged. This gave us an opportunity to see some fine old churches and monuments of the Basilicata region. So the conference was both intellectually and culturally stimulating.

My talk was on Saturday, September 5, in the concluding session. I spoke about my work with Basil Gordon on a generalization and refinement of the two Andrews hierarchies of theorems to moduli $2^k - 1$, for $k \geq 3$ emanating from Schur's classic theorem which is the case $k = 3$. Unlike the other theorems we had proved using the method of weighted words, this was done without a series = product *key identity*. For the partitions given by difference conditions, Gordon and I showed that the generating functions amalgamated nicely to lead to the product representation directly. On hearing my talk, Dominique Foata asked whether a *key identity* of the type we had for the Schur and Göllnitz theorems, could be found? This key identity has remained elusive to this day!

George Andrews gave the concluding talk of the conference entitled "Debts I owe". With his characteristic humility, he acknowledged the help and encouragement of many mathematicians whose work had led him to a variety of projects such as P. A. MacMahon, Issai Schur, G. N. Watson, Nathan Fine, and of course Ramanujan. After Andrews' lecture, Herb Wilf and I left Maratea in his rental car and spent the night at a Holiday Inn in Naples. The next morning I left Naples by British Airways for London Gatwick Airport where I connected to a Virgin Atlantic Boeing 747 for Orlando. Virgin is known for its superior service, and the thing I remember most was the "English afternoon tea" served at 3:30 pm as we were flying over the infamous Bermuda triangle! The highlight of tea were the scones with Scottish clotted cream, and this was simply divine. It was a sweet ending to a most memorable trip.

The refereed proceedings of the conference edited by the organizers Dominique Foata and Guo-Niu Han was brought out by Springer in 2001. I regretted not having contributed to the volume because I was very busy with administrative work in 1998–99 during my first year as Chair. But I made it up for his 65th birthday conference by not only contributing a paper, but also in editing the proceedings.

The 60th birthday conference, Penn State, Oct 22–23, 1998

Penn State University wanted to hold its own conference for Andrews' 60th birthday, and so they had it in late October. The Dean opened this conference, following which, as at Maratea, Richard Askey gave the first lecture. I was asked to be the Chair for the next talk by Ken Ono. In the afternoon session, there were plenary talks by Barry McCoy and Herb Wilf. That night there was a conference banquet at the famous Nittany Lion Inn, a landmark of the Penn State campus. Andrews gave a fine speech in which he thanked all participants, his colleagues at Penn State, and the administration for their support of his efforts over the years.

My hour talk at the conference on the theme "Weighted partition identities and applications" was in the morning session of the second day. I chose this topic because the first set of weighted partition identities I proved were at Penn State when I visited in Fall 1994. The conference concluded with the afternoon plenary talks by Gian-Carlo Rota (perhaps the greatest combinatorialist of his time), and Steve Milne (Ohio State).

The Andrews-60 conference at Penn State was followed by an AMS Meeting the next two days (Oct 24–25). Many speakers of the Special Session on q-series and partitions at the AMS Meeting arrived earlier to attend the Andrews-60 conference. In those days I always travelled with my tennis racquet, and I had a competitive game with James Haglund (AMS special session speaker) in the indoor courts of the Penn State Club.

Conference for the 65th birthday and Election to The National Academy of Sciences, April 1, 2004

On August 21, 2003, I received the pleasant message from James Sellers of Penn State University, that George Andrews had been elected to The National Academy of Sciences. A fitting recognition indeed. The letter said that Penn State University wanted to celebrate this in a grand way, but needed time to prepare. So it was decided that instead of having the 65th birthday conference in late Fall 2003, to combine the birthday conference with the celebration for the election to the National Academy of Sciences with a one day conference on April 1, 2004 — no April fool's joke!

The one day conference had only four talks, all one hour. The speakers were Ken Ono (who had moved to the University of Wisconsin), Ae Ja Yee (Penn State), George Andrews, and me. I was honored to be included in this group of four. In view of the election to the National Academy, the conference was well attended by Penn State faculty, but there were also participants from outside the University, like Peter Paule from Austria.

The conference opened at the Hentzel Union Building with Ono's lecture followed by the talk of Ae Ja Yee, and this completed the morning session. Following this, all participants were treated to a fine lunch at the stately Nittany Lion Inn, a campus icon. My talk was the first in the afternoon and it was on the theme "Andrews, Ramanujan, and partitions". I was pleased that the distinguished statistician C. R. Rao, Eberly Professor at Penn State, attended my lecture. There was a Reception after my talk, following which George Andrews gave the concluding lecture entitled "Further debts I owe", in continuation of what he had said in his talk at Maratea!

The one day event ended with a grand banquet at the Scanticon (Conference Center), now known as the Penn Stater. I was asked to speak at the banquet along with Ken Ono, and Peter Paule. We all expressed our appreciation to Andrews for being an inspiring leader in our field. There were of course speeches by the Dean and Andrews' colleagues including

Gary Mullen and James Sellers, who thanked him for his substantial service and leadership at Penn State University.

The 70th birthday conference, Penn State, Dec 5–7, 2008

As early as July 2007, I received a warm letter from James Sellers inviting me to give one of the hour lectures of the George Andrews 70th Birthday Conference "Combinatory Analysis - partitions, q-series and applications", to be held at Penn State in December 2008. I was doubly honored when I was also invited to be a Co-Organizer of the conference along with James Sellers, Ae Ja Yee (both at Penn State), and Peter Paule (Director of the Research Institute of Symbolic Computation in Linz, Austria). The real work for the conference was done by the local organizers Sellers and Yee, with Paule and me giving suggestions and comments from time to time.

In the Fall of 2008, as the conference neared, I was also invited to address the Algebra and Number Theory Seminar on the day prior to the conference. So I arrived in State College on December 3 afternoon at 3:30 pm. I took a taxi to the Atherton Hotel where I was put up, and at 4:00 pm immediately after I had checked in, Andrews picked me up to take me to the lecture of Richard Askey in the College of Education. Askey gave a beautiful lecture — as his lectures always were — on some classical problems in Geometry. I was at that time preparing a volume entitled "The Legacy of Alladi Ramakrishnan in the mathematical sciences", in memory of my father who had passed away in June 2008. My father had a deep interest in geometry and had given novel proofs of certain theorems in Euclidean geometry which even the great geometer H. S. M. Coxeter had appreciated. Askey knew my father well, and I had invited him to contribute to that memorial volume. When I said that I loved his lecture and that my father had a passionate interest in geometry, he immediately responded saying that he would contribute a paper on that topic to the volume in honor of my father, which he did.

After the Dean inaugurated the conference on December 5 morning, the opening plenary lecture was given by the great Richard Stanley of MIT on "Hook lengths and contents of partitions". This set the tone for the conference. Stanley was followed by Jim Lepowsky (Rutgers) who gave a talk entitled "Partition theory as an inspiration for vertex operator algebras", emphasizing the central role of the Rogers-Ramanujan identities to the modulus 5, and the Andrews-Gordon identities to all higher odd moduli. Mine was the concluding talk of the first session. I spoke on "Two fundamental partition statistics - hook lengths and successive ranks", both of which had figured in my work on weighted partition identities.

The post-lunch session opened with Ken Ono's talk on the role of Ramanujan's mock theta functions in the differential topology of 4-manifolds. My colleague Frank Garvan, a former PhD student of Andrews and an authority on partition congruences, gave a talk entitled "Yet even more partition congruences" in which he discussed congruences arising from Andrews' famous function $spt(n)$, which is the number of occurrences of the smallest part among all partitions of the integer n.

The first hour lecture of the second day was by Kimmo Eriksson of Malardalen University, Sweden. Eriksson is a brilliant researcher who along with Bosquet-Melou had found a deep refinement of Euler's celebrated theorem by introducing the concept of Lecture Hall Partitions. The paper by Bosquet-Melou and Eriksson appeared in the very first issue of The Ramanujan Journal in 1997, and has been very influential. Eriksson also coauthored a lovely book on partitions with Andrews. Subsequently, he said goodbye to pure mathematics and started work at the Interdisciplinary Center for the Study of Cultural Evolution in Stockholm. He said in his lecture that he thought he would never see integer partitions again in his research, but he was wrong. They entered unexpectedly in his work at the Center! Eriksson is a hilarious speaker and in his talk, full of mathematical content, he sent the audience roaring in laughter.

On the night of the second day, there was a Banquet at the Penn Stater Conference Center. I was honored again to be asked to be one of the six banquet speakers. I compared Andrews to Paul Erdős in terms of the influence they have had in their fields, their encyclopaedic knowledge, and their warm and kind hearted nature that was so encouraging to so many young researchers. I concluded by pointing out that I am unique in having collaborated with both Andrews and Erdős, and by having them as my mentors — Erdős in the first part of my career, and Andrews in the second.

After Bruce Berndt's first plenary talk of the third day, the rest of the morning was devoted to parallel sessions of contributed talks. The conference concluded on the third day afternoon with Askey's talk entitled "Problems for George Andrews and others", and Andrews' final lecture of the conference on parity in partition identities.

As I said earlier, much of the work before and during the conference was done by the local organizers Sellers and Yee. My work really started AFTER the conference. My proposal to publish the refereed proceedings of the conference as a Special Volume of The Ramanujan Journal was accepted by the organizers. I was assisted by Sellers, Paule, and Yee, who were co-editors with me. This four hundred page volume appeared toward the end of 2010 and it gave me a sense of satisfaction that The Ramanujan Journal had honored George Andrews appropriately.

The 75th birthday conference, Tianjin, China, Aug 2–4, 2013

One of the most energetic and influential figures in combinatorics and in the theory of partitions is William (Bill) Chen in Tianjin, China. He dons several hats simultaneously as Director for the Center for Combinatorics in Tianjin University, as Provost of the University, as a member of the political party in China, as Editor-in-Chief (until December 2019) of the journal Annals of Combinatorics published by Birkhauser, and as mentor for dozens of PhD students and post-docs at the Center for Combinatorics. He considers Doron Zeilberger and George Andrews as his two gurus. Andrews was an Editor for the Annals of Combinatorics, and is an Advisor for the Center. So it was natural for him to organize a conference in August 2013 in honor of Andrews for his 75th birthday. It was gracious of Bill to invite me to give one of the plenary talks, and to serve on the Organizing Committee along with Peter Paule, James Sellers and Ae Ja Yee. Having enjoyed Chen's lavish hospitality in 2010 at the Zeilberger 60 Conference in Tianjin, and observed the perfection of his arrangements, I readily accepted the invitation, more so because it was to honor George. Since I had formed such a positive impression on my first visit to Tianjin in 2010, I told Mathura that she should accompany me on this trip, and she did. This made it even more enjoyable for me.

The 13-hour flight from Detroit to Beijing on the Delta Boeing 777, was a sheer delight. I had time to work on the flight and have a restful sleep. We arrived quite fresh in Beijing on the afternoon of July 31, and checked in at the modern Hilton at Beijing's Capital Airport for a one-night stay before proceeding to Tianjin. Our spacious corner room commanded a fine view of Beijing Airport and the runway, and for an aviation enthusiast like me, it was the very definition of heaven, for I could see giant aircraft of various international airlines landing so gracefully one behind the other.

Bill Chen sent a car to pick us up at 10:00 am the next morning. Tianjin is two hours away from Beijing, and maybe considered as a sister city. The two cities are connected by a super highway, and we reached the Huigao Garden Hotel at noon (where we were accommodated along with other participants), in time for a welcome lunch in the company of Frank Garvan, Richard Stanley, and of course, George Andrews.

The Conference on the Combinatorics of Partitions and q-series opened the next morning (Aug 2) at 8:20 am with Bill Chen's welcome. I was asked to speak at the inaugural ceremony. What I had done since 2008 was to also get a book form of the Andrews-70 volume of The Ramanujan Journal published in my book series Developments in Mathematics (DEVM). This book entitled *Combinatory Analysis* had just appeared, and so I presented it to Bill Chen and to Andrews, and thus had it officially "released" in Tianjin. Following the inauguration, Andrews gave the Opening Lecture of the conference on "A new perspective on fifth and seventh order mock theta functions". Peter Paule followed Andrews with a talk on "Partition analysis and partition congruences". I gave my lecture "Andrews and the Göllnitz theorem" as the final talk before lunch.

The afternoon session lasted until 5:00 pm, and so while I was attending it, Mathura and Larissa Berkovich were shown around Tianjin by one of Bill Chen's cheerful assistants by name Sarah. That night there was a magnificent banquet in honor of George. Speakers at the banquet waxed eloquent about Andrews and appropriately so. I was also asked to speak at the banquet. I not only thanked George for being our inspiring leader but said that 2013 was very special for me as the Centenary of Paul Erdős and the 75th birthday year of Andrews — my two gurus!

The opening talk of the second day was by the eminent Richard Stanley. Although it was titled "Three enumerative titbits", like any other Stanley talk, important results flowed relentlessly like a gushing waterfall. I also enjoyed the afternoon talk of my colleague Frank Garvan, who, true to his nature, presented high quality results quietly and modestly. On the night of the second day, we were treated to a delightful magic show and a dinner at a fancy downtown restaurant.

On the third and last day, I was asked to chair the session in which Bill Chen spoke. He was followed by my good friend Mourad Ismail. The concluding talk of the conference was by Richard Askey with an interesting title: "Some theorems looking for applications and some problems looking for solutions". It was a fitting finale for the conference program following which there was yet another fabulous banquet, this one attended by the President of Nankai University. After that we enjoyed a breathtaking river cruise and since it was a clear night, Tianjin was dazzling.

Ever since my boyhood days of international travel, I have always enjoyed going up a tower for a view of the city. Our hotel was very close to the Tianjin Tower, and so before we checked out on August 5 morning, Mathura and I went up the Tower accompanied by Frank Garvan. Mornings are the best time to go up towers because the air is clear. We did get a splendid view of the city. While I was exuberantly expressing my admiration of the view, Garvan was quietly smiling in approval.

Bill Chen was at the hotel to personally wish all of us goodbye, and we too could thank him for his unmatched hospitality. Mathura and I were taken by car to the Hilton Hotel at Beijing airport where we rested one night before departing for Hong Kong the next day.

From Beijing to Hong Kong we flew by an Airbus A-330 of Dragonair a subsidiary of Cathay Pacific. Even though it was a subsidiary, the service on board was superb. We arrived at 4:00 pm at the new Hong Kong International Airport, a technological marvel. Hong Kong is composed of several islands, and this sprawling airport has been built on one of the outer islands. It was a thirty minute drive to Central Hong Kong where we checked in at the ultra-modern Conrad Hong Kong Hotel and were given a room on the 60th floor that commanded a breathtaking view of Hong Kong harbor, and of Kowloon across the causeway. I have wonderful memories of visits to enchanting Hong Kong since my boyhood days, but every time we stayed on the Kowloon side. This was the first time we were staying in Central Hong Kong, and it was wonderful. The view of Hong Kong by night, of the boats and the Star Ferry plying across the causeway, excites our senses. The special sightseeing trip on this visit for us was to the gigantic Buddha atop a hill on Lantau island at the recommendation of my daughter Amritha who visited it a year before. This is the island on which the new Hong Kong Airport is located, and the cable car ride over several hill tops first provides jaw dropping views of the airport. Then as we near the last hill, the giant Buddha looms in full majesty against the sky. From the cable car station at the top, it is a stiff walk to the base of the Buddha, from where one has to ascend two hundred steps to reach the Buddha. But the climb is well worth it, not just to see the Buddha up close, but also to enjoy the panoramic view of several islands of Hong Kong.

After spending three memorable nights in Hong Kong, Mathura and I returned to Beijing by a Cathay Pacific Boeing 777 enjoying their renowned Marco Polo service. We caught a connecting Delta flight at Beijing to return to Florida via Detroit. Bill Chen send me a file of fantastic photographs, as an ideal memento of the wonderful conference.

The 80th birthday conference, Penn State, June 21–24, 2018

To celebrate the 80th birthday of George Andrews, the conference *Combinatory Analysis 2018: Partitions, q-Series, and Applications* was held at his home turf — The Pennsylvania State University. Once again, it was an honor to be invited to give a plenary talk, and to be a Co-Organizer of the conference along with James Sellers, Ae Ja Yee, Peter Paule, Bruce

Berndt, and William Chen, with Sellers and Yee taking on the main organizational aspects at Penn State. I felt triply honored because I was the Opening Speaker at 8:30 am on June 21. My talk was not on partitions and q-series; instead it was on "The local distribution of the number of small prime factors - a variation of the classical theme". This was in analytic number and was joint work with my recent PhD student Todd Molnar. It was considerate on the part of the organizers and Andrews to allow me to speak on this topic; they felt it would be a refreshing change. I was followed by my colleague Frank Garvan, who spoke about the higher order mock theta conjectures. This was very relevant because it was Andrews and Garvan who had originally formulated the mock theta conjectures.

One of the highlights of the conference was the announcement of the solution to the famous Borwein Conjecture by a young Chinese mathematician Chen Wang working in the group of Christian Krattenthaler at the University of Vienna. Wang was originally scheduled to give a contributed talk, but in view of this spectacular achievement, he was asked to give the first lecture on the second day. The Borwein conjecture had resisted all attempts for a few decades, including assaults by Andrews himself, and so this talk by Wang was a fine gift to Andrews.

Kimmo Eriksson was back at Penn State for this 80th birthday conference. As in the case of the 70th birthday conference at Penn State, Eriksson gave another hilarious lecture. His talk had two themes. The first was about an iterative process for partitions called Belgian solitaire. The second was about what we can learn about effective mathematics education from the analysis of big data. This also was relevant to the conference in view of Andrews' views on mathematics education and the data deluge.

There was an impressive book exhibit both by Springer and World Scientific at the Macalester Building, the home of the Penn State Mathematics Department. On June 22, I had meetings with Rochelle Kronzek (World Scientific) and Elizabeth Loew (Springer) to discuss my projects with these two publishing houses.

On the night of June 23, there was the Conference Banquet at the Penn Stater Conference Center. Andrews' wife Joy, and his daughter Amy, gave touching speeches describing Andrews as an ideal husband and father. Bill Chen who could not be present at the conference, sent an exquisite Chinese vase as a gift; this was carefully brought all the way from China by two of his PhD students. I was asked to speak at the banquet, and I described my long and fruitful association with Andrews which began with his first visit to India in the Fall of 1981 when my father hosted him in Madras.

Our friendship grew over the years, and I have had the pleasure of enjoying his hospitality at Penn State on numerous occasions, just as I have enjoyed hosting him in Florida annually, and in India many times. Of course, during these four decades, I have had the privilege of collaborating with him in the theory of partitions and q-series. After the speeches were over, Andrews and his son Derek entertained us on the piano, and their music was a sweet ending to a most memorable evening.

I was asked to be the Chair for the final session on June 24, with George Andrews giving the final lecture entitled "Dyson's most beautiful identity, sequences in partitions, and mock theta functions". Interestingly, Andrews wore a Ramanujan tie, so beautifully designed by his son Derek.

I suggested that the refereed proceedings of the conference be published both as a journal volume and as a book, and the organizers endorsed this suggestion enthusiastically. The research papers appeared in 2020 in a special volume of *Annals of Combinatorics*, edited by William Chen, and published by Birkhauser. The book, also published by Birkhauser, contains all the papers in the journal volume together with some survey articles, personal recollections, and photographs; Peter Paule took the lead in editing this book. My article "My association and collaboration with George Andrews" is in this book.

George Andrews remains youthful at 80 — both in his attitude towards life, and in his research. So we expect more milestone birthday conferences. We wish him many more years of good health so that he can enjoy the fine things in life with family and friends, and continue to be our inspiring leader in the theory of partitions and q-series.

11.5   The Erdős Memorial and Centenary Conferences (1999, 2013)

Paul Erdős (1913–96) was the most influential mathematician of the twentieth century owing to the number of areas to which he made path-breaking contributions, the hundreds of young mathematicians he encouraged and groomed, and the multitude of problems he posed, that has engaged, and continues to engage, the attention of researchers since the 1930s. Naturally for someone so eminent and influential, both the memorial conference (1999) and the centenary conference (2013) attracted luminaries and young mathematicians from around the world to pay homage to him in his country of birth, Hungary. I had the privilege of attending both these conferences in Budapest and pay tribute to my great mentor.

The Erdős Memorial Conference, July 4–11, 1999

Although Erdős died in September 1996, it took nearly three years before this conference was held because a lot of planning had to go into it. The conference was the collective effort of the Hungarian mathematical community led by Professor Vera T. Sos, a noted combinatorialist who was a close friend of Erdős and the wife of the late Paul Turan, one of the great Hungarian number theorists, and a close collaborator of Erdős.

In Fall 1998, I received an invitation from Andras Sarkozy, to speak in the Number Theory Session of the conference. I was in India with my family in the summer of 1999, and so from India, I made a one week trip to Hungary. I flew by a Northwest Airlines DC-10 from Bombay to Amsterdam where I connected to a KLM flight to Budapest. However, just as we pulled out of the terminal in Bombay, the Captain noticed that the aircraft was overweight due to cargo, and that some of it had to be off loaded. We were not asked to get out of the aircraft; instead a delicious Indian dinner was served as we waited for the procedure to be completed. Owing to the delay in the departure from Bombay, I missed my connection in Amsterdam to Budapest, but KLM had kindly re-booked me on the next flight.

I checked in at the magnificent Hyatt Regency which had an enviable location on banks of the great Danube, and right across from the Hungarian Academy of Sciences where the conference was held. The Hyatt and the Academy were separated by a lovely park, all on the Pest side of Budapest. Between the Hyatt and the Academy, is the famous chain bridge, one of many connecting Pest with Buda on the other side of the Danube. I made it on time to Budapest to attend the Conference Welcome Reception at the Hungarian Academy of Sciences at 7:00 pm on Sunday, July 4.

Sitting in the palatial Formal Hall of the Hungarian Academy of Sciences during the opening ceremony on July 5 morning, surrounded by 500 mathematicians, my mind was transported to December 1987, when mathematical luminaries gathered in Madras to pay homage to Ramanujan for his centennial. The splendid opening lecture by Bela Bollobas (himself a discovery of Erdős) of Cambridge University entitled "Paul Erdős - from prodigy to grand old man, the mathematical journey", began with a slide containing the names of Hardy and Ramanujan. Bollobas was referring to the fundamental work of Hardy and Ramanujan on the number of prime factors of integers which led to the celebrated Erdős-Kac Theorem and the birth of probabilistic number theory. Bollobas concluded his lecture by recalling Einstein's great tribute to Mahatma Gandhi that "succeeding generations will scarce believe that such a person walked this earth". Bollobas said that those very words applied equally well to Paul Erdős.

Another plenary speaker at the conference was Professor Tim Gowers of Cambridge University, winner of the Fields Medal in 1998. He was a former PhD student of Bollobas and hence a grand student of Erdős. He spoke on "Arithmetic progressions in sparse sets". Gowers began his lecture by describing the fundamental theorems proved, and questions raised, by Erdős in this area, and concluded with his own research work that won him the Fields Medal. Gowers alluded to a famous $3,000 problem of Paul Erdős (unsolved to this day in full generality), that if $\{a_n\}$ is an infinite sequence of positive integers such that

$$\sum_{n=1}^{\infty} \frac{1}{a_n} = \infty,$$

then there are arbitrarily long arithmetic progressions in the sequence. Gowers expressed the view that notoriously hard problems like the prime k-tuples conjecture would be solved before this conjecture of Erdős is settled!

An example of a mathematician who was encouraged by Erdős in his teens, and who is now at the pinnacle, is Prof. Lovasz, who spoke at this conference. Lovasz had just received the Wolf Prize of Israel for lifelong contributions to mathematics. Erdős had won this prize in 1983. While introducing Lovasz, Vera Sos remarked that had Erdős been alive, he would have been the happiest person to see his disciple win the Wolf Prize.

The organizers did a wonderful job in inviting very senior mathematicians who had known Erdős for several decades as well as young researchers working on the latest advances. Some of the senior mathematicians who introduced the plenary speakers were asked to give brief reminiscences. Professor Alan Baker of Cambridge University who won the Fields Medal in 1970 introduced Professor Peter Elliott who spoke about the work of Erdős in Probabilistic Number Theory. Baker alluded to the visits of Erdős to Cambridge over the years, and his close interaction with Harold Davenport, Richard Rado, and Louis Mordell.

Other plenary speakers included Ronald Graham, former President of the AMS, who gave a magnificent lecture on Erdős and Ramsey Theory, Peter Borwein who described Erdős' work on polynomials, and Joel Spencer who spoke about Erdős and random graphs. The subject of random graphs was initiated by the fundamental work of Erdős and Alfred Renyi. Another plenary lecture I attended was that of Carl Pomerance on arithmetical functions. In his talk, Pomerance referred to my joint work with Erdős which I did as an undergraduate student on the arithmetical function $A(n)$, the sum of the prime factors of an integer n.

In addition to the plenary talks, there were special sessions talks in the areas number theory, combinatorics, graph theory, set theory, geometry, analysis, computer science, and probability, where Erdős had a strong influence. These sessions were organized by Hungarian mathematicians who were experts in those areas and who had collaborated with Erdős.

The talks of the special session on number theory were all held in the Formal Hall. This session featured talks by leading number theorists such as Kubilius (one of the four architects of Probabilistic Number Theory), and Hillel Furstenberg (who had given a proof of Szemeredi's famous theorem using ergodic theory, and who later won the Abel Prize). Aleksandar Ivic who spoke in this session, started his talk quoting my joint work with Erdős on $A(n)$. My lecture in the special session was on the morning of the final day, Saturday July 10. I spoke on "A variation of a theme of Sylvester - a smoother road to the Göllnitz' big partition theorem".

Wednesday, July 7 afternoon was free for sightseeing. Right after lunch, I took a bus tour of Budapest and saw (from the outside), the War Memorial, the Opera House, the Grand Palace, and the magnificent Parliament Building. I was back just in time to join the Danube Boat Trip (4:00–6:30 pm) that was arranged for all participants.

The formal conference banquet was on Thursday, July 8, at the Hungarian Academy of Sciences, in and around the Formal Hall. On Tuesday evening, there was a special showing of the documentary film "*N* is a number - a portrait of Paul Erdős" produced by George P. Csicsery.

The conference concluded around 1:00 pm on Saturday, July 10, with a talk by George Szekeres (Sydney) entitled "An old comrade looks back (Esther's problem)".

It was nice to renew contact with senior mathematicians I had first met when I was an undergraduate — like Szekeres, who had invited me to talk at the University of New South Wales in October 1973, and J. P. Kahane, whom I met at MATSCIENCE in January 1973.

After one week of intense mathematics at the conference, I was back in Madras on July 12.

The Erdős Centenary Conference, June 30–July 5, 2013

The Erdős Centenary Conference was just as grand as the memorial conference, and attracted just as many mathematicians from around the world. On my part, I had proposed to the AMS that I could edit a feature article on Erdős for his centenary for the Notices of the AMS, and the Editor of the Notices, Steven Krantz, agreed to this suggestion. There were to be about ten contributors to the feature article and I utilized the conference to talk to some of the contributors. I was invited to speak in the Session on Number Theory and my trip was supported by the mathematics department and the College of Liberal Arts and Sciences.

I departed Florida on June 29 for New York's JFK Airport where I connected to an Air France Airbus A-330 flight to Paris. At Charles de Gaulle Airport, Paris, where I arrived the next morning, I took another Air France flight to Budapest. That flight was delayed by one hour because one passenger did not show up! So I arrived in Budapest at 4:00 pm and went straight to the Hyatt, which now had been changed to The Sofitel. The hotel was just as immaculate as it was under Hyatt management, except that everything now had a French touch to it. I was in time to attend the Welcome Reception at the Hungarian Academy of Sciences that evening.

The conference opened on Monday, July 1, at the Formal Hall of the Academy. As in the case of the memorial conference, Bela Bollobas was the Opening Plenary Speaker. It was fourteen years after the memorial conference, and several new stars had appeared on the mathematical firmament. So the second plenary talk was by Ben Green of Cambridge University who, in his PhD thesis, had solved the famous Cameron-Erdős Conjecture.

Like the 1999 conference, there was a plenary talk by Fields Medalist Tim Gowers. I also enjoyed the plenary talk of Fields Medalist Terence Tao, probably the most influential young mathematician of our generation. Tao began his lecture with a picture of him as a young boy talking to Erdős. Tao was a child prodigy and Erdős had spotted him.

Some of the other plenary speakers were Noga Alon (Tel Aviv University), Fields Medalist Elon Lindenstrauss (Hebrew University, Jerusalem), and Carl Pomerance (Dartmouth University). From the Hungarian side, one of the plenary talks was given by Janos Pintz of the Alfred Renyi Institute, who in collaboration with Dan Goldston and Yildirim had proved the sensational result that

$$\liminf_{n\to\infty}\frac{p_{n+1} - p_n}{\log n} = 0,$$

where p_n is the n-th prime.

The number theory session featured talks by Harold Diamond (Urbana, Illinois), Robert Tijdeman (Leiden, Holland), Helmut Maier (Ulm), Kevin Ford (Urbana, Illinois), Gerald Tenenbaum (Nancy, France), to name a few. Tijdeman opened his lecture by saying that in one's mathematical career, the first few papers that are written by the individual determine that person's trajectory. He mentioned this in connection with the fact that Erdős' first paper was on a new proof of Bertrand's postulate that there is always a prime number in every interval $[n, 2n]$. This propelled him into prominence and it is this proof that is given in all textbooks today. In one of the talks at this session, Bob Hough, a PhD student of Kannan Soundararajan at Stanford, outlined his proof of the famous Covering Congruences Conjecture of Erdős.

My talk was on the last day. I spoke on the theme "Multiplicative functions and small divisors", which was my joint work with Erdős and Jeffrey Vaaler (University of Texas, Austin) in the 1980s. I also mentioned further work on this problem by Soundararajan, Ritabrata Munshi, and others. Just as I was about to start my talk, Aleksandar Ivic in the audience exclaimed: "Krishna, you divorced yourself from analytic number theory a quarter century ago and moved to partitions and q-series. It is nice to hear you speak now on analytic number theory." I was also asked to chair this final session on number theory.

The conference banquet was on the boat Europa. We departed at around 5:00 pm and returned at 9:00 pm. As we rode up and down the Danube, Budapest looked magnificent both in daylight and at night.

One evening George Csicsery showed his film on Erdős at 100. I spoke to him and suggested that he should add a small segment on Erdős and Ramanujan. After I returned to Florida, I sent Csicsery an article on Erdős and Ramanujan that I had written in December 1996 (the year of the passing of Erdős). I do not know whether my suggestion was adopted.

It was pleasure to reconnect with all my old Hungarian mathematical friends — Vera Sos, Andras Sarkozy, Imre Katai, and Janos Pintz. Sos and Sarkozy contributed to the feature article on Erdős that I edited for the AMS Notices. One evening Sarkozy and his son Gabor, graciously hosted a classy dinner for me, Cameron Stewart, Harold Diamond, and Carl Pomerance, at a traditional French restaurant in Budapest.

I departed Budapest on July 6, and on the trans-atlantic leg, I flew for the first time on the mammoth Airbus A-380 double decker super jumbo. The Air France flight from Paris Charles de Gaulle Airport to New York Kennedy Airport was silken smooth. I could not even feel the giant aircraft leave the ground at take-off! An interesting feature of the A-380 is that there are three cameras — one under the nose, one on top of the tail, and one outside the belly underneath, so that one could get different views while flying. I chose the view from the tail camera because this way we see the entire aircraft and the view in front. Watching the landing at Kennedy Airport from the tail camera was awesome.

Back in Florida, my duty was to complete the Erdős 100 feature for the Notices of the AMS. This feature appeared in two parts:

PART I (Feb, 2015)

Krishnaswami Alladi and Steven Krantz, "One of the most influential mathematicians of our time"

Lazlo Lovasz and Vera Sos, "The Erdős Centennial"

Ronald L. Graham and Joel Spencer, "Ramsey theory and the probabilistic method"

Jean-Pierre Kahane, "Bernoulli convolutions and self-similar measures"

Melvyn B. Nathanson, "Paul Erdős and additive number theory"

PART II (March 2015)

Noga Alon, "Paul Erdős and the probabilistic method"

Dan Goldston, "Erdős' work on primes"

Andras Sarkozy, "Erdős and sequences"

Gerald Tenenbaum, "Paul Erdős and the divisors"

Stephan Ramon Garcia, and Amy L. Shoemaker, "Wetzel's problem, Paul Erdős, and the Continuum Hypothesis: A mathematical mystery"

I had a feeling of satisfaction for having done something to preserve the memory of my illustrious mentor.

The President of the Norwegian Academy of Sciences, proposing a toast during the Abel Banquet. Seated next to him are the 2008 Abel Laureates — Jacques Tits (left) and John Thompson (right). Oslo, May 19, 2008.

In memory of Paul Erdős

L to R - Mathura, Ron Graham (UC San Diego), Vera Sös (Alfred Renyi Institute, Budapest), Jean-Louis Nicolas (Univ. Lyon), Krishna, and his parents, during a party at the Alladi House, Gainesville, in honor of Graham who delivered the First Erdős Colloquium at the University of Florida, March 15, 1999.

Cam Stewart (Univ. Waterloo), Andras Sarkozy (Alfred Renyi Institute, Budapest), Krishna, Carl Pomerance (Dartmouth), and Harold Diamond (Univ. Illinois), in front the Hungarian Academy of Sciences, Budapest, during the Erdős Centennial Conference, July 2, 2013.

The success of the SASTRA Ramanujan Prize

The SASTRA Ramanujan Prize with an age limit of 32 for eligibility, recognizes outstanding mathematicians very early in their careers. When the SASTRA laureates subsequently win prizes with a hallowed tradition, it is a testimony to the high standards of the SASTRA Prize. Since its inception in 2005, five SASTRA prize winners have later won Fields Medals. Of the four Fields Medals given in 2022, two went to former SASTRA Prize Winners — James Maynard and Maryna Viazovska. In the first picture below, we see Krishna Alladi with James Maynard (Oxford University) on the grounds of Sterling (now Indeco) Resorts near Kumbakonam (Ramanujan's hometown), Dec 21, 2014. In the second photograph, you see Maryna Viazovska (Ecole Polytechnique Lausanne), sitting on a window sill in Ramanujan's humble home in Kumbakonam, Dec 20, 2017. Ramanujan used to sit on this window sill and work out his "sums" as he watched the passers-by on the street.

Chapter 12

Fostering the legacy of Ramanujan

NOTE: In §12.2, I talk about the Ramanujan Journal, and in §12.3, I discuss the SASTRA Ramanujan Conferences and Prizes. Descriptions of both of these are spread over several sections of the book, and discussed as part of the events of various academic years. However, for those who would be specifically interested in knowing about the Ramanujan Journal or the SASTRA Ramanujan Prize, I have briefly discussed these topics collectively in this chapter even though there would be some overlap with what was described in different sections of earlier chapters.

Inspired by the gathering of mathematical luminaries in India to pay homage to Ramanujan for his centennial in 1987, I seriously began to work since then in various ways to foster Ramanujan's legacy, and here I shall give an account of my efforts. This has been a major component of my service to the profession, and also can be viewed as my service to India.

12.1 Annual articles for Ramanujan's birthday

When I took to a research career in mathematics in the seventies, my father told me that in addition to doing research, I should also be active in informing the general public about the great developments in mathematics. Since Ramanujan is a hero to every Indian, and any Ramanujan related news attracts the Indian public, I decided to write articles of general interest to be published in India around Ramanujan's birthday each year in December starting from 1987. In order to properly understand Ramanujan's work or to get an idea of the place he occupies in the world of mathematics, we need to compare his life and work with other mathematical luminaries whose life and work have things in common with Ramanujan. Thus after writing the first article "Ramanujan - an estimation" in 1987 for the Centennial, and following it with an article "Ramanujan - the second century" in which

I discussed briefly the impact that his research will have in the following decades, I wrote a series of articles on the lives and works of mathematical giants who had strong links with Ramanujan. These articles all appeared in "The Hindu", India's National Newspaper, annually in December. I felt that these articles would not only inform students and the lay public about the contributions of many great mathematicians in history, but also enable them to appreciate Ramanujan better by comparing him with other legendary mathematicians.

From time to time, in addition to writing about mathematicians, I also wrote about certain aspects of Ramanujan's mathematics — such as his work related to the number π, his work on partitions, and on the solution by Manjul Bhargava and Jonathan Hanke to the problem of determining all universal quadratic forms — a problem stemming from Ramanujan's work. These also appeared in The Hindu.

Periodically, I was asked to review books on Ramanujan, such as the printed version of Ramanujan's Lost Notebook that was released on Ramanujan's 100th birthday, the classic biography of Ramanujan entitled "The Man Who Knew Infinity" by Robert Kanigel, and the two books by Bruce Berndt and Robert Rankin, on Ramanujan's Letters, and on Essays and Surveys pertaining to Ramanujan. These appeared in *The Hindu*, *The American Scientist* and in *The American Mathematical Monthly*.

The collection of my Ramanujan related articles of general interest was published by Springer as a book entitled "Ramanujan's place in the world of mathematics" [B1] and released at the Ramanujan 125 Conference at SASTRA University in 2012. George Andrews encouraged me to publish this book, and graciously wrote a Foreword to it. An enlarged Second Edition containing six more articles was published by Springer in 2021.

12.2 The Ramanujan Journal

The Ramanujan Centennial was indeed a magnificent celebration, but a Centennial is an event that happens once. I wanted something that would not only be a permanent memorial for Ramanujan, but would also relate to emerging trends. So I had the idea to launch *The Ramanujan Journal* that would be devoted to all areas of mathematics influenced by Ramanujan. The name Ramanujan brings to mind a whole range of topics including: (i) hypergeometric and q-hypergeometric series, (ii) partitions, compositions and combinatory analysis, (iii) circle method and additive number theory, (iv) elliptic and theta functions, (v) modular forms and automorphic functions, (vi) special functions and definite integrals, (vii) continued fractions, (viii) diophantine analysis, (ix) number theory, (x) Fourier analysis, and (xi) connections between Lie algebras and q-series — areas that have felt his magic touch and some that he has gloriously transformed. Thus the journal would simultaneously have a sense of focus and a broad scope. I proposed this idea to George Andrews and Bruce Berndt in 1993 at the Joint Annual Meeting of the American and Canadian Mathematical Societies in Vancouver and they liked the idea. In the Fall of 1993, based on a suggestion of Bruce Berndt, John Martindale, Editor of the Science and Technology Division of Kluwer Academic Publishers of The Netherlands, called me and expressed interest in starting a new journal along the lines I proposed. I then consulted Basil Gordon, Jon Borwein and Peter Borwein, and they heartily endorsed this proposal. They felt that the journal would have the positive effect of bringing together papers in the above areas which otherwise would be scattered in the literature.

Encouraged by this support, I sent out a proposal in March 1994 to about 100 top mathematicians worldwide. Many responded with enthusiasm and so in March 1995, David Larner of the Kluwer head office in Dordrecht in The Netherlands, and John Martindale of the Kluwer office in Norwell, Massachusetts, came to Florida for discussions, invited me to be the Editor-in-Chief of the new journal, and asked me to form the Editorial Board. I was able to get 26 leading mathematicians representing the above areas to join the Editorial Board (for the list of Founding Editors, see §7.8). I must mention that the late Marvin Knopp called me by telephone and expressed his support and great appreciation of the idea to launch the journal. Indeed he was one of the Founding Editors.

I have served as Editor-in-Chief of the journal since its inception in 1997. In 2004, Kluwer and Springer merged, and adopted the more widely known Springer name. The subsequent Springer Editors Ann Kostant, Elizabeth Loew and Marc Strauss, who have managed the journal in succession, have been extremely supportive. Indeed, the journal which started with publishing one volume per year of four issues with about 100 pages per issue when it was published by Kluwer, has quintupled in size since Springer took over in 2004; in the last five years, the Journal has been publishing three volumes with three issues per volume of at least 250 pages per issue. But from 2022, the journal started publishing 12 issues per year in three volumes The journal has established itself as the main venue for papers in Ramanujan related areas, and the increase in size reflects both the importance of the journal and the explosion of activities in the above fields with a significant portion of the work having its roots in the writings of

Ramanujan. By launching this journal and serving as its Editor-in-Chief, I feel a sense of satisfaction in fostering the legacy of Ramanujan, furthering the worldwide pursuit of research in Ramanujan related mathematics, and serving the profession.

12.3   The SASTRA Ramanujan Conferences and the SASTRA Ramanujan Prizes

One of the major forces to the enter the world of Ramanujan at the start of the new millennium was SASTRA University, a new university launched in the state of Tamil Nadu in South India in the eighties. SASTRA, an acronym for Shanmugha, Arts, Science, Technology, Research Academy, is a private university with its main campus in Tanjore (now Tanjavur). SASTRA grew by leaps and bounds and in 2003 purchased Ramanujan's home in Kumbakonam (near Tanjore and in the Tanjore district) to refurbish it and maintain it as a museum. This home was in a dilapidated condition, but owing to the purchase by SASTRA, we now have the active involvement of university administrators and academicians in the preservation of Ramanujan's legacy for posterity. In connection with the purchase of this home, and seeing the need for expansion, SASTRA also opened a branch campus in Kumbakonam, and named it the Srinivasa Ramanujan Centre. Also, in connection with the opening of the Srinivasa Ramanujan Centre, SASTRA decided to hold a three day international conference at the Centre to conclude on Ramanujan's birthday, December 22, 2003, and had succeeded in getting the acceptance of Dr. Abdul Kalam, the President of India, to inaugurate this conference. The story as to how I was approached by S. Swaminathan, son the SASTRA Vice-Chancellor, to be the lead organizer of this conference is described in §9.6. So here I only say that I brought a team of leading researchers to the conference, including the great Ramanujan expert George Andrews, who gave both the opening technical lecture of the conference as well as the concluding Ramanujan Commemoration (Public) Lecture. At the valedictory function, the participants suggested that such conferences should be conducted by SASTRA annually in December. At that point, Dean Vaidhyasubramaniam (the elder son of Vice Chancellor Sethuraman) invited me to help organize these annual conferences, and I have been doing this every year since 2003.

For the second conference in 2004, Vice Chancellor Sethuraman invited me to inaugurate the conference and also deliver the concluding Ramanujan Commemoration Lecture. Just before the start of the inaugural ceremony, the Vice Chancellor told me that he would like to set apart $10,000 annually for a worthwhile cause in the name of Ramanujan. He said that I should announce this during my inaugural speech and also specify how it should be used. When I spoke, I announced this "annual gift" from the Vice Chancellor and said that one should utilize this to launch the SASTRA Ramanujan Prize, an annual prize of $10,000 to be given to a mathematician not exceeding the age of 32, for outstanding contributions to areas influenced by Ramanujan (see §9.8 for more details). I said that age limit is 32 because Ramanujan lived only for 32 years and so the challenge to the candidates for the prize is to show what they have accomplished in that time frame. I pointed out that the Fields Medal has an age limit of 40, and by having an age limit of 32 for the SASTRA Prize, we will be recognizing talented mathematicians much earlier. The Vice Chancellor endorsed my suggestion, and requested me to Chair the prize committee! That is how I got involved with the SASTRA Ramanujan Prize. The prize is given annually at SASTRA University, Kumbakonam (Ramanujan's hometown) around December 22 (Ramanujan's birthday).

Since 2005, the prize has been awarded to the following brilliant mathematicians: 2005 - Manjul Bhargava (Princeton) and Kannan Soundararajan (Michigan) - two full prizes, 2006 - Terence Tao (UCLA), 2007 - Ben Green (Cambridge, UK), 2008 - Akshay Venkatesh (Stanford), 2009 - Kathrin Bringmann (Cologne), 2010 - Wei Zhang (Columbia), 2011 - Roman Holwinsky (Ohio State), 2012 - Zhiwei Yun (Stanford), 2013 - Peter Scholze (Bonn), 2014 - James Maynard (Oxford), 2015 - Jacob Tsimerman (Toronto), 2016 - Kaisa Matomaki (Turku, Finland) and Maksym Radziwill (Montreal) - shared, 2017 - Maryna Viazovska (Ecole Polytechnique, Lauzanne), 2018 - Yifeng Liu (Yale) and Jack Thorne (Cambridge, UK) - shared, 2019 - Adam Harper (Warwick), 2020 - Shai Evra (Princeton and the Hebrew University), 2021 - Will Sawin (Columbia).

The prize has established itself as one of the most prestigious and coveted owing to caliber of the winners. This has been due to (i) the support of the international mathematical community in making excellent nominations and writing letters on the work of the candidates, (ii) the willingness of leading mathematicians to serve on the prize committee, and (iii) the publicity given by professional bodies like the American Mathematical Society and the European Mathematical Society.

An account of the path-breaking work of each of these winners and a description of the events relating to the SASTRA Conferences is given in the individual sections in Chapter 10. I have published two important articles ([A5], [A6]) on the work of the SASTRA Prize winners.

The Fields Medals with an age limit of 40 were instituted with two lofty goals in mind: (i) to recognize pioneering work by brilliant young mathematicians, and (ii) to encourage these young researchers to continue to influence the growth of mathematics. The SASTRA Ramanujan Prize has a more stringent age limit of 32. The winners of the SASTRA Ramanujan Prize have also shaped the development of mainstream mathematics and will continue to do so in the years ahead; the fact that the SASTRA laureates have subsequently won major prizes with a hallowed tradition, is a testimony to this. Indeed, SASTRA Prize Winners Bhargava, Venkatesh, Scholze, Maynard and Viazovska went on to win Fields Medals.

The SASTRA Ramanujan Prize is one of the finest ways to preserve and honor Ramanujan's legacy. It has given me pride and satisfaction to have played a role in launching the prize and serving as Chair of the prize committee since the inception of the prize in 2005 — satisfaction in identifying and recognizing brilliant young mathematicians, and pride in serving India and the mathematical community in this noble effort.

Krishna Alladi receiving the Honorary Doctorate of Science (honoris causa) from India's Union Minister of Education Mr. Dharmendra Pradhan, on September 18, 2022 during the 36th Convocation of SASTRA University. Others in the photo are (L to R): Dean S. Swaminathan, Mr. Chamu Krishna Sastry (Recipient of the Honorary Doctorate of Letters), Minister of State Murugan, Chancellor R. Sethuraman, and Vice-Chancellor S. Vaidhyasubramaniam.

Chapter 13

Collaboration with Erdős, Andrews, and Gordon

I have had the good fortune of collaborating with three giants of the mathematical world — Paul Erdős, George Andrews, Basil Gordon,

The great mathematician G. H. Hardy once said that he had done something that no one else could claim, namely that he had "collaborated with Littlewood and Ramanujan in something like equal terms". I am no Hardy, but I can claim with pride that I am perhaps the only one to have collaborated, and interacted closely with, both Paul Erdős and George Andrews, two of the most eminent mathematicians of our times. I shall first describe my collaboration with Erdős and Andrews which molded my career, and conclude by describing how equally significant my interaction with the "Great Guru" Basil Gordon was. Here I describe the working habits of the three mathematicians and my collaboration with them in general terms; details of my interaction and specifics of collaboration over the years are given in various sections of this book. So there is some overlap between this chapter and certain earlier sections, but I hope the reader will find it convenient that I have described separately here my collaborations with Erdős, Andrews and Gordon.

Just as Erdős was like a mentor to me in the early part of my career when I worked in analytic and elementary number theory, George Andrews acted (and continues to act) as a guide in the latter part of my career in the theory of partitions and q-hypergeometric series. My close mathematical interaction with Erdős was during the years 1973 to 1989, and with Andrews from 1990 onwards.

13.1 Collaboration with Paul Erdős

Erdős had encyclopedic knowledge of several areas in which he worked and this was immensely helpful to me in my research in number theory. We wrote five joint papers. The work for the first three was done when I was an undergraduate student in India and a graduate student at UCLA. The last two were written during the period when I was transitioning from India to take up my permanent position in Florida.

Any one who spent time discussing seriously with Erdős had a good chance ending up as his collaborator. Those who have collaborated with Erdős — and there are hundreds of them — have expressed both their pleasure and pride in this accomplishment. I have described in §2.7 how I first met Erdős in Madras in 1975, and how our collaboration started. It is a charming story that reveals his greatness as a human being, and how he sought talented young mathematicians and encouraged them. It was due to that initial meeting with Erdős and our mathematical interaction, that I went to UCLA to do my PhD. Our collaboration always started with me asking him a question on a problem that I was thinking about. This usually was when he visited a university where I was working, or from a letter I wrote to him. He was my guest when I was visiting: (i) the Institute for Advanced Study in 1981–82, (ii) the University of Texas, Austin in 1982–83, and (iii) the University of Hawaii in 1984–85. My last two papers with Erdős (also with Jeffrey Vaaler as a collaborator) was based on work on multiplicative functions and small divisors that I was doing between 1982 and 85. Erdős' visits were always brief — a week or two at most. So the discussion on problems of interest to both of us would continue through correspondence. Not only was Erdős the greatest collaborator in history, he was also the ultimate in correspondence, writing about a dozen letters to mathematicians every day. He would promptly reply to questions you asked or comment on ideas you presented, often providing key references and directions in which you could proceed further. It was he who informed me about the fundamental papers of de Bruijn on the functions $\Phi(x, y)$ (the number of integers up to x all of whose prime factors are at least y) and the dual function $\Psi(x, y)$ (the number of integers up to x whose prime factors do not exceed y), when I was an undergraduate student in India starting to think about arithmetic functions. I studied these papers of de Bruijn and investigated weighted versions of these functions in my PhD thesis (1975–78) and later at the University of Michigan (1978–81). With regard to my work on multiplicative functions and small divisors, when I showed him a conjecture which I could not prove, he told me to my surprise, that my conjecture was not strong enough! He said that a stronger conjecture could be formulated and proved by an induction argument. Vaaler and I subsequently found two proofs of this stronger conjecture. This is what led to my last two joint papers with Erdős and Vaaler.

It was primarily due to Erdős' support that I was given a permanent position at the University of Florida that started in January 1987. Erdős was an annual visitor to Florida spending two weeks each spring. He was constantly on the move but always made sure to visit places when the temperatures there were balmy. In an article I wrote for his 80th birthday, I said, "Somehow like migrating birds, he always hovered around the isotherm 70 degrees Fahrenheit!" Thus he was in Calgary and Boulder in the summers, Florida in the Spring, and UCLA in December. Although he visited Florida annually, I never collaborated with him during my tenure in Florida, and I feel a trifle guilty about this. My interest had shifted to the theory of partitions and q-series and that was why I did not work with Erdős after moving to Florida. He expressed his disappointment by saying that he was not good at identities, but preferred inequalities. The theory of partitions and q-series is all about identities, whereas, the study of arithmetic functions involves estimates which means either bounds for the main term or bounds for the error terms, and these are inequalities. Not only has it been a privilege to have collaborated with Erdős, one of the legendary figures in the world of mathematics, it was education in many ways which widened my mathematical horizon considerably.

13.2 Collaboration with George Andrews

My shift to the theory of partitions after arriving in Florida was much influenced by the work of George Andrews. I had always wanted to understand the partition and q-hypergeometric underpinnings in Ramanujan's spectacular work, and so after the Erdős 70th birthday conference in the lovely hill station of Ootacamund (Ooty) that I conducted for MATSCIENCE in January 1984, I wrote to Andrews requesting some of his papers because he is the king in the worlds of partitions and Ramanujan. Within a month, I received two large parcels of about 100 reprints of his papers. So with these, and his book on partitions [B3] as the basis, I gave a series of lectures at MATSCIENCE on partitions and q-series, the notes of which I still use today at the University of Florida. The best way to learn any subject, is to lecture on it — the preparation forces you to learn the subject properly, and on your own. Even though I lectured on this topic, I was not sure whether I would do research in the area. This desire to do research strengthened during the Ramanujan Centennial in 1987 in India when Andrews, Askey and Berndt were the stars of the event. I was most influenced by the lectures of Andrews, and my collaboration with him which started in 1990 was facilitated after my initiation into partitions and q-series due to Basil Gordon's mentorship.

The collaboration with Andrews started in a most amazing manner. I have described it at length in §7.4, but here I will just say that, Gordon and I had found a remarkable multi-parameter q-hypergeometric identity for a generalization of a deep partition theorem of Göllnitz. We arrived at this identity by combinatorial reasoning, but were unable to prove it. When Andrews arrived in November 1990 in Gainesville to deliver the Frontiers of Science Lecture at the University of Florida, I showed him the identity. He said it was remarkable, focused on it non-stop for the three days he was in Gainesville, and handed me an eight page proof at the airport before his departure! That led to my first paper with Andrews for which Gordon was a co-author.

Over the next few years, I worked on reformulations of Göllnitz's partition theorem and published some papers on weighted partition versions of the Göllnitz theorem. In that process I discovered by combinatorial arguments on separate occasions, two simpler but equivalent versions (cubic and quartic) of the original multi-parameter *key identity* that Gordon and I had found in 1990. I described these to Andrews by email and over the phone, and on both occasions, he responded within a few days with remarkable q-hypergeometric proofs of these identities. This resulted in two papers with Andrews on the combinatorial and q-hypergoemetric approaches to the cubic and quartic key identities for Göllnitz' theorem. Andrews may not write a dozen letters per day like Erdős did, but he is very prompt in replying when you ask him a question.

I spent my first sabbatical at Penn State University for the full academic year 1992–93 at his kind and generous invitation. It was the most productive year in my life. Inspired by his presence, and owing to frequent discussions I had with him, I wrote five papers in the theory of partitions and q-series, one of which was jointly with him and Gordon on generalizations and refinements of Capparelli's partition theorems. The story of how Capparelli's conjectures were first proved by Andrews is just as amazing as to how he proved the key identity for Göllnitz' theorem constructed by Gordon and me, and this is described in detail in §7.5. I will just emphasize here that one of the benefits of being in close proximity with the giants of the field is that you get to learn about the most exciting developments. In my case, being at Penn State with Andrews in 1992–93 got me involved in the study of the Capparelli theorems, and the outcome was that Andrews, Gordon and I wrote a paper generalizing and refining the Capparelli theorems using the method of weighted words that Gordon and I had introduced earlier in connection with Schur's partition theorem.

Andrews and I work well as collaborators because we take complementary approaches — mine being combinatorial and Andrews' being q-hypergeometric. Andrews has spent the entire spring semester in Florida since 2005, and so I have had an opportunity to interact with him closely. This has benefited me immensely in my own work. Andrews is to theory of partitions and q-series what Erdős is to elementary number theory: they both have encyclopedic knowledge of their fields. If I come up with an idea in the theory of partitions, I talk to Andrews to find out if it is new or if it has been considered in some form by others. But Andrews' stay in Florida has actually resulted only in one joint paper for us. It evolved from a talk he gave in our Number Theory Seminar in which he announced and proved a new dual of the Göllnitz theorem. He proved the dual by a study of certain generating functions of partitions by classifying these partitions as per the largest part — a technique he has used with great effect over the decades. When I saw this dual theorem, I asked myself what its key identity would be, and I showed combinatorially that it is a different version of the key identity for the Gollnitz theorem that Gordon and I had obtained. So we joined forces to write a paper on the dual of the Göllnitz theorem. I should add here that Andrews, Berkovich, and I wrote an important paper providing a four parameter extension of the deep theorem of Göllnitz; this settled in the affirmative a 35 year old problem of Andrews whether such an extension exists. For this paper which appeared in *Inventiones Mathematicae*, the collaboration with Andrews was through correspondence.

My joint work with Andrews is among my most significant researches. It has been a pleasure and a learning experience working with him.

13.3 Collaboration with Basil Gordon

Basil Gordon (1931–2012) was a towering figure in combinatorics and number theory. He made fundamental contributions to several areas, such as the theory of partitions, modular forms, mock theta functions, and coding theory. He was one of the few who was at home with both combinatorial and modular form techniques. He was one of the leaders in the world of Ramanujan's mathematics. Two of his most significant and appealing results are: (i) a generalization of the celebrated Rogers-Ramanujan partition theorem to all odd moduli greater than on equal to 5, and (ii) the Gollnitz-Gordon identities which are to the modulus 8 what the Rogers-Ramanujan identities are to the modulus 5. Owing to the breadth of his expertise, and his genial temperament, he guided several PhD students in a wide spectrum of fields. I was not his PhD student at UCLA, nor was he on my committee, but I interacted with him at the number theory seminars. Being aware of his stature, I invited him to be speaker in December 1987 for a conference I was organizing in Madras at Anna University for the Ramanujan Centennial. That is how my close friendship with Gordon began.

It came as a pleasant surprise, when one day in the spring of 1989, I received a phone call from him saying that he had a fully paid sabbatical in the Fall of 1989 and that he would like to visit the University of Florida. This was indeed a gift from heaven for me, because I was seriously thinking of moving into the realm of partitions as my new research area, and Gordon would be an ideal person to facilitate my entrance into this exciting domain. Gordon spent two months in Gainesville, and he and I worked together on two separate problems: (i) understanding a continued fraction of Ramanujan combinatorially, and (ii) to develop a generalization of Schur's famous partition theorem of 1926. This resulted in two papers, of which our work on generalizations of Schur's partition theorem turned out to have more significant impact.

I call Gordon as *The Great Guru*, because not only did he have a wealth of ideas, but he had a philosophy as to how to approach certain questions, and he shared his ideas generously. He had this idea of generalizing partition identities at the "base level" and lifting them up in levels by dilations and translations to get well known partition theorems as consequences. I worked with him on developing this technique and that is how what is now known as *the method of weighted words* evolved, with the first significant application being to Schur's partition theorem. It was clear that the method would be more widely applicable, and indeed it has, as testified by the dozens of papers that have been written by various active researchers.

In the next few years, I visited Los Angeles regularly to work with Gordon. In ancient Hindu culture, there is the *gurukula* style of learning. The student lives with the guru and observes him in close quarters as he is practicing the art. Working with Gordon was like undergoing this gurukula training. On my visits to Los Angeles, I stayed at his stately home in Santa Monica a few times. But whether I stayed with him or elsewhere, our discussions would start at his home in the morning. We would go out to lunch, then come back to his home to work until we would go out again for dinner. All day at his home, and even at lunch and dinner, the talk was about mathematics. It was "immersion" in the true sense, and I learnt an enormous amount by this close contact with him. On one of these visits to Los Angeles, he told me that we should apply the method of weighted words to the deep theorem of Göllnitz. We worked hard on it and constructed the key identity combinatorially; it was this identity which Andrews proved q-hypergeometrically during his visit to Florida in November 1990.

During my sabbatical at Penn State, I used to call Gordon regularly during mornings his time. He gave his unlisted phone number, which is very unusual. But every time I called him, I made sure I had a mathematical idea to discuss. During one of these phone calls he suggested that we should approach the Capparelli partition theorems by the method of weighted words. We discussed this over several phone calls and constructed a key identity that gave a generalization and refinement of Capparelli's theorems. On seeing this, Andrews improved on his earlier proof of the Capparelli conjectures and that is how the Alladi-Andrews-Gordon paper on generalizations and refinements of the Capparelli theorems came about. I submitted it to the Journal of Algebra because the Capparelli conjectures which came out a study of vertex operators in Lie algebras, was published in that journal.

I am truly grateful to Gordon who helped me enter the fascinating world of partitions and q-series, following which, I was smitten with that incurable, but wonderful, q-disease!

Left to right: Alexander Berkovich, George Andrews, and Krishna Alladi solved the problem of finding the four dimensional extension of Göllnitz' (Big) Partition Theorem in 2000. This picture was taken at the Millennial Number Theory Conference, University of Illinois, Urbana, May 22, 2000.

Some academic international travel pictures

Krishna, Mathura and their host Prof. Peter Paule (Director, Research Institute of Symbolic Computation, Linz, Austria) relax at a restaurant atop a mountain in the lovely Salzkammergut region of Austria where Paule took them on an excursion - April 29, 2016.

L to R: front row - Gaurav Bhatnagar (then at Univ. Vienna), Christian Krattenthaler (Univ. Vienna), Krishna, and Peter Paule, outside the Research Institute of Symbolic Computation, Linz, Austria, after Krishna's colloquium - May 3, 2016.

Shigeru Kanemitsu translating Krishna's talk into Japanese as Krishna gives his lecture to bright high school students at the University of Kinki, Iizuka, Japan, March 15, 2001.

L to R - Lalitha, Mathura, and Amritha with Prof. Hata, at *Kinka Kuji*, the Golden Temple in Kyoto, Japan - July 17, 2002.

L to R - Ali Nesin, Krishna, Mathura, Gabriela Nesin, and Ali Bulent (Krishna's host in Turkey), at the *Nesin Mathematics Village* in Turkey, August 10, 2014. Krishna and Gabriela (daughter of Ali Nesin) wrote a Feature Article on the Nesin Mathematics Village in the Notices of the AMS in 2015. Ali Nesin was recognized with the Leelavati Prize at the International Congress of Mathematicians in 2018.

Krishna with Hershel Farkas on the campus of the Hebrew University in Jerusalem, April 30, 2013.

Indian classical music and dance in the Alladi family

Many mathematicians have deep interest in art and classical music, and my strong interest is in Carnatic music and Bharathanatyam

Semmangudi Srinivasier, the doyen of Carnatic music, at Ekamra Nivas with the Alladi family (August 6, 1996). His vibrant music is a passion for all of us.

My mother Lalitha Ramakrishnan pursued Carnatic music vocal since her late teens, and has given several concerts over the decades. She was a graded artiste of the All India Radio during the seventies.

Mathura is an accomplished Carnatic vocalist, Bharathanatyam dancer and teacher. As Director and Guru of the JATHISWARA School of Dance and Music in Florida, she has trained about 100 students of whom 16 have had their Arangetrams.

L to R - Amritha and Lalitha, trained by Mathura, during their Bharathanatyam Arangetram at the Madras Music Academy in July 1999.

The author and his family with WSPC staff

Rochelle Kronzek (Editor, World Scientific), Alfred Hales (Emeritus, UCLA), and Krishna, at the World Scientific booth, AMS Annual Meeting, Baltimore, Jan 17, 2020.

Krishna with Max Phua (Executive Director, World Scientific), and editorial staff Lai Fun and Rok Ting, at the World Scientific Offices in Singapore, May 8, 2015.

Krishna with World Scientific Staff at the World Scientific headquarters in Singapore on July 18, 2022. L to R - Liu Nijia, Rok Ting, Krishna and Rajesh Babu.

L to R - Rochelle Kronzek (Editor, World Scientific) with Mathura, Amritha and Lalitha during a party at the Alladi House, Gainesville, for the Alladi-60 Conference, March 18, 2016.

Some important letters and papers

Letter from Professor Helmut Hasse

Prof. Dr. H. Hasse
2070 Ahrensburg
Hagener Allee 35
Tol. 20 87

1.11.1976

Dear Professor Ramakrishnan,

Thank you for your explanatory letter of Oct.25, and for letting me have the right paper, viz., that of your son in collaboration with Professor Erdős.

First of all let me say, that your son may be very proud and grateful for having found the interest and collaboration of such a distinguished mathematician. What else can you expect under such circumstances than results of great importance.

I read the paper with great interest, finding the results a welcome supplement to our knowledge of prime numbers and prime decomposition, items that have occupied the mathematical giants dating from Euclid's time.

With kind regards, very sincerely

Yours

H. Hasse

Helmut Hasse was one of the greatest mathematicians of the 20th century. He had written a letter to me in 1972 (upon seeing my fledgling research on the Farey-Fibonacci sequence) saying that it was original and also stating, "Let not excessive reading spoil your innate originality". After my collaboration with Erdős started in 1975, my father informed Hasse about that and sent him my work. This response of 1976 from Hasse to my father is about my joint work with Erdős. I was then a graduate student at UCLA.

Letter from Nobel Laureate Hans Bethe

CORNELL UNIVERSITY
LABORATORY OF NUCLEAR STUDIES
ITHACA, N. Y. XXXXX 14853

December 23, 1974

Professor Alladi Ramakrishnan
MATSCIENCE
Madras 20
India

Dear Professor Ramakrishnan:

Thank you very much for your letter of 7 November.

Professors Littauer and Fuchs, as well as I, expect to be in Cornell next May. There is only one short period, probably May 15-18, when I expect to be absent. We shall all be happy to see you here.

When talking to Professor Fuchs, he told me that he was much impressed by your son Krishna's papers. It is remarkable that an undergraduate can already publish so much. Professor Fuchs said that some of the ideas were quite original.

With best wishes for the New Year.

Yours sincerely,

Hans A. Bethe

HAB:vhr

Hans Bethe (Cornell University) won the 1967 Nobel Prize in Physics for his work on stellar nucleosynthesis. My father knew Prof. Bethe well and so he sent Bethe some of my papers that I wrote as an undergraduate, requesting him to get an assessment from mathematicians at Cornell. In this letter, Prof. Bethe says that mathematics professor Wolfgang Fuchs, a reputed complex analyst, expressed a positive opinion of my work as an undergraduate. In 1974 when this letter was written, I was in the final (third) year of my BSc course at Vivekananda College of the University of Madras, India.

Letter from UCLA Chancellor Charles Young

UNIVERSITY OF CALIFORNIA, LOS ANGELES

BERKELEY • DAVIS • IRVINE • LOS ANGELES • RIVERSIDE • SAN DIEGO • SAN FRANCISCO  SANTA BARBARA • SANTA CRUZ

OFFICE OF THE CHANCELLOR
LOS ANGELES, CALIFORNIA 90024

March 11, 1975

Mr. Krishnaswami Alladi
27 Luz Church Rd.
Madras, Tamil Nadu 600004   INDIA

Dear Mr. Alladi:

It is my pleasure to inform you that upon the recommendation of the Department of _____ Mathematics _____ and of the Committee on Fellowships of the Graduate Council you have been selected for a Chancellor's Intern Fellowship tenable at the University of California, Los Angeles, for the academic year 1975-76. This award represents the most prestigious award the University may bestow on an entering graduate student.

While a Chancellor's Intern Fellow, you are entitled to a first-year Fellowship, two years as a Teaching or Research Assistant and, if advanced to candidacy for the PhD degree within three years of initiation of your tenure, a one-year Dissertation Fellowship.

Please indicate your acceptance or declination of this award on the enclosed form and return two copies to us as soon as possible and no later than April 15, 1975. A copy of the formal award transmittal will be forwarded to your major department upon receipt of your acceptance.

"As a member of the Association of Graduate Schools This institution adheres to the policy concerning the 15 April deadline for binding acceptance of a stipend offer. It should be pointed out, however, that a multiplicity of offers held by a single student causes hardships for other students who may be deprived of receiving any offer at all. The AGS institutions therefore strongly encourage early declinations of any offers you may have received, in which you do not have a genuine interest."

May I offer my congratulations and my best wishes for a successful graduate academic career.

Sincerely yours,

Charles E. Young
Chancellor

encl
cc: Department of Mathematics

The Chancellor's Fellowship, the most prestigious at UCLA, is a four year fellowship. Charles Young was Chancellor (= President) at UCLA from 1968 to 1997. He then served as the President of the University of Florida during 1999–2004, and I had occasion to meet him during my tenure as Chair.

Alladi's first joint paper with Paul Erdős

PACIFIC JOURNAL OF MATHEMATICS
Vol. 71, No. 2, 1977

ON AN ADDITIVE ARITHMETIC FUNCTION

K. Alladi and P. Erdős

We discuss in this paper arithmetic properties of the
function $A(n) = \sum_{p^a\|n} \alpha p$. Asymptotic estimates of $A(n)$
reveal the connection between $A(n)$ and large prime factors
of n. The distribution modulo 2 of $A(n)$ turns out to be an
interesting study and congruences involving $A(n)$ are considered. Moreover the very intimate connection between
$A(n)$ and the partition of integers into primes provides a
natural motivation for its study.

0. **Introduction.** Let a positive integer n be expressed as a
product of distinct primes in the canonical fashion $n = \prod_{i=1}^{r} p_i^{a_i}$. Define
a function $A(n) = \sum_{i=1}^{r} \alpha_i p_i$.

(i) The function $A(n)$ is not injective. In fact for a fixed
integer m, the number of solutions in n to $A(n) = m$, is the number
of partitions of m into primes.

(ii) $A(n)$ fluctuates in size appreciably. It is easily seen that
$A(n) = n$ when n is a prime, while $A(n) = O(\log n)$ when n is a power
of a small prime. Actually the "average order" of $A(n)$ turns out
(as a corollary to Theorem 1.1) to be $\pi^2 n/6 \log n$. The term average
order is defined below.

(iii) The function $A(n)$ is additive and one can expect it to take
odd and even values with equal frequency.

The term "average order" calls for some explanation. We follow
the usage in Hardy and Wright [6]. If $f(n)$ is a function defined
on the positive integers we consider

$$F(x) = \sum_{n \leq x} f(n) .$$

Usually F can be expressed in terms of well behaved functions like
polynomials or exponentials and the like. That is we seek an asymptotic estimate for F in terms of these functions. Then we seek a
similar well behaved function g so that

$$F(x) = \sum_{n \leq x} f(n) \sim \sum_{n \leq x} g(n) .$$

The function g may be thought of as the average order of f. For
instance if φ is the Euler function then

$$F(x) = \sum_{n \leq x} \varphi(n) = \frac{3x^2}{\pi^2} + O(x \log x) \sim \sum_{n \leq x} \frac{6n}{\pi^2}$$

so the average order of $\varphi(n)$ is $6n/\pi^2$.

*This is my first joint paper with the legendary Paul Erdős on the additive arithmetic
function $A(n)$, extending considerably my initial observations on $A(n)$ as an undergraduate in India in 1973, when I had called $A(n)$ as a "new logarithmic function". This
joint paper was published in the Pacific Journal of Mathematics in 1977 when I was doing my PhD at UCLA. It is perhaps my most quoted paper. I thank the Pacific Journal
of Mathematics for giving permission to reproduce the first page of the paper.*

Letter of acceptance of Alladi-Erdős paper I

UNIVERSITY OF SOUTHERN CALIFORNIA
UNIVERSITY PARK
LOS ANGELES, CALIFORNIA 90007

DEPARTMENT OF MATHEMATICS

November 9, 1976

Professor K. Alladi
Department of Mathematics
University of California
Los Angeles, California 90024

Dear Professor Alladi:

It is a pleasure to report that your paper, with Erdos

has been accepted for publication in the Pacific Journal.

Your manuscript is being sent to our central Editorial Offices;
I estimate that you will receive the galleys in about *10* months.

The Pacific Journal is cooperating with the Zentralblatt, in asking
that the author of each accepted article send a review of the article
directly to them.

Yours sincerely,

J. Dugundji, Editor
Pacific Journal of
    Mathematics

JD:HD

*Letter of acceptance of my first joint paper with Paul Erdős, from Professor Dugundji,
Editor, Pacific Journal of Mathematics.*

My Mathematical Universe

A letter from Paul Erdős to Krishna Alladi

1975 I 24

THE UNIVERSITY OF NEW SOUTH WALES

P.O. BOX 1 · KENSINGTON · NEW SOUTH WALES · AUSTRALIA · 2033

TELEPHONE 663 0351
EXTN.

PLEASE QUOTE

Dear Mr Alladi,

Let $m = \prod p_i^{d_i}$, $L(m) = \sum d_i \cdot p_i$, assume m is not a prime and $m \equiv 0 \pmod{L(m)}$. I can prove your conjecture $\sum \frac{1}{m} < \infty$. In fact denote by $N(x)$ the number of integer $m_i < x$, $m_i \equiv 0 \pmod{L(m_i)}$, $m_i \neq p$. I hope to prove

(1) $$N(x) < x \exp\left(c (\log x \, \log\log x)^{1/2}\right)$$

The method is similar to my old paper "on primitive abundant numbers", J. London Math Soc 1935 see also Remarks on 2 - abundant numbers, Acta Arithmetica 5 (1958), 25-33

To prove (1) we distinguish several cases. Let $P(m)$ be the greatest prime factor of m. A theorem of N. G. de Bruijn states that the number of integer $m < x$ with $P(x) < x^{1/u}$ is less than $\frac{x}{u!}$. (Indigationes Math 1950 or 51). Thus we can assume

(2) $$P(m_i) > \exp\left(\log x \, \log\log x\right)^{1/2}$$

for if (2) is not satisfied $u = \left(\frac{\log x}{\log\log x}\right)^{1/2}$ and

$$\frac{x}{u!} < x \exp - c(\log x \, \log\log x)^{1/2}$$

thus the numbers not satisfying (1) can be neglected.

Similarly we can assume ($d(m)$ is the number of divisors of n)

(3) $$d(m) < \exp \frac{1}{10} \left(\log x / \log\log x\right)^{1/2}$$

numbers but this would need more care and unexpected
difficulties may arise (but I do think all is well)

Please let me know if this is correct + clear - if you want
more explanation please return this letter and indicate
the places which are not O.K.

To get a genuine asymptotic formula for $N(x)$ is probably
hopeless.

Kind regards to you, your parents + colleagues

P. Erdős

After the middle of next month my adress is: Univ of Florida
Math Dept Gainesville Florida

I have dozens of letters that Erdős wrote to me over the years from all parts of the globe. I provide here a sample, and that too only the first and last page of an important letter in which he outlined a proof of a certain conjecture I made and introduced me to certain fundamental ideas relating to integers without large prime factors. Note that in concluding the letter he says where he will be next so that I could be in touch with him as he was criss-crossing the world.

Letter of offer of the
Hildebrandt Research Assistant Professorship

The University of Michigan

DEPARTMENT OF MATHEMATICS

ANN ARBOR, MICHIGAN 48109

3 February 1978

(313) 764-0335

Mr. Krishnaswami Alladi
Department of Mathematics
University of California
Los Angeles, California 90024

Dear Mr. Alladi:

I am writing to follow up our telephone conversation of this afternoon and officially offer you an appointment as a T. H. Hildebrandt Research Assistant Professor at the University of Michigan. The appointment is for three years, non-renewable, and the teaching duties consist of two courses one term and one course the other. The salary for the first year will be $14,500 and not less than that for the other two.

We will include you in a proposal to the National Science Foundation for summer research money, and there is no doubt in my mind that you will be awarded summer research money, in the amount of 2/9 of your academic year salary. If this fails, we will be glad to offer you summer teaching.

There are numerous fringe benefits associated with an appointment at the University, including medical insurance, group life insurance (an amount three times one's academic salary), and a retirement plan (TIAA-CREF) in which the University pays ten percent and the individual five percent of his salary. In addition, your moving expenses in coming to Ann Arbor will be reimbursed at the rate of seventeen cents per mile for transportation, plus up to one-half the cost of packing and freighting household goods, but not to exceed $500.

I am required to state that the above offer has not been approved by the Regents of the University of Michigan. That group meets infrequently to give formal approval to all of the University's activities, and it is our custom not to request prior approval in matters of offers of employment. However, I know of no case where the Regents have not given subsequent approval of an appointment when requested, and hence this matter need cause you no concern.

- 2 -

Let me say once again how glad I and my colleagues are that you will be coming to Michigan. We all look forward to a very interesting and profitable three years with you here.

Sincerely yours,

Fred Gehring

F. W. Gehring
Chairman

FWG:er

P.S. Please reply with your official acceptance of this offer as soon as possible so that we may notify the Administration.

Beginning from 1978, when I received the offer, the Hildebrandt Research Assistant Professorship, was converted from a two-year position to a three-year position. The starting salary then was $14,500 for the first academic year; this gives an idea what the salaries were like in those days.

Offer from the Institute for Advanced Study, Princeton

THE INSTITUTE FOR ADVANCED STUDY

HARRY WOOLF
Director

February 26, 1981

Dr. Krishnaswami Alladi
Department of Mathematics
University of Michigan
Ann Arbor, Michigan 48109

Dear Dr. Alladi:

On the recommendation of the Faculty in the School of Mathematics, I am pleased formally to offer you a membership in the Institute for Advanced Study for the academic year 1981-82. The term dates are: first term, Monday, September 21, 1981 to Friday, December 18, 1981; second term, Monday, January 4, 1982 to Friday, April 2, 1982. In general, members are expected to be in residence during term time except for short absences; absences of more than a few days should be discussed with the appropriate officer of the School and me.

To help defray the expenses of your stay in Princeton, we can make available to you a grant-in-aid of $14,500 for the year.

If your response to this invitation is favorable, and if you are interested in Institute housing, would you be kind enough to fill out and return the enclosed housing reservation form? Please do so as soon as possible for the number of units is limited and we would like to be able to provide you with suitable accommodations while you are here.

We all look forward with pleasure to having you with us, and we would find it most helpful to have a response to our invitation no later than one month after receipt of this letter.

Sincerely yours,

Harry Woolf

Enclosures

Within a few weeks of receiving the offer from the University of Michigan, the Institute for Advanced Study (IAS) in Princeton sent me a letter offering me a Visiting Membership for the Academic Year 1978–79, which I could not accept because I had just accepted the offer from Michigan. Upon completion of my term at Michigan, I re-applied to the IAS, and Prof. Harry Woolf, the Director at IAS, again made me this offer in 1981, which I accepted.

Letter from Professor Robert Steinberg

UNIVERSITY OF CALIFORNIA, LOS ANGELES UCLA

BERKELEY • DAVIS • IRVINE • LOS ANGELES • RIVERSIDE • SAN DIEGO • SAN FRANCISCO SANTA BARBARA • SANTA CRUZ

DEPARTMENT OF MATHEMATICS
405 HILGARD AVENUE
LOS ANGELES, CALIFORNIA 90024-1555
March 25, 1994

Dear Alladi:

This is to inform you that your very interesting paper with Andrews and Gordon has been accepted for publication in the Journal of Algebra. I'm sorry for the delay, caused mostly by a slow referee.

Please send in the enclosed form, filled in as indicated.

With best wishes,

Robert Steinberg

Professor Robert Steinberg, one of my teachers at UCLA, was a world renowned algebraist. He was an Editor of the Journal of Algebra. One of my joint papers with George Andrews and Basil Gordon was on generalizations of the Capparelli partition theorems, which has connections with the theory of Lie algebras. So I submitted our paper to Prof. Steinberg for publication in the Journal of Algebra. I cherish this letter from Steinberg in his beautiful handwriting, accepting our paper.

Letter from Professor Hugh Montgomery
after the Alladi-60 Conference

THE UNIVERSITY OF MICHIGAN
DEPARTMENT OF MATHEMATICS

EAST HALL, 530 CHURCH STREET
ANN ARBOR, MI 48109-1043
PHONE: 734 764-0335   FAX: 734 763-0937
http://www.math.lsa.umich.edu/

May 27, 2016

Dear Krishna:

I enjoyed the Gainesville conference very much. As you are probably aware, Doron Zeilberger responded to my talk with work on the Shapiro polynomials. Perhaps you have not heard that Brad Rodgers (a former student of Terry Tao who is at Michigan now) has given a complete proof of Saffari's conjecture on the distribution of $|P_k(x)|^2$. He uses the representation theory of $SU(2)$, so his proof is quite different from Zeilberger's approach. The two have now met, and may start to collaborate.

I am now writing up my talk as a paper for the conference proceedings. I should have it done shortly.

After so much talk about Ramanujan in Gainesville, on my return to Ann Arbor I looked in my offprint collection to see if I had anything by Ramanujan. I found the three papers that you find enclosed, and I would like you to have them. I'm sorry that many years ago I wrote my name on them. It seems like a blemish now. These papers came from Davenport's collection. Perhaps Hardy gave them to Davenport, or maybe they were among Hardy's papers, and Littlewood gave them to Davenport after Hardy's death. In the next few years I shall be donating the bulk of my offprint collection, including papers that came to me from Davenport and from Ingham, to the AIM reprint collection.

Best wishes,

Hugh

After the conference in March 2016 for my 60th birthday, Professor Montgomery sent this letter, and along with it, three reprints of Srinivasa Ramanujan's papers as a gift for me. He mentions in this letter, that he wrote his name on the reprints and feels that it is a blemish. In thanking him for these reprints, I said that I was honored to have his name on the reprints because he (Montgomery) was my link with Trinity College.

Cover of a reprint of Srinivasa Ramanujan's paper

ON THE EXPRESSION OF A NUMBER IN THE
FORM $ax^2 + by^2 + cz^2 + du^2$

BY

S. RAMANUJAN, B.A., Trinity College

REPRINTED FROM THE
PROCEEDINGS OF THE CAMBRIDGE PHILOSOPHICAL SOCIETY, Vol. XIX. Part 1

CAMBRIDGE
AT THE UNIVERSITY PRESS
January 1917

This is one of three reprints of Srinivasa Ramanujan's papers that Professor Montgomery sent me as a gift. It is the problem on universal quadratic forms stemming from this paper that Manjul Bhargava and Jonathan Hanke solved in 2005.

Letter from Fields Medalist Gerd Faltings

Jan-10-2003 11:23am From-MATH DEPT PENN STATE UNIVERSITY +8148653795 T-270 P.002/002 F-856

Inventiones

mathematicae

Springer International

Professor Dr. G. Faltings
MPI für Mathematik
Postfach 7280
53072 Bonn

Professor Andrews (Georg)
Department of Mathematics
Eberly College of Science
The Pennsylvania State University
218 McAllister Building
University Park, PA 16802
USA

03/01/2003

Dear Colleague,

The editors are pleased to inform you that your manuscript *(joint paper with Alladi + Berkovich)*
A NEW FOUR PARAMETER Q-SERIES IDENTITY AND ITS PARTICITION IMPLICATIONS

has been accepted for publication and is being forwarded to the publisher. You will receive
the page proofs in due course. Please note that only typographical errors should be corrected,
as it is possible that you will be invoiced for any other changes made.

ZENTRALBLATT publishes abstracts of mathematical papers.
You are invited to send an abstract of your paper to the following address :
Zentralblatt für Mathematik
Franklinstraße 11
10587 Berlin, Germany

Any further correspondance concerning the manuscript should be addressed to:
Springer-Verlag
Journal Production Mathematics
Attn.: Jean-Marie Lerner
Postfach 10 52 80
D - 69042 Heidelberg, Germany
Tel.: +49 6221/487 8963, Fax: +49 6221/487 8688
e-mail: Lerner@springer.de

If possible, please submit the text of your paper by e-mail, using Springer's TEX macros,
and indicate the format of each file. Please ensure that the e-mail version is identical
with the hard copy.

Sincerely yours,

Gerd Faltings

*In 2000, Alexander Berkovich, George Andrews and I found a four parameter extension
of a deep theorem of Göllnitz, and thereby settled a question that Andrews posed three
decades earlier. Andrews submitted our joint paper to Inventiones Mathematicae, a very
prestigious journal. This letter of acceptance was sent to us by the Fields Medalist Gerd
Faltings, who was Editor-in-Chief of that journal.*

References

Books

B1) K. Alladi, *Ramanujan's Place in the World of Mathematics*, Springer, New York, Edition 1 (2013); Edition 2 (2021).

B2) K. Alladi, J. R. Klauder, C. R. Rao (Eds.), *The Legacy of Alladi Ramakrishnan in the Mathematical Sciences*, Springer, New York (2010).

B3) G. E. Andrews, *The Theory of Partitions*, in Ency. of Math. Vol. II, Addison-Wesley, Reading (1976).

B4) G. E. Andrews and F. G. Garvan (Eds.), *Analytic Number Theory, Modular Forms, and q-Hypergeometric Series* (Proceedings of the Number Theory Conference in honor of Krishna Alladi's 60th birthday at the University of Florida) Springer, New York (2018).

B5) A. Baker, *Transcendental Number Theory*, Cambridge Univ. Press, London (1975).

B6) Peter L. Duren (Ed.), *A Century of Mathematics in America*, Parts I, II, III, AMS, Providence (1988–89).

B7) P. D. T. A. Elliott, *Probabilistic Number Theory, Vols. I and II*, Springer, New York (1980).

B8) A. Flexner, *The Usefulness of Useless Knowledge* (with a companion essay by Robbert Dijkgraff), Princeton Univ. Pres, Princeton (2017 - reprinted).

B9) H. Halberstam and H.-E. Richert, *Sieve Methods*, Acad. Press, New York (1974).

B10) G. H. Hardy, *Ramanujan - Twelve Lectures inspired by his work*, Chelsea (reprinted), New York (1961).

B11) P. Hoffman, *The Man Who Loved Only Numbers - The Story of Paul Erdős and the Search for Mathematical Truth*, Hyperion, New York (1998).

B12) Alladi Ramakrishnan, *The Alladi Diary* (Edited by Krishnaswami Alladi), World Scientific, Singapore (2019).

B13) S. Ramanujan, *Collected Papers*, Chelsea (reprinted), New York (1962).

B14) B. Schechter, *My Brain is Open - The Mathematical Journeys of Paul Erdős*, Simon and Schuster, New York (1998).

Articles of general interest

A1) K. Alladi, The discovery and rediscovery of mathematical genius - a Review of "The Man Who Knew Infinity: A life of the genius Ramanujan" (book by Robert Kanigel), *The American Scientist*, **80** (1992), 388–389.

A2) K. Alladi, "Institute visit inspired creation of the Institute of Mathematical Sciences in Madras - Alladi Ramakrishnan (1923–2008)", *The Institute Letter*, Princeton (Spring 2009), p. 7.

A3) K. Alladi, "Touched by a Goddess" - Review of the Movie "The Man Who Knew Infinity" *Inference - International Review of Science*, **2** (2016).

A4) K. Alladi, "Review of the movie on the Mathematical Genius Ramanujan", *Asia Pacific Mathematics Newsletter* (World Scientific), **6** (2016), 29–41.

A5) K. Alladi, "The SASTRA Ramanujan Prize - its Origins and its Winners", *Notices AMS*, **66** (2019), 64–72.

A6) K. Alladi, "Ramanujan's Legacy - the work of the SASTRA Prize Winners", *Phil. Trans. Royal Soc. London*, **378** (2020), 1–19.

A7) D. Goldfeld, "The Elementary Proof of the Prime Number Theorem: an Historical Perspective", in *Number Theory - New York 2003*, Springer, New York (2004), 179–192.

A8) J. Spencer and R. L. Graham, "The Elementary Proof of the Prime Number Theorem", *Math. Intelligencer*, **31** (2009), 18–23.

Index

Springer and WSPC Editors at mathematics events in Florida

Krishna has been active in publishing and has had a close association with Springer and World Scientific, whose Editors have regularly attended, and have had book exhibits at, events at the University of Florida arranged by Krishna, especially the Ramanujan Colloquium sponsored by George Andrews and The Pennsylvania State University. In this picture we have three Springer Editors and Rochelle Kronzek of WSPC at a dinner in honor of Alex Lubotzky (Hebrew University) for his Ramanujan Colloquium (Mar 11, 2019). From L to R: Ann Kostant (Springer), Alex Lubotzky, Rochelle Kronzek (WSPC), Elizabeth Loew (Springer), Marc Strauss (Springer), Mathura Alladi and George Andrews.